Plastic Packaging

Edited by
Otto G. Piringer and Albert L. Baner

Further Reading

Elias, H.-G.

Macromolecules
Volume 1: Chemical Structures and Syntheses

2005
ISBN: 978-3-527-31172-9

Elias, H.-G.

Macromolecules
Volume 2: Industrial Polymers and Syntheses

2007
ISBN: 978-3-527-31173-6

Elias, H.-G.

Macromolecules

Volume 3: Physical Structures and Properties
2008
ISBN: 978-3-527-31174-3

Elias, H.-G.

Macromolecules
Volume 4: Applications of Polymers

2009

ISBN: 978-3-527-31175-0

Brennan, J. G.

Food Processing Handbook

2006
ISBN: 978-3-527-30719-7

Plastic Packaging

Interactions with Food and Pharmaceuticals

Edited by
Otto G. Piringer and Albert L. Baner

Second, Completely Revised Edition

WILEY-VCH Verlag GmbH & Co. KGaA

The Editors

Dr. Otto G. Piringer
FABES ForschungsGmbH
Schragenhofstr. 35
80992 München
Germany

Dr. Albert L. Baner
Nestle Purina Petcare
Checkerboard Square
St. Louis, MO 63164
USA

■ All books published by Wiley-VCH are carefully produced. Nevertheless, authors, editors, and publisher do not warrant the information contained in these books, including this book, to be free of errors. Readers are advised to keep in mind that statements, data, illustrations, procedural details or other items may inadvertently be inaccurate.

Library of Congress Card No.: applied for

British Library Cataloguing-in-Publication Data
A catalogue record for this book is available from the British Library.

Bibliographic information published by the Deutsche Nationalbibliothek
Die Deutsche Nationalbibliothek lists this publication in the Deutsche Nationalbibliografie; detailed bibliographic data are available in the Internet at <http://dnb.d-nb.de>

© 2008 WILEY-VCH Verlag GmbH & Co. KGaA, Weinheim

All rights reserved (including those of translation into other languages). No part of this book may be reproduced in any form – by photoprinting, microfilm, or any other means – nor transmitted or translated into a machine language without written permission from the publishers. Registered names, trademarks, etc. used in this book, even when not specifically marked as such, are not to be considered unprotected by law.

Typesetting Thomson Digital, Noida, India
Printing betz-druck GmbH, Darmstadt
Book Binding Litges & Dopf GmbH, Heppenheim
Cover Design Anne Christine Keßler, Karlsruhe

Printed in the Federal Republic of Germany
Printed on acid-free paper

ISBN: 978-3-527-31455-3

Contents

Preface *XV*
List of Contributors *XVII*

1	**Preservation of Quality Through Packaging** *1*	
	Albert Baner and Otto Piringer	
1.1	Quality and Shelf-Life *1*	
1.2	Physical and Chemical Interactions Between Plastics and Food or Pharmaceuticals *4*	
1.3	The Organization of this Book *5*	
	Further Reading *12*	
2	**Characteristics of Plastic Materials** *15*	
	Johannes Brandsch and Otto Piringer	
2.1	Classification, Manufacture, and Processing Aids *15*	
2.1.1	Classification and Manufacture of Plastics *16*	
2.1.1.1	Raw Materials and Polymerization Processes *17*	
2.1.1.2	Addition Polymerization *18*	
2.1.1.3	Condensation Polymerization *19*	
2.1.1.4	Synthesis of Copolymers, Block, and Graft Copolymers *19*	
2.1.1.5	Polymer Reactions *20*	
2.1.1.6	Plastic Processing *22*	
2.1.2	Processing Aids *23*	
2.1.2.1	Initiators and Crosslinkers *24*	
2.1.2.2	Catalysts *25*	
2.2	Structure and States of Aggregation in Polymers *26*	
2.2.1	Structure *26*	
2.2.2	States of Aggregation *29*	
2.3	The Most Important Plastics *32*	
2.3.1	Thermoplastics *32*	
2.3.1.1	Polyethylene *32*	

2.3.1.2	Polypropylene	34
2.3.1.3	Polybutene-1	36
2.3.1.4	Polyisobutylene	36
2.3.1.5	Poly-4-methylpentene-1 (P4MP1)	36
2.3.1.6	Ionomers	36
2.3.1.7	Cyclic Olefin Copolymers (COC)	37
2.3.1.8	Polystyrene	38
2.3.1.9	Polyvinyl Chloride	39
2.3.1.10	Polyvinylidene Chloride	41
2.3.1.11	Thermoplastic Polyesters	41
2.3.1.12	Polycarbonate	42
2.3.1.13	Polyamide	43
2.3.1.14	Polymethylmethacrylate	44
2.3.1.15	Polyoxymethylene or Acetal Resin	45
2.3.1.16	Polyphenylene Ether (PPE)	45
2.3.1.17	Polysulfone	45
2.3.1.18	Fluoride Containing Polymers	46
2.3.1.19	Polyvinylether	46
2.3.2	Thermosets	46
2.3.2.1	Amino Resins (UF, MF)	47
2.3.2.2	Unsaturated Polyester (UP)	47
2.3.3	Polyurethanes	48
2.3.4	Natural and Synthetic Rubber	49
2.3.5	Silicones	51
2.3.6	Plastics Based on Natural Polymers Regenerated Cellulose	54
2.3.6.1	Biodegradable Polymers	54
2.3.7	Coatings and Adhesives	55
2.3.7.1	Lacquers	56
2.3.7.2	Plastic Dispersions	57
2.3.7.3	Microcrystalline Waxes	57
2.3.7.4	Temperature-Resistant Coatings	58
2.3.7.5	Printing Inks and Varnishes	59
	References	60
3	**Polymer Additives**	**63**
	Jan Pospíšil and Stanislav Nešpůrek	
3.1	Introduction	63
3.2	Antifogging Agents	64
3.3	Antistatic Agents	65
3.4	Blowing Agents	65
3.5	Colorants	66
3.6	Fillers and Reinforcing Agents	66
3.7	Lubricants	67
3.8	Nucleating Agents	67
3.9	Optical Brighteners	68

3.10	Plasticizers	68
3.11	Stabilizers	70
3.11.1	Antiacids	71
3.11.2	Antimicrobials	72
3.11.3	Antioxidants	72
3.11.3.1	Chain-Breaking Antioxidants	73
3.11.3.2	Hydroperoxide Deactivating Antioxidants	74
3.11.4	Dehydrating Agent	75
3.11.5	Heat Stabilizers	75
3.11.6	Light Stabilizers	76
3.11.6.1	Light Screening Pigments and UV Absorbers	76
3.11.6.2	Photoantioxidants	77
3.12	Transformation Products of Plastic Stabilizers	78
3.12.1	Transformation Products from Phenolic Antioxidants and UV Absorbers	79
3.12.2	Transformation Products from Hydroperoxide Deactivating Antioxidants	83
3.12.3	Transformation Products from Hindered Amine Stabilizers	84
3.12.4	Transformation Products from Heat Stabilizers for PVC	85
3.13	Conclusions	86
	References	86

4   Partition Coefficients   89
Albert Baner and Otto Piringer

4.1	Experimental Determination of Polymer/Liquid Partition Coefficients	89
4.2	Thermodynamics of Partition Coefficients	90
4.2.1	Equilibrium Between Different Phases in Ideal Solutions	91
4.2.1.1	Partitioning in Ideal Solutions: Nernst's Law	92
4.2.2	Equilibrium Between Different Phases in Nonideal Solutions	93
4.2.2.1	Partition Coefficients for Nonideal Solutions	94
4.2.3	Partition Coefficients for Systems with Polymers	96
4.2.4	Relationship Between Partition Coefficients and Solubility Coefficients	98
4.3	Estimation of Partition Coefficients Between Polymers and Liquids	99
4.3.1	Additive Molecular Properties	99
4.3.2	Estimation of Partition Coefficients Using QSAR and QSPR	102
4.3.3	Group-Contribution Thermodynamic Polymer Partition Coefficient Estimation Methods	102
4.3.3.1	Estimation of Partition Coefficients Using RST	104
4.3.3.2	Estimation of Partition Coefficients Using UNIFAC	104
4.3.3.3	Estimation of Partition Coefficients Using Group-Contribution Flory Equation-of-State	108

4.3.3.4	Estimation of Partition Coefficients Using Elbro Free Volume Model 108
4.3.3.5	Comparison of Thermodynamic Group-Contribution Partition Coefficient Estimation Methods 108
4.3.4	Vapor Pressure Index Partition Coefficient Estimation Method 109
4.3.4.1	Examples of Vapor Pressure Index Values 112
	References 118

5	**Models for Diffusion in Polymers** 123
	Peter Mercea
5.1	Diffusion in Polymers – The Classical Approach 125
5.1.1	Diffusion in Rubbery Polymers 126
5.1.1.1	Molecular Models 126
5.1.1.2	The Molecular Model of Pace and Datyner 129
5.1.1.3	Free-Volume Models 131
5.1.1.4	The Free-Volume Model of Vrentas and Duda 133
5.1.2	Diffusion in Glassy Polymers 135
5.2	Diffusion in Polymers – The Computational Approach 140
5.2.1	Molecular Dynamics 142
5.2.2	The Transition-State Approach 150
5.3	Conclusions 154
	References 158

6	**A Uniform Model for Prediction of Diffusion Coefficients with Emphasis on Plastic Materials** 163
	Otto Piringer
6.1	Introduction 163
6.2	Interaction Model 166
6.2.1	Model Assumptions 166
6.3	Prerequisites for Diffusion Coefficients 168
6.3.1	Critical Temperatures of *n*-Alkanes 168
6.3.2	Melting Temperatures of *n*-Alkanes 170
6.3.3	Melting Temperatures of Atom Clusters 173
6.3.4	Critical Compression Factor 175
6.3.5	The Entropy of Evaporation 175
6.3.6	The Reference Temperature and the Reference Molar Volume 176
6.4	The Diffusion Coefficient 178
6.4.1	Diffusion in Gases 178
6.4.2	Diffusion in the Critical State 181
6.4.3	Diffusion in Solids 181
6.4.3.1	Self-diffusion Coefficients in Metals 181
6.4.3.2	Self-Diffusion Coefficients in Semiconductors and Salts 183
6.4.3.3	Self-Diffusion Coefficients in n-Alkanes 184
6.4.4	Diffusion in Liquids 184
6.4.4.1	Self-Diffusion Coefficients in Metals 184
6.4.4.2	Self-Diffusion Coefficients in n-Alkanes 185

6.4.5	Diffusion in Plastic Materials	188
6.4.5.1	Diffusion Coefficients of n-Alkanes in Polyethylene	188
6.4.5.2	Diffusion Coefficients of Additives in Polymers	191
	References 193	

7 Transport Equations and Their Solutions   195
Otto Piringer and Titus Beu

7.1	The Transport Equations	195
7.1.1	The Terminology of Flow	196
7.1.2	The Differential Equations of Diffusion	197
7.1.3	The General Transport Equations	200
7.2	Solutions of the Diffusion Equation	201
7.2.1	Steady State	202
7.2.2	Nonsteady State	202
7.2.3	Diffusion in a Single-Phase Homogeneous System	203
7.2.3.1	Dimensionless Parameters and the Proportionality of Mass Transfer to the Square Root of Time	209
7.2.3.2	Comparison of Different Solutions for the Same Special Cases	212
7.2.4	Diffusion in Multiphase Systems	213
7.2.4.1	Diffusion in Polymer/Liquid Systems	213
7.2.4.2	Influence of Diffusion in Food	224
7.2.5	Surface Evaporation	225
7.2.6	Permeation Through Homogeneous Materials	227
7.2.7	Permeation Through a Laminate	228
7.2.8	Concentration Dependence of the Diffusion Coefficient	228
7.2.9	Diffusion and Chemical Reaction	229
7.3	Numerical Solutions of the Diffusion Equation	230
7.3.1	Why Numerical Solutions?	230
7.3.2	Finite-Difference Solution by the Explicit Method	231
7.3.2.1	von Neumann Stability Analysis	236
7.3.2.2	The Crank–Nicholson Implicit Method	237
7.3.3	Spatially Variable Diffusion Coefficient	240
7.3.4	Boundary Conditions	241
7.3.5	One-Dimensional Diffusion in Cylindrical and Spherical Geometry	243
7.3.6	Multidimensional Diffusion	245
	References 246	

8 Solution of the Diffusion Equation for Multilayer Packaging   247
Valer Tosa and Peter Mercea

8.1	Introduction	247
8.2	Methods for Solving the Diffusion Problem in a Multilayer (ML) Packaging	248
8.3	Solving the Diffusion Equation for a Multilayer Packaging in Contact with a Foodstuff	251

8.4	Development of a User-Friendly Software for the Estimation of Migration from Multilayer Packaging *256*	
	References *261*	
9	**User-Friendly Software for Migration Estimations** *263*	
	Peter Mercea, Liviu Petrescu, Otto Piringer and Valer Tosa	
9.1	Introduction *263*	
9.2	MIGRATEST©Lite – A User-Friendly Software for Migration Estimations *266*	
9.2.1	Basic Features of MIGRATEST©Lite and Input Data Menus *266*	
9.2.2	Estimation of Migration with MIGRATEST©Lite *276*	
9.2.3	Output Information Delivered by MIGRATEST©Lite *278*	
9.2.4	Case Examples Computed with MIGRATEST©Lite *278*	
9.2.5	Migration Estimations with the MIGRATEST©EXP Software *281*	
9.2.6	Case Examples Computed with MIGRATEST©EXP *287*	
	References *296*	
10	**Permeation of Gases and Condensable Substances Through Monolayer and Multilayer Structures** *297*	
	Horst-Christian Langowski	
10.1	Introduction: Barrier Function of Polymer-Based Packaging *297*	
10.2	Permeation Through Polymeric Materials *302*	
10.2.1	Substance Transport Through Monolayer Polymer Films *303*	
10.2.2	Substance Transport Through Multilayer Polymer Films (Laminates) *305*	
10.2.3	Units for Different Parameters *307*	
10.3	Substance Transport Through Single and Multilayer Polymer Substrates Combined with One Inorganic Barrier Layer *307*	
10.3.1	Numerical Modeling *307*	
10.3.2	Simplification: Barrier Improvement Factor *311*	
10.3.3	Multilayer Polymer Substrates Combined with One Inorganic Layer *313*	
10.3.4	Polymer Substrates Combined with an Inorganic Barrier Layer and Other Polymer Layers on Top of the Inorganic Layer *314*	
10.3.5	Temperature Behavior of the Structures Shown Above *316*	
10.3.6	Substance Transport Through Thin Polymer Layers Having Inorganic Layers on Both Sides *317*	
10.5	Substance Transport Through Polymers Filled with Particles *320*	
10.6	Experimental Findings: Polymer Films and One Inorganic Barrier Layer *321*	
10.6.1	Structures and Defects in Inorganic Barrier Layers on Polymer Substrates *323*	

10.6.2	Comparison of Model Calculations and Experimental Results for Combinations of Polymer Films and One Inorganic Barrier Layer *324*	
10.6.3	Apparent Additional Transport Mechanisms for Water Vapor *327*	
10.6.4	Properties of Systems with at least One Inorganic Layer Embedded Between to Polymer Layers or Films *332*	
10.7	Experimental Findings: Combinations of Polymer Films and More Than One Inorganic Barrier Layer *332*	
10.8	Experimental Findings: Polymers Filled with Platelet-Shaped Particles *333*	
10.9	Experimental Findings: Permeation of Flavors Through Mono- and Multilayer Films and Combinations with Inorganic Barrier Layers *338*	
10.10	Conclusions *342*	
	References *342*	
11	**Migration of Plastic Constituents** *349*	
	Roland Franz and Angela Störmer	
11.1	Definitions and Theory *349*	
11.1.1	Migration, Extraction, and Adsorption *349*	
11.1.2	Functional Barrier *350*	
11.1.3	Legal Migration Limits and Exposure *350*	
11.1.4	Parameters Determining Migration *352*	
11.2	Indirect Migration Assessment *354*	
11.2.1	Worst-Case (Total Mass Transfer) Assumption *355*	
11.2.2	General Considerations: Taking Solubility and/or Low Diffusivity of Certain Plastics into Account *357*	
11.2.3	Migration Assessment of Mono- and Multilayers by Application of Complex Mathematical Models *359*	
11.2.4	Multilayers *359*	
11.3	Migration Experiment *361*	
11.3.1	Direct Migration Measurement in Conventional and Alternative Simulants *361*	
11.3.2	Accelerated Migration Tests: Alternative Migration Tests *362*	
11.3.3	Choice of Appropriate Test Conditions *365*	
11.3.3.1	Food Simulants *365*	
11.3.3.2	Time–Temperature Conditions *367*	
11.3.3.3	Surface-to-Volume Ratio *369*	
11.3.3.4	Migration Contact *370*	
11.4	Analysis of Migration Solutions *372*	
11.4.1	Overall Migration *372*	
11.4.1.1	Aqueous and Alternative Volatile Simulants *372*	
11.4.1.2	Olive Oil *372*	
11.4.1.3	Modified Polyphenylene Oxide (Tenax®) *373*	
11.4.2	Specific Migration *374*	
11.4.2.1	Vinyl Chloride EU Directives *374*	

11.4.2.2	EN 13130 Series	374
11.4.2.3	Further Standard Methods	375
11.4.2.4	Methods of Analysis in Petitions to the European Commission	376
11.4.2.5	Methods in Foods (Foodmigrosure Project)	377
11.5	Development of Methods, Validation, and Verification	378
11.5.1	Establishing (Juristically) Valid Performance of Methods	378
11.5.2	A Practical Guide for Developing and Prevalidation of Analytical Methods	380
11.5.3	Validation Requirements for EU Food Contact Petitions and US FDA Food Contact Notifications	387
11.5.4	Determination of the Detection Limit	387
11.5.5	Analytical Uncertainty	389
11.5.6	Use of the Precision Data from Fully Validated Methods	390
11.6	Sources of Errors	394
11.6.1	Highly Volatile Migrants	394
11.6.2	Reaction with Food/Simulant Constituents	395
11.6.3	Migrants in Reactive Processes (e.g., Primary Aromatic Amines from Adhesives)	397
11.7	Migration into Food Simulants in Comparison to Foods	400
11.8	Consideration of Non Intentionally Added Substances (NIAS) and Other not Regulated Migrants	407
	References	409
12	**US FDA Food Contact Materials Regulations**	**417**
	Allan Bailey, Layla Batarseh, Timothy Begley and Michelle Twaroski	
12.1	Introduction	417
12.2	Regulatory Authority	417
12.2.1	Federal Food, Drug and Cosmetic Act (FFDCA)	417
12.2.2	National Environmental Policy Act (NEPA)	421
12.3	Premarket Safety Assessment	422
12.3.1	Introduction	422
12.3.2	Chemistry Information	422
12.3.2.1	Migrant Levels in Food	423
12.3.2.2	Packaging Information	423
12.3.3	Toxicology Information	426
12.3.3.1	Safety Assessment	427
12.3.3.2	General Considerations	431
12.3.4	Environmental Information	432
12.3.4.1	Claim of categorical exclusion	432
12.3.4.2	Environmental Assessment (EA)	434
12.3.4.3	Polymeric Food Packaging Materials	435
12.3.4.4	Inadequacies in EAs	436
12.4	Final Thoughts	437
12.5	Conclusions	438
	References	438

13	**Community Legislation on Materials and Articles Intended to Come into Contact with Foodstuffs** 441
	Luigi Rossi
13.1	Introduction 441
13.2	Community Legislation 442
13.2.1	Directives/Regulations Applicable to all Materials and Articles 442
13.2.1.1	Framework Directives/Regulation 442
13.2.1.2	Regulation on Good Manufacturing Practice 445
13.2.2	Directives Applicable to One Category of Materials and Articles 446
13.2.2.1	Directive on Regenerated Cellulose Film 446
13.2.2.2	Directive on Ceramics 447
13.2.2.3	Directive on Plastics Materials 448
13.2.2.4	Field of Application 448
13.2.2.5	EU List of Authorized Substances 449
13.2.2.6	Restricted Use of Authorized Substances (OML, SML, QM, and QMA) 450
13.2.2.7	Authorization of New Substances 451
13.2.2.8	Directives on the System of Checking Migration 452
13.2.2.9	Functional Barrier 454
13.2.2.10	Fat (Consumption) Reduction Factors 455
13.2.2.11	Declaration of Compliance 456
13.2.2.12	Specific Rules for Infants and Young Children 457
13.2.2.13	Special Restrictions for Certain Phthalates now Authorized at EU Level 457
13.2.2.14	Simulant for Milk and Milk Products 458
13.2.2.15	Other Complementary Community Initiatives 458
13.2.3	Directives Concerning Individual or Groups of Substances 459
13.2.3.1	Directives on Vinyl Chloride 459
13.2.3.2	Directive on MEG and DEG in Regenerated Cellulose Film 459
13.2.3.3	Directive on Nitrosamines in Rubber Teats and Soothers 459
13.2.3.4	Regulation on the Restriction of Use of Certain Epoxy Derivatives 459
13.2.3.5	Directive on the Suspension of the Use of Azodicarbonamide as Blowing Agent in Plastics 460
13.2.3.6	Regulation on Some Plasticizers in Gaskets in Lids 460
13.3	National Law and European Mutual Recognition 460
13.3.1	Future Commission Plans 462
13.4	National Legislations and Council of Europe Resolutions 462
13.5	Conclusions 462
14	**Packaging Related Off-Flavors in Foods** 465
	Albert Baner, Francois Chastellain and André Mandanis
14.1	Introduction 465
14.2	Sensory Evaluation 466
14.3	Identification of Off-Flavor Compounds 468

14.4	Physical Chemical Parameters Determining Off-Flavors	*469*
14.5	Derivation of Threshold Concentrations of Sensory-Active Compounds	*474*
	References	*494*

15 Possibilities and Limitations of Migration Modeling *499*
Peter Mercea and Otto Piringer

15.1	Correlation of Diffusion Coefficients with Plastic Properties	*501*
15.2	The Partition Coefficient	*511*
	References	*521*

Appendices *523*

Appendix I *525*
Peter Mercea
References *552*

Appendix II *557*
References *589*

Appendix III *591*
A Selection of Additives Used in Many Plastic Materials *591*

Index *607*

Preface

This second edition confirms that in an active field like the interaction of plastic packaging with food and pharmaceuticals such a book can only be a work in progress. In just seven years after the first edition was published enough significant new research and learning has taken place to necessitate an update. Most chapters in the second edition have been rewritten to reflect advances in the estimation of physical and chemical interaction parameters like diffusion and partition coefficients as well as new developments and methods for estimating diffusion and migration with user friendly software. We have also incorporated new state of the art material on permeation, migration testing, regulatory development and off-flavors.

The goal of the second edition remains to provide a practical and accessible treatment of plastic packaging interactions with food and pharmaceuticals that fills the gap between the many general food packaging books and ones that are very mathematical and theoretical.

The interaction between plastics and foods and pharmaceuticals remains a very active field and recent trends continue to shape research and development in this area. This makes it more important than ever to understand the interactions between food/pharmaceuticals and plastic packaging as plastic packaging usage and range of application continues to expand every year.

At the same time more plastics are being used there is increased regulatory scrutiny of all chemicals in general including those used in food packaging. No longer are chemicals being monitored and regulated just on the bases of their acute and chronic toxicities but also their environmental and endocrine disruption activities at trace levels far below those previously evaluated. Regulatory activity and chemical monitoring has been made easier and more widely available by affordable and increasingly powerful analytical techniques with the ability to detect ever lower levels of substances. These factors amount to more stringent control and monitoring of potential migration of substances from plastic packaging into foods and pharmaceuticals. Economic trends such as the growing global trade in packaging where some packaging is coming from markets with little food packaging chemical safety regulation and/or enforcement requires increased vigilance and monitoring of packaging sourced from these areas.

All these regulatory and economic trends are against a backdrop of increasing environmental, health and safety awareness among media savvy consumers. Today's consumers have many product choices available to them and consumer product companies must ensure the quality and safety of their food products or risk losing their business. Finally, there is an increasing desire for companies and consumers alike to operate and live in a more sustainable manner so that both are looking for ways to reduce, recycle and reuse plastic packaging and to substitute traditional petrochemical based plastics with newer biopolymer based plastics. All of these trends make an understanding of the interactions between plastics and food and pharmaceuticals critical to their optimal use and safety as packaging materials.

This book is surely not the last word on the subject of plastic packaging interaction and largely reflects the point of view of its authors. We do hope that this work will be of practical use to people concerned with plastic packaging interactions as well as providing a starting point and stimulation for continued research in this field.

December 2007

A. L. Baner
St. Louis, MO (USA)

O. G. Piringer
Munich, Germany

List of Contributors

Allan B. Bailey
Food and Drug Administration
Center for Food Safety and Applied Nutrition
5100 Paint Branch Parkway
College Park
Maryland 20740
USA

Albert L. Baner
Nestle Purnia Petcare PTC
Checkerboard Square
St. Louis, MO 63164
USA

Layla Batarseh
Food and Drug Administration
Center for Food Safety and Applied Nutrition
5100 Paint Branch Parkway
College Park
Maryland 20740
USA

Timothy H. Begley
Food and Drug Administration
Center for Food Safety and Applied Nutrition
5100 Paint Branch Parkway
College Park
Maryland 20740
USA

Titus A. Beu
University Babes-Bolyai
Faculty of Physics
Kogalniceanu 1
3400 Cluj-Napoca
Romania

Johannes Brandsch
Fabes Forschungs-GmbH
Schragenhofstraße 35
80992 München
Germany

Francois Chastellain
Nestec S.A.
Avenue Nestle 55
1800 Vevey
Switzerland

Roland Franz
Fraunhofer-Institut für Verfahrenstechnik und Verpackung
Abteilung Produktsicherheit und Analytik
Giggenhauserstraße 35
85354 Freising
Germany

Plastic Packaging. Second Edition. Edited by O.G. Piringer and A.L. Baner
Copyright © 2008 WILEY-VCH Verlag GmbH & Co. KGaA, Weinheim
ISBN: 978-3-527-31455-3

List of Contributors

Horst-Christian Langowski
Technische Universität München
Wissenschaftszentrum Weihenstephan
für Ernährung, Landnutzung und
Umwelt
Weihenstephaner Steig 22
85350 Freising-Weihenstephan
Germany

André Mandanis
Nestle Research Center
Vers-chez-les-Blanc
1000 Lausanne 26
Switzerland

Peter Mercea
Filderstraße 17
70180 Stuttgart
Germany

Stanislav Nešpůrek
Academy of Sciences of the Czech
Republic
Institute of Macromolecular Chemistry
Department of Polymer Materials
Heyrovsky Square 2
162 06 Prague 6
Czech Republic

Liviu Petrescu
Fabes Forschungs GmbH
Schragenhofstraße 35
80992 München
Germany

Otto Piringer
Fabes Forschungs GmbH
Schragenhofstraße 35
80992 München
Germany

Jan Pospíšil
Academy of Sciences of the Czech
Republic
Institute of Macromolecular Chemistry
Department of Polymer Materials
Heyrovsky Square 2
162 06 Prague 6
Czech Republic

Luigi Rossi
European Commission
Office 4/17
Rue de la Loi (B232 4/17)
1049 Brussels
Belgium

Angela Störmer
Fraunhofer-Institut für
Verfahrenstechnik und Verpackung IVV
Giggenhauserstraße 35
85354 Freising
Germany

Valer Tosa
National Institute for Research and
Development of Isotopic and
Molecular Technologies
P.O. Box 700
400293 Cluj-Napoca
Romania

Michelle Twaroski
Food and Drug Administration
Center for Food Safety and Applied
Nutrition
5100 Paint Branch Parkway
College Park
Maryland 20740
USA

1
Preservation of Quality Through Packaging
Albert Baner and Otto Piringer

Plastics are defined as processable materials based on polymers. These materials can be transformed into finished products, such as bottles, containers, films, hoses, coatings, lacquers, etc. As a result of today's multitude of plastic applications there is a corresponding enormous variety of plastic materials. The polymer matrix as well as the incorporated plastic additives can be made to differ in such a variety of ways with respect to their chemical composition and structure that one finds or can develop a tailor made product for every application.

Packaging is one major field of application for plastic materials. The development of self-service stores with their large variety of products is unimaginable without plastics. The most important function of a packaging material is the quality preservation of the packed goods. Among these goods, foods hold a place of special importance due to their principal chemical instability. This instability is also the characteristic for other products containing active substances, in particular pharmaceuticals.

In order to fulfill the task of quality assurance of the packed product with minimal impact both on the product and on the environment, the packaging must be optimized by taking into consideration various criteria. This book provides assistance in package optimization functions. Special emphasis is given for mass transport between plastic materials and packed goods and the consequences of such interaction for quality assurance and legislation.

1.1
Quality and Shelf-Life

Products being offered on the market can, thanks to the currently available manufacturing and preservation methods as well as the various transportation modes, come from all regions of the country, continent and other continents together.

Many products consist of numerous ingredients which have a relatively low chemical stability. Such labile goods are exposed to numerous spoilage possibilities and one of the most important factors leading to longer shelf-lives is their packaging.

Plastic Packaging. Second Edition. Edited by O.G. Piringer and A.L. Baner
Copyright © 2008 WILEY-VCH Verlag GmbH & Co. KGaA, Weinheim
ISBN: 978-3-527-31455-3

In order to describe what a product shelf-life is or what it means in terms of quality retention and measurement, the word "quality" must be defined. Whatever from a legal standpoint in different countries is used as definition, the quality (Q) determining properties of a product are in principle based on the product's components. Thus Q can be described as a function of the chemical composition of the product:

$$Q = f(c_1, c_2, \ldots, c_i, \ldots, c_n) \tag{1.1}$$

Let c_i designate the concentration of a specific component i in the product and n the number of different components. If Q_j is defined as a function of the concentration of component i, then the change in quality ΔQ_j over the time interval Δt becomes a function of the concentration change Δc_i in this time interval. In this case it is not necessary to know the change in concentration of all n ingredients and their change with time. If for example the change in concentration with time of ingredient i can be measured, then maybe this variation can be correlated with a quality change (Figure 1.1). Even though at constant concentrations (curve 1) there is no quality change taking place with respect to i, an increase in concentration (curve 2), for example resulting from mass transport of a plastic component into the product, leads to quality loss. There are of course cases where an increase in ingredient concentration during storage can lead to improvement in quality, for example, during the ripening processes of cheeses or alcoholic beverages. A reduction in quality also takes place through the loss of an ingredient (curve 3), for example diffusion of aromatic compounds through the packaging and into the atmosphere.

For various product ingredients or undesirable foreign substances, limits can be assigned (shaded field in Figure 1.1) outside which a significant quality reduction can occur compared to the initial quality. The importance of individual ingredients for product quality can vary considerably and therefore also the width of the allowable concentration. The importance and allowable concentration range are determined by the component's chemical structure.

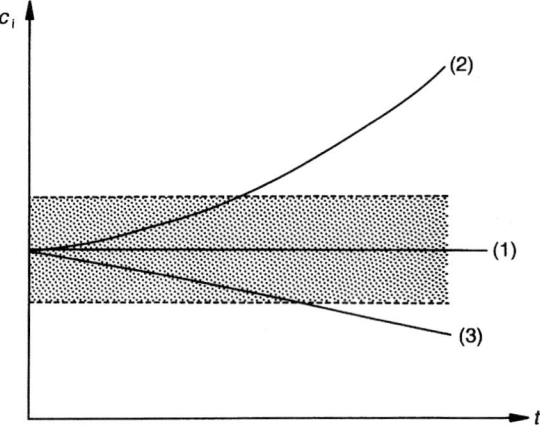

Figure 1.1 Concentration variation with time of food ingredients.

Quality preservation through packaging means therefore to maintain as long as possible a particular concentration c_i within a certain value range. The time interval within which the product quality remains completely unchanged can be very short. It is therefore more important in practice to define the shelf-life over a time interval up until the limit where the most important product quality characteristics just still remain. This means, amongst other things, that during this time in the product neither undesirable compounds that have health significance nor odor or taste is allowed to occur. This requirement has two important consequences: first, the necessity of an objective quality evaluation for changes in quality and second, the adaptation of packaging to this requirement resulting from the product shelf-life. The solution of both problems has to meet the legal requirements.

The quality requirements as well as requirements derived from them are subject to change over time. Besides objective criteria that result from technical advances there are also subjective, political, and media generated emotional criteria that also play important roles.

One goal of the present technological development is the production of food that still possesses as many quality attributes of the raw materials as possible. This leads to products which may still contain many naturally occurring, chemically unstable materials that are preserved by gentle processing methods. These types of product require a much higher initial quality compared to other foods manufactured or treated under harsher processing conditions. One consequence is that it is possible to have a more rapid quality decrease for the product with high initial value (1) than for the product with lower initial quality (2), both having ideal packaging (Figure 1.2). Q_{min} designates the minimal acceptable quality where sufficient or adequate product-specific characteristics are still maintained.

Unpackaged products show a faster time-related quality decrease (left sides of the hatched triangles) than ideally packaged products (right sides of triangles). The area

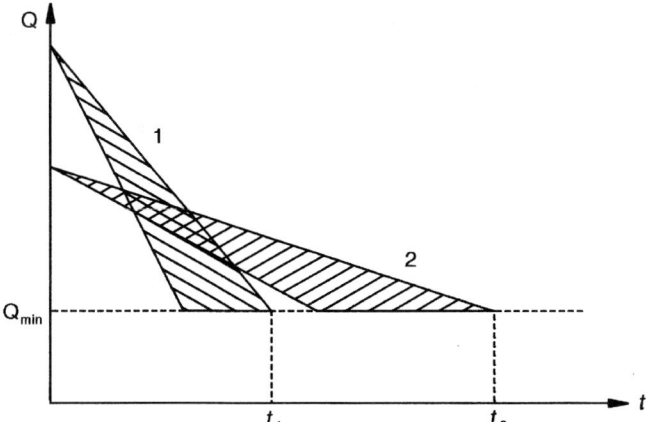

Figure 1.2 Quality loss over time of two foods (1 and 2) each having different initial quality. The left straight line shows the unpackaged and the right one an ideally packaged condition.

that can be influenced by packaging lies inside the hatched field for a given product. The straight line representation is a simplification because quality losses do not necessarily have to be linear. The conclusions are thus in some cases foods having high initial quality but shorter shelf-life can use lower quality packaging than products with lower initial quality and longer shelf-lives.

The shelf-life of milk is given here as an example. For the 6- to 7-day stability of high quality fresh pasteurized milk a relatively simple polyethylene coated carton package is satisfactory. However, the much longer stability of a lower quality aseptic milk requires a sophisticated package that includes for example an additional barrier layer.

There are of course areas in which a very long shelf-life is preferred over a high initial quality product. Examples of this are the establishment and maintenance of emergency reserves and the supplying of remote regions, some of which having high temperatures. The packaging requirements for these cases are particularly high. In general however, the trend today arising from higher product quality consciousness is away from product "mummification" and toward "fresh" appealing goods.

1.2
Physical and Chemical Interactions Between Plastics and Food or Pharmaceuticals

If one has knowledge of specific sensitivities of a food or the properties of another product, one can derive the necessary packaging requirements. The most essential requirement today compared to previous requirements is the simultaneous optimization with respect to several criteria. For example, these optimization criteria could include a protective function, material and energy expenditures during manufacture, as well as disposability and other environmental considerations. Such optimization is always a compromise between different solutions which can lead to the appearance of new problems. With reference to several criteria, optimization generally means the reduction of safety margins in reference to a certain criterion. Fulfilling for example the criterion of packaging minimization, the permeability is increased to the allowable maximum, that may mean that exceeding or falling short of a packaging specification value by even a small amount might lead to a significant change in the quality of a packaged product.

In future package development, optimization from an ecological viewpoint will play an especially important role and minimization of packaging will help make this possible. One should never forget however that quality assurance of the packaged product and therefore the guarantee of consumer safety will always have priority and must remain the most important criterion for optimization. The fulfillment of these requirements assumes complete knowledge of possible interactions between packaging and product during their contact time. In this respect the properties of both parts of package, the packaging material and the product, must be coordinated with one another. Here possible interactions between the two parts play an important role in the quality assurance of the product.

The term interaction encompasses the sum of all mass transports from the package into the product as well as mass transport in the opposite direction (Figure 1.3). The

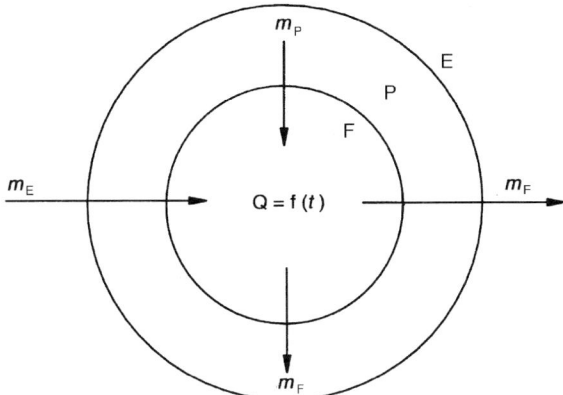

Figure 1.3 Mass transports in the packaged food. m_E and m_P represent the mass transport from the environment E of the package P and from P into the food F. m_F represents the mass transfer from F into P and E.

mass transfers, often coupled with chemical reactions, lead to quality, Q, changes in the product and packaging material.

Mass transport is understood to mean the molecular diffusion in, out and through plastic materials like that shown schematically in Figure 1.3. This figure represents most applications where there is a layer of plastic material separating an external environmental medium from an inner product medium. The product can be a sensitive medium with a complex chemical composition, e.g., food, that must be protected from external influences such as oxygen and contaminants. It can also be an aggressive chemical that must not escape into the surrounding environment. Because this plastic material barrier layer usually includes low molecular weight substances incorporated into the polymer matrix, there are many applications in which the transport of these substances into the product and environment must be minimized.

The mass transport of package components to the product is known as migration, and the mass transport of product components to the package is known as scalping. Permeation means the mass transport of components through the package in both directions.

1.3
The Organization of this Book

Chapter 2

The goal of Chapter 2 is to draw the attention of the reader to the enormous variety of plastic materials which results from their different chemical structures, synthetic routes, and contents of additives. The permeability of plastics to low molecular compounds is often uniquely obtained via specific processing routes. In order to select the appropriate material for specific applications some knowledge about the chemical composition, structure, and corresponding properties of the plastics is necessary. Knowledge of composition is also an indispensable requirement for successful search of potential migrants from plastic materials into a product with the aim of its quality assurance. The many chemical compounds mentioned in this chapter may help to identify an analytically separated compound as a rest monomer, oligomer, degradation product, or impurity of the investigated plastic.

From a short overview of the principal manufacturing procedures, the raw materials and processing aids, much useful information can be obtained especially concerning the permeability (functional barrier properties) of the plastic and potential migrants having relevant toxicological or sensorial properties.

Despite the enormous number of potential starting substances in practice only a finite number of basic polymeric structures with well-defined interaction and transport processes form the majority of practical applications. This is of great help for making theoretical estimations of transport properties. Nevertheless, it must not be forgotten that even for well-defined basic structures, e.g. polyethylene, there are hundreds of grades of polyethylenes available which differ more or less in their composition and structure. As a consequence the transport properties (diffusion coefficients of low molecular components) tend to scatter around an average value even for a well-defined plastic material over a more or less wide range, as can be seen from the values listed in Appendix I. This means one can predict transport processes and partition behaviors of such materials only within a limited range of precision. But this precision improves rapidly as more details about the composition and structure of the materials are known.

Chapter 3

The characteristic functions and the representative structures of plastic additives used to make marketable and durable materials are included in this chapter. In comparison to the polymeric matrix, the additives are in general low molecular compounds and the stabilizers in particular are much more reactive than polymers.

Due to the high reactivity of the important category of stabilizers, many reactions can occur in the polymeric matrix. As a result a variety of degradation products appear, a fraction of which are able to migrate into the product in contact with plastic while a fraction can remain immobilized in the polymer matrix. Both the chemical nature of the degradation products and their concentrations are of great importance for the quality assurance of the product in contact with plastic. Estimation of migration of the additives themselves or their degradation products is possible only if the mass balance of these products can be predicted or measured and their chemical nature known.

The formation of various transformation products from the stabilizers cannot be avoided. The sacrificial fate of stabilizers is part of their activity mechanism which is providing protection to plastics against degradation. In elucidating transport phenomena in commodity polymers, the presence of combinations of stabilizers along with varying amounts of their transformation products with sometimes very different molecular parameters, has to be taken into account.

Chapter 4

One of the two fundamental material constants which govern the mass transfer of a compound between two contacting phases, e.g., a plastic P in the liquid L or gas G, is the partition coefficient of the compound between the two phases. This chapter deals with the thermodynamic basis fundamentals of partition and some of the methods that can be used to estimate its magnitude.

Different estimation methods based on additive molecular properties are described. The oldest and best known treatment is based on the so-called regular solution theory. This method although widely used qualitatively has a very limited application range for quantitative calculations. Methods for estimating partitioning of almost any chemical structure based on structural increments (group contributions) are commonly used in chemical engineering. UNIFAC, one of the oldest and most comprehensive methods that can be used for polymers, is presented here as a typical example. Due in part to the extremely large variation range of partition values extending over many orders of magnitude, the precision of estimation with the UNIFAC method is in general within one order of magnitude which is sufficient for most practical applications. A serious drawback of the method is its rather complicated handling requiring programmable calculators or computer programs.

Estimation using quantitative structure activity relationship (QSAR) and quantitative structure property relationship (QSPR) is a field of computational chemistry.

In order to offer a simple procedure for practical applications an additional new method, the vapor-pressure-index (VPI) estimation method, is introduced. This method is easy to use and with the linear relationships and data from Appendices II and III the partition coefficients, especially between ethanol/water mixtures and polyolefins, can be estimated with reasonable agreement with experimental values.

Chapter 5

In addition to the partition coefficients discussed in the preceding chapter, the second fundamental material constant which governs the mass transfer of a compound i from a plastic P into a liquid L or gas G is the diffusion coefficient $D_{P,i}$ of i in the matrix of P. A brief review of the most frequently cited and used models for diffusion in polymers is presented in this chapter. The chapter discusses some "classical" approaches for analyzing and quantifying diffusion processes in polymers. It is pointed out that although some of these models can lead to quite remarkable agreements between theory and experiment, none of them is a truly predictive diffusion model.

A review is given of the more recent computational approaches describing the process of diffusion in polymers and the $D_{P,i}$ values estimated from them. These approaches have a true *ab initio* predictive character. At the same time these models are not yet capable of estimating diffusion coefficients for the complex polymer–migrant systems usually found in food packaging applications.

Chapter 6

An original deduction of an equation for diffusion coefficients of substances in plastic materials is presented in this chapter. The development uses a uniform model which is applicable to all aggregation states. One goal of this chapter is to demonstrate the reasonable agreement between calculated and measured diffusion coefficients in gases, liquids, and solids, with special emphasis of plastics. The model is based on assumptions about interaction of the molecules in a macroscopic system, starting in its critical state. The only needed parameters for estimations are the critical temperature, critical volume, and critical pressure of the compounds involved, as well as the glass and melting temperatures and molecular weights of the plastic matrices.

Chapter 7

The starting point for a mathematical treatment of all specific cases of interactions between packaging and product is a general mass transport equation. This partial differential equation has analytical solutions only for special cases. For solutions involving complicated cases, simplifying approximations are used or numerical solutions are carried out.

In order to understand the literature on this subject it is necessary to know how the most important solutions are arrived at, so that the different assumptions affecting the derivation of the solutions can be critically evaluated.

The selection of different equation solutions included here are diffusion from films or sheets (hollow bodies) into liquids and solids as well as diffusion in the reverse direction, diffusion controlled evaporation from a surface, influence of barrier layers and diffusion through laminates, influence of swelling and heterogeneity of packaging materials, coupling of diffusion and chemical reactions in filled products as well as permeation through packaging.

Despite the large number of analytical solutions available for the diffusion equation, their usefulness is restricted to simple geometries and constant diffusion coefficients. However, there are many cases of practical interest where the simplifying assumptions are introduced when deriving analytical solutions are unacceptable. This chapter also gives an overview of the most powerful numerical methods used at present for solutions of the diffusion equation.

Chapter 8

The principles of a numerical method to solve the diffusion equation for a monolayer packaging in contact with a liquid F are presented in Section 7.2. In the following this topic will be extended to the one-dimensional (1D) diffusion problem for multilayer (ML) materials in contact with various types of foods. In this respect a brief presentation of the main numerical approaches developed to solve this mass transfer problem will be made. Then the presentation will be focused on a numerical method developed to solve the transport equations for a ML packaging in contact with any type of homogenous foodstuffs, F. This method is based on a finite difference technique and was developed in 1D for the general case in which the transport processes are controlled not only by the diffusion coefficients (D_i) in the packaging and foodstuff (D_F) but also by partition coefficients (K_{ij}) between any two adjacent layers i and j of the packaging as well as partition coefficients between the packaging and foodstuff (K_{pf}). The numerical algorithms of this FD method were then implemented into a computer program which can be run on a regular PC. A major concern with this computer program was to check if it produces correct results. For this a series of test were designed/conducted and the results are presented.

Chapter 9

The application of the methods described in Chapters 7 and 8 in practice needs the use of adequate computer programs (software). Two user-friendly programs developed by FABES GmbH are described in this chapter.

The aim in developing MIGRATEST©Lite was to provide to a large spectrum of potential users from industry and research and development as well as from the enforcement laboratories a user-friendly tool for a quick and easy estimation of migration of substances from plastic (polymeric) films into foods and food simulants. A special emphasis was to conceive the software in such a manner to include the actual aspects and data from the EU documents related to migration regulation.

The second program for migration estimations is MIGRATEST©EXP. This user-friendly software is based on a numerical solution of the differential migration equation as described in Chapter 8. A series of examples from practice are described together with the principal operation steps. This program is especially adapted for multilayer plastics in contact with liquid, highly viscous and solid products.

Chapter 10

Sensitive foods and encapsulated technical products are generally sensitive to their surroundings, in particular to oxygen and water vapor. As the permeabilies of favorably priced commodity polymers (for food packaging) and also more expensive specialty polymers (for encapsulation of technical devices) to water vapor, oxygen, and other substances are far too high for most applications, a thorough understanding of the permeation processes is essential. To improve the barrier properties of single polymers, the following strategies are pursued: production of polymeric multilayered structures under inclusion of barrier polymers, multilayered structures which incorporate one, sometimes even several inorganic layers and hybrid structures where polymeric matrices are filled with inorganic particles.

For purely polymeric monolayer and multilayer structures, the permeation process can be represented mathematically via the one-dimensional form of the related transport equations.

In the case of nontrivially shaped inorganic particles, single or multiple thin inorganic layers and even thicker inorganic foils, all of them embedded in polymeric matrices, the whole three-dimensional geometry of the samples has to be taken into account on the microscopic scale: for particles incorporated in polymers, their size, shape and orientation have to be regarded. For inorganic layers, numbers of inevitable defects and their size distribution play the decisive role.

All these parameters can be combined to specific geometry factors, which, in combination with the coefficients of diffusion and solution of the polymeric matrix or of the polymeric substrate film and optional further polymeric top layers, determine the final permeation properties. This concept has been verified in many different cases, mostly for oxygen as the permeating substance.

An exception from this concept occurs when the permeation of condensable substances such as water vapor or flavors is involved. In such cases, much higher values of permeability are often observed than to be expected from the considerations mentioned above. It is to be expected – although it has still not been proven unambiguously – that substances may condense in inorganic structures, leading to much higher local concentrations and thus to a higher permeability.

In most technically relevant cases, however, especially in the case of the favored polymeric multilayers, the concept described in this chapter gives a sufficiently accurate quantitative description of the transport phenomena.

Chapter 11

This chapter provides a critical review of modern food packaging migration testing by addressing both the test requirements and their availability and the practicality of different migration assessment schemes and analytical methods. In order to enable the reader to select and tailor his own specific migration test approach the first section of Chapter 11 starts with an introduction to the principles of migration testing and the primary factors controlling migration processes. After that an efficient schematic for food law compliance testing is presented covering modern indirect, semidirect, and direct migration tests. A major focus in the second section is given to the analytical aspects of specific migration testing. After discussion of the general requirements for analysis methods, detailed and practical guidance is given on how to develop, validate, and document analytical methods that are suitable for compliance testing that fulfills food contact material legal requirements. An overview of existing methods currently used in Europe provides the necessary information to complete this topic. Additionally, practical examples of specific migration test methods are given along with their related difficulties and specific problems.

An important aspect discussed in this chapter is the recent results obtained with an EU project concerning migration into food.

Chapter 12

During the past decade regulatory processes in the United States have changed significantly with regard to components of food contact materials, allowing industry a variety of options for obtaining authorization for their safe use. A thorough understanding of the US regulatory processes for the substances, presently referred to as food contact substances (FCS), allows industry to determine the most appropriate regulatory option based on the intended use. In this chapter the US

Food and Drug Administration's (FDA) regulatory authority and premarket safety evaluation of FCS is discussed.

FDA's safety assessment relies on evaluating probable consumer exposure to an FCS, including all constituents or impurities, as a result of the proposed use and other authorized uses, and ensuring that such exposures are supported by the available toxicological information. A general discussion of the recommended chemistry, toxicology, and environmental information for a submission relating to an FCS follows.

FDA's approach to the safety assessment of the substances is exposure driven, in that it is specific to the intended use and the resultant dietary exposure, which determines the amount of toxicological data consistent with the tiered requirements.

Chapter 13

In order to harmonize the legislation in the European Community a broad program of action started in 1972. The Community legislation has established rules for the most complex and important area of packaging, that of plastic materials. The Commission of the Community is currently preparing a series of texts which should make it possible for legislation on plastics to be fully harmonized at the Community level by the year 2010.

This chapter describes the main aspects of current Community legislation on materials and articles intended to come into contact with foodstuffs.

Chapter 14

Off-flavors represent one of the major quality issues in the food industry and can result in significant economic damage to a company. Even if they may not represent any health risk, they can seriously damage the quality image of a brand, and the confidence that the consumer has in it. By law, for example in the European Union as well in the United States, packaging materials or substances transferring from the package to the foodstuff are not allowed to impart unacceptable changes to the organoleptic characteristics of the packed foodstuffs.

This chapter focuses on the off-flavors which are associated with packaging materials. Because of the number of raw materials, additives, adhesives, inks, solvents, and other chemicals used in the food packaging industry, and the number of suppliers/converters implicated in the manufacture of finished printed materials, many different sources of contamination are possible. The origin of the problems can be divided into three main categories: migration of odorous substances from package to food and to package headspace; inadequate protection of food from environmental influences; reaction of substances in packaging material with each other or with food components.

Sensory evaluation remains the most widely used method to assess the sensory quality of packaging materials. It represents the starting point in off-flavor investigations.

Sensory analysis is also a starting point for subsequent analytical work to identify the cause of the off-flavor and for taking corrective action. Since many different types of off-flavor contamination exist, specific and accurate descriptors are needed to characterize the problem with precision at an early stage of the off-flavor investigation.

Chapter 15

Worldwide investigations over the last 50 years have demonstrated that interactions between plastics and foodstuffs or other products occur as foreseeable physical processes. Standardization of migration measurements is based on this knowledge. However, the variety of substances occurring in interaction processes and the necessary time and cost requirements to carry out all the analysis for a complete quality assurance for consumer safety necessitate additional tools in order to fulfill this task. Such a tool is a recently finished EU Project for evaluation of migration models in support of Directive 2002/72/EC.

Beyond the characterization of the polymer and food (simulant), the key input parameters for the use of a migration model are the diffusion coefficient, D_P, of the migrant in the plastic material P, as well as the partition coefficient, $K_{P/L}$, of the migrant between P and the product (e.g. food, liquid) L. As already shown in Chapter 6 a considerable improved estimation of diffusion coefficients is now possible.

An improved estimation method for partition coefficients described in Chapter 4 has been found especially useful for migration modeling of additives from plastics into various products. The use of certain solvents as food simulants for controlling migration from plastics is widely practiced and allowed by food regulations. Nevertheless, there is a real danger in some cases when the food simulant has a significantly lower migration value compared to the food. Finding the correct food simulant is of great practical importance. In principle, ethanol/water mixtures are very appropriate for this application.

Possibilities and limitations of the actual knowledge in these fields are discussed.

Further Reading

Asby, R., Cooper, I., Harvey, S. and Tice, P. (1997) *Food Packaging Migration and Legislation*, Pira International, Leatherhead.

Barnes, K.A. and Sinclair, C.R. (eds) (2007) *Chemical Migration and Food Contact Materials*, CRC Press, Woodhead Publishing Limited, Cambridge.

Bart, J.C.J. (2005) *Additives in Polymers. Industrial Analysis and Applications*, Wiley, New York.

Brody, A.L. and Marsh, K. (eds) (1997) *The Wiley Encyclopedia of Packaging Technology*, 2nd edn., Wiley, New York.

Brown, W.E. (1992) *Plastics in Food Packaging: Properties, Design, and Fabrication*, Dekker, New York.

Brydson, J.A. (1995) *Plastics Materials*, Butterworth-Heinemann, Oxford.

Hanlon, J.F., Kelsey, R.J. and Forcinio, H.E. (1998) *Handbook of Package Engineering*, Technomic Publishing, Lancaster, PA.

Heiss, R. (1970) *Principles of Food Packaging. An International Guide*, Keppler Verlag, Germany.

Hernandez, R.J. (1997) Food packaging materials. in *barrier properties, and selection. Handbook of Food Engineering Practice* (eds E. Rostein, R.P. Singh and K.J. Valentas). CRC Press, Boca Raton, FL, Chapter 9.

Hernandez, R.J. and Giacin, J.R. (1997) *Factors affecting permeation, sorption, and migration in package-product systems.* in: *Food Storage Stability*, (eds I.A. Taub and R.P. Singh). CRC Press, Boca Raton, FL, Chapter 10.

Hotchkiss J.H. (eds) (1988)Food and Package Interaction. ACS Symposium Series, 365, American Chemical Society, Washington, DC.

Hotchkiss J.H. and Risch S.J. (eds) (1991) Food and Packaging Interactions II. ACS Symposium Series, 473, American Chemical Society, Washington, DC.

Katan, L.L. (1996) *Migration from Food Contact Materials*, Aspen Publishers, Frederick, MD.

Piringer, O.G. (1993) *Verpackungen für Lebensmittel. Eignung, Wechselwirkungen, Sicherheit*, VCH Verlagsgesellschaft mbH, Weinheim, Germany.

Piringer, O.G. and Baner, A.L. (eds) (2000) *Plastic Packaging Materials for Food. Barrier Function, Mass Transport, and Legislation*, Wiley-VCH, Weinheim, New York.

Risch, S.J. (1999)New Developments in the Chemistry of Packaging. ACS Symposium Series, American Chemical Society, Washington, DC.

Robertson, G.L. (2006) Food Packaging. *Principles and Practice*, 2nd edn., CRC Taylor & Francis, Boca Raton, London, New York.

Vergnaud, J.M. (1991) Liquid Transport Processes in Polymeric Materials. *Modeling and Industrial Applications*, Prentice-Hall, Englewood Cliffs, NJ

Zweifel, H. (ed) (2001) *Plastics Additives Handbook*, 5th edn., Hanser Publishers, Munich.

2
Characteristics of Plastic Materials

Johannes Brandsch and Otto Piringer

2.1
Classification, Manufacture, and Processing Aids

A common feature of all plastic materials is their backbone made of polymers that means natural or synthetic macromolecules composed of thousands of atoms and having correspondingly high molecular masses. A polymer molecule is built up by repetition of small, simple chemical units which are connected with covalent bonds. In addition to the primary structure of the polymer, determined by the strong intramolecular covalent bonds, there are weak intermolecular interactions of secondary van der Waals and/or hydrogen bonds which are responsible for the high variety of materials obtained after compounding and processing polymers to plastics. These materials are examples where neither the primary chemical structure nor the technological design solely determines the product's properties. As a consequence, in plastic technology the complete chain of knowledge is explored, from pure synthesis in organic chemistry, via catalysis, chemical engineering, and polymer design, to the development of new polymeric systems and the final mechanical engineering steps for product design. Only the organic nature, also built up from polymers, uses much more advanced synthesis routes with enzymes as catalysts, is mainly based on hydrogen bonding, and makes maximum use of the self-organizing possibilities of macromolecules.

A new polymer which cannot be processed is only of a limited value. During processing most of the final material properties are obtained, or the materials themselves can be polymerized, given the large flexibility of the intermediate fluid state of matter that is often present at intermediate temperatures during processing (Meijer 2005). The ultimate properties of polymers in terms of modulus, strength, impact resistance and, last but not the least, their permeability to low molecular compounds are often uniquely obtained via specific processing routes, and clearly illustrate this particular behavior and the opportunities that polymers provide.

Plastics are a twentieth-century discovery, the first plastics being derived from high molecular weight natural raw materials, e.g., regenerated cellulose (cellophane) from cellulose around 1910. Plastics were first seen as replacements for natural raw

Plastic Packaging. Second Edition. Edited by O.G. Piringer and A.L. Baner
Copyright © 2008 WILEY-VCH Verlag GmbH & Co. KGaA, Weinheim
ISBN: 978-3-527-31455-3

materials during times of shortage, e.g., synthetic rubber during the First World War. However, since the Second World War a new class of useful materials has been developed whose properties can be tailored through control of their syntheses and processing to fit every desired application. With a yearly production of over 200 million tons, plastics form a pillar of the economy without which today's standard of living would not be attainable.

The importance of plastics is attested by the abundance of scientific and technical literature on the subject. For several reasons a short discussion on the manufacture, structure, and properties of plastics is necessarily a brief introduction to the following problems treated in the other chapters in this book:

1. Measurable residual amounts or conversion products from the many different raw materials and processing aids used in various plastic synthesis processes can remain in the finished material. Knowledge of these materials is indispensable for the toxicological evaluation of the plastics and their analysis. The same applies for the many chemically different additives which are incorporated into the polymer matrix to allow better processing, to increase stability, and to give material-specific properties.

2. The rate of migration of low molecular weight molecules from plastics into foods or other products and interactions of the plastics with product components depends on the molecular structure and the macroscopic (aggregate) nature of the plastic material. In order to perform useful estimations of mass transfers, for example from plastic into its environment, a basic knowledge of the structure of the plastic components and their influences on this phenomenon is necessary.

In view of the enormous abundance of data and knowledge about plastics, only a few representative examples from a multitude are presented in the following. When searching for solutions to special cases of interaction, one should always try to learn as much as possible about the manufacture, composition, and properties of the packaging as well as about their environment. This is necessary in order to evaluate the possible reactions that could occur, to make estimations of migration, and to make comparisons with the actual interaction problem at hand.

2.1.1
Classification and Manufacture of Plastics

Plastics can be classified according to whether they are made from converted natural products (regenerated cellulose) or from completely synthetic products. They can then be further classified according to their manufacturing method in terms of their polymerization reactions, either condensation or addition reactions. They are then further divided, according to their physical properties, into thermoplastics, elastomers, and thermosets.

The combining of carbon atoms in an unlimited number through covalent bonding leads to the synthesis of macromolecules. Depending on the way that covalent bonding occurs, heteroatoms besides carbon such as oxygen, nitrogen, and sulfur can be included.

Thermoplastics are composed of threadlike chain molecules tangled together. This group of plastics takes its name from the properties resulting from such structures. Thermoplastics soften with increasing temperature, which allows them to be formed and then become hard again as they cool. A network of covalent bonds crosslinking the polymer chains leads to the formation of elastomers. Thermosets are composed of networks of primary valence covalently crosslinked molecular structures. The crosslinking step occurs during forming, and afterward the plastics are heat stable (not thermoplastic).

Thermoplastics are delivered in the form of granules and powders to production sites that are separated from the polymer synthesis. After the addition of the necessary additives, e.g., plasticizers, and after additional processing steps, the final material is referred to as a plastic.

The synthesis steps that occur, for example in forming the thermoset coatings used in food contact materials and articles, takes place in the final production phases. This difference in processing compared to the processing of thermoplastics is not insignificant with regard to quality considerations of the finished products when one considers the possible interactions from a food regulation viewpoint. The following synthesis paths deal mainly with thermoplastics, but apply as well to the preliminary steps in thermoset production. The thermosets are then crosslinked or hardened at their point of application. The most important crosslinking reactions will be briefly discussed at the end of this section.

The simplest chemical compounds used directly in synthesis reactions and which are incorporated into the macromolecular chain as a structure sequence are called monomers. Monomers are either unsaturated, that is, they have one or more double bonds, or are bifunctional compounds. The corresponding polymer is produced by a technical polymerization reaction of either a free radical chain reaction (unsaturated monomers) or an intermolecular condensation reaction (bifunctional).

2.1.1.1 Raw Materials and Polymerization Processes

Fossil-based raw materials, mainly oil, gas, and occasionally coal, are used almost exclusively for the manufacture of monomers. Plant materials, the so-called renewable resources, have been used earlier and can become more significant once again in the future. Although the plastics in these cases are obtained by direct polymerization of their monomers, the synthesis of the monomers themselves often requires several intermediate steps. The multifunctional multiple intermediate compounds in the plastic synthesis steps cannot be clearly defined as monomers in every case. The polycondensation reaction of terephthalic acid with ethylene glycol, for example, leads directly to polyethylene (PE) terephthalate. However, the reactions of the well-defined chlorosilanes require several intermediate steps to form silicon polymers. The chlorosilanes which can be recognized as forming repeating segments in the silicon chain fulfill the above definition for monomers. In general, the corresponding residual monomers can be found in the finished polymer material. However, in the case of silicon no chlorosilanes or residuals from the intermediate steps can be found in the finished material. The chlorosilanes themselves are not directly used in the synthesis of silicon, but rather the siloxanes from the intermediate step.

To avoid any misunderstanding in the definition of a monomer, the substances that are used directly in the synthesis of plastics are designated here as starting materials. These starting materials can be "real" monomers, e.g., ethylene, or a mixture of intermediates, e.g., siloxanes. Whereas it is assumed that residual starting materials will remain in the finished plastic, the raw materials of the starting materials are assumed to be completely converted, i.e., decomposed, so that they are not detectable.

2.1.1.2 Addition Polymerization

The most important bulk plastics, e.g., the polyolefins, are produced using addition polymerization processes. The molecules of the starting materials contain double bonds which are broken with the help of initiators or catalysts. The resulting free radicals then undergo a chain reaction to form a macromolecule. In practice, there are numerous processes with different reaction conditions. The start of chain reactions requires a radical produced as a rule by the disintegration of initiator substances, usually peroxide.

The finished plastic, usually in the form of granules, can contain small amounts of undestroyed residual initiator and/or other disintegration products, residual monomers, and low molecular weight polymerization products (oligomers) as well as residuals of other processing aids. Oxidation reactions resulting from traces of unsaturated compounds, present during the processing of the plastic material, can lead to the formation of sensory active compounds. Some of the necessary additives for further converting to the packaging material may already be added to the plastic granules (see Chapter 3).

If it is possible to trace the nature of the plastic and some of the substances contained in it back to the manufacturing process, then this can be useful for solving product problems related to migration or the formation of off-odors (see Chapter 14). The same goes for the knowledge of package material converting processes and the additives used in them.

The monomer can be polymerized either directly, that is undiluted (block or substance polymerization), or in the presence of a nonpolymerizable solvent (solvent polymerization). In the first case, there is a problem with dissipating the localized heat of a reaction (traces of decomposition products result from overheating) and, in the second case, the solvent must be completely removed. Other possibilities are to suspend the monomers in dispersions, e.g., in water (suspension or pearl polymerization), or eventually to use an emulsifier (emulsion polymerization). Emulsifiers and dispersants are considered to be processing aids for the production of polymers.

In addition to the use of radical-producing initiators, other catalysts can also be used for ionic addition polymerization reactions. Compared to low density polyethylene (LDPE) produced by radical polymerization, the use of metal oxide catalysts produces higher density PEs. A further possibility for the synthesis of such a high density polyethylene (HDPE) is metal complex polymerization with coordination or Ziegler–Natta catalysts (metal alkyl–metal halide catalysts) under low pressure. This type of polymerization is particularly important because it is stereo selective (see polypropylene). In this connection, the metallocene catalyst systems must be mentioned. These systems have high activities and polymerization rates, and the level of isotacticity can be controlled.

Another plastic addition polymerization synthesis possibility is ionic polymerization (cationic and anionic).

2.1.1.3 Condensation Polymerization

Starting materials with two different reactive functional groups can polymerize without any further external assistance with the help of an initiator or a catalyst. Another direct polymerization possibility exists between two different starting materials (monomers), having each two identical functional groups. These reactions are usually subdivided into three groups: polycondensation, polyaddition (not to be confused with radical addition polymerization), and ring opening reactions.

A typical example of a condensation polymerization reaction is the reaction between poly functional alcohol (e.g., glycol) and dicarboxylic acid (e.g., terephthalic acid).

Condensation polymerizations are equilibrium reactions, which means they eventually stop reacting when small molecular weight reaction products like water are no longer removed from the system. These characteristics of the condensation polymerization reaction also have an effect on the chemical properties of such plastics. In the presence of water, particularly at high temperatures, PE terephthalate begins to hydrolyze and low molecular weight oligomers are produced which can be transferred into food in contact with the plastic.

2.1.1.4 Synthesis of Copolymers, Block, and Graft Copolymers

Polymerization involving a single type of monomer produces a homopolymer (Figure 2.1(a)), while a mixture of different monomers leads to a mixed polymer or copolymer. Copolymerization offers the possibility to tailor-make a number of different structures which differ from one another in terms of solubility, reactivity, and many other properties. The different monomers can alternate with one another in the polymer chain (Figure 2.1(b)) or be randomly distributed (random copolymer, Figure 2.1(c)). Random copolymers are obtained by radical addition polymerization whereby the statistical distribution of the medium length chains of a given monomer is relatively short. This means a given sequence in a random copolymer consists of 1–10 monomer units. For polymerization degrees ranging from 1000 10 000, these sequences alternate with one another up to 100–1000 times.

In block copolymers (BCPs), a targeted distribution of the different monomers leads to sequences containing many monomers of a single type (Figure 2.1(d)). A magnitude of 100 to 1000 monomer molecules per chain means the polymer chain contains only two to four sequences. Graft copolymers (GCP) are relatively the same only where the sequences of one monomer are attached to a chain made up of the other monomers (Figure 2.1(e)).

Block copolymers are fundamentally different from polymer mixtures in that they comprise only one type of molecule containing sequences of unlike monomer units

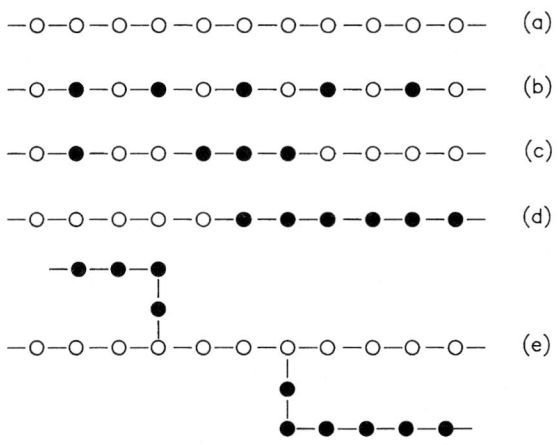

Figure 2.1 (a) Homopolymer, (b) and (c) copolymer, (d) block copolymer, (e) graft copolymer.

covalently bonded together. When a homogeneous melt of BCPs is cooled, unlike interactions will become unfavorable and unlike monomer units will try to segregate so that an apparent upper-critical solution point is observed. Phase separation per se is impossible and a local ordering process will occur. A typical example is the polyolefin BCP which has alternating polystyrene-polybutadiene blocks.

2.1.1.5 Polymer Reactions

In order to obtain materials with certain properties, polymers can be modified using different types of chemical reactions. The crosslinking reactions used for the manufacture of elastomers and thermosets referred to earlier are the most important types of chemical reaction. Preformed polymer chains are bound together at a later time using a built-in reactive functional group that is activated, for example, through heating. An example of such a process is the epoxy lacquers. Through an addition reaction between a diphenol (bisphenol A) and epichlorohydrin, an intermediate for the following crosslinking step is formed. In the intermediate step, for example bisphenol-A-diglycidyl ether (BADGE) as well as polycondensation, high molecular weight products based on BADGE are formed. Following the addition of a hardener, e.g., polyamine, the crosslinking reaction leading to the formation of the three-dimensional thermoset takes place.

Other substances are also used as polymerization processing aids, like solvents (e.g., benzyl alcohol) and accelerators (e.g., nonylphenol). These processing aids are not significantly or not to a measurable degree chemically incorporated into the crosslinked polymer. Under the conditions of the epoxy thermoset reaction the epichlorohydrin, for example, is completely decomposed. Under the current, state-of-the-art hardening technology, practically no epichlorohydrin can be detected in the finished product.

It is especially difficult to make a definitive comprehensive list of all starting materials (positive list) used to make polymeric package materials by three-dimensional crosslinking, which could be transferred into a product. In addition to the numerous oligomers coming from the combined intermediate steps, there is a variety of combination possibilities and mixtures of polymer starting materials, together with the corresponding processing aids (catalysts, crosslinkers) and additives (stabilizers, plasticizers), which further complicate the inclusion of all these compounds into food regulations and quality analysis systems (see Chapters 11–13 and 15).

Crosslinked polymers or polymer networks include a wide range of polymeric materials with specific processing and material properties. The crosslinked materials range from vulcanized elastomers to thermosetting materials, and a considerable number of adhesives and coatings, soft gels, etc. The advantages of crosslinked polymers are dimensional stability, higher thermal and chemical resistance, reversible rubber elasticity, and generally their ability to store information characterizing their birth and growth processes. Almost all reactions between groups producing bonds can be used for the preparation of polymer networks. The condition is that some of the constituent units must participate in three or more bonds by which they are linked to neighboring units. A few examples are as follows.

Polyester networks. Triol or tetrol (e.g., trimethylolpropane or pentaerythritol) are added as branching agents.

Polyurethane networks. It includes a variety of systems obtained by reactions of polyisocyanate compounds. The urethane, urea, and isocyanurate groups are the most frequent types of stable bonds used in network formation.

Cured epoxy resins. Networks prepared by the reaction of polyepoxides with polyamines represent the most important group of cured epoxy resins.

Network involving silicon-containing groups. Poly(dimethylsiloxane) networks have been one of the most frequently studied model networks. One main reaction involves the formation of $-Si-O-Si-$ bond by crosslinking hydroxy-terminated poly(dimethylsiloxanes) with polyfunctional alkoxysilanes, e.g., with tetraethoxysilane (Dusek, 2005).

The reactive processing of polymers and composites involves the simultaneous development of the polymer structure and materials shaping to produce polymer artifacts with final properties suitable for particular applications. The methods used in reactive processing come under the general headings of casting, coating, and molding, and involve the use of viscous (as opposed to viscoelastic) materials, which include liquid monomers, prepolymers, polymer solutions, dispersions, and (low molar mass) polymer powders.

The simplest reactive polymer process is perhaps monomer casting, and examples include monomers such as methylmethacrylate, styrene, caprolactame, and N-vinyl carbazole.

Coating is a surface modification method. Reactive molding processes include compression (CM), resin transfer (RTM), and reaction injection molding (RIM) (Stanford et al. 2005).

2.1.1.6 Plastic Processing

Melt Blending In polymer processing, for example, melt blending of homopolymers is a common route toward materials with specific properties that are superior to those of the constituents. Most polymers are thermodynamically immiscible in the molten state. Hence, the blending process yields a heterogeneous morphology that is characterized by the shape, size, and distribution of the constituting domains. At the end of the melt blending process, the morphology is frozen-in in the solid state (Janssen, 2005).

Structure Formation by Orientation Phenomena Long chain molecules lend themselves to the attainment of oriented products with associated advantageous properties. There are principally two different routes toward this goal: (1) to draw or to otherwise orient an initially random crystalline solid and (2) to orient chains in their random state (solution or melt) first and "set" this orientation by subsequent crystallization. In this respect, a distinction needs to be made between rigid rod molecules and flexible chains. To achieve orientation, rigid rods only need to be aligned, while flexible chains need both extending and aligning, the two occurring simultaneously in the course of a normal orientation process (Keller 2005).

Reactive Extrusion For years, both chemistry and engineering practice involving polymer systems used inert liquid carriers to avoid complications caused by the high viscosity of undiluted bulk systems; these systems are otherwise difficult to handle in classical reactors of a stirred-port version, such as beakers and tanks. When a polymeric system was too viscous, it was simply diluted with more liquid effluents.

During the last three decades and due to ever-increasing concerns and tougher government regulations about the excessive use of solvents, reactions have more and more been run in solventless polymer melts using polymer processing machines as chemical reactors, e.g., batch mixers and screw extruders. However, the absence of a solvent may raise other complications: high temperature, heterogeneity of the reacting media, low mass and heat transfer, low polarity, etc., rendering the proper control of the selectivity and the rate of the reaction difficult. Resin suppliers, specialty compounders, and even end-product users are now trying to diversify the existing polymers by chemically modifying or reactively blending them inside a screw extruder. Reactive extrusion (REX) may be defined as a manufacturing process that combines the traditionally separated polymer chemistry (polymerization or chemical modification) and extrusion (blending, compounding, structuring, devolatilization, and shaping) into a single process carried out in a screw extruder. A disadvantage may be possible side reactions which are difficult to control (Hu and Lambla 2005).

Compatibilization Processes Most polymer blends are phase-separated systems with some dispersion of one polymer in the other. The properties of the polymer blends deeply depend on the morphology. Generally, some 1 µm or even smaller particles are required to get good synergy of the properties of the component polymers. The mean particle size in polymer blends is the result of a balance between mechanical and interfacial forces: the lower the interfacial torsion, the smaller the particles. A fine and stable morphology and good interfacial adhesion can be reached by effective control of the interfacial properties through adequate interfacial modification. These processes of interfacial modification resulting in lower interfacial tension, inhibited coalescence, and improved adhesion are generally allied "compatibilization processes." Interfacial modification can be brought about by the addition of a BCP or GCP at the interface. However, the synthesis of premade BCP is often difficult and/or expensive. Therefore, in many cases, the BCP or GCP is produced *in situ* by blending reactive materials (Inoue and Marechal 2005).

2.1.2
Processing Aids

The processing aids necessary for polymerization can be divided into two groups: substances which directly influence the manufacturing process and substances which provide an adequate medium for the polymerization process to take place (e.g. solvents and emulsifiers).

Even though some processing aids could be chemically incorporated into the polymer, they do not appear systematically in the repeating units of the polymer chains and therefore cannot be considered to be monomers or starting materials. In general, processing aids are used only in small amounts and after the synthesis is completed they practically exert no influence on the finished polymer. Thus, they are very different in nature from additives which are also often used sparingly but which exert a much greater influence in the finished plastic.

Processing aids are used to directly influence the synthesis process function as reaction controllers. Depending on their chemical state they can function as reaction accelerators (the actual catalysts and starters or initiator substances), crosslinkers and/or hardeners, reaction inhibitors or catalyst deactivators, molecular weight controllers, chain splitters, or lengtheners. From a chemical standpoint (structure and method of function) the radical builders, mainly peroxides and azo compounds, are treated separately from the catalysts which are mainly metals, metal oxides, salts (redox systems), and organo-metal compounds. The carrier substances, promoters, and deactivators are placed in the catalyst class of substances.

The second category of processing aids contains substances which function as working media to assist initiators and catalysts. These medium-forming substances are, in general, much less reactive or are nonreactive compounds, most of which can be found as additives in further converting and polymer use applications. These processing aids are mainly solvents, dispersants, emulsifiers, precipitants, antifoaming and degassing agents, pH controllers, stabilizers, germinating agents, blowing agents for foams, and others.

Substances that are also used as additives for plastic processing and applications will be treated in Chapter 3. In the following, the two most important substance classes in the first category are discussed.

Table 2.1 Classes of organic peroxides.

Alkylhydroperoxide
Dialkylperoxide
Peroxy carbonic acid
Diacylperoxide
Peroxy carbonic acid ester
α-Oxyperoxide
α-Aminoperoxide

2.1.2.1 Initiators and Crosslinkers

Organic peroxides are used as initiators or starter substances for many polymerization reactions because they easily decompose to form radicals. There are presently about 50 technically important organic peroxides. These can be formally described as derivatives of hydrogen peroxide (HOOH). Through substitution of one or both of the hydrogen atoms by alkyl, aralkyl, or acyl groups, a series of compound classes is obtained whose reactivity is heavily structure dependent (Table 2.1). The stability of the peroxide in every substance class increases with the number of carbon atoms. Along with radical formation, oxidation is the other important peroxide property. The oxidation potential decreases according to the order: peroxy carbonic acid hydroperoxide > diacylperoxide > peroxy carbonic acid esters > dialkylperoxide.

Some of the alkylhydroperoxides, e.g., cumylhydroperoxide, are used for the polymerization of diolefin (butadiene) with comonomers (e.g., styrene) at low temperatures (5–20°C).

Alkylhydroperoxides are also interesting because of their formation in natural products. Unsaturated fatty acids and their esters (plant oils) are oxidized in air to peroxides (autooxidation). This leads to a stepwise breaking of the double bonds leading to the formation of aldehydes, ketones, and fatty acids, all of which make their presence known through their strong odor intensities (e.g., rancid oil).

Dialkylperoxides are used as high temperature catalysts for suspension and bulk polymerization as well as hardeners for unsaturated polyester resins and for crosslinking polymers because of their good thermal stability. While the liquid di-*tert*-butylperoxide is relatively volatile at application temperatures, dicumylperoxide is much less volatile and has the disadvantage of forming decomposition products with intense odors (acetophenone) (Chapter 14).

Diacylperoxide is used mainly as an initiator for radical polymerizations as well as for hardening and crosslinking polyester resins as a result of its ability to easily thermally decompose.

It can be helpful to consider the reaction products of the peroxides used in terms of their eventual interactions between the plastic and the filled product.

Peroxycarbonic acid esters play the most important role in vinyl, ethylene, and styrene polymerizations. In particular, mixtures of the *tert*-butylesters of the peroxybenzoic acids and the peroxyaliphatic fatty acids are used.

In addition to the above-mentioned homolysis, peroxides can also be ionically decomposed leaving behind by-products. Primary and secondary peroxyesters decompose into the corresponding carbonic acids and carbonyl compounds. α-Oxyperoxides are used as crosslinkers and hardeners. Adding transition metal salts, a wider useable temperature range can be achieved for warm hardening of polyester resins.

In order to distribute initiators, which are often solids, throughout the reaction medium, nonreactive processing aids such as phthalates are often used.

Inhibitors, which act as radical absorbers, are used to slow down peroxide-controlled polymerizations. Most of these compounds are chinones, aromatic nitrogen compounds, or aromatic amines. Impurities can lead to lengthened inhibition times and subsequently an uncontrolled course of polymerization so that only the highest purity starting materials are used in the polymerization.

Along with the peroxides, other compounds are used as radical formers. The azo-fatty acid nitriles play an important role here. Azo-isobutyric acid nitrile or 2,2'-azo bis (2-methylpropionitrile) decomposes under heat to nitrogen and two radicals, which can start a polymerization chain reaction.

Finished products are not allowed to show any positive reactions for peroxide on their surfaces.

2.1.2.2 Catalysts

Substances which increase the rate of a chemical reaction without themselves being used up or incorporated into the finished product are called catalysts. In heterogeneous catalysts, the reaction takes place on the surface of a solid support. The activity of the catalyst in this case is determined by the structure and size of the surface area as well as by the way the catalyst is produced. Catalysts are not limited to immobilization on solids; they can also be introduced as homogeneous catalysts in the solution. Between heterogeneous and homogeneous catalysts, there is a possibility of evenly distributing small particles (dispersions) of the catalyst in a liquid phase.

Heterogeneous catalysts are mostly metals or metal oxides. They can be used as they are or together with inert carriers, which themselves are usually metal oxides. Homogeneous catalysts are as rule cations of certain metals or complexes of metal atoms with an organic molecule (ligands).

With regard to the possibility of a residual catalyst in the plastic being transferred into the filled product, the following can be said from the above considerations: residual immobilized solid heterogeneous catalysts, provided that they cannot be separated from the polymer, can only play an interactive role on the surface of the material in contact with the filled product. The migration of metals and metal oxides from within the plastic has practically no significance because compared to organic compounds they are not dissolved in the plastic and therefore are not subject to diffusion in the form of molecules and ions. However, cases where traces of a catalyst have some residual activity can have negative effects on the properties and stability of the plastic over a period of time. This can happen when reactions occur with substances diffusing through the plastic, e.g., oxygen, that subsequently lead to the decomposition of the plastic structure.

When dispersion systems are used, some residual dispersants can remain in the plastic and then migrate out. In the case of homogeneous catalysts, migration can also occur. In particular, this applies to the residual organic parts of the organo-metal compounds which remain as breakdown products in the plastic after completion of polymer synthesis and destruction of the catalyst. This also applies to residual solvents that are not fully evaporated.

The heterogeneous catalyst systems containing a mixture of the elements Ca, Mg, Al, Si, Ti, Cr, V, and Zr are most important for polymer synthesis.

Polymerizates intended for use as food contact materials may not contain a total of more than 0.1% of these catalysts in the form of oxides.

Further residual metal oxides from catalysts used for polymer synthesis, e.g., polyterephthalic acid diol ester, are oxides of antimony, gallium, germanium, cobalt, manganese, zinc, and titanium. Residual amounts of these oxides in the polymer are also restricted.

2.2
Structure and States of Aggregation in Polymers

2.2.1
Structure

The overall nature and properties of plastic are determined by its chemical structure, macromolecule mass, and the additives compounded into it.

The polymer molecular chains form the backbone of the thermoplastics' polymer structure. The nature and orientation of the monomer units in the chain determine the primary structure of the polymer. This primary structure of the chain can be differentiated into three groups as follows.

1. *Pure carbon chains*. The chains can be unsubstituted (e.g., PE) or contain single or multiple substitutions (polyvinyl compounds).
2. Chains that in addition to carbon contain heteroatoms such as O, N, P, or S. Here the nature of the segment, R, can vary as well as the nature of the bound hetero groups, X, in the segment: $-R-X-R-X-R-X-R-X-$.
3. Chains that are exclusively composed of heteroatoms. The most important representatives of this group are the silicons.

The nature of the elements found in the polymer chain and the covalent bonds that exist between them allow one to predict the properties of the corresponding thermoplastic. The primary valence bonds determine the secondary valence bonds such as the van der Waals forces, polar bonds, and hydrogen bonds occurring between the polymer chains. The intermolecular forces existing in thermoplastics composed of carbon and hydrogen are due to weak van der Waals attractions. These forces rapidly decrease with increasing temperature. The thermoplastics containing heteroatoms have comparatively much stronger polar attractions. Atoms such as Cl, F and the atomic groups OH, CN and COOR create dipoles, which increase the attractive forces between the chain molecules and cause the thermoplastic to be much stronger. One type of dipole attraction that is particularly strong is the hydrogen bond

```
          R   R   R   R   R
          |   |   |   |   |
isotactic C - C - C - C - C - C - C - C - C - C
          |   |   |   |   |   |   |   |   |   |


          R           R           R
          |           |           |
syndio-   C - C - C - C - C - C - C - C - C - C
tactic    |   |   |   |   |   |   |   |   |   |
              R           R


          R   R       R
          |   |       |
atactic   C - C - C - C - C - C - C - C - C - C
          |   |   |   |   |   |   |   |   |   |
                  R           R
```

Figure 2.2 Polymer chain configurations of atactic, isotactic, and syndiotactic polymers.

such as occurs between OH and NH groups and O atoms on other chains. The hydrogen bonds are responsible for the strength and stiffness of polymers like polyamides for example, as well as their normally undesirable affinity for absorbing water.

Unsubstituted polymer chains cannot form different stereo isomers, while substituted polymers can have a large number of different possible isomeric forms. As a result, it is possible to have various configurations for substituted polymers. For example, polystyrene produced by radical polymerization is atactic, which means the phenyl groups bound to every second C-atom are randomly distributed on both sides of the polymer chain. Polymers produced using Ziegler catalysts, made from monomers such as styrene, propene, and others, are isotactic and/or syndiotactic (Figure 2.2):

The regularity of the polymer chain structural group locations is an ideal representation. In practice, the chains can have more or less branching depending on the manufacturing process. As an example, the branching occurring in PE can be schematically represented in Figure 2.3.

PE–LD — high pressure polymerization

PE–HD — low pressure polymerization

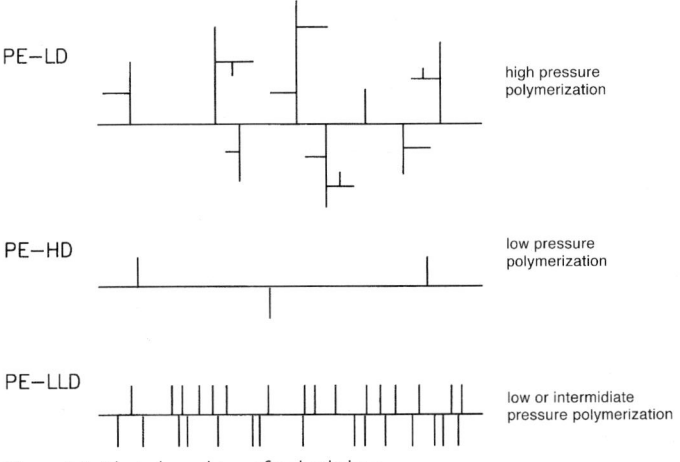

PE–LLD — low or intermidiate pressure polymerization

Figure 2.3 Chain branching of polyethylene.

Side groups in atactic structures and chain branching hinder crystallization of the thermoplastic polymers in contrast to unbranched and isotactic configurations, which lead to increased crystallinity. With increasing crystallinity the density, strength, and stiffness are increased but the transparency and processability of the plastic decrease.

The extreme hardness and lack of formability of thermoset plastics is due to the extremely large number of primary valence bonds between the plastic's atoms. In an extreme case of crosslinking with covalent bonds, practically no weak covalent bonds exist that can be loosened by increasing the temperature. This means that no thermoplastic processing of the polymer is possible. A thermoset plastic can be depicted as being a huge single molecule. The removal of these intramolecular bonds by increased temperature leads to destruction of the polymer. In practice, there exists a range of plastics having different degrees of crosslinking, from those of the uncrosslinked thermoplastics to the completely crosslinked thermosets. With increasing degree of crosslinking the strength, stiffness, and thermal resistance of the plastics increase. By varying the degree of crosslinking, the elastic behavior of elastomers can be established over a wide range. The elasticity of an elastomer is, therefore, determined by both the primary valence and the covalent bonds between the molecule chains.

The properties of plastics are also determined by the chain length and the distribution of different chain lengths. The relative molecular mass of a macromolecule, M_r, has a large influence on the polymer's flow properties, its glass transition temperature, T_g, and its mechanical properties.

A pure natural polymer, e.g., a protein, is monodisperse, meaning that it has a single, definite mass. A synthetic polymer, however, is polydisperse. That means the sample is a mixture of molecules with various chain lengths and molar masses. The various techniques that are used to measure molar masses result in different types of mean values of polydisperse systems.

The mean obtained from the determination of molar mass by osmometry is the number-average molar mass, $M_n = \frac{1}{N}\Sigma_i N_i M_i$, which is the value obtained by weighting each molar mass by the number of molecules of that mass present in the sample. N_i is the number of molecules with molar mass M_i and N is the number of all molecules. The number of molecules is proportional to the amount of substance, which means the number of moles. Light scattering experiments give the weight-average molar mass, $M_w = \frac{1}{m}\Sigma_i m_i M_i$. This is the average calculated by weighting the molar masses of the molecules by the mass of each one present in the sample. In this expression, m_i is the total mass of molecules of molar mass M_i and m is the total mass of the sample. The ratio M_w/M_n is called the heterogeneity index or polydispersity index. A synthetic polymer normally spans a range of molar masses, and the different averages yield different values. Typical synthetic materials have $M_w/M_n \approx 4$. The term monodisperse is conventionally applied to synthetic polymers in which the index is less than 1.1. Commercial PE samples might be much more heterogeneous, with a ratio $20:50$ for LDPE and $4:15$ for HDPE. One consequence of a narrow molar mass distribution for synthetic polymers is often a higher degree of a three-dimensional long-range order in the solid and therefore higher density and melting point. An important influence of the heterogeneity index is observed in the diffusion behavior of the polymer matrix. A high index value, which means a high proportion of polymer molecules with relative low masses in the matrix, increases the diffusion coefficients of additives solved in the polymer (see Chapters 6 and 15). The spread of values is controlled by the choice of a catalyst and reaction conditions. In practice, it is found that a long-range order is determined more by structural factors (branching) than by molar mass (Atkins and de Paula 2006).

All synthetically produced thermoplastics exhibit a characteristic distribution of the macromolecule's length and mass. With increasing degree of polymerization the tensile strength, tear resistance, hardness, strain at break, impact resistance, and melt viscosity all increase. At the same time the tendency to crystallize, flowability, swelling, and stress-cracking tendency decrease (Elias, 2005).

2.2.2
States of Aggregation

A low molecular weight substance can exist in one of three aggregate states: solid, liquid, or gas independent of temperature or pressure. The transition from one state into another is sharp, and the corresponding melting and boiling temperatures are characteristic properties of the substance. These small molecules are usually arranged in orderly crystals in the solid state. Macromolecules are for the most part irregular amorphous structures. A large difference exists between low molecular weight substances and most polymers, in that polymers can show a coexistence of crystalline and amorphous regions. In amorphous polymers, changes of state are less well defined and may occur over a finite temperature range. The polymer chain mobility resulting mainly from the primary polymer structure is responsible for this characteristic. Here it is understood that polymer chain mobility means its freedom of movement, i.e., the rotation of certain chain segments, and not translational and rotational movements of the whole polymer. A measure of the chain mobility is the glass transition temperature or freezing temperature T_g (Table 2.2). Above this temperature, the polymer chains can move freely and the polymer is rubbery or plastic. Below the glass transition temperature, the chain mobility stops and the polymer becomes glassy and hard (glassy state). This difference between the glassy and rubbery states of a plastic has a big influence on the values of the diffusion coefficients (see Chapter 6).

When heating an amorphous polymer, it eventually reaches a temperature designated as its flow temperature. This is a very viscous transition state and further heating leads to a viscous melt. Such a transition again changes significantly the diffusion behavior of the plastic (Chapter 6).

If a sample of an amorphous polymer is heated to a temperature above its glass transition point and then subjected to a tensile stress, the molecules will tend to align themselves in the general direction of the stress. If the mass is then cooled below its transition temperature while the molecule is still under stress, the molecules will

Table 2.2 Abbreviations, densities, and glass transition temperatures of some important thermoplastics.

Polymer	Abbreviation	Density (g/cm^3)	Glass transition temperature (°C)
Low-density polyethylene	LDPE	0.915–0.94	−30 ± 15
High-density polyethylene	HDPE	0.945–0.964	−30 ± 15
Linear low-density polyethylene	LLDPE	0.90–0.935	−30 ± 15
Poly-4-methylpentene-1	P4MP1 (PMP)	0.83	55
Polypropylene	PP	0.90–0.91	−17 ± 5
Polystyrene	PS	1.04–1.12	80–100
Acrylonitrile-butadiene-styrene polymer	ABS	1.03–1.07	–
Polymethyl methacrylate	PMMA	1.18–1.24	99–104
Polyvinyl acetate	PVAC	1.19	28–31
Polyvinyl alcohol	PVAL (PVA)	1.19–1.27	70–85
Ethylene-vinyl acetate copolymer	EVA	0.91–0.97	–
Polyvinyl chloride	PVC	1.39–1.43	80–100
Polytetrafluoroethylene	PTFE	2.28–2.30	115–125
Polyethylene terephthalate	PETP (PET)	1.37	67–81
Polybutylene terephthalate	PBTP (PBT)	1.3–1.5	48–55
Polycarbonale	PC	1.20–1.24	120–150
Polyoxymethylene	POM	1.42–1.435	188–199
Polyamide	PA	1.12–1.14	50–60

become frozen in an oriented state. Such an orientation can have significant effects on the properties of the polymer mass. The polymer is thus anisotropic.

In addition to the deliberate monoaxial or biaxial orientation carried out to produce an oriented filament or sheet, orientation will often occur during polymer processing, whether desired or not. Thus in injection molding, extrusion or calendering the shearing of the melt during flow will cause molecular orientation.

If a polymer molecule has a sufficiently regular structure, it may be capable of some degree of crystallization. Crystallization is limited to certain linear or slightly branched polymers with a high structural regularity. Well-known examples of crystalline polymers are PE, acetal resins, and polytetrafluoroethylene.

There are substantial differences between the crystallization of simple molecules such as water and copper sulfate and of polymers such as PE. For example, the lack of rigidity of PE indicates a much lower degree of crystallinity than in the simple molecules. In spite of this the presence of crystalline regions in a polymer has a large effect on such properties as density, stiffness, and clarity.

The essential difference between the traditional concept of a crystal structure and crystalline polymers is that the former is a single crystal, whilst the polymer is polycrystalline. A single crystal means a crystalline particle grown without interruption from a single nucleus and relatively free from defects. The term "polycrystallinity" refers to a state in which clusters of single crystals are involved, developed from the more or less simultaneous growth of many nuclei.

There are two principal theories of crystallization in polymers. The fringed micelle theory considers that the crystallinity present was based on small crystallites of the order of a few hundred Angström units in length. This is very much less than the length of a high polymer molecule, and it was believed that a single polymer molecule actually passed through several crystallites. The crystallites thus consisted of a bundle of segments from separate molecules which had packed together in a highly regular order. Between the crystallites, the polymer passed through amorphous regions in which molecular disposition was random. Thus, there is the general picture of crystallites embedded in an amorphous matrix.

This theory helps to explain many properties of crystalline polymers, but it was difficult to explain the formation of larger structures such as spherulites.

The lamellae formation theory is based on studies of polymer single crystals. It was found that under many circumstances the polymer molecules folded upon themselves at intervals of about 100 A to form lamellae, which appear to be the fundamental units in a mass of crystalline polymer. Crystallization spreads by the growth of individual lamellae as polymer molecules align themselves into position and start to fold. For a variety of reasons, such as a point of branching or some other irregularity in the structure of the molecule, growth would then tend to proceed in many directions. In effect, this would mean an outward growth from the nucleus and the development of spherulites. In this concept, it is seen that a spherulite is simply caused by growth of the initial crystal structure, whereas in the fringed micelle theory, it is generally postulated that the formation of a spherulite requires considerable reorganization of the disposition of the crystallites.

Both theories are consistent with many observed effects in crystalline polymers. The closer packing of the molecules causes an increased density. The decreased intermolecular distances will increase the secondary forces holding the chain together and increase the value of properties such as tensile strength, stiffness, and softening point.

The properties of a given polymer very much depend on the way in which crystallization has taken place. A polymer mass with relatively few large spherulitic structures will be very different in its properties to a polymer with far more, but smaller, spherulites. A polymer crystallized under conditions of high nucleation/growth ratios, with smaller structures, is generally more transparent.

Homogeneous nucleation occurs when, as a result of statistically random segmental motion, a few segments adopt the same conformation as they would in a crystallite.

High nucleation rates can be achieved together with high growth if *heterogeneous nucleation* is employed. In this case, nucleation is initiated by seeding with some foreign particle. This can be of many types but is frequently a polymer of similar cohesive energy density to that being crystallized, but of a higher melting point. Nucleating agents are now widely used in commercial products. They give a high degree of crystallization and good clarity in polymer films.

In later discussions in this book, it should be noted that migration behavior is strongly dependent on the given polymer sample. This dependence comes from the variety of plastics that exist with respect to their chemical nature, structure, molecular mass distribution, manufacturing and processing conditions.

2.3
The Most Important Plastics

2.3.1
Thermoplastics

2.3.1.1 Polyethylene

Polyethylene is the most widely used mass–produced plastic. The worldwide production capacity of PE in 2007 is estimated to 79×10^6 metric tons per year. Of this amount 21×10^6 tons are LDPE, 22×10^6 tons linear LDPE (LLDPE), and the rest of 36×10^6 tons HDPE (Beer et al. 2005).

The development of PE began in 1936 with the introduction of the high pressure polymerization process of ethylene to LDPE (0.915–0.94 g cm^{-3}), which produced a relatively low molecular weight polymer. The manufacture of HDPE by low pressure polymerization first began after the discovery of the Ziegler catalysts in 1953. The HDPE produced using this process has a medium density (0.945 g cm^{-3}). The Philips and Standard Oil process was also developed in the 1950s and produces HDPE with the highest density (0.96 g cm^{-3}). LLDPE is the copolymer of ethylene with about 8% 1-butene, 1-hexene, or 1-octene. Copolymers with 10–20% 1-octene are "plastomers." LLDPE is denoted as linear, because the ramifications are already contained in the comonomers and not produced as transfer reactions. With these short side chains, LLDPE has a density range from 0.900 g cm^{-3} for very LDPE (VLDPE) to 0.935 g cm^{-3} for the octene-ethylene copolymer (Brydson, 1995).

Metallocene polyethylene (mLLDPE) is obtained from ethylene with metallocene catalysts. They are characterized by relative narrow distribution of their molecular masses, by low extractable contents and good organoleptic properties.

The number average molecular weight of LDPE with a density of 0.92 g cm^{-3} (high pressure polymer) is $10\,000 < M_n < 40\,000$. The relative molecular masses of all PE families are generally smaller than $M_r = 3 \times 10^5$.

All polyethylenes are semicrystalline. Their densities and melting temperatures decrease with the increase of ramification. The crystalline melting temperatures are about 108°C for LDPE and LLDPE, 135°C for HDPE and 144°C for the ideal crystallites of linear PE. The softening temperatures of the wax-like thermoplastic are between 80 and 130°C. The glass temperatures are between −80 and −30°C.

At present, many hundreds of grades of PE, most of which differ in their properties in one way or other, are available.

PE possesses good chemical stability. The mechanical properties are dependent on the molecular weight and degree of chain branching. PE can be easily heat sealed, is tough and has high elasticity. It has good cold resistance properties and is a good water vapor barrier. However, LDPE has low barrier properties to gases, aromas, and fats. With increasing density, all the barrier properties increase as well as the stiffness, hardness, and strength, as a result of the higher crystallinity. At the same time, there is a decrease in the impact resistance, toughness, resistance to stress cracking, cold resistance, and transparency.

Since PE is a crystalline hydrocarbon polymer incapable of specific interaction, there are no solvents at room temperature. Materials of similar solubility parameters and low molecular weight will however cause swelling, the more so in low density polymers. LDPE has a gas permeability in the range normally expected with rubbery materials. HDPE has a permeability of about one-fifth that of LDPE.

The processing of PE is normally carried out at temperatures between 150 and 210°C. However, temperatures as high as 300°C can be reached during paper coating. PE is stable at these high temperatures under inert atmospheres and, when being processed under these conditions, the oxygen concentration in and around the plastic should be kept as low as possible.

The chemical stability of PE is comparable to paraffin. It is not affected by mineral acids and alkalis. Nitric acid oxidizes PE and halogens react with it by substitution mechanisms. By chlorination in the presence of sulfur dioxide, chlorine groups and sulfonyl chloride are incorporated and an elastomer is formed. Oxidation of polyethylene which leads to structural changes can occur to a measurable extent at temperatures as low as 50°C. Under the influence of ultraviolet (UV) light, the reaction can occur at room temperature.

In order to obtain sufficient adhesion of printing inks on PE surfaces, oxidation of the surface must take place. This can be accomplished either with flame or by corona treatment. Significant off-odors can be produced as a result of the oxidation process. In particular, unsaturated ketones and aldehydes are implicated in these off-odors.

LDPE is used mostly in the form of films over thicknesses ranging from 15 to 250 μm. Coextrusions, laminates, shrink films, films for the building industry and for agricultural purposes, shopping bags, trash bags, and household films are all made from LDPE. The coating of cartons and paper with LDPE using an extrusion process makes the packaging (milk cartons and paper bags) water tight and heat sealable. Films of polyethylene terephthalate (PET), PP, PA, and other plastic substrates are extrusion coated with PE and, in so doing, are converted into water, gas and aroma–tight, hot-fill and aseptic packaging.

Blown containers from LDPE are used as packaging in the pharmaceutical and cosmetic industries as well as for foods, toys, and cleaning agents. Injection-molded LDPE is used to make buckets and various household and kitchen containers.

The most important application area of HDPE is the production of containers and injection-molded articles. Bottles for detergents, gasoline cans, and heating oil tanks are some examples. The most common use of HDPE for injection-molded articles is for the production of storage and distribution containers, like buckets and bottle cases. However, processing into films and pipes has become increasingly more common. Films made out of HDPE possess high fat resistance (as wrappers for meat) and have better aroma barrier properties compared to lower density PE materials.

Copolymers of PE with vinyl acetate (EVA), acrylic acid ester, and methyl acrylic acid increase the heat sealability, adhesion to other materials, and seal strength, and they improve the polymer's cold resistance and transparency. EVA in the form of shrink films are well suited for meat packaging because of their relatively high gas permeabilities.

EVA-copolymers are used as sealants. With vinyl acetate contents ranging from 15 to 40%, these copolymers are particularly applicable for the production of hot melts because of their good compatibility with filled products and in mixtures with other plastics. Ethylene vinyl alcohol copolymer (EVOH) is a plastic with exceptional barrier properties. It is manufactured by the saponification of EVA.

2.3.1.2 Polypropylene

Since approximately 1986, polypropylene (PP) has ranked third in the bulk plastic production after PE and polyvinyl chloride (PVC), with an estimated annual production of 40×10^6 tons per year. PP is composed of linear hydrocarbon chains and therefore its properties quite closely resemble those of PE. PP and its copolymers can be classified into three categories: monophasic homopolymer (h-PP), monophasic random copolymer (r-PP), and heterophasic copolymer (heco-PP). The h-PP can be either isotactic, syndiotactic, or atactic.

Isotactic PP is obtained with propylene and transition metal catalysts or metallocene catalysts. Syndiotactic PP is also obtained with metallocen catalysts. Atactic PP, which in reality is highly ramificated PP, is an adhesive and not a typical polymer. Statistical copolymers and BCPs of propylene with ethylene or 1-butene are also denoted as technical polypropylene. The properties of isotactic PP are particularly useful. The stereo regularity of the macromolecule chain construction and the related high crystallinity give PP its outstanding characteristics. Large scale commercially produced PP is up to 95% isotactic in nature.

Homopolymer PP is one of the lightest thermoplastics, having a density ranging from 0.90 to 0.91 g cm^{-3}. Pure isotactic PP has a melting temperature of 176°C. In general, the melting temperature of commercial materials is around 150–170°C with melting beginning around 140°C, which is much higher than PE. The molecular weights M_r of the atactic PP are between 20 000 and 80 000 and for the isotactic PP between 2×10^5 and 5×10^5.

The chemical compatibility of PP is similar to that of HDPE. PP can be swelled by aromatic and chlorinated hydrocarbons and dissolved in them at higher temperatures. The tertiary C atoms reduce the chemical inertness of PP and make it, above all, more sensitive to oxidation. This sensitivity to oxidation must be compensated for by the addition of antioxidants.

PP possesses good water vapor barrier and fat resistance properties. Normal PP films have limited food packaging applications (e.g., packaging of bread) because of their low cold temperature resistance. Copolymer mixtures with ethylene are used to improve cold resistance and heat sealability as well as material strength and, above all, seal strength.

The statistical distribution of the monomer units leads to a significant decrease in the degree of crystallinity with small amounts of ethylene (4–15%). These products result in better strength and transparency. The melting range is also broader and begins at a lower temperature.

The large availability of nucleating agents based on sorbitol enables the fabrication of clear PP products, with transparencies between 89% (homopolymer) and 96% (metallocene random copolymers).

PP is an excellent material for injection and extrusion processes. The packaging containers, in particular bottles, made using these processes should be mentioned. PP bottles maintain their shape well at high temperatures which allows them to be hot-filled. Improved properties are obtained by using polymer mixtures with ethylene as mentioned before. Injection-molded containers are used for frozen foods, e.g., ice cream. In addition, steam sterilizable containers and dishes can be produced for heating in microwaves. New PP packaging developments are multiple-layer bottles and cans with inner barrier layers, which can be hot-filled or sterilized in an autoclave as well as directly steam sterilized. PP packaging can be filled with liquids that are surface active because of its good stress-cracking resistance (Brydson, 1995).

Over 40% of PP produced in Europe is used to make films. Random copolymers with ethylene are superior with regard to toughness, transparency, shrink characteristics, and sealability. These films exhibit low stiffness, strength and hardness. The thickness of these films lies between 12 and 125 µm.

By stretching, usually biaxially, under their melting temperature range, PP films can be orientated. Orientation can be used to improve properties like strength, cold stability (down to $-50°C$), and heat resistance. Heat sealable biaxially orientated PP (BOPP) is used for the packaging of confectionery products, baked goods, snack products, pasta, potato products, and dried fruits. In addition, BOPP films play an important role in cigarette packaging as well as in cosmetic and pharmaceutical packaging.

OPP films serve as carriers in laminates. The shrink characteristics of PP are greatly increased by biaxial stretching. After unwrapping an article, the film shrinks back to its unoriented state upon nearing the crystalline melting point. The disadvantage of orientation is that it is not particularly suitable for heat sealing. To overcome this disadvantage, the BOPP is coated with a heat sealable layer having a low melting temperature. By coextruding a core layer of homopolymer PP along with two heat sealable layers and subsequently stretching it, the thickness of the heat seal layers can be reduced to 1 µm.

For the packaging of sensitive foods, PP films are coated with polyvinylidene chloride, polyvinyl acetate, EVA–copolymers, polyacrylates, styrene-butadiene copolymers, LDPE, poly-1-butene, or random copolymers of propene with ethylene and 1-butene. By using these various coatings, PP has recently sharply reduced the use of regenerated cellulose (cellophane), the previous market leader in this area.

Polymer chain segments of pure PP and pure PE placed one after the other form BCPs that have an increased degree of crystallinity. Depending on the manufacturing process, copolymers with an ethylene fraction of up to 30% can also be noncrystalline, thus forming an ethylene–propylene elastomer.

Products with similar properties are also obtained from mixtures of homo-PP with PE as well as with ethylene–propylene elastomers. Such mixtures are interesting because of their high flexibility at low temperatures. These mixtures, which may contain up to 50% elastomer, are designated as elastomer–modified thermoplastics. These materials with elastomer content over 50% are referred to as thermoplastic rubbers.

2.3.1.3 Polybutene-1

The manufacturing process and properties of polybutylene-1 are comparable to PP. Compared to the aliphatic hydrocarbons, this material is not as inert as PE and PP. Its high burst strength and tear strength are very advantageous for the manufacture of hot water pipes (resistant up to 95°C).

Good flexibility is a characteristic of polybutylene films, which makes it a good plastic coating for paper and aluminum because of its high tear strength.

Butene-1 can be copolymerized with ethylene as well as with higher α-olefins like propylene and 4-methyl-pentene.

2.3.1.4 Polyisobutylene

The rubbery character and particular physical and chemical properties of polyisobutylene stem from its paraffinic origins. Its outstanding properties are its low glass transition temperature, very low water vapor permeability, and resistance to many chemicals. At room temperature polyisobutylene is resistant to dilute and concentrated mineral acids and bases, as well as hydrogen peroxide.

Low molecular weight polyisobutylene is used both as the soft component and as the adhesive component in glues and sealants. The high molecular weight polymers are quite similar to vulcanized rubber.

Polymer mixtures of isobutylene and polyolefins are used for the manufacture of lacquers. Polymer mixtures with styrene also have different applications for impregnation compounds, glues, etc.

Copolymers of isobutylene with styrene ($<10\%$) and isoprene ($<3\%$) may be used for the manufacture of food contact materials. The following polymerizates and polymer mixes can be added to these polymerizates: PE, PP, styrene–acrylonitrile mixed polymers, mixed polymers of ethylene, propylene, butylene, vinyl esters, and unsaturated aliphatic acids as well as salts and esters and polybutene-1.

2.3.1.5 Poly-4-methylpentene-1 (P4MP1)

Poly-4-methylpentene-1 possesses a largely isotactic structure with over 40% crystallinity. Its characteristic properties are high transparency and a very low density of $0.83\,g\,cm^{-3}$, as well as its applications at high temperatures up to 150°C.

4-Methylpentene-1 can be copolymerized with n–alkenes (C2–C5; $<10\%$, C6–C14; $<5\%$).

2.3.1.6 Ionomers

Polymers having carboxyl groups in their ionized form are obtained through high pressure polymerization of ethylene with 1–10% of an unsaturated organic acid and treatment with compounds of magnesium, zinc, calcium, or sodium (e.g., magnesium acetate). These polymers have a type of ionic crosslinking, which is stable at normal temperatures but can be reversed at higher temperatures. By these means, ionomers possess increased stiffness while still having sufficient flexibility at high temperatures. Ionomer films, also known as Surlyn (DuPont), have high water vapor permeability because of their low crystallinity compared to PE. At the same time, ionomers show exceptional oil and fat resistance. A particular advantage of this material is its good adhesion to other substrates. Thus, it is used in laminations because of its good resistance to delamination. Disadvantages compared to PE are its high cost and low application temperatures.

The maximum zinc content of the finished ionic crosslinked mixed polymerization material is about 3.5%. Polymer mixtures of ethylene, propylene, butylene, vinyl esters, and unsaturated aliphatic acids as well as salts and esters can be used in the manufacture of ionomer food contact materials. The materials are then differentiated according to whether they contain no crosslinking or have ionic or physical cross-linked materials, using peroxide.

For the manufacture of uncrosslinked ionomer polymer mixtures ethylene, butene-1, isobutylene, vinyl chloride, vinylidene chloride, aliphatic carboxylic acids of vinyl esters (C2–C18), aliphatic unsaturated mono and dicarboxylic organic acid esters (C3–C8) with mono aliphatic saturated alcohols (C2–C12), and unsaturated aliphatic mono and dicarboxylic organic acids (C3–C8) can be used as raw materials.

Ionomer polymer mixtures can be cut with various products. The materials used to cut the ionomer are paraffins, microcrystalline waxes, plasticizer-free vinyl chloride, polymer mixtures of PE, PP as well as natural and synthetic rubber. Finished materials made from uncrosslinked ionomer mixtures may not be used for contact with fatty foods.

Peroxide crosslinked ionomer mixtures made from the above–mentioned raw materials may be used in contact with fatty foods under certain limited conditions, according to their raw materials and additives. Vinyl chloride and vinylidene chloride should not be used in materials having either ionic (ionomer) or physical (e.g., through electron beam irradiation) crosslinks. Crosslinked PE can be used for the manufacture of food contact materials, e.g., for drinking water pipes and fittings. The crosslinking can be done using either peroxide or electron beam irradiation.

2.3.1.7 Cyclic Olefin Copolymers (COC)

Cyclic olefin copolymers are characterized by high clarity, transparency, rigidity, strength and heat resistance, good stability, and low moisture absorption. These amorphous plastics are related to PE and PP in terms of their chemical structure. Norbornene, a cyclic olefin used as a comonomer with ethylene, makes COC very rigid and hard as compared with polyolefins. The new metallocene catalyst technology makes it possible to produce a wide range of grades with different glass transition temperatures from 70 to 180°C.

Typical applications include pharmaceutical and food primary packaging. The material can be added to standard plastics such as PE and PP to improve the property profile. Meat, cheese, and sausage benefit from higher performance films. In complex multilayer solutions, COC improves not only the barrier properties but

2.3.1.8 Polystyrene

With an annual production of over 12×10^6 tons, polystyrene (PS) occupies the fourth place behind PE, PP, and PVC on the bulk polymer list. PS has been commercially produced since 1930. As a thermoplastic plastic, it can be processed between temperatures of 150 to 300°C. At higher temperatures, depolymerization takes place by splitting out the styrene. Products formed from PS are hard and transparent, with high brilliance and resistance to many chemicals. Its disadvantages are its brittleness and sensitivity to stress cracking. Because of its high permeability to gases and vapors, it is mainly used as a material for products requiring short shelf-lives, usually refrigerated and not having too high a fat content, such as yoghurt, ice cream, fresh cheese, and coffee cream. PS is also used as a divider or organizer for fruits, eggs, baked goods, and sweets. In the past few years, PS has been increasingly replaced in many application areas by the less expensive PP.

Polystyrene has little strength if its molecular weight is below 50 000 but increase rapidly, with molecular weight up to 100 000. An increase above 100 000 has a little further effect on the tensile strength, but continues to have an adverse effect on the ease of flow. The melting temperature of the isotactic PS is around $T_m = 230°C$ and for the syndiotactic PS is around $T_m = 270°C$.

Styrene can be copolymerized with many monomers. The following monomers can be used along with styrene in the manufacture of food contact materials: α-methylstyrene, vinyltoluene, divinylbenzene, acrylonitrile, ethyleneoxide, butadiene, fumaric and maleic acid esters of the monofunctional saturated aliphatic alcohols C1–C8, acrylic acid ester and methacrylic acid, maleic acid anhydride, methylacrylamide-methylol ether, vinylmethyl ether, vinylisobutyl ether. Styrene and/or α-methylstyrene and/or vinyltoluene should be the main mixture component in every case.

Through the polymerization of a styrene rubber solution, one obtains styrene-butadiene (SB) mass. SB forms a two-phase system in which the styrene is the continuous phase and the rubber, usually a butadiene base, is the discontinuous phase. The rubber phase also contains pockets of styrene. The SB polymer, because of its properties, is also known as impact resistant or high impact PS (HIPS).

The combination of high transparency, toughness and high impact resistance of thermoplastic SB polymers (Styrolux from BASF), and their miscibility with PS over a wide range opens a broadly adjustable characteristics spectrum for blends (Schwaben 2005). Another possibility is a mixture of PS and SB with a thermoplastic elastomer that likewise consists of styrene and butadiene. Such a plastic with the trade name Styroflex can be further combined with PS/SB or HIPS with advantageous properties for applications at low temperatures (e.g., for ice packing).

When the continuous phase is formed by a copolymer of styrene and acrylonitrile, then one obtains a material known as acrylonitrile-butadiene-styrene (ABS). The acrylonitrile part improves the stress–crack resistance of the polymer.

Although the absorption of water vapor by PS is very low, it cannot withstand boiling water. PS is soluble in aromatic and halogenated hydrocarbons as well as in ethers, esters, and ketones. Even though many single substances do not attack PS, one can observe obvious synergistic effects with respect to interactions. Essential oils and various cosmetics and medical preparations attack PS and lead to stress cracking among other problems.

PS is resistant to salt solutions, bases, and dilute acid solutions. Oxidizing acids cause oxidative degradation of PS. In the presence of oxygen, UV light leads to yellowing and brittleness. The chemical stability of PS is on the whole lower than PE.

Packaging for food and pharmaceuticals as well as household appliances and containers like drinking cups and disposable dishes are all made from PS and SB. Oriented PS films are used for packaging milk products and cigarettes. The low heat conductance and high impact resistance at low temperatures are advantageous properties for the use of PS as a packaging material for freezing.

Of the styrene copolymers used for food packaging, the styrene-acrylonitrile copolymer known as SAN still needs to be mentioned. SAN copolymers possess better mechanical properties and better resistance to oils and aroma compounds than PS. Copolymers with acrylonitrile fractions of 20–35% find uses as household and camping dishes. Copolymers with a higher acrylonitrile content (>60%) have earned particular importance as barrier plastics. With an increasing acrylonitrile fraction, the gas permeability decreases sharply.

Copolymers with Methacrylic Acid Esters are also used in Packaging.

The breakthrough to bulk plastic was made possible by the development of ABS polymers. This copolymer mixture (acrylonitrile, butadiene, and styrene) leads to a combination of technologically important properties that have allowed its use for many diverse applications.

The continuous phase in the ABS polymer is responsible for most of its chemical properties. Because of the presence of only C–C binding in the polymer chain, no hydrolytic reactions can take place. ABS polymers are in general resistant to aqueous salt, base or acid solutions and are not dissolved by paraffinic hydrocarbons. Depending on the type and amount of the rubber phase, a weight gain due to uptake of hydrocarbons can take place. The copolymer is also resistant to fat and various cosmetic creams. This results from increased polarity due to acrylonitrile. Halogenated hydrocarbons, aromatics, esters, and ketones however dissolve the SAN phase. Oxidative substances, in particular acids, destroy the polymer chain. The good cold resistance of this material (to −40°C) deserves a mention.

Finally, styrene homopolymers with the addition of ∼6% of a low molecular weight hydrocarbon, for example pentane, are the most important starting materials for the manufacture of PS hard foams.

2.3.1.9 Polyvinyl Chloride

The ability of vinyl chloride to polymerize was first observed over 150 years ago. Polyvinyl chloride has been industrially manufactured since approximately 1930. Even though pure PVC is fairly unstable, the manifold applications are made possible by the discovery of effective stabilizers and other additives for the polymer.

At the beginning of the 1970s, the annual production figures for PVC and PE were quite similar. However, since 1971 serious health problems and complications have been seen in persons exposed to vinyl chloride in the air. The formation of liver angiosarcomas in workers in the PVC industry marked a turning point in PVC production. The residual concentration of the vinyl chloride monomer in PVC that was approximately 300–400 ppm (mg kg^{-1}) in the 1960s was reduced to 2–5 ppm in 1976 and presently is well under 1 ppm. The additional technological effort needed to remove the remaining residual monomer and the decreasing acceptance by consumers of this plastic in the meantime led to a relative decrease in the use of PVC compared to PE. This decrease has been particularly noticeable for food packaging. Additionally, PVC can be replaced by PP in various applications. Nevertheless, PVC has still maintained a leading position among the bulk plastics today because of its low price and numerous application possibilities. The global annual capacity for PVC production is about 30×10^6 tons.

PVC is resistant to nonpolar (hydrocarbon) and strongly polar substances (water, inorganic acid). Middle polarity compounds such as cyclohexanone, dimethylformamide, acetone, chlorinated hydrocarbons, tetrahydrofuran, and phenol all either swell PVC or dissolve it. This behavior can easily be attributed to the slightly polar structure of the PVC macromolecule.

When PVC is pyrolyzed, the main decomposition product is hydrochloric acid, along with small amounts of saturated and unsaturated hydrocarbon side products. PVC is easily degraded through the effect of heat, light and mechanical energy. In order to improve the low stability of this plastic, a series of additives are incorporated into the PVC melt. The most important additives for the processing of PVC are the plasticizers, which may be incorporated at elevated temperatures to give mixtures stable at room temperature.

Due to its particularly good polymer characteristics, PVC has an enormously wide spectrum of applications. Blow-molded containers for packaging liquid products (beverages, edible oils, detergents, cosmetics, and pharmaceuticals) receive special consideration, as do dishes for fatty foods (highly stable against low polarity substances) and films (such as soft PVC films with high gas permeability) for fresh meat packaging. Soft PVC is also used as a component in seals.

Vinyl chloride can be copolymerized with a series of monomers: vinylidene chloride, *trans*-dichloroethylene, vinylesters of aliphatic carboxylic acid (C_2–C_{18}), acrylic acid esters, methyl acrylic, and/or maleic acid as well as fumaric acid with monofunctional aliphatic saturated alcohols (C_1–C_{18}), monofunctional aliphatic unsaturated alcohols (C_3–C_{18}), vinyl ethers from monofunctional aliphatic saturated alcohols (C_1–C_{18}), propylene, butadiene, maleic acid, fumaric acid, itaconic acid, acrylic acid, methyl acrylic acid (total <8%) and *n*-cyclohexylmaleic acid (<7%).

PVC can be blended with numerous other polymers to give it better processability and impact resistance. For the manufacture of food contact materials, the following polymerizates and/or polymer mixtures from polymers manufactured from the above–mentioned starting materials can be used: chlorinated polyolefins; blends of styrene and GCPs and mixtures of polystyrene with polymerizate blends; butadiene-acrylonitrile-copolymer blends (hard rubber); blends of ethylene and propylene, butylene, vinyl ester, and unsaturated aliphatic acids as well as salts and esters; plasticizer–free blends of methyl acrylic acid esters and acrylic acid esters with monofunctional saturated alcohols (C_1–C_{18}) as well as blends of the esters of methyl acrylic acid; butadiene and styrene as well as polymer blends of acrylic acid butyl ester and vinylpyrrolidone; polyurethane manufactured from 1,6-hexamethylene diisocyanate, 1,4-butadiol, and aliphatic polyesters from adipic acid and glycols.

For unplasticized chlorinated PVC, unplasticized chlorinated polymer blends of vinyl chloride and mixtures of these copolymers with other polymer blends, the following starting materials can be used: PVC (homopolymer); polymer blends of vinyl chloride, vinylidene chloride *trans*-dichloroethylene, ethylene, propylene, butylene, maleic acid, fumaric acid, itaconic acid, acrylic acid, methacrylic acid as well as chlorine.

Unplasticized PVC and polymer blends of vinyl chloride can be added to chlorinated polymers manufactured using the above starting materials.

Because of the increasing amount of criticism from consumer groups due to the formation of hydrochloric acid during burning and because of plasticizer migration from soft PVC films, PVC is continually being replaced by other plastics. The strongest competition exists in the container market where it is being replaced by PET for beverage packaging. Soft PVC films can also be replaced by other polyolefin-based systems.

2.3.1.10 Polyvinylidene Chloride

Homo and copolymers of vinylidene chloride (VDC) possess extremely high barrier properties to gases, water, and aromas as well as good resistance to water and solvents. The barrier properties of polyvinylidene chloride (PVDC) come from the dense packing of its polymer chains (without voids or branching), which are crystalline in their stable form. The chlorine content of the high density polymer is 73% (1.80–1.97 g cm^{-3}, crystalline).

PVDC dissolves at room temperature only in polar solvents like hexamethylphosphoric acid amide or tetramethylene sulfoxide. Amorphous PVDC can also be dissolved in tetrahydrofuran. Above 125 °C PVDC decomposes by giving off hydrochloric acid. Under the influence of high energy irradiation, basic compounds and heavy metals cause decomposition. Stability with respect to decomposition can be increased by copolymerization with vinyl chloride, acrylonitrile, methyl acrylate, and others. At the same time, copolymerization decreases crystallinity and increases the gas permeability.

In packaging, thin PVDC films are used as barrier layers in laminates. PVDC dispersion coatings provide very good barrier properties on paper, regenerated cellulose, OPP, and other plastic films. The coatings can also be manufactured so that they are heat sealable. Because of their heat sealability, fat and oil resistance, and good flexibility, PVDC polymers are exceptional packaging materials.

2.3.1.11 Thermoplastic Polyesters

The most important representatives of this group are PET and polybutylene terephthalate (PBT). Even though the cost of these plastics is presently in the medium price range, one can count on a reduction in their price in the future due to their widespread use.

The following starting materials can be used for the manufacture of polyesters for food contact: terephthalic acid, isophthalic acid (<25%), adipic acid, azelaic acid, sebacic acid, ethylene glycol, butanediol-1,4, 1,4-dihydroxymethyl cyclohexane, terephthalic acid methyl ester, azelaic acid dimethyl ester, sebacic acid dimethyl ester, oligomeric diglyceride ether of 4,4'-dioxidiphenyl-2,2-propane (a.k.a. disphenol-A-diglyceride ether (>2%), n-decane dicarboxylic acid-1,10 (>15%), polyethylene glycol (<0%)). PE (<5%) or PP (<5%) may be added to polyesters manufactured from the above starting materials.

The linear saturated polyesters are hard, semicrystalline thermoplastics that are impact resistant even at low temperatures, smooth, and have good wear resistance. Their glass temperatures are around 67–80°C and the melting temperature $T_m = 255°C$. The molecular weights of commercial polymers are between 10 000 and 20 000.

The barrier properties of PET are good with respect to gases, aromas, and fats and have slightly lower barrier properties against water vapor. Because of its partial crystallinity, PET has a high strength at short–time load over a wide temperature range from $-60°C$ to over 200°C.

The glassy clearness and strength are improved by stretching the plastic. Biaxially stretched PET films with thicknesses around 12 µm are important substrates in barrier laminates and have a wide spectrum of uses, especially for longer times at high temperatures over 150°C.

Cardboard baking dishes coated with crystalline PET or crystalline PBT ($T_m = 227°C$) can be used in convection ovens up to temperatures of 200–220°C. Single portion dishes made from heat-formed films find wide applications in microwave ovens. Biaxially stretched PET covers an important application area of bottles, wide-mouth jars, and cans. These containers are particularly well suited for carbonated beverages, edible oils, and spirits. The gas barrier properties can be improved by coextrusion with a barrier layer such as polyamide. Another good barrier polymer is polyethylenenaphthalat (PEN), which is obtained from ethyleneglycol and dimethyl-2,6-naphthalene-dicarboxylate. The melting temperature of PEN is $T_m = 266°C$. Coating of PET and the use of oxygen absorbers are also methods for oxygen reduction. With such improved barrier properties, PET can also be used for beer and wine.

2.3.1.12 Polycarbonate

Polycarbonate (PC) is a high value plastic with high strength and hardness along with good toughness. PC has a high resistance to heat, up to 135°C, as well as to cold, down to $-90°C$. The relatively expensive glass-clear amorphous plastic, however, possesses relatively high permeability to gases and water vapor which makes it necessary to combine it with a suitable barrier layer. PC is well suited for the manufacture of dishes and kitchen utensils, coffee filters and machines, baby bottles, and other containers.

For food contact articles, the following starting materials can be used: 4,4'-dioxy-diphenyl-2,2-propane (bisphenol A), 4,4'-dioxy-diphenyl-1,1-cyclohexane, 2,6-bis-(2'-hydroxy-5'-methylbenzyl)-4-methyl phenol (<1%), 1,4-bis-(4',4'-

dihydroxytriphenyl-methyl)-benzol (<1%), diphenyl carbonate, phosgene, terephthalic acid dichloride, isophthalic acid dichloride, 4,4'-dioxy-diphenyl-3,3'-oxindol (<1%), 3,3-bis-(3-methyl-4-hydroxyphenyl)-2-indolinone (<1%).

The following substances for limiting polymer chain length in the manufacture of PC can be used: phenol (<2%), tertiary butylphenol (<3%), or 4-(1,1,3,3-tetramethyl-butyl)-phenol (<5%). These substances are incorporated into the polymer macromolecule. PC can be mixed with copolymers of styrene, butadiene, and acrylonitrile where the PC forms the bulk of the total mixture.

2.3.1.13 Polyamide

The manufacture of the large variety of polyamides (commonly referred to as nylons) occurs through polycondensation of amino carboxylic acids (or functional derivatives of them, e.g., lactams) and from diamines and dicarboxylic acids. Labeling the amino groups with A and the carboxyl groups with B allows differentiation of the different chemical structures between the two types: AB (from amino carboxylic acids) and AA-BB (from diamines and dicarboxylic acids). The number of C atoms in the monomers acts as a code number for the identification of the polyamides. The polycaprolactam manufactured from caprolactam (type AB) is then called polyamide 6 (PA 6). The number of carbon atoms in diamine is given first for type AA-BB followed by the number of atoms in the dicarboxylic acid, e.g., PA 66 for polyhexamethylenediadipic amide from hexamethylenediamine and adipic acid. For copolymers the components are separated by a slash, e.g., PA 66/6 (90 : 10) is a copolymer composed of 90 parts PA 66 and 10 parts PA 6.

For food contact articles, the following can be used as starting materials: straight chain ω-amino acids (C6–C12) and their lactams; adipic acids, azelaic acid, sebacic acids, dodecane dicarboxylic acids, and heptadecanedicarboxylic acids of hexamethyldiamine; isophthalate acid, bis(4-aminocyclohexyl)-methane, 2,2-bis(4'-amino-cyclohexyl)-propane, 3,3'-dimethyl-4,4'-diaminodicyclohexyl-methane, terephthalic acid or its methylester, 1,6-diamino-2,2,4-trimethylhexane, 1,6-diamino-2,4,4-trimethylhexane, 1-amino-3-amino-methyl-3,5,5-trimethylhexane.

Polymer blends having PA as the bulk phase can contain ethylene, propylene, butylene, vinylesters, and unsaturated aliphatic acids as well as their salts and esters. PA 6, PA 12, PA 6/66, PA 6/12, and PA 11 all have particular importance in the packaging area. Because of the strong polar nature of the CONH group, hydrogen bonds are formed between neighboring macromolecules. As a result PA is hard, temperature resistant, and some types are highly crystalline. The shorter the segment between the amide groups, the more water the polar groups can absorb. The absorption of water increases the strength while at the same time decreasing the stiffness. Air-moisture-conditioned samples are not in a danger of stress cracking. Except for the chlorinated hydrocarbons, polyamides are resistant to most solvents, fats, oils, alkalis, and acids. They can be dissolved in concentrated sulfuric acid, phenol, and m-cresol.

Even though PA have good barrier properties to gases and aromas, the barrier properties against water are only mediocre. The melting points vary between 175 and 262°C ($T_m = 215°C$ for PA 6 and $T_m = 262°C$ for PA 66) and the molecular weight between 9000 and 15 000. PA is also applicable for use at low temperatures ranging from $-50°C$ to $-70°C$.

The main application of this relatively expensive plastic in the packaging area is in laminates. Its good barrier properties are improved by combining it with PE, which has good water vapor barrier and heat sealability properties. Thermoformable laminates are used for vacuum or inert gas packed meat products, fish, and cheese. Biaxial stretching of PA (OPA) improves its stiffness and leads to its use as a carrier film together with a heat sealable layer in laminates. Vacuum and inert gas packing of coffee, milk powder, and meat products are some of the many examples of such applications. These PA laminates are also used in the inner bags for "bag-in-a-box" liquid packages.

Future trends in connection with polyamides used as barrier polymers are combinations with nanotechnology. Lanxess has developed a material with nanoscale fillers that shows a significantly increased oxygen barrier. The material in question is PA 6, which is modified with 1 nm silicate platelets during manufacture. These act as tiny obstacles that considerably delay the progress of oxygen molecules through the film (see also Chapter 10). The new film material combines the advantages of low-cost PA 6 with those of the more impermeable ethylene vinyl alcohol (EVOH) copolymer (Joachimi, Kunststoffe). As a result of the so-called nano-effect, very small additions of nanoparticles can achieve what is otherwise only possible with large quantities of conventional mineral fillers.

2.3.1.14 Polymethylmethacrylate

In comparison to bulk plastics, thermoplastic polymethylmethacrylate (PMMA) is much more expensive. Its particular characteristics are clarity, hardness, low absorbance, and resistance to aqueous solutions, acids, alkalis, carbon dioxide, and fat. It is attacked or dissolved by polar organic solvents. Typical food contact articles are dishes, cups, and silverware. In addition it has orthopedic and denture uses.

For food contact articles made from acrylic and methacrylic acid ester copolymers and their polymer blends, the following starting materials can be used:

- esters of methacrylic acid and acrylic acid with mono- and multifunctional saturated aliphatic alcohols C_1–C_{18}, dimethylamino ethanol, cyclohexylamino ethanol, trimethyl ammonium ethanol chloride, ether alcohol, phenol, benzyl alcohol,
- styrene and α-methylstyrene,
- acrylic acid, methyl acrylic acid, maleic acid, itaconic acid,
- amides of acrylic and methacrylic nitrile,
- butadiene,
- vinylidene chloride,
- vinyl and allyl esters of acrylic and methyacrylic acid, and
- triallylcyanurate.

The esters of methacrylic acid and acrylic acid should form the largest fractions in the finished product.

The addition of different starting materials to methylmethacrylate modifies the properties of the PMMA melt. For example, by lowering the working temperature or increasing the warm form stability (through incorporation of maleic acid and styrene) the plastic can be made impact tough down to −40°C and is suitable for the manufacture of blown containers and deep-drawn packaging for food and medicines.

2.3.1.15 Polyoxymethylene or Acetal Resin

Polyoxymethylene (POM) plastics are highly crystalline thermoplastics that are obtained by polymerization of formaldehyde and can also be in the form of trioxymethylene oligomers (trioxane). Polyacetals are primarily engineering materials being used to replace metals.

In food contact articles ethylene oxide, butane diolglyceride ether, butane diolformal, 1,3-dioxane, 1,3-dioxolane (total < 6%) can be used as comonomers. Amines, triphenylphosphine, borotrifluoride, and others can be used as catalysts (total < 0.1%) and various polymerization regulators and polymerization inhibitors can be used.

Because of their high crystallinity POM polymers are white opaque, have high strength, and toughness even at low temperatures (usable from −40°C to 100°C and for short times at 150°C). They are resistant to alcohols, esters, hydrocarbons, and weak alkalis, but are not resistant to acids (pH < 4). The copolymers are more resistant to hot water.

2.3.1.16 Polyphenylene Ether (PPE)

The amorphous thermoplastic poly-2,6-dimethylphenylene ether or PPE is characterized by a very high glass transition temperature (215°C) and a low density compared to other engineering thermoplastics. The polymer is also characterized by superior resistance to hydrolysis and moisture uptake. It was introduced commercially in 1964 under the trade name PPO.

PPE is capable of forming unusual blends with PS. Normally, they are designated as modified PPE resins and combine the best properties of PPE and PS.

Food packaging represents a special segment among extrusion grades. Modified PPE provides the required impact strength, resistance to hydrolysis, dimensional stability, and melt strength in applications such as microwaveable frozen food packing and single-serve portions. Compared to PP, PPE + PS blends are break-resistant at frozen-food temperatures and guarantee good stiffness at considerable thinner walls. Their stiffness exceeds that of PP or PET even at elevated temperatures (Peters and Parthasarathy 2005).

2.3.1.17 Polysulfone

As a result of incorporating benzene rings in the molecular chain, the temperature resistance of thermoplastic polysulfone (PSU) is quite high and lies around 130°C. Better temperature resistance at higher temperatures can be obtained by bonding the benzene rings using oxygen, sulfur, or sulfur groups as well as nitrogen–containing imide groups. Polysulfone (PSU) and polyethersulfone (PES) are two amorphous polar thermoplastics having lower application temperatures from −70 to −100°C and constant use temperatures in air from 150–170°C (PSU), 200°C (PES), and for short times up to 200°C (PSU) and 260°C (PES).

These plastics have high strength and stiffness and are particularly suitable for the manufacture of microwave dishes, hot water containers, and other household articles. The uptake of water by polysulfones influences the mechanical properties in a similar manner to PA. One obtains particularly good heat resistance with polyimide plastic dishes.

2.3.1.18 Fluoride Containing Polymers

Even though these plastics are used mainly as coatings in the manufacture of food contact materials, they are treated in this section as thermoplastics.

The most important representative of this group is polytetrafluorethylene (PTFE), which is synthesized by the radical polymerization of tetrafluoroethylene. This high molecular weight, crystalline, linear polymer is exceptionally resistant to solvents and other chemicals. PTFE melts around 320–345 °C and can be used continuously between temperatures of -200 and 260 °C. Because of its nonwetability, PTFE has exceptional antiadhesive properties and in addition shows good slip characteristics. For these reasons, the polymer is used as a raw material for temperature-resistant nonstick coatings for frying pans, pots, and other cooking pots and utensils.

Plastics with similar properties and applications can be found in the form of polymer mixtures of PTFE with other polymers (see Section 2.3.7).

2.3.1.19 Polyvinylether

Polyvinylethers form a further group of thermoplastics, which are not used as containers or packaging films. They are atactic polymers forming oils, sticky soft resins, or nonsticky rubber elastic materials according to their molecular weight and composition. All polyvinylethers are very resistant to saponification by dilute acids and alkalis. They can subsequently be used as unsaponifiable polymer plasticizers and for the manufacture of glues.

The following starting materials can be used for the manufacture of polyvinylether polymers: vinylmethylether, vinylethylether, vinylisobutylether, vinyloctylether, vinyldecylether, and vinyloctadecylether.

Polymerization of polyvinylethers is initiated using cationic initiators like borotrifluoride, and the finished product may not contain more than 0.4% borane and 0.3% fluoride as decomposition products. Tertiarybutylphenol disulfide (<0.15%) can be used as a stabilizer as well as a polymerization regulator.

The solubility of the polymer in various media is dependent on the nature of its vinyl groups. Polyvinylmethyl ether dissolves in aromatic hydrocarbons, esters, ketones, alcohols, and cold water. However, polyvinylbutylether is soluble in aliphatic hydrocarbons but not in water, methanol, and ethanol. Polyvinyloctadecyl ether possesses a wax-like consistency. The polymers can be mixed with one another as well as with many natural resins and plastics.

2.3.2
Thermosets

The most important thermosets are:

- phenolic resins (PF)
- urea-formaldehyde resins (UF)

- melamine-formaldehyde resins (MF)
- unsaturated polyester resins (UP)
- epoxide resins (EP).

2.3.2.1 Amino Resins (UF, MF)

Melamine resins are used from this group of thermosets for the manufacture of food contact materials. Melamine can be used in mixtures with urea and in some applications with phenol (<1%). The polymerization process is catalyzed in the presence of organic acids (e.g., acetic acid, lactic acid, tartaric acid, citric acid), hydrochloric acid, sulfuric acid, phosphoric acid, sodium and potassium hydroxide, ammonia, calcium, or magnesium hydroxide as well as salts of these substances (total < 1%) which cause the elimination of water and lead to a cured resin system. Stearic acid can be used as a lubricant as can zinc, calcium, and magnesium salts; esters of montanic acid with ethandiol and 1,3-butandiol as well as silicone oil (total <1%).

Dishes and cups for eating and drinking are manufactured by molding and can be recognized by their light, nonfading color. These articles are resistant to hot water, organic solvents, oils, fats and alcohols. The polymers containing phenol may only be used for household and kitchen utensils and equipment (intended only for short food contact times).

2.3.2.2 Unsaturated Polyester (UP)

As a preliminary step in the manufacture of unsaturated polyester thermoset plastic, one uses low molecular weight linear polyester ($M_r \ll 10\,000$) obtained by a polycondensation of polyglycols with saturated and unsaturated dicarboxylic acids. The precondensate can then be dissolved and stored in the stabilized comonomer, e.g., styrene, with which it will be crosslinked later. The crosslinking polymerization reaction between the polyester chains and the styrene bridges is initiated with the help of organic peroxides, which are added dispersed in plasticizers. The reaction begins at 60–90°C and then proceeds exothermally. In addition to this, a cold hardening reaction can also be carried out. For this reaction, cold accelerators are necessary, e.g., tertiary amines or cobalt naphthenate.

Food contact materials must be hardened in a hot air oven (80–100°C) for a couple of hours. After the oven treatment, the material must be washed for 1–2 h with hot water (80°C) in order to remove volatile and water-soluble components.

The following substances are recommended starting materials for unsaturated polyesters: fumic acid, maleic acid, methyl acrylic acid, adipic acid, phthalic acid, resinic acid, isophthalic acid, terephthalic acid, hydrated or halogenated phthalic acids, aliphatic and substituted aliphatic single and multifunctional alcohols up to C_{18}, alkoxylated and hydrated phenols and bisphenols, styrene, vinyltoluene, acrylic acid and methyl acrylic acid esters of the C_1–C_4 alcohols, and tricyclodecane dimethanol.

The hardened, light stable, lightly pigmented food contact materials are not susceptible to stress cracking, but are resistant to alcohols, ethers, hydrocarbons, fats, and can be used with weak acids and boiling water. Typical applications are beverage containers and silos for foodstuffs.

Phthalate-based polycondensate resins, modified with unsaturated fatty acids and styrene as well as with vinyltoluene have many applications in the form of paints and lacquers as "alkyd resins."

2.3.3
Polyurethanes

This plastic includes a large group of polyaddition polymers which are formed through the reaction of bifunctional or trifunctional alcohols with di- or polyisocyanates. By varying the starting materials, linear as well as crosslinked macromolecules with correspondingly different properties are formed. Alcohols with three functional groups and/or triisocyanate are used to make crosslinked polyurethane (PUR) elastomers.

The market is dominated by flexible foam and rigid and semirigid foam. Other materials are reaction injection molding (RIM) products and thermoplastic rubbers, surface coatings, sealants, adhesives, and synthetic leathers.

With respect to food packaging applications, there are many different cases. Most of the food contact articles as well as coatings do not come into complete contact with foods for long time.

Examples are storage containers, container coatings, seals having large surface areas and packaging. The following raw materials can be used: adipic acid, carbon dioxide, 1,2-ethanediol, 1,2-propanediol, glycerin, 1,6-hexanediol, 1,4-butanediol, 2,2-dimethylpropanediol, 2,3-butanediol, diethyleneglycol, propoxylated trimethylpropane, diethoxyhydrochinone as well as polyesters of these diols and acids with hydoxyl end groups. In addition, polyethers based on ethylene dioxide, propylene dioxide, and/or tetrahydrofuran with free hydroxyl groups can be used.

The following isocyanates are mentioned: hexamethylene-1,6-di-isocyanate (HDI), bis-(4-isocyanatocyclohexyl)methane (H12MDI), isophorone diisocaynyte (IPDI), diphenylmethane-2,4'-di-isocyanate and diphenylmethane-4,4'-di-isocyanate (MDI), 2,4-toluene di-isocyanate, and 2,6-toluene diisocyanate (TDI).

Crosslinked polyurethanes are used as glue layers in food packaging. As opposed to glued seams, which when properly used have no or very minute food contact surface areas, the glue layer is a very thin coating covering the complete package surface area. But it has no possible direct contact with the food due to a covering material. Nevertheless, it cannot be ignored that some migration may take place from the polyurethane layer through the cover material layer. There should be no migration of residual reactive low molecular weight starting material components found in the glue during the manufacture of the laminate.

The following starting materials can be used for these applications:

- polyesterpolyole (polyesters containing hydroxyl end groups) based on carboxylic acid, adipic acid, phthalic acid, trimellitic acid (benzene-1,2,4-tricarboxylic acid), sebacic acid, malic acid, ethanediol, propanediol-1,2, butanediol-1,3, butanediol-1,4, 2,3-dimethyl propanediol-1,3, 1,1,1-trimethylol propane, diethylene glycol, hexanediol-1,6, glycerin, dimethylpropanediol-1,3, glycerin,

- polyetherpolyole (polyether containing hydroxyl end groups),
- reaction products with hydroxyl end groups manufactured from 1 and 2 polyoles with diisocyanates mentioned above,
- reaction products with end isocyanate groups manufactured from the above-mentioned isocyanates and polyoles.

As accelerants, caprolactam and, as chain lengtheners, tri-isopropanol amine are recommended.

After the removal of solvents, not more than 10 g of polyurethane should be applied per 1 m^2. Residual solvents and amines may not be detected in foods or food simulants coming in contact with the laminate films.

Linear polyurethane is used for coating paper. The above-mentioned diols and polyethers based on ethylene oxide and propylene oxide with free hydroxyl groups, adipic acid, phthalic acid, and the isocyanates indicated in (3) may be used as starting materials.

The main type of water-borne, dry bond laminating adhesives used for laminating food contact materials are the epoxy urethane adhesives. As there are no reactive isocyanate groups in epoxy urethane adhesives, there is no risk of the generation of aromatic amines during use. However, BADGE and bisphenol-A can be potential migrants from epoxy urethanes.

As emulsifiers Na and K salts of glycine, lysine, and taurines (total <0.8%) as well as N-(2-amino-ethyl)-3-amino propane sulfonate and/or N-(2-amino-ethyl)-2-amino ethane sulfonate (total <5.0%) are recommended. These emulsifiers are incorporated into the macromolecules and therefore can be found only in very small amounts in the finished product. Esters of montanic acid with ethanediol and/or 1,3-butanediol and mixtures of these esters with unesterified montanic acid as well as Ca salts (total <1.5%) can be used as lubricants and mold release agents.

Linear polyurethane for paper coating can be mixed with plastic dispersions where the polyurethane forms the bulk of the coating.

Polyurethane is also used as a foam, mostly in a sheet form as an underlay or a middle layer, for example in fruit bins. The following starting materials for polyurethane foam can be used: polyester with hydroxyl end groups made from adipic acid, diethylene glycol, trimethyl propane (methylol) as well as polyether based on ethylene oxide and/or propylene oxide with free hydroxyl groups in combination with 2,4-toluene diisocyanate and 2,6-toluene diisocyanate. Stabilizers, dispersants, and amines (as catalysts in amounts up to 1.2%) can be used.

2.3.4
Natural and Synthetic Rubber

The consumption of classical elastomers in 1997 was about 16×10^6 tons, of what 60% was synthetic rubber: styrene-butadiene rubber (SBR) 53%, butadiene rubber (BR) 18%, isoprene rubber (IR) 2%, chloroprene rubber (CR) 3%, ethylene-propylene-diene rubber (EPDM) 8%, and acrylonitrile-butadiene rubber (NBR) 4%.

In food contact materials and articles that contain rubber, only a portion of the material's surface area contacts the food and for only a short time. For this reason, articles containing rubber can be divided into five categories corresponding to the intended application of the rubber–containing materials and articles. These categories are as follows (Franck et al., 2006):

Category 1. Food contact materials and articles which according to their intended use come in contact with food for 24 h to several months. Examples of such applications are storage containers, protective container coatings, seals with large surface areas, and sealing rings for cans, jars, bottles, and similar articles.

Category 2. Food contact materials and articles which according to their intended use come in contact with food for a maximum of 24 h. Examples are hoses for pumping food, stoppers and caps for bottles, sealing rings for pressure cookers, hoses for coffee machines, lid seals for milk containers, and valve balls.

Category 3. Food contact materials and articles which according to their intended use come in contact with food for a maximum of 10 min (short time contact). Examples are shaker beakers, milking machine hoses, seals for milk processing equipment, membranes, pistons, faucets, pump rotors, roller coverings, and conveyors for fatty foods as well as gloves and aprons worn when handling food.

Category 4. Food contact materials and articles from which, according to their intended use conditions, no mass transfer to food could occur. Examples are conveyor belts and roller covers, suction and pressure lines for filling or emptying tanks, seals for pipes, pumps, faucets, and valves for drinking water.

Special category. Food contact materials and articles which are intended to be used during the consumption of food which, according to their intended use, will be put in the mouth. Examples are toys, balloons, nipples, pacifiers, teething rings, and bite protectors.

Solid rubber, latexes, and rubber dispersions may be used for the manufacture of food contact materials. The starting materials used for solid rubber are natural rubber, GCPs of natural rubber with acrylic or methylacrylic acid esters of monofunctional C_1–C_4 alcohols; butadiene and isoprene polymers; polymers and polymer mixtures of mono-, di-, and trichlorobutadiene including styrene or acrylonitrile (chloroprene rubber); polymer mixtures of butadiene as well as isoprene and styrene and/or acrylonitrile (nitrile rubber) and/or divinyl benzene and/or acrylic as well as methacrylic acid; polymer mixtures of isobutylene and isoprene (butyl rubber); polymer mixtures of ethylene and α-olefin with chain length C_3–C_4 and/or acyclic or cyclic double unsaturated monomers (ethylene-propylene-rubber); chlorosulfonated PE for rubberized materials and clothing.

These starting materials can be used alone and in combination with various polymers or polymer mixtures. The starting materials for solid rubber may also be used for the manufacture of latexes and rubber dispersions.

In addition to the mentioned starting materials many processing aids, stabilizers, and other additives are used which are potential migrants into the product coming in contact with the rubber.

2.3.5
Silicones

The term silicone, or more precisely polyorganosiloxane, is used to describe a group of plastics whose backbones are composed of an inorganic chain of alternating silicon and oxygen atoms. The silicon valences not bound to oxygen are saturated using organic groups (R):

$$-\underset{R}{\overset{R}{Si}}-O-\underset{R}{\overset{R}{Si}}-O-\underset{R}{\overset{R}{Si}}-O-$$

With this structure, the silicones occupy a special classification between the inorganic silicates and organic polymers. This structure leads to a series of very useful material characteristics which no other product group has. The numerous possible compounds with linear or crosslinked spatial structures can lead to oils, greases, rubbers, or resins that are used in every conceivable technology.

Uses of silicone fluids: polish additives, release agents, water-repellent applications, lubricants, and greases

Uses of silicone resins: laminates, molding compositions

Uses of silicone rubbers: car uses (sealing rings, O-rings, gaskets, ignition cables, etc.), cable insulations

Diverse applications: blood transfusion tubing, antibiotic container closures, domestic refrigerators, nonadhesive rubber-covered rollers for handling different materials, potting and encapsulation, medical applications (particularly for body implants).

The building blocks of the polymer are the silicon–containing siloxane units. Their composition comes from the fact that every oxygen atom acts as a bridge between two silicon atoms and only half of the silicon atom is needed for bonding. The number of oxygen bonds determines the functionality of a siloxane unit (Table 2.3). The use of the symbols found in the table simplifies the description of the chemical composition of the macromolecule, usually polymethyl siloxane. If the siloxane methyl groups are partly or completely replaced by other groups, then a corresponding symbol is used like D^{Ph2}, T^{Ph}, D^{Vi}, M^H (Ph = C_6H_5–, Vi = CH_2=CH–, H = hydrogen). As an example, one obtains the simplified notation:

$$(CH_3)_3Si-O-[-\underset{CH_3}{\overset{CH_3}{Si}}-O]_x[-\underset{CH_3}{\overset{C_6H_5}{Si}}-O]_y-\underset{CH_3}{\overset{CH_3}{Si}}-CH=CH_2 \quad : MD_x D_y^{Ph} M^V$$

The most important starting materials (>95% by weight) for silicone synthesis are methyl chlorosilanes. Next to them are phenyl chlorosilanes, methyl phenyl chlorosilanes, methyltrifluoropropyl chlorosilanes, and a variety of silanes containing organic functional groups such as hydroxyl, amine, epoxy, acrylate, and carboxyl which are responsible for the different products.

Table 2.3 Labeling of siloxane unit structures.

Unit of siloxane	Formula	Symbol
CH$_3$–Si(CH$_3$)(CH$_3$)–O–	$(CH_3)_3SiO_{1/2}$	M
CH$_3$–Si(CH$_3$)(O)–O–	$(CH_3)_2SiO_{2/2}$	D
–O–Si(CH$_3$)(O)–O–	$CH_3SiO_{3/2}$	T
–O–Si(O)(O)–O–	$SiO_{4/2}$	Q

The intermediate steps leading to the formation of polydimethyl siloxanes are composed of linear and/or cyclic dimethyl siloxanes formed by hydrolysis or methanolysis of dimethyl dichlorosilane:

$$n\,Cl-Si(CH_3)(CH_3)-Cl + (n+1)\,H_2O \rightarrow HO\,D_n\,OH + 2n\,HCl$$

$$n\,(CH_3)_2\,SiCl_2 + n\,H_2O \rightarrow D_n + 2n\,HCl$$

$$n\,(CH_3)_2\,SiCl_2 + 2n\,CH_2OH \rightarrow D_n + 2n\,CH_3Cl + n\,H_2O$$

The hydrochloric acid by-product can be used again for the formation of silanes. The same can happen with the methylene chloride produced by methanolysis. If the dimethyldichlorosilane used contains residual tri- or tetrafunctional silanes or other impurities, a clean-up step of polydimethyl siloxane is conducted using a ring closing reaction with the help of alkali. The synthesis of linear polydimethyl siloxane for silicone oil and elastomers (raw materials for silicone rubber) takes place either through acid or basic ring opening and polymerization, or through polycondensation of linear dihydroxymethyl siloxane with hydroxyl end groups. The average chain length and subsequent viscosity of silicone oil is adjusted by the addition of chain-building siloxane types such as MD_nM, M_2, or water.

The branched polyorgano-siloxanes present in silicone resins are manufactured in an analogous manner. The intermediates from the siloxane oligomers are the starting materials for the polycondensation reaction. The resulting product can contain OH− or alkoxy groups and can be used for further processing to make silicone–based resins. These resins can be pure silicones or copolymers with polyethers.

The details of silicone synthesis mentioned here show the complicated problem of assigning starting materials for these plastics, which can be the object of mass transfer processes. From the analytical and toxicological viewpoints, these materials are completely altered during the course of the reaction by, for example, ring formation and alteration in the presence of alkali. The only detectable substances that can be considered are the residual low molecular weight oligomer mixtures. A listing and description of individual components is just not possible.

The technically most important silicone oils are of the type MD_xM. They are resistant to high temperatures up to 250°C in the presence of air. The products containing phenyl groups are the most stable. The low temperature dependence of their viscosity makes their use at very low temperatures (−60°C) possible.

Their packaging uses are usually in the form of emulsions. They function as coatings and coverings, for example as lubricants for corks and plastic casings or as antistick coatings for rubber stoppers and seals. The addition of thickening fillers (e.g., calcium salts or straight saturated monofunctional carboxylic acids C_{10}–C_{20}) to silicone oil produces greases as well as pastes.

Silicone elastomers (silicone rubber) are marked by their high stability and constant mechanical properties over a wide temperature range (−50 to +200°C and for short times up to 300°C). They can be manufactured using a two–component system that becomes crosslinked at room temperature (RTV 2 K) or through formation of crosslinks at high temperatures (hot vulcanizable, HV). The RTV 2 K elastomers are crosslinked either with polydimethyl siloxane with hydroxyl end groups in the presence of a catalyst (Di-*n*-octyl-tin-dilaurate) or with polydimethyl siloxane with dimethylvinyl siloxy end groups with the help of a platinum catalyst. The crosslinking of HV products takes place between the polymethyl siloxane containing vinyl groups with the help of organic peroxide. Platinum catalysts can also be used for crosslinking.

Silicone elastomers can be used for paper coatings, nipples, teething rings, seals for baking ovens, refrigerators, etc.

Silicone resins can be made from pure polyorganic siloxanes with a high percentage of tri- and tetrafunctional siloxy groups, which are crosslinked in the presence of a catalyst by heating at 250°C for a long time. They can also be made from a mixture of polyorganic siloxanes with polyesters.

One differentiates resins into those for direct food contact use like lacquers for baking forms and paper coatings and those resins for indirect food contact use like grills, warming plates, and others.

2.3.6
Plastics Based on Natural Polymers Regenerated Cellulose

The group of plastics known as regenerated celluloses forms a transition from the natural polymers to the completely synthetically produced plastics. As a converted natural product, regenerated cellulose foil (cellophane) has formed the basis for the first transparent, flexible packaging for food and tobacco which contributed significantly to the development of the self–service shops since the beginning of the 1920s.

Regenerated cellulose, as its name implies, is manufactured from cellulose. The raw material is transformed to sodium-cellulose-xanthogenate (viscose) with sodium hydroxide and carbon disulfide. With this process, the cellulose chains composed of over 1000 glucose units are reduced to approximately 400 units. The very viscous viscose is passed through a die under pressure into a sulfuric acid bath, where it coagulates and a regeneration of cellulose takes place. The soft, regenerated cellulose is then passed through a series of baths where it is further strengthened, bleached, and conditioned. During these processes, plasticizers and moisturizing agents are worked into the regenerated cellulose film.

In order to make the regenerated cellulose film heat sealable, it is lacquered. The resulting different film types are then described according to an international standard nomenclature. With a three place number, the weight-to-surface area ratio is given. With a combination of letters, the following properties are labeled: unlacquered (P), nitrolacquered (M), PVDC lacquered (X), heat sealable (S), suitable for folding and twist wrapping (F), and high water vapor permeability (D).

The glass-clear, shiny and tear-resistant regenerated cellulose is permeable to water vapor, but resistant to contact with fats. P-regenerated cellulose foil is used for the packaging of baked goods and dough products for short storage times, and also used for candies. The permeability of the lacquered films to water vapor, oxygen, and aromas depends on the type of lacquer. X-regenerated cellulose foil is particularly impermeable. Lacquered regenerated cellulose is used for the packaging of bread and other baked goods, particularly for fatty products, for spices, nuts, dried fruits, pralines, foods with intensive odors, and cigarettes. The outstanding properties of regenerated cellulose as a packaging material are its machinability and thick seal seams, because cellulose does not melt when the heat seal temperature is set too high.

For economic reasons, regenerated cellulose has been replaced recently or is in serious competition with PP films in many applications. Regenerated cellulose has the advantage that it biodegrades well in composting.

2.3.6.1 Biodegradable Polymers

Polysaccharides One of the industrially available natural polymers for the manufacture of biodegradable plastic materials is starch. Starch is produced and stored by various types of plants. The differing properties of various starches result from their different proportions of amylose (linear starch chains) and amylopectin (branched starch chains).

The solid polymeric starch granules with their defined structures and crystallinities can be incorporated as biodegradable filler into nonbiodegradable synthetic polymers like PE. Starch and PE mixtures can be extruded. Until now, shopping bags and compost sacks have been manufactured and various application possibilities tested in the packaging area.

Starch is suitable, in principle, for processing as a thermoplastic. Water retained in the starch can function as a plasticizer. Starch has also been used as a raw material for producing foams.

By fermentation of the starch one obtains pullalane, which belongs to the polymaltoses group. The expensive material is not extrudable, but rather suitable for molding processes and can be used for water soluble packaging.

Polyesters Various procaryotic microorganisms can produce polyhydroxyalkanoates, using regenerable carbon sources. This polymer is a storage material and can make up to 90% of the dried cell weight. The most widely researched material in this group up till now is the poly-D(−)-3-hydroxybutyric acid (PHB).

ICI has developed a fermentation process for PHB having various levels of polyhydroxyvalerate (PHV) as a copolymer. This polymer is marketed under the product name BIOPOL. The material is extrudable and can be used as bottles for packaging cosmetics. Because of the high migration rate of the triacetine plasticizer used (see Chapter 4), it is not suitable for food use. Currently, efforts are being made to manufacture this material as a film.

While PHB and PHV are not considered true plastics, another biodegradable polymer polycaprolactone (PCL) is a plastic material because its monomer ε-caprolactone is obtained on an industrial scale from petrochemical products (cyclohexanone and peroxyacetic acid). This synthetic plastic with its low melting point is easily extrudable, and applications in the packaging area are envisioned.

In view of the necessity for getting waste disposal under control coupled with limited fossil raw material resources, biodegradable polymer and in particular polymers from renewable resources will gain importance in the future. In the most sensitive application area, food contact materials, and articles, it is possible initially to use these materials in very limited amounts. The easy decomposition of these packaging materials is at variance with the inertness needed to protect packaged food. These polymers are particularly sensitive to moisture. By finishing operations such as surface treatments, one could improve the inertness of these polymers. However, the degradability would be diminished by such processes.

2.3.7
Coatings and Adhesives

Most of the plastics discussed here now could be used, not just in the manufacture of different materials and articles, but also as finishing materials in the form of coatings on other substrates or as adhesives between two layers. By using coatings, paper for example could be made impermeable to water or fat, tin cans protected against corrosion, kitchen articles and appliances made nonstick, etc. Whereas films, containers, and other plastic materials and articles are manufactured from polymerized materials, one often uses partly polymerized mixtures as coatings and adhesives. These mixtures are then chemically crosslinked or thermally treated to bind them to the substrate material to make the finished product. The variety of substances used in the coating and adhesive mixtures makes the study of mass transfer processes often difficult. These substances, for example, can be solvents, crosslinking agents, or certain starting materials. In addition, the coating formulations are composed of many low molecular weight polymers whose properties have been selected for a particular application. Coated articles can also come in contact with food.

2.3.7.1 Lacquers

The manufacture of lacquers with good adhesion, independent of the intended application, requires numerous chemical substances. These can be broadly divided into volatile and nonvolatile components. Solvents belong to the volatile group and binding agents and additives belong to the nonvolatile group.

Binding agents are the main coating components and principal film formers. Film formation can take place using the following techniques:

- application of a solution or dispersion of macromolecular binding agent onto a substrate and evaporation of the solvent; in order to reduce the amount of organic solvent, water based solutions and dispersions are being used more frequently,
- application of the binding agent in molten form,
- film formation by chemical reaction.

The physically dried binding agents include acrylic resins, polyesters, silicones, cellulose derivatives, and others. Polyamides and polyolefins are used as raw material for powder lacquers. Differing from physically dried lacquers, chemically dried lacquers contain crosslinked macromolecules. The raw materials used as binding agents can, in this case, be used without solvents because of their low viscosities and react with the substrate by one of the above-mentioned polymerization reactions.

One can further divide lacquers according to their drying processes into oxidative drying, cold hardening, irradiation hardening, and oven drying.

The additives used for lacquer manufacture are composed of processing aids like hardening accelerators, flow agents, crosslinking agents, and others like plasticizers, fillers and, if necessary, pigments.

Natural products and transformed natural products can also be used as binding agents for packaging applications. These are shellac, dammar gum, dried unsaturated oils (linseed oil), rosin and hydrated esters of rosin, cellulose acetate and propionate acetate, cellulose acetobutyrate, cellulose nitrate, and ethyl cellulose.

The following polycondensation, polymerization, and polyaddition products can be used as binding agents: alkyd and polyester resins, epoxy compounds, phenol-formaldehyde resin, urea and/or melamine-aldehyde resin, cyclic urea resin, carbamide acid ester formaldehyde resin, ketone formaldehyde resin, polyurethane, polyvinylester, polyvinyl acetate, polyvinyl chloride and polymer mixtures, PE, polystyrene, styrene mixtures and GCPs, polyamide, polycarbonate, polyvinyl ether, polyacrylic and methacrylic acid esters, polyvinyl flouride, polyvinylidene chloride copolymers, UV, and/or electron-irradiated lacquers.

2.3.7.2 Plastic Dispersions

A dispersion is a system consisting of two or more phases in which a material is finely distributed throughout another. Dispersions can be liquid/liquid (emulsions) or solid/liquid (suspensions). Polymer dispersions consist of an internal phase – the polymer or copolymer and an external phase – water.

The consumption of polymer dispersions in 1997 was 10×10^6 tons. The market is divided among styrene-butadien dispersions (35%), dispersions containing vinyl acetate (32%), styrene and styrene-acrylate dispersions (25%), and others in minor quantities. They have many applications: coatings and paints, adhesives, textile finishing, paper coatings, and others. When used as coatings, the dispersions should be suitable for food contact. Many substances can be used as monomers:

- acrylic acid and methacrylic acid esters of C_1–C_8 monofunctional aliphatic saturated alcohols and ether alcohols (RO–$(CH_2)_x$–OH),
- vinyl esters of C_1–C_8 aliphatic saturated carboxylic acids,
- vinyl chloride, vinylidene chloride,
- acrylonitrile and methacrylonitrile,
- ethylene, butadiene, isoprene, isobutylene, propylene, 2-chlorobutane, 2,3-dichlorobutadiene, tetrafluoroethylene, styrene,
- malic acid and fumaric acid esters from C_1–C_8 monofunctional aliphatic saturated alcohols or C_3–C_{18} monofunctional aliphatic unsaturated alcohols, esters of C_3–C_{12} aliphatic carboxylic acids with C_3–C_{18} unsaturated alcohols, esters of unsaturated dicarboxylic acid with polyethylene glycol and/or PP glycol,
- vinyl ethers of C_1–C_{18} monofunctional aliphatic saturated alcohols,
- acrylic acid, methacrylic acid, crotonic acid, malic acid, fumaric acid, itaconic acid, vinyl sulfonic acid, styrene sulfonic acid, half esters of malic and fumaric acid, itaconic acid with C_1–C_{18} monofunctional aliphatic saturated alcohols as well as their alkali and ammonium salts, vinyl pyrolidone, amides of acrylic and methacrylic acids and N-methylolamide of acrylic and methacrylic acid as well as their ethers, N-vinyl-N-methylacetamide,
- acrylic acid ester of diethylaminoethanol and/or methacrylic acid ester of dimethylaminoethanol,
- acrylic acid and methacrylic acid esters of C_2–C_{18} aliphatic dialcohols, divinyl and diallyl esters of C_3–C_{18} saturated aliphatic dicarboxylic acids, vinyl and allyl esters of acrylic acid and crotonic acid triallyl cyanurate.

Some of these dispersions can be crosslinked by vulcanization.

2.3.7.3 Microcrystalline Waxes

The substances discussed here are used mostly for coating, impregnation and lamination of paper packaging. They can be divided into four groups:

1. Hard paraffins having natural origins. Included in this group are mixtures of solid, pure, mostly straight-chain hydrocarbons coming from oil, brown coal, and shale tar oils. The solidification temperature must lie between 43°C and 75°C. For liquid packages, in particular milk, this temperature may not be below 52°C. Further conditions (kinetic viscosity, iodine color number) and purity requirements are given. In particular these materials may not contain any fluorescence-quenching substances.

2. Microcrystalline wax. These are mixtures of solid, purified, mainly branched, saturated microcrystalline hydrocarbons made from oil with solidification temperatures between 50 and 90°C. They must also meet specific purity requirements.

3. Synthetic hard paraffin. This includes mixtures of high molecular weight, solid, purified, mainly straight-chain hydrocarbons with solidification temperatures between 92 and 105°C.

4. Low molecular weight PP with a relative molecular mass between 2500 and 6000.

The above-mentioned materials can be mixed with one another. A series of other polymers and resins can also be added if the substances listed in 1 to 4 form the bulk of the material. Additional materials are: PE, PP, low molecular weight polyolefins, polyterpenes (mixtures of aliphatic and cycloaliphatic hydrocarbons produced by polymerisation of terpene hydrocarbons), polyisobutylene, butyl rubber, dammar gum, glycerine and pentaerythrite esters of rosin acid and their hydration products, polyolefin resins, hydrated polycyclopentadiene resin (substance mixtures manufactured by thermal polymerization of a mixture mainly composed of di-cyclopentadiene with methylcyclopentadiene, isoprene and piperylene which is then hydrogenated).

The finished products coated with substances or substance mixtures listed above may not be used as food contact materials for fats and oils or for foods containing fat in which the fat forms an external phase.

2.3.7.4 Temperature-Resistant Coatings

Fluoropolymers are used for the manufacture of coatings for frying pans, pots, fryers, and other cooking equipment and utensils. Polytetrafluoroethylene with a melting point of approximately 327°C is mostly used, but polymer mixtures with perfluoroalkylvinyl ether and hexafluoropropylene can also be used.

According to the intended conditions of use, these types of cooking equipment can reach temperatures of 320–340°C for short times (<15 min) in places not covered by food (e.g., foods like sausages, pancakes).

In addition to the polymer substrate, a series of binding resins made from polyamide and polyimide, polyphenylsulfide (PPS), polyether sulfone (PES), and/or silicone resin are necessary for applying the coating. Such substances like laminating agents (lithium polysilicate, aluminum phosphate, and phosphoric acid) and various additives are also used. Included in these additives are emulsifiers and further processing aids (e.g., silicone oil).

Under the present state of the art technology with correctly applied sintering of the coating, the total amount of substance released during heating of the cooking item is well under $1\,\mathrm{mg\,dm^{-2}}$.

2.3.7.5 Printing Inks and Varnishes

Printing inks, including varnishes, are products manufactured from pigments, binders, solvents, and additives. Many of the substances mentioned in this chapter and especially in the above sections are also used as components of printing inks and varnishes. There are solvent-based, waterborne, oleo-resinous, or energy-curing (UV or electron beam) systems. They are applied as flexography, offset, gravure printing, and roller varnishing.

Pigments used in printing inks include both inorganic pigments (e.g., carbonblack, titanium dioxide) and insoluble organic pigments prepared often from azo, anthraquinone, and triarylmethane dyes and phthalocyanines The manufacture of inks consists of dissolving or dispersing resins in organic solvents or oils to produce the vehicle (varnish), mixing and dispersing the pigment into the vehicle, and then introducing the necessary additives. In lithography and letterpress, where inks are dried by absorption and oxidation, vehicles are generally mixtures of mineral and vegetable oils and resins. Flexograpic inks, which are designed to dry quickly by evaporation, can be based on water or organic solvents such as ethanol, ethyl acetate, n-propanol, or isopropanol, with a wide variety of resins. Vehicles for gravure inks may contain aromatic or aliphatic hydrocarbons and ketones as solvents. Inks for screen printing use organic solvents that are somewhat less volatile than those used for flexography or gravure (e.g., higher glycol ethers and aromatic and aliphatic hydrocarbons) (Robertson 2006). UV radiation-cured inks, commonly based on acrylates, are used in all of the printing processes to varying degrees. For UV curing, a variety of photoinitiators are used which may produce a series of reaction products. Traces of photoinitiators and their degradation products can be transported into foods by diffusing through the packaging layer and/or by set-off. Set-off can occur in the stack or on the reel subsequent to printing, when ink transfers from the print to the reverse of the adjacent sheet.

Binders are the film-forming components of inks in which the coloring material is finely dispersed. After the drying of the print, binders serve to adhere the ink film to the printed surface and contribute to functional properties.

1. Binders for conventional drying are:
    - rosin-derived resins obtained from rosin, maleic acid, fumaric acid, glycerol, pentaerythritol, isophthalic acid, formaldehyde;
    - hydrocarbon resins and phenolics made from cresol, cyclopentadiene, formaldehyde, phenol, and 4-*tert*-butyl phenol;
    - cellulose resins: nitrocellulose, ethylcellulose, cellulose acetate butyrate (CAB), cellulose acetate propionate (CAP), alcohol soluble propionates (ASP);
    - acrylic acid: styrene copolymer obtained from acrylic acid and its methyl, ethyl and butyl esters, methacrylic acid and its methyl, ethyl, butyl and ethyl-hexyl esters, styrene and α-methyl-styrene;
    - vinyl resins: polyvinyl acetate, PVC, polyvinylbutyral, polyvinylether;

- polyamide resins: PA 66, 610, 11, 6, 12;
- polyurethane resins;
- urea and melamine formaldehyde resins;
- oxidative oils, alkyds, and fatty acid monoalkylesters obtained from soybean oil, linseed oils, sunflower oil, palmitic acid, isophthalic acid, phthalic anhydride, benzoic acid, adipic acid, maleic anhydride, trimethylol propane, glycerol, pentaerythritol;
- cyclohexanone formaldehyde copolymer;
- polyester resins from terephthalic, isophthalic and adipic acid, phthalic anhydride, azelaic acid, hexahydrophthalic anhydride, cyclohexane dicarboxylic acid, 2,2-dimethyl-1,3-prpanediol, trimethylol propane, diethylene glycol, ethylene glycol, 1,6-hexanediol,1,2-propanediol;
- epoxy resins from BADGE, bisphenol-A, epichlorohydrin, cyclo aliphatic epoxides.

2. Binders for energy curing packaging inks are:
   - stenomeric acrylates: tripropylene glycol diacrylate (TPGDA), dipropylene glycol diacrylate (DPGDA), 1,1,1-trihydroxymethylpropyl triacrylate (TMPTA), pentaerythritol tri-tetraacrylate (PETA), glycerol propoxylate triacrylate (GPTA), polyoxyethylpentaerythritol tetraacrylate (PPTTA);
   - eurymetric acrylates (products with a rather high molecular weight and a wide molecular weight distribution): aromatic and aliphatic urethane acrylates, polyester acrylates, polyether acrylates, epoxy acrylates.

Additives are used in small quantities and determine the technical properties of the printing inks and overprint varnish. They are often specific to certain ink types: acid catalyst, adhesion promoter, amine solubilizer, antifoam agent, antimist, antiatatic, biocide, chelating agent, flow agent, gallant, ink stabilizer, optical brightener, photoinitiator, slip agent.

References

Atkins, P. and de Paula, J. (2006) *Physical Chemistry*, 8th Edn. Oxford University Press, Oxford.

Beer, G., Wiesecke, J., Mannebach, G., Auffermann, J., Vogt, H., Bisson, P. and Marczinke, B. (2005) Low-density polyethylene (PE-LD/LLD). *Kunststoffe Plast Eur.*, **95** (10), 54–59. (translated from Kunststoffe).

Brydson, J.A. (1995) *Plastics Materials*, 6th Edn. Butterworth-Heinemann, Oxford.

Elias, H.G. (2005) *Macromolecules*, Vol. 1: *Chemical Structures and Syntheses*, Wiley-VCH, Weinheim.

Dusek, K. (2005) Network formation. In: *Materials Science and Technology*, Vol. 18 (eds. R.W. Cahn, P. Haasen and E.J. Kramer), Wiley-VCH, Weinheim. pp. 402–427.

Franck, R., Wieczorek, H. and Otto, U. (eds) (2006) *Kunststoffe im Lebensmittelverkehr*, Carl Heymanns Verlag KG, Köln.

Hu, G.-H. and Lambla, M. (2005) Fundamentals of reactive extrusion: an overview. In: *Materials Science and Technology*, Vol. 18 (eds. R.W. Cahn, P. Haasen and E.J. Kramer), Wiley-VCH, Weinheim. pp. 345–400.

Inoue, T. and Marechal, P. (2005) Reactive processing of polymer blends: polymer-polymer interface aspects. In: *Materials Science and Technology*, Vol. 18 (eds. R.W. Cahn, P. Haasen and E.J. Kramer), Wiley-VCH, Weinham. pp. 429–463.

Janssen, J.M.H. (2005) Emulsions: the dynamics of liquid–liquid mixing. In: *Materials Science and Technology*, Vol. 18 (eds. R.W. Cahn, P. Haasen and E.J. Kramer), Wiley-VCH, Weinheim. pp. 154–188.

Keller, A. (2005) Flow-induced orientation and structure formation. In: *Materials Science and Technology*, Vol. 18 (eds. R.W. Cahn, P. Haasen and E.J. Kramer), Wiley-VCH, Weinheim. pp. 189–268.

Meijer, HEH. (2005) Processing for properties. In: *Materials Science and Technology*, Vol. 18 (eds. R.W. Cahn, P. Haasen and E.J. Kramer), Wiley-VCH, Weinheim. pp. 3–75.

Peters, E.N. and Parthasarathy, M. (2005) Polyphenylene ether (PPE). *Kunststoffe Plast Eur.*, **95** (10), 108–113. (translated from Kunststoffe).

Robertson, G.L. (2006) *Food Packaging. Principles and Practice*, 2nd Edn., CRC Taylor and Francis, Boca Raton, FL.

Schwaben, H.-D. (2005) Polystyrene (PS). *Kunststoffe Plast Eur.*, **95** (10), 68–75. (translated from Kunststoffe).

Sparenberg, B. (2005) Cyclic olefin copolymers (COC). *Kunststoffe Plast Eur.*, **95** (10), 156–160. (translated from Kunststoffe).

Stanford, J.L., Ryan, A.J. and Elwell, M.J.A. (2005) Structure development in reactive systems. In: *Materials Science and Technology*, Vol. 18 (eds. R.W. Cahn, P. Haasen and E.J. Kramer), Wiley-VCH, Weinhiem. pp. 466–512.

3
Polymer Additives

Jan Pospíšil and Stanislav Nešpůrek

3.1
Introduction

When considering mass transport phenomena in plastics used as packaging materials in contact with fresh or processed food, bottles for drinks, cosmetics, pharmaceuticals, as greenhouse or mulching films in agriculture and various articles or implants in human and veterinary medicine (for example, bags for blood transfusion and dialysis, coverings for facial or limb corrections, surgical fibers or supports, syringes), specific attention has to be paid to a possible transfer of low molecular weight compounds (residual monomers, oligomers, stabilizers, polymer degradation products). Commodity plastics used for the above-mentioned applications, i.e., high density polyethylene (HDPE), low density polyethylene (LDPE), linear-low density polyethylene (L-LDPE), isotactic polypropylene (PP), poly(vinyl chloride) (PVC), poly(ethylene terephthalate) (PET), polyamide (PA), and polystyrene (PS), are a commercial failure without additives. The latter are organic or inorganic chemicals enabling processing of plastics, shaping their use, and enhancing end-use performance. Additives influence the future development of plastics and make them value-added materials under increased environmental regulations at regional difference. They are used at levels of 0.05 wt.% up to about 20 wt.%. Their market is very cost competitive, standardized across the globe, and corresponds to the world growth of plastics' production. Additives are classified by function and not chemistry (Table 3.1), and are used under stringent legislation and environmental rules. The legal bans against particular chemicals are global selective. Chemical legislation proposed by the European Commission as Registration-Evaluation and Authorisation of Chemicals (REACH) (Bennemann, 2005) may affect additive application rules.

About 75% of all additives have been consumed in PVC. Additives modifying plastics' properties constitute about 70% of the totally consumed amount in comparison to 23% of property extenders and 7% of processing aids (Pfaendner, 2006).

Some additives or their residues remaining in trace amounts in the plastics, such as processing aids, are not generally declared in commercial articles. The presence of

Table 3.1 Additives used in plastics for contact with food, pharmaceuticals and medical applications.

Additive	Polymer							
	HDPE	LDPE	L-LDPE	PP	PS	PA	PET	PVC
Antifogging agent	+	+	+				+	+
Antistatic agent	+	+	+	+	+	+	+	+
Blowing agent				+	+			+
Colorants	+	+	+	+	+	+	+	+
Fillers, reinforcing agents	+	+	+	+	+	+	+	+
Lubricants	(+)	+	+	+	+			+
Nucleating agents	+			+		+	+	
Optical brighteners	+	+	+	+	+	+	+	+
Plasticizers								+
Stabilizers								
Antiacid	+	+	+	+			+	+
Antimicrobials	+	+	+	+			+	+
Benzofuranone (lactone)	+	+	+	+	(+)			
Dehydrating agent							+	
Dialkylhydroxylamine				+				
Heat stabilizers (organotins, metal soaps)							+	+
Hindered amine stabilizers	+	+	+	+	+	+	(+)	
Organic phosphate (not as flame retardants)							+	+
Organic phosphite, phosphonite	+	+	+	+	+		+	+
Phenolic antioxidants	+	+	+	+	+	+	+	+
Thioethers (thiosynergists)				+	+			
UV absorbers	+	+	+	+	+	+	+	

additives having properties of stabilizers and plasticizers must be declared. Their suppliers must provide appropriate details on the regulation status of each additive with respect to industrial hygiene, the environment, and indirect contact with food. A part of the additives, stabilizers in particular, is very reactive and is present in the plastic matrix in a chemically transformed form (see Section 3.12).

Characteristic functions of plastic additives providing marketable packaging materials and their representative structures are outlined in this chapter. Details may be found in specialized monographs or reviews, for example, Pospíšil (1995), Pospíšil and Nešpůrek (2000a), Pritchard (1998), and Zweifel (2001).

3.2
Antifogging Agents

The optical clarity of packaging materials containing food with high water content (vegetables, fruits, various cheeses, fresh meet) or greenhouse films can be impaired after cooling below the dew point by the formation of small discrete water droplets on the inner surface of the films making the content invisible. In applications where perforated films having effective water vapor transmissions are undesirable, antifogging agents with properties of surface active substances, such as poly(oxy ethylene), sorbitan monooleate, or polyglycerol are effectively used (Wylin, 2001).

3.3
Antistatic Agents

Static electricity is a considerable problem in highly insulating plastics. Ionogenic and nonionogenic additives reducing the chargeability of plastics are applied either from solutions on the plastic surfaces ("external" antistatics) or mixed into the plastic masses during processing (migratory "internal" antistatics decreasing transparency). A semi-conducting layer dissipating charges is formed on the surface of the plastic (Clint, 1998). Hydrophobic heads of the molecule of antistatic agents are fixed to the surface of the macromolecule and separate it from the direct contact with air humidity. The hydrophilic tails absorb humidity, increase electrical conductivity of the surface, and help to prevent electrostatic build-up. Spark discharges capable of igniting flammable gases are prevented, and dust pickup at surfaces of films or containers is eliminated.

Cationactive agents (quarternary ammonium, phosphonium, or sulfonium salts, for example (3-dodecanoylamidopropyl) trimethylammonium methylsulfate (**1**)) and anionactive agents (for example, alkylsulfonate (**2**) or salts of alkylated benzenesulfonic acid) are typical ionic external antistats. Nonionic agents (such as poly(ethylene glycol) monoethers (**3**) or ethoxylated fatty alkylamines or amides, glycerol, poly(ethylene glycols)) are used as internal antistatics. Their molecular architecture is important in terms of the compatibility with the plastics and the migration rate to the polymer surface. On the one hand, the solubility in the plastic must not be too good so that there is an adequate tendency for it to migrate to the surface. On the other, the solubility must not be too low so that blooming is too fast. The solubility of external antistatics plays no role in their activity mechanism. Ionic antistatics are of choice for PVC and PS. Nonionic antistatics are excellent for polyolefins (PO).

Concentration of antistatics in plastics is usually from 0.1 to 2 wt.%. Special grades of electro-conducting (EC) carbon black are used in PO at levels higher than 10 wt.%. Other conducting fillers incorporating antistatic effects, such as metals or organic semiconductors (for example, polypyrrole or polyaniline), are not commonly used in plastics for contact with food.

$$[(CH_3)_3N^+(CH_2)_3NHCOC_{11}H_{23}]\ CH_3OSO_3^- \quad\quad [O=SO_2C_{12}H_{25}]^-\ Na^+ \quad\quad HO[(CH_2)_2O]_8C_{12}H_{25}$$

$$\textbf{1} \quad\quad\quad\quad\quad\quad \textbf{2} \quad\quad\quad\quad\quad\quad \textbf{3}$$

3.4
Blowing Agents

Expanded PVC or PS packaging materials are widely used in the catering industry. Various chemical blowing agents generating thermally inert gases are used (Hurnik,

2001), for example 2,2′-azobis(isobutyronitrile), *p*-toluenesulfonylhydrazide, *p*-toluenesulfonylsemicarbazide, diisopropylhydrazodicarboxylate, or 5-phenyltetrazole. Low molecular weight and migrating transformation products of these additives remain in the final expanded polymer matrix.

3.5
Colorants

Medium-soluble dyes having most different structures (derivatives of anthraquinone, quinophthalone, perinone, methine, azine, furanone) and essentially medium-insoluble pigments are used to give plastic articles market appeal or functional demand (light screening, conductivity). They also change transparency and weathering resistance (Christie, 1993; Pospíšil and Nešpůrek, 2000b). Pigments surpass dyes in light fastness, heat, and migration resistance. Colorants may be either mixed into the polymer mass or applied as printing inks on plastic surfaces. Some processing aids such as dispersants, binding agents (acrylic, alkyd, polyester, or melamine resins), or solvents have to be used together with colorants.

Inorganic pigments, such as titanium dioxide, zinc oxide, or carbon black, also act as efficient light screens and, in this way, protect plastics against photodegradation (Pospíšil and Nešpůrek, 2000b). Iron oxides and structurally very different organic pigments (such as various azo compounds or metal phthalocyanines) act primarily as colorants. Pigments containing cadmium and chromium(VI) are of environmental concern for food packaging. Various "effect" pigments (powdered metals, fluorescent, or perlescent pigments) are available. The surface adsorption of migrating stabilizers on pigments has to be taken into account when considering transport processes in plastics (Pospíšil and Nešpůrek, 2000b).

3.6
Fillers and Reinforcing Agents

Fillers are mostly powdered inorganic additives, such as calcium carbonate, talc (hydrated magnesium silicate), kaolin (hydrated aluminum silicate), mica (complex potassium/aluminum silicate), or silica (silicon dioxide), used to increase bulk and improve mechanical (impact resistance) and physical (heat and flame resistance) properties of plastics (Hohenberger, 2001). Glass, carbon, and polyester fibers are used as specific reinforcing additives in the manufacture of large rigid containers. Particular fillers are mostly coated to improve surface properties and compatibility with the polymer matrix. Polarity of fillers and their surface properties affect the adsorption of some migrating additives (Pospíšil and Nešpůrek, 2000b), such as hindered amine stabilizers (HAS) on acid fillers or phenolic antioxidants and acid products arising from sulfur-containing hydroperoxide-decomposing antioxidants or heat stabilizers on basic fillers. Transport processes of additives in plastics are influenced by this way.

3.7
Lubricants

Melt rheology of plastics is affected, and the processing above the glass transition temperature is improved by additives reducing the external friction on plastics/processing equipment interfaces and protecting from sticking to the mold of the machinery ("outer" lubricants) and internal friction on macromolecule/macromolecule interfaces improving the movement of polymeric chains ("inner" lubricants added into the polymer mass). Plasticizers reduce the danger of thermomechanical degradation (melt fracture) and improve the dispergation of components in multiphase systems (for example, polymer/pigment, polymer/filler, compatibilized blend). Finished articles have smoother surfaces due to the mold release and slip effects of lubricants.

Fatty alcohols C_{12}–C_{22}, fatty acids C_{14}–C_{18}, their esters with fatty alcohols, glycerol, or pentaerythritol, amides (4) or diamides (5) and metallic soaps, calcium stearate in particular, acids C_{28}–C_{31} from montan wax and their esters, esters of phthalic acid (6, R=C_8H_{17}), paraffin wax C_{20}–C_{70}, PE waxes C_{125}–C_{700} or their oxidized (polar) grades containing hydroxyl and carbonyl groups are typical lubricants (Richter, 2001). Some polymeric processing aids, for example fluoropolymers or silicones used in PO, are related to lubricants. The nonpolar lubricants (waxes) are readily soluble in PO. On the contrary, fatty acids and their esters are insoluble and function as outer lubricants. The latter come easier in contact with food than the inner lubricants. The solubility or compatibility of lubricants such as 4 or 5 is temperature dependent.

$$C_{15\text{-}17}H_{31\text{-}35}\overset{O}{\overset{\|}{C}}NH_2 \quad \left[C_{15\text{-}17}H_{31\text{-}35}\overset{O}{\overset{\|}{C}}NHCH_2\right]_2 \quad \underset{(6)}{\text{phthalate with COR groups}}$$

(4) (5) (6)

Lubricants are used in rigid PVC (PVC-U) in amounts of 0.3–0.8 wt.% (metallic soaps acting primarily as heat stabilizers are used at levels up to 1.5 wt.%). In PO, calcium or zinc stearates used as antiacids provide the lubricating effect at 0.1 to 0.2 wt.% levels. Fine particle fillers, such as silica gel or chalk, called antiblocking agents, can also be classified among external lubricants. They make the separation of rolled films easier.

3.8
Nucleating Agents

To achieve consistent properties and morphology of semicrystalline plastics, nucleating agents promoting or control formation and size of sphaerulites in crystallizable polymers are used. They provide sites for adsorbing some segments of the polymer chain and formation of a prerequisite for nuclei. Nucleation can occur on the surface of some impurities in the polymer (residues of polymerization catalysts in PO) or on fillers. Special additives controlling the process are preferred (Kurja and Mehl, 2001). They form either a physical gelation network within the polymer or provide single nucleation sites dispersed within the plastic mass. Sodium salts of organophosphates, salts of benzoic or phthalic acid, some organic compounds, for example sorbitol bis(3,4-dimethylbenzylidene diacetal) (7) or finely ground fillers (clays, silica), are used in amounts of 0.1–0.3 wt.%.

3.9
Optical Brighteners

Fluorescent agents improve the whiteness or brilliant white appearance of finished articles from thermoplastics and mask discoloration accounting for polymer degradation. The additives used up to 0.05 wt.% absorb light having wavelengths lower than 400 nm and re-emit it as visible blue or violet light (Christie, 1993). Derivatives of benzoxazole, phenylcoumarine (8), or stilbene (9) are mostly used.

3.10
Plasticizers

Plasticizers are additives gelling the polymer, improving processibility and flexibility or stretchability of plastics by decreasing their melt viscosity, glass transition temperature, and modulus of elasticity of the final product without alteration of the chemical character of the polymer (Rahman and Brazel, 2004). Plasticizers account for about one-third of the global additive market. More than 80% of them are used in PVC. The recent plasticizer development has focussed on material challenges improving resistance to leaching, migration, and health risk upon chronic exposure according to safety regulations. The aspects of toxicity during oral, dermal, or intraperitonal exposures have been strictly considered. Extraction resistance is the main criterion to avoid an indirect contact with human body.

Plasticizers are generally differentiated as external and internal. The former are bound to the plastics' macromolecule as a solvent, by van der Waals forces. Even small amounts of most plasticizers have noticeable effects on plastics. The role of diffusion and volatility of external plasticizers increases rapidly with temperature. Relatively low molecular weight external plasticizers used in polymer at high concentration levels become the most problematic extractable additives in plastics. Consequently, high molecular weight external plasticizers with limited migration rates are preferred together with plasticizer-functionalized polymers.

Various organic plasticizers having boiling points over 300 °C, such as di(2-ethylhexyl) phthalate (**6**, R=C_8H_{17}); di(2-ethylhexyl) adipate (**10**); oligomers with molecular weights ranging up to 3500 g mol^{-1} formed from dicarboxylic acids and diols; esters of trimellitic, citric, lactic, or benzoic acids; epoxidized fatty acids or organic phosphates, such as tris(2-ethylhexyl) phosphate (**11**) or (2-ethylhexyl) diphenyl phosphate (**12**). Most phosphates act at the same time as flame retardants. Some of them, such as **12**, are smoke suppressants as well. Phthalates, **6**, R=C_8H_{17} in particular, are the most worldwide-used plasticizers (Rahman and Brazel, 2004). Esters of dicarboxylic acids, such as **10**, impart very good flexibility at lower temperatures. Aliphatic oligomeric esters are excellent fat-resistant plasticizers. Epoxidized plasticizers provide some costabilizing effect to heat stabilizers in PVC and improve oil and water resistance. Blends of plasticizers are commonly used. This indicates the diversity in plasticizers that may be present in plastics.

$$\left[-(CH_2)_2\overset{O}{\overset{\|}{C}}OC_8H_{17}\right]_2 \qquad O=P(C_8H_{17})_3 \qquad O=P(OC_8H_{17})(OC_2H_5)_2$$

$$\text{10} \qquad\qquad\qquad \text{11} \qquad\qquad \text{12}$$

Due to high application levels of plasticizers, their behavior with respect to weathering, microbial sensitivity, and impact on optical properties of the polymer matrix have to be considered.

The plasticizer content in the plastics plays a key role in the evaluation of food contact materials and articles. As long as the danger that the plasticizer can easily transfer to food exists, the application in plastics is fundamentally undesirable. According to recommendations, which are valid in many countries, the application of external plasticizers is limited to tubes or hoses for liquid foods, coatings, seals, and sealants. A total plasticizer content of 35 wt.% may not be exceeded in films, coatings and tubes made from soft PVC, vinyl-chloride-based copolymers, and chlorinated PE-based blends with PVC. These materials in the form of powder or fine particles having a size lower than 3 mm and their final articles should not come into contact with food containing fat, waxed or paraffin-coated foods, milk and milk products including cheeses, and foods containing alcohol or essential oils. High standards are set for the application of articles from plasticized PVC in medicine and cosmetics. Soft PVC films with high oxygen permeability for packing fresh meat may not contain phthalates and phosphates, and the total amount of plasticizers may not exceed 22 wt.%. Specific attention has to be paid to the application of water nonextractable plasticizers for chlorinated rubber used as a coating for drinking water containers. Particularly, phthalates (**6**) are currently the focus of attention regarding their negative impact on the environmental human health. However, based on many toxicological investigations and their evaluation, it can be concluded that their application has no relevance to human health: they do not induce carcinogenity, reproductive effects, or endocrine modulations (Cadogan, 2002). The information obtained from tests with rodents is not transferable to human beings.

Internal plasticizers are nonextractable copolymers of suitable monomers. They have mostly substantially lower glass transition temperatures due to the presence of plasticizing ("soft") segments (such as poly(ethylene-co-vinyl acetate) with approximately 45 wt.% vinyl acetate content, ethylene-vinyl acetate-carbon monoxide terpolymer or chlorinated PE) and worse mechanical properties. They are designed for rather special applications in medicinal articles. A combination with external plasticizers may provide an optimal balance of effects. For example, for food contact products made from poly(vinylidene chloride) a combination of 5 wt.% of citrate or sebacate esters can be used with ca 10 wt.% of polymeric plasticizers.

3.11
Stabilizers

All plastics used in packaging materials gradually degrade during their lifetime (processing, storage, application) by combined attacks of chemical deteriogens (oxygen and its active forms, atmospheric pollutants such as NO_x or SO_2), harmful physical effects of the environment (tropospheric solar radiation, heat, and mechanical stress), high-energy radiation in sterilization processes, and microorganisms. The relevant degradation processes are classified as melt (processing) degradation, thermal degradation, long-term heat aging (thermal oxidation), weathering (including photo-oxidation) and biodeterioration. Some of the processes are catalyzed by traces of metallic impurities and are sensitized by polymer inherent impurities. Different deteriogens attack plastics either in concerted or consecutive processes.

Plastics are rather different as far as their inherent sensitivity to degradation is concerned. Differences in the sensitivity are due to the chemical structure, i.e., presence of degradation-sensitive moieties in polymer construction units; defect structures (structural inhomogeneities) present in unpredictable amounts and/or formed as a consequence of adventitious oxidation during manufacture, storage, and shipping; and sensitizing degradation during processing and subsequent application. The progressive degradation is catalyzed or sensitized by nonpolymeric impurities, such as various metal contaminants (including residues of polymerization catalysts) or photoactive dyes or pigments (Pospíšil et al., 1997). The level of structural inhomogeneities sensitizing degradation of plastics increases gradually during the polymer lifetime. Chain scission, branching, or crosslinking and formation of new functional groups, such as olefinic unsaturation C=C and polymer-bound oxygenated groups (hydroxyl, carbonyl, carboxyl), are chemical consequences of degradation (Pospíšil et al., 1998). They are accompanied by changes of polymer appearance (discoloration, loss of gloss or transparency, surface cracks) and undesirable changes in mechanical properties (elongation, tensile and impact strengths).

Plastics have to be stabilized to withstand environmental stresses during different phases of their lifetime. Stabilizers for packaging materials are commercialized under different trade names. According to their activity mechanisms, they are classified as antiacids, antimicrobials, antioxidants, dehydrating agents, light and heat stabilizers (Pospíšil and Nešpůrek, 2000a, 2000b; Zweifel, 2001). Other stabilizers, such as metal deactivators, antiozonants, and fire retardants, are not used in plastics for contact with food and related applications and are not mentioned here.

Molecular architecture, including functional moieties determining the stabilization mechanism and structural modifications (sterical and polar effects of substituents) optimizing the activity, is the principal factors governing the long-term inherent chemical activity of stabilizers (Pospíšil and Nešpůrek, 2000a). The final effect is obtained using combinations of stabilizers acting by different mechanisms. Structural modifications optimizing physical relations between the stabilizer and the polymer matrix (solubility, migration, compatibility), efficiency spectrum (bifunctional stabilizers), and assuring physical persistence (resistance to volatilization, blooming, and leaching into contact environments) by increasing the molecular weight of stabilizers and synthesis of stabilizer-functionalized polymers (Pospíšil, 1991; Pospíšil and Nešpůrek, 2000a) improve integral properties.

Stabilizers characteristic of plastics for packaging materials are briefly outlined. The market offers under different trade names many stabilizers having an identical chemical structure.

3.11.1
Antiacids

Acid scavengers (antiacids) are stabilizers contributing significantly to the performance of PO, PVC, PA, or PET containing acid contaminants. They are commonly used in the base stabilization packages (Thürmer, 1998) and contribute significantly to the overall performance of the stabilizer in polymers containing halogenated flame retardants, prevent depletion of HAS by acid deposits characteristic of pesticide vapors attacking stabilized greenhouse films or acid dew and rain, prevent corrosion of metallic parts of the equipment, reduce discoloration by chelating residual amounts of titanium and aluminum catalysts in PO, and deactivate hydrochloric acid arising from the residues of Ziegler–Natta catalysts in PO or from the thermodegrading PVC. Moreover, they have a complementary lubricating and weak nucleating effect. There is a good selection of antiacids: calcium or zinc salts of weak organic acids (for example, stearic or lactic) and inorganic acids, epoxidized oils (soybean, linseed, sunflower) or esters of oleic acid (e.g., **13**), and fillers such as synthetic hydrotalcite $Mg_{4.5}Al_2(OH)_3CO_3 \cdot 3.5H_2O$ or hydrocalumite $[Ca_2Al(OH)_6]CO_3$. Very beneficial complementary effects are imparted in blends of antiacids consisting of hydrotalcite, zinc oxide, and calcium stearate.

3.11.2
Antimicrobials

Most plastics are resistant to biodegradation by bacteria, fungi, and molds. They may suffer by biodeterioration due to the growth of microorganisms on surfaces contaminated with nutrients or damages in polymers plasticized by esters of fatty acids (phthalates and phosphates are more resistant). This results in a loss of flexibility and light transparency, development of discoloration, and odor. Proliferation of pathogenic microbes is critical in food storage or medical applications (Sawan, 2000).

Antimicrobials (biocides) prevent polymers against growth of microorganisms consuming parts of the material (plasticizers) as a nutrient (Nichols, 2004). Modern packaging applications are designed so that the strictly regulated substance with a migration limit to food <5 mg kg^{-1} food has a direct antimicrobial effect on the food itself. The inhibitory activity is affected by the ultraviolet (UV) stability of the additive.

The range of antimicrobials designed to meet needs of plastics developed significantly in recent years. There are different biocides of strict environmental concern available (Sawan and Manivannan, 2000). 10,10-Oxybisphenoxazine (**14**) accounts for about one-half of the worldwide market. It is used at a 0.6–5.0 wt.% level. There are some metal-free replacements, for example N-octyl-4-isothiazolin-3-one (**15**). Polymer-anchored films with built-in biocidal moieties protect against direct contamination with microbes. Strict requirements dealing with toxicity and ecotoxicity of antimicrobials vary by country.

3.11.3
Antioxidants

Protection of plastics against oxygen-triggered degradation is a key condition on how to prolong their service life. Polymers are oxidized either in thermal processes during processing and long-term heat aging (LTHA) or in processes triggered by solar UV radiation in the range of 295–400 nm during weathering. Alkylperoxyradicas POO˙ (Eq. (3.2)) and alkylhydroperoxides POOH (Eq. (3.3)) arise as primary oxidation products of polymers PH in chain propagation steps after thermal, mechanochemical, catalytic, or radiative initiation accounting for macroalkyls P˙ (Eq. (3.1)). Homolysis of POOH accounts for alkoxy radicals PO˙ (Eq. (3.4)), a source of carbonyl compounds >CO (chromophoric impurities) formed by β-scission (Eq. (3.5)).

$$PH \xrightarrow{\Lambda, \text{shear}, h\nu, M^{n+}} P^\cdot \qquad (3.1)$$

$$P^\cdot + O_2 \rightarrow POO^\cdot \qquad (3.2)$$

$$POO^\cdot PH \rightarrow POOH + P^\cdot \qquad (3.3)$$

$$POOH \xrightarrow{\Lambda, h\nu, M^{n+}} PO^\cdot + HO^\cdot \qquad (3.4)$$

$$PO^\cdot \xrightarrow{\Lambda, h\nu} {>}CO \qquad (3.5)$$

Two classes of antioxidants are effective in thermal oxidation. Chain-breaking or "primary" antioxidants limit the rate of the chain propagation steps by trapping carbon- or oxygen-centered free radicals. Hydroperoxide decomposing or "secondary" antioxidants prevent chain initiation by deactivating POOH. Photoantioxidants deactivate P˙, POO˙, and POOH in plastics exposed to photo-oxidation.

3.11.3.1 Chain-Breaking Antioxidants

Scavenging of POO˙ by hindered phenols, aromatic amines (not used in plastics and not mentioned here), and dialkylhydroxylamines is the principal mechanism protecting plastics against thermal oxidation. Substituted benzo[b]furan-2-one, a processing stabilizer for PO, is considered as a scavenger of alkyl radicals. This mechanism is also assumed as a supporting process in dialkylhydroxylamines and is one of the key steps in the activity of HAS (Pospíšil and Nešpůrek, 2000a).

Hindered phenols are used preferentially in processing and LTHA of PO where their low-discoloring properties are advantageous (Pospíšil and Nešpůrek, 2000a; Zweifel, 1998). They are also effective for the thermal stabilization of styrene-based polymers, PET, and aliphatic PA. The common concentration of phenols in plastics ranges between 0.025 and 0.3 wt.%.

Phenols may be used as single stabilizers. Their efficiency is very effectively improved in combinations with organic compounds of trivalent phosphorus and benzofuranone (in melt stabilization of PO), activated sulfides or HAS (in long-term heat stabilization), and HAS and UV absorbers (in weathering/photostabilization) (Zweifel, 1998). Low polymer discoloration accounts for some phenol transformation products, salts with transition metals, and gas fading by NO_x (Pospíšil and Nešpůrek, 2000a).

Effective but volatile 2,6-di-*tert*.butyl-4-methylphenol (**16**) was replaced for most applications by less volatile analogs, for example **17**, **18**. Synthetic DL-α-tocopherol (**19**) was introduced for effective melt stabilization of packaging materials (Laermer and Zambetti, 1992). This antioxidant may be listed among generally recognized as safe (GRAS) stabilizers.

Di(octadecyl)hydroxylamine (**20**) was introduced for processing and LTHA stabilization of PP. It is used in concentrations of 0.04–0.06 wt.% in combination with aromatic phosphites and HAS and may potentially replace a part of phenolic antioxidants in applications where discoloration due to antioxidant transformation products is extremely restricting (CIBA, 1996).

5,7-Di-*tert*.butyl-3-(3,4-dimethylphenyl)-3*H*-benzo[b]furan-2-one (**21**, a blend with 10 wt.% of 2,3-dimethylphenyl isomer) was introduced for high-temperature melt stabilization of PO (Voigt and Todesco, 2002). It provides in an oxygen-deficient atmosphere an outstanding color maintenance and is used in amounts of 0.01–0.02 wt.% in blends with hindered phenols **17** or **18** and aromatic phosphites (total amount of the stabilizer blend is approximately 0.1 wt.%). Scavenging of P· radicals limiting crosslinking in PO is assumed in the activity mechanism of **21** as well (Pospíšil and Nešpůrek, 2000a).

3.11.3.2 Hydroperoxide Deactivating Antioxidants

Alkylhydroperoxides arise according to Eq. (3.3) and homolyze to radicals. The latter act as thermo- and photoinitiators in plastics' degradation. Additives reducing POOH to alcohols compete with the homolysis and, consequently, with the chain initiation and transfer. Organic compounds of sulfur and trivalent phosphorus are used as hydroperoxide deactivating antioxidants in plastics (Pospíšil and Nešpůrek, 2000a; Zweifel, 1998).

Sulfur containing stabilizers, such as dialkyl thiodipropionate (**22**, $R=C_{12}-C_{18}$), are classified as thiosynergists and are used in combination with phenolic antioxidants in the long-term heat stabilization of PP (Zweifel, 1998). Thiosynergists are conventionally used in excess to phenolics (the ratio is 1 : 3–5); the total concentration of the stabilizer combination is usually not greater than 0.75 wt.%. An antagonistic effect may arise between HAS and acid transformation products of thiosynergists (Pospíšil, 1995; Pospíšil and Nešpůrek, 2000a).

Organic compounds of trivalent phosphorus are excellent processing stabilizers for PO and are commonly used in combinations with phenolic antioxidants, benzofuranone (**21**), and hydroxylamine (**20**) (Zweifel, 1998). They reduce POOH to alcohols POH, potentially react directly with oxygen or deactivate radicals PO· or POO·, and inhibit a photocatalytic effect of residues of polymerization catalysts (Pospíšil and Nešpůrek, 2000a). Phosphites reduce phenol consumption during melt processing of PO and improve color of final products, stabilize PET by chelating transition metals from residues of polymerization catalysts, and stabilize PVC by deactivating labile chlorine atoms by acting as costabilizers with metal soaps and organotin stabilizers. The activity of organophosphorus compounds and their hydrolytic stability are structure dependent (Tocháček and Sedlář, 1995). Aliphatic phosphites, such as tris(dodecyl) phosphite or mixed aliphatic-aromatic phosphites, for example **23**, are used in PVC. Aromatic phosphite **24**, cyclic phosphite **25** [$R=C_{18}H_{37}$, 2,4-di-*tert*.butylphenyl,2,4-bis-(α,α-dimethylbenzyl)], and phosphonite **26** are effective in PO (Zweifel, 1998). Phosphite **24** is also used in PET. Application levels of phosphites are between 0.05 and 0.3 wt.% in PO, PS, or PET. A higher concentration, up to 1 wt.%, may be used in PVC.

[ROC(O)(CH$_2$)$_2$]$_2$S   [Ph–O–POC$_{10}$H$_{21}$]$_2$ [Ph–O]$_3$P

22   23   24

[ROP(O–)(O–)]$_2$C   [Ph–O–P(Ph–O)]$_2$ Ph–N=C=N–Ph

25   26   27

3.11.4
Dehydrating Agent

PET undergoes during melt processing acid-catalyzed hydrolytic degradation. The content of humidity must be kept below 0.005 wt.%. This may be achieved by addition of bis(2,6-di-*tert*.butylphenylcarbodiimide) (**27**) (Karayannidis *et al.*, 1998).

3.11.5
Heat Stabilizers

PVC is the worldwide major bulk plastic with an outstanding market profile and a range of applications. It suffers by a low processing stability. The degradation is manifested by the formation of light-absorbing conjugated sequences of double bonds and deep discoloration of PVC. Photodegradation takes place in PVC outdoor application. Heat stabilizers (HS) having a preventive or curative character control thermal degradation (Bacalogulu *et al.*, 2001). Some chemically different groups of HS have been used. Food contact, medical applications, toys, and water pipes/containers are of principal concern (Dave, 2004).

Metal carboxylates (barium, calcium, lead, or zinc soaps of fatty acids **28**, R=C$_{11}$–C$_{17}$, M=Ba, Ca, Pb, Zn; synergistic binary salts Ba/Zn or Ca/Zn are mostly used) and various organotin compounds, for example dibultin maleate (**29**) or dioctyltin bis(iso-octylthioglycolate) (**30**), are typical HS. Concentration levels of heat stabilizers in various PVC products are in the range of 1.5–3.0 wt.% for metal soaps and of 0.3–2.5 wt.% for organotin stabilizers. There are selective global bans against compounds of heavy metals, mainly in Europe.

(RCO$_2$)$_2$M^{2+} (C$_4$H$_9$)$_2$Sn(O–C(O)–CH=CH–C(O)–O)   (C$_8$H$_{17}$)$_2$Sn(SCH$_2$CO–O–C$_8$H$_{17}$)$_2$

28   29   30

[Ph(SH)–C(O)O(CH$_2$)$_2$–O]$_2$ CH$_3$-N-C(O)-N(CH$_3$)-C(NH)=C-CH$_3$ (**32**)   (C$_4$H$_9$O)$_3$P=O

31   32   33

The new generation of totally metal-free HS (mostly considered as costabilizers) shapes future trends in PVC stabilization. Promising results were obtained with ester thiol (**31**) or pyrimidinedione derivative (**32**) (Hopfmann et al., 1998). These compounds synergize with common HS. Similar supporting effects have different other stabilizers, mainly aliphatic or mixed aliphatic/aromatic phosphites (**23**), phenolic antioxidants, antiacids, polyols (pentaerythritol or sorbitol), esters of β-aminocrotonic acid, derivatives of dihydropyridine, dehydroacetic acid, dibenzoylmethane, or pyrazolone (Dave, 2004).

PET degrades thermally during processing at 280–300°C. The chain scission is due to transesterification catalysts (zinc, manganese or cobalt acetates, alkyl titanates). Tributyl phosphate (**33**) or triphenyl phosphate, considered as heat stabilizers and metal deactivators, is used (Karayannidis et al., 1998). Besides HS, commercial PET is doped by an antioxidant combination of hindered phenols with aliphatic or aromatic phosphites and, in some cases, by a dehydrating agent.

3.11.6
Light Stabilizers

Commodity aliphatic carbon chain polymers theoretically do not absorb the actinic part (295–400 nm) of the terrestrial sunlight. The situation is dramatically influenced by the presence of trace amounts of polymer-bound UV light-absorbing impurities, such as polymeric hydroperoxides or carbonyl compounds (internal chromophores) or nonpolymeric photosensitizing impurities, such as some ions of transition metals or colorants (admixed chromophores) (Pospíšil and Nešpůrek, 2000a, 2000b). Chromophores absorb actinic solar radiation and are transformed in excited states. They sensitize the formation of macroalkyls from the matrix polymer, accelerate the homolysis of POOH, and generate singlet oxygen from the ground state oxygen. Consequently, outdoor application of any polymeric material is connected with irreversible weathering/photodegradation. Polymers differ in their inherent resistance to phototriggered processes. Their long-term outdoor application without photostabilization is impossible. Preventive protection is achieved with UV absorbers and light screening pigments. Curative stabilization involves quenching of excited species, deactivation of peroxidic species by HAS and some metal thiolates (Pospíšil and Nešpůrek, 2000a, 2000b). The latter as well as quenchers cannot be used in food-contact applications due to ecological objections and are not notified here.

3.11.6.1 Light Screening Pigments and UV Absorbers

For some applications, white pigments (titanium dioxide in particular), inorganic colored pigments, or carbon black are used to screen harmful solar radiation (Pospíšil and Nešpůrek, 2000b). Organic UV absorbers absorb the harmful solar radiation preferentially to the polymer (Pospíšil and Nešpůrek, 2000a, 2000b). Ultraviolet absorbers can also act as filters screening radiation penetrating through the packaging material and protect the inside material by this way. Intramolecular radiative and radiationless processes deactivate the absorbed energy. The ideal UV absorber is expected to absorb all terrestrial UV-A and UV-B radiation, but no radiation having wavelengths longer than 400 nm. Different classes of commercialized UV absorbers are used. Those acting by the excited state intramolecular proton transfer (ESIPT) mechanism (Pospíšil and Nešpůrek, 2000a, 2000b) include phenolic derivatives of benzophenone (**34**), benzotriazole (**35**), and 1,3,5-triazine (**36**). Nonphenolic UV absorbers are represented by oxanilide **37**, malonate **38**, and bifunctional malonate/HAS **39**.

34   **35**   **36**   **37**

38   **39**

Relatively high concentration levels of UV absorbers (0.25–0.5 wt.%) are used in plastics containing conventional basic stabilization (i.e., a combination of a phenolic antioxidant, organic phosphite, and calcium stearate) (Zweifel, 1998). Combinations of UV absorbers with HAS or benzoate-based phenols (**40**) are beneficial in pigmented PO.

3.11.6.2 Photoantioxidants

Conventional phenolic antioxidants, such as **16** to **18**, are sensitive to photolysis and have a reduced efficiency in photostabilization of polymers (Pospíšil, 1981). Hydroxybenzoates, for example **40**, have a higher photoresistance. The photoantioxidant problem has been solved by HAS, a prominent class of light stabilizers (Pospíšil and Nešpůrek, 2000a, 2000b). Optimization of their inherent chemical efficiency, resistance to acid environment and physical persistence, and application of blends of HAS differing in molecular weight substantially increased the application spectrum of HAS in plastics. HAS are effective as long-term heat stabilizers at a temperature range up to about 110 °C and photoantioxidants (Pospíšil, 1995; Pospíšil and Nešpůrek, 2000a). They are also active as stabilizers in γ-rays sterilized items. The mechanism of HAS (represented as >NH, Scheme 3.1) includes deactivation of hydroperoxides in the first step resulting in some intermediates (>NOH, NOP) oxidizing in the respective nitroxide >NO·, the key intermediate in the cyclic HAS mechanism. Nitroxide scavenges alkyl radicals P·. Formed O-alkylhydroxylamine >NOP reacts with POO· radicals and regenerates the nitroxide. The cyclic regenerative mechanism explains the high efficiency of HAS (Pospíšil, 1995; Pospíšil and Nešpůrek, 2000b).

Commercialized HAS contain 2,2,6,6-tetramethylpiperidine moiety and differ in molecular architecture. (A piperazinone moiety is exceptional in commercialized

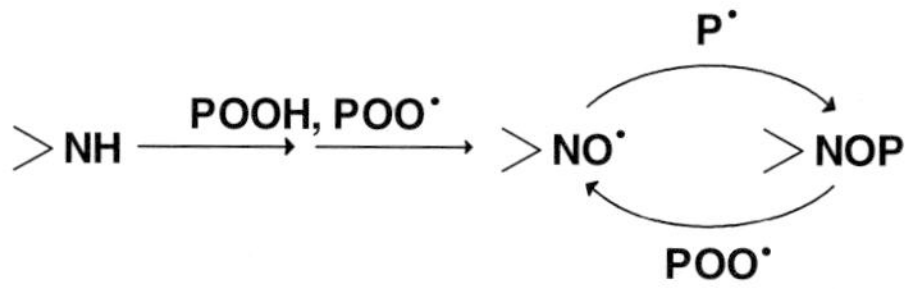

Scheme 3.1 Mechanism of photoantioxidant activity of HAS.

HAS.) Hindered secondary >NH and tertiary >NR (R is mostly methyl) amino groups, O-alkylhydroxylamine group >NOR, or acylamido group >NCOCH$_3$ are the functional moieties in HAS. Compounds **41–44** are examples of commercialized HAS. The piperidine moiety can be combined in one molecule with phenolic or UV absorbing moieties (e.g., **39**) (Pospíšil and Nešpůrek, 2000a, 2000b).

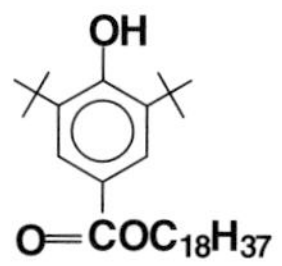

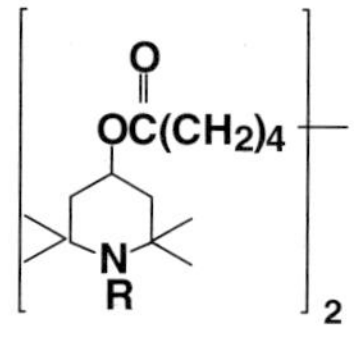

The optimum effectiveness in plastics' protection has been obtained in cooperative combinations of HAS with phenolic antioxidants, dialkylhydroxylamine, organic phosphites, and/or UV absorbers. The common concentration of HAS in plastics ranges generally from 0.1 to 1 wt.% (Zweifel, 1998). Very impressive results were obtained using combinations of two HAS, such as **41** (R=H)/**43**, **42**/**44**, or **43**/**44**. HAS-functionalized PO prepared by reactive processing (Al-Malaika, 1999) or photografting (Avár and Bechtold, 1999) are extraction resistant systems. Application of secondary or tertiary HAS with sulfur-containing stabilizers should be avoided due to potential antagonistic effects. HAS containing O-alkylhydroxylamine (e.g., **41**, R=OC$_8$H$_{17}$) or acylamide moieties are more resistant to acids (Pospíšil, 1995).

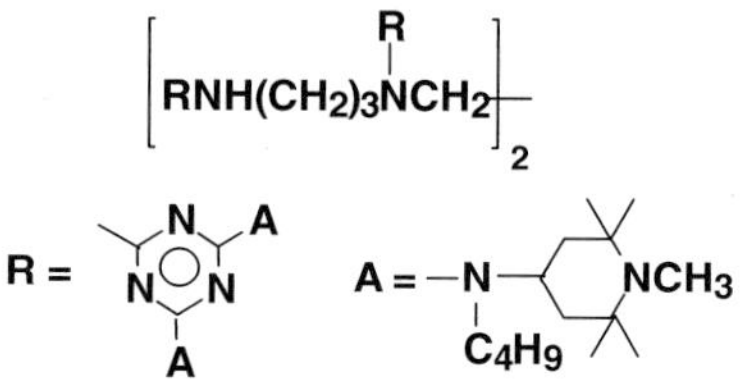

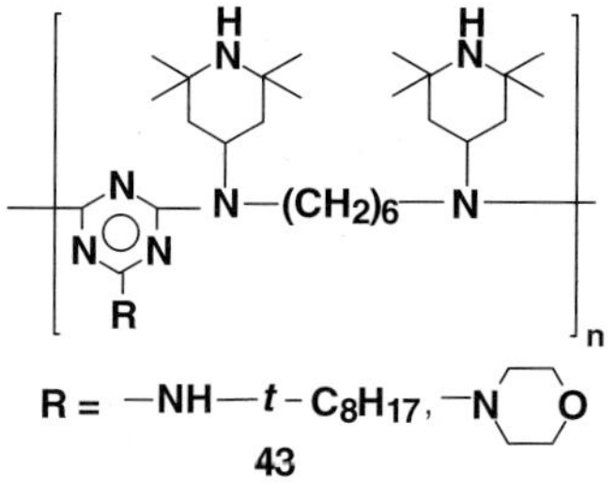

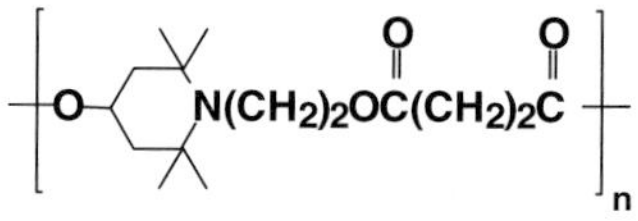

3.12
Transformation Products of Plastic Stabilizers

Plastic stabilizers are chemically more reactive compounds than the protected polymer matrix. Consequently, their application is connected with the transformation of the original structure of the stabilizer and is regarded as sacrificial stabilizer consumption. Chemical transformation is a consequence of the deactivation of polymer-borne free-radical intermediates participating in initiation or propagation steps of chain degradation (alkyls, alkoxyls, alkylperoxyls, acylperoxyls), of oxidation with hydroperoxides or singlet oxygen, or as a consequence of deactivation of the harmful solar radiation or excited chromophores. The sacrificial consumption of the original structure of the stabilizer proceeds during all phases of the polymer lifetime: processing, thermal aging, or weathering. The consumption has a heterogeneous character with characteristic concentration gradients, and proceeds preferentially in randomly distributed sites with high concentration of oxidized species and in polymer surface layers (Pospíšil and Nešpůrek, 2000a; Pospíšil et al., 2005, 2006). In addition to the sacrificial consumption, oxidation by atmospheric pollutants (nitrogen oxides, ozone), chelating with metal impurities (including residues of polymerization catalysts), acid environment, or sensitized photolysis deplete the stabilizers. New compounds formed from the originally added stabilizer accumulate gradually in the polymer matrix. They have different structures, properties, and molecular weights from the original compounds. As a consequence, some transformation products discolor or stain polymers, have an antagonistic effect on other additives, differ in environmental solubility or volatility, and can cause organoleptic problems. The consumption of stabilizers results in the ultimate phase in reduction of the effective concentration of stabilizers under a level no more prone to protect the degrading polymer against loss of serviceable properties. Extensive product studies deciphering chemical reactions of some stabilizers in degrading polymers and describing structures of principal transformation products were performed (Pospíšil, 1995; Pospíšil et al., 1996; Pospíšil and Nešpůrek, 2000a, 2000b). Products are present in the aged polymer matrix in trace amounts; some of them are formed only transiently and their determination is difficult. Typical structures of transformation products of stabilizers are outlined in this chapter.

3.12.1
Transformation Products from Phenolic Antioxidants and UV Absorbers

Compounds containing phenolic moieties form a great family of chain-breaking antioxidants and UV absorbers. Their durability is affected by reactions of the phenolic part. The chain-breaking activity of hindered phenols is based on hydrogen transfer from the phenolic hydroxy group to alkylperoxyl derived from the oxidizing polymer (Pospíšil and Nešpůrek, 2000a). Transiently formed phenoxyls, exemplified by **45** derived from the simplest antioxidant **16**, are the primary free-radical intermediates arising from phenols. They react in mesomeric cyclohexadienonyl form, for example **46**, or rearrange into phenolic benzyl radicals **47**. Radicals **46**, **47** are precursors of all isolated transformation products (Pospíšil 1981; Pospíšil et al.., 1996; Pospíšil and Nešpůrek 2000a).

Phenoxyls are short-living species detectable in the polymer matrix by electron spin resonance spectroscopy. Phenoxyls (and their mesomeric forms) participate in autorecombination (coupling) processes, recombination with alkylperoxyls, and disproportionation reactions. The transformation of individual nuclei in polynuclear phenols (e.g., **18**) proceeds independently. This accounts for a rather complicated mixture of structures in one molecule. Steric effects of substituents influence recombination reactions. Fully hindered phenoxyls (derived from **16–18**, containing tertiary alkyls in both ortho positions to the phenolic hydroxy group) are prone to react preferentially in the less hindered position para. The coupling of benzyl-type radicals takes place at sites of high concentration of phenoxyls in the degrading polymer and accounts for dimeric products, for example **48**, having higher molecular weight and, consequently, lower migration rates and extractability than the parent phenol. The coupling is structure dependent and can be restricted in phenoxyls derived from phenols bearing on the para-methyl group voluminous substituents.

Transformation products having structures of alkylated 4-alkylperoxycyclohexa-2,5-dien-1-ones (**49**) arise from cyclohexadienonyl radicals **46** and alkylperoxyls ROO· in sites of high concentration of both radical species (Pospíšil et al., 1999; Pospíšil and Nešpůrek, 2000a). Alkylperoxycyclohexadienones, analogous to the simple **49**, were isolated in model experiments from other phenols. In polynuclear phenols, for example **18**, the reaction proceeds gradually and products with partially or fully transformed molecules were isolated.

During weathering, phenolic antioxidants are photooxidized into hydroperoxycyclohexadienones, such as **50**. The presence of peroxidic moieties in **49** and **50** together with their conjugation with the carbonyl moiety in one molecule renders them thermolabile at temperatures exceeding 100°C and photolysable upon solar UV radiation (Pospíšil et al., 1999, 2005; Pospíšil and Nešpůrek, 2000a). Both processes account for homolysis of the peroxidic bonds. Free-radical fragments formed accelerate the oxidative degradation of the polymeric matrix. Low molecular weight products of homolysis, such as **51** to **53**, were detected in low amounts.

3.12 Transformation Products of Plastic Stabilizers

The oxidative transformation of phenolic antioxidants by atmospheric nitrogen oxides (NO_x) takes place in the PO matrix and accounts for discoloring cyclohexadienones (**54** is an example) in a process called gas fading (Pospíšil and Nešpůrek, 2000a).

Hindered phenoxyls derived from **16–18** containing on the α-carbon atom of the substituent in position four at least one hydrogen atom disproportionate to the parent phenol and relevant quinone methide (QM). The disproportionation competes with the free-radical recombination of phenoxyls. QMs are considered as principal transformation products of phenols. 3,5,3′,5′-Tetra-*tert*.butylstilbene-4,4′-quinone (**55**), the strongly discoloring dimeric QM (Pospíšil *et al.*, 1999, 2002a), is an example. QM **55** is formed by oxidation of **16** to phenoxyl **45**, disproportionation into unstable "monomeric" QM, and its dimerization. Phenoxyl derived from propionate-type phenol octadecyl 3-(3,5-di-*tert*.butyl-4-hydroxyphenyl)propionate (**17**) bearing bulky substituent in the para position disproportionates to a primary QM **56**. The latter dimerizes by C–C coupling in the β-position to the QM ring into unconjugated (**58**) and strongly discoloring conjugated (**60**) QM (Pospíšil *et al.*, 1996, 2002a; Pospíšil and Nešpůrek, 2000a). The transformation mechanism is complicated and involves aromatization of **56** and **58** to phenolic cinnamates **57** or **59** (R in **56** to **60** is $C_{18}H_{37}$). The process explains the final low discoloring properties of propionate-type phenolics.

56 57 58 59 60

The cascade regenerating phenolic antioxidants from their quinone methinoide transformation products, shown for the QM **56**, is also considered for other QM derived from propionate-type phenols, for example **18**, and explains their excellent antioxidant activity and low discoloration in plastics. Dimerization of the cinnamate **57** via the β-position to the aromatic nucleus also enhances the physical persistence of the quinone methinoide transformation products in comparison with the parent phenols.

QMs are the final stable transformation products of phenolic antioxidants. Their formation cannot be avoided in phenols containing a methyl or substituted methyl in position two or four to the phenolic hydroxy group. QMs absorb visible light at $\lambda_{max} = 420$ to 463 nm (Pospíšil *et al.*., 1999). As a consequence, the sacrificial transformation of phenols results in discoloration of the polymer (Pospíšil *et al.*, 2002a). However, the stabilization power is not completely lost after QM formation. This is due to the mentioned regeneration and the ability of QM to trap alkyl radicals and react with organic phosphites during processing of PO (reactive bleaching) or with HAS during PO weathering (Pospíšil *et al.*, 1996; Pospíšil and Nešpůrek, 2000a).

Moreover, conjugated QMs (e.g., **55**, **60**) photobleach reversibly in the PO matrix upon UV-A radiation (Pospíšil et al., 2002b).

In polynuclear phenols, the individual phenolic nuclei are oxidized (consumed) independently. Consequently, compounds are formed having in one molecule phenolic groups in various degrees of transformation. For example, oxidized tris-nuclear phenol 1,3,5-tris(3,5-di-*tert*.butyl-4-hyroxybenzyl)-2,4,6-trimethylbenzene forms a mixture of mono-, bis-, and tris-alkylperoxycyclohexadienones, a product having in one molecule quinone methinoide and alkylperoxycyclohexadienoide moieties (**61**), and a compound **62** with aldehydic groups. Volatile 2,6-di-*tert*.butyl-1,4-benzoquinone (**52**) is split off.

61   **62**

Great attention has been paid to DL-α-tocopherol (**19**), a synthetic analog of the natural antioxidant of plant origin (Laermer and Zambetti, 1992). Its application in packaging materials in contact with food was studied intensively. A detailed elucidation of the sacrificial consumption of α-tocopherol in thermal oxidation and photo-oxidation of organic materials indicated transformation products, having structures analogous to those of synthetic hindered phenolic antioxidants (Al-Malaika et al.., 2001; Pospíšil and Nešpůrek, 2000a). Alkylperoxycyclohexadienone **63** (R=$C_{12}H_{25}$), hydroperoxycyclohexadienone **63** (R=H), reactive QM **64**, complicated oligomers arising from **64**, and α-tocopherylquinone **65** were identified. Compounds **64** and **65** discolor PO doped with synthetic α-tocopherol.

63   **64**   **65**

Long-term model experiments indicate a possibility of hydrolysis of ester-type phenolic antioxidants **17**, **18** to free acids and the respective alcohols (Bartoldo and Ciardelli, 2004). Hydrolysis in the PO matrix seems to be, however, improbable.

Phenolic moieties differing in the steric hindrance are present as key functional groups conditioning the ESIPT stabilizing mechanism in UV absorbers **34** to **36**

(Pospíšil and Nešpůrek, 2000a, 2000b). Oxidation with alkylperoxyls proceeds by a mechanism analogous to this characteristic of phenolic antioxidants. Phenolic groups in UV absorbers are, fortunately, more resistant to oxidative transformations. It is because a nonoxidizing zwitterionic nonphenolic form with an intramolecular hydrogen bond arises reversibly during the ESIPT mechanism. Nevertheless, after a long-term service, a small part of the phenolic moieties of UV absorbers is oxidized by the phototriggered free-radical assisted process (Pospíšil and Nešpůrek, 2000b), breaking the H-tunneling from the phenolic group to carbonyl or nitrogen functions in the UV absorber. The molecule with transformed hydroxy groups is no more prone to act via the ESIPT mechanism, and the photostabilizing function is lost. Substituted 1,4-benzoquinone **66** ($R=C_8H_{17}$) and alkylperoxycyclohexadienone **67** are formed from 2-hydroxybenzophenone **34** and 2-(3,5-di-*tert*.butyl-2-hydroxyphenyl)benzotriazole (**35**), respectively. Quinone methide **68** arises from benzotriazole containing propionate-type phenolic moiety and is expected to isomerize in a phenolic cinnamate in a mechanism analogous to the rearrangement of **56** to **57**. The ESIPT mechanism may be thus formally reestablished to some extent. The service lifetime of UV absorbers is prolonged in combination with phenolic antioxidants and HAS protecting the ESIPT mechanism from depletion.

66   **67**   **68**

3.12.2
Transformation Products from Hydroperoxide Deactivating Antioxidants

Hydroperoxide deactivating antioxidants (HD) avoid harmful homolysis of POOH. This accounts for reduction of POOH into alcohols POH and a gradual transformation of the HD in oxidized products, generally $HD(O)_n$ (Eq. (3.6))

$$HD + nPOOH \rightarrow HD(O)_n + nPOH \tag{3.6}$$

Sulfur-containing antioxidants (e.g., **22**) react with ROOH and yield in the first stoichiometric step product $HD(O)_n$ ($n=1$), i.e., the corresponding sulfoxide **69** ($R=C_{12}-C_{18}$, $n=1$) or thiosulfinate **70** ($n=1$) (Pospíšil and Nešpůrek, 2000a). The primary products **69**, **70** ($n=1$) deactivate another molecule of POOH in the second stoichiometric step. This accounts for $HD(O)_2$, i.e., sulfone **69** ($R=C_{12}-H_{18}$, $n=2$) and thiosulfonate **70** ($n=2$). The primary oxidation products $HD(O)$ are thermolabile and decompose to peroxidolytic products in a very intricate mechanism involving free-radical species. From **69**, **70** ($n=1$), sulfenic **71** ($n=0$), and sulfoxylic acid **72** ($n=0$) are formed and remain in the polymer matrix together with sulfurless fragments, such as **73**. Acids **71**, **72** ($n=0$) act as "peroxidolytic" catalysts of decomposition of hydroperoxides and are responsible for the high (overstoichiometric) activity of sulfidic antioxidants. In final transformation steps, higher oxidized acids, such as sulfinic **71** ($n=1$), sulfonic **71** ($n=2$), thiosulfurous **72** ($n=1$), or thiosulfuric **72** ($n=2$), are formed in trace amounts together with sulfur oxides (SO_2, SO_3). Analogous transformations of the sulfur moieties take place in phenolic sulfides.

$$\left[\underset{O}{ROC(CH_2)_2}\text{―}S(O)_n\right]_2 \qquad C_{18}H_{37}SS(O)_nC_{18}H_{37} \qquad \underset{O}{ROC(CH_2)_2S(O)_nOH}$$

          69   **70**   **71**

$$C_{18}H_{37}SS(O)_nOH \qquad \underset{O}{ROCCH\!=\!CH_2} \qquad O\!=\!P(H)(OAr)_2$$

          72   **73**   **74**

Volatile products may be sources of undesirable organoleptic problems. This limits the use of organic thiocompounds in odor-sensitive applications. Polar organic S-protonic acids **71**, **72** deactivate basic stabilizers (HAS) (Pospíšil et al., 1995). The peroxidolytic effect of **71**, **72** is reduced in the presence of some antiacids or fillers, for example calcium carbonate (Pospíšil and Nešpůrek, 2000a).

The reaction of organic compounds of trivalent phosphorus in nonradical decomposition of hydroperoxides (Eq. (3.6)) is structure dependent, and the rate decreases in the series arylphosphonite > alkylphosphite > hindered arylphosphite. The stoichiometric phase of their sacrificial transformation results in structurally relevant compounds of pentavalent phosphorus, for example **11**, **12**, **33** (Habicher et al., 2005; Pospíšil and Nešpůrek, 2000a). Aliphatic open chain and cyclic phosphites are sensitive to humidity and hydrolyze during storage and handling. Sterically hindered aromatic phosphites are more resistant to hydrolysis. Products of hydrolysis (e.g., hydrogen phosphate **74**) may be present in stored phosphites used for stabilization. Analogous hydrogen phosphates and products of deeper hydrolysis retain properties of HD antioxidants and melt stabilizers (Tocháček and Sedlář, 1995). The hydrolysis is catalyzed by acid impurities arising from residues of polymerization catalysts in PO or from degrading PVC, and accounts for phosphorous acid in the ultimate phase.

Acid products arising either during storage or during processing from hydrolysis-sensitive phosphites can corrode the processing equipment and form sticky cakes. To minimize the hydrolytic instability, basic additives such as tris(2-hydroxypropyl) amine are added to phosphates or the amine moiety can be build-in into the phosphate molecule (Pospíšil and Nešpůrek, 2000a).

3.12.3
Transformation Products from Hindered Amine Stabilizers

A complex of reversible transformations is characteristic of the integral stabilizing mechanism of HAS. Nitroxides > NO·, O-alkylhydroxylamines > NOP, and hydroxylamines > NOH are formed as stabilizing active forms from HAS during the cyclic regenerative process (Pospíšil et al., 1995; Pospíšil and Nešpůrek, 2000a). Nitroxides, the key sacrificial intermediary products, arise in very low concentrations in the polymer matrix and can be detected by high-sensitive spectral methods (Marek et al., 2006). Some irreversible chemical transformations account for the loss of HAS activity either due to the formation of species unable to regenerate nitroxides, such as salts of strong organic or mineral acids (**75**, R = H, methyl; X = residue of an acid, e. g., **71**, **72**, H_2SO_4), N-acyloxy derivative **76** arising by recombination of nitroxide with acyl radicals generated by Norrish photolysis of polymer-bound ketones, or volatile cyclic and open-chain products **77** to **80** arising due to photolysis or ozonolysis of nitroxides derived from various HAS (Pospíšil, 1995).

3.12.4
Transformation Products from Heat Stabilizers for PVC

As preventive stabilizers, HS deactivate initiation sites in the PVC backbone by elimination or complexation of reactive chlorine atoms or by chemical binding of hydrogen chloride released from the degrading PVC. The curative mechanism of heat stabilizers accounts for the reduction of the rate of formation of polyene sequences $-[CH=CH]_n-$ and deactivation of hydroperoxides in PVC (Bacalogulu et al., 2001). Transformation products of heat stabilizers resulting in their sacrificial action are represented by the compound **81**, a metal-free fragment **82**, various alkyltinchlorides $R_{4-n}SnCl_n$ and tin tetrachloride formed from organotinthioglycolate **30** (Burley, 1987), metal chlorides (e.g., calcium or zinc dichlorides) and the corresponding free fatty acids generated from metal soaps **28**, and chlorohydrin **83** arising from **13** (Bacalogulu et al., 2001). Thiol **82** was reported to stop the growth of polyenes by addition to the C=C bond in PVC. Organotin stabilizers containing sulfur (e.g., **30**) or their transformation products **81**, **82** are also considered as HD antioxidants (Pospíšil and Nešpůrek, 2000a). This activity accounts for oxidation of the sulfur moiety. For example, thiol **82** is transformed to disulfide **84** and sulfenic acid **85** having peroxidolytic properties.

$(C_8H_{17})_2SnCl(SCH_2\underset{\underset{O}{\|}}{C}O-iC_8H_{17})$ $HS CH_2\underset{\underset{O}{\|}}{C}O-iC_8H_{17}$

81   82

$CH_3(CH_2)_7\underset{OH}{\underset{|}{C}}H\ \underset{Cl}{\underset{|}{C}}H(CH_2)_7\underset{\underset{O}{\|}}{C}OC_8H_{17}$ $\left[SCH_2\underset{\underset{O}{\|}}{C}OC_8H_{17}\right]_2$ $HO(O)SCH_2\underset{\underset{O}{\|}}{C}O-iC_8H_{17}$

83   84   85

3.13
Conclusions

Additives are organic and inorganic chemicals enabling processing of plastics, shaping their use, and enhancing end-use performance. They are classified by function and not chemistry as (a) modifiers introducing new effects (blowing agents, clarifying agents, nucleating agents, plasticizers, fillers, and reinforcing agents), (b) processing aids (lubricants), and (c) property extenders (antistatic agents, antifogging agents, colorants, optical brighteners, stabilizers). Formation of various transformation products from stabilizers added to plastics cannot be avoided because of their high chemical reactivity. The sacrificial fate of stabilizers is in agreement with their activity mechanism accounting for protection of plastics against degradation. In elucidation of transport phenomena in commodity polymers, the presence of combinations of originally added stabilizers with varying amounts of their transformation products having sometimes very different molecular parameters has to be taken in account.

Acknowledgments

The work was supported by the Ministry of Industry and Trade of the Czech Republic (grant FTA-TA 2/018) and the Grant Agency of the Academy of Sciences of the Czech Republic (project AV0Z 405005005). Technical assistance of Mrs D. Dundrová is appreciated.

References

Al-Malaika, S. (1999) The good, the bad and the ugly in the science and technology of antioxidant grafting on polymers. In *Chemistry and Technology of Polymer Additives* (eds S. Al-Malaika, A. Golovoy and C.A. Wilkie), Blackwell, Oxford, pp. 1–20.

Al-Malaika, S., Issenguth, S. and Burdich, D. (2001) The antioxidant role of vitamin E in polymers: V. Separation of stereoisomers and characterization of the oxidation products of DL-α-tocopherol formed in polyolefins during melt processing. *Polym. Degrad. Stab.*, **73**, 491–503.

Avár, L. and Bechtold, K. (1999) Studies on the interaction of photoreactive light stabilizers and UV absorbers. *Prog. Org. Coat.*, **35**, 11–17.

Bacalogulu, R., Fisch, M.H., Kaufhold, J. and Sander, H.J. (2001) PVC stabilizers. in *Plastics Additives Handbook*, 5th Edition, (ed. H. Zweifel), Hanser Publishers, Munich, pp. 427–483.

Bartoldo, M. and Ciardelli, F. (2004) Water extraction and degradation of sterically hindered phenolic antioxidants in polypropylene films. *Polymer*, **45**, 8751–8759.

Bennemann, R. (2005) The plastics industry and REACH.Proceedings of the 11th International Plastics Additives and Modifiers Conference Addcon World 2005.Rapra Technology Shrewsbury, pp. 2/1–8.

Burley, J.W. (1987) Mechanistic aspects of the thermal stabilization of PVC by organotin compounds. *Appl. Organometal Chem.*, **1**, 95–113.

Cadogan, D. (2002) Plasticizers for PVC: Health and environmental impact, Proceedings of the 8th International Plastics Additives and Modifiers Conference Addcon World 2002, Rapra Technology Shrewsbury. pp. 33–44.

Christie, R.M. (1993) Pigments: structures and synthetic procedures. *Rev. Prog. Color*, **23**, 1–18.

CIBA Specialty Chemicals (1996) Fiberstab systems Information A 737A3M26.

Clint, J.H. (1998) Surfactants: application in plastics. in *Plastics Additives, An A-Z Reference*. (ed. G. Pritchard), Chapman and Hall, London, pp.604–612.

Dave, T. (2004) New generation of PVC stabilizers – sustainability beyond 2010. Proceedings of the 10th International Plastics Additives and Modifiers Conference Addcon World, Rapra Technology, Shrewsbury. pp. 115–121.

Habicher, W.D., Bauer, I. and Pospíšil, J. (2005) Organic phosphites as polymer stabilizers. *Macromol. Symp.*, **225**, 147–164.

Hohenberger, N. (2001) Fillers and reinforcements/coupling agents. in *Plastics Additives Handbook*, 5th Edition, (ed. H. Zweifel), Hanser Publishers, Munich, pp. 901–948.

Hopfmann, T., Wehner, W., Ryningen, A., Stoffelamo, J.U., Clucas, P. and Pfaendner, R. (1998) Organic based stabilizer systems for poly(vinyl chloride) pipes – new stabilizer generation. *Plast. Rubber Compos. Process. Appl.*, **27**, 442–446.

Hurnik, H. (2001) Chemical blowing agents. in *Plastics Additives Handbook*, 5th Edition (ed. H. Zweifel), Hanser Publishers, Munich, pp. 699–724.

Karayannidis, G.P., Sideridon, I.D. and Zamboulis, D.X. (1998) Antioxidants for poly (ethylene terephthalate). in *Plastics Additives, An A-Z Reference*, (ed. G. Pritchard), Chapman and Hall, London, pp. 95–107.

Kurja, J. and Mehl, N.A. (2001) Nucleating agents for semicrystalline polymers. In *Plastics Additives Handbook*, 5th Edition (ed. H. Zweifel), Hanser Publishers, Munich, pp. 949–971.

Laermer, S.F. and Zambetti, P.F. (1992) Alpha tocopherol (vitamin E) – the natural antioxidants for polyolefins. *Plast. Film Sheeting*, **8**, 228–248.

Marek, A., Kaprálková, L., Schmidt, P., Pfleger, J., Humlíček, J., Pospíšil, J. and Pilař, J. (2006) Spectral resolution of the degradation process in stabilized polystyrene and polypropylene plaques exposed to accelerated photodegradation or oven test. *Polym. Degrad. Stab.*, **91**, 444–458.

Nichols, D. (2004) Commonly used biocides in plastics and a variety of test methods available to demonstrate their efficiency. Proceedings of the 10th International Plastics Additives and Modifiers Conference Addcon World 2004, Rapra Technology Shrewsbury. pp. 51–67.

Pfaendner, R. (2006) How will additives shape the future of plastics. *Polym. Degrad. Stab.*, **91**, 2249–2256.

Pospíšil, J. (1981) Photooxidation reactions of phenolic antioxidants. In *Developments in Polymer Photochemistry-2*, (ed. N.S. Allen), Applied Science Publishers, London, pp. 53–132.

Pospíšil, J. (1991) Functionalized oligomers and polymers as stabilizers for conventional polymers. *Adv. Polym. Sci.*, **101**, 65–167.

Pospíšil, J. (1995) Aromatic and heterocyclic amines in polymer stabilization. *Adv. Polym. Sci.*, **124**, 87–189.

Pospíšil, J., Horák, Z., Kruliš, Z. and Nešpůrek, S. (1998) The origin and role of structural inhomogeneities and impurities in material recycling of plastics. *Macromol. Symp.*, **135**, 247–263.

Pospíšil, J. and Nešpůrek, S. (2000a) Highlights in the inherent chemical activity of polymer stabilizers. in *Handbook of Polymer Degradation*.2nd edn. (ed. H.S. Hamid,), Dekker, New York, pp. 191–276.

Pospíšil, J. and Nešpůrek, S. (2000b) Photostabilization of coating. Mechanisms and performance. *Prog. Polym. Sci.*, **25**, 1261–1335.

Pospíšil, J., Nešpůrek, S., Pfaendner, R. and Zweifel, H. (1997) Material recycling of plastics waste for demanding applications: upgrading by restabilization and compatibilization. *Trends Polym. Sci.*, **5**, 294–300.

Pospíšil, J., Nešpůrek, S. and Zweifel, H. (1996) The role of quinone methides in thermostabilization of hydrocarbon polymers: I. Formation and reactivity of quinone methides. *Polym. Degrad. Stab.*, **54**, 7–14.

Pospíšil, J., Nešpůrek, S. and Zweifel, H. (1999) Formation and role of conjugated cyclic dienones in polymer stabilization. In *Chemistry and Technology of Polymer Additives*. (eds S. Al-Malaika, A. Golovoy and C.A. Wilkie), Blackwell, Oxford, pp. 36–61.

Pospíšil, J., Habicher, W.D., Pilař, J., Nešpůrek, S., Kuthan, J., Piringer, G.O. and Zweifel, H. (2002a) Discoloration of polymers by phenolic antioxidants. *Polym. Degrad. Stab.*, **77**, 531–538.

Pospíšil, J., Nešpůrek, S., Zweifel, H. and Kuthan, J. (2002b) Photobleaching of polymer discoloration caused by quinone methides. *Polym. Degrad. Stab.*, **78**, 251–255.

Pospíšil, J., Pilař, J., Billingham, N.C., Marek, A., Horák, Z. and Nešpůrek, S. (2006) Factors affecting accelerated testing of polymer photochemistry. *Polym. Degrad. Stab.*, **91**, 417–422.

Pospíšil, J., Pilař, J., Nešpůrek, S., Kruliš, Z. and Habicher, W.D. (2005) Effect of weathering components on efficiency and durability of photostabilizers in polymers. In *Natural and Artificial Ageing of Polymers* (ed. T. Reichert), Gesellschaft für Umweltsimulation, Pfinztal, pp. 175–184.

Pritchard, G. (ed.),(1998) *Plastics Additives, An A-Z Reference*, Chapman and Hall, London

Rahman, M. and Brazel, C.S. (2004) The plasticizer market: an assessment of traditional plasticizers and research trends to meet new challenges. *Prog. Polym. Sci.*, **29**, 1223–1248.

Richter, E. (2001) Lubricants. *Plastics Additives Handbook*, 5th Edition (ed. H. Zweifel), Hanser Publishers, Munich, pp. 511–552.

Sawan, S.P. and Manivannan, G. (2000) *Antimicrobial/Antiinfective Materials. Principles, Applications, Devices*. Technomic Publishers, Lancaster.

Thürmer, A. (1998) Acid scavengers for polyolefins. In *Plastics Additives, An A-Z Reference* (ed. G. Pritchard), Chapman and Hall, London, pp. 43–48.

Tocháček, J. and Sedlář, J. (1995) Hydrolysis and stabilization performance of bis(2,4-di tert.butylphenyl)pentaerythrityl diphosphate in polypropylene. *Polym. Degrad. Stab.*, **50**, 345–352.

Voigt, W. and Todesco, R. (2002) New approaches to melt stabilization of polyolefins. *Polym. Degrad. Stab.*, **77**, 397–402.

Wylin, F. (2001) Antifogging additives. in *Plastics Additives Handbook*, 5th Edition (ed. H. Zweifel), Hanser Publishers, Munich, pp. 609–626.

Zweifel, H. (1998) *Stabilization of Polymeric Materials*, Springer, Berlin

Zweifel, H. (ed.),(2001) *Plastics Additives Handbook*, 5th Edition, Hanser Publishers, Munich.

4
Partition Coefficients
Albert Baner and Otto Piringer

Partition coefficients, K, are fundamental physicochemical parameters describing the distribution of a solute between two contacting phases at equilibrium. They are material constants required when solving practical mass transport problems (i.e., migration calculations). In packaging/product systems, the concentrations of most migrateable package components are present in the dilute concentration range. At very low concentration the behavior of the solute molecule is isolated in solvent from other solute molecules and the solute molecule exhibits maximum nonideality.

This chapter addresses estimation methods for solubility values and partition constants at infinite dilution which are used to describe the behavior of solutes in typical food and pharmaceutical packaging polymer materials and their food and pharmaceutical or simulant contact phases.

4.1
Experimental Determination of Polymer/Liquid Partition Coefficients

In most cases the necessary material constant can be determined by direct measurement. Appendix II lists examples of experimentally determined partition coefficients taken from the literature. There are several methods used in the literature for experimentally determining partition coefficients. The "plaque sorption method" used by Becker *et al.* (1983) and Koszinowski and Piringer (1986) exposes several discs of plastic film to a dilute mixture containing several different solutes in a liquid food phase. After reaching equilibrium the concentrations of solutes in the contacting liquid phase and the sorbed solutes, which are extracted from the polymer phase using a suitable organic solvent, are then both quantified using gas chromatography (GC). Direct static headspace gas chromatography (e.g., Kolb *et al.* 1992) and indirect static headspace methods such as the PRV phase ratio variation) method developed by Ettre *et al.* (1993) are widely used. Athes *et al.* (2004) found the PRV method to be more accurate. Fukamachi *et al.* (1996) used an adsorptive column method to recover solutes

Plastic Packaging. Second Edition. Edited by O.G. Piringer and A.L. Baner
Copyright © 2008 WILEY-VCH Verlag GmbH & Co. KGaA, Weinheim
ISBN: 978-3-527-31455-3

4.2
Thermodynamics of Partition Coefficients

In practice because of the time and cost required to measure the numerous types and combination possibilities of plastics and contacting media, only a limited selection of such experimental constants is available. Consequently, one cannot practically avoid using estimated partition coefficient values. Such estimations are possible within a degree of accuracy adequate for practical purposes when the chemical structures of the migrating substance, the polymer, and the contacting media are known. Thermodynamic terms are used to characterize the equilibrium distribution of a diffusant substance between plastic (P) and contacting media (e.g., a liquid L which is also used in this text to represent food or pharmaceuticals). The most important of these terms is the chemical potential μ.

During a spontaneously occurring process at constant temperature T and constant volume V, there is a decrease in the free energy A. For a spontaneous process at constant temperature and constant pressure p, a decrease in the free enthalpy G takes place. Because most spontaneous processes occur at constant pressure, the free enthalpy is particularly important for describing such processes.

When energy, e.g., in the form of heat, is supplied to the system under constant pressure, only a fraction of this energy serves to increase the internal energy of the system, U. The remainder of the energy goes for expansion work (volumetric work) against the external pressure. The sum of U and the volumetric expansion work $p \cdot V$ is the enthalpy H. With entropy of a system defined as the ratio of the amount of heat q and temperature, $S = q/T$, the two quantities $A = U - TS$ and $G = H - TS$ are thus defined.

The quantities U, H, S, A, G, q, and V are extensive and p and T intensive quantities. When an extensive quantity is related to the amount of material in a mole, then it becomes a molar and therefore specific quantity with which the properties of the material under consideration can be described.

The molar free enthalpy G_m, in analogy with a mechanical system, is called the chemical potential and is designated with μ ($\mu = G_m$). In a mechanical system, a body moves in the direction of decreasing potential (e.g., an object falls to earth or a ground state) and from this comes the analogy with the chemical potential.

Designating a standard pressure of 1 bar with $p^\ominus$, the chemical potential of a system that behaves as an ideal gas is given as

$$\mu = \mu^\ominus + RT \ln\left(\frac{p}{p^\ominus}\right) \tag{4.1}$$

where $\mu^\ominus$ is the standard chemical potential at $p^\ominus$. Equation (4.1) allows the evaluation of μ for a perfect gas at any given pressure and temperature.

For the room temperature and atmospheric pressure conditions coming into question in this book, most gases behave like a perfect gas, which means that their chemical potential can be described using Eq. (4.1).

A gas phase can be composed of several gaseous elements or compounds. If one labels the mole amounts of components 1, 2, ... with $n_1, n_2, \ldots$, then the free enthalpy is a function of p, T and the quantities $n_1, n_2, \ldots$ so that $G = G(p, T, n_1, n_2, \ldots)$. The complete differential of this function is

$$dG = \left(\frac{\partial G}{\partial p}\right)_{T,n_1,n_2,\ldots} dp + \left(\frac{\partial G}{\partial T}\right)_{p,n_1,n_2,\ldots} dT + \left(\frac{\partial G}{\partial n_1}\right)_{p,T,n_2,\ldots} dn_1 \\ + \left(\frac{\partial G}{\partial n_2}\right)_{p,T,n_1,\ldots} dn_2 + \cdots. \qquad (4.2)$$

From the definition of the chemical potential, the case for a single-component gas having n moles results in $G = n \cdot G_m = n \cdot \mu$ whereby the partial derivative of n at constant temperature T and p is given as $(\partial G/\partial n)_{p,T} = \mu$. One can consequently hold all other variables constant and define the partial derivatives of G for $n_1, n_2, \ldots$ as the chemical potentials of the single components, e.g.,

$$\mu_1 = \left(\frac{\partial G}{\partial n_1}\right)_{p,T,n_2,\ldots} \qquad (4.3)$$

4.2.1
Equilibrium Between Different Phases in Ideal Solutions

Component a making up a liquid phase (L) in contact with a gas phase (G) forms a two-phase system. In the equilibrium state, the chemical potentials of component a in the gas and contacting phases are equal. The equilibrium saturated vapor pressure of the pure component a in the gas phase over the pure liquid phase a can be designated with p_a^*. Using the expression for a perfect gas, Eq. (4.1), for the chemical potential of a, one gets an expression of the chemical potential of component a in liquid L, $\mu_a^*(L)$, in the equilibrium state:

$$\mu_a^*(L) = \mu_a^*(G) = \mu_a^\ominus + RT \ln(p_a^*/p^\ominus). \qquad (4.4)$$

For a two-component liquid phase composed of a and b whose partial pressures in the gas phase at equilibrium with the liquid phase are p_a and p_b, one can write

$$\mu_a(L) = \mu_a(G) = \mu_a^\ominus + RT \ln(p_a/p^\ominus) \\ \mu_b(L) = \mu_b(G) = \mu_b^\ominus + RT \ln(p_b/p^\ominus). \qquad (4.5)$$

From Eqs. (4.5) and (4.4) one gets equations that are analogous to Eq. (4.4) for components a and b:

$$\mu_a(L) = \mu_a^*(L) + RT \ln(p_a/p_a^*) \\ \mu_b(L) = \mu_b^*(L) + RT \ln(p_b/p_b^*). \qquad (4.6)$$

In a so-called ideal solution, that is a mixture of several components with very similar properties, the ratio of p_a/p_a^* is equal to the mole fraction x_a of component a in the liquid phase. Thus Raoult's law is valid:

$$p_a = x_a p_a^*. \tag{4.7}$$

Consequently Eq. (4.6) can be expressed in the form

$$\begin{aligned}\mu_a(L) &= \mu_a^*(L) + RT \ln(x_a) \\ \mu_b(L) &= \mu_b^*(L) + RT \ln(x_b).\end{aligned} \tag{4.8}$$

If two pure liquids, one composed of only n_a moles of component a and the other of only n_b moles of component b, are mixed and the total free enthalpy of the two liquids in separate initial states, G_A, is considered, then the enthalpy of mixing G_E for an assumed ideal solution can be calculated. Because $n_a + n_b = n$, $n_a/n = x_a$, and $n_b/n = x_b$, Eq. (4.8) results in

$$\begin{aligned}\Delta_L G_M = G_E - \pi G_A &= n_a[\mu_a^*(L) + RT \ln x_a] + n_b[\mu_b^*(L) + RT \ln x_b] \\ &\quad - n_a \mu_a^*(L) - n_b \mu_b^*(L) = nRT[x_a \ln x_a + x_b \ln x_b].\end{aligned} \tag{4.9}$$

Because $x_a < 1$ and $x_b < 1$ the enthalpy of mixing, $\Delta_L G_M$, is always negative: $\Delta_L G_M < 0$.

The same result is obtained for mixtures of two perfect gases. Because there are no interactions between the individual particles of perfect gases (atoms or molecules) the decrease in the free enthalpy during mixing can be traced back to the increase in entropy. Mixing increases the disorder of the system.

In ideal solutions there exist interactions between the individual particles. However, because the molecular properties of components a and b are very similar, the interactions between a and b in the mixture can be assumed to be on average the same as those between a and a as well as those between b and b in pure liquids. Consequently in this case $\Delta_L G_M$ is the increase in entropy due to mixing.

For very dilute solutions of component a in component b, all the neighboring molecules of a are b molecules. Thus the partial pressure above the solution can be expressed as a constant, h, thus giving an expression which is Henry's law (Denbigh 1981):

$$p_a = h \cdot x_a \tag{4.10}$$

4.2.1.1 Partitioning in Ideal Solutions: Nernst's Law

Given two partially immiscible liquids b and c, consider a third component (subscript a) which is present in the two liquid layers. If this substance is sufficiently dilute in each layer, it may behave individually as an ideal solute in both of them even though the system as a whole is nonideal. When this condition is satisfied the result is equilibrium:

$$\mu_a^b = \mu_a^c \tag{4.11}$$

which may be replaced by

$$\mu_a^{*b} + RT \ln x_a^b = \mu_a^{*c} + RT \ln x_a^c \tag{4.12}$$

and therefore

$$\ln K' = \ln \frac{x_a^b}{x_a^c} = \frac{\mu_a^{*c} - \mu_a^{*b}}{RT} \tag{4.13}$$

where the partition coefficient K' is defined in terms of the mole fraction ratio to be $K' = x_a^b/x_a^c$, which is independent of composition. The ratio x_a^b/x_a^c is independent of the individual values of x_a^b and x_a^c in the region where each solution is ideal. This partition coefficient is also equal to the ratio of the Henry's law coefficients in the two solvents:

$$K' = \frac{h_a^c}{h_a^b} \tag{4.14}$$

4.2.2
Equilibrium Between Different Phases in Nonideal Solutions

Nonideal solutions deviate as a rule from Raoult's law. One can however still retain the form of the equation derived for ideal solutions if, instead of the mole fraction x_a, the activity a_a is used:

$$a_a = \gamma_a x_a \tag{4.15}$$

With the help of the activity coefficient, γ_a, in Eq. (4.15), all deviations from the ideal state can be taken into consideration. Instead of Eq. (4.8) for the chemical potential of component a in the liquid mixture one now gets

$$\mu_a(L) = \mu_a^*(L) + RT \ln x_a + RT \ln \gamma_a \tag{4.16}$$

One can now define an existing liquid phase as the standard state so that γ_a goes to 1 when $x_a \to 1$. All deviations from ideal behavior in a mixture ($x_a \neq 1$) are then included in the term $RT \ln(\gamma_a)$.

Because the chemical potential of component a in the liquid phase and that in the contacting gas phase are equal in equilibrium, it is possible to determine the partition coefficient for component a between the liquid and gas phases with the help of thermodynamic quantities.

For the equilibrium state for a substance, a, distributed between the gas (G) and liquid (L) phases then

$$\mu_a(G) = \mu_a^\ominus + RT \ln (p_a/p^\ominus) = \mu_a(L) = \mu_a^*(L) + RT \ln x_a + RT \ln \gamma_a \tag{4.17}$$

and because

$$\mu_a^*(L) = \mu_a^\ominus + RT \ln (p_a^*/p^\ominus) \tag{4.18}$$

then

$$RT \ln \frac{x_a}{p_a} = \mu_a^\Theta - \mu_a^\Theta - RT \ln p_a^* + RT \ln p^\Theta - RT \ln p^\Theta - RT \ln \gamma_a = RT \ln \frac{1}{p_a^* \gamma_a} \quad (4.19)$$

and consequently

$$\frac{x_a}{p_a} = \frac{1}{p_a^* \gamma_a} \quad \text{or} \quad p_a = \gamma_a \cdot x_a \cdot p_a^* \quad (4.20)$$

which is a form of Raoult's law for nonideal solutions where $a_a = p_a/p_a^*$.

The thermodynamic characteristics of solutions are often expressed by means of excess functions. These are the amounts by which the free energy, entropy, enthalpy, etc. exceed those of a hypothetical ideal solution of the same composition (Denbigh 1981). The excess free energy is closely related to the activity coefficients. The total free enthalpy of a system is

$$G = \sum n_i \mu_i \quad (4.21)$$

Substituting from Eq. (4.16) one obtains

$$G = \sum n_i \mu_i^* + RT \sum n_i \ln x_i + RT \sum n_i \ln \gamma_i \quad (4.22)$$

If the solutions were ideal the last term of course would be zero. The excess free energy, G^E, is thus defined as

$$G^E = H^E - TS^E = RT \sum n_i \ln \gamma_i \quad (4.23)$$

Differentiating the above expression at constant temperature and then applying conditions of constant temperature and pressure gives (Denbigh 1981)

$$\underline{G}_i^E = \underline{H}_i^E - T\underline{S}_i^E = RT \ln \gamma_i = \left(\frac{\partial G^E}{\partial n_i} \right)_{T,p,n_i} \quad (4.24)$$

where $\underline{G}_i^E$ is the excess free energy of mixing per mole, $\underline{H}_i^E$ is the excess enthalpy of mixing per mole, and $\underline{S}_i^E$ is the excess entropy of mixing per mole. Note that the excess free energy of mixing is also referred to as the excess chemical potential μ_i^E in some notations. A regular solution is a special case where solutions of similar sized molecules are completely randomly oriented in solution (i.e., no attractive forces other than dispersion forces), such that the volume change on mixing is quite small and the excess entropy per mole of mixture is essentially zero. For regular solutions then $\underline{S}_i^E = 0$ but $\underline{H}_i^E \neq 0$. Another special case, athermal solutions, is assumed to have zero (or negligibly small) enthalpy of mixing.

4.2.2.1 Partition Coefficients for Nonideal Solutions

For partitioning substance a between two nonideal liquid solutions with superscripts b and c one gets

$$\mu_a^b = \mu_a^c \quad \text{so that} \quad \mu_a^{*b} + RT \ln x_a^b + RT \ln \gamma_a^b = \mu_a^{*c} + RT \ln x_a^c + RT \ln \gamma_a^c \quad (4.25)$$

Solving it, one gets

$$\ln\frac{x_a^b \cdot \gamma_a^b}{x_a^c \cdot \gamma_a^c} = \frac{\mu_a^{*c} - \mu_a^{*b}}{RT} \tag{4.26}$$

where the partition coefficient K' for nonideal solutions is now defined to be the ratio

$$K' = \frac{x_a^b \cdot \gamma_a^b}{x_a^c \cdot \gamma_a^c} \tag{4.27}$$

Note that for ideal solutions $\gamma_a = 1$ and Eq. (4.27) is equal to Eq. (4.13).

Experimentally it is often more convenient to describe solute concentrations in terms of molar (mole/volume) or weight/volume concentration quantities instead of mole fractions. A partition coefficient K can be defined to be a ratio of concentrations:

$$K = \frac{c_a^b}{c_a^c} \tag{4.28}$$

The relationship between a mole fraction partition coefficient (K') and a molar concentration partition coefficient (K') for dilute solutions is

$$K = \frac{c_a^b}{c_a^c} = \frac{x_a^b \cdot \underline{V}_c}{x_a^c \cdot \underline{V}_b} = K' \cdot \frac{\underline{V}_c}{\underline{V}_b} \tag{4.29}$$

In order to get the molar concentration $c_{L,a}$ (e.g., moles/volume) of a in the liquid phase at equilibrium, one relates the quantity of material $n_a(L)$ mol in the liquid phase to the volume of the liquid, V_L, to get

$$c_{L,a} = \frac{n_a(L)}{V_L} = \frac{n_L \cdot x_a^L}{V_L} = \frac{x_a^L}{\underline{V}_L} \text{ (mol/volume)} \quad c_{L,a} = \frac{x_a^L \cdot M_a}{\underline{V}_L} \text{ (mass/volume)}$$
$$\tag{4.30}$$

where n is the total number of moles in the system and $\underline{V}_L$ is molar volume (e.g., volume/mole) of the liquid phase. With the molar mass M_a the concentration $c_{L,a}$ is expressed in mass/volume. For dilute solutions (the case most applicable for food packaging systems) one can make a simplifying assumption and use the molar volume of the pure bulk liquid phase instead of the molar volume of the liquid solution.

From the ideal gas law

$$PV = nRT \tag{4.31}$$

one obtains for the gas phase $V_G/n_a(G) = \underline{V}_{G,a} = RT/p_a$ and with it the molar concentration of a in the gas phase $c_{G,a}$:

$$c_{G,a} = \frac{n_a(G)}{V_G} = \frac{p_a}{RT} \text{ (mol/volume)} \quad c_{G,a} = \frac{p_a \cdot M_a}{RT} \text{ (mass/volume)} \tag{4.32}$$

where $\underline{V}_G$ is the molar volume of the ideal gas. For $T = 298.15$ K ($25°$C) and $p = 1$ Pa $= 1$ N m$^{-2} = 1$ kg m^{-1} s^{-2}, one obtains the value for an ideal gas with $R = 8.31451$ J K^{-1} mol^{-1} for $\underline{V}_G = 2478.94$ m^3 mol^{-1} and for $p = 1$ atm $= 101\,325$ Pa, $\underline{V}_G = 24.47$ dm^3 mol^{-1}.

The dimensionless partition coefficient $K_{G/L}$ for equilibrium of a dilute solution between an ideal gas and liquid is obtained by combining Eq. (4.28) with (4.20), (4.30), and (4.32) to give

$$K_{G/L}(a) = \frac{c_{G,a}}{c_{L,a}} = \frac{V_L}{x_a} \cdot \frac{p_a}{RT} = \frac{p_a^* \gamma_{L,a} V_L}{RT} \tag{4.33}$$

Using the index P for a polymeric material phase one obtains an equation analogous to Eq. (4.33) for molar concentrations:

$$K_{G/P}(a) = \frac{c_{G,a}}{c_{P,a}} = \frac{p_a^* \gamma_{P,a} V_P}{RT} \tag{4.34}$$

From the ratio of $K_{G/L}$ to $K_{G/P}$ the partition coefficient between a polymer and a liquid can be calculated for molar concentration ratios ($\gamma_a^b = \gamma_{L,a}$; $\gamma_a^c = \gamma_{P,a}$):

$$K_{P/L}(a) = \frac{K_{G/L}(a)}{K_{G/P}(a)} = \frac{c_{P,a}}{c_{L,a}} = \frac{\gamma_{L,a} V_L}{\gamma_{P,a} V_P}. \tag{4.35}$$

Note that molar concentration partition coefficients K are the same as if the partition coefficient were defined as a ratio of weight/volume concentrations due to the canceling out of the relative molecular weights of the solute in the numerator and denominator.

With Eqs. (4.33) to (4.35) the first goal of describing partition by using parameters resulting from thermodynamic state functions is reached even though $\gamma_{L,a}$ and $\gamma_{P,a}$ are related to the corresponding chemical potential as well as the free enthalpy.

4.2.3
Partition Coefficients for Systems with Polymers

A complication in estimating the partition of some solute a in systems containing polymers occurs due to the large, in general nonuniform, molar volumes of polymers. It is usually more convenient to work with weight/volume concentrations, weight fractions (molal concentrations) or volume fractions. Instead of mole fractions one can define analogous weight fractions using the mass g of the system's i components:

$$w_i = \frac{g_i}{\sum_{i=1}^{j} g_i} \tag{4.36}$$

with analogous equations for Raoult's law:

$$p_a = w_a \cdot p_a^* \tag{4.37}$$

for very dilute ideal solutions and

$$p_a = \Omega_a \cdot w_a \cdot p_a^* \tag{4.38}$$

for nonideal solutions where Ω_a is referred to as the weight fraction (molal) activity coefficient. For dilute liquid solutions having components with well-defined molar weights the relationship between the molal activity coefficient and molar activity coefficient is

$$\Omega_{L,a} = (M_L/M_a)\cdot\gamma_{L,a} \tag{4.39}$$

where M_i is the molecular mass. For dilute concentrations of substance a in a polymer, $c_{P,a}$ can be approximated using weight fractions as

$$c_{P,a} \cong w_a \cdot \rho_P = \frac{p_a}{p_a^* \cdot \Omega_{P,a}} \cdot \rho_P = \frac{a_a}{\Omega_{P,a}} \rho_P. \tag{4.40}$$

For dilute concentrations of a in a liquid the molar concentration in terms of the weight fraction is

$$c_{L,a} \cong w_a \cdot \rho_L = \frac{p_a}{p_a^* \cdot \Omega_{L,a}} \cdot \rho_L = \frac{a_a}{\Omega_{L,a}} \rho_L \tag{4.41}$$

where $a_a = p_a/p_a^*$ is the activity.

Combining Eqs. (4.39) and (4.40) one can derive an expression for partition coefficients of dilute solutions calculated using molal activity coefficients. For $K_{G/L}$ one combines Eqs. (4.38), (4.33), the ideal gas law equation (4.31) and Eq. (4.41) to get

$$K_{G/L}(a) = \frac{\Omega_{L,a} \cdot p_a^* \cdot M_a}{RT \cdot \rho_L} \tag{4.42}$$

$$K_{G/P}(a) = \frac{\Omega_{P,a} \cdot p_a^* \cdot M_a}{RT \cdot \rho_P}. \tag{4.43}$$

Combining Eqs. (4.42) and (4.43) one gets $K_{P/L}$:

$$K_{P/L}(a) = \frac{K_{G/L}(a)}{K_{G/P}(a)} = \frac{c_{P,a}}{c_{L,a}} = \frac{\Omega_{L,a} \cdot \rho_P}{\Omega_{P,a} \cdot \rho_L}. \tag{4.44}$$

In some cases it is more convenient to define a molal activity coefficient for the solute in the polymer phase and a molar activity coefficient for the liquid phase. Combining Eqs. (4.33) and (4.43) one gets

$$K_{P/L} = \frac{\gamma_{L,a} \cdot V_L \cdot \rho_P}{\Omega_{P,a} \cdot M_a}. \tag{4.45}$$

Sometimes volume fractions, Φ_a, are used which can be defined as

$$\Phi_a = \frac{V_a}{\sum_j V_a} \tag{4.46}$$

as well as volume fraction activity coefficients, γ_a^Φ:

$$p_a = \Phi_a \cdot \gamma_a^\Phi \cdot p_a^* \tag{4.47}$$

with related partition coefficient expressions for dilute solutions (e.g., $\Phi_{L,a} \cong V_a/V_L$ and $c_{L,a} \cong \Phi_{L,a} \cdot \rho_a$) based on volume fraction activity coefficients and assuming ideal gas law behavior (Eq. (4.31)):

$$K_{G/L}(a) = \frac{\gamma_{L,a}^{\Phi} \cdot p_a^* \cdot M_a}{RT \cdot \rho_a} \tag{4.48}$$

$$K_{G/P}(a) = \frac{\gamma_{P,a}^{\Phi} \cdot p_a^* \cdot M_a}{RT \cdot \rho_a} \tag{4.49}$$

$$K_{P/L}(a) = \frac{K'_{G/L}(a)}{K'_{G/P}(a)} = \frac{c_{P,a}}{c_{L,a}} = \frac{\gamma_{L,a}^{\Phi}}{\gamma_{P,a}^{\Phi}}. \tag{4.50}$$

Using the above expressions for partition coefficients defined on the basis of weight fraction and volume fraction activity coefficients one avoids the difficulty of trying to define what exactly the molar volume of a polymer is.

Due to the difficulties caused by using polymer molar volumes, it is common in some estimation methods to use the molar volume of the polymer's repeating structural unit (i.e., molar volume of monomer) instead of the actual molar volume when estimating the values of physical properties. This molar volume is designated $\underline{V}_P$ and one can use Eqs. (4.33) and (4.35) without difficulty because the relative molar mass of the monomeric structural unit M_P is known and the density of the polymer ρ_P is easy to determine ($\underline{V}_P = M_P/1000\,\rho_P$ dm^3 mol^{-1}). Taking for example polyethylene with a density of $\rho_P = 0.93$ g cm^{-3} at 25°C and the monomeric structural unit $-CH_2-CH_2-$, with $M_P = 28$ g mol^{-1}, then $\underline{V}_P = 0.030$ dm^3 mol^{-1}. For calculating the partitioning of a substance i between polyethylene and water according to Eq. (4.35), $\underline{V}_L = 18/1000 = 0.018$ dm^3 mol^{-1} is used for water.

In general the corresponding molar volume for nonpolymeric liquids can be obtained without difficulty, because the densities are available from tables or can easily be determined experimentally.

4.2.4
Relationship Between Partition Coefficients and Solubility Coefficients

The partition coefficient describing solute partitioning between air and polymer is often referred to as a solubility coefficient, S. The solubility coefficient can be expressed in terms of the molal activity coefficient for polymers using Eq. (4.40):

$$S = \frac{c_{P,a}}{p_a} = \frac{\rho_P}{\Omega_{P,a} \cdot p_a^*}. \tag{4.51}$$

Solubility has CGS units of g cm^{-3} Pa^{-1}, where p_a is the solute partial pressure and p_a^* is the solute saturated vapor pressure at the temperature of the system. Note that at ambient temperature and pressure one can assume ideal gas behavior. Henry's constant for a solute in a polymer is a special solubility coefficient case where $c_{P,a} \propto h \cdot p_a^*$.

4.3
Estimation of Partition Coefficients Between Polymers and Liquids

The majority of partition coefficient estimation methods are based on thermodynamic models that estimate the activity coefficient of a solute in a polymer and in a liquid phase. There has been significant effort devoted to activity coefficient estimation methods in the chemical engineering, environmental, and pharmaceutical research fields because the necessary experimental data for many substances are not available and are difficult to measure. In addition partition coefficients in polymer/liquid systems have been modeled using quantitative structure activity relationships (QSAR) (Tehrany *et al.* 2006).

All current partition coefficient/activity coefficient estimation models are by necessity semiempirical in nature or are statistical based (e.g., QSAR) because still too little is known about solution theory for outright estimation. Chemical modeling is not readily available and is not far enough developed to do these types of calculations. The constants required by the models must be estimated using either experimental data points (e.g., an infinite dilution activity coefficient or a molar volume) or by using group contributions derived from experimental data (e.g., interaction constants, molecular volumes, and surface areas).

4.3.1
Additive Molecular Properties

The physical properties of a substance are dependent on the nature of the atoms found in it and the type of bonds between them. The size and shape of the molecules from which a substance is composed determine the aggregate state and all related specific properties such as melting point, vapor pressure, density, viscosity, and solubility in various media. This also includes the number value of the partition coefficient. Fundamentally, there are two completely different ways to determine the number values of specific quantities. The first and most exact way is by direct experimental measurement. This requires significant effort and is not always feasible for two main reasons. First, there is a multitude of system combinations coming into question; and second the measurement techniques themselves in most cases are not simple. Numerous sources of error in the various experimental methods can make the use of published data difficult because it is not always possible to critically evaluate them.

The second way to obtain numeric values for specific quantities is a purely deductive approach where calculation of a number value is attempted with the help of theoretical derivations from quantum mechanics and statistical mechanics. Even with modern computing facilities, it is not yet possible to carry out these calculations without a series of simplifying assumptions. Such simplifications can nevertheless have large negative effects on the precision of the calculated results.

In practice one is obliged to use yet a third way that goes between the two extremes in order to get useable number values for material-specific properties (Chapter 15). This way touches on the estimation of values with the help of experimentally determined data collections that allow the estimation of values using a series of simplifications and more or less theoretically based assumptions. In view of the often large simplifications used, the results can be astoundingly reliable. Even when estimations obtained in a semiempirical way are considered only as approximations, they are in practice extremely useful and in many cases the only useful estimation of a property. In the next section it will be shown that even very empirical approximation methods are ultimately based on theoretical foundations.

One of the reasons why estimated data can be so reliable is the experience that values for a given substance within a homologous series show a very slow change in their properties. A substance class is designated a homologous series when its individual members have the same structural characteristics and are differentiated only through the number of a repeating structural unit. The simplest example of homologous series is the class of straight chain saturated hydrocarbons, the so-called normal or n-alkanes, with the elementary composition C_zH_{2z+2}, where z is the number of carbon atoms in a molecule in this series. Specific properties such as melting point and boiling point increase after a certain z value in a predictable way. This behavior is based on the similarity of the repeating structural unit which is a CH_2 or methylene group. There are deviations for small z values which are caused by the different behavior of the end CH_3 or methyl group. Because of the very slow change of the value within a homologous series, values of individual members that contain errors can be discovered easily and missing values can be estimated by interpolation or extrapolation. Homologous series, and in particular the homologous series of n-alkanes, serve as the backbone for estimating values for various specific properties of an organic substances (see Chapter 6).

A further reason for the reliability of estimated values is the relative independence of various structural units within a molecule as long as these units or functional atomic groups are not placed too near to one another. Using the basis of this relative independence, the possibility exists to simply sum the individual contributions of the characteristic atom groups to give a value for the properties of the substance. However, complications exist due to influences from neighboring groups, which must be taken into account to improve the estimation in the so-called second approximation. As an example, take the di-ketone compound $CH_3COCH_2CH_2CH_2CH_2COCH_3$, with two identical carbonyl groups in positions 2 and 7. For estimating the value of a specific property, one can in general add the contributions from the carbonyl groups, and the methylene and the two methyl end groups to account for the complete structure. If both of the CO groups are neighbors, as for example in the molecule $CH_3CH_2CH_2COCOCH_2CH_2CH_3$, then there must be a contribution added or subtracted to account for this, because the two groups influence one another. Despite the complications inherent in adding together structural group contributions of a molecule, which are referred to as structural increments in the following text, these methods are practical and useful for the estimating the values of specific properties.

The simplest example of a quantity obtained by the addition of structural increments is the relative molecular mass M_r (which is dimensionless and is commonly but not correctly referred to as molecular weight). The relative molecular mass is made up by strict addition of the single atomic masses (atomic weights), A_r, and can be calculated using the elemental formula of the substance. The influence of neighboring groups takes no place in calculating molecular mass, because such influences are based on the binding relationships between the atoms and these have no influence on the masses. The relative molecular mass is a dimensionless quantity because it comes from the ratio of the actual molecular mass to the 1/12 atomic mass of the carbon ^{12}C isotope. This is comparable to the molar mass or mole mass M which has the dimensions $g\,mol^{-1}$.

A further specific quantity that can be calculated by adding together contributions from individual building blocks is the molar volume $\underline{V}$, usually expressed in $cm^3\,mol^{-1}$. This is the volume that the mole of a given substance occupies. Compared to mole mass the molar volume is dependent on the physical state of the substance as well as pressure and temperature. As a result there are several difficulties in summing the individual contributions to give molar volume, because along with the volume of individual molecules the spatial volume between molecules must also be considered. The spatial volume is dependent on the state of the system, which means the opposing spatial ordering of molecules in a crystal lattice or liquid. For this reason it is not so easy to give such an exact value for a specific property such as density (specific weight) from M and $\underline{V}$ as it is for M and M_r.

For the following applications, additive quantities are needed that can reproduce the intermolecular binding relationships, e.g., partition coefficients are based on such relationships.

For all additive mole constants there are two generally valid rules:

(1) The additive structural increments in either atomic constants or in binding constants can be determined purely by calculation. Secondly, additional values or structural increments can be added for certain structural elements e.g., ring systems, branching, and others.

(2) The structural increments are considered to be purely calculated quantities for which no simple physical meaning can be established.

When putting together a table of additive structural increments, attention must be given that it does not contain too many values that unnecessarily complicate its application. The most sensible method is to combine several atoms in groups, e.g., into functional groups, for which a constant value can be used. By doing this, the table becomes more convenient to use. The additive principle makes it possible to simply sum the group contributions with the help of the following simple formula:

$$E = \sum_a v_a E_a \tag{4.52}$$

where E represents the sum of all molecular increments, v_a is the number of a given type of increment, and E_a is the value of the structural increment a.

There already exists a substantial literature devoted to the estimation of various material properties with the help of additive structural increments (Reid et al. 1987; Van Krevelen 1990).

4.3.2
Estimation of Partition Coefficients Using QSAR and QSPR

Quantitative structure activity relationship (QSAR) and quantitative structure property relationship (QSPR), a field of computational chemistry, are widely used to predict the biological activity of molecules for drug discovery, pesticide development, molecular synthesis, and environmental toxicity. Quantitative structure relationship techniques attempt to find consistent relationships between the variations in the values of molecular properties and their activity (QSAR) or property (QSPR) for a series of compounds so these "rules" can be used to evaluate new chemical entities (Richon and Young 2007). Selassie (2003) published a review of the history, tools, techniques, and applications of QSAR. A QSAR generally takes the form of a linear equation (Richon and Young 2007):

$$\text{activity/property} = \text{constant} + (c_1 \cdot P_1) + (c_2 \cdot P_2) + (c_3 \cdot P_3) + \ldots + (c_n \cdot P_n) \quad (4.53)$$

where the parameters P_n are molecule physiochemical or structural properties and the coefficients c_n are calculated by fitting variations in the parameters and the activity/property. Parameters used in QSAR include electronic (e.g., Hammett constants, quantum mechanical indices), hydrophobicity (e.g., octanol/water partition coefficient), steric (e.g., molecular weight, molecule size, and shape), indicator variables (e.g., LUMO – lowest unoccupied molecular orbital), and molecular structure descriptors (Selassie 2003). To develop a QSAR statistical linear regression methods are usually used to fit a set of experimental training data. As a result the quality of QSAR is only as good as the quality of the experimental data used to derive the model. Numerous computer-based programs have been developed using different parameters to build QSARs (Richon and Young 2007).

There has been some initial limited work calculating solute (six solvents) partition coefficients between polymers (polyamide and polyester) and aqueous (water, 10 and 95% ethanol, 3% acetic acid) food stimulant mixtures using QSPR (Tehrany et al. 2006). Molecular descriptors were obtained from a molecular modeling database and then the equation fitting was carried out using a math software program. For experimental polymer/liquid partition coefficients larger than 5, four parameters were used: polarity polymer, polarity of the food, molecular weight, and lowest unoccupied molecular orbital. For experimental partition coefficients smaller than 5, two additional parameters for the molecular weight of the stimulant and the hydrophilic–hydrophobic balance were used. The QSPR approach shows promise but has so far only been applied to a limited number of solute/polymer/food simulant mixtures.

4.3.3
Group-Contribution Thermodynamic Polymer Partition Coefficient Estimation Methods

The estimation of activity coefficients in polymer systems presents special problems for modeling and there are several activity coefficient estimation models that have been proposed. There are thermodynamic estimation methods based on the regular solution theory (RST) with empirical corrections (Baner and Piringer 1991), equations of state (Flory 1970; High and Danner 1990; Chen et al. 1990; Bogdanic and Fredenslund 1994 – GCFLORY), statistical mechanics (Oishi and Prausnitz 1978 – UNIFAC-FV), quantum mechanical (Mavrovouniotis 1990; Canstantinou et al. 1993), and free volume models (Kontogeorgis et al. 1993).

For many thermodynamic estimation models the activity coefficient is commonly described as being roughly composed of two or three different components. These components represent combinatorial contributions (γ_i^c) which are essentially due to differences in size and shape of the molecules in the mixture, residual contributions (γ_i^r) which are essentially due to energy interactions between molecules, and free volume (γ_i^{fv}) contributions which take into consideration differences between the free volumes of the mixture's components:

$$\ln \gamma_i = \ln \gamma_i^c + \ln \gamma_i^r + \ln \gamma_i^{fv}. \tag{4.54}$$

The combinatorial and free volume contributions can be thought of as being roughly analogous to the excess entropy of mixing (see Eq. (4.23)). The residual contribution is roughly analogous to the excess enthalpy of mixing. In some models the free volume contribution is treated in a separate term (e.g., UNIFAC-FV and GCFLORY) and in others (e.g., Elbro et al. 1990 – ELBRO-FV) the combinatorial and free volume contributions are essentially combined into the free volume term thus eliminating the combinatorial term. The RST considers essentially only a residual contribution.

One of the central problems with estimating activity coefficient in polymer systems is that general observations made for low molecular weight component systems are no longer valid for polymers. It is observed that the solution-dependent properties are no longer directly proportional to the mole fraction of solute in the polymer at dilute concentrations. For example, the solute partial pressure in a system containing a polymer is no longer directly proportional to its mole fraction which is an apparent deviation from Raoult's law.

Some estimation models work around the difficulty of using mole fractions for polymers by using weight fractions (molal concentration). This avoids the problem of defining what is a mole of polymer. However, even using weight fractions and defining the estimated activity coefficients on a molal basis is not enough to overcome the differences between what is expected from a low molecular weight liquid system and what is experimentally observed in a polymer system. This difference is treated in most estimation models by describing the difference as being due to free volume differences between the polymer and liquid. The free volume concept was specifically developed to describe the variation of polymer system properties from those of a liquid system. The concept of free volume varies on how it is defined and used but is generally acknowledged to be related to the degree of thermal expansion of the molecules. When liquids with different free volumes are mixed that difference contributes to the excess functions (Prausnitz et al. 1986). The definition of free volume used by Bondi (1968) is the difference between the hard sphere or hard core volume of the molecule ($\underline{V}^W$ = van der Waals volume) and the molar volume at some temperature:

$$\underline{V}^{fv} = \underline{V}_i - \underline{V}^W. \tag{4.55}$$

A modern way of estimating activity coefficients and thus partition coefficients is through the use of group-contribution parameters. Methods have been developed to allow chemical engineers to estimate activity coefficients in liquid and polymeric systems. The principle of the method is to start with a thermodynamic framework describing how molecules interact in solutions and then use additive molecular increments (i.e., molecular groups such as methyl groups CH_3, hydroxyl groups OH) to obtain the combinatorial, residual, or free volume contribution estimation of how a molecule will interact with other molecules of the solvent or polymer. While combinatorial and free volume parameters are based on well-known measures of molecular size and dimensions, residual group-contribution parameters are developed for each model by fitting a set of experimental data.

In the following sections, taken from Baner (2000), the accuracies of four of the currently available group-contribution thermodynamic partition coefficient estimation methods that are suitable for systems involving polymers are compared in Tables 4.1 and 4.2.

4.3.3.1 Estimation of Partition Coefficients Using RST

The RST in combination with additive structural increments has a wide application for estimating the relative solubilities of organic substances in polymers and the solubility of polymers in various solvents (Barton 1983; Van Krevelen 1990; Brandrup *et al.* 2003). The RST uses solubility parameters to estimate activity coefficients and was one of the first thermodynamic-based estimation methods. It has been successful in giving relatively good activity coefficient estimations for hydrocarbons. RST as its name implies is for regular solutions. It uses the geometric mean assumption which assumes that interactions between different molecules in a mixture are similar to those the molecules experience between themselves in a pure mixture. As a result its best estimations are for mixtures of similar sized nonpolar molecules. When estimating partition coefficient values, one is quickly confronted with this method's application limits, particularly with polar and nonpolar structures. The accuracy of RST is not good for estimations of very dilute concentrations of organic solute partition coefficients between polyolefins and alcohols (Baner and Piringer 1991). They estimated activity coefficients using an empirical correlation along with solubility parameters estimated by the group-contribution method of Van Krevelen (1990).

4.3.3.2 Estimation of Partition Coefficients Using UNIFAC

The UNIFAC (unified quasi chemical theory of liquid mixtures functional-group activity coefficients) group-contribution method for the prediction of activity coefficients in nonelectrolyte liquid mixtures was first introduced by Fredenslund *et al.* (1975). It is based on the unified quasi chemical theory of liquid mixtures (UNIQUAC)

Table 4.1 Partition coefficients of 13 aroma compounds between 100% ethanol and polyethylene.

Aroma compound	Experiment[a] $K_{P/L}$	UNIFAC-FV	GCFLORY (1990)	GCFLORY (1993)	ELBRO-FV	UNIFAC (L) GCFLORY (P) (1990)	UNIFAC (L) GCFLORY (P) (1993)	UNIFAC (L) ELBRO-FV (P)	Regular solution theory
D-limonene	0.67	2.18	4.3	82	0.92	0.37	0.22	1.9	0.3
Diphenylmethane	0.27	1.09	0.040	36	0.27	0.096	1.4	1.2	0.17
Linalylacetate	0.058	0.31	1.5	20	0.15	0.18	0.60	0.34	0.074
Camphor	0.062	0.36	0.029	2.3	0.21	0.012	0.097	0.38	0.16
Diphenyloxide	0.23	0.68	0.52	13	0.20	0.48	0.48	0.75	0.027
Isoamylacetate	0.0464	0.27	0.30	1.6	0.16	0.26	0.41	0.26	0.053
Undelactone	0.052	0.66	0.20	3.4	0.32	0.20	0.32	0.63	–
Eugenol	0.0181	0.00093	0.0051	0.034	0.00051	0.0019	0.0041	0.0010	0.008
Citronellol	0.025	0.018	0.046	0.39	0.019	0.052	0.15	0.22	–
DMC	0.0113	0.0091	–	0.22	0.0082	–	0.12	0.012	–
Menthol	0.02	0.019	0.0088	0.24	0.017	0.0097	0.081	0.23	0.027
PEA	0.0097	0.0067	0.0037	0.13	0.0076	0.010	0.11	0.010	0.0029
cis-3-Hexenol	0.0139	0.014	0.034	0.16	0.015	0.040	0.15	0.016	0.0029
(Undelactone)	0.052	–	–	2.8	–	–	0.47	–	–
(Limonene)	0.67	–	0.011	64	–	1.2	0.60	–	–
(Camphor)	0.062	–	0.25	1.8	–	0.23	0.21	–	–
(Menthol)	0.02	–	0.012	0.17	–	0.017	0.11	–	–
AAR		5.0	5.6 (11)	67 (63)	4.7	3.5 (3.2)	6.5 (6.8)	4.89	2.91
AAR s.d.		5.2	6.4 (17)	90 (89)	9.0	2.2 (2.3)	3.3 (3.4)	4.77	2.15
AAR c.v. %		103	115 (156)	134 (141)	193	63 (70)	52 (50)	97.5	73.9

[a] Experimental data from Koszinowski and Piringer (1989) and Baner (1992).
[b] () calculated using cyclic group contribution parameters (GCFLORY only). Calculated using polymer density = 0.85, (L) = liquid phase activity coefficient, (P) = polymer phase activity coefficient. – not possible to calculate a result. AAR = average absolute ratio: for calculated (calc.) > experimental (esp.) = calc./exp.; for calc. < exp. = exp./calc. AAR s.d. = standard deviation of absolute ratio. AAR c.v. % = % coefficient of variation absolute ratio.

Table 4.2 Average absolute ratios of estimated aroma compound partition coefficients between polymers and solvents.

Partition system	UNIFAC-FV	GCFLORY (1990)	GCFLORY (1993)	ELBRO-FV	UNIFAC (L) GCFLORY (P) (1990)	UNIFAC (L) GCFLORY (P) (1993)	UNIFAC (L) ELBRO-FV (P)	Regular solution theory
PE ($\rho_P = 0.918$)/13 aromas/100% ethanol[a]	5.0 (5.2)	5.6 (6.4) c:11 (17)	67 (90) c:(63) (89)	4.7 (9.0)	3.5 (2.2) c:3.2 (2.3)	6.5 (3.3) c:6.8 (3.4)	4.89 (4.8)	2.9 (2.2)
PE ($\rho_P = 0.918$)/13 aromas/100% water[b]	17.7 (30.3)	8060 (24 300) c:2120 (607)	653 000 (1.9E+6) c:653 000 (1.9E+6)	4.2E+19 (1.1E+20)	10.2 (10.0) (6.2) (3.9)	45 (75) c:43 (75)	18 (29)	—
PP ($\rho_P = 0.902$)/13 aromas/100% ethanol[d]	5.8 (6.8)	9.2 (15) c:11 (21)	133 (147) c:115 (130)	5.7 (12)	5.5 (6.6) c:4.6 (6.3)	9.9 (6.3) c:11 (6.8)	5.8 (6.6)	—
PVC ($\rho_P = 1.36$)/13 aromas/100% ethanol[b]	14 (27)	59 (122) c:62 (122)	73 (181) c:68 (178)	24 (53)	122 (334) c:131 (347)	16 (21) c:14 (17)	13 (26)	—
PET ($\rho_P = 1.37$)/7 aromas/100% methanol[c]	2.9 (1.8)	1380 (2210) c:922 (2240)	2.6E+6 (3.2E+6) c:3.6E+6 (4.3E+6)	4.4 (3.1)	620 (1220) c:4.8 (4.6)	3.3 (3.3) c:3.4 (3.2)	4.1 (3.8)	—
PET ($\rho_P = 1.37$)/7 aromas/100% 1-propanol[c]	12 (11)	73 (85) c:27 (49)	127 (199) c:127 (199)	10 (12)	250 (480) 4.6 (4.0)	15 (18) 15 (18)	18 (22)	—
PA ($\rho_P = 1.14$)/6 aromas/100% methanol[c]	4.5 (2.9)	—	—	7.9 (3.5)	—	—	6.6 (5.0)	—
PA ($\rho_P = 1.14$)/7 aromas/100% 1-propanol[c]	6.1 (2.7)	—	—	7.0 (3.8)	—	—	12 (8.2)	—
Average AAR:	8.5	1598	100	9.1	169	16	10.3	

c denotes calculated values using cyclic group contribution parameters (GCFLORY only). — not possible to calculate a result. AAR = average absolute ratio: for calculated (calc.) > experimental (exp.) = calc./exp.; for calc. < exp.: = exp./calc.() = AAR s.d. = standard deviation of absolute ratio.
[a] Experimental data from Koszinowski and Piringer (1989) and Baner (1992).
[b] Experimental data from Koszinowski and Piringer (1989).
[c] Experimental data from Franz (1990).
[d] Experimental data from Becker et al. (1983).

(Abrams and Prausnitz 1975) which is a statistical mechanical treatment derived from the quasi chemical lattice model (Guggenheim 1952). UNIFAC has been extended to polymer solutions by Oishi and Prausnitz (1978) who added a free volume contribution term (UNIFAC-FV) from the polymer solution equation-of-state equation of Flory (1970). The UNIFAC activity coefficient estimation model uses the form of Eq. (4.54).

The UNIFAC method has a useful temperature range of −23 to 152°C. The accuracy of infinite dilution estimations for relatively low molecular weight organic molecules ($M_i < 200$) averages from 20.5% for 3357 low molecular weight compounds (Thomas and Eckert 1984), 21% for 6 compounds in three classes of solvents (Park and Carr 1987) and 21.1% for 791 series of measurements containing 1773 data points (Larsen et al. 1987). Several modifications have been proposed to the UNIFAC model in addition to the free volume contribution already mentioned. Proposed modifications to the combinatorial and residual terms that improve estimations by about 11% (Weidlich and Gmehling 1987; Larsen et al. 1987) but cannot be used for polymers (Fredenslund 1989). Other UNIFAC modifications have been developed (Bastos et al. 1988; Iwai et al. 1985; Doong and Ho 1991) that attempt to increase the predictions made by the original model for infinite dilutions, hydrocarbon solubility, and swelling/dissolving of semicrystalline polymers.

Mixtures of hydrocarbons are assumed to be athermal by UNIFAC meaning there is no residual contribution to the activity coefficient. The free volume contribution is considered significant only for mixtures containing polymers and is equal to zero for liquid mixtures. The combinatorial activity coefficient contribution is calculated from volume and surface area fractions of the molecule or polymer segment. The molecule structural parameters needed to do this are the van der Waals or hard core volumes and surface areas of the molecule relative to those of a standardized polyethylene methylene CH_2 segment. UNIFAC for polymers (Oishi and Prausnitz 1978) calculates in terms of activity (a_i) instead of the activity coefficient and uses weight fractions (Eq. (4.36)) instead of mole fractions.

Tables 4.1 and 4.2 show estimated molar activity coefficients from UNIFAC and molal (weight fraction) activity coefficients from UNIFAC-FV to calculate a liquid activity coefficient of a test set of 13 aroma compounds partitioned between various polymers and alcohol solvents using Eq. (4.45). The UNIFAC estimations of activity coefficients were estimated using a BASIC computer program based on a program (Sandler 1989) modified for calculations for polymers (UNIFAC-FV) to carry out the free volume correction using weight fraction based activity coefficients (Oishi and Prausnitz 1978). The program was rewritten using published algorithms (Goydan et al. 1989) for binary polymer/solute solutions and the interaction parameters were updated using the UNIFAC 5th revision interaction parameters (Hansen et al. 1991). In UNIFAC calculations the polymer monomer repeat unit is used to represent the chemical structure and molecular weight of the polymer in the polymer activity coefficient calculations. For polyethylene (PE) polymers the amorphous PE density ($\rho_P = 0.85$) gave the most accurate estimations.

4.3.3.3 Estimation of Partition Coefficients Using Group-Contribution Flory Equation-of-State

The group-contribution Flory equation-of-state (GCFLORY) developed by Chen *et al.* (1990) and later revised (Bogdanic and Fredenslund 1994) is a group-contribution extension of the Flory equation of state (Flory 1970). The equation is similar to the Holten–Anderson model and incorporates a correlation for the degree of freedom parameter. The method has combinatorial and free volume contributions to the activity coefficient similar to UNIFAC and uses the UNIFAC surface area and volume functional group parameters. The model estimates activities and uses weight fractions.

Copies of the FORTRAN versions of the GCFLORY programs were obtained from the authors (POLGCEOS March 5, 1991, Chen *et al.* 1990 and GC-FLORY EOS April 28, 1993, Bogdanic and Fredenslund 1994). An average number molecular weight of 30 000 was used in the calculations for all polymers. The model's resulting activity coefficient estimations are relatively insensitive to variations in polymer molecular weights.

4.3.3.4 Estimation of Partition Coefficients Using Elbro Free Volume Model

The Elbro free volume model (ELBRO-FV) is based on the free volume term proposed by Elbro *et al.* (1990). The ELBRO-FV model uses only the free volume and residual activity coefficient contribution terms. The residual term is taken from the UNIFAC model and is thus equal to zero for athermal mixtures. For the activity coefficient estimation calculations the residual portion of the activity coefficient (Eq. (4.54)) was calculated in a simple spreadsheet and added to the interaction activity coefficient portion calculated using UNIFAC. An average number molecular weight of 30 000 was used for the polymers. The van der Waals volumes were calculated using the group-contribution method of Bondi (1968).

4.3.3.5 Comparison of Thermodynamic Group-Contribution Partition Coefficient Estimation Methods

A test set of 6 to 13 aroma compound partition coefficient between different food contact polymers (low density polyethylene (LDPE), high density polyethylene (HDPE); polypropylene (PP), polyethylene terephthalate (PET), poly amide (PA)) and different food simulant phases (water, ethanol, aqueous ethanol/water mixtures, methanol, 1-propanol) were taken from the literature (Koszinowski and Piringer 1989; Baner 1992, 1993; Franz 1990, 1991; Koszinowski 1986; Piringer 1993). In Table 4.1 the experimental data were compared to estimations using the different thermodynamic group-contribution partition coefficient estimation methods described above.

Table 4.1, with partition coefficient estimation results for 13 aroma compounds partitioned between polyethylene (PE) and ethanol, is an example of the estimation accuracy one can expect using these methods. In order to compare the different estimation methods, average absolute ratios of calculated to experimental values were calculated for the partitioned substances. For calculated values greater than experimental values the calculated value is divided by the experimental value. For calculated values less than the experimental values the inverse ratio is taken. Calculating absolute ratios gives a multiplicative factor indicating the relative differences between values of the experimental and estimated data. A ratio of 1 means the experimental value is equal to the estimated value (Baner et al., 1994).

The liquid phase and polymer phase activity coefficients were combined from the different methods to see if better estimation accuracy could be obtained since some estimation methods were developed for estimation of activity coefficients in polymers (e.g., GCFLORY, ELBRO-FV) and others have their origins in liquid phase activity coefficient estimation (e.g., UNIFAC). The UNIFAC liquid phase activity coefficient combined with GCFLORY (1990 and 1993 versions) and ELBRO-FV polymer activity coefficients was shown to be the combinations giving the best estimations out of all the possible combinations of the different methods.

Table 4.2 shows examples of average absolute ratios (estimated values to experimental values) and their corresponding standard deviations for several other polymer/aroma/solvent systems. The results are representative of the estimation behavior of these models and do not include all data tested to date. Baner (1992) tested the partition coefficient estimations for these aroma compounds and polymer additives between aqueous ethanol solutions and polyethylene with similar results. UNIFAC-FV was found to be the most consistent, most widely applicable and overall gives the most accurate partition coefficient estimations of all the models. However, the UNIFAC model partition coefficient estimations here are less accurate than the 20% expected variation reported in the literature (Thomas and Eckert 1984; Park and Carr 1987) for solute in solvent activity coefficient data. GCFLORY (Chen et al. 1990) could not estimate water and polar polymers (PVC, PET, no amide group for polyamide) containing systems well. GCFLORY by Bogdanic and Fredenslund (1994) is very similar to GCFLORY by Chen et al. (1990) but made worse estimations because it was not intended to be used for estimating activity coefficients in low molecular weight liquids. When the GCFLORY polymer activity coefficient was used with the liquid phase activity coefficient from UNIFAC, partition coefficient estimations were significantly improved. The GCFLORY models had cyclic and aliphatic group-contribution terms. Calculations were made for four ring-containing aroma compounds using both cyclic and aliphatic groups to see whether there was any advantage of one set of groups over the other for these estimations. In most cases the results were very similar either slightly better or slightly worse.

ELBRO-FV cannot model water as a solvent; otherwise it was often better than UNIFAC-FV in accuracy. For water-containing systems UNIFAC liquid activity coefficient estimations should be used with ELBRO-FV polymer activity coefficients. The scope of application of the RST is limited only to solute partitioning in PE/ethanol systems. Their better accuracy is not surprising since they are essentially correlations of this same experimental data.

4.3.4
Vapor Pressure Index Partition Coefficient Estimation Method

Given the inaccuracies of the RST and the large amount of effort required to use the UNIFAC and similar thermodynamic group-contribution methods (all of which still have inherent errors), a third estimation, based on additive molecular properties (increments), can be presented. This approach leads to sufficiently accurate values by using simple calculations for the most important practical partitioning cases. The method referred to here as the vapor pressure index method (VPIM) is based on a recognized fact in gas chromatography that the partitioning of a substance between a gas and a polymer liquid can be estimated based on its structural increments and these can be used as characteristic quantities for identification.

As mentioned previously the main disadvantage of the regular solution theory is its inability to model polar systems, e.g. aqueous solutions, which are very important for food and pharmaceutical packaging. This limits the range of application of the RST. The main problems when applying UNIFAC and similar type of methods to food and pharmaceutical packaging are the lack of basic empirical data and limited models for interpreting systems with large molecular weights such as plastics and additives. In addition these estimation methods require specific computer programs to calculate the results. For this reason the vapor pressure index estimation method was developed. The method is easy to use, the input parameters required are readily available or can be estimated and the method can be applied to several different applications. The insight leading to the development of the method starts with the field of gas chromatography (GC) where retention indexes have been successfully used for years to identify separated substances.

Starting from the well-known definition of a retention index in GC (Kovats 1958), a molecular retention index based on partition coefficients of a substance a and two homologous n-alkanes with j and $j+1$ carbon atoms, between a polymeric separation phase and a gas, has been defined (Piringer et al. 1976; Piringer 1993):

$$M_e = 14 \cdot \frac{\log(K_j) - \log(K_a)}{\log(K_j) - \log(K_{j+1})} + M_j. \tag{4.56}$$

In the following procedure the homologous series of n-alkanes is used as a reference compound class. For these substances it is possible to calculate the equilibrium saturated vapor pressure p_j^* of a pure n-alkane with j carbon atoms at a given temperature T by using a generalized Antoine equation:

$$\log(p_j^*) = A - \frac{B}{\frac{w}{w_{j,e}} \cdot \frac{T}{T_{j,\infty}} + C - \frac{j}{0.036 \cdot j^2 + 100}} \tag{4.57}$$

with $A = 9.4278$ $T_{j,\infty} = 1036.2\,\text{K}$ $j = (M_j - 2)/14$ $B = 2.8586$ $w_{j,e} = (1 + 2\pi/j)^{j/e}$, $C = 0.015$ $w = e^{2\pi/e}$.

In Chapter 6 the meaning of the so-called interaction functions $w_{j,e}$, and w and of the limited temperature $T_{j,\infty}$ used in this equation is explained.

The corresponding saturated vapor pressure p_a^* of a pure substance a can be calculated with the same equation (4.57) if $j = (M_r + W_a - 2)/14$ is used. This vapor

4.3 Estimation of Partition Coefficients Between Polymers and Liquids

pressure represents the vapor pressure of a hypothetical n-alkane with the relative molecular mass $M_j = M_r + W_a$. The vapor pressure index W_a is a dimensionless group-contribution element with the same unit as the relative molecular mass M_r of the substance, that can be added to M_r. The value of W_a depends on the structure of the molecule a. Polar functional groups contribute with positive values, whereas branching in the molecule give negative contributions. The n-alkanes have by definition $W_a = 0$.

The partial vapor pressure $p_{L,a}$ of a substance a over a solution containing a in liquid L or the partial vapor pressure $p_{P,a}$ over the polymer P is the product between its saturated vapor pressure p_a^* over the pure liquid a, the mole fraction $x_{L,a}$ or $x_{P,a}$, and the activity coefficient $\gamma_{L,a}$ or $\gamma_{P,a}$ (Eq. (4.20)). The product $\gamma_{L,a} p_a^* = h_{L,a}$ or $\gamma_{P,a} p_a^* = h_{P,a}$ is the corresponding Henry constant of a in L or P. The Henry constant $h_{L,a}$ can also be calculated with Eq. (4.57) whereby $j = (M_r + W_a + G_L - 2)/14$ is once again the number of carbon atoms in a hypothetical n-alkane which has a saturated pure vapor pressure corresponding to a relative molecular mass $M_j = M_r + W_a + G_L$. The same procedure is used for $h_{P,a}$. The vapor pressure index $G_{L,a}$ is an additional dimensionless group-contribution element. Its value is a function of the chemical nature of the substance a and the liquid phase L.

From the above description one can resume that the Henry constant $h_{L,a}$ or $h_{P,a}$ can be estimated in the form of a function $h = f(M_r + W_a + G_L)$ in which the argument is obtained from the relative molecular mass M_r in addition to two structure increments, W_a and G_L or G_P, which take into account the specific structures of the substance a and L or P. The index values are very characteristic for a given chemical structure as will be shown below.

The dimensionless partition coefficients, $K_{G/L}(a)$, of substance a between a gas G and the liquid L and $K_{G/P}(a)$ between G and the polymer P are given in Eqs. (4.33) and (4.34). In this manner one obtains the partition coefficient $K_{P/L}(a)$ of a between P and L:

$$K_{P/L} = \frac{K_{G/L}(a)}{K_{G/P}(a)} = \frac{h_{L,a} \cdot \underline{V}_L}{h_{P,a} \cdot \underline{V}_P}. \tag{4.58}$$

The molar volume $\underline{V}_L$ of the liquid L is easy to determine. As mentioned in Section 4.1.3 it is difficult to use correct molar volumes $\underline{V}_P$ of the polymer. By convention in many cases the molar volumes of the corresponding monomers are used and these have an order of magnitude about 30–100 cm³ mol⁻¹. In the following estimation method a value of $\underline{V}_P = 100 \, \text{cm}^3 \, \text{mol}^{-1}$ is used per definition.

The estimation of partition coefficients will require a corresponding collection of index values. To obtain such collections, one can calculate the W_a, G_L as well as G_P indices using experimentally measured vapor pressure values or from corresponding literature data. The relative vapor pressure index W_a is calculated from the pure vapor pressure p_a^* and the vapor pressures of two homologous n-alkanes with j and $j+1$ carbon atoms in the molecule:

$$W_a = 14 \cdot \frac{\log(p_j) - \log(p_a*)}{\log(p_j) - \log(p_{j+1})} + 14 \cdot j + 2 - M_r \quad \text{and}$$

$$G_L = 14 \cdot \frac{\log(p_{j,a}) - \log(h_{L,a})}{\log(p_{j,a}) - \log(p_{j+1,a})} + 14 \cdot j + 2 - (M_r + W_a). \tag{4.59}$$

To calculate W_a one starts with the above equation for W_a by using $j = (M_r - 2)/14$ in the first step for the calculation of $\log(p_j)$ and $\log(p_{j+1})$ with Eq. (4.57). The W_a-value obtained is added to M_r and the above Eq. (4.57) is used again but with $j = (M_r + W_a - 2)/14$. The new value of W_a is used for the collection.

The corresponding index value, G_L or G_P, is calculated in the same manner whereby the corresponding Henry constant $h_{L,a}$ or $h_{P,a}$ is used instead of p_a*. In the first step $j = (M_r + W_a - 2)/14$ is used for calculation of $\log(p_{j,a})$ with Eq. (4.57). In the second step, as shown above, the calculation of G_L is repeated by using $j = (M_r + W_a + G_L - 2)/14$ in Eq. (4.57).

4.3.4.1 Examples of Vapor Pressure Index Values

The aliphatic hydrocarbons, the alcohols, and acids are substance classes with extreme different polarities. The vapor pressure index W_a for the straight chain n-alkanes reference series is by definition equal to zero. For alcohols and acids the corresponding values are particularly large. However, the index values decrease with increasing molecular weight throughout the homologous series as the relative contribution of the OH group to interactions between the alcohol molecules in the liquid decreases. A good estimation of the W_a-values is obtained with the following linear functions of their relative molecular mass:

$W_a = 65.5 - 0.166\, M_r$ primary aliphatic alcohols
$W_a = 67.7 - 0.295\, M_r$ secondary aliphatic alcohols
$W_a = 37.24 - 0.152\, M_r$ tertiary aliphatic alcohols
$W_a = 63 - 0.068\, M_r$ organic acids.

For each additional branching, for example a methyl group, a contribution $\Delta W_a = -4.5$ must be added.

The values within a homologous series can be well represented by a linear equation. For a given size molecule there is a pronounced decrease in the W_a-values in the alcohol series primary > secondary > tertiary. Single branched molecules, e.g. a CH_2 group, can be assigned a negative value of -4.5 and molecules with two branches have a negative value of -9, etc. By definition the branching next to the OH group present in tertiary alcohols is already taken into account. A direct consequence of this is that the pure vapor pressure for tertiary alcohols is much higher compared to a primary alcohol with the same molecular mass.

For organic compounds with less polar groups in comparison with alcohols and acids, the decrease of the W_a-values with the increase of their molecular mass is less pronounced. For the carbonyl group in aldehydes and ketones, as an example, one can use a mean value of $W_a = 17$ and for an aromatic ring, $W_a = 13$. Esters and ethers have smaller W_a-values, as well as halogen derivatives.

The G_L values for the aliphatic alcohols over water ($L = W$) can also be calculated as a linear function of molecular mass M_r:

$G_W = 3.19 - 1.107\ M_r$ primary aliphatic alcohols (Figure 4.1)
$G_W = 43.3 - 1.17\ M_r$ secondary aliphatic alcohols
$G_W = 59 - 1.17\ M_r$ tertiary aliphatic alcohols.

For each additional branching, for example a methyl group, a contribution ΔW_a +4.5 must be added.

Compared to the W_a-values, molecular branching is accounted for by adding positive values. This means that the water molecules have easier access to the immediate vicinity of OH groups in molecules opened up by branching (i.e., interaction of water with the OH group is shielded from the effect of the hydrocarbon

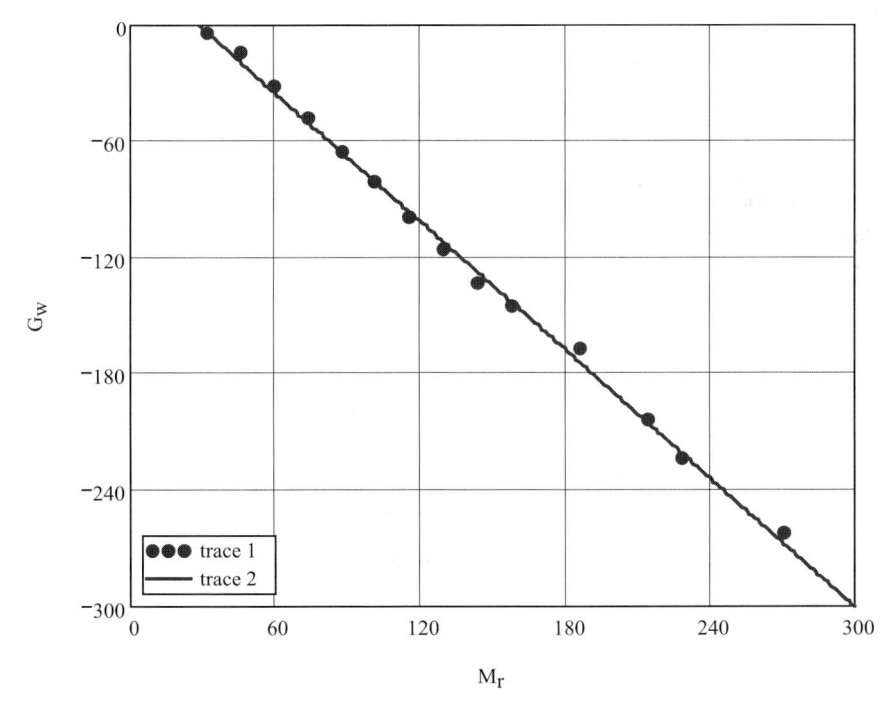

Figure 4.1 The G_w values for the primary aliphatic alcohols over water (trace 1) as a linear function of the relative molecular mass, $G_w = 31.9 - 1.107\ M_r$ (trace 2).

tail of the molecule). The effect of a positive contribution from an index is an increase in solubility in the corresponding liquid.

Up to a relative molecular mass of approximately $M_r = 250$ the W_a-values are positive for a substance and become larger the more polar the molecules. At the same time negative W_a-values are a sign of branched molecules. As molecules become larger the W_a-values become negative as can be seen in Figure 4.2, where the W_a-values for the additives from Appendix III are correlated with their molecular weights. This apparently peculiar result can be explained by the fact that large molecules with several functional groups and branching, e.g. polymer additives, are not as tightly packed solids as the unbranched n-alkane reference molecules. Important for practical applications however is the tendency for W_a to decrease linearly as a function of M_r.

This makes it possible in general to estimate W_a-values using linear relationships in the case of complex molecules whereby along with an equation for average values it allows maximum and minimum limits to be assigned. This is particularly important in all cases in which a "worst case" estimation is made (e.g., for regulatory or safety purposes).

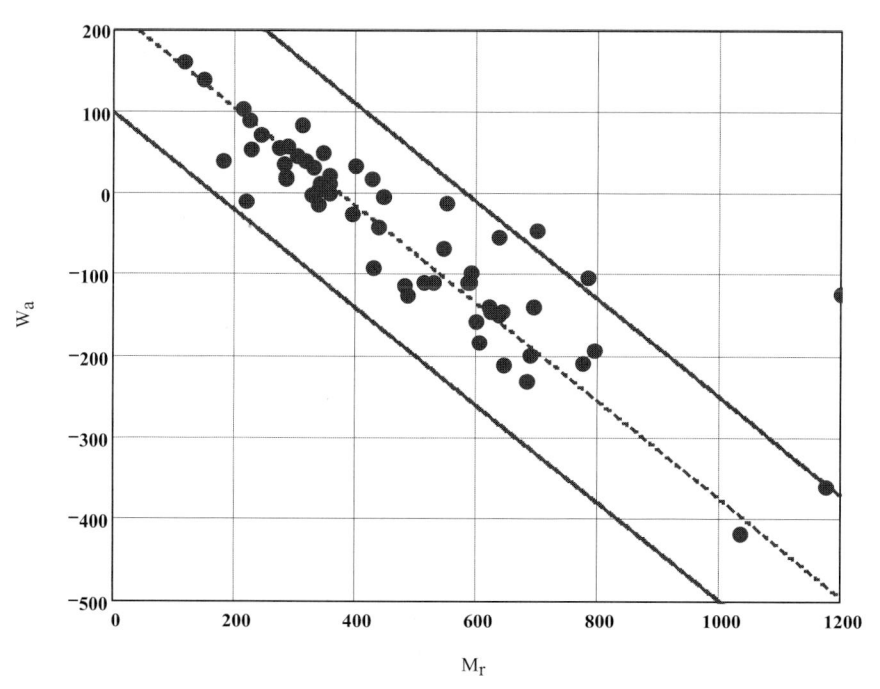

Figure 4.2 The W_a-values of the additives listed in Appendix III in correlation with their molecular weights. The three lines are the corresponding upper limit, $W_{a,max} = 350 - 0.6\, M_r$, the main value function, $\underline{W}_a = 225 - 0.6\, M_r$, and the lower limit, $W_{a,min} = 100 - 0.6\, M_r$, respectively.

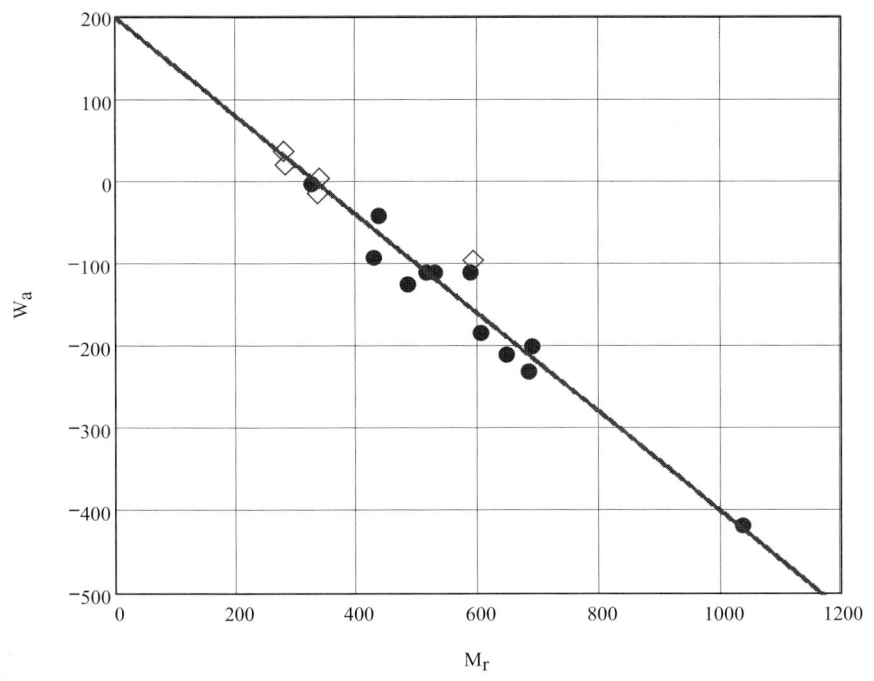

Figure 4.3 The W_a-values of antioxidants with long aliphatic chains and phosphites (•) and of aliphatic amides (◇) from Appendix III in correlation with the linear function $W_a = 200 - 0.6\,M_r$.

Chemical families with specific structures can be well correlated with linear functions as exemplified in Figure 4.3 for some antioxidants and amides from Appendix III.

The same tendencies are observed for $G_L = G_W$ values over water ($L = W$) for all additives (Figure 4.4):

$$W_{a,min} = 100 - 0.6\,M_r \quad G_{W,min} = -M_r$$

$$\underline{W}_a = 225 - 0.6\,M_r$$

$$W_{a,max} = 350 - 0.6\,M_r \quad G_{W,max} = -0.182\,M_r.$$

As mentioned above the alkanes and alcohols form two chemical families with extremely different polarities. This results in a much lower volatility for alcohols compared to alkanes with the same mass. The very high Henry constant values for hydrocarbons over water produce consequently an extremely low solubility of these molecules in water.

As with the W_a-values exemplified in Figure 4.3, chemical families with specific structures can be correlated well with certain linear functions, as shown in Figure 4.5.

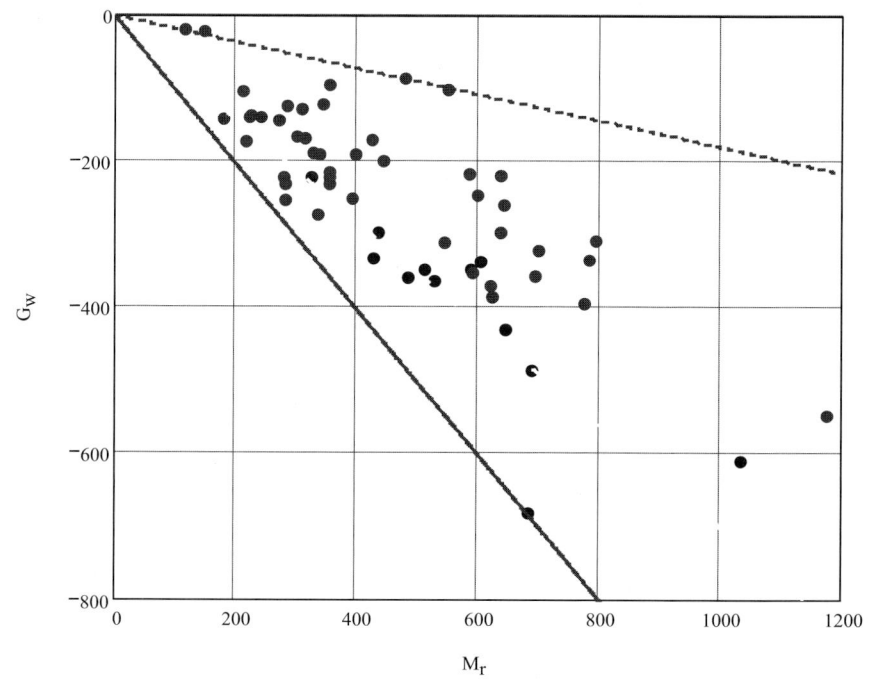

Figure 4.4 The G_W-values of the additives listed in Appendix III in correlation with their molecular weights. The two lines are the corresponding upper limit, $G_{W,max} = -0.182\, M_r$ and the lower limit, $G_{W,min} = -M_r$, respectively.

Several additional linear relationships for estimating the G_L values of additives in ethanol, methanol, and polyethylene glycol as well as G_P values for polyolefins are given:

$G_L = -(M_r - 4)$ for n-alkanes and aliphatic hydrocarbons in water

$G_L = -20 - 0.2\, M_r$ for additives in ethanol

$G_L = -7.3 - 0.41\, M_r$ for additives in methanol and polyethyleneglycols

$G_P = -14 - 0.24\, M_r$ for additives in polyolefins.

Similar relationships can be created for any class of substances and media. The main advantage of linear relationships between vapor pressure unit contributions is that they are easy to calculate by hand or in programs. Similar to the estimation of diffusion coefficients shown in Chapter 6, the equations for estimating partition coefficients can be incorporated into a standard mathematical computer calculation

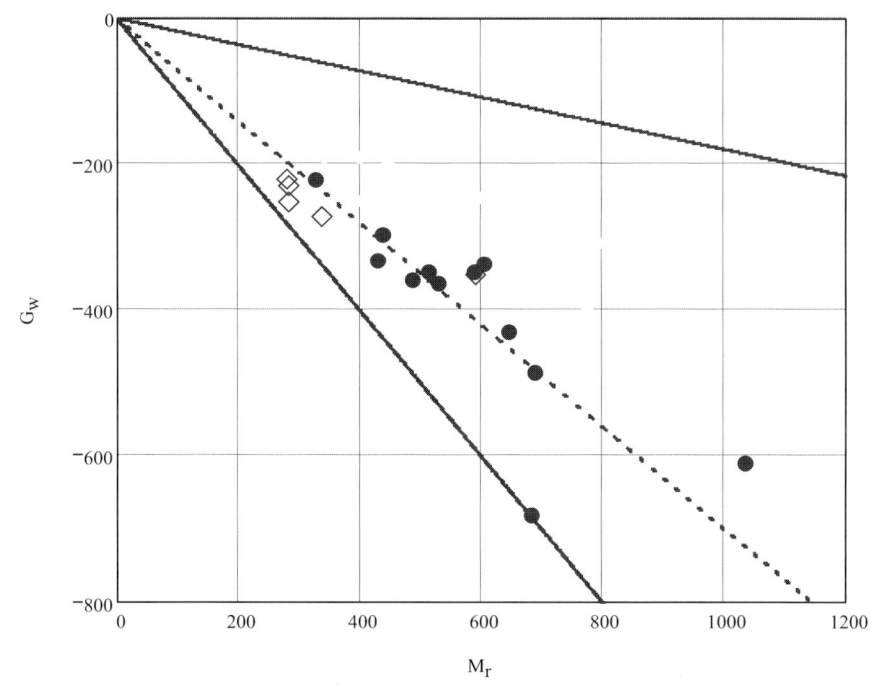

Figure 4.5 The G_W-values of antioxidants with long aliphatic chains and phosphites (●) and of aliphatic amides (◇) from Appendix III in correlation with the linear function $G_W = -0.7\, M_r$.

program (e.g., Mathsoft or Excel). By doing so the field of application for migration estimation will be expanded considerably and many of the present cases with extremely overestimated values will be now more realistic.

An important class of systems for which partition coefficients, $K_{P/L}$, are needed in migration modeling is polyolefins in contact with mixtures of ethanol/water. One can assume that practically all foods and pharmaceuticals in contact with nonpolar plastics, like polyolefins, can be simulated with ethanol/water mixtures (Figure 4.6).

It is possible to estimate the vapor pressure index G_L of ethanol/water mixtures with the following relation:

$$G_L = (G_E - G_W) \cdot \frac{c_E}{100} + G_W$$

where c_E is the concentration of ethanol in % (w/w). G_E is the index for pure ethanol and G_W is the index for water. The molar volume $\underline{V}_L$ of the ethanol/water mixture can be estimated with $\underline{V}_L \simeq 17 + 1.037^c$, where $c = c_E$.

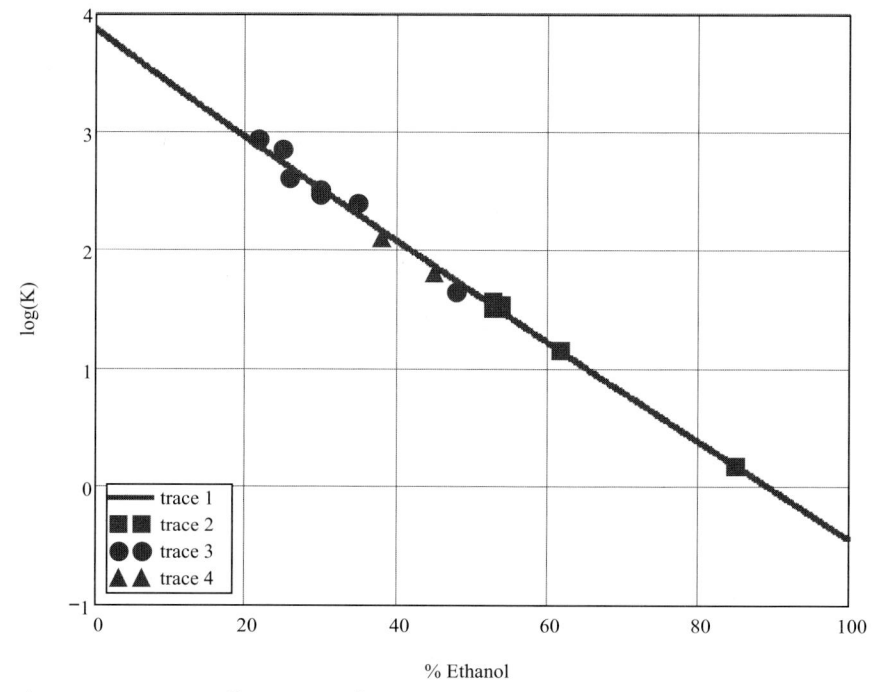

Figure 4.6 Partition coefficients $K_{P/L}$ of CHIMASORB 81 between PE and cheese, mayonnaise, butter, chocolate (trace 2), meet with fat content between 0 and 50% (trace 3), soft drinks (trace 4), and ethanol/water mixtures calculated with Eq. (4.58) (trace 1).

References

Abrams, D.S. and Prausnitz, J.M. (1975) Statistical thermodynamics of liquid mixtures: A new expression for the excess Gibbs energy of partly or completely miscible systems. *AICHE J.*, **21**(1), 116–128.

Athes, V., Peña, Y., Lillo, M., Bernard, C., Pdrez-Correa, R. and Souchon, I. (2004) Comparison of experimental methods for measuring infinite dilution volatilities of aroma compounds in water/ethanol mixtures. *J. Agric. Food Chem.*, **52**, 2021–2027.

Baner, A.L. (1992) Partition coefficients of aroma compounds between polyethylene and aqueous ethanol and their estimation using UNIFAC and GCFEOS Ph.D. Dissertation, Michigan State University, E. Lansing.

Baner, A.L. (2000) The estimation of partition coefficients, solubility coefficients, and permeability coefficients for organic molecules in polymers. Chapter 5 in *Food Packaging Testing Methods and Applications* (S.J. Risch ed.). ACS Symposium Series 753.

American Chemical Society, Washington, DC.

Baner, A.L. (1993) Unpublished data.

Baner, A.L. and Piringer, O. (1991) Prediction of solute partition coefficients between polyolefins and alcohols using the regular solution theory and group contribution methods. *Ind. Eng. Chem. Res.*, **30**, 1506–1515.

Barton, A.F.M. (1983) *CRC Handbook of Solubility Parameters and Other Cohesion Parameters*, CRC Press, Boca Raton, FL.

Bastos, J.C., Soares, M.E. and Medina, A.G. (1988) Infinite dilution activity coefficients predicted by UNIFAC group contribution. *Ind. Eng. Chem. Res.*, **27**, 1269–1277.

Becker, K., Koszinowski, J. and Piringer, O. (1983) Permeation von riech- und aromastoffen durch polyolefine (Permeation of flavor and aromas through polyolefins). *Deutsche Lebensmittel-Rundschau.*, **79** (8), 257–266.

Bogdanic, G. and Fredenslund, A. (1994) Revision of the group-contribution Flory equation of state for phase equilibria calculations in mixtures with polymers: 1. Prediction of vapor-liquid equilibria for polymer solutions. *Ind. Eng. Chem. Res.*, **33**, 1331–1340.

Bondi, A. (1968) *Physical Properties of Molecular Crystals, Liquids, and Glasses*, Wiley, New York.

Brandrup, J., Immergut, E.H. and Gruelke, E.A. (2003) *Polymer Handbook*, 4th edn., Wiley-Interscience, New York.

Canstantinou, L., Prickett, S.E. and Mavrovouniotis, M.L. (1993) Estimation of thermodynamic and physical properties of acyclic hydrocarbons using the ABC approach and conjugation operators. *Ind. Eng. Chem. Res.*, **32**, 1734–1746.

Chen, F., Fredenslund, A. and Rasmussen, P. (1990) Group-contribution Flory equation of state for vapor-liquid equilibria in mixtures with polymers. *Ind. Eng. Chem. Res.*, **29**, 875–882.

Denbigh, K. (1981) *The Principles of Chemical Equilibrium*, 4th edn., Cambridge University Press, Cambridge.

Doong, S.J. and Ho, W.S.W. (1991) Sorption of organic vapors in polyethylene. *Ind. Eng. Chem. Res.*, **30**, 1351–1361.

Elbro, H.S., Fredenslund, A. and Rasmussen, P. (1990) A new simple equation for the prediction of solvent activities in polymer solutions. *Macromolecules*, **23** (21), 4707.

Ettre, L.S., Welter, C. and Kolb, B. (1993) Determination of gas-liquid partition coefficients by automatic equilibrium headspace-gas chromatography utilizing phase ratio variation methodology. *Chromatographia*, **35**, 73–84.

Flory, P.J. (1970) Thermodynamics of polymer solutions. *Discuss. Faraday Soc.*, **48** (7), 7–29.

Franz, R. (1991) Fraunhofer-Institut für Lebensmitteltechnologie und Verpackung ILV Annual Report (now Fraunhofer IVV, Freising, Germany), Munich, 1991, p. 47.

Franz, R. (1990) Personnel communication. Fraunhofer-Institut IVV, Freising, Germany, 1990.

Fredenslund, A. (1989) UNIFAC and related group-contribution models for phase equilibria. *Fluid Phase Equilibria*, **52**, 135–150.

Fredenslund, A., Jones, R.L. and Prausnitz, J.M. (1975) Group-contribution estimation of activity coefficients in nonideal liquid mixtures. *AICHE J.*, **21** (6), 1086–1099.

Fukamachi, M., Matsui, T., Hwang, Y., Shimoda, M. and Osajima, Y. (1996) Sorption behavior of flavor compounds into packaging films from ethanol solution. *J. Agric. Food Chem.*, **44**, 2810–2813.

Goydan, R., Reid, R.C. and Tseng, H. (1989) Estimation of the solubilities of organic compounds in polymers by group-contribution methods. *Ind. Eng. Chem. Res.*, **28**, 445–454.

Guggenheim, E.A. (1952) *Mixtures*, Clarendon Press, Oxford.

Hansen, H.K., Rasmussen, P., Fredenslund, A., Schiller, M. and Gmehling, J. (1991) Vapor-liquid equilibria by UNIFAC group contribution: 5. Revision and extension. *Ind. Eng. Chem. Res.*, **30**, 2352–2355.

High, M.S. and Danner, R.P. (1990) Prediction of solvent activities in polymer solutions. *Fluid Phase Equilibria*, **55**, 1–15.

Holten-Anderson, J., Rasmussen, P., Fredenslund, A. (1987) Phase equilibria of polymer solutions by group contribution. 1. Vapor liquid equilibria. *Ind. Eng. Chem. Res.*, **26** (7), 1382–1390.

Iwai, Y., Ohzono, M. and Aria, Y. (1985) Gas chromatographic determination and correlation of weight-fraction Henry's constants for hydrocarbon gases and vapors in molten polymers. *Chem. Eng. Commun.*, **34**, 225–240.

Kolb, B., Welter, C. and Bichler, C. (1992) Determination of partition coefficients by automatic equilibrium headspace gas chromatography by vapor phase calibration. *Chromatographia*, **34**, 235–240.

Kontogeorgis, G.M., Fredenslund, A. and Tassios, D.P. (1993) Simple activity coefficient model for the prediction of solvent activities in polymer solutions. *Ind. Eng. Chem. Res.*, **32**, 362–372.

Koszinowski, J. (1986) Diffusion and solubility of n-alkanes in polyolefins. *J. Appl. Polym. Sci.*, **31**, 1805–1826.

Koszinowski, J. and Piringer, O.G. (1986) Influence of the packaging material and the nature of the packed goods on the loss of volatile organic compounds – permeation of aroma compounds. *Food Packaging and Preservation. Theory and Practice*, Elsevier, London, pp. 325–340.

Koszinowski, J. and Piringer, O. (1989) The influence of partition processes between packaging and foodstuffs or cosmetics on the quality of packed products. *Verpackungs-Rundschau: Technische-wissenschaftliche, Beilage*.**40** (5), 39–44.

Kovats, E. (1958) *Helv. Chim. Acta*, **41**, 1915.

Larsen, B.L., Rasmussen, P. and Fredenslund, A. (1987) A modified UNIFAC group-contribution of phase equilibria and heats of mixing. *Ind. Eng. Chem. Res.*, **26**, 2274–2286.

Mavrovouniotis, M.L. (1990) Estimation of properties from conjugate forms of molecular structures: the ABC approach. *Ind. Eng. Chem. Res.*, **29**, 1943–1953.

Oishi, T. and Prausnitz, J.M. (1978) Estimation of solvent activities in polymer solutions using a group-contribution method. *Ind. Eng. Chem. Process Res. Dev.*, **17** (3), 333–339.

Park, J.H. and Carr, P.W. (1987) Predictive ability of the MOSCED and UNIFAC activity coefficient estimation methods. *Anal. Chem.*, **59**, 2596–2602.

Piringer, O. (1993) Verpackungen für Lebensmittel: Eignung. *Wechselwirkungen, Sicherheit*, VCH, Weinheim, Germany, p. 258, Chapter 6.1.

Piringer, O., Jalobeanu, M. and Stanescu, U. (1976) *J. Chromatogr.*, **119**, 423.

Prausnitz, J.M., Lichtenthaler, R.N. and Gomes de Azevedo, E. (1986) *Molecular Thermodynamics of Fluid-Phase Equilibria*, Prentice-Hall, Englewood Cliffs, NJ. p. 193, Chapter 6.

Reid, R.C., Prausnitz, J.M. and Poling, B.E. (1987) *The Properties of Gases and Liquids*, McGraw-Hill, New York.

Richon, A.B. and Young, S.S. (2007) An introduction to QSAR methodology. http://www.netsci.org/science/compchem/feature19.html.

Sandler, S.I. (1989) *Chemical and Engineering Thermodynamics*, Wiley, New York.

Selassie, C.D. (2003) History of quantitative structure-activity relationships. *Burger's Medicinal Chemistry and Drug Discovery*, 6th edn., Vol. 1. Wiley, New York, p. 1–48. Chapter 1.

Tehrany, E.A., Fournier, F. and Desorbry, S. (2006) Simple method to calculate partition coefficient of migrant in food stimulant/polymer system. *J. Food Eng.*, **77** (1), 135–139.

Thomas, E.R. and Eckert, C.A. (1984) Prediction of limiting activity coefficients by a modified separation of cohesive energy density model and UNIFAC. *End. Eng. Chem. Process Des. Dev.*, **23**, 194–209.

Van Krevelen, D.W. (1990) *Properties of Polymers: Their Correlation With Chemical Structure; Their Numerical Estimation and Prediction from Additive Group Contributions*, 3rd edn., Elsevier, Amsterdam.

Weidlich, U. and Gmehling, J. (1987) A modified UNIFAC model: 1. Prediction of VLE, h^E, and γ^∞. *Ind. Eng. Chem. Res.*, **26**, 1372–1381.

5
Models for Diffusion in Polymers
Peter Mercea

The literature contains a very large amount of both experimental and theoretical information on the diffusion of small molecules, especially gases, in polymers. Since the pioneering works (Mitchell 1983; Graham 1866) on the diffusion of gases through rubber septa interest in diffusion phenomena in polymers have continuously increased and diversified. The considerable interest and the concentrated academic and industrial research efforts in the study of diffusion in polymers arise from the fact that important practical applications for these materials depend to a great extent on diffusion phenomena. In the last five decades, a series of classic books and reviews has been devoted to the presentation of the main topics, experimental results, theories, and applications for the diffusion of small molecules in polymers (Barrer 1951; Tuwiner 1962; Rogers 1965; Crank and Park 1968; Stannett et al. (1972, 1979); Hwang and Kammermeyer 1975; Vieth et al. 1976; Meares 1976; Mason and Lonsdale 1990; Frisch and Stern 1983; Rogers and Machin 1972; Vieth 1991; Koros 1990; Stern 1994; Aminabhavy et al. 1988; Paul and Yampol'skii 1994). An interested reader will most certainly find in one of these publications information, which apply to her/his special area of interest.

To produce today an exhaustive review about the field of mass transport (diffusion) in polymers would require an ample amount of work and space, which is beyond the aim of this book. Therefore, in this chapter, we will present concisely only some outlines of the models currently found in the literature for the interpretation of the diffusion of small molecules in polymers (Crank and Park 1968; Hwang and Kammermeyer 1975; Stannett et al. 1979; Frisch and Stern 1983; Vieth 1991; Paul and Yampol'skii 1994). An important aspect in this presentation will be to show to which extent the mathematical formalism developed in the framework of a diffusion model allows accurate predictions of the diffusion coefficients, D_p, of small molecules in a given polymer. Namely, the knowledge of D_p is essential for the theoretical modeling of substance migration from polymeric (packaging) materials into various contact media (foods, food simulants, and/or pharmaceuticals).

In many publications, the term "small molecules" denotes permanent or rare gases, hydrocarbons from C1 to C4, some hexafluorides and simple organic vapors.

However, from the point of view of the practical applications of polymers, especially in the packaging sector, the term "small molecules" includes also a wide range of organic substances, other than macromolecules, with a relative molecular weight, M_w, up to about 2500 g/mol (Koros 1990; Aminabhavy et al. 1988; Schlotter and Furlan 1992; Piringer 1993; Comyn 1985; Clough et al. 1985; Vergnaud 1991; Scott 1980; Hauschild and Spingler 1988; Stastna and DeKee 1995; Crompton 1979; Piringer and Baner 2000).

Many theoretical models have been proposed in the last six decades to describe the diffusion of small molecules in polymer matrices. According to the distance–time scale at which the basic physicochemical processes involved in these models have been set one could classify them in two main categories, namely, "microscopic" and "atomistic" diffusion models.

The "microscopic" models have been developed since the late 1930s, and therefore, can be coined as "classical," too. They attempted to demonstrate that experimental results, i.e., the dependence of diffusion coefficients on temperature, nature of the small molecule, and its concentration in the polymer, can be put into correspondence with some "microscopic" structural or energy parameters of the polymer matrix and/or of the diffusing species. Very often at the basis of these classical models phenomenological "heuristic" reasonings were made and used to develop eventually mathematical formulae capable to quantify a diffusion processes investigated experimentally. The "microscopic" parameters involved in the development of the classical diffusion models are most often average data on the structural geometry and dimensions of the polymer macromolecules and/or diffusant molecule, i.e., the length of the polymer chains and chain segment angles, estimated average spacing between the polymer chains, uniformly distributed average free volumes, diffusant collision diameters, etc. In fact, most of these models are not truly microscopic because the mathematical formulae developed in their framework rely on macroscopic properties of the polymer, such as thermal expansion, compressibility, viscosity, and/or density.

The following section is devoted to the presentation of some of these classical "microscopic" diffusion models. As already special emphasis is to discuss how the mathematical formulae of these models can correlate with experimental data and predict diffusion coefficients. This latter aspect is of great interest not only from a fundamental point of view but also in many practical fields where the possibility to predict a diffusion process might be a more economic alternative to its experimental investigation.

Since about two decades, with the advent of powerful computers and development of appropriate software, works are published in which diffusion of some small molecules in selected polymeric matrices is described by means of computer simulations. The theoretical models use in this respect can be coined as "atomistic" ones. In principle the development of such a "computational approach" starts from very elementary physical–chemical data – also called "first principles" – on the small molecule–polymer, SM–P, system. The dimensions of the atoms, the interatomic distances and molecular chain angles, the potential fields acting on the atoms and molecules, and other local parameters are used to generate a computer polymer structure, to insert the small molecules in its free volumes and then to simulate their motion in the polymer matrix. Determining the size and rate of these motions makes it possible to calculate the diffusion coefficient and characterize the diffusional mechanism.

Nowadays the "atomistic" simulation of diffusional processes in polymers is a domain of vivid research activity, which already produced remarkable results. There are expectations that "the computational approach" will become one day a practical and powerful tool to predict diffusion phenomena in complex SM–P systems. The second section of this chapter is devoted to a concise presentation of this approach.

In recent years nonparametric models (schemes) were developed for the estimation of D_p based on hierarchical classification of experimental results available from the literature (Vitrac et al. 2006). Such a model is, in fact, not describing either at a "microscopic" or at an "atomistic" level the diffusion process taking place in a SM–P system, but diffusion coefficients in a given polymer, known from literature, are classified by using "descriptors" which reflect certain individual properties of the small molecules. How such a hierarchical classification can then be used to estimate the D_p for another small molecule will be presented briefly in Section 5.3.

5.1
Diffusion in Polymers – The Classical Approach

Historically most of the "microscopic" diffusion models of the "classical approach" were formulated for amorphous polymer structures and are based on concepts derived from diffusion in simple liquids. An amorphous polymer can often be regarded with good approximation as a homogeneous and isotropic structure. In semicrystalline polymers the crystalline regions can be considered in a first approximation as impenetrable obstacles in the path of the diffusion process and sources of heterogeneous properties for the SM–P system. The effect of polymer crystallites on the mechanism of substance diffusion in a semicrystalline polymer has often been analyzed from the point of view of barrier property enhancement in polymer films (Cussler et al. 1988; Cussler 1990).

In approaching the problem of modeling diffusion in polymers, regardless if in a "classical" or "computational" manner, an important feature must be highlighted, namely that markedly different diffusion mechanisms operate at temperatures above and below the glass transition temperature, T_g, of the polymer. This is due, in principle, to the fact that polymers at temperatures $T > T_g$, so-called rubbery polymers, respond rapidly to changes in their physical condition. Therefore, the SM–P system adjusts immediately to a new equilibrium when a diffusing species is absorbed by and transported through the rubbery matrix of the polymer. The time taken to reach the new state of equilibrium is very much shorter than the characteristic time involved in the diffusion of the molecule through the matrix of the polymer (Crank and Park 1968; Stannett et al. 1979; Frisch and Stern 1983). In rubbery polymers the diffusion of small molecules is generally Fickian, i.e., it follows Fick's laws of diffusion (Crank and Park 1968), except special cases in which, for example, sorption equilibrium is not achieved at the polymer interfaces (Frisch 1962).

In contrast, for "glassy" polymers, i.e., when $T < T_g$, the motion of the polymer chains is not sufficiently rapid to completely homogenize the small molecules' environment. Therefore, below T_g the polymer is not in a true equilibrium within the time scale of a conventional diffusion or sorption experiment (Crank and Park 1968; Vieth et al. 1976; Stannett et al. 1979; Frisch and Stern 1983; Vieth 1991). Because of that in glassy polymers the diffusion of small molecules is a much more complex process, see Section 5.2

According to the physicochemical parameters and the type of mathematical formalism used in the development of "microscopic" diffusion models, one could classify them into: molecular, free volume, and respectively, hybrid models.

This classification should, in principle, be valid for both rubbery and glassy polymers. However, until now more detailed and true "microscopic" treatments have mainly been models for diffusion in rubbery polymers. An explanation for this may be the difficulty to develop consistent "microscopic" diffusion models for the much more complex diffusion process occurring in glassy polymers (Vieth et al. 1976; Frisch and Stern 1983; Saxena and Stern 1982; Paul and Koros 1976; Stern and Saxena 1980).

5.1.1
Diffusion in Rubbery Polymers

5.1.1.1 Molecular Models

Molecular models to describe diffusion in polymers began to be developed with the pioneering work published more than six decades ago by Barrer (1939, 1942). The attempt was to model the diffusional process by analyzing specific motions of diffusing molecules and the surrounding polymer chains relative to each other and taking into consideration the pertinent molecular forces acting between them. Earlier molecular models used statistical–mechanical concepts, which were greatly oversimplified and can only roughly be labeled as "microscopic." The mathematical formulae developed in their frameworks expressed most often only the dependence of the diffusion activation energy, E_d, on certain parameters of the penetrant–polymer system.

In the "activated zone" model given by Barrer 1942, it was assumed that diffusion of a small penetrant in a polymer matrix occurs when a total energy equal to or greater than a critical energy is absorbed in the region surrounding the penetrant molecule. The energy of activation for diffusion appears in any region of the polymer as a result of thermal fluctuations in it. This energy is shared by the penetrant molecule and adjacent polymer segments and is stored in a so-called zone of activation (Barrer (1939, 1942)). The diffusional jump occurs within the lifetime of the energy fluctuations with the additional requirement for cooperation and synchronization between segmental rotations and intermolecular vibrations within the zone of activation. In the framework of this model an expression for calculating the diffusion coefficient was developed too. It was used to correlate experimental results obtained for the diffusion of simple gases and organic vapors in elastomers (Barrer 1939). The use of this expression is not trivial because it requires among others the knowledge of a series of data on the viscosity of the polymer as well as "heuristic" assumptions on certain parameters included in the expression.

A somewhat closer phenomenological tie to the "microscopic" structure and motions in the penetrant–polymer system was proposed by Meares 1954. The main assumption made there was that diffusion of a small penetrant in the matrix of a polymer takes place by "jumps" between "holes" or "vacancies" created by the thermal fluctuation of the polymer chains. It was further assumed that the unit diffusion step for small penetrants in polymers is governed by the energy required to separate the polymer chains surrounding a molecule of penetrant and to give a space of sufficient cross section for the molecule to pass to another "hole." Thus, above T_g the activation process of diffusion is breaking off van der Waals bonds between polymer chains and creating voids of "cylindrical" form. For very small penetrant molecules, i.e., He, Ne, or N_2, complete separation of the segments beyond the limit of van der Waals interactions would not be required but for larger penetrants this will be the case. Applying the theory of absolute reaction rates (Glasstone et al. 1941) to diffusion, an expression for calculating the diffusion coefficients was derived by Meares 1954 too. The diffusion coefficients are assumed to increase with the square of the length, λ^*, of the assumed "cylindrical" void and depend on the temperature, T too. For simple gases diffusing through rubbery polymers the model produced reasonable correlations of the experimental results measured at different temperatures, T (Meares (1954,1993) and allowed thus the estimation of λ^* in the investigated polymers. However, the model was not used/tested later to see if it held for the diffusion of more complex organic vapors in rubbery polymers.

A further attempt to correlate the, E_d, with "microscopic" structural features of the polymer matrix was proposed by Brandt 1959. This classical diffusion model is constructed on the assumption that, in order to create a passageway for a penetrant molecule, the energy of activation is used in part to bend the molecular chain segments which surround the penetrant – intramolecular energy E_b – and in part to overcome the attractive forces between the segments – intermolecular energy E_i. In principle, in this diffusion model a semipredictive calculation of D_p was developed but the procedure relies on the knowledge of a series of additional parameters, such as internal pressure in the polymer, average geometric parameters of the polymer chains which must be evaluated from structural, density, specific volume data on the polymer, and other parameters. The discrepancies between theory and experiment were ascribed to limitations resulting from the oversimplified mechanism of diffusion assumed in the development of the model. Moreover, the validity of the model was tested only for the diffusion of very simple penetrants (permanent gases) in rubbery polyethylene.

Nevertheless, taking into account the simplicity of the motions assumed to be responsible for the diffusional jump and the crude method of estimating some of the structural and energy parameters involved, it might be reasonable to consider that such estimates will at best give information on the order of magnitude of the calculated diffusion coefficients.

The molecular model concept presented by Brandt 1959 was further refined and developed by DiBenedetto (1963a, 1963b), which are often cited in the literature. In this statistical molecular model it is assumed that the whole structure of the polymer matrix consists of a random distribution of identical *n*-center segments in cylindrical cells. The bundle of parallel chains creates a cylindrically symmetric force field, which influences the motions of the central chain. During the course of normal thermal vibrations and rotations of the polymer chains, the unit cell expands and contracts. The motion of the segments can be coordinated to produce a cylindrical void adjacent to a molecule absorbed into the polymer, which is then able to "diffuse" into this void. The oscillatory movement of the polymer segments is several orders of magnitude slower than the translational rate of the penetrant molecule. Therefore, the diffusional process consists of a "normal" state in which four parallel chains are separated only by the mean intermolecular distance. In the "activated" state, the polymer chains are separated and the volume of the unit cell consists of the average free volume plus a cylindrical void large enough to accommodate a penetrant molecule of diameter d, Figure 5.1.

By analyzing the interaction between a chain center and the centers from adjacent chains, a relation for the potential interaction energy in both the normal and activated state was obtained (DiBenedetto 1963a). This formula was eventually used to develop a relationship for the calculation of E_d. To use this relation knowledge of parameters determined from thermodynamic and molecular data on the penetrant–olymer system is required. Unfortunately, the model presented (DiBenedetto 1963a) has

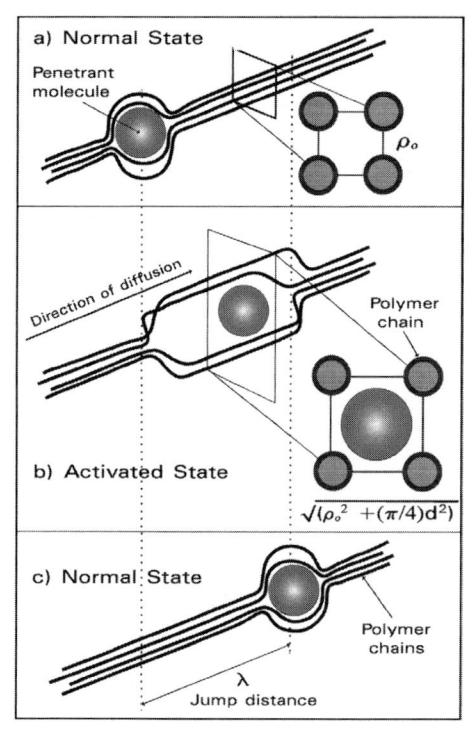

Figure 5.1 The activation process of diffusion (DiBenedetto 1963a).

not been developed by its author to address the problem of diffusion coefficient calculation.

A process or manufacturing engineering concerned with the type of problems highlighted in this book might be interested not only in a general presentation of the one or the other of the diffusion models but also to what extent such a model can be used to calculate D_p for a small penetrant in a given polymer. Inspecting from this point of view the "microscopic" models presented so far one finds out that their mathematical formulae for D_p require one or more adjustable parameters which can be determined only by fitting theoretical curves to experimental results. Once these parameters are determined the expressions can be used to interpolate and/or extrapolate the experimental data in a certain range. When using such expression to extrapolate experimental data one should take into account at least two aspects: (i) not to perform calculations far beyond the range investigated in the experiments and (ii) to make sure that at the level where the calculations are performed the penetrant–polymer system did not change its properties, for example, because of a phase transition or swelling.

One can state that a diffusion model, which allows only such calculations and extrapolations of diffusion coefficients exhibits only "correlative" and "semi-predictive" capabilities. All "classical" diffusion models presented in this section fall into this category.

5.1.1.2 The Molecular Model of Pace and Datyner

One of the most popular and detailed statistical molecular models for the diffusion of simple penetrants in amorphous rubbery polymers was proposed by Pace and Datyner (1979a, 1979b). This model is based on features taken from some of the models presented above.

To construct the model, it has been assumed that the amorphous polymer regions possess an approximately paracrystalline order with chain bundles locally parallel. A penetrant molecule may diffuse through the matrix of the polymer by two modes of motion, Figure 5.2.

First, similar to the model described by DiBenedetto (1963a, 1963b), the penetrant molecule is allowed to move along the axis of a "tube" formed by adjacent parallel chains. This is called "longitudinal movement" and is regarded in a first approximation as requiring no activation energy. Second, as in the model given by Brandt 1959, the molecule may move perpendicular to this axis – "transverse movement" – when two adjacent polymer chains separate sufficiently to permit its passage. The longitudinal movement occurs much more rapidly than the transverse. It was estimated by Pace and Datyner 1979a that the longitudinal motion occurs at least three orders of magnitude faster than the macroscopically determined diffusion rates. Hence, the penetrant molecule will move backward and forward in the "tube" formed by the parallel chains, Figure 5.2, many times before it may progress further by moving, through a transverse jump, to an adjoining "tube." The two motions occur in series and the observed E_d is required to separate the polymer chains, which then allows a transverse motion. Therefore, this model ascribes to the transverse motions the role of rate determining step in diffusion of the penetrant through polymer matrix. In this

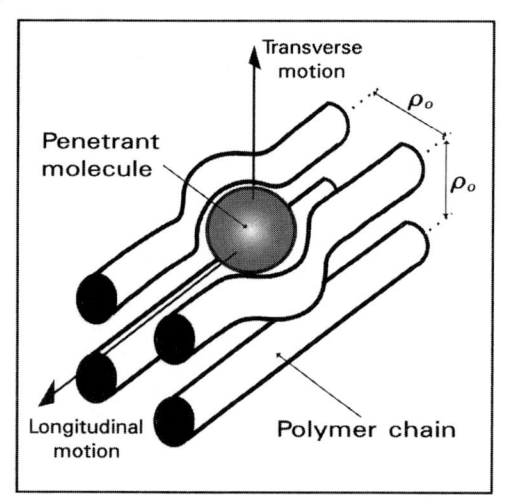

Figure 5.2 Proposed polymer microstructure with locally parallel chains and possible motions of spherical penetrant.

phenomenological picture, the average diffusional jump length is then given by the length of the "tube" while the jump frequency is monitored by the frequency of chain separations that allows penetrant passage between the two adjacent "tubes." The E_d is evaluated from the energy needed for a symmetric separation of two polymer chains to allow a penetrant of diameter d to pass through. For "nonspherical" penetrant molecules the model uses, instead of d, an effective diameter d_{eff} defined by the relation $d_{eff} = d^2/d_l$, where d_l is the longest molecular dimension determined from models of the molecule. The model allows a reasonable correlation of the experimental E_d with diameters of the diffusant species (Pace and Datyner (1979b, 1979c). The main advantage of this "classical" molecular model is that E_d can be estimated without using any adjustable parameters derived from correlation with experimental diffusion data. However, the equation for E_d contains a series of material, structural, and thermodynamic parameters of the host polymer as well as the dimension and shape of the penetrant molecules (Pace and Datyner (1979a, 1979b). The determination of these data, as far as they have not already been tabulated in some publication, depends on the availability of certain experimental data and is not always a simple task.

As for the problem of a formula for D_p this molecular model has to resort to the theory of stochastic processes (Chandrasekar 1943). In a homogeneous medium in which a penetrant may jump with equal probability in all the directions D is given by

$$D = \frac{1}{6} \lambda^2 \nu \tag{5.1}$$

where λ^2 is the mean-square "jump" distance and ν the average jump frequency of the diffusing molecule. This frequency can be equated with that of the chain openings that permit passage of the penetrant molecule and was calculated in the framework of the model by using statistical mechanical relations for the probability density of energy distribution in multicomponent systems (Wheeler 1938). Eventually, the relationship derived for D was (Pace and Datyner 1979a)

$$D = \left(\frac{\lambda}{\zeta}\right)^2 \left(\frac{\varepsilon^*}{\rho^*}\right)^{5/4} \left(\frac{\sqrt{\beta}}{m^*}\right)^{1/2} \frac{\delta^l}{\partial \Delta E/\partial d} \exp\left(-\frac{E_d}{RT}\right) \qquad (5.2)$$

where $\delta^l = d + \rho^* - \langle \rho \rangle$ and β is a chain-bending modulus. Within the limits of the simplifying assumptions made by Pace and Datyner 1979a the numerical factor Θ was found to be equal with 9.0×10^{-4}. Let us assume that for a given penetrant–polymer system ζ, ε^*, m^* and ρ^* are known or can be determined from appropriate experimental data reported in the literature. Now assuming that E_d has been estimated theoretically with the model by substituting the above parameters into Eq. (5.2) one could, in principle, calculate a diffusion coefficient D. The problem is that both Eqs. (5.1) and (5.2) contain the mean-square "jump" distance, λ, which is generally not known or very difficult to determine experimentally. Thus, in order to calculate a D a "shrewd guess," based on similitude with other SM–P polymer, or a special experiment (for example, using a positron annihilation techniques (Yampol'skii et al. 1994), is needed to obtain λ. In conclusion the mathematical formalism developed in this model for D is not truly predictive. This means that for a given polymer–penetrant system D cannot be estimated using only readily available "first principles," i.e., thermodynamic, molecular, and structural data on the penetrant–polymer system.

5.1.1.3 Free-Volume Models

The "free-volume" models are by far the most widely used theories for interpreting and predicting the diffusion of small penetrants in polymers (Crank and Park 1968; Stannett et al. 1979; Frisch and Stern 1983; Fujita 1961; Guo et al. 1992; Vrentas and Duda 1986). The beauty of these models is that they describe the very complex process of diffusion in polymers in a way that is based on realistic concepts, relying on parameters that have some relevant physical significance. The basic assumption of the free-volume models is that the mobility of both polymer segments and penetrant molecules is primarily determined by the available free volume in the penetrant–polymer system. The free volume of the polymer is regarded as an "empty" volume between the chains of the polymer, while the penetrant free volume is the volume not occupied between the molecules of the penetrant. It is assumed that the diffusion of a small molecule in the polymer depends on two factors. First, the molecule should obtain sufficient energy to overcome attractive forces and second, it should be surrounded by free volume.

Most free-volume models for diffusion in polymers follow the phenomenological basis set by Cohen and Turnbull 1959 where the self-diffusion of an ideal liquid of hard spheres ("molecules"), which behave according to the van der Waals physical model (Hirschfelder et al. 1954), has been analyzed. These molecules are confined – for most of the time – in a "cage" formed by their immediate neighbors. A local fluctuation in density may open a "hole" within a cage, large enough to permit a considerable displacement of the sphere contained by it. This displacement gives rise to diffusion only if another sphere jumps into the "hole" before the first sphere returns to its initial position. Diffusion occurs not as a result of an activation process in the ordinary sense but rather as a result of the redistribution of the free volume within the liquid of hard spheres.

The functional relationships derived in the model of hard-spheres have been reinterpreted over course of the time, leading to a series of more sophisticated free-volume diffusion models. Some of these models are presented briefly below. One of the first attempts to correlate experimental diffusion data with free volume has been made by Meares 1958a. The experiments showed that above the glass transition temperature of the polymer, $T > T_g$, the diffusion is Fickian and that measured average diffusion coefficient steeply increases with the concentration, c_s, of penetrant in the polymer. This finding was quantified with an empirical relation proposed earlier by Kokes et al. 1952 for the intrinsic diffusion coefficient D^+:

$$D^+ = D_{c \to 0} \exp(w_{c_s}) \tag{5.3}$$

where w is a parameter. Equation (5.3) was then refined by deriving, in terms of a theory for polymer segmental mobility (Bueche 1953), a relationship between D^+ and its concentration, expressed as solvent volume fraction, v_s. To calculate D^+ with this formula, some thermodynamic and free-volume parameters for the penetrant–polymer system must be calculated from data given in the literature and two adjustable parameters must be determined by fitting the theoretical curves to experimental diffusion data (Meares 1958a). With these data the formula for D^+ showed an excellent fit over the concentration range which covered a 1000-fold increase of D^+ (Guo et al. 1992; Meares 1958a) thus, awarding in principle the model correlative and semipredictive capabilities for the estimation of the diffusion coefficient of a solvent in a polymer above T_g. Though the basic assumptions of this "free-volume" model cannot describe the much more complex diffusion processes occurring in migration of complex organic molecules from polymeric packaging materials.

One of the simplest early free-volume diffusion models was formulated by Fujita 1961 and Fujita et al. 1960. The concept of this model was considered as an advance, because some of the parameters required to describe the concentration-dependence of the diffusion coefficient could be obtained from the physical–chemical properties of the polymer and penetrant. The relation proposed for the calculation of the thermodynamic diffusion coefficient, D_T, was (Fujita 1961; Fujita et al. 1960)

$$D_T = A_d\, RT \exp(-B_d/V_f) \tag{5.4}$$

where V_f is the average fractional free volume. The proportionality coefficient A_d is considered to be dependent primarily upon the size and shape of the penetrant, while B_d is a parameter, which is independent of temperature and penetrant concentration. To work effectively with Eq. (5.4) the magnitude of its parameters must be determined. For this the free volume of the penetrant–polymer system must be evaluated from viscosity data. Eventually, the two adjustable parameters A_d and B_d must be calculated by fitting appropriate experimental diffusion data. For the diffusion of organic vapors in rubbery polymers, the correlation between theoretical curves and experimental data is often acceptable. In such cases the model can be used in a semipredictive manner in order to estimate diffusion coefficients beyond the penetrant concentration and/or temperature range where experimental results were collected. As already mentioned, the model includes in its formulae the adjustable coefficients A_d and B_d which cannot be determined from "first principles." Hence, one cannot ascertain true predictive capabilities from the model and thus, it is of little effective help for the type of practical diffusion coefficient estimations which are needed for the estimation of migration from packaging polymers.

5.1.1.4 The Free-Volume Model of Vrentas and Duda

In the last three decades, Vrentas, Duda, and their coworkers have published a substantial number of papers on the free-volume model of diffusion in polymer–solvent systems they developed in the late 1970s (Vrentas et al. 1988a, 1985, 1987, 1991; Vrentas and Vrentas 1994a, 1994b; Ganesh et al. 1992; Vrentas and Duda 1977a, 1977b, 1977c, 1977d, 1978; Duda et al. 1982). This model, which is often cited and used in the literature, underwent a number of modifications over the years and appears to apply well to the diffusion of organic solvents in rubbery and glassy polymers.

In order to develop a consistent free-volume diffusion model, there are some issues which must be addressed, namely: (i) how the currently available free volume for the diffusion process is defined, (ii) how this free volume is distributed among the polymer segments and the penetrant molecules, and (iii) how much energy is required for the redistribution of the free volume. Any valid free-volume diffusion model addresses these issues both from the phenomenological and quantitative points of view such that the diffusion process is described adequately down to the "microscopic" level. Vrentas and Duda stated that their free-volume model addresses these three issues in a more detailed form than previous diffusion models of the same type. Moreover, it was stated that the model allows the calculation of the absolute value of the diffusion coefficient and the activation energy of diffusion mainly from parameters, which have physical significance, i.e., so-called first principles. In the framework of this model the derivation, of the relation for the calculation of the self-diffusion coefficient of the solvent D_{1s} is not a trivial task. A relation can obtained which gives the dependence of D_{1s} on the nature of the penetrant and its concentration in the polymer–solvent system, the temperature and on the molecular weight of the polymer. For a rubbery polymer a condensed form of this relation, valid also for low-penetrant concentration levels, can be cited from Vrentas and Vrentas 1994a:

$$D_{1s} = \bar{D}_o \exp\left(-\frac{E^+}{RT}\right) \exp\left\{-\gamma\left(\frac{\bar{V}_s^* \chi_s + \bar{V}_p^* \chi_p \xi}{\bar{V}_{FH}}\right)\right\} \tag{5.5}$$

For the definition of all parameters involved in the above relation see (Vrentas et al. 1985; Vrentas and Vrentas 1994a). The explicit form of Eq. (5.5) contains 15 parameters of which 13 can be determined from thermodynamic and molecular data of the penetrant and polymer. Parameters include two specific hole-free-volumes for the components, free-volume parameters for the penetrant and polymer, the thermal expansion coefficient of the polymer, free-volume overlap factors, glass transition temperatures, the fractional composition of the system, etc. For a noninitiated reader, the procedures followed to determine these 13 parameters are not quite simple, although the authors of the model state that the data needed for this purpose are generally available in the literature. In the scheme for the estimation of these parameters presented by Vrentas and Vrentas 1994a one can see that in order to perform calculations with the model, two parameters must be calculated by fitting the theoretical curves to experimental results obtained in the so-called zero-penetrant concentration limit. Thus, it is stated that using a nonlinear regression analysis "...all of the parameters of the theory can be determined in general with as few as two diffusivity data points" (Vrentas and Vrentas 1994a). The results obtained with this complex, but straightforward procedure, have shown that the model provides excellent correlations for diffusivity data in several polymer–solvent systems (see Figure 5.3).

Having mentioned the correlative capabilities of this model, one can consider its semipredictive abilities. It was mentioned that a number of diffusion data taken from a limited range of penetrant concentration are required to calculate two of the parameters of the model. Once these parameters have been determined, one can make theoretical predictions for diffusion coefficients over a wider range of penetrant concentration or temperature variation. This is a critical test for any theoretical model, since a useful model should have at least an established semipredictive capability. These results are encouraging evidence that the proposed model is a suitable tool for a more accurate description of the diffusion process in rubbery polymers. The model was most often tested by its authors for polymer–solvents systems like: polystyrene (PS), poly(methylacrylate) (PMA), poly(ethylmethacrylate) (PEM), and poly(vinylacetate) (PVAc); and for toluene, benzene, ethylbenzene as solvents. The experimental test conditions reported by Vrentas et al. 1988a 1985, 1987, 1991, Vrentas and Vrentas

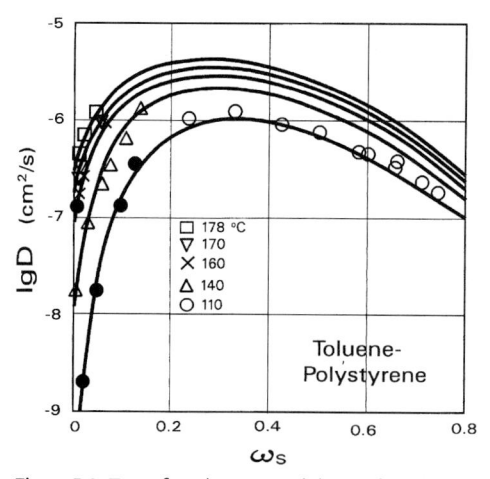

Figure 5.3 Test of predictive capabilities of proposed free-volume model using data for the toluene polystyrene system. Only data points represented by solid symbols were used to obtain free-volume parameters.

1994a, 1994b Ganesh et al. 1992, and Vrentas and Duda 1977a 1997b, 1977c, especially, for high concentration of solvent in the polymers, often differ considerably from what is generally of interest when these polymers are used in the packaging sector. Therefore, to assess the potential use of this free-volume diffusion model in the field of small substance migration in polymeric food packaging, the model must be tested for penetrant–polymer systems, which are specific for this field. Moreover, it is to mention that, because the model contains two parameters that cannot be determined from "first principles" but only by fitting a limited amount of experimental data, one cannot ascribe true predictive capabilities of the model.

To conclude this section, it may be interesting to mention what was said recently by Stern 1994 on the future of the free-volume diffusion models: "... However, phenomenological transport models based on free-volume concepts are likely to become obsolete during the coming decade, due to the development of computational techniques of simulating polymer microstructures..." The development of such techniques and their results are discussed in Section 5.2.

5.1.2
Diffusion in Glassy Polymers

The process of diffusion of small molecules in glassy polymers is a much more complex than that in rubbery ones. In principle, one can classify the experimental diffusion results obtained with glassy polymers in three categories:

- Case I (or Fickian) diffusion, when the rate of diffusion is much less than that of the relaxation of the SM–P system,
- Case II diffusion, Super Case II diffusion when the diffusion process is very rapid in comparison with the relaxation processes of the SM–P system, and
- Anomalous diffusion, when the diffusion and relaxation rates are comparable. In this case, it is assumed that the diffusion process is affected by the presence of pre-existing "holes" or "microcavities" in the structure of the polymer matrix.

The complexity of diffusion below T_g results, at least in part, because the free rotation of the polymer chains is restricted below T_g. Thus, it was assumed that fixed microcavities or "holes" of various sizes result throughout the matrix of the polymer below T_g. These "holes" are "frozen" into the polymer as it is quenched from the rubbery state (Haward 1973). The concept that two mechanisms of sorption may be implicated in the diffusion and behavior of small molecules in amorphous glassy polymers was first suggested by Meares 1957. Here and later by Meares 1958b, it was speculated that below T_g the "holes" may act to immobilize a portion of the molecules by binding them at high-energy sites at the periphery of the "holes" or by entrapment in the "holes." Based on this concept it has been suggested (Barrer et al. 1958) that the sorption of organic vapors in a glassy polymer is due to two concurrent mechanisms: (i) ordinary dissolution in the matrix of the polymer (so-called Henry's law sorption) and (ii) a "hole"-filling process obeying Langmuir's law. This phenomenological model was accompanied, for the sorption of simple gases and organic vapors, by the equation (Barrer et al. 1958)

$$C = k_D p + \frac{a_1 p}{1+bp} \tag{5.6}$$

where a_1, b, and k_D are adjustable coefficients and p is the pressure of the gaseous penetrant. It has been reasoned that a_1 and b are given approximately by statistical thermodynamic treatment of Langmuir's isotherm (Fowler and Guggenheim 1949) and k_D by the lattice theory of penetrant polymer solutions (Barrer 1947a). Later it was postulated that in Eq. (5.6) one may equate $a_1 = c'_H{}^{2/3}$ and designate b and c'_H as "hole affinity" and "hole saturation" constants, respectively (Michaels et al. 1963). This quantitative description of the solution of a simple penetrant in a glassy polymer is known today as the dual sorption theory (with total immobilization) (DST). The problem is that the basic assumptions of DST cannot be justified *a priori* (Vieth et al. 1976). The possibility that penetrant molecules adsorbed in "holes" may not be completely immobilized is one of these problems and has been addressed (Paul 1969; Petropoulos 1970). If that is the case, both the normally dissolved penetrant molecules (according to Henry's law) and the partially immobilized ones could diffuse through the matrix of the polymer and contribute to the diffusional flux. Moreover, in order to better describe real systems, another key postulate from the initial DST should be relaxed, namely, that the normally dissolved species and those adsorbed into the "holes" are always in local equilibrium (Petropoulos 1970). That means the diffusion model should incorporate some kinetics for the immobilization process. There will be cases where the diffusion and immobilization proceed at comparable rates; and limiting cases where one of the two processes predominates. The phenomenological sorption theory which resulted from taking into account of these assumptions is known as dual sorption (with partial immobilization) theory.

Because of the assumed dual sorption mechanism present in glassy polymers, the explicit form of the time-dependent diffusion equation in these polymers is much more complex than that for rubbery polymers (Petropoulos 1970; Tshudy and von Frankenberg 1973; Crank 1975; Wang et al. 1969; Fredrickson and Helfand 1985). As a result exact analytical solutions for this equation can be found only in limiting cases (Crank 1975; Wang et al. 1969; Vieth and Sladek 1965). In all other cases, numerical methods must be used to correlate the experimental results with theoretical estimates. Often the numerical procedures require a set of starting values for the parameters of the model. Usually, these values are "shroud guessed" in a range where they are expected to lie for the particular penetrant–polymer system. Starting from this set of arbitrary parameters, the numerical procedure adjusts the values until the best fit with the experimental data is obtained. The problem, which may arise in such a procedure (Toi et al. 1983), is that the numerical procedures may lead to excellent fits with the experimental data for quite different starting sets of parameters. Of course the physical interpretation of such a result is difficult.

However, the mathematical formulae of DSTs satisfactorily present the dependence of the solubility and diffusion coefficients for simple gases and organic vapors on the concentration of the penetrant in the glassy polymer (Vieth et al. 1976; Stannett et al. 1979; Frisch and Stern 1983; Vieth 1991; Stern 1994; Frisch 1962; Saxena and Stern 1982; Chern et al. 1983).

From the point of view of earlier discussions, namely, the true prediction of diffusion coefficients for volatile and nonvolatile organic penetrants in glassy polymers, the diffusion equations derived in the framework of DSTs have only a limited usefulness. This means that, because the parameters of the DST models are not directly related to "first principles," the equations can be used with success to correlate experimental results, but not to predict diffusion coefficients.

One possible solution to this problem is to develop "microscopic" diffusion models for glassy polymers, similar to those already presented for rubbery polymers. In Pace and Datyner 1980 some of the results obtained with the statistical model of penetrant diffusion in rubbery polymers, Section 5.1.1 are combined with simple statistical mechanical arguments to devise a model for sorption of simple penetrants into glassy polymers. This new statistical model is claimed to be applicable at temperatures both above and below T_g. The model encompasses dual sorption modes for the glassy polymer and it has been assumed that "hole"-filling is an important sorption mode above as well as below T_g. The sites of the "holes" are assumed to be fixed within the matrix of the polymer. Starting from these assumptions and using elementary statistical mechanical arguments, the authors of the model estimated the values of parameters approximately, which were then included in relation to the solubility coefficient (Pace and Datyner 1980). For a series of simple gases diffusing in some glassy polymers, solubility data calculated with the model were compared with the experimental sorption data. Semiquantitative to qualitative agreements between theory and experiment were found. Unfortunately, for the scope of the present book, the model was not developed for estimating of diffusion coefficients of small organic molecules in glassy polymers.

Local density fluctuations occur in penetrant–polymer systems both above and below T_g. It is then reasonable to expect that a free-volume diffusion model should also provide an adequate description of the diffusion of small penetrants in glassy polymers. To reach this goal the free-volume model for diffusion of small penetrants in rubbery polymers, Section 5.1.1 was modified to include transport below T_g (Vrentas et al. 1980, 1987, 1988b; Vrentas and Vrentas 1992, 1992b; Vrentas and Duda 1978).

In principle, the diffusion process in a penetrant–polymer system can be characterized by determining the mutual diffusion coefficient and its dependence on temperature, penetrant concentration, pressure, and polymer molecular weight. When molecular relaxation in the polymer–solvent system is much faster than the diffusive transport, the conformational changes in the polymer structures appear to take place instantaneously. The diffusional transport is comparable in such cases to the transport observed in simple liquids. This type of transport mechanism is considered to characterize quite well polymer solvent systems for $T > T_g$. As the temperature decreases toward T_g the probability that a local fluctuation in density will produce a "hole" of sufficient size so that a polymer jumping unit or a penetrant molecule can move in decreases. When $T < T_g$ the "hole"-free-volume which can be redistributed with no energy change in the penetrant–polymer system becomes very small. Below T_g the motions of the polymer are so hindered that, for a given penetrant concentration, significant movements do not occur at the time scale of the diffusion experiment. Moreover, at a very low-penetrant mass fraction, the structure of the glassy polymer is essentially unaffected by the presence of the penetrant and the diffusion process is Fickian (Vrentas et al. 1988a; Vrentas and Duda 1978; Vrentas and Vrentas 1992). The diffusion process under such conditions has been denoted as an elastic diffusion process (Vrentas et al. 1988a; Vrentas and Duda 1977d), which can be analyzed using the classical theory of diffusion. In the limit of zero penetrant mass fraction, these phenomenological assumptions were included into the relations of a mathematical formalism which led eventually to an expression for the dependence of the mutual diffusivity on temperature (Vrentas and Duda 1978):

$$\ln \frac{D(T)}{D(T_{g2})} = \frac{\gamma \xi}{K_{12}} \overline{V_p} * \frac{T - T_{g2}}{K_{22}\left[\frac{K_{22}}{9^+} + T - T_{g2}\right]} \tag{5.7}$$

where the parameter 9^+ describes the character of the change of the volume contraction which can be attributed to the glass transition. For glassy polymers, $T < T_{g2}$, the temperature dependence of D at zero penetrant concentration can be described by an E_d (Vrentas and Duda 1978; Vrentas et al. 1980):

$$E_d = \frac{RT^2(\gamma * \overline{V_p} * \xi / K_{12}^+)}{K_{22}\left[\frac{K_{22}}{9^+} + T - T_{g2}\right]} \tag{5.8}$$

The temperature dependence of D for the n-pentane–polystyrene system both above and below T_{g2} has been calculated using the formulae of this free-volume model (Vrentas et al. 1987). The results obtained are shown in Fig. 5.4 along with a few experimental data (Holley and Hopfenberg 1970) for the same system at three temperatures below T_{g2}.

Similarly to Fig. 5.4 for other glassy polymer–solvent systems also the predictions of this free-volume theory are in general agreement with experimental data on the

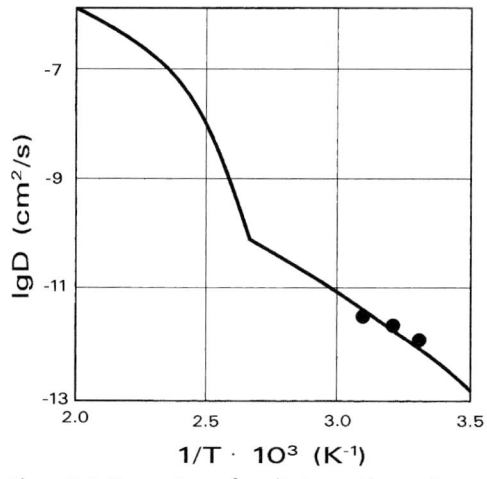

Figure 5.4 Comparison of predictions with experiment for the n-pentane–polystyrene system (Vrentas et al. 1987; Holley and Hopfenberg 1970).

temperature dependence of D in the vicinity of T_{g2}. In particular, the theory predicts a step change in E_d at T_{g2}, and this is consistent with most experimental investigations of polymer–solvent diffusion at temperatures just above and below the glass transition temperature (Crank and Park 1968; Stannett et al. 1979; Vieth 1991). Vrentas, Duda, and their coworkers refined in recent years their free-volume model for diffusion in glassy polymers to address also the problem of Fickian diffusion at finite solvent concentrations (Vrentas et al. 1987; Vrentas and Vrentas 1992, 1994b). For this the free volume and thermodynamical parameters involved in Eq. (5.7), which give the solvent self-diffusion coefficient D_{1s} in a rubbery polymer, were adapted to describe adequately the phenomenology of diffusion below the glass transition temperature, T_{gm}, of the polymer–solvent mixture at a particular solvent mass fraction. A series of assumptions on the structure, properties and sample history, and the introduction of an additional expansion coefficient were necessary (Vrentas and Vrentas 1994b) to express the behavior of the free-volume parameters below T_{gm}. Eventually, a set of equations was obtained and it was stated that using them calculation of D_{1s} for glassy polymers is no more difficult than computing D_{1s} for rubbery polymer–solvent systems (Vrentas and Vrentas 1992, 1994b). However, it was emphasized that the predictions of the model are sensitive to the sample preparation history, which means reasonably good agreement between theory and experiment will be obtained only for sample preparation histories that are similar to the one used in the model. Anyway one can see that up to 19 parameters are needed to express, with this free-volume model, the concentration and temperature dependence of D_{1s} in a glassy polymer (Vrentas and Vrentas 1992, 1994b). It is stated in these publications that all these parameters except two can be estimated from physicochemical data generally available in the literature. To determine the remaining parameters a small, amount of experimental diffusion data is needed.

The reasonably good agreement between theory and experiment shown by this free-volume model (Vrentas and Vrentas 1994b) recommends it as an interesting tool to model the diffusion in glassy polymers used in the packaging sector. However, the problem is that the correlative, semipredictive and predictive capabilities of this model do not address exactly the type of diffusion coefficient prediction, which is of interest for the estimation of many migration processes in polymeric packaging. When we state this we are thinking not only on how difficult it would be to specify all the parameters of the model for a complex penetrant like an antioxidant or stabilizer but even more if the model is still valid for this type of polymer–penetrant systems.

The above sections have presented models that link the process of diffusion of small penetrants in polymers to "microscopic" features of the penetrant–polymer system. Strictly speaking the type of diffusion models presented above are not truly "microscopic" because they actually describe average and not truly local "microscopic" properties of the penetrant–polymer system. Sometimes even excellent correlations of experimental data offered by these models are due to the fact that the experimental methods used to determine the diffusion coefficients are in turn probing the penetrant–polymer system over "nonmicroscopic" distances and comparatively long times.

Somewhat closer to the designation of a "microscopic" model are those diffusion theories, which model the transport processes by stochastic rate equations. In the most simple of these models a unique transition rate of penetrant molecules between smaller "cells" of the same energy is determined as a function of gross thermodynamic properties and molecular structure characteristics of the penetrant–polymer system. Unfortunately, until now the diffusion models developed on this basis also require a number of adjustable parameters without precise physical meaning. Moreover, the problem of these later models is that in order to predict the absolute value of the diffusion coefficient at least a most probable "average length" of the elementary diffusion jump must be known. In the framework of this type of "microscopic" model, it is not possible to determine this parameter from "first principles."

To conclude one can state that in the framework of the "classical" "microscopic" diffusion models more or less complex mathematical formulae have been developed with the aim of interpreting experimental data and even offering an insight on the mechanics of diffusion. The mathematical relations for the diffusion coefficient rely on parameters that must be determined from given physicochemical and structural data about the penetrant–polymer system. But, almost without exception, these models also include a number of adjustable parameters, which can be determined only by fitting experimental data to theoretical curves. In some models, the physical meaning of these adjustable parameters is quite unsubstantiated. Moreover, among the earlier "classical" diffusion models some "shrewd guessing" of some model parameters is needed. Therefore, one can state that the main limitation of all these phenomenological models is that they cannot truly predict diffusion coefficients only from "first principles."

5.2
Diffusion in Polymers – The Computational Approach

It was shown in the above section that as a rule, at the base of the "classical" or "microscopic" diffusion models, there are ad hoc (heuristic) assumptions on a certain molecular behavior of the polymer–penetrant system. The fact that the mathematical formulae developed on such basis often lead to excellent correlations and even semipredictions of diffusion coefficients must be acknowledged. It is true that the "classical" models are not capable to predict diffusion coefficients only from "first principles" but this is often not an obstacle to hinder their use in certain types of investigations. Therefore, we are quiet sure that this type of diffusion models will certainly be used in the future too for the interpretation of diffusion experiments.

The problem of diffusion modeling in polymers changes to some degree when one envisages to develop a really atomistic model, with truly predictive capabilities and without making any ad hoc assumption on the molecular behavior and/or motions in the polymer–penetrant system. In principle, a possibility to develop such diffusion modelings, is to simulate theoretically the process of penetrant diffusion in a polymer matrix by computer calculations.

For this one starts by considering only an appropriate set of "first principles" which describe at a truly atomistic level, the polymer and the penetrant. Then, these data about the atoms and molecules of the polymer are used to generate, by some means, a polymeric structure that has the "microscopic" and "macroscopic" properties of the true polymer, i.e., a low-energetic state, an appropriate distribution of torsional angles, a physically acceptable distribution of unoccupied volume, density, and so on (Catlow et al. 1990; Roe 1991; Gusev et al. 1993b; Müller-Plathe 1994; Theodorou 1996). Once this structure is generated a number of penetrant molecules are randomly "inserted" in it (where enough unoccupied volume is available). Then, the system is left to pursue its "molecular dynamics," i.e., the atoms and molecules of the system are allowed to move in the force fields and under the interactions acting inside the system, over a certain time interval. During this process there is no interference from the outside and, in particular, no heuristic assumptions are made about the molecular motions. If the process is simulated consistently for enough time, by observing, for example, the average displacement of the penetrant species, one can eventually calculate their diffusion coefficient (Müller-Plathe 1994). Though, this scheme sounds very elegant and attractive its practical achievement is a complex and demanding task. Because of that computer simulation, as a method for the estimation of the diffusion coefficients in polymers, has only lately become a practicable approach.

The prerequisites that make the development of "atomistic" simulations of diffusion in polymers possible are the development of powerful methods for the simulation of polymer microstructures and dynamics and also great computation capabilities of the computers.

The first attempts in the direction of simulating theoretically at an atomistic level the diffusion of simple gas molecules in a polymer matrix were made at the beginning of the 1970s (Jagodic et al. 1973). But, the systematic development of *ab initio* computer simulations of penetrant diffusion in polymeric systems dates only from the late 1980s (Rigby and Roe 1988; Boyd and Pant 1989; Shah et al. 1989; Trohalaki et al. 1989). At the beginning of the 1990s it was achieved to simulate some qualitative aspects, such as the diffusion mechanism, temperature, and pressure dependence of diffusion coefficients (Takeuchi and Okazaki 1990; Takeuchi 1990; Takeuchi et al. 1990; Boyd and Pant 1991; Müller-Plathe 1991a). The polymers chosen for investigation mainly fell into two categories, either they were easily described (model elastomers or polyethylene) or they were known to have, for simple permanent gases like H_2, O_2, N_2, H_2O or CH_4, large diffusion coefficients (polydimethylsiloxane (PDMS), Sok et al. ; Tamai et al. 1994, 1995 and atactic polypropylene (aPP), Müller-Plathe 1992a). The advantage of simulation at room temperature, for example, the diffusive motion of H_2 in aPP (D about 10^{-6} cm^2/s), is that the diffusive motion of the hydrogen molecules can already be sampled in relatively short simulations, about 0.5 ns (Müller-Plathe 1992a).

Based on these encouraging achievements, since the mid-1990s, the interest of the researchers shifted from easy-to-compute polymer–penetrant systems to those which have interesting technological potentials in such fields as: gas barriers (Müller-Plathe 1993; Gusev et al. 1993a; Han and Boyd 1996; Bharadwaj and Boyd 1999), gas or liquid separation processes (Niemelä et al. 1996; Fritz and Hofmann 1997; Hofmann et al. 1997; Charati and Stern 1998; Fried and Goyal 1998), molded objects (packagings, for example) (Hedenqvist et al. 1998), or swelling of polymers by solvents (Tamai et al. 1996; Müller-Plathe 1996a, 1998; Müller-Plathe and van Gunsteren 1997).

Trying to model, theoretically, the transport of small penetrants in polymer matrices, one realizes that the characteristic length and time scales vary greatly and depend on the polymer morphology (Müller-Plathe 1994). Most of the polymers used technologically are either amorphous or partially crystalline. From experimental results obtained over the past five decades it is commonly assumed that both diffusion and sorption in crystalline polymers are orders of magnitude smaller than in amorphous ones (Stannett et al. 1979; Frisch and Stern 1983; Koros 1990; Stern 1994; Piringer and Baner 2000; Vitrac et al. 2006). These facts determine that different theoretical and computational techniques will be appropriate for modeling the diffusion in different polymer–penetrant systems (Müller-Plathe 1994). For the diffusion of small penetrants, i.e., simple gases and vapors of water and/or simple organic substances, in purely amorphous polymers the computational techniques of choice will be molecular dynamics, MD (Gusev et al. 1993b; Müller-Plathe 1994; Theodorou 1996; Allen and Tildesley 1987; van Gunsteren and Weiner 1989; Baranyai 1994) or the transition-state approach, TSA (Gusev et al. 1993a; Gusev and Suter 1992, 1993a, 1993b). In a semicrystalline polymer a similar task can be approached, for example, by a Monte Carlo two-phase model (Müller-Plathe 1991b). So far, the "atomistic" modeling of diffusion of small penetrants in polymers was predominantly done for amorphous polymers and using the MD or TSA techniques, which will be presented briefly in the next section.

5.2.1
Molecular Dynamics

Because time is explicitly present in the formulations of MD, this technique is the most straightforward way of computer simulating the motion of penetrant molecules in amorphous polymer matrices (Gusev et al. 1993b; Müller-Plathe 1994; Theodorou 1996). The MD method allows one to look at a truly "atomistic" level within the system as it evolves in time. Excellent reviews on the use of MD for simulating penetrant diffusion in polymers have been published in (Roe 1991; Gusev et al. 1993b; Müller-Plathe 1994; Theodorou 1996). A summary of the basic concepts and some relevant results obtained so far with MD will be presented below.

To start a MD simulation of a diffusional process an amorphous polymer structure of the host material must first be theoretically generated. This structure must be low in energy and have the known physical properties of the polymer: chain length and distribution of torsional angles of polymer chains, density, distribution of free volume, etc. The origins of the MD approach to the problem of generating polymer structures lie in works done in the late 1970s to investigate theoretically amorphous bulk polymers (de Vos and Bellemans 1974; Skvortsov et al. 1977; De Santis and Zachmann 1978; Wall and Seitz 1977; Bishop et al. 1980; Weber and Helfand 1979). A MD approach to the problem of modeling the structure of amorphous polymers was introduced by Theodorou and Suter 1985a and a few years later developed by Theodorou and Suter 1985a and Müller-Plathe 1993 to allow a detailed description of such systems. An overview of the various MD methods used to generate amorphous polymer structures can be found by Hofman et al. 1996. The principal methods are (i) structural generation methods which in an ideal case are used to generate a structure which needs no further refinement, (ii) structural refinement methods that ideally are so efficient that the starting structure can be arbitrary, and (iii) coarse-graining methods in which the atomistic model of a polymer is mapped into a coarse representation of several atoms or even monomers.

To generate a polymer structure theoretically its matrix is presented as an ensemble of microscopic structures, which satisfies the requirements of detailed mechanical equlibria (Theodorou and Suter 1985b). For every atom its initial position and velocity have to be specified. Chain bond lengths and bond angles are fixed. Molecular movements are allowed to occur exclusively through rotations around the skeletal bonds of macromolecules. A polymer chain meeting these assumptions is built in vacuum by an iterative process that is started from an initially guessed "parent" structure which is then relaxed to a state of minimum potential energy (Theodorou and Suter 1985b). The density of the structure obtained must be eventually equal that of the simulated polymer. The free volume in the polymer can be estimated from the generated structure. To obtain a statistical average of this free volume, a number of structures are generated starting from different "parent" chain configurations. Once the host structure was generated the next step in the MD simulation of diffusion is to "place" (insert) the diffusant molecules into the computed structure. The condition for inserting the penetrant molecules into the structure is to find "free-volumes" where the energy is below a certain threshold and that any two of the penetrant molecules are separated by some minimum distance. Then the penetrant and polymer molecules are allowed to interact with each other and move within the limits of constrains they are subjected to. The straightforward technique is now to follow by computer simulation the displacement of the penetrants into the potential field of the system and eventually to estimate the mean-square displacement (MSD) of the penetrant species.

Among the first remarkable results of MD simulations was the finding that diffusion of small molecules in amorphous polymeric structures proceeds by "hopping" (jumping) motions (Takeuchi 1990). From a phenomenological point of view this is not a new result if one takes into account that such a mechanism was intuitively assumed in some "microscopic" diffusion models long before the development of computer simulation techniques, see the preceding section. The new aspect is that the computational approach has led to this picture of the diffusion mechanism starting from true "first principles" of the penetrant–polymer system and not on the basis of "shrewd guesses." To illustrate this type of motion in Figure 5.5, a typical trajectory of a water molecule through an amorphous elastomer (PDMS) is presented (Fritz and Hofmann 1997; Hofmann et al. 1997).

From this figure, one can discern that the voids forming the free volume of the rubbery polymer are clearly separated from each other and that there are two types of motion of the penetrant molecule:

1. For a relatively long period of time (typically a few 100 ps) the penetrant molecule stays confined in certain small regions of space, the "cavities" of the polymer

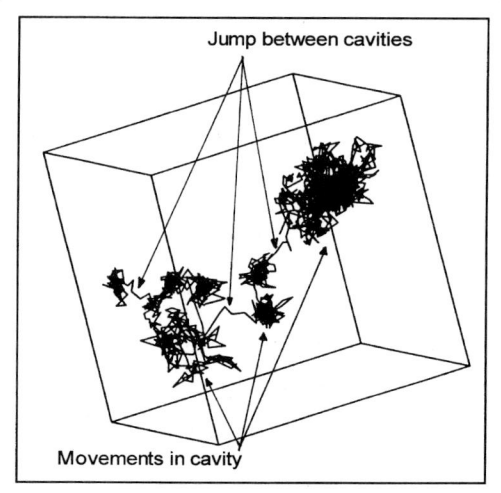

Figure 5.5 A typical trace of the center of mass of one representative water molecule in a PDMS matrix (Fritz and Hofmann 1997).

matrix. It explores the cavity thoroughly without being able to move beyond the confines of the volume it resides in. Thereby, the penetrant is reflected by the polymer matrix about every few picoseconds (Müller-Plathe 1994; Fritz and Hofmann 1997; Hofmann et al. 1997);

2. The quasistationary period is interrupted by quick leaps from one such cavity to another close by. The jump between the two neighboring cavities is preceded by the formation of a channel between them. Under favorable circumstances (right momentum) the penetrant then slips through this opening, essentially without activation energy or more exactly surpassing a small energy barrier, due to the fact that the channels are on average narrow (Müller-Plathe 1994). The jump duration is short, compared to the residence time in the cavities.

A "hopping" event in a polymer matrix, as found typically in MD simulations is presented in Figure 5.6 (Takeuchi 1990).

As announced above these findings are in astonishing agreement with the "heuristic" pictures of the diffusion mechanism discussed in the framework of some "microscopic" diffusion models. But, besides being free of the conceptual drawbacks (the ad hoc assumptions) of the "classical" diffusion models the MD method of computer simulation of diffusion in polymers makes it possible to get an even closer look at the diffusion mechanism and explain from a true atomistic level well-known experimental findings. For example, the results reported by Fritz and Hofmann 1997 and Hofmann et al. 1997 on the "hopping" mechanism reveal the following additional features, namely.

In a rubbery polymer with flexible macromolecular chains (PDMS, for example) the cavities forming the free volume are clearly separated from each other. The detailed evaluation of the movement of a penetrant particle from cavity (1) to the

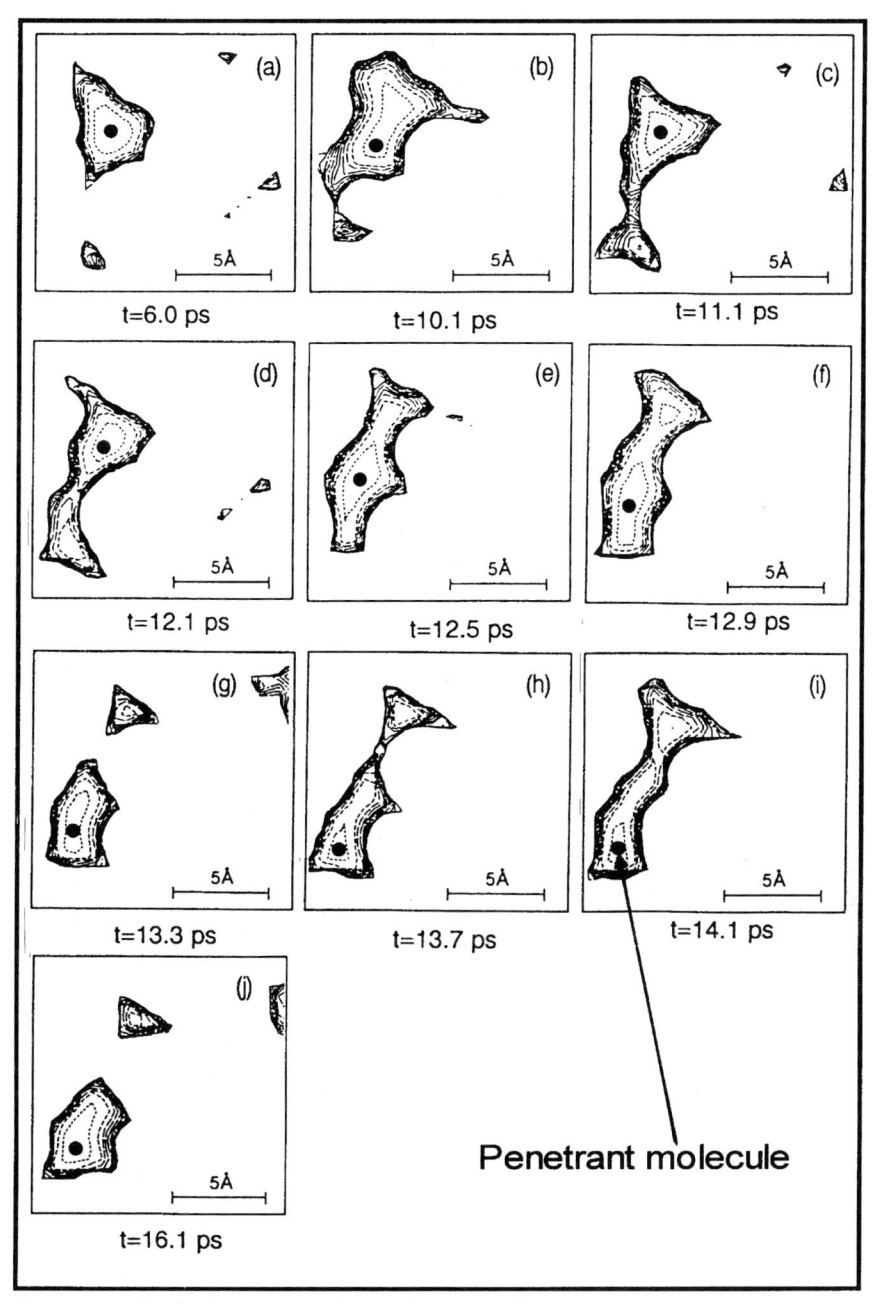

Figure 5.6 Molecular dynamics simulation of a "jump" of an O_2 molecule in a glassy – CH_2 – matrix (Takeuchi 1990).

neighboring (2), did not show any immediate back jumps (2) → (1). This is mainly due to the fact that the channel between (1) and (2) closes quiet quickly. In a polymer with stiff chains (a glassy polyimide, PI, for example) the individual cavities are closer to each other and a rather large number of immediate back jumps occurred during the time interval simulated (Hofmann et al. 1997). This indicates that once a channel between two adjacent cavities in a stiff chain polymer is formed it will stay open for some 100 ps. This makes the back jump (2) → (1) of the penetrant more probable than a jump to any other adjacent hole (3), a process which seems to be one cause for the general tendency that the diffusion coefficient of small penetrants in stiff chain glassy polymers is smaller than in flexible chain rubbery polymers.

The results of MD simulations will be useful if they are able to reproduce diffusion coefficients measured experimentally with sufficient accuracy. Given the scatter between the results of different experiments reported in the literature, a computational method can be considered accurate enough if, for absolute diffusion coefficients, it reproduces the experimental values within one order of magnitude. Such results are presented in Table 5.1.

The results given in Table 5.1 show that the agreement between the diffusion coefficients, predicted from MD simulations and experimental ones, ranges from reasonable to excellent. At temperatures around 300 K this is found both for polymers which are above their glass transition temperature, T_g (PDMS, PIB, PE, and aPP) and for polymers which are below T_g (PET, PS, PTMSP, PI, and PAI). As a trend one can notice, and this not only from Table 5.1 but also from other works published in the last decade, that the agreement between MD simulations of diffusion and solvation of small penetrants in polymers and experiment steadily improved. These are encouraging developments, showing that modern software and powerful computers are today capable to model and predict diffusional processes in a certain types of polymer–penetrant systems. To perform a consistent MD simulation the general condition is tied to the ability of the MD procedure to simulate a polymer–penetrant system large enough to sample the configurational statistics of the polymer sufficiently well. For a simple polymer like linear polyethylene with flexible chains one may need a few hundred [–CH_2–] repeat units of a few hundred to a few thousand atoms (Müller-Plathe 1994). To generate a bulk PDMS structure in which three water molecules are "inserted" 220 monomer units [–$Si(CH_3)_2$–O–], i.e., 2238 atoms, were, for example, used by Fritz and Hofmann 1997. One might expect that many more repeat units are needed if the polymer has stiff chains (Müller-Plathe1994). However, it should be noted that it is the number of flexible bonds in a chain and not just the number of repeat units that is a decisive parameter for the achievable quality of the amorphous polymeric structure generated from a chain (Sok et al. 1992). Other factors determining the range of application of the MD method arise from the mobility of the penetrant itself. To be sufficiently precise with the computer simulation one needs to observe, say, 10 jump events for every single penetrant (which is probably the bare minimum). At equilibrium and assuming hopping motion the diffusion coefficient can be given by Eq. (5.1), where λ is now the mean-square "jump" distance and ν^{-1} the average residence time between jumps. Hence for a D of about 5×10^{-6} cm^2/s (a comparatively high-diffusion coefficient for

Table 5.1 Diffusion coefficients calculated by molecular dynamics and from experiment.

Polymer	Diffusant	D_{calc} (cm^2/s) × 10^6 (K)	Reference	D_{exp} (cm^2/s) × 10^6 (K)	Reference
Polydimethylsiloxane (PDMS)	He	180.0 (300)	Sok et al. 1992	100.0 (300)	Barrer and Choi 1965
	CH$_4$	21.0 (300)	Sok et al. 1992	20.6 (300)	Stern et al. 1987
	CO$_2$	11.0 (300)	Charati and Stern 1998	26.0 (308)	Charati and Stern 1998
	N$_2$	12.0 (300)	Charati and Stern 1998	39.0 (308)	Charati and Stern 1998
	O$_2$	18.0 (300)	Charati and Stern 1998	41.0 (308)	Charati and Stern 1998
	H$_2$O	20.0 (300)	Fritz and Hofmann 1997	14.5 (298)	Okamoto et al. 1988
	Ethanol	4.4 (300)	Fritz and Hofmann 1997	4.5 (298)	Okamoto et al. 1988
Polypropylmethylsiloxane (PPMS)	CH$_4$	12.0 (300)	Charati and Stern 1998	11.0 (308)	Charati and Stern 1998
Polytriflouropropylmethylsiloxane (PTFPMS)	CO$_2$	13.0 (300)	Charati and Stern 1998	8.1 (308)	Charati and Stern 1998
	CH$_4$	2.3 (300)	Charati and Stern 1998	5.6 (308)	Charati and Stern 1998
Polyphenylmethylsiloxane (PPhMS)	CO$_2$	3.2 (300)	Charati and Stern 1998	5.3 (308)	Charati and Stern 1998
	CH$_4$	0.5 (300)	Charati and Stern 1998	1.2 (308)	Charati and Stern 1998
Polyisobutylene (PIB)	CO$_2$	0.8 (300)	Charati and Stern 1998	2.0 (308)	Charati and Stern 1998
	He	30.0 (300)	Müller-Plathe et al. 1993	5.96 (300)	van Amerongen 1950
	H$_2$	9.0 (300)	Müller-Plathe et al. 1993	1.52 (300)	van Amerongen 1950
	O$_2$	0.169 (300)	Müller-Plathe et al. 1992a	0.081 (300)	van Amerongen 1950
	CH$_4$	0.63 (300)	Pant and Boyd 1993	1.7 (300)	Lundberg et al. 1969
Polyethylene (PE)	CH$_4$	1.12 (300)	Pant and Boyd 1993	0.54 (298)	Michaels and Bixler 1961
	CH$_4$	1.6 (300)	Tamai et al. 1994	0.54 (298)	Michaels and Bixler 1961

Table 5.1 (Continued)

Polymer	Diffusant	D_{calc} (cm^2/s) × 10^6 (K)	Reference	D_{exp} (cm^2/s) × 10^6 (K)	Reference
	C$_{10}$H$_{16}$ (Limonene)	4.0 (350)	Karlsson et al. 2002	0.45 (350)	Karlsson et al. 2002
atacticPolypropylene (aPP)	H$_2$	44.0 (300)	Müller-Plathe 1992b	4.9 (300)	Mercea 1996
	O$_2$	4.0 (300)	Müller-Plathe 1992b	0.95 (300)	Mercea 1996
	CH$_4$	0.48 (300)	Müller-Plathe 1992b	0.24 (300)	Mercea 1996
Polyamidimide (PAI)	H$_2$	0.97 (300)	Hofman et al. 1996	1.3 (300)	Fritsch and Peinemann 1995
Polyimide (PI)	N$_2$	0.28 (300)	Hofman et al. 1996	0.52 (300)	Fritsch and Peinemann 1995
	O$_2$	0.74 (300)	Pant and Boyd 1993	2.69 (300)	Fritsch and Peinemann 1995
Poly[1-(trimethylsilyl)-1-propyne] (PTMSP)	He	465 (300)	Fried and Goyal 1998	316 (300)	Masuda et al. 1988
	O$_2$	23.8 (300)	Fried and Goyal 1998	30.0 (303)	Okamoto et al. 1988
	N$_2$	20.5 (300)	Fried and Goyal 1998	36.0 (298)	Srinivasan et al. 1994
	CH$_4$	16.7 (300)	Fried and Goyal 1998	22.0 (303)	Okamoto et al. 1988
	CO$_2$	4.0 (300)	Fried and Goyal 1998	19.5 (303)	Okamoto et al. 1988
Polyethyleneterephthalate (PET)	CH$_4$	~0.01 (extrapolated to 340)	Bharadwaj and Boyd 1999	0.0092 (340)	Michaels et al. 1963
Polystyrene (PS)	CH$_4$	~0.056 (extrapolated to 325)	Han and Boyd 1996	0.0338 (323)	Barrie et al. 1980

packaging applications) and a "jump distance" of about 0.5 nm – see Bharadwaj and Boyd 1999, for example – one finds that in a 1-ns simulation one will encounter about 12 jumps on average. It is interesting to notice that if the MD simulation is done in steps of 1 fs = 10^6 time steps must be computed to complete a 1 ns simulation (Fried and Goyal 1998). To simulate with MD slower diffusion processes, i.e., smaller D, one must either extend the duration of the simulation (and hence the computing time and costs) or to "insert" several penetrants at the same time in the generated polymer structure and thereby improve the quality of the sampling (Müller-Plathe 1994; Bharadwaj and Boyd 1999; Fritz and Hofmann 1997; Hofmann et al. 1997). However, the later option is valid only if the diffusion coefficient is not very sensitive to the penetrant concentration. With nowadays software and computers MD simulations can be extended to about 10 ns, which brings the D_s of about 5×10^{-7} cm^2/s within reach of the method. Diffusion processes which evolve at a rate of 5×10^{-7} cm^2/s or faster are typical for:

- the diffusion, at very low concentrations, of small penetrants (simple gases or vapors) in low-barrier polymers, i.e., siloxanes (Tamai et al. 1994; Fritz and Hofmann 1997), polypropylenes (Stern et al. 1987; Müller-Plathe 1992b), polybutylenes (Müller-Plathe et al. 1992, 1993; Pant and Boyd 1993) and polyethylenes (Tamai et al. 1994; Stern et al. 1987; Pant and Boyd 1993) at room temperature or polystyrene (Han and Boyd 1996), polyethylenetherephthalate (Bharadwaj and Boyd 1999) well above room temperature;
- the diffusion, at room temperature, of simple gases and vapors through glassy polymers with large interchain regions, i.e., poly[1-(trimethylsilyl)-1-propyne] (Niemelä et al. 1996).

However, in the packaging sector the large majority of the diffusion processes in polymers imply penetrants with a relative molecular weight, M_w, ranging between 100 and 1200 g/mol and have often quite complex structures. Recently, a MD simulation of limonene, an small organic substance with $M_w = 136.2$ g/mol, diffusion in PE between 77 and 227°C was reported in Karlsson et al. 2002. A diffusion process ranging from 2 to 100 ns was simulated and the obtained D_p – ranging from about 6×10^{-6} to 2×10^{-5} cms^{-1} – was within 30% of the experimental values. However, from experiments one knows that for organic molecules with $100 < M_w < 1200$ g/mol, in most polymer used in the packaging sector, D_p ranges from 10^{-9} to 10^{-14} cm^2/s or even lower levels (see Appendix I). Müller-Plathe 1994 stated that to study with MD techniques polymer–penetrant systems in which the diffusion coefficients are that small, is certainly out of reach for several generations of supercomputers to come. As for personal computers, PCs, Karlsson et al. 2002 stated that with today's PC power a MD run to simulate a diffusional process of more than 100 ns takes too long a time for practical reasons.

The possibility of extending MD to slower diffusion processes has been discussed (Müller-Plathe 1994; Karlsson et al. 2002). But applying such algorithms has a tradeoff on the overall quality of the computational approach. To perform calculations at time scales beyond those accessible to MD is possible nowadays by using the transition state approach (TSA) proposed by Gusev et al. 1993a,1993b and Gusev and Suter 1993b. This method will be presented briefly below.

5.2.2
The Transition-State Approach

As already mentioned in Section 5.1.1.1 one of the early theoretical models of gas diffusion in solid polymers (Barrer 1937, 1939, 1951) was based on the transition-state theory (TST) (Glasstone et al. 1941). More than 60 years ago it was assumed ad hoc that gas molecules move through a dense polymer in a series of activated "jumps" between "holes" which exist in the polymer matrix. Fortunately, results of *ab initio* MD simulations, Section 5.2.1, demonstrate that the computed trajectories of small penetrants in atomistic structures of dense polymers are consistent with the "heuristic" picture of this early "classical" model. In its framework it was estimated, from solubility data, that at room temperature the vibrational frequency v_o of the gas molecule trapped by the surrounding chains is of about 10^{12} s^{-1} model (Barrer 1947b). This finding is also in reasonable agreement with the bouncing frequency of a small gas molecule inside a "cavity" of the polymer matrix, as found in MD simulations. These results indicate that the "jumps" of a penetrant in a dense polymer could be treated as an elementary process, thus justifying the use of TST for developing a computer simulation technique to evaluate the rate of the penetrant jumps and out of this the diffusion coefficients. The development of a TSA based on a simplified description of thermal motions in the host matrix and stochastic methods in treating the penetrant dynamics, promises to allow much longer simulation intervals than MD can practically achieve nowadays (about 100 ns). This feature is important because: (i) the occurrence, in some polymer–penetrant systems, of anomalous diffusion (Gusev et al. 1993b; Gusev and Suter 1993a) leads to the necessity of carrying out very long MD simulation runs for penetrants to enter the Einstein diffusive regime (Gusev et al. 1993b; Müller-Plathe 1994) and (ii) unpracticably long MD simulations would be needed to simulate and predict slower diffusion processes, Section 5.2.3 and Appendix I.

In the development of the TSA besides the "jumping" mechanism already mentioned another fundamental mechanistic features assumed that the penetrant dynamics is coupled to the elastic motion of the polymer chains, but, to a first approximation, is independent from the structural relaxations of the matrix (Gusev et al. 1993a, 1993b; Gusev and Suter 1993a, 1993b).

The thermal motion causes the polymer matrix to move in its configurational space. At short times the vibrational modes of motion dominate: vibration of chemical bonds or bond angles, small-amplitude rotations of side groups or wiggling of torsion angles. As time goes by, the system tends to perform structural relaxation, for example, through torsional transitions in the main chain or in side groups. Using MD to simulate an appropriate penetrant trajectory one can specify an upper bound for times at which the system at hand can be treated as essentially executing elastic motions (Gusev et al. 1993b). Elastic motion implies that the atoms of the matrix fluctuate about their equilibrium positions. Allow now a small dissolved molecule to reside in the system and suppose that one can neglect the correlation between the structural relaxation of the matrix and the dynamics of the penetrant. In this case one can write a penetrant distribution function $\rho(r)$ which is obtained by integrating over all possible values of the deviations of the host atoms the result of the potential energy of interaction between the dissolved molecule and the host atoms and a normalized probability density, $W(\{\Delta\})$, describing the elastic fluctuations (Gusev and Suter 1993b). The function $\rho(r)$ is related to the Helmholtz energy, $A(r)$, of the dissolved molecule at location r according to a general equation given by Barrer 1947b. The TST can be used then for describing the spatial movement of a dissolved molecule as a series of activated jumps between adjacent local minima of $A(r)$ (Gusev and Suter 1993b). The rate constants R_{ij} for the penetrant transition from site i to site j can be written as (Glasstone et al. 1941):

$$R_{i,j} = k^* \frac{kT}{h} \frac{Q_{i,j}}{Q_i} \tag{5.9}$$

where $Q_{i,j}$ and Q_i denote partition functions of penetrant on the crest surface between sites i and j and in site i, respectively. In Eq. (5.9), k^* is a transmission factor taken to be about 0.5 (Gusev and Suter 1993b). It was shown that in the quasiclassical case one can link $Q_{i,j}$ and Q_i to the function $\rho(r)$ (Landau and Lifshitz 1967). Hence, specification of the elastic fluctuations of the atoms of the host matrix through the probability density $W(\{\Delta\})$ yields $\rho(r)$ which, in turn, yield the transition rates R_{ij} (Gusev and Suter 1993a). If the network of local minima of $A(r)$ with the associated R_{ij} are known, one can use stochastic methods to evaluate the correlation function describing the penetrant dynamics (Gusev and Suter 1992, 1993b). The procedures of simulating the dynamics of guest molecules on the network of sites and of evaluating $W(\{\Delta\})$ were described by Gusev and Suter 1992, 1993b, respectively. An important parameter in these procedures is the mean-square deviation $\langle \Delta_\alpha^2 \rangle$ of host atom α from its average position. The values of $\langle \Delta_\alpha^2 \rangle$ are expected to depend on the time scale of averaging: for very short times $\langle \Delta_\alpha^2 \rangle$ increases with time approaching then, by definition of elastic motion, an asymptotic value. Using atomistic short-scale trajectories calculated with MD and specifying an averaging time one can calculate for $\langle \Delta_\alpha^2 \rangle$ a "smearing" factor $\langle \Delta^2 \rangle$ and use it in the TSA simulations (Gusev et al. 1993b; Gusev and Suter 1993b). Another possible way to evaluate $\langle \Delta^2 \rangle$ is to match the short-time region of the mean-square displacement $\langle r^2 \rangle$ of the penetrant versus time curves obtained from TSA with those from MD calculations (Gusev et al. 1993b).

To actually carry out a TSA simulation a three-dimensional grid, with grid interval of about 0.2 (5×10^6 equispaced points by Gusev and Suter 1993b) is built and the Helmholtz energies at all grid points are computed. Before this can be done in practice, a value for $\langle \Delta^2 \rangle$ must be found. Then, local minima and the crest surfaces must be found, using the procedures given by Gusev and Suter 1991, 1992, 1993b. To study the dynamics of the penetrant molecules on the network of sites a Monte Carlo procedure is employed (Gusev et al. 1993b).

Eventually, the stochastic trajectory of a dissolved molecule is obtained and subsequently, by averaging a large number of such trajectories, about 10^3 in Gusev

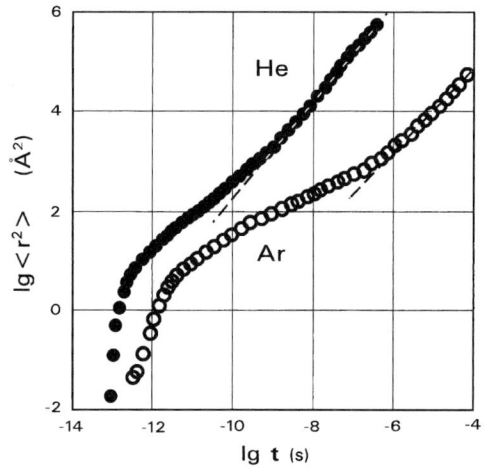

Figure 5.7 Computed dynamics of He and Ar in polycarbonate at 300 K (Gusev and Suter 1993b).

and Suter 1993b, the diffusion coefficient D of the penetrant in the polymer can be calculated from the plot of $\langle r^2 \rangle$ versus the time t (using for the linear portions (Einstein diffusion) of the curves a simple equation similar to Eq. (5.3)). In Figure 5.7 the $\langle r^2 \rangle$ of He and Ar in glassy PC at 300 K, as calculated with TSA, are shown. The results plotted in this figure represent averages over 500 independent simulation paths. The simulations presented in Fig. 5.7 show a region of "anomalous diffusion" of the penetrant He for $\langle r^2 \rangle$ smaller than $\sim 10^3$ (simulation interval of $\sim$0.5 ns). This is similar to those reported by Müller-Plathe 1992b, 1994 on the MD simulation of He diffusion in rubbery polyisobutylene, where the transition to normal diffusion was captured at around $\langle r^2 \rangle \sim 10^{-2}$ and a simulation interval of $\sim$0.1 ns. It is believed that this anomalous behavior is caused by a separation of time scales consistent with the jumping pattern (Müller-Plathe 1994).

The very fast motions of the penetrant molecules inside the cavities (timescales of several 100 ps) are determined by the shape of these cavities. Therefore, these motions do not have a random-walk-like behavior and consequently it is not appropriate to use the Einstein equation, i.e., $D = \langle r^2 \rangle \, 6t$ (which similar to Eq. (5.1)), to calculate D. In fact the Einstein equation holds true if the slope of the $\log \langle r^2 \rangle$ versus $\log t$ plot is equal to one. A direct consequence of this fact is that, in order to predict diffusion coefficients, a MD or TSA computation must simulate a time interval long enough to get fulfilled the above requirement. For some polymer–penetrant systems, this means already, the need to carry out simulations over time intervals that are out of reach of the MD method ($t > 100$ ns) (Hofmann et al. 1997; Gusev and Suter 1992, 1993b). In these cases, the method of choice will be the TSA.

In Table 5.2 a comparison between diffusivities obtained with the TSA method and experimental D_s is presented. From this table one can see that, in all cases computed D_s agree with experimental data to within an order of magnitude. Moreover, most of these D_s are considerably smaller than the 5×10^{-7} cm^2/s lower threshold assumed to

Table 5.2 Diffusion coefficients calculated with the transition-state approach and from experiment.

Polymer	Diffusant	$D_{calc.}$ (cm^2/s) $\times 10^7$ (K)	Reference	$D_{exp.}$ (cm^2/s) $\times 10^7$ (K)	Reference
Polycarbonate (PC)	He	35.0 (300)	Gusev and Suter 1993b	64.6 (308)	Muruganandam et al. 1987
	O$_2$	0.10 (300)	Gusev and Suter 1993b	0.56 (308)	Muruganandam et al. 1987
	N$_2$	0.09 (300)	Tamai et al. 1994	0.18 (308)	Muruganandam et al. 1987
Polyamidimide (PAI)	H$_2$	15.9 (300)	Hofmann et al. 1997	9.4 (300)	Hofmann et al. 1997
	O$_2$	0.40 (300)	Hofmann et al. 1997	0.30 (300)	Hofmann et al. 1997
	N$_2$	0.20 (300)	Hofmann et al. 1997	0.10 (300)	Hofmann et al. 1997
Polyimide	O$_2$	15.3 (300)	Hofmann et al. 1997	7.40 (300)	Hofmann et al. 1997
	N$_2$	2.6 (300)	Hofmann et al. 1997	2.80 (300)	Hofmann et al. 1997
Polyvinylchloride (PVC)	He	17.0 (318)	Gusev et al. 1993b	40.0 (318)	Okamoto et al. 1988, Barrer et al. 1967, Tikhomirov et al. 1968,
	Ne	2.0 (318)	Gusev et al. 1993b	4.0 (318)	Okamoto et al. 1988, Barrer et al. 1967, Tikhomirov et al. 1968,
	Ar	0.04 (318)	Gusev et al. 1993b	0.05 (318)	Okamoto et al. 1988, Barrer et al. 1967, Tikhomirov et al. 1968,
	Kr	0.003 (318)	Gusev et al. 1993b	0.01 (318)	Okamoto et al. 1988, Barrer et al. 1967, Tikhomirov et al. 1968

be in reach of nowadays MD simulations Section 5.3. This is an encouraging sign that computer simulations of diffusional processes are already able to predict, with a reasonable accuracy and for small and simple penetrants, D's around $10^{-10}\,\text{cm}^2/\text{s}$. From the point of view the packaging sector it would be interesting to learn if and when further theoretical developments of the TSA method will be able to simulate (predict) such slow diffusional processes for organic penetrants with a much more complex structure, see Appendix I.

Two "atomistic" approaches have been presented briefly above: molecular dynamics and the transition-state approach. They are still not ideal tools for the prediction of diffusion constants because: (i) in order to obtain a reliable chain packing with a MD simulation one still needs the experimental density of the polymer and (ii) though TSA does not require classical dynamics it involves a number of simplifying assumptions, i. e., duration of jump mechanism, elastic polymer matrix, size of smearing factor, that impair to a certain degree the *ab initio* character of the method. However, MD and TSA are valuable achievements, they are complementary in several ways and can be used to predict the diffusion coefficients of small penetrants (so far simple gases and simple organic vapors) in both rubbery and glassy amorphous polymers. These computational methods can be used to understand the behavior of small penetrants in the matrix of a polymer starting from an "atomistic level" and without ad hoc assumptions on the movements of the polymer chains. In this respect, MD is the less coarse grained of the two methods. The main drawback of MD is the computational cost that nowadays prohibits simulations beyond 100 ns, which are still being far from routine. The TSA is well suited to extend the time-scale of simulations, bringing new phenomena within reach. In this respect, it is important to use MD and TSA in conjunction. The limitations of the TSA, as developed so far, are evident when one intends to simulate penetrant–polymer systems where there is a strong interaction between the penetrant and the host atoms, or where larger penetrant molecules require a deformation of the polymer structure for their passage. In such systems as well as in systems where the penetrant induces a swelling of the polymer matrix MD seems to be the method of choice to properly simulate the diffusion mechanism (Müller-Plathe and van Gunsteren 1997; Müller-Plathe 1996b, 1998).

5.3
Conclusions

A process or manufacturing engineer is often confronted with the difficult and expensive task of measuring experimentally the migration (diffusion) of rather complex organic molecules in rubbery or glassy semicrystalline polymer matrices. For such systems the knowledge/prediction of diffusion coefficients would be crucial for the theoretical estimation of substance transfer, for example, from a polymeric packaging into the wrapped good (foodstuff, medicine, etc.). Therefore, a theoretical method/model for performing the true prediction of diffusion coefficients for small organic penetrants in polymers would be of great help to reduce the costs and work time nowadays spent in the field of polymer packaging research and law enforcement. The problem is that ideally such a theoretical method/model should be (i) as simple as possible, (ii) rely on parameters which are well known and easily available, and (iii) at last but not least, the use of the method to predict diffusion processes should not consume more time and resources than the direct migration/diffusion experiments. If a given diffusion model cannot meet the one or the other of these requirements, from a purely pragmatic point of view, a process engineer or law enforcer may not see incentives to use the theoretical approach instead of a well-established experimental one.

Unfortunately, it seems that none of the diffusion models presented in the above sections meets completely these practical goals.

It is beyond any question that the type of "classical" diffusion models presented in Section 5.1 were, at the time of their conceivement, important steps for the qualitative understanding of the phenomenology of penetrant diffusion in polymers. Moreover, some of these models are very successful in rationalizing average experimental diffusion coefficients with macroscopic parameters as temperature and penetrant concentration. Trying to use these models for predicting diffusion coefficients for penetrant–polymer systems, which are specific in the packaging sector one is confronted with several problems.

With no exception, in all "classical" diffusion models one or more adjustable parameters enter in the formula of the diffusion coefficient. To calculate the magnitude of this/these parameter/s a number of diffusion experiments must be performed, preferably with the very penetrant–polymer system which one intends to simulate theoretically. In one of the most widely used "classical" models – the free-volume models of Fujita, Vrentas, Duda, and their alternatives (Kosiyanon and McGregor 1981; Paul 1983; Doong and Ho 1992; Lodge et al. 1990; Frick et al. 1990) – more than a dozen structural and physical parameters are needed to calculate the free volume in the penetrant–polymer system and subsequently, the diffusion coefficients. This might prove to be a relatively simple task for simple gases and some organic vapors, but not for the nonvolatile organic substances (rest-monomers, additives, stabilizers, fillers, plasticizers) which are typical for polymers used in the packaging sector.

Once the adjustable parameters of the diffusion model are known its expressions can be used to make predictions of the diffusion coefficient and further use them to estimate a migration process. However, in order to make realistic migration estimations, the parameters of the simulated migration process – temperature, migrant concentration, pressure, degree of polymer swelling – should not differ too much from those of the said diffusion experiment.

The question now is: "How will a practitioner from the field of polymer packaging prefer to establish the level of migration from his samples?"

- Perform direct experimental migration tests with his samples. or
- Perform first a series of diffusion experiments (which often require quite sophisticated equipment), derive from the obtained results, with the aid of a "classical model," the parameters to calculate the diffusion coefficient (the theoretical schemes to do this are often nontrivial) and then use another theoretical formalism to estimate the migration.

Most likely the practitioner will chose scheme (a) unless the theoretical estimation of the migration cannot be done in a much simpler manner than scheme (b). A prerequisite for this would be a simple and reliable scheme to obtain the diffusion coefficients needed in the migration estimation. Recently an attempt in this direction was reported for some selected polymers by Vitrac et al. 2006. Classification and regression tree (CART) algorithms were used to build up a scheme to predict diffusion coefficients in PEs and PPs. For this more than 650 diffusion coefficients reported in the literature were classified by using three molecular descriptors: the van der Waals volume of the diffusant, its gyration radius, and a dimensionless shape parameter. A characteristic CART structure, as shown in Figure 5.8 for a certain class of polyethylenes, can be reached by a stepwise division of the population (diffusion coefficients) at a node into branches in such a way that the subpopulations were internally as homogenous and externally as heterogeneous as possible with respect to the classification criterion. The algorithms used to obtain such a CART diagram for a certain polymer and diffusion temperature can be used, in principle, to predict the diffusion coefficient for molecules not included in the database with which the diagram was generated. For this one determines the magnitude of the three molecular descriptors mentioned above and implements them into the CART algorithms. The CART calculations will eventually associate to the descriptors a diffusion coefficient D_p which can then be used for migration estimations.

However, the usefulness of this scheme to estimate D_p is limited. To build up a CART scheme, as presented in detail in Vitrac et al. 2006, is not a trivial task. One must develop for this a special software and one needs a comprehensive D_p database for each type of polymer and diffusion condition.

Based on the progress which was recorded in recent years in the computer simulation of diffusion processes in polymers one may be confident that these

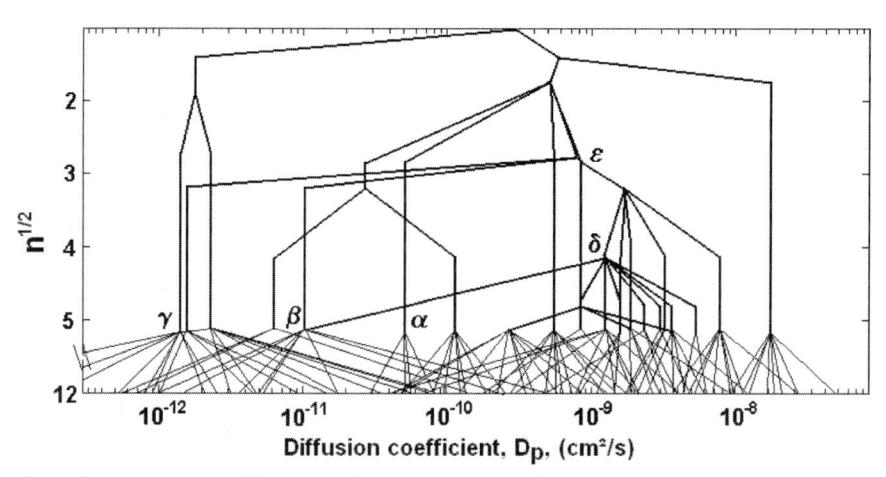

Figure 5.8 Clustering of diffusion coefficient according to the pruning level of the full classification tree for medium and high-density polyethylene at 23°C (Vitrac et al. 2006).

methods will one day able to provide a true prediction of diffusion coefficients needed for migration estimations from polymeric packaging materials. However, to reach this goal a series of problem must yet be solved.

As mentioned in Section 5.2, a MD simulation is better suited to describe at a true atomistic level the host matrix and the dynamics of the penetrants. However, most of the MD performed so far are dealing with purely amorphous polymers and with very simple penetrants. Such MD simulations use cells with a length of a few nm. In the packaging practice, however, most of the polymers are partly crystalline and the penetrants are often complex organic molecules. Crystallite dimensions may range from several 10 nm to several micrometers and crystallites often aggregate to form larger domains of macroscopic dimensions (Paul and Koros 1976; Stern and Saxena 1980). In (Müller-Plathe 1994) it was mentioned that a straightforward atomistic MD simulations of a semicrystalline material is not yet achievable. Moreover, at an atomic level, the interaction of most such organic molecules with the host matrix is much stronger and difficult to quantify that the interaction of the simple and relatively inert molecules investigated so far with MD simulations.

Even assuming that further developments of the MD and/or Monte Carlo two-phase techniques will be able to cope with these problems, a question remains: how long will be the time interval simulated? According to those mentioned in Section 5.2.1 simulations of a few 100 ns to a few μs are needed to predict a D in the range of 10^{-10} to 10^{-11} cm^2/s, which is often found in technical applications of polymers. From today's perspective the computer time (and costs) needed for a MD simulation of such duration are out of practical reach.

With the TSA developed by Gusev et al. 1993a and Gusev and Suter 1992, it was possible to almost reach simulation intervals of 1 ms. This makes, in principle, possible to predict D_s as small as a few 10^{-12} cm^2/s. Therefore, the TSA seems to be a good choice to predict D_s for the type of diffusion processes encountered in packaging applications. But for this, the actual TSA algorithms must be developed to take into account also strong interactions between the penetrant and the host atoms, and the deformation of the polymer structure at the passage of complex penetrant molecules.

To evaluate if the "atomistic" computer simulation of diffusion processes may help to reduce the considerable volume and costs of experimental migration testing performed nowadays, it is necessary to consider also the following aspects.

How much software development and computing time will be needed to predict D for a penetrant–polymer system? Hofmann et al. 1997 stated that even rather the fast TSA simulation technique will presumably not lead to a fast predictability of transport parameters for large numbers of hypothetical polymers in the near future. This was mainly attributed to the fact that the construction of well-equilibrated polymer packing models is still demanding large amounts of computer time (not to mention the much longer time needed to effectively develop the appropriate algorithms).

Then an important aspect is how precise the predicted D will be? So far an agreement within one order of magnitude between an experiment and an atomistic simulation is considered to be a good achievement. Certainly, such a *D* prediction precision would be of practical value for migration estimations too. However, a question remains: "To what sophistication must be developed the computer simulation approaches to meet this requirement for the type of penetrant–polymer systems which are usual in packaging sector?"

In the end it is legitimate to mention that for a considerable number of process engineers and law enforcement personnel the material costs of using an atomistic computational approach to predict a *D* and subsequently, uses it in a migration estimations will also play an important role. Pragmatically speaking one can expect that somebody interested to reduce its expenses for migration testing from polymeric packaging, will not have too much interest to replace these tests with much more expensive and less precise theoretical simulations.

References

Allen, M.P. and Tildesley, D.J. (1987) *Computer Simulation of Liquids and Solids*, Oxford University Press, Oxford.

Aminabhavy, T.M., Shanthamurthy, U. and Shukla, S.S. (1988) *J. Macromol. Sci., Rev. Macromol. Chem. Phys.*, **C28**, 421.

Baranyai, A. (1994) *J. Chem. Phys.*, **101**, 5070.

Barrer, R.M. (1937) *Nature*, **140**, 106.

Barrer, R.M. (1939) *Trans. Faraday Soc.*, **35**, 262 & 644.

Barrer, R.M. (1942) *Trans. Faraday Soc.*, **38**, 321.

Barrer, R.M. (1947a) *Trans. Faraday Soc.*, **43**, 3.

Barrer, R.M. (1947b) *Trans. Faraday Soc.*, **39**, 3.

Barrer, R.M. (1951) *Diffusion in and Through Solids*, Cambridge University Press, Cambridge.

Barrer, R.M., Barrie, J.A. and Slater, J. (1958) *J. Polym. Sci.*, **27**, 177.

Barrer, R.M. and Choi, H.T. (1965) *J. Polym. Sci.*, **10**, 111.

Barrer, R.M., Mallinder, R. and Wong, P.S.-L. (1967) *Polymer*, **8**, 321.

Barrie, J.A., Williams, M.J.L. and Munday, K. (1980) *Polym. Eng. Sci.*, **20**, 20.

Bharadwaj, R.K. and Boyd, R.H. (1999) *Polymer*, **40**, 4229.

Boyd, R.H. and Pant, K. (1989) *Polym. Preprints*, **30**, 30.

Boyd, R.H. and Pant, K. (1991) *Macromolecules*, **24**, 6325.

Bishop, M., Ceperley, D., Frisch, H.L. and Kales, M.H. (1980) *J., Chem. Phys.*, **72**, 322.

Brandt, W.W. (1959) *J. Phys. Chem.*, **63**, 1080.

Bueche, F. (1953) *J. Chem. Phys.*, **21**, 1850.

Catlow C.R.A., Parker S.C., and Allen M.P. (eds)(1990) *Computers Simulation of Fluids, Polymers and Solids*, Kluwer, Dordrecht.

Chandrasekar, S. (1943) *Rev. Mod. Phys.*, **15**, 1.

Charati, S.G. and Stern, S.A. (1998) *Macromolecules*, **31**, 5529.

Chern, R.T., Koros, W.J., Sanders, E.S., Chen, S.H. and Hopfenberg, H.B. (1983) *Am. Chem. Soc., Symp. Ser.*, **223**, 47.

Clough R.L., Billingham N.C., and Gillen K.T. (eds), (1985) Polymer Durability: Degradation, Stabilization and *Lifetime Predictions*, ACS Adv. Chem. Ser. 249.

Cohen, M.H. and Turnbull, D. (1959) *J. Chem. Phys.*, **31**, 1164.

Comyn, J. (1985) *Polymer Permeability*, Elsevier, New York.

Crank, J. and Park, G.S. (1968) *Diffusion in Polymers*, Academic Press, London.

Crank, J. (1975) *Mathematics of Diffusion*, 2nd edn. Claredon Press, Oxford.

Crompton, T.R. (1979) *Additive Migration from Plastics into Food*, Pergamon Press, Oxford.

Cussler, E.L. (1990) *J. Membr. Sci.*, **52**, 275.

Cussler, E.L., Huges, S.E., Ward, W.J. and Aris, R. (1988) *J. Membr. Sci.*, **38**, 161.

de Vos, E. and Bellemans, A. (1974) *Macromolecules*, **7**, 812.

De Santis, R. and Zachmann, H.G. (1978) *Prog. Colloid. Poly. Sci.*, **64**, 281.

DiBenedetto, A.T. (1963a) *J. Polym. Sci.*, **A1**, 3459.

DiBenedetto, A.T. (1963b) *J. Polym. Sci.*, **A1**, 3477.

Doong, S.J. and Ho, W.S.W. (1992) *Ind. Eng. Chem. Res.*, **31**, 1050.

Duda, J.L., Vrentas, J.S., Ju, S.T. and Liu, H.T. (1982) *AIChEJ*, **28**, 279.

Fredrickson, G.H. and Helfand, E. (1985) *Macromolecules*, **18**, 2201.

Frick, T.S., Huang, W.J., Tirrell, M. and Lodge, T.P. (1990) *J. Polym. Sci., Polym. Phys. Ed.*, **28**, 2067.

Fried, J.R. and Goyal, D.K. (1998) *J. Polym. Sci., Part B. Polym. Phys.*, **36**, 519.

Frisch, H.L. (1962) *J. Chem. Phys.*, **37**, 2408.

Frisch, H.L. and Stern, S.A. (1983) Diffusion of small molecules in polymers, *CRC Crit. Rev. Solid State Mater. Sci.*, **11**, 123.

Fritsch, D. and Peinemann, K.-V. (1995) *J. Membr. Sci.*, **99**, 29.

Fritz, L. and Hofmann, D. (1997) *Polymer*, **38**, 1035.

Fowler, R.H. and Guggenheim, E.A. (1949) *Statistical Thermodynamics*, Cambridge University Press, Cambridge.

Fujita, H. (1961) *Fortschr. Hochpolym. Forsch.*, **3**, 1.

Fujita, H. (1993) *Macromolecules*, **26**, 643 & 4720.

Fujita, H., Kishimoto, A. and Matsumoto, K. (1960) *Trans. Faraday Soc.*, **58**, 424.

Ganesh, K., Narajan, R. and Duda, J.L. (1992) *Ind. Eng. Chem. Res.*, **31**, 746.

Graham, T. (1866) *Phil. Mag.*, **32**, 401.

Glasstone, S., Laidler, K.J. and Eyring, H. (1941) *The Theory of Rate Processes*, McGraw-Hill, New York, p. 524.

Guo, C.J., DeKee, D. and Harrison, B. (1992) *Chem. Eng. Sci.*, **47**, 1525.

Gusev, A.A., Arizzi, S., Suter, U.W. and Moll, D.J. (1993a) *J. Chem. Phys.*, **99**, 2221.

Gusev, A.A., Müller-Plathe, F., van Gunsteren, W.F. and Suter, U.W. (1993b) *Adv. Polym. Sci.*, **116**, 207.

Gusev, A.A. and Suter, U.W. (1991) *Phys. Rev.*, **A43**, 6488.

Gusev, A.A. and Suter, U.W. (1992) *Polym. Preprints*, **16**, 631.

Gusev, A.A. and Suter, U.W. (1993a) *Makromol. Chem., Macromol. Symp.*, **69**, 229.

Gusev, A. A. and Suter, U. W. (1993b) *J. Chem. Phys.*, **99**, 2228.

Han, J. and Boyd, R.H. (1996) *Polymer*, **37**, 1797.

Hauschild, G. and Spingler, E. (1988) *Migration bei Kunststoffverpackungen*, Wissenschaftliche VGmbH, Stuttgart.

Haward R.N. (ed.) (1973) *The Physics of Glassy Polymers*, Applied Science Publ. London.

Hedenqvist, M.S., Bharadwaj, R. and Boyd, R.H. (1998) *Macromolecules*, **31**, 1556.

Hofmann, D., Fritz, L., Ulbrich, J. and Paul, D., (1997) *Polymer*, **38**, 6145.

Hofman, D., Ulbrich, J., Fritsch, D. and Paul, D. (1996) *Polymer*, **37**, 4773.

Holley, R.H. and Hopfenberg, H.B. (1970) *Polym Eng. Sci.*, **10**, 376.

Hirschfelder, J.O., Curtis, C.F. and Byrd, B.R. (1954) *Molecular Theory of Gases and Liquids*, Wiley, New York.

Hwang, S.-T. and Kammermeyer, K. (1975) *Membranes in Separations*, Wiley Interscience, New York.

Jagodic, F., Borstnik, B. and Azman, A. (1973) *Makromol. Chem.*, **173**, 221.

Karlsson, G.E., Johansson, T.S., Gedde, U.W. and Hedenqvist, M.S. (2002) *Macromolecules*, **35**, 7453.

Kokes, R.J., Long, F.A. and Hoard, J.L. (1952) *J. Chem. Phys.*, **20**, 1711.

Koros W.J. (ed.) (1990) *Barrier Polymers and Structures, ACS Symp. Ser. 423*, American Chemical Society, Washington.

Kosiyanon, R. and McGregor, R. (1981) *J. Appl. Polym. Sci.*, **26**, 629.

Landau, L.D. and Lifshitz, E.M. (1967) *Physique Theorique*, Mir, Moscow.

Lodge, T.P., Lee, J.A. and Frick, T.S. (1990) *J. Polym. Sci., Polym. Phys. Ed.*, **28**, 2067.

Lundberg, J.L., Mooney, E.J. and Rogers, C.E. (1969) *J. Polym. Sci.*, **7**, 947.

Mason, E.A. and Lonsdale, H.K. (1990) *J. Membr. Sci.*, **51**, 1.

Masuda, T., Iguchi, Y., Tang, B.-Z. and Higashimura, T. (1988) *Polymer*, **29**, 2041.

Meares, P. (1954) *J. Am. Chem. Soc.*, **76**, 3415.

Meares, P. (1957) *Trans. Faraday Soc.*, **53**, 101.

Meares, P. (1958a) *J. Polym. Sci.*, **27**, 391 & 405.

Meares, P. (1958b) *Trans. Faraday Soc.*, **54**, 40.

Meares, P. (1976) *Membrane Separation Processes*, Elsevier, New York.

Meares, P. (1993) *Eur. Polym. J.*, **29**, 237.

Mercea, P. (1996) Fraunhofer-Institutre IVV, Report 1/97, Munich.

Michaels, A.S., Vieth, W.R. and Barrie, J.A. (1963) *J. Appl. Phys.*, **34**, 1 & 13.

Michaels, A.S. and Bixler, H.J. (1961) *J. Polym. Sci.*, **6**, 413.

Mitchell, J.K. (1831) *Philadelphia J. Med. Sci.*, **13**, 36.

Müller-Plathe, F. (1991a) *J. Chem. Phys.*, **94**, 3192.

Müller-Plathe, F. (1991b) *Chem. Phys. Lett.*, **177**, 527.

Müller-Plathe, F. (1992a) *J. Chem. Phys.*, **94**, 3192.

Müller-Plathe, F. (1992b) *J. Chem. Phys.*, **96**, 3200.

Müller-Plathe, F. (1993) *J. Chem. Phys.*, **98**, 9895.

Müller-Plathe, F. (1994) *Acta Polym.*, **45**, 259.

Müller-Plathe, F. (1996a) *Macromolecules*, **29**, 4782.

Müller-Plathe, F. (1996b) *Macromolecules*, **29**, 4728.

Müller-Plathe, F. and van Gunsteren, W.F. (1997) *Polymer*, **38**, 2259.

Müller-Plathe, F. (1998) *J. Chem. Phys.*, **108**, 8252.

Müller-Plathe, F., Rogers, S.C. and van Gunsteren, W.F. (1992a) *Macromolecules*, **25**, 6722.

Müller-Plathe, F., Rogers, S.C. and van Gunsteren, W.F. (1992b) *Chem. Phys. Letts.*, **199**, 237.

Müller-Plathe, F., Rogers, S.C. and van Gunsteren, W.F. (1993) *J. Chem. Phys.*, **98**, 9895.

Muruganandam, N., Koros, W.J. and Paul, D.R. (1987) *J. Polym. Sci., Polym. Phys. Ed.*, **25**, 1999.

Niemelä V.S., Lepänen, J. and Sundholm, F. (1996) *Polymer*, **37**, 4155.

Okamoto, K., Nishioka, S., Tsuru, S., Sasaki, S., Tanaka, K. and Kita, H. (1988) *Kobunshi Ronbushu*, **45**, 993.

Pace, R.J. and Datyner, A. (1979a) *J. Polym. Sci., Polym. Phys. Ed.*, **17**, 437.

Pace, R.J. and Datyner, A. (1979b) *J. Polym. Sci., Polym. Phys. Ed.*, **17**, 453.

Pace, R.J. and Datyner, A. (1979c) *J. Polym. Sci., Polym. Phys. Ed.*, **17**, 465.

Pace, R.J. and Datyner, A. (1980) *J. Polym. Sci., Polym. Phys. Ed.*, **18**, 1103.

Pant, P.V.K. and Boyd, R.H. (1993) *Macromolecules*, **26**, 679.

Paul, D.R. (1969) *J. Polym. Sci. A-2*, **7**, 1811.

Paul, C.W. (1983) *J. Polym. Sci., Polym. Phys. Ed.*, **21**, 425.

Paul, D.P. and Koros, W.J. (1976) *J. Polym. Sci., Polym. Phys. Ed.*, **14**, 675.

Paul, D.R., Yampol'skii and Yu, P. (1994) *Polymeric Gas Separation Membranes*, CRC Press, Boca Raton, FL.

Petropoulos, J.H. (1970) *J. Polym. Sci., A-2*, **8**, 1797.

Piringer, O. (1993) *Verpackungen für Lebensmitteln*, VCH, München.

Piringer O-.G. and Baner A.L. (eds) (2000) *Plastic Packaging Materials for Food*, Wiley-VCH, Munich.

Rigby, D. and Roe, R.-J. (1988) *J. Chem. Phys.*, **89**, 5280.

Roe, R.-J. (1991) *Computer Simulations of Polymers*, Prentice-Hall, Englewood Cliffs, NJ.

Rogers, C.E. (1965) Solubility diffusivity, in: *Physics Chemistry of the Organic Solid State*, (eds D. Fox, M.M. Labes and A. Weissberger), Interscience, New York. p. 509.

Rogers, C.E. and Machin, D. (1972) *CRC Crit. Rev. Macromol. Sci.*, 245.

Saxena, V. and Stern, S.A. (1982) *J. Membr. Sci.*, **12**, 65.

Schlotter, N.E. and Furlan, P.Y., (1992) *Polymer*, **33**, 3323.

Scott G. (ed.) (1980) *Developments in Polymer Stabilisations-3*, Elsevier, London.

Shah, V.M., Stern, S.A. and Ludovice, P.J. (1989) *Macromolecules*, **22**, 4660.

Skvortsov, A.M., Sariban, A.A. and Birshtein, T.M. (1977) *Vysokomol. Soeden.*, **25**, 1014.

Sok, R.M., Berendsen, H.J.C. and van Gunsteren, W.F. (1992) *J. Chem. Phys.*, **96**, 4699.

Srinivasan, R., Auvil, S.R. and Burban, P.M. (1994) *J. Membr. Sci.*, **86**, 67.

Stannett, V.T., Hopfenberg, H.B. and Petropoulos, J.H. (1972) *MTV Int. Rev. Sci., Macromol. Sci.*, **8**, 329.

Stannett, V.T., Koros, W.J., Paul, D.R., Lonsdale, H.K. and Baker, R.W. (1979) *Adv. Polym. Sci.*, **32**, 71.

Stastna, J. and DeKee, D. (1995) *Transport Properties in Polymers*, Technomic, Lancaster.

Stern, S.A. (1994) *J. Membr. Sci.*, **94**, 1.

Stern, S.A. and Saxena, V. (1980) *J. Membr. Sci.*, **7**, 47.

Stern, S.A., Shah, V.M. and Hardy, B.J. (1987) *J. Polym. Sci., Polym. Phys. Ed.*, **25**, 1263.

Takeuchi, H. (1990) *J. Chem. Phys.*, **93**, 2062 & 4490.

Takeuchi, H. and Okazaki, K. (1990) *J. Chem. Phys.*, **92**, 5643.

Takeuchi, H., Roe, R.-J. and Mark, J.E. (1990) *J. Chem. Phys.*, **93**, 9042.

Tamai, Y., Tanaka, H. and Nakanishi, K. (1994) *Macromolecules*, **27**, 4498.

Tamai, Y., Tanaka, H. and Nakanishi, K. (1995) *Macromolecules*, **28**, 2544.

Tamai, Y., Tanaka, H. and Nakanishi, K. (1996) *Macromolecules*, **29**, 6750 & 6761.

Theodorou, D.N. (1996) in: *Diffusion in Polymers*, (ed P. Neogi), Marcel Decker, New York.

Theodorou, D.N. and Suter, U.W. (1985a) *Macromolecules*, **18**, 1206 & 1467.

Theodorou, D.N. and Suter, U.W. (1985a) *J. Chem. Phys.*, **82**, 955.

Tikhomirov, B.T., Hopfenberg, H.B., Stannett, V.T. and Williams, J.L. (1968) *Die Makromoleculare Chem.*, **118**, 177.

Toi, K., Maeda, Y. and Tokuda, T. (1983) *J. Appl. Polym. Sci.*, **28**, 3589.

Trohalaki, S., Rigby, D., Kloczkowsi, A., Mark, J.E. and Roe, R.-J. (1989) *Polym. Preprints*, **30**, 23.

Tshudy, J.A. and von Frankenberg, C. (1973) *J. Polym. Sci., A-2*, **11**, 2077.

Tuwiner, S.B. (1962) *Diffusion and Membrane Technology*, ACS Monographies, Reinhold, New York.

van Amerongen, G.J. (1950) *J. Polym. Sci.*, **5**, 307.

van Gunsteren W.F. and Weiner P.K. (eds) (1989) *Computer Simulation of Biomolecular Systems*, ESCOM, Leiden.

Vergnaud, J.M. (1991) *Liquid Transport Processes in Polymeric Materials*, Prentice-Hall, Engelwood Cliffs, NJ.

Vieth, W.R. (1991) *Diffusion in and Through Polymers*, Hanser, München.

Vieth, W.R., Howell, J.H. and Hsieh, J.H. (1976) *J. Membr. Sci.*, **1**, 177.

Vieth, W.T. and Sladek, K.J. (1965) *J. Colloid Sci.*, **20**, 1014.

Vitrac, O., Lézervant, J. and Feigendbaum, A. (2006) *J. Appl. Polym. Sci.*, **101**, 2167.

Vrentas, J.S. and Duda, J.L. (1977a) *J. Polym. Sci., Polym. Phys. Ed.*, **15**, 403.

Vrentas, J.S. and Duda, J.L. (1977b) *J. Polym. Sci., Polym. Phys. Ed.*, **15**, 417.

Vrentas, J.S. and Duda, J.L. (1977c) *J. Polym. Sci., Polym. Phys. Ed.*, **15**, 441.

Vrentas, J.S. and Duda, J.L. (1977d) *J. Appl. Polym. Sci.*, **21**, 1715.

Vrentas, J.S. and Duda, J.L. (1978) *J. Appl. Polym. Sci.*, **22**, 2325.

Vrentas, J.S. and Duda, J.L. (1986) in: *Encyclopedia of Polymer Science and Engineering*, 2nd edn., (Vol. 5) Wiley, New York.

Vrentas, J.S., Duda, J.L. and Hou, A.-C. (1987) *J. Appl. Polym. Sci.*, **33**, 2581.

Vrentas, J.S., Duda, J.L. and Ling, H.-C. (1988a) *Macromolecules*, **21**, 1470.

Vrentas, J.S., Duda, J.L. and Ling, H.-C. (1988b) *J. Polym. Sci., Polym. Phys. Ed.*, **26**, 1059.

Vrentas, J.S., Duda, J.L. and Ling, H.-C. (1991) *J. Appl. Polym. Sci.*, **42**, 1931.

Vrentas, J.S., Duda, J.L., Ling, H.-C. and Hou, A.-C. (1985) *J. Polym. Sci., Polym. Phys. Ed.*, **23**, 275 & 289.

Vrentas, J.S., Liu, H.T. and Duda, J.L. (1980) *J. Appl. Polym. Sci.*, **25**, 1297.

Vrentas, J.S. and Vrentas, C.M. (1992) *J. Polym. Sci., Polym. Phys. Ed.*, **30**, 1005.

Vrentas, J.S. and Vrentas, C.M. (1994a) *Macromolecules*, **27**, 4684.

Vrentas, J.S. and Vrentas, C.M. (1994b) *Macromolecules*, **27**, 5570.

Wall, F.T. and Seitz, W.A. (1977) *J. Chem. Phys.*, **67**, 3722.

Wang, T.T., Kwei, K.T. and Frisch, H.L. (1969) *J. Polym. Sci.*, A-2, **7**, 879.

Weber, T.A. and Helfand, E. (1979) *J. Chem. Phys.*, **71**, 4760.

Wheeler, T.S. (1938) *Trans. Nat. Sci. India*, **1**, 333.

Yampol'skii, Yu. P., Bespalova, N.B., Finkel'shtein, E.Sh., Bondar, V.I. and Popov, A.V. (1994) *Macromolecules*, **27**, 2872.

6
A Uniform Model for Prediction of Diffusion Coefficients with Emphasis on Plastic Materials

Otto Piringer

6.1
Introduction

Diffusion is a mass transport process resulting from random molecular motions. Such molecular motions occur in gases and condensed phases and can be described in principle as using the commonly held theoretical picture of "random walk." This means the particles (molecules, atoms) move in a series of small random steps and gradually migrate from their original positions. Each particle can jump through a distance λ in time τ. But the direction of each step may be different, and the net distance traveled must take the changing directions into account. The coefficient of diffusion D is related to λ and τ in the Einstein–Smoluchowski equation:

$$D = \frac{\lambda^2}{2\tau} \tag{6.1}$$

If $\lambda/\tau = \bar{c}$, and λ are interpreted as the mean speed of the particle and the mean free path, then Eq. (6.1) has the same structure as the following equation obtained from the kinetic theory of gases:

$$D_G = \frac{1}{3}\lambda\bar{c} = \frac{1}{3}\left(\frac{kT}{2^{1/2}\sigma p}\right)\left(\frac{8kT}{\pi m}\right)^{1/2} \tag{6.2}$$

where k is the Boltzmann constant, $\sigma = \pi d^2$ is the collision cross-section of a particle with diameter d, $m = uM_r$ is the mass of a particle with its relative molecular weight M_r and atomic mass unit u, p is the pressure, and T the absolute temperature. This implies that the diffusion of a perfect gas is a random walk with an average step size equal to the mean free path. The value of the self-diffusion coefficient D_G can be obtained with two specific parameters, σ and m, for the diffusing particle.

Plastic Packaging. Second Edition. Edited by O.G. Piringer and A.L. Baner
Copyright © 2008 WILEY-VCH Verlag GmbH & Co. KGaA, Weinheim
ISBN: 978-3-527-31455-3

From a detailed analysis of molecular motion in dilute gases, it is observed that a much better prediction of diffusion coefficients results with the Chapman–Enskog equation. This equation, which describes a mixture of two solutes A and B (binary gas system), is

$$D_{G,AB} = \frac{3}{16}\sqrt{\frac{4\pi kT}{M_{AB}}}\frac{1}{n\pi\sigma_{AB}^2 \Omega_D} f_D \qquad (6.3)$$

where $M_{AB} = 2[(1/M_A) + (1/M_B)]^{-1}$, with molecular weights M_A, and M_B, σ_{AB} is a characteristic length, n is the number density of molecules in the mixture, Ω_D is the diffusion collision integral, and f_D is the correction term with the order of unity.

The collision integral for diffusion depends upon the choice of the intermolecular force law between colliding molecules and is a function of temperature. It is observed that the characteristic length depends upon the intermolecular force law selected. In comparison with the simple Eq. (6.2) for perfect gases, Eq. (6.3) takes into account the interactive forces between real molecules. But, while in the first case only two specific parameters are needed, the diffusion collision integral, Ω_D, in the second case, is a complicated function of several parameters.

Whereas λ and τ are characteristic parameters of microscopic particle motion, the diffusion coefficient D is a specific macroscopic parameter. The main difficulty in predicting D for condensed systems lies in the derivation of correct expressions for the microscopic parameters.

The most common basis for estimating diffusion coefficients in liquids is the Stokes–Einstein equation:

$$D_L = \frac{kT}{6\pi\eta a} \qquad (6.4)$$

in which D_L is related to another macroscopic parameter, the viscosity η of the medium. The solute radius is denoted by a. It must not be forgotten, however, that Eq. (6.4) was derived for a very special situation, in which the solute is much larger than the solvent molecule. Nevertheless, many authors have used the form of Eq. (6.4) as a starting point in developing empirical predictions. A significant difference in Eq. (6.4) compared to Eqs. (6.2) and (6.3) for gases lies in the effect of temperature on the diffusion coefficient in liquids. This can be assumed via the viscosity η to be an exponential function of the form $A\exp(-B/T)$.

Diffusion coefficients in solids are commonly expressed as exponential functions:

$$D_S = D_0 \exp\left(-\frac{E_A}{RT}\right) \qquad (6.5)$$

where the pre-exponential athermal factor D_0 and the molar activation energy E_A are estimated empirically (Murch 1995).

As shown in Chapter 5, the diffusion behavior in polymers lies between that of liquids and solids. As a consequence, the models for diffusion coefficient estimation are based on ideas drawn from diffusion in both liquids and solids.

In general, diffusion coefficients in gases can often be predicted accurately (Kestin et al. 1984). Predictions of diffusion coefficients in liquids are also possible using the Stokes–Einstein equation or its empirical parallels. On the contrary in solids and polymers, models allow coefficients to be correlated but predictions are rarely possible (Cussler 1997).

The aim of this chapter is the development of a uniform model for predicting diffusion coefficients in gases and condensed phases, including plastic materials. The starting point is a macroscopic system of identical particles in the gaseous state. All particles in the gas are treated as monoatomic particles. However, the main distinction of this treatment from the kinetic theory of perfect gases lies in taking interactions between particles into consideration from the beginning. The difference from the treatment of perfect gases is accounted for by two parameters, the critical temperature, T_c and the critical molar volume, V_c. The self-diffusion of noble gases is used for validation of this model.

Liquid systems occupy a special place between gaseous and solid phases. The high mobility of particles in a liquid and the values of diffusion coefficients falling in a narrow range, support the assumption of a nearly constant free volume, as a prerequisite for the manifestation of the typical properties of a liquid phase. Equations for diffusion coefficients in the liquid state can be achieved from both sides, the gaseous and solid state.

The self-diffusion of metals in the solid state refers also to a system with monoatomic particles, but in another extreme situation regarding the interaction processes. For diffusion in solid states, the melting temperature, T_m, of the condensed phase is a basic specific parameter instead of the critical temperature for gases.

Whereas the noble gases and metals are simple systems with monoatomic particles, polymers are much more complicated in structure. An amorphous polymer phase is considered to be a pile having maximum disorder amongst the particles. Such a close-packed assemblage has some similarity with liquids. An amorphous solid phase with interacting particles, having a certain degree of mobility, is considered to be essential for the diffusion process in plastic materials. For these systems, the homologous series of n-alkanes is used as a reference class of chemical compounds, which asymptotically reach an unlimited molecular chain in the form of polymethylene. The structures of the individual members of this series do not deviate significantly from one another. Out of all known classes of chemical compounds–the saturated open-chain and nonbranched (normal) hydrocarbons–the n-alkanes or n-paraffins, represent the most frequently studied homologous series with the largest number of available members in pure form. The number i of carbon atoms in n-alkane or more exactly, the i structural methylene groups including the two methyl end groups, can be interpreted in such comparisons as playing the role of i identical subunits, which compose the molecule. The properties of a macroscopic n-alkane sample can then be described as a function of i subunits making up every molecule of the macroscopic system. The homologous series of n-alkanes is used as the reference backbone. Each member of the series represents a macroscopic system of identical particles (the n-alkane molecules) and the particles of the system themselves represent subsystems of i uniform structural units. At room temperature, these systems in all physical states occur among the series of n-alkanes.

The self-diffusion coefficients in solid and liquid alkanes are also derived theoretically, as well as the diffusion coefficients of additives in polyethylene, which is considered as the reference structure for all polymers.

6.2
Interaction Model

As mentioned in the introduction, the discussion starts from a macroscopic system composed of $n \gg 1$ identical particles. Among the particles there exists an attractive interaction that is responsible for the formation of condensed phases. The particles on the other hand possess a certain degree of freedom of motion in any direction within the system.

The background process in all interactions is an energy exchange between the n particles of the system.

6.2.1
Model Assumptions

The following assumptions apply to systems of n identical particles (molecules, atoms) that leave their chemical identities essentially unchanged. Above a certain number of particles in the system, the sum of all interactions on a single particle by the other particles of the system becomes independent of the number of particles. This allows the identification of a macroscopic system by its specific properties, e.g., its melting point.

1. The interactions between the n particles are based on an exchange of finite values $\varepsilon_r = \varepsilon/\varepsilon_0$ of energies, where ε is relative to a reference amount ε_0. The consequence of this exchange is a relative density of interaction energy, $q_{r,n} = (1 + \varepsilon_r/n)^n$, in the form of a n-fold product with the limit value $q_r = \exp(\varepsilon_r)$, for $n \to \infty$. The exponential expression is assumed because:

    - it reaches a constant value, independent of the number of particles in a macroscopic system with $n \gg 1$.
    - the exponential function represents a mathematical order of magnitude which is higher than that of any other power of ε_r, this means $\exp(\varepsilon_r/\varepsilon_r^a) \to \infty$ for $a > 1$ and $\varepsilon_r \to \infty$.

The relative density q_r is responsible for specific properties of a macroscopic system, e.g., its melting point and critical temperature and again is considered as a starting point for additional dynamic processes occurring in the system. Self-diffusion of the particles is an example of such a process. Based on the same mathematical assumption, the magnitude of the diffusion coefficient $D = D_u \exp(q_r)$ is derived as an exponential function of q_r with a unit amount D_u. This assumption is further supported by many empirically established equations describing dynamic properties of macroscopic systems.

2. The exchange of energy between the interacting particles is propagated by periodical processes. The proportional factor 2π takes this into account, where the relative energy transferred from one particle during an interaction step is written as $\varepsilon_r = \varepsilon/\varepsilon_0 = 2\pi(\varepsilon_0 + a)/\varepsilon_0 = 2\pi + \alpha$, with a specific parameter α of the system.

3. Mathematically, the exchange of energy between n particles can be treated as a permutation. The whole process can be understood formally as a result of $n!$ consecutively occurring in the individual interaction steps. Each individual step represents a transport of one energy quantum from one particle to another and is interpreted mathematically as one change of places between the two particles. The total number of such place exchanges equals $n!$. The relative number, p_n, of exchanges related to $n!$ in which no particle remains in its starting position is then:

$$p_n = \frac{1}{2!} - \frac{1}{3!} + \cdots + (-1)^n \frac{1}{n!} \qquad \lim_{n \to \infty} p_n = p_e = \frac{1}{e} \qquad (6.6)$$

The limit value $p_e = 1/e$ for $n \gg 1$ is designated as the maximal probability of a place exchange in a macroscopic system

With these assumptions a common characteristic of all macroscopic particle systems can be expressed as $q_r = \exp(\varepsilon_r) = \exp(2\pi + \alpha)$ (assumptions 1 and 2). However, taking into consideration a diminution of ε_r, which is proportional to the maximum probability p_e of place exchange (assumption 3), the value $q_r = \exp(\varepsilon_r p_e) = \exp[(2\pi + \alpha)/e] = e^{\alpha/e} e^{2\pi/e} = C^{1/e} w$ becomes the specific relative density of interaction energy with the specific parameter $C^{1/e} = e^{\alpha/e}$ for a system and with $w = e^{2\pi/e}$.

4. A special case occurs whenever the n particles are molecules from a homologous series of chemical compounds, e.g., n-alkanes. In such a case the specific parameter C can be enlarged into $C = C_0(1 + 2\pi/i)^i$, where i represents the number of carbon atoms in an unbranched alkane chain and $C^{1/e} = C_0^{1/e}(1 + 2\pi/i)^{i/e} = Kw_{i,e}$ results with a common specific parameter $K = C_0^{1/e}$ for all members of the homologous series. The product $(1 + 2\pi/i)^i$ is a consequence of assumptions 1 and 2 and increases asymptotically to the limit $e^{2\pi}$ for very long molecular chains ($i \gg 1$). The i substructures in the form of methylene groups $-CH_2-$, including the two methyl groups $-CH_3$ at the ends of a n-alkane molecule, do manifest their relative individualities in the form of a multiplicative effect on the interaction intensity between the molecules of the macroscopic system.

Summarizing the above results, the relative density $q_{r,i}$ of interaction energy between the particles of a macroscopic system of n-alkane with i carbon atoms in the molecular chain is

$$q_{r,i} = Kw_{i,e}w = C_0^{1/e}\left(1 + \frac{2\pi}{i}\right)^{i/e} e^{2\pi/e} \qquad (6.7)$$

If a specific property $f(i)$ of the macroscopic system can be correlated with $q_{r,i}$ in the form of a direct proportionality, then a simple dimensionless relationship between the values of this property for one member i and a member with $i \gg 1$ of the homologous series results from Eq. (6.7):

$$\frac{f(i)}{f(\infty)} = \frac{q_{r,i}}{q_{r,\infty}} = \frac{(1+2\pi/i)^{i/e}}{e^{2\pi/e}} = \frac{w_{i,e}}{w} \tag{6.8}$$

The number $w = e^{2\pi/e}$ derived from the above assumptions of the model is the common limit value of the two power sequences:

$$w_n = \left(1 + \frac{2\pi}{n}\right)^{\frac{n}{(1+1/n)^n}} \quad \text{and} \quad w_{i,e} = \left(1 + \frac{2\pi}{i}\right)^{i/e}$$

$$\text{with} \quad \lim_{n \to \infty} w_n = \lim_{n \to \infty} w_{i,e} = e^{2\pi/e} = w \tag{6.9}$$

These power sequences, designated as interaction functions, represent the mathematical backbone of the model described in this chapter.

6.3
Prerequisites for Diffusion Coefficients

This subchapter may be skipped by readers interested only in the resulting equations for diffusion coefficients. But it is important for understanding the ideas which have led to these equations.

In order to support the above assumptions and the interaction functions, Eq. (6.9) was developed from the model. The next three sections contain direct correlations of the relative density of interaction energy, q_r, with specific properties of macroscopic systems.

6.3.1
Critical Temperatures of *n*-Alkanes

The critical temperature may be considered to be a measure of the intensity of interaction between n particles of a system, as produced by van der Waals forces. Critical temperatures are especially suitable for the comparison of numerical values within a homologous sequence because at these temperatures the systems are in corresponding states. The essential characteristic of a particle system in its liquid state is the unordered translational movement of the particles in comparison to the particle vibrations in the x, y, and z directions in a solid. In a liquid system there is a continuous exchange of places which occurs between the system's particles and there are no particular spatial orientations. These considerations greatly simplify the problem of modeling liquid systems.

Although the critical temperature for $n \gg 1$ is practically independent of the number of particles, there exists a possibility for estimating the influence of the number of i structural subunits composing a particle based on the value of the critical temperature of a macroscopic system. If $T_{c,i}$ represents the critical temperatures of an *n*-alkane containing i carbon atoms, we may tentatively let the dimensionless ratio $T_{c,i}/T_{c,\infty}$ be equal to the ratio of the two corresponding interaction functions $w_{i,e}$ and w in Eq. (6.8), due to the common specific parameter $K = C_0^{1/e}$ for all members of the homologous series:

$$\frac{T_{c,i}}{T_{c,\infty}} = \frac{\left(1+\frac{2\pi}{i}\right)^{i/e}}{e^{2\pi/e}} = \frac{w_{i,e}}{w} \tag{6.10}$$

Experimental values for the critical temperatures of n-alkanes are known up to eicosane ($i = 20$) (Reid et al. 1987). For longer molecular chains, the experimental determination of the critical temperature is not possible with sufficient accuracy due to the onset of thermal decomposition.

By means of Eq. (6.10) it is possible to calculate, starting from each experimental value corresponding to i carbon atoms, a limit value $T_{c,\infty}$, (Table 6.1). Due to the fact that the terminal methyl groups in the initial members of the n-alkane represent an important deviation from a system containing only methylene groups, it is more convenient to use alkanes having chains as long as possible for the determination of $T_{c,\infty}$. As seen in Table 6.1 these deviations become unimportant after $i = 9$. This is because the individual $T_{c,\infty}$ values are irregularly distributed for the 12 longest chains ($i = 9 - 20$). The mean limit value obtained from Table 6.1 is $\bar{T}_{c,\infty} = T_c = 1036.2$ K. Figure 6.1 shows the estimated curve for $T_{c,i}$ from Eq. (6.10) using $T_{c,\infty} = T_c = 1036.2$ K as well as the experimental values of $T_{c,i}$ for $1 \leq i \leq 20$.

The remarkable coincidence between the ratios of the critical temperatures, $T_{c,i}/T_c$ within the homologous series and the ratios of the corresponding values of the interaction function $w_{i,e}/w$ supports the interpretation that this function is a measure of the energy density of interaction.

Due to the translation and rotation of particles in the liquid state of a macroscopic system, the value of the interaction function may be assumed to be independent of

Table 6.1 Critical temperatures of n-alkanes.

i	T_c, (K)	$T_{c,\infty}$ (K)	$T_c - T_{c,\infty}$
9	594.6	1039.1	−2.9
10	617.7	1036.8	−0.6
11	638.8	1035.5	+0.7
12	658.2	1034.9	+1.3
13	676	1034.8	+1.4
14	693	1036.0	+0.2
15	707	1034.7	+1.5
16	722	1036.7	−0.5
17	733	1034.4	+1.8
18	748	1039.2	−3.0
19	756	1035.4	+0.8
20	767	1036.8	−0.6

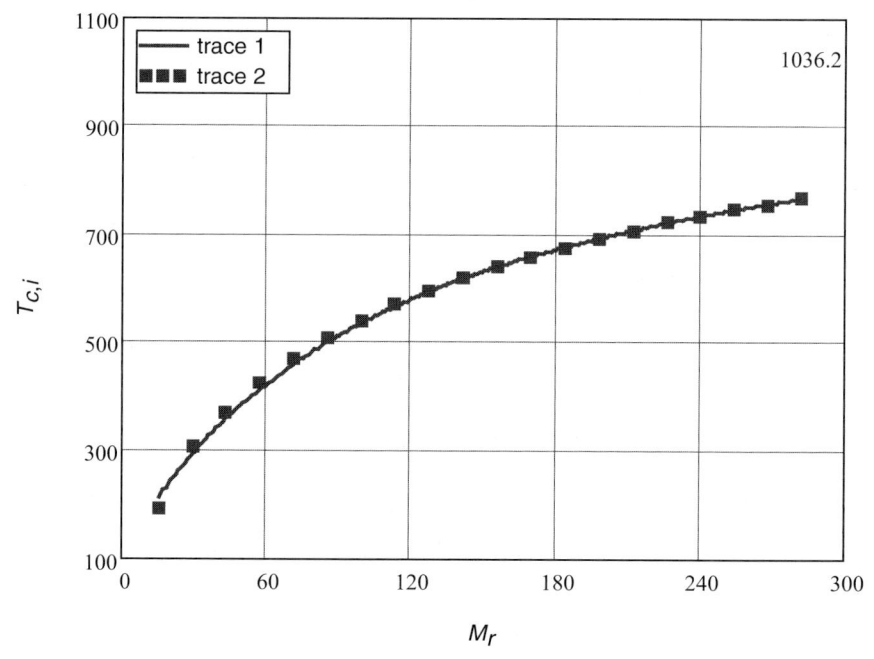

Figure 6.1 Critical temperatures, $T_{c,i}$ of n-alkanes as a function of the relative molecular mass, $M_r = (14i + 2)$. Calculated values using Eq. (6.10) (trace 1), experimental values (trace 2).

the configuration of the particles within the system. Therefore, there is no need for data related to orientation. This is also valid for the i chainlike subunits of an alkane molecule. Due to the possibility of a free rotation of any of the i subunits around the bond axis with the neighboring subunits, a relative motion of segments of several subunits is also possible.

6.3.2
Melting Temperatures of n-Alkanes

The melting point of a pure substance is one of the most important specific macroscopic properties for its characterization and identification. It is also a specific parameter for crystalline polymers. In a macroscopic crystal composed of n-alkane molecules the individual alkane chains lie in arrangements laid out next to one another. Very long chains can be folded but still have chain segments lying next to one another. The result of such an arrangement is a structure composed of lamellae lying on top of one another. The thickness of the lamellae is determined by the chain length. Compared to its thickness, the length and width of lamellae in principal can be of any size. The crystal in the direction of the extended chains is therefore composed of any number of layers (lamellae).

When the amplitude of vibration of the particles reaches a certain fraction of the distance from their neighbors, the crystal melts. Since there are different orientations along the coordination axis, one can assume that the vibrations in the x, y, and z directions are not equal. This propagation of interactions in the solid state is significantly different from the case in liquids.

The melting points of macroscopic alkane crystal samples are determined mainly by the chain length of the molecules, e.g., by the number i of the methylene groups (particles) including the two methyl end groups. If one lays an alkane chain along the x-axis of the coordination system, then one can express the x-direction material specific constant $C_x^{1/e}$ with the help of an analogue of the product in the form $C_x^{1/e} = C_{0,x}^{1/e}(1+2\pi/i)^{i/e} = K_x w_{i,e}$. The material constant K_x is a specific parameter for the alkane homologous series.

The molecule chains, each of which are composed of i particles are assumed to be oriented along the x-axis and are bound to one another head to head by van der Waals forces. A different orientation exists in the other two coordinate axis directions where van der Waals forces exist between the methylene group (particle) of one molecule to that in the corresponding neighbor particle.

If the melting point of a macroscopic alkane sample is determined solely by the lamellae thickness of its crystal, then one can set the melting point, $T_{m,i}$, of an alkane with i C atoms proportional to the value of the x-component $q_{r,i}(x) = K_x w_{i,e} w$ Eq. (6.7). However, the interactions occurring in both the y and z axes must also be taken into account. The strength of these interactions is comparable to those in a system composed of particles similar in size and structure to the methylene groups, that means for particles with $i = 1$. But whereas $K_x w_{i,e}$ with $i > 1$ refers to molecules with covalent bonds between the subparticles i in the x-direction, no such bonds exist in the y and z-directions. Therefore the value of the constant K_x must not be the same as $K_y = K_z$. Taking this into account, a relative interaction contribution in the y- and z-axis, respectively, of $q_{r,i}(y) = q_{r,i}(z) = (K_x/w)w_{1,e}w$ is assumed as an approximation. For the relative contribution $K_y = K_z = K_x/w$ of the interactions in the y- and z-directions the w-value for an infinite methylene chain is chosen as reference. Even the constant K_x may increase slightly with i at the beginning of the homologous series, but reaches a constant value for $i \gg 1$. Then the ratio of the melting point for an alkane with i C atoms, $T_{m,i}$ and the limit value, $T_{m,\infty}$ for $i \gg 1$ can be represented by the following equation:

$$\frac{T_{m,i}}{T_{m,\infty}} = \frac{2 \cdot (K_x/w) \cdot w_{1,e} \cdot w + K_x \cdot w_{i,e} \cdot w}{2 \cdot (K_x/w) \cdot w_{1,e} \cdot w + K_x \cdot w \cdot w} = \frac{2(w_{1,e}/w) + w_{i,e}}{2(w_{1,e}/w) + w} \quad (6.11)$$

In order to determine the melting point $T_{m,\infty}$, an accurate melting point of a known alkane is necessary. The first few members of the homologous series do not show a regular increase in their melting points with carbon number. This irregularity is caused by crystal structural differences between molecules with even and odd i values as well as by the methyl end groups. To avoid these effects one should select as large an alkane molecule as possible as a reference. On the other hand shorter chains have sharp melting points, i.e., the transition from solid to liquid state occurs within a small temperature interval, e.g., approximately 0.25 K for $i = 44$. A slower, more gradual melting process is observed for longer chains which can lead to large measurement errors when not taken adequately accounted for. In a related systematic study (Stack et al. 1989) a melt temperature of $T_{m,i} = 400.65$ K was determined at the equilibrium between the melted and crystalline state for synthesized straight chain alkane, $C_{192}H_{386}$. With the value for $i = 192$ as reference a value of $T_{m,\infty} = 415.14$ K was calculated using Eq. (6.11). In Table 6.2 and Figure 6.2, melting temperatures $T_{m,i}$ of the unbranched alkanes, calculated with Eq. (6.11) are compared with experimentally measured values (Broadhurst 1962). The reason for the deviation of the measured melting points to smaller values than the calculated temperatures for short chains ($i < 24$) are due to the above mentioned structural differences and the slightly increase of K_x with i.

In a model (Hagemann and Rothfus 1979), which incorporates the possible crystal systems into which alkanes can crystallize (hexagonal, orthorhombic, monoclinic or triclinic), the possible interactions between all atoms of an alkane were calculated and summed using the known distances between atoms and bond angles between H–C and C–C. In this way results are expressed as a function of chain length for every crystal system. Using nonlinear (parabolic) curve fitting onto the experimental melting point curves for the triclinic system in the range from $26 \leq i \leq 100$ the limit value for an infinitely long alkane chain was found to be $T_{m,\infty} = 415.55$ K. This value is used in Table 6.2 as the limit melting temperature for unfolded polymethylene.

Table 6.2 Comparison of calculated and experimental melting temperatures, $T_{m,i}$ for n-alkanes with $i > 23$ carbon atoms.

i	$T_{m,i}$ (K) calc	$T_{m,i}$ (K) exp	ΔT calc–exp
24	324.6	324.0	+0.6
25	327.4	326.7	+0.7
28	334.7	334.6	+0.1
29	336.9	336.9	0.0
31	341	341	0.0
32	342.8	343.2	−0.4
34	346.3	346	+0.3
35	347.9	347.9	0.0
37	350.9	350.9	0.0
38	352.3	352.2	+0.1
40	355	354.7	+0.3
43	358.5	358.3	+0.2
44	359.6	359.5	+0.1
46	361.7	362.2	−0.5
50	365.4	365.3	+0.1
58	371.5	371.4	+0.1
64	375.2	375.2	0.0
66	376.3	376.8	−0.5
67	376.8	377.3	−0.5
70	378.3	378.4	−0.1
82	383.3	383.5	−0.2
94	387.1	387	+0.1
100	388.7	388.4	+0.3
192	401.0	400.7	+0.3
720	411.6	411.7	−0.1

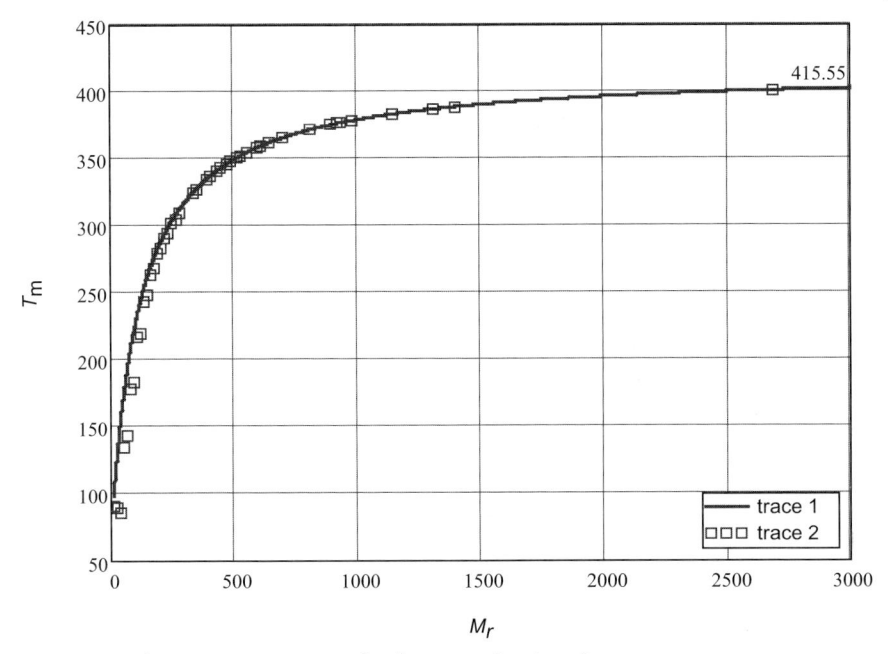

Figure 6.2 Melting temperatures $T_{m,i}$ of n-alkanes as a function of the relative molecular mass, $M_r = (14i + 2)$. Calculated values with Eq. (6.11) (trace 1) and experimental value (trace 2).

When alkane chains become long enough ($i > 200$) they start to fold. The layer thickness of a lamella is then no longer proportional to the total chain length but rather proportional to the length of the chain segment between a CH_3 end group and a folding point or the segment between two folding points. Because the thickness of the lamella and consequently the number of C atoms in a chain segment is responsible for the melting process of an alkane, the melting point can lie below the corresponding i value of the alkane. A melt temperature of $T_{m,\infty} = 411.65$ K was measured for a particularly carefully crystallized polyethylene (Flory and Vrij 1963). A segment length of approximately 720 C atoms was found from the measured layer thickness in these samples. With Eq. (6.11) one obtains a melting point of $T_{m,i} = 411.57$ for $i = 720$, with $T_{m,\infty}$ 415.55 K. However, the melting point temperature curve for large i values changes very little with each additional chain segment, e.g., 0.1 K for $\Delta i = 20$ (Table 6.2).

6.3.3
Melting Temperatures of Atom Clusters

Spherical particles crystallize preferably into one of the densest spherical packing, e.g., argon and gold atoms are cubic close-packed (ccp). The ccp structure gives rise to a face-centered unit cell, so may also be denoted cubic F (or fcc, for face-centered cubic). In a cluster, assumed to be approximately spherical and composed of N spherical particles, the diameter of the cluster behaves as though $d = N^{1/3} d_0$, that means it is composed of approximately $N^{1/3}$ particles with diameter d_0 arranged along one coordination axis. In a ccp system each atom is surrounded by 12 atoms, which means $12 + 1$ atoms are in direct contact but only $13/3$ along one axis. Because each atom has two neighbors along one axis, the total number of interacting atoms along one axis is $n = (13/6)N^{1/3} = 2.17 N^{1/3}$. If the difference in interactions between particles on the surface and inside the cluster is neglected, then it can be assumed in the case of spherically shaped particles that no anisotropy occurs in the properties of the cluster.

When the amplitude of vibration of the particles reaches a certain fraction of the distance from their neighbors, the crystal melts. Since there is no preferred orientation, one can assume that the vibration in the x, y, and z directions are equal. When one disregards any material specific characteristics in the first approximation so that the specific parameter $C = 1$, then all systems composed of spherical particles can be treated in a similar manner. The relationship of the melting point temperature $T_{m,N}$ of a cluster composed of N spherical particles to the melting point T_m of a macroscopic sample composed of the same type of particles can be correlated with a relation similar to Eq. (6.8), where instead of the interaction function $w_{i,e}$ for macroscopic systems, the interaction function w_n defined in Eq. (6.9) for n particles interacting along one coordination axis is used:

$$\frac{T_{m,N}}{T_m} \cong \frac{3w_n}{3w} \quad \text{with} \quad w_n = \left(1 + \frac{2\pi}{n}\right)^{\frac{n}{(1+1/n)^n}} \quad \text{and} \quad n = 2.17 N^{1/3} \quad (6.12)$$

The factor 3 indicates the contributions of the interactions along the three coordination axis. To test the application of the expression in Eq. (6.12), published data (Buffat and Borel 1976) measuring the melting of gold clusters ranging in diameter from 2 to 25 nm can be used. The number of atoms (particles) N in such large sited clusters lies between 300 to over 2 million. Even the melting points of such large clusters are significantly lower than the melting point of macroscopic gold samples where $T_m = 1336$ K. With an atomic diameter of gold, $d_0 = 0.288$ nm, the particle number n used in Eq. (6.12) ranges from 8 to 80. In Table 6.3 the average values of published

Table 6.3 Calculated and experimental melting points of atomic gold clusters as a function of the particle number and in comparison with the melting point of gold, $T_m = 1336$ K.

d (nm)	$N^{1/3}$	$T_{m,N}$ (K) calc	$T_{m,N}$ (K) exp
2.5	8.69	1025	500
5.0	17.37	1155	1088
7.5	26.06	1208	1177
10.0	34.75	1237	1240
12.5	43.43	1256	1256
15.0	52.12	1268	1268
17.5	60.81	1277	1277
20.0	69.49	1284	1284
22.5	78.18	1290	1290

experimental melting point data obtained for different cluster diameters d are presented in comparison with the calculated $T_{m,N}$ values. At cluster diameters below 5 nm the differences between calculated and measured temperatures become significant due to additional effects, e.g., a different behavior of atoms at the surface or errors in the temperature measurements of very small particles.

The following three sections contain direct correlations between certain values of the interaction functions defined in Eq. (6.9) and other important parameters which characterize macroscopic systems.

6.3.4
Critical Compression Factor

The first term $w_1 = (1 + 2\pi)^{1/2}$ of the power series w_n defined in Eq. (6.9) plays a special role within the interaction model. It can be assumed to represent the relative molar energy of a perfect gas phase in which no attractive interaction between the particles occur. But as a member of the series this value is related to the other members and especially to the limit value w. Then the limit value w represents the relative molar energy of the same system in the condensed state, resulted from the attractive interactions between the particles. If V_c, p_c, T_c, and R represent the critical molar volume of a compound, the critical pressure and critical temperature of the system and the gas constant, respectively, then the ratio w_1/w can be identified as the critical compression factor, Z_c:

$$\frac{p_c V_c}{RT_c} = Z_c = \frac{w_1}{w} = \frac{(1+2\pi)^{1/2}}{e^{2\pi/e}} = 0.2675 \qquad (6.13)$$

Taking into account an empty (free) volume fraction in the critical state, the critical molar volume is written as $V_c = w_1 V_0 = 2.7 V_0$, where V_0 represents the molar volume of the particles in the system which is comparable with the constant b in the van der Waals equation, with $V_c = 3b$.

From a data collection with 349 experimental values for the critical compression factor (Reid et al. 1987) obtained with organic and inorganic compounds and elements, a mean value of $Z_c = 0.2655$ is obtained with a standard deviation of $\sigma = 0.0346$.

The critical constants obtained from the van der Waals equation, $V_c = 3b$, $p_c = a/27b^2$ and $T_c = 8a/27Rb$ deliver a critical compression factor $Z_c = 3/8 = 0.375 > 0.2675$, which is higher than the mean experimental value.

6.3.5
The Entropy of Evaporation

Systems with comparable amounts of disorder are especially important for developing a common basis for relationships between diffusion coefficients. Such a comparable amount of disorder is generated when any liquid evaporates and becomes a gas. According to Trouton's rule the entropy of evaporation has values around 85 J K^{-1} mol^{-1} for many liquids at their boiling point T_b at a standard pressure of 1 bar. This rule was modified by Hildebrand in 1915 (Hildebrand et al. 1970). According to Hildebrand, the value of the molar entropy of evaporation, ΔS_v, for many substances is nearly the same at temperatures where their molar vapor volumes are equal to the standard value of $24.8\,\mathrm{dm^3\,mol^{-1}}$ at 25 °C. The validity of this rule extends over boiling points ranging over three orders of magnitude and for classes of substances as different as monoatomic noble gases, high boiling metals, and compounds with polyatomic molecules with complex structures. The deviation from the mean value of $84.9\,\mathrm{J\,K^{-1}\,mol^{-1}}$ does not exceed $1.5\,\mathrm{J\,K^{-1}\,mol^{-1}}$ with few exceptions when using the Hildebrand correction.

As a conclusion from the Hildebrand/Trouton rule, the definition of a standard vapor phase in a standard state with a well-known amount of disorder, related to the corresponding liquid phase, can be made. This definition can be used as a starting point for modeling diffusion coefficients of gases.

The change in entropy ΔS for a reversible isothermal expansion of an ideal gas from its initial volume V_1 to a volume V_2 is $\Delta S = R \ln(V_2/V_1)$ and therefore $V_2/V_1 = \exp(\Delta S/R)$. By setting V_2/V_1 equal to the ratio between the molar volume $V_G^\circ = 24.8\,\mathrm{dm^3\,mol^{-1}}$ of an ideal gas under standard conditions ($T = 298.15$ K, $p = 1$ bar) and assigning a volume V_L° to one mole of a liquid at T_h, then $V_G^\circ/V_L^\circ = \exp(\Delta S_v/R) = \exp(\Delta H_v/RT_h)$. The value ΔH_v stands for the molar enthalpy of evaporation at the Hildebrand temperature, T_h, and ΔS_v is the molar entropy of evaporation. By using $\Delta S_v = 84.9\,\mathrm{J\,K^{-1}\,mol^{-1}}$ the value $V_L^\circ = 0.91\,\mathrm{cm^3\,mol^{-1}}$ is obtained. An interpretation of the Hildebrand/Trouton Rule is that this "free" volume, V_L°, allows for the freedom of movement of molecules (particles) necessary for the liquid state at the temperature T_h. The explanation of the constant entropy of evaporation is that it takes into account only the translational entropy of the vapor and the liquid. It has to be pointed out that V_L° does not represent the real molar volume of a liquid, but designates only a fraction of the corresponding molar volume of an ideal gas V_G°, derived from the entropy of evaporation. The real molar volume V_L of the liquid contains in addition the molar volume occupied by the molecules V_0. As a result the following relations are valid: $V_L = V_L^\circ + V_0$ and $V_G = V_G^\circ + V_0$. However while $V_L^\circ < V_0$ and V_L is practically independent of the pressure, $V_0 \ll V_G^\circ$ in the gaseous phase. Only in the critical state does $V_G/V_L = 1$ and the entropy difference between the two phases vanishes.

In the next section the correlation of the molar entropy of evaporation and the molar volumes with the interaction functions will be shown.

6.3.6
The Reference Temperature and the Reference Molar Volume

Now let us consider a macroscopic system in which the particles are members of a homologous series of the kind defined above in the 4th assumption. In this case the particles themselves represent subsystems made of chains with i interacting particles, e.g., atoms or group of atoms. The relative density of interaction energy, $q_{r,i}$ in such a system is defined in Eq. (6.8). But as already mentioned in Section 6.3.2 the first term $w_{1,e}$ plays an outstanding role in the series $w_{i,e}$ because no chains with covalent bonds between the subparticles i exist for $w_{1,e}$. This situation is similar to the outstanding role of the first term w_1, which represents perfect gases without any interactions, in contrast with the other terms in the series w_n. But w is the limit value of both the series w_n and $w_{i,e}$. The difference $w_1 - w_{1,e}$ between the two first terms of the two interaction series is interpreted to be a pure translational contribution to the energy of the system and can therefore be related to its temperature. In order to establish a connection between the difference $w_1 - w_{1,e}$ and SI units let us start with $n = 1000$ mol of hypothetical particles with the atomic mass unit $u = 1.66054 \times 10^{-27}$ kg in a volume $V = 1$ m^3. This system with $N = nN_A = 10^3$ (mol) 6.02214×10^{23} (mol^{-1}) particles defines the mass unit of $Nu = 1$ kg. The total energy of the system is $E = \sum n_i \varepsilon_i$ where n_i particles are in the state with energy ε_i. Using the Boltzmann distribution and $E = (N/q)\Sigma \varepsilon_i \exp(-\beta\varepsilon_i)$ with $\beta = 1/kT$, the following relationship between the internal energy of the system and the two terms w_1 and $w_{1,e}$ is postulated to be:

$$\frac{E}{\sum_i \varepsilon_i e^{-\beta\varepsilon_i}} = \frac{N}{q} = w_1 - w_{1,e} = (1+2\pi)^{1/2} - (1+2\pi)^{1/e} = 0.6227236 \quad (6.14)$$

With the translational partition function $q = V/\lambda^3$ and $\lambda^3 = h^3/(2\pi ukT)^{3/2}$ a value for the temperature $T = 2.98058$ K results, when using the values for N, V, u, the Planck constant $h = 6.62608 \times 10^{-34}$ J s and the Boltzmann constant $k = 1.38066 \times 10^{-23}$ J K^{-1}.

The self-diffusion of particles and the entropy of the system are both a result of random particle motions. With the Sackur–Tetrode equation the molar entropy, S_m, of the above system can be calculated at temperature T and pressure p:

$$S_m = R \ln\left(\frac{e^{5/2}kT}{p\lambda^3}\right) \quad (6.15)$$

As shown in the previous section a common feature of all systems in the liquid state is their molar entropy of evaporation at similar particle densities at pressures with an order of magnitude of one bar. Taking this into account a reference temperature, T_r, will be selected for systems at a standard pressure, $p° = 10^5$ Pa = 1 bar, having the same molar entropy at the pressure unit, $p_u = 1$ Pa at $T = 2.98058$ K. As can easily be verified, the same value of molar entropy and consequently the same degree of disorder results at $p°$ if a 100-fold value of the above T-value is used in Eq. (6.15). This value denoted as $T_r = 298.058$ K is the reference temperature for the following model for diffusion coefficients. The coincidence of T_r with the standard temperature $T° = 298.15$ K is pure chance. With T_r a reference molar volume of gas, $V_{G,r} = RT_r/p° = 24.782$ dm^3 mol^{-1} is defined, which by chance is very close to the value of the standard volume $V_G° = RT°/p° = 24.79$ dm^3 mol^{-1}. But due to these coincidences, the values of the standard temperature and standard volume are used in the following equations for the reference values: $T_r \simeq T° = 298.15$ K and $V_{G,r} \simeq V_G° = 24.79$ dm^3 mol^{-1}.

With the reference temperature and the limit value w of the interaction function w_n for a macroscopic system we can also define the "free volume" $V°_L$ mentioned in Section 6.3.5 and a reference critical molar volume.

Assuming that the molar volume V_L° discussed in the previous section is a consequence of interaction between the particles in a condensed macroscopic system, a reference molar volume, $V_{L,r}$, is defined using the reference volume $V_{G,r}$ and w:

$$\frac{V_{G,r}}{V_{L,r}} = \exp\left(\frac{\Delta S_r}{R}\right) = \exp\left(\frac{wR}{R}\right) = e^w = 24078.88 \tag{6.16}$$

$V_{L,r} = V_{G,r}/e^w = 1.03 \text{ cm}^3 \text{ mol}^{-1}$. The product $wR = 10.0891 \times 8.31447 = 83.8854$ J K^{-1} mol^{-1} defines a reference entropy of evaporation, ΔS_r, and lies within the limits of deviation from the values of the Hildebrand entropy of evaporation. The effect of the interaction function lies in an isothermal contraction of the gaseous phase proportional to $1/e^w$.

If we further consider an ideal gas phase at unit pressure, $p_u = 1$ Pa, at temperature $T_0 = 1$ K and with an energy amount $R_0 = 1$ J K^{-1} mol^{-1} instead of $R = 8.314$ J K^{-1} mol^{-1} then the molar volume of this hypothetical gas phase is $V_{0,G} = 1$ m^3 mol^{-1}. With the same application of the interaction function a reference critical molar volume results: $V_{c,r} = V_{0,G}/e^w = 4.153 \times 10^{-5}$ m^3 mol^{-1}.

6.4
The Diffusion Coefficient

The interaction functions w_n, and $w_{i,e}$ with the limit value w defined in Eq. (6.9) have been correlated directly with different properties of macroscopic systems. These numbers can now be considered again as starting points for dynamic processes occurring in macroscopic systems. In all aggregate states diffusion is considered to be a consequence of interactions between the particles that are in conformity with the first assumption of the model. This means the diffusion coefficient can be described as an exponential function of an interaction energy. The uniform structure of this function for all aggregate states is

$$D = D_0 \exp(w_n - A/A_0 - E_A/RT) \tag{6.17}$$

The first term in the exponent, w_1 for monoatomic gases, $w_{1,e}$ for monoatomic particles in condensed states and $w_{i,e}$ for chains made of i particles in condensed states, represents an ideal system in conformity with Eq. (6.8) and the limit value w. The second and third term in the exponent refer to real systems. A is proportional to the molar cross-section of the diffusing particle, whereas E_A represents an activation energy proportional to the critical temperature, T_c, for gases and to the melting temperature, T_m, for condensed states, respectively. These last two terms act as brakes of the diffusion process.

6.4.1
Diffusion in Gases

At constant temperature the ratio $D_{G,2}/D_{G,1}$ of the diffusion coefficients in a gas at two states 1 and 2 equals the ratio V_2/V_1 of the system volumes at these states. Then together with the first assumption a starting point for modeling diffusion coefficients will be the relation $D_{G,2} = (V_2/V_1)D_{G,1} = (V_2/V_1)D_u\exp(q_r)$ with the unit value $D_u = 1\,\text{m}^2\,\text{s}^{-1}$ and $q_r = w_1$ for a perfect gas. By selecting $V_1 = 10^5\,V_G^\circ = RT^\circ/p_u$ at $p_u = 1\,\text{Pa}$ as a reference volume and with $V_2 = RT/p$, the following equation results for a self-diffusion coefficient, D_G, in a perfect gas:

$$D_G = \frac{T \cdot p_u}{T^0 p} D_u \exp(w_1) \tag{6.18}$$

This equation is typical for an ideal gas in which neither an interaction term in the form of activation energy nor the size of the diffusing particles are considered. At $p = 1\,\text{bar}$ and $T = 273\,\text{K}$ for example, $D_G = 1.36\,\text{cm}^2\,\text{s}^{-1}$ is obtained using Eq. (6.18) compared with $1.4\,\text{cm}^2\,\text{s}^{-1}$ measured in He (Landolt–Börnstein 1969).

For real gases two additional terms in the exponent of Eq. (6.18) must be introduced, acting as brakes:

1. A molar activation energy E_A of diffusion is defined as the product $E_A = w_{1,e}RT_c$, with the critical temperature, T_c, of the system. This definition takes into account the connection between the interaction terms of the model, w_1 and $w_{1,\,e}$ and IS units, as shown in Section 6.3.6. At the critical temperature, T_c, a pure translational amount, $w_1 - w_{1,e}$, of the relative energy density is responsible for the magnitude of D_G at pressures $p < p_c$.

2. The second term takes into account the size magnitude of the diffusing particle. Let $A_m = V_0^{2/3}\,\text{m}^2\,\text{mol}^{-1}$ be the molar cross-sectional area of the diffusing particle. When moving through the system a particle with this cross-sectional area must overcome the force exerted on it by the other particles in the system while they are moving about in a disordered fashion. The magnitude of this force F divided by the unit area $A_u = 1\,\text{m}^2$ to which the force is being applied defines a pressure, $p = F/A_u$. The product $pA_m d = (A_m/A_u)Fd = E_{m,A}\,\text{J}\,\text{mol}^{-1}$ represents the necessary molar work to overcome the resistance of the matrix by moving A_m along the distance d in a direction perpendicular to it. Referred to the corresponding work for moving A_u along the same distance, $E_{u,A} = Fd\,\text{J}$, we get $E_{m,A}/E_{u,A} = \varepsilon_{m,A} = A_m/A_u = A_m\,\text{mol}^{-1}$. On the other hand, the pressure p can be expressed as $p = m(N/V)\langle v^2\rangle$, where N/V is the density of the moving matrix particles, $m = M_r u$ is the particle mass and $\langle v^2\rangle$ is the mean value of the square of the particle velocity. With the atomic mass unit u, the relative particle mass M_r, the amount of particles n (mol), the Avogadro constant N_A and the volume V, we get $p = M_r u(nN_A/V)\langle v^2\rangle = uN_A \langle v^2\rangle M_r/(V/n)$. The product $p(V/n) = uN_A M_r \langle v^2\rangle = 10^{-3}M_r\langle v^2\rangle = E_{m,k}\,\text{J}\,\text{mol}^{-1}$ defines the molar kinetic energy of the moving particles. Referred to the corresponding energy $E_{u,k} = m_u M_r \langle v^2\rangle\,\text{J}$, using the mass unit $m_u = 1\,\text{kg}$ instead of the molar mass unit, $M_u = uN_A = 10^{-3}\,\text{kg}\,\text{mol}^{-1}$, we get $E_{m,k}/E_{u,k} = \varepsilon_{m,k} = 10^{-3}\,\text{mol}^{-1}$. From the above considerations a dimensionless value, $\varepsilon_m = \varepsilon_{m,A}/\varepsilon_{m,k} = 1000 A_m = 1000 V_0^{2/3} = 1000(V_c/w_1)^{2/3}$ results as relative measure of the resistance against the movement of the diffusing particle.

Collecting the above results, the following final equation can be established for the self-diffusion coefficient in a macroscopic system in the gaseous state:

$$D_G = \frac{T}{T^o} \frac{p_u}{p} D_u \exp\left(w_1 - 10^3 \left(\frac{V_c}{w_1}\right)^{2/3} - w_{1,e} \frac{RT_c}{RT}\right) \qquad (6.19)$$

With Eq. (6.19) a system (0) can be formally defined, for which $D_{G,0} = D_u = 1\, m^2 s^{-1}$. This state is obtained with a hypothetical gas having $T_c = T_r = 298.15$ K. At $p = p_u = 1$ Pa and $T = T_r$ the unit value for $D_{G,0}$ is obtained if $w_1 - w_{1,e} = 10^3 (V_{c,0}/w_1)^{2/3} = 10^3 V_0^{2/3}$. From this condition a critical molar volume value $V_{c,0} = 4.19376 \times 10^{-5}\, m^3\, mol^{-1}$ results, which is very close to the reference critical molar volume, $V_{c,r} = 1/e^w = 4.153 \times 10^{-5}\, m^3\, mol^{-1}$ defined above. Then an expansion coefficient, α, must be considered for the molar volume $V_0 = V_{c,0}/w_1$ of the particles in the system at $T = T_r$. Taking this into account one can write $V_{c,r}(1 + \alpha T_r) = (1/e^w)(1 + \alpha T_r) = V_{c,0} = V_0 w_1$ and in the above relation, $w_1 - w_{1,e} = 10^3[(e^w w_1)^{-1}(1 + \alpha T_r)]^{2/3}$, a value $\alpha = 3.29 \times 10^{-5}\, K^{-1}$ results, which is typical for expansion coefficients in solids. For example, iron's expansion coefficient is $\alpha = 3.54 \times 10^{-5}\, K^{-1}$. These considerations are used further in Section 6.4.2 for the diffusion in the solid state.

Using measured values of critical molar volumes and critical temperatures, the self-diffusion coefficients of substances in the gas phase at pressures $p < p_c$ significantly below their critical values can be calculated using Eq. (6.19). In Table 6.4 the self-diffusion coefficients of noble gases at 1 bar, calculated with Eq. (6.19) are compared with the experimental values (Kestin et al. 1984). Helium behaves like a perfect gas at $T \gg T_c$ (Eq. 6.18) without interaction ($T_c = 0$) and atomic size ($V_c = 0$). Molecular hydrogen has the same behavior.

Table 6.4 Self-diffusion coefficients of noble gases at 1 bar.

	T_c (K)	$V_c \times 10^5$ (m^3 mol^{-1})	T (K)	D_G (cm^2 s^{-1}) calc	D_G (cm^2 s^{-1}) exp
He	5.19	5.74	50	0.093	0.089
			100	0.21	0.29
			200	0.44 (1.0)[a]	0.92
			300	0.67 (1.5)[a]	1.82
Ne	44.44	4.16	50	0.021	0.021
			100	0.11	0.08
			200	0.34	0.27
			300	0.59	0.53
Ar	150.8	7.49	200	0.083	0.087
			300	0.21	0.19
			373	0.32	0.28
Kr	209.4	9.12	200	0.040	0.045
			300	0.12	0.10
			373	0.20	0.15
Xe	289.7	11.84	300	0.058	0.058
			373	0.11	0.09

[a] Values obtained for a perfect gas with $T_c = 0$ and $V_c = 0$.

6.4.2
Diffusion in the Critical State

Equation (6.19) can be easily adapted for the critical state if p is substituted with the critical pressure, p_c, obtained from Eq. (6.13) and if w_1 in the exponent is substituted with $w_{1,e}$. This takes into account the absence of a pure translational energy contribution in the critical state. On the contrary, an additional negative term, the critical compression factor $Z_c = w_1/w$, is introduced in the exponent, taking into account the decrease in diffusion velocity caused by attraction between the particles. As a result the following equation gives the coefficient of self-diffusion in the critical state:

$$D_c = \frac{V_c p_u}{RT^o} \frac{w}{w_1} D_u \exp\left(-10^3 \left(\frac{V_c}{w_1}\right)^{2/3} - \frac{w_1}{w}\right)$$

$$= \frac{T_c}{T^o} \frac{p_u}{p_c} D_u \exp\left(-10^3 \left(\frac{V_c}{w_1}\right)^{2/3} - \frac{w_1}{w}\right) \tag{6.20}$$

The diffusion coefficients at the critical state are the upper limits for the diffusion coefficients in liquids. Therefore Eq. (6.20) will be used later for the development of equations for the liquid state.

6.4.3
Diffusion in Solids

6.4.3.1 Self-diffusion Coefficients in Metals

In comparison with the behavior of a gas, the expansion of the system is neglected and the ratio of the volumes at two states is $V_2/V_1 \cong 1$. Then together with the first assumption a starting point for modeling diffusion coefficients in the solid state is $D_S = D_u \exp(w_{1,e})$ with the unit value $D_u = 1$ m^2/s and the interaction function $w_{1,e}$ for a macroscopic system made of monoatomic particles. But as for real gases, two additional terms must be introduced in the exponent acting as brakes and the following equation results finally for the self-diffusion coefficient, D_S, in solid metals:

$$D_S = D_u \exp\left(w_{1,e} - 10^3 w \left(\frac{1}{e^w}(1+\alpha T)\right)^{2/3} - w w_{1,e}^{2/3} T_m R/RT\right) \tag{6.21}$$

Whereas the diffusing particles in the gas phase move through a matrix in which all particles are in the same condition of movement, the diffusing atoms in the solid metal move through a rigid matrix made of immovable particles. This different situation is taken into account with the limit value w of the interaction function as a proportionality factor in the second term of the exponent. The bracket at exponent 2/3 contains the critical molar volume, $V_{c,0}$, at temperature T, as defined in Section 6.4.1. In contrast to the gaseous state where the molar volume $V_0 = V_c/w_1$ of the diffusing particles is used, in the above equation the value $V_{c,0}$ is introduced, because the molar cross-sectional area of the moving particles is larger than the area resulting from the diameters of the atoms in the metallic bonds. The difference between the particle radius which results from $V_{c,0}$ and the particle radius from V_0 is about 0.072 nm. This value corresponds to the difference between a van der Waals radius and an atomic radius in a covalent bond. The jump of an atom from an interstitial place in the matrix to another place needs the bigger cross-sectional area resulting from the van der Waals radius.

The factor w in the third term in the exponent of Eq. (6.21) plays the same role as $w_{1,e}$ in the equation for the diffusion in gases. But instead of the critical temperature, T_c, the melting temperature, T_m, of the solid phase is introduced. In addition the factor $w_{1,e}^{2/3}$ is introduced. This takes into account the contribution of the interaction energy, relative to w, which results from the interaction between the moving diffusing particle (its cross-sectional area) with the solid matrix.

In Table 6.5 the calculated self-diffusion coefficient values for 10 metals at different temperatures are listed together with the corresponding experimental values collected from the literature (Landolt–Börnstein 1991). In addition the empirically determined pre-exponential factor values, D_0, and activation energies of diffusion, E_A, in Eq. (6.5) are given, as well as the calculated activation energies, $E_A = w(w_{1,e})^{2/3}T_m R$, from Eq. (6.21). In the temperature range between 300 K and 3000 K the D_S values are obtained between 10^{-8} and 10^{-23} cm^2 s^{-1}. The ratio between calculated and experimental values in most cases are in the range 0.3–3. The values and the remarkable

Table 6.5 Comparison of self-diffusion coefficients for 10 metals calculated with Eq. (6.21) and experimental values.

	T_m (K)	E_{exp} (kJ mol^{-1})	E_{calc} (kJ mol^{-1})	$D_{0\,exp}$ (cm^2 s^{-1})	T (K)	D_{exp} (cm^2 s^{-1})	D_{calc} (cm^2 s^{-1})
Mg	923	135	126	1	740	3.0×10^{-10}	4.7×10^{-10}
					910	1.8×10^{-8}	2.0×10^{-8}
Zn	693	96	95	0.18	513	3.0×10^{-11}	9.2×10^{-11}
					690	9.7×10^{-9}	2.6×10^{-8}
Cd	594	82	81	0.18	400	3.5×10^{-12}	1.0×10^{-11}
					590	9.9×10^{-9}	2.5×10^{-12}
Y	1825	252	249	0.82	1100	8.9×10^{-13}	4.9×10^{-13}
					1600	4.9×10^{-9}	2.1×10^{-8}
Al	933	126	127	0.18	300	2.1×10^{-23}	2.7×10^{-23}
					480	3.5×10^{-15}	5.4×10^{-15}
V	2192	308	299	0.36	1100	8.5×10^{-16}	2.0×10^{-15}
					1600	3.2×10^{-11}	5.0×10^{-11}
Nb	2741	396	374	0.52	1300	6.4×10^{-17}	2.9×10^{-16}
					2600	5.8×10^{-9}	6.8×10^{-9}
Ta	3269	424	446	0.21	1200	7.4×10^{-20}	1.2×10^{-20}
					2900	4.8×10^{-9}	1.9×10^{-9}
Pt	2042	279	279	0.22	1500	4.2×10^{-11}	5.9×10^{-11}
					2000	1.1×10^{-8}	1.4×10^{-8}
Au	1336	176	182	0.12	900	7.3×10^{-12}	9.1×10^{-12}
					1200	2.6×10^{-9}	3.7×10^{-9}

agreement between the calculated and measured self-diffusion coefficients result mainly from the activation energies. The similarity between the empirical activation energies and the values $E_A = w(w_{1,e})^{2/3} T_m R$ supports the assumption which led to Eq. (6.21). The influence of the melting temperature in the exponent is much bigger than the pre-exponential factor D_0 (Murch 1995). In Eq. (6.21) the pre-exponential factor results from the nearly constant value $10^3 w (e^{-w}(1 + \alpha T))^{2/3}$. Values calculated for additional 20 metals lie generally within one order of magnitude with the experimental data. The 10 metals shown in Table 6.5 crystallize in cubic close-packed structures (Al, Pt, Au), in hexagonally close-packed structures (Mg, Zn, Cd, Y) and cubic I (body-centered cubic) structures (V, Nb, Ta).

6.4.3.2 Self-Diffusion Coefficients in Semiconductors and Salts

It is interesting to compare the diffusion coefficients in the metallic state with the self-diffusion coefficients in nonmetallic state, but in important electronic materials as silicon and germanium. The three-dimensional structure based on covalent bonds between the atoms with a tetrahedric orientation suggests that the diffusing particles are not monoatomic as in the metallic state. If instead of $w_{1,e}$ the value $w_{6,e}$ is introduced in Eq. (6.21) than the self-diffusion coefficient, D_S of the silicon-type is

$$D_S = D_u \exp\left(w_{6,e} - 10^3 w \left(\frac{1}{e^w}(1+\alpha T) \right)^{2/3} - w w_{6,e}^{2/3} T_m R / RT \right) \qquad (6.22)$$

With this equation a very good agreement with the measured data occurs (Gösele and Tan 1995). The same good agreement is obtained for germanium as shown in Table 6.6.

A different behavior results for diffusion in crystallized salts. For the self-diffusion of the Cl^- anion in NaCl, $w_{2,e}$ must be used in Eq. (6.21), whereas w_1 is held for the diffusion of the cation Na^+ (Laurent and Benard 1957).

Table 6.6 Comparison of self-diffusion coefficients in semiconductors and salts calculated with Eq. (6.22) and Eq. (6.21) with experimental values.

	T_m (K)	E_{exp} (kJ mol^{-1})	E_{calc} (kJ mol^{-1})	$D_{0\,exp}$ (cm^2 s^{-1})	T (K)	D_{exp} (cm^2 s^{-1})	D_{calc} (cm^2 s^{-1})
Si	1683	425	405	20	1100	1.3×10^{-19}	3.1×10^{-19}
					1670	1.0×10^{-12}	9.8×10^{-13}
		396	405	8	1100	1.3×10^{-18}	3.1×10^{-19}
					1670	3.3×10^{-12}	9.8×10^{-13}
Ge	1220	300	294	21.3	770	9.5×10^{-20}	7.0×10^{-20}
					1200	1.9×10^{-12}	8.6×10^{-13}
NaCl Cl$^-$	1081	215	182	110	1073	3.8×10^{-9}	9.9×10^{-10}
					873	1.5×10^{-11}	9.7×10^{-12}
NaCl Na$^+$	1081	155		0.5	1073	1.4×10^{-8}	2.2×10^{-8}
					873	2.7×10^{-10}	5.2×10^{-10}

6.4.3.3 Self-Diffusion Coefficients in n-Alkanes

Let us now consider a macroscopic system made of n-alkanes. In the following section an equation for the self-diffusion coefficient of a n-alkane with a number i of carbon atoms in the molecule will be derived in a manner analogous to that for metals. The starting point is Eq. (6.21) for metals. The first term in the exponent, $w_{1,e}$, must be replaced with $w_{i,e}$ to take into account the polyatomic chain with i methylene groups, inclusive of the two methyl end-groups. The same substitution is necessary in the third term.

Using the homologous series of n-alkanes as a reference structure series a relation of the molar cross-sectional area $10^3 V_0^{2/3} = 10^3 (V_c/w_1)^{2/3}$ to the specific properties of n-alkanes is derived in the following manner. In the special situation of a n-alkane with repeating $-CH_2-$ groups in the molecular chain, a constant value of the ratio between the critical volume and its relative molecular mass, $V_{c,i}/M_{r,i}$ can be expected for the homologous series except the first member, methane. For the n-alkanes with $i = 2 - 18$ a mean value $V_{c,i}/M_{r,i} = 4.4594 \times 10^{-6}$ m^3 mol^{-1} is obtained (Reid et al. 1987). With this value $10^3(V_{c,i}/w_1)^{2/3} = 10^3(4.4594 \times 10^{-6} M_{r,i}/w_1)^{2/3} = 0.14 M_{r,i}^{2/3} = 0.14(14i+2)^{2/3}$ results because the relative molecular mass of an n-alkane with i C atoms is $M_{r,i} = 14i + 2$. But as in Eq. (6.21) for solid metals, the critical volume is used instead of the molar volume and consequently $0.14(w_1 M_{r,i})^{2/3} = 0.14 w_1^{2/3}(14i+2)^{2/3}$ replaces the product $10^3 V_{c,0}$ in the second term. Also w in this term is substituted with $w_{i,e}$ because the brakes which the diffusing particle must overcome are the molecular chains with van der Waals interactions in the matrix in comparison with the much stronger metallic bonds. The resulting equation for self-diffusion coefficients in crystallized n-alkanes is

$$D_{S,i} = D_u \exp(w_{i,e} - w_{i,e} \cdot 0.14 w_1^{2/3}(14i+2)^{2/3} - w w_{i,e}^{2/3} T_{m,i} R/RT) \quad (6.23)$$

Taking n-heptane as an example, a value $D_{S,i} = 7.7 \times 10^{-21}$ cm^2 s^{-1} in the solid phase near the melting temperature $T_{m,i} = 183$ K results with the above equation. Although no experimental data for this case are available, the order of magnitude at that temperature seems not unrealistic in comparison with the data for metals in the solid state. The above equation is important as a starting point for treating diffusion coefficients in liquids, especially in plastic materials.

6.4.4
Diffusion in Liquids

6.4.4.1 Self-Diffusion Coefficients in Metals

When the temperature reaches the melting point, the rigid matrix changes into the liquid state, in which the metallic particles can move much easily. This is taken into account by introducing the molar volume $V_0 = V_{c,0}/w_1$ in Eq. (6.21) instead of $V_{c,0}$ and the diffusion coefficient for liquid metals, D_L at the temperature T is

$$D_L = D_u \exp\left(w_{1,e} - 10^3 w \left(\frac{1}{e^w \cdot w_1}(1+\alpha T) \right)^{2/3} - w w_{1,e}^{2/3} T_m R/RT \right) \quad (6.24)$$

This equation can be considered as a reference for all metals, because at $T = T_m$ the value of D_L depends only slowly from T and has the typical value of about $1 \times 10^{-5}\,\text{cm}^2\,\text{s}^{-1}$ for the liquid state.

6.4.4.2 Self-Diffusion Coefficients in n-Alkanes

Equation (6.23) can be adapted for the self-diffusion-coefficient, $D_{Lm,i}$ in liquid n-alkanes at the melting temperature $T_{m,i}$ in the following manner:

$$D_{Lm,i} = D_u \exp(w_{i,e} - w_{i,e} \cdot 0.14(14j+2)^{2/3} - ww_{j,e}^{2/3} T_{m,i} R/RT) \qquad (6.25)$$

with

$$j = (M_{r,i} - 2)/14)^{1/3}.$$

The first term in the exponent, $w_{i,e}$, remains unchanged. In the second term w_1 disappears because the molar volume replaces the bigger critical molar volume, as in the liquid metals and in the gas phase. The product $0.14(14j+2)^{2/3}$ in the second term represents the relative molar cross-sectional area of the diffusing particle. As shown in Section 6.4.1 this area, $1000 A_m = 1000 V_0^{2/3} = 1000 (V_c/w_1)^{2/3}$ results as a measure of the relative resistance against the movement of the diffusing particles. In the special situation of a n-alkane $10^3(V_{c,i}/w_1)^{2/3} = 0.14 M_{r,i}^{2/3} = 0.14(14i+2)^{2/3}$ results as shown above. But in contrast to the crystalline state, the diffusing molecules in the liquid state can be considered in a first approximation to be present in the form of spherical coils, which offer a smaller cross-sectional area, than the sum of i methylene groups. As a consequence, a relative diameter of $(M_{r,i} - 2)/14)^{1/3} = j$ results in comparison to i and the factor $(14j+2)$ is used instead of $14i+2$. With j, a corresponding interaction function, $w_{j,e} = (1 + 2\pi/j)^{j/e}$ results, which is used in the third term of the exponent in Eq. (6.25).

The change from a linear orientation to spherical coils in the whole matrix can be understood as a gradual process, which occurs only to a small degree at very low temperatures, e.g., near the melting point, with short chains of small molecules. With increasing temperature and longer chains the transformation to spherical coils takes places to a much higher extent. In any case the diffusing particles can be assumed to exist in all situations in a more or less well coiled form. The consequence of these considerations is a gradual change from the use of $w_{i,e}$ in the second term to $w_{j,e}$ for the matrix molecules. The limit value w remains in the third term as in the solid state, because it refers to this state, from which the activation energy is derived together with the melting temperature. The above picture of the liquid phase near the melting temperature is naturally an extreme simplification. Taking again n-heptane as an example, with Eq. (6.25) a value $D_{Lm,i} = 4.14 \times 10^{-6}\,\text{cm}^2\,\text{s}^{-1}$ results in the liquid phase at the melting point of 183 K. The small molecules of the heptane matrix at this low temperature can be assumed to exist mainly in a crystalline-like state and thus the coil formation, as explained above, is done if at all, only to a small extent. This means $w_{i,e}$ with the real carbon number i in the second term of the exponent in Eq. (6.25), can be used. Due to the gradual transformation of the matrix molecules into coils as described above, the calculated diffusion coefficients for bigger alkanes remain smaller than the experimental values if $w_{i,e}$ is used for the matrix. But with long chain alkanes ($i > 18$) the calculated values fit well with experimental data if $w_{j,e}$ instead of $w_{i,e}$ is used in the second term in Eq. (6.25)

The melting of n-alkanes with very long chains occurs slowly and at the corresponding high temperatures one can assume that most of the molecules at the melting point are in spherical shapes. The relative movement of the matrix molecules has the consequence that instead of w the interaction function $w_{i,e}$, or even $w_{j,e}$ at high temperatures, must be used in the last term of Eq. (6.25). Alkanes with long chains have already a behavior similar to polymers.

It is now possible to arrive in the critical state starting from the liquid state at the melting temperature. An analogous equation for the critical state as Eq. (6.20) is used with the only difference in the first term in the exponent. Based on the above consideration about the coil formation at high temperatures a corresponding expression for the first term in the exponent of Eq. (6.20) can be given. It represents the relative cross-sectional area for an n-alkane with i carbon atoms in the critical state. Together with the critical temperature $T_{c,i} = 1036.2(w_{i,e}/w)$ from Eq. (6.10) and the critical pressure, $p_{c,i}$ from Eq. (6.13) the following formula results from Eq. (6.20) for the self-diffusion coefficient of an n-alkane with i carbon atoms, in the critical state:

$$D_{c,i} = \frac{1036.2 \cdot (14i+2)}{298.058 \cdot 5.155 \times 10^8} \exp\left(-w_{j,e} \cdot 0.14(14j+2)^{2/3} - \frac{w_1}{w}\right) \quad (6.26)$$

The lower limit for the diffusion in liquids is at the melting point. An equation for the self-diffusion coefficient of n-alkanes in the liquid state between the melting and critical temperature can now be written. The upper limit of the self-diffusion coefficient in the liquid phase is obtained at the critical temperature, $T_{c,i}$ with Eq. (6.26) for the n-alkane series and the lower limit at the melting temperature, $T_{m,i}$, with Eq. (6.25). It can be assumed that the self-diffusion coefficient $D_{L,i}$ at a temperature T between $T_{m,i}$ and $T_{c,i}$ follows the exponential function $D = ae^{-b/T}$. Collecting all results from the above steps and writing $D_{Lm,i} = a\exp(-b/T_{m,i})$ and $D_{c,i} = a\exp(-b/T_{c,i})$, the following equation is obtained for the diffusion coefficient $D_{L,i}$:

$$D_{L,i} = a \exp\left(-\frac{bR}{RT}\right) \quad (6.27)$$

with

$$b = \frac{\ln\left(\frac{D_{Lm,i}}{D_{c,i}}\right)}{\frac{1}{T_c}\frac{w}{w_{i,e}} - \frac{1}{T_{m,i}}} \quad \text{and} \quad a = D_{c,i}\exp\left(\frac{b}{T_{c,i}}\right) = D_{c,i}\exp\left(\frac{b}{1036.2}\frac{w}{w_{i,e}}\right)$$

Taking into account that for the reference homologous series of n-alkanes the relative molecular mass of the member i in the series is $M_{r,i} = 2 + 14i$, the self-diffusion coefficient $D_{L,i}$ can be calculated with Eqs. (6.25) and (6.26). This can be done by using only two values based on experimental results, the limit value of

6.4 The Diffusion Coefficient

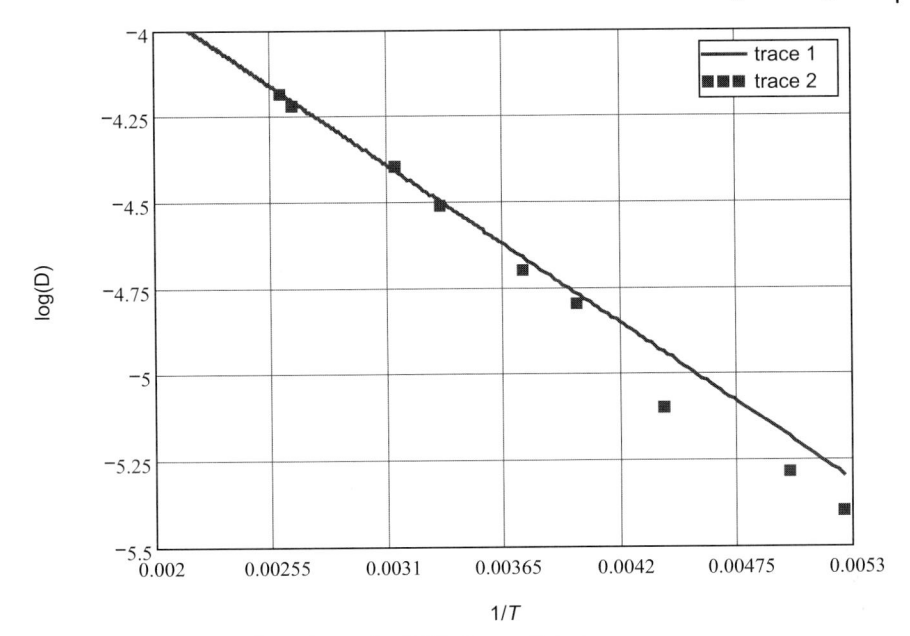

Figure 6.3 Temperature dependence of self-diffusion coefficients of n-heptane. Calculated with Eq. (6.27) (trace 1) and experimental values (trace 2).

the critical temperature, $T_c = 1036.2$ K and the mean value for the ratio, $V_{c,i}/M_{r,i} = 4.4594 \times 10^{-6}$ m^3 mol^{-1}.

Table 6.7 and Figure 6.3 show a comparison between experimental (Landolt-Börnstein 1969) self-diffusion coefficients and calculated values obtained with Eq. (6.27).

Table 6.7 Comparison of experimental self-diffusion coefficients in n-alkanes with values calculated with Eq. (6.27).

i	M_r	T (°C)	T (K)	$D_{L,i} \times 10^5$ (cm^2 s^{-1}) calc	$D_{L,i} \times 10^5$ (cm^2 s^{-1}) exp	Remarks for use of Eq. (6.25)
7	100	25	298	3.2	3.1	with $w_{i,e}$ [a]
8	114	25	298	1.4	2.8	with $w_{i,e}$ [a]
		84	357	3.0	4.5	with $w_{i,e}$ [a]
9	128	25	298	0.9	1.7	with $w_{i,e}$ [a]
		84	357	2.1	4.0	with $w_{i,e}$ [a]
10	142	25	298	0.4	1.3	with $w_{i,e}$ [a]
		84	357	1.1	3.7	with $w_{i,e}$ [a]
12	170	60	333	0.2	1.4	with $w_{i,e}$ [a]
18	254	50	323	2.2	0.5	with $w_{j,e}$ [a]
32	450	100	373	0.5	0.3	with $w_{j,e}$ [a]
82	1150	120	393	0.04	0.1	with $w_{j,e}$ [a] and $w_{i,e}$ [b]
94	1318	120	393	0.015	0.055	with $w_{j,e}$ [a] and $w_{i,e}$ [b]

[a] $w_{i,e}$, or $w_{j,e}$ in the second term;
[b] $w_{i,e}$ instead of w in the third term.

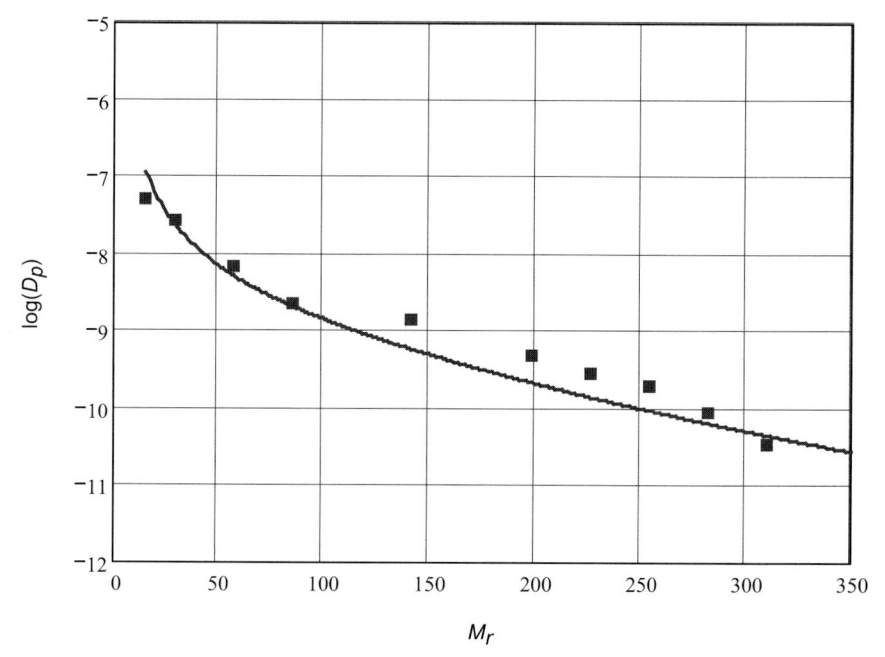

Figure 6.4 Comparison of measured diffusion coefficients (cm²/s) at 25 °C for 10 n-alkanes with calculated values (curve) using Eq. (6.28) for HDPE with $T_{m,p} = 110\,°C$ and $M_{r,p} = 50{,}000$.

6.4.5
Diffusion in Plastic Materials

The diffusion behavior in polymers lies between that of liquids and solids. As a consequence, the models for diffusion coefficient estimation are based on ideas drawn from diffusion in both liquids and solids.

6.4.5.1 Diffusion Coefficients of n-Alkanes in Polyethylene

The self-diffusion in n-alkanes can now be used as basis for modeling the diffusion coefficients, D_p, of additives in polyethylene. This polymer is the matrix which results asymptotically from the homologous series of n-alkanes for $i \gg 1$.

From the above discussion one can assume that the matrix molecules are to a high extent, spherical coils above the glass temperature, T_g. The same assumption can be made for diffusing n-alkanes. This means the polymer matrix is similar to the matrix of an n-alkane in the liquid state at the melting point. The following equation follows directly from Eq. (6.25) for the diffusion coefficient of an n-alkane with i carbon atoms and the relative molecular mass $M_{r,i} = (14i + 2)$ in polyethylene with the relative molecular mass $M_{r,p}$ and the melting temperature $T_{m,p}$:

$$D_{P,i} = D_u \exp(w_{i,e} - w_{p,e} \cdot 0.14(14j+2)^{2/3} - ww_{j,e}^{2/3} T_{m,p} R/RT) \tag{6.28}$$

with $i = (M_{r,i} - 2)/14$

$w_{i,e} = (1 + 2\pi/i)^{i/e}$, $j = (i^{1/3})$, $w_{j,e} = (1 + 2\pi/j)^{j/e}$, $p = (M_{r,p}/14)^{1/3}$, $w_{p,e} = (1 + 2\pi/p)^{p/e}$

The only difference from Eq. (6.25) lies in the use of the interaction function $w_{p,e} = (1 + 2\pi/p)^{p/e}$, where $p = (M_{r,p}/14)^{1/3}$, with the relative molecular mass of the polymer and the melting temperature, $T_{m,p}$ K. The numbers $i, j, w_{i,e}$, and $w_{j,e}$ refer to the diffusing molecule and have the same meaning as defined above.

Figure 6.4 shows a comparison between calculated and experimental (Koszinowski 1986 and Appendix I) values of diffusion coefficients for n-alkanes in high density polyethylene (HDPE) at 25 °C.

In Figure 6.5 a comparison between calculated and experimental values for the diffusion of n-alkanes in low-density-polyethylene (LDPE) are shown. From Appendix I the values of diffusion coefficients at 40 °C from all n-alkanes are considered as far as possible. With Eq. (6.28) an upper limit with $M_{r,p} = 3000$ and $T_{m,p} = 353$ K (80 °C) and a lower limit with $M_{r,p} = 100{,}000$ and $T_{m,p} = 393$ K (120 °C) are calculated, respectively. A polymer like LDPE is polydisperse in the sense that a sample spans a large range of molar masses. From the diffusion point of view, the smaller molecules probably play a relatively more important role in comparison with the bigger molecules due to the easier movement of the diffusing solute through a matrix with a large portion of small molecules. The molecular mass range between 3000 and 100,000 for $M_{r,p}$ and the temperature range between 80 and 120 °C for $T_{m,p}$, correlate

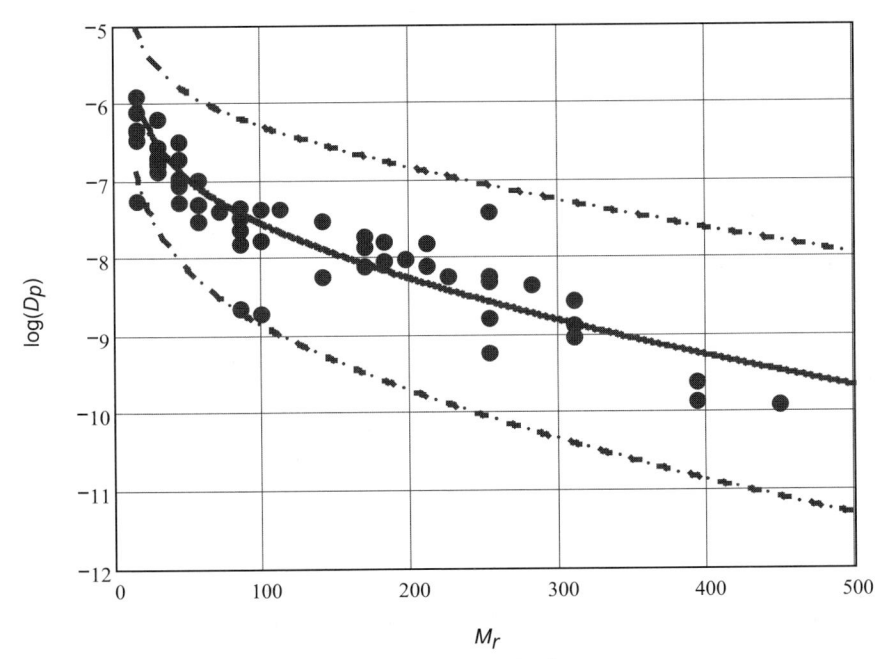

Figure 6.5 Comparison of the diffusion coefficients (cm² s⁻¹) at 40 °C of the n-alkanes from Appendix I in LDPE with calculated values (curve) using Eq. (6.28), with $T_{m, p} = 363$ K (90 °C) and $M_{r,p} = 30{,}000$. An upper limit with $M_p = 3000$ and $T_{m, p} = 353$ K (80 °C) and a lower limit with $M_p = 100{,}000$ and $T_{m, p} = 393$ K (120 °C) are also given (dashed curves).

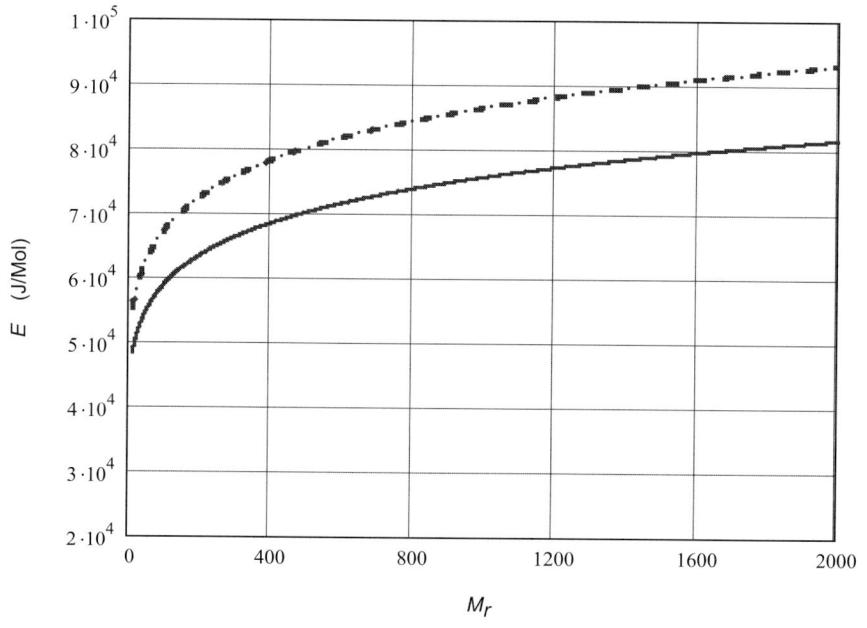

Figure 6.6 The activation energies in PE obtained from Eq. (6.28) with $T_{m,\,p} = 353$ K (80 °C, lower limit, continuous curve) and $T_{m,\,p} = 403$ K (130 °C, upper limit, dashed curve).

well with the properties of LDPE. The experimental values are from many different sources of low-density polyethylene samples with probable different ranges of molar masses.

In Figure 6.6 the activation energy, $E_A = w \cdot w_{j,e}^{2/3} \cdot T_{m,p} \cdot R$ as a function of $M_{r,i}$ is given for two melting points of PE, $T_{m,p} = 353$ K (80 °C) and $T_{m,p} = 403$ K (130 °C).

At temperatures $T > T_{m,p}$ the interaction function $w_{p,e}$ instead of w will be used in the third term: This term represents the interaction between the relative cross-section area of the diffusing particle expressed as $w_{j,e}^{2/3}$, and the matrix particles. In the molten matrix the interaction takes place mainly with individual spherical coils instead with the whole macroscopic phase expressed as w. The diffusion coefficient of the alkane i in PE above the melting temperature, $T > T_{m,p}$ is therefore

$$D_{P,i} = D_u \exp(w_{i,e} - w_{p,e} \cdot 0.14(14j+2)^{2/3} - w_{p,e} w_{j,e}^{2/3} T_{m,p} R / RT) \qquad (6.29)$$

In Figure 6.7 the log($D_{p,i}$)-curve calculated with Eq. (6.29) is shown, where $M_{r,p} = 15{,}000$, $T_{m,p} = 363$ K (90 °C) and $T = 450$ K (177 °C) are used (trace 1). Four experimental values (trace 2) were used, obtained with deuterated polyethylene segments (Gell et al. 1997).

In a first approximation one can use Eq. (6.28) or (6.29) for all additives dissolved in polyethylene, where $i = (M_{r,a} - 2)/14$ and $M_{r,a}$ represents the relative molecular mass of the diffusing additive. Only the adequate values for $M_{r,p}$ and $T_{m,p}$ must be known.

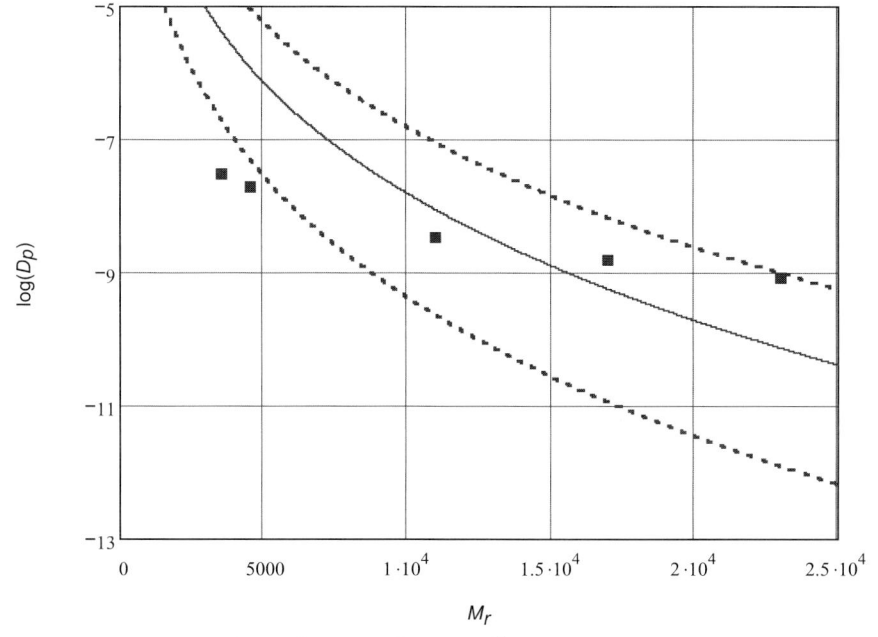

Figure 6.7 Comparison of the diffusion coefficients (cm²/s) of 5 deuterated PE-samples at 177 °C with calculated values (middle curve) calculated with Eq. (6.29), with $M_p = 15{,}000$, $T_{m,\,p} = 363$ K (90 °C). An upper limit with $M_{r,\,p} = 10{,}000$, $T_{m,\,p} = 353$ K (80 °C) and a lower limit with $M_{r,\,p} = 30{,}000$, $T_{m,\,p} = 373$ K (100 °C) are also given.

6.4.5.2 Diffusion Coefficients of Additives in Polymers

Due to the very complex structures of many plastic materials, with many unknown parameters, it is practically not possible to predict with sufficient precision the values of diffusion coefficients without additional empirical parameter values. If this additional empirical information obtained for a plastic material sample from migration measurement is denoted with $A_P = A'_P - \tau/T$, with the athermal, dimensionless number A'_P and the parameter τ with the dimension of a temperature, respectively, then a general applicable equation (see Eq. (6.28)) can be written as follows:

$$D_P = D_u \exp(w_{i,e} + A_P - w_{p,e} \cdot 0.14(14j+2)^{2/3} - ww_{j,e}^{2/3} T_{m,p} R / RT) \qquad (6.30)$$

Both values, A'_P and τ_P can be obtained from two diffusion measurements at different temperatures, using a reference solute in the polymer matrix (see Chapter 15).

Diffusion coefficients of additives at temperatures below the glass point T_g of polyethylene are not known due to the low value of these temperatures. But for polyethyleneterephthalate (PET) with $T_g \approx 80$ °C values of D_p exist below and above this temperature. Although the matrix structure of PET differs significantly from that of PE, one can use in a first approximation the above equations for estimation of the diffusion coefficients in PET. For $T < T_g$ an equation analogous to Eq. (6.23) for n-alkanes in the solid state can be used:

$$D_{P,i} = D_u \exp(w_{i,e} - w_{p,e} \cdot 0.14 \cdot w_1^{2/3}(14j+2)^{2/3} - ww_{i,e}^{2/3} T_g R/RT) \tag{6.31}$$

with $i = (M_{r,i} - 2)/14$, $w_{i,e} = (1 + 2\pi/i)^{i/e}$, $p = (M_{r,p}/14)^{1/3}$, $w_{p,e} = (1 + 2\pi/p)^{p/e}$ and T_g instead of $T_{m,p}$.

At temperatures near T_g the above equation is used without the factor w_1 because the molar volume instead of the critical volume in the second term in the exponent marks the transition from the glassy to the rubbery state of the polymeric matrix:

$$D_{P,i} = D_u \exp(w_{i,e} - w_{p,e} \cdot 0.14 \cdot (14j+2)^{2/3} - ww_{i,e}^{2/3} T_g R/RT) \tag{6.32}$$

At temperatures $T \gg T_g$ but below the melting point $T_{m,p}$ Eq. (6.28) for the rubbery phase is used.

In Figure 6.8, a comparison between the diffusion coefficients obtained for additives in PET with the three Eqs. (6.31) at 40 °C, (6.32) at 80 °C and (6.28) at 175 °C with

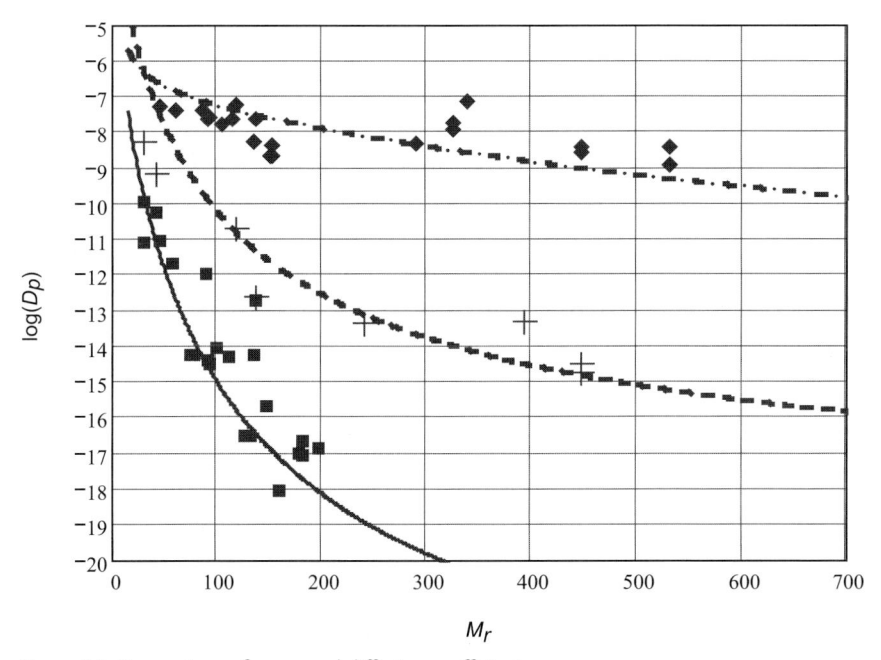

Figure 6.8 Comparison of measured diffusion coefficients (cm^2/s) of several compounds in PET at
a) 40 °C with calculated values (lower, continuous curve) using Eq. (6.31) with $M_{r,p} = 10{,}000$, $T_g = 80\,°C$;
b) 80 °C with calculated values (middle, dashed curve) using Eq. (6.32) with $M_{r,p} = 10{,}000$, $T_g = 80\,°C$;
c) 175 °C with calculated values (upper, dashed curve) using Eq. (6.28) with $M_{r,p} = 10{,}000$, $T_{m,p} = 528\,K\ (255\,°C)$.

experimentally obtained data (Begley et al. 2005) are shown. In these equations, $T_g = 353$ K (80 °C), $T_{m,p} = 528$ K (255 °C) and $M_{r,p} = 10,000$ are used (see Chapter 2).

Due to the big range over more than ten orders of magnitude, a good agreement between experimental and calculated values results from this comparison.

References

Begley, T., Castle, L., Feigenbaum, A., Franz, R., Hinrichs, K., Lickly T., Mercea P., Milana, M., O'Brien, A., Rebre, S., Rijk, R. and Piringer, O. (2005) Evaluation of migration models that might be used in support of regulations for food-contact plastics. *Food Addit. Contam.*, **22** (1), 73–90.

Broadhurst, J.W. (1962) *J. Research Natl. Bur. Standards A*, **66**, 241.

Buffat, P.H. and Borel, J.P. (1976) Size effect on the melting temperature of gold particles. *Phy. Rev.*, **13** (6), 2287–2298.

Cussler, E.L. (1997) *Diffusion. Mass Transfer in Fluid Systems*, 2nd edn. Cambridge University Press, Cambridge.

Flory, P.J. and Vrij, A. (1963), Melting points of linear-chain homologs: The normal paraffin hydrocarbons. *J. Amer. Chem. Soc.*, **85**, 3548–3553.

Gell, C.B., Graessley, W.W. and Fetters, L.J. (1997) *J. Polym. Sci B Polym. Phys.*, **35**, 1933.

Gösele, U.M. and Tan, T.Y. (1995) Equilibria, nonequilibria, diffusion and precipitation, in: W. Schröter (ed.), Electronic Structure and Properties of Semiconductors, Vol. 4, pp. 198–243, in R. W., P. Haasen, E. J. Kramer (eds.), Materials. Science and Technology: A Comprehensive Treatment, Wiley-VCH, Wienhiem.

Hagemann, J.W. and Rothfus, J.A. (1979) Computer modeling for stabilities and structural relationships of n-hydrocarbons. *J. Am. Chem. Soc.*, **56**, 1008–1013.

Hildebrand, J.H. (1915) *J. Am. Chem. Soc.*, **37**, 970.

Hildebrand, J.H., Prausnitz, J.M. and Scott, R.L. (1970) *Regular and Related Solutions*, van Nostrand Reinhold Company, New York.

Kestin, J., Knierim, K., Mason, E.A., Najafi, B., Ro, S.T. and Waldman, M. (1984) Equilibrium and transport properties of the noble gases and their mixtures at low density. *J. Phys. Chem., Ref. Data*, **13** (1), 229–303.

Koszinowski, J. (1986) *J. Appl. Polym. Sci.*, **31**, 1805.

Landolt–Börnstein (1969) *Zahlenwerte und Funktionen, II. Band, 5. Teil, Bandteil a Transportphänomene*, Springer, Berlin, New York.

Landolt-Börnstein (1991) Numerical data and functional relationships in science and technology, new series, in: H. Mehrer, N. Stolica, N.A. Stolwuk, Self-diffusion in solid metallic elements, in H. Mehrer (ed.), Diffusion in Solid Metals and Alloys, Vol. 26, Springer, Berlin, Heidelberg, New York, London.

Laurent, J.F. and Benard, J. (1957) Autodiffusion des ions dans les cristaux uniques des halogenures de potassium et des chlorures alcalins. *J. Phys. Chem. Solids* (Pergamon Press), **3**, 7–19.

Murch, G.E. (1995) Diffusion in crystalline solids, in Haasen P (ed), *Phase Transformations in Materials*, Vol. 5, pp. 77, in: R. W Cahn,, P. Haasen, E.J. Kramer (eds.), Materials Science and Technology: A *Comprehensive Treatment*, Wiley-VCH, Wienhiem.

Reid, R.C., Prausnitz, J.M. and Poling, B.E. (1987) *The Properties of Gases and Liquids*, 4th edn. McGraw-Hill, New York.

Stack, G.M., Mandelkern, L., Kröhnke, C. and Wegner, G. (1989) Melting and crystallization kinetics of a high molecular weight *n*-alkane: $C_{192}H_{386}$. *Macromolecules*, **22** (11), 4351–4361.

7
Transport Equations and Their Solutions
Otto Piringer and Titus Beu

7.1
The Transport Equations

Interactions between packaging and product are always connected with transport processes occurring within the packaging system. A transport process is understood to be a general movement of mass, energy, or other quantity from one location to another. An example of mass transport in packed liquid products is the convection that occurs during the heating or shaking of the package. Macroscopic regions of the liquid move with different speeds, relative to one another, and cause mixing to occur. With heating, a simultaneous transport of heat takes place along with mass transport. The convection of mass and energy takes place in liquid products during distribution of the packaging from the manufacturer to its final storage destination and during heating and cooling of the package.

Mixing by convection in viscous and solid packed products has very little or no practical significance. A special case is the mixing of particulate products by shaking, which gives results similar to convection. The most important transport processes in solid-, viscous-, and liquid-filled products during the storage period are diffusion and thermal conductance. Mass transport by diffusion and energy transport by conductance have a common molecular basis. They are both affected by the unordered movement of molecules in the medium in which transport takes place. It is the vibration of atoms and groups of atoms transmitted to neighboring atoms which is responsible for conductance in solids. Unordered collisions between the mobile molecules of a liquid or gas are also a source of mass transport by diffusion (Chapters 5 and 6).

A further example of energy transport through packaging into the filled product is electromagnetic radiation. This radiation in the form of light can start chemical reactions or, in the case of microwaves, be transformed into heat and then further distributed through the packaging system by conduction or convection.

In addition to mass and energy, other quantities can also experience transfer. Flowing layers with different flow rates in a convection stream can influence one another. The slower flowing layer acts as a brake on the faster layer, while at the same

Plastic Packaging. Second Edition. Edited by O.G. Piringer and A.L. Baner
Copyright © 2008 WILEY-VCH Verlag GmbH & Co. KGaA, Weinheim
ISBN: 978-3-527-31455-3

time the faster layer acts to accelerate the slower one. The cause of this behavior is the inner friction of the liquid appearing as a viscosity difference, which is a consequence of the attractive forces between the molecules. Viscosity can be explained as the transport of momentum. The viscosity of different media can be very different and thus plays an important role in transport processes.

7.1.1
The Terminology of Flow

For the mathematical description and understanding of transport processes, it is advantageous to have several common characteristics, regardless of the nature of the transport quantity, to allow them to be treated in a similar manner. Without knowledge of their fundamental causes at the molecular level, which corresponds to their historical development, transport processes can be described with help from quantities that can be quantitatively measured on a macroscopic level. One such quantity is that of flux.

The flux J is understood to be the amount of a quantity transported per unit time through a unit surface area.

Flux is a vector for which a direction must be specified in addition to the quantity or contribution J. This is accomplished with the help of the unit vector e giving

$$J = Je = J_x + J_y + J_z = J_x i + J_y j + J_z k \tag{7.1}$$

J_x, J_y, and J_z are the vector components in the x-, y-, and z-axis directions of the coordinate system, J_x, J_y, and J_z are their contributions, and i, j, and k are the corresponding unit vectors. Given a mass quantity m that is transported during time t through an area A, then let J represent the contribution of the mass flux. For energy transport, then J is the contribution of the energy flux with the dimensions $J/m^2 s$ (J represents Joule).

In a very general sense, the flux of a quantity G is proportional at a given location to the gradient of the scalar field produced by the flux $a(x, y, z)$. Mathematically, one obtains the contributions of the three components with the gradient of a (grad a) from the partial derivative of a at the coordinates x, y, and z which for the flux G results in

$$J(G) = -b \operatorname{grad} a = -b\left(\frac{\partial a}{\partial x} i + \frac{\partial a}{\partial y} j + \frac{\partial a}{\partial z} k\right) \tag{7.2}$$

The location-independent proportionality factor is designated as b. The minus sign in Eq. (7.2) shows that the flux goes in the direction of decreasing a values. This means that the quantity G "flows" down the gradient.

The usefulness of the flow terms as common characteristics for transport processes allows them to illustrate such seemingly diverse processes as convection, momentum transport (viscosity), diffusion, and heat conductance. To simplify the written expression, the flux components of the four processes are expressed in Eq. (7.3) in the direction of one axis of the coordinate system whereby, instead of the partial derivative for the function, a variable and useful form of the derivative expression is used

(a) $J_x(\text{mass, convection}) = \rho \dfrac{\mathrm{d}x}{\mathrm{d}t} \boldsymbol{i}$

(b) $J_z(\text{momemtum in } x \text{ direction}) = -\eta \dfrac{\mathrm{d}v_x}{\mathrm{d}z} \boldsymbol{k}$

(c) $J_x(\text{mass, diffusion}) = -D \dfrac{\mathrm{d}c}{\mathrm{d}x} \boldsymbol{i}$

(d) $J_x(\text{energy, conduction}) = -\kappa \dfrac{\mathrm{d}T}{\mathrm{d}x} \boldsymbol{i}$

(7.3)

In Eq. (7.3a), ρ and $\mathrm{d}x/\mathrm{d}t$ are the contributions of the density and the velocity of the liquid in the x direction, respectively. The material-specific constants η, D, and κ are for the viscosity, diffusion, and thermal conductivity coefficients, respectively. The derivatives in the z and x directions, $\mathrm{d}v_x/\mathrm{d}z$, $\mathrm{d}c/\mathrm{d}x$, and $\mathrm{d}T/\mathrm{d}x$ are for the velocity components (in the x direction), the concentration, and temperature, respectively. A comparison of the four equations in Eq. (7.3) shows the similarities between the expressions.

With respect to their individual historical development, the four expressions are quite separate. While this representation of momentum can be traced back to Newton, the expression for heat conductance was first derived by the mathematician and physicist Joseph Fourier at the beginning of the 19th century. The physiologist Adolf Fick, who was concerned with measuring the transport of oxygen in blood, recognized the analogy of diffusion to heat conductance and published in 1855 the diffusion equation now known as Fick's first law Eq. (7.3c). The relationships between the different processes at the molecular level was first recognized by Einstein and other physicists and led to quantitative relationships between material-specific constants, in particular between D and η, which are important for calculating their respective contributions (see Chapters 5 and 6).

7.1.2
The Differential Equations of Diffusion

During a diffusion process, e.g., the migration of an additive from a plastic into the atmosphere, a change in the concentration of the diffusing substance takes place at every location throughout the plastic. The mass flux caused by diffusion is represented by a vector quantity, whereas the concentration c and its derivative of time t is a scalar quantity and connected by the flux with the help of the divergence operator. The following example serves to emphasize this relationship.

In a body with any given shape, e.g., a piece of soap, there is an aroma compound which is initially uniformly distributed throughout the entire body. During storage without any packaging, a decrease in concentration takes place due to diffusion into the atmosphere, particularly in the outer layers of the soap. The resulting scalar concentration field with the levels $c_1 > c_2 > c_3$ (Figure 7.1) forms a gradient field that describes the external direction of the aroma compound flux.

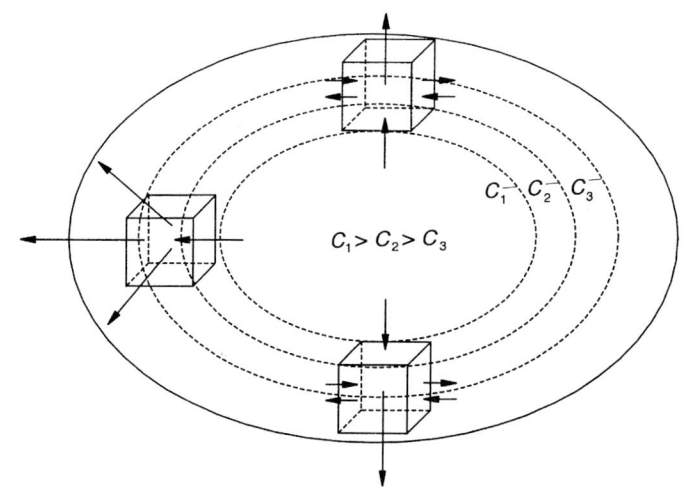

Figure 7.1 Diffusion and the divergence operator.

Now consider only a suitably small section of the soap in the form of a cube with side lengths of Δx, Δy, and Δz (Figure 7.2).

The aroma compound will diffuse in as well as out of the cube because of its perpendicular-side surface areas. Due to the greater decrease in the aroma near the soap's external surface, the flux out of the side of the cube closer to the surface is greater than the flux into the side of the cube that lies deeper in the soap. The difference between the aroma diffusing in and out will be positive which means one can consider the cube as an aroma source. As a consequence of the flux out of the cube, the concentration in the cube decreases with time. The concentration is also a function of time, $c = c(x, y, z, t)$, and its decrease with time, i.e., the partial derivative

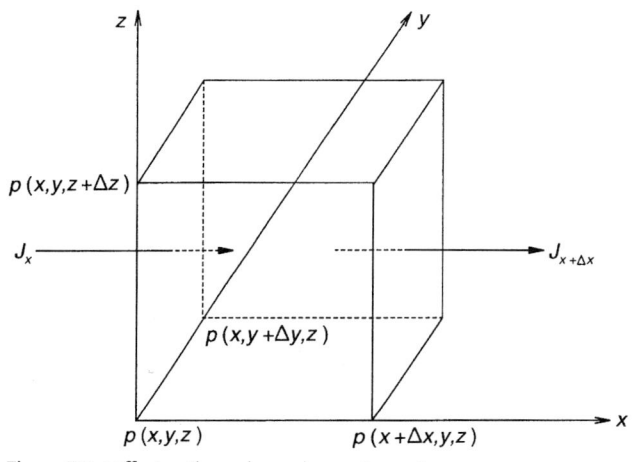

Figure 7.2 Diffusion through a volume element.

$-\partial c/\partial t$ in the cubic volume $\Delta V = \Delta x\,\Delta y\,\Delta z$, represents the net flux out of the cube and designated div $\boldsymbol{J}$ the divergence of the flux.

Mathematically, the divergence is obtained as the sum of the differences between flux components in and out of the cube in the coordinate axis direction with respect to the cube's volume. Placing the coordinate axis parallel to the corners of the cube as a helpful construction (Figure 7.2), one can label the incoming flux component contributions through the side walls $\Delta y\,\Delta z$ at the location x with $J_x(x)$ and the outgoing component through the opposite side wall at $x + \Delta x$ on the x-axis with $J_x(x+\Delta x)$. When this is done in the same manner for the other components, then one gets

$$-\frac{\partial c}{\partial t} = \frac{[J_x(x+\Delta x) - J_x(x)]\,\Delta y\,\Delta z}{\Delta x\,\Delta y\,\Delta z} + \frac{[J_y(y+\Delta y) - J_y(y)]\,\Delta x\,\Delta z}{\Delta x\,\Delta y\,\Delta z} + \frac{[J_z(z+\Delta z) - J_x(z)]\,\Delta x\,\Delta y}{\Delta x\,\Delta y\,\Delta z} \quad (7.4a)$$

By letting the length of the cube's sides Δx, Δy, and Δz become infinitely small, and then the differences on the right side of Eq. (7.4a) become the partial derivatives of the flux component contributions at the location $P(x, y, z)$ and one obtains

$$\frac{\partial c}{\partial t} = -\left(\frac{\partial J_x}{\partial x} + \frac{\partial J_y}{\partial y} + \frac{\partial J_z}{\partial z}\right) = D\left(\frac{\partial^2 c}{\partial x^2} + \frac{\partial^2 c}{\partial y^2} + \frac{\partial^2 c}{\partial z^2}\right) \quad (7.4b)$$

Then the contributions of the diffusion flux in the direction of the three coordinate axis, according to Eqs. (7.2) and (7.3a), are

$$J_x = -D\,\partial c/\partial x \quad J_y = -D\,\partial c/\partial y, \text{ and } J_z = -D\,\partial c/\partial z. \quad (7.5)$$

With help from the divergence and gradient, one obtains the same result in the form of the following expression:

$$\frac{\partial c}{\partial t} = -\operatorname{div} \boldsymbol{J} = D\operatorname{div}\operatorname{grad} c = D\nabla^2 c = D\left(\frac{\partial^2 c}{\partial x^2} + \frac{\partial^2 c}{\partial y^2} + \frac{\partial^2 c}{\partial z^2}\right) \quad (7.6)$$

The mathematical operator ∇, called Nable or del, appearing in Eq. (7.6), has the following structure:

$$\nabla = \frac{\partial}{\partial x}\boldsymbol{i} + \frac{\partial}{\partial y}\boldsymbol{j} + \frac{\partial}{\partial z}\boldsymbol{k} \quad (7.7)$$

When del is applied to concentration c, $\nabla c = \operatorname{grad} c$, and to the vector of the diffusion flux $\boldsymbol{J} = -D\operatorname{grad} c$, it gives $\nabla \boldsymbol{J} = -\operatorname{div}\boldsymbol{J} = D\operatorname{div}\operatorname{grad} c = D\nabla^2 c$. The application of the del operator twice leads to a scalar, to a vector, and once again to a scalar, then $\boldsymbol{i}\cdot\boldsymbol{i} = \boldsymbol{j}\cdot\boldsymbol{j} = \boldsymbol{k}\cdot\boldsymbol{k} = 1$ and $\boldsymbol{i}\cdot\boldsymbol{j} = \boldsymbol{i}\cdot\boldsymbol{k} = \boldsymbol{j}\cdot\boldsymbol{k} = 0$, and subsequently to

$$\nabla^2 = \left(\frac{\partial^2 c}{\partial x^2} + \frac{\partial^2 c}{\partial y^2} + \frac{\partial^2 c}{\partial z^2}\right) \quad (7.8)$$

Equations (7.5) and (7.6) are known as Fick's second law for the case where the diffusion has a constant diffusion coefficient.

The immediate result of the above discussion is that the diffusion equation can be transformed into the differential equation for heat conduction by substitution of c by T and D by κ. This analogy has the consequence that practically all mathematical solutions of the heat conductance equation are applicable to the diffusion equation. The analogy between diffusion and conductance should be kept in mind in the following discussion although the topic here will be mainly the treatment of the diffusion equation, which represents the most important process of mass transport.

7.1.3
The General Transport Equations

If diffusion and convection currents are similar in magnitude, then the total transport is the sum of all the individual contributions. While convection currents caused by mild shaking of low-viscosity liquids lead to a much faster mixing than by diffusion processes, the influence of convection decreases with increasing viscosity (e.g., mayonnaise).

A decrease in concentration in addition to physical transport effects can also be the consequence of a chemical reaction taking place. The concentration decrease per unit of time caused by chemical reaction is defined as the rate of reaction r and is a function of the concentrations present at the reaction site:

$$r = \frac{dc}{dt} = k\,c^n \qquad (7.9)$$

The proportionality factor k is the reaction rate constant. The exponent n, usually 1 or 2, specifies the order of the reaction.

The simultaneous occurrence of reaction and transport processes can be represented by adding the contributions together, and for the total concentration decrease over time at a given point $P(x,y,z)$ in the media considered by the general transport equation one obtains

$$-\left.\frac{\partial c}{\partial t}\right|_{total} = \operatorname{div} \boldsymbol{J}\ (\text{Diffusion}) + \operatorname{div} \boldsymbol{J}\ (\text{Convection}) + r\ (\text{reaction})$$

$$= -D\left(\frac{\partial^2 c}{\partial x^2} + \frac{\partial^2 c}{\partial y^2} + \frac{\partial^2 c}{\partial z^2}\right) + \left(v_x \frac{\partial c}{\partial x} + v_y \frac{\partial c}{\partial y} + v_z \frac{\partial c}{\partial z}\right) + k\,c^n \qquad (7.10)$$

A typical example of transport and reaction occurring during storage of a package is the spoilage of fat-containing food by oxidation with oxygen transported from the atmosphere through the packaging.

Equation (7.10) is a mass balance. At every location, a decrease in concentration of substance i takes place by transport and chemical reaction. Thus the total decrease $-\partial c/\partial t$ is equal to the amount of substance leaving the location, which includes the changes due to diffusion and convection plus the loss due to chemical reaction. By this, the description of the location where the processes take place is properly described as the source of substance i.

7.2
Solutions of the Diffusion Equation

For interactions between packaging and product, the above descriptions of both material transport processes by diffusion and convection as well as the simultaneous chemical reactions come into consideration. The general transport equation (7.10) is the starting point for solutions of all specific cases occurring in practice. Material loss through poorly sealed regions in the package can be considered as convection currents and/or treated as diffusion in the gas phase.

A solution of the general equation delivers the concentration contribution at every point in time and at every location throughout the volume considered and thus $c = c(x, y, z, t)$. The general form of the transport equation as a second-order partial differential equation has no solution. Analytical solutions are given, however, for numerous special cases. For solutions involving complicated cases, simplifying approximations are used or numerical solutions are carried out. Since the general equation (7.10) represents a starting point not only for interesting interactions but also for the complete chemical reaction technology, there are numerous solutions described in the literature which can be applied to interaction problems. The usefulness of analogous considerations was already mentioned in the comparison of diffusion and heat conductance.

Since Eq. (7.10) is composed of the sum of its members, it is logical to consider next the contribution of each individual component. The fastest step in a group of simultaneous overlapping processes is the most important. If the overall process is the result of a series of processes taking place one after another, for example as a consequence of transport processes through one or more boundary surfaces, then the slowest step of the process determines the rate of the overall process. Mass transport by diffusion is without doubt the most important process throughout the storage of packed products. The discussion of the solution begins then with the diffusion equation (7.5) or (7.6). In order to start with the most general case in which the diffusion coefficient D is not constant, one can also write

$$\frac{\partial c}{\partial t} = \nabla^2 D c = \frac{\partial^2 D c}{\partial x^2} + \frac{\partial^2 D c}{\partial y^2} + \frac{\partial^2 D c}{\partial z^2} \tag{7.11}$$

While numerical methods come into question for solutions involving variable D, D can be assumed to be constant or practically constant for most cases of practical interest. In addition, simplified solutions for diffusion along the x-axis can be used instead of the general solution, except for some particular cases which will be pointed out later. This greatly simplifies presentation of the problem and the resulting equation for diffusion is

$$\frac{\partial c}{\partial t} = D \frac{\partial^2 c}{\partial x^2} \tag{7.12}$$

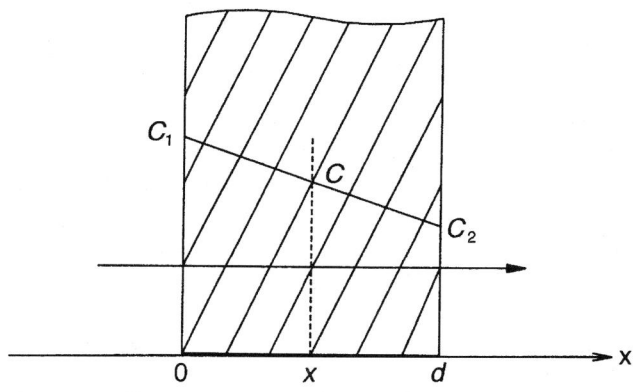

Figure 7.3 Diffusion (permeation) through a film at steady state.

7.2.1
Steady State

The simplest case to solve is when the concentration stays constant over time in the polymer. If diffusion occurs only along the direction of the x-axis, then

$$D \frac{\partial^2 c}{\partial x^2} = 0 \qquad (7.13)$$

This particular case exists, for example, in the diffusion of a substance through a film with thickness d (Figure 7.3), if the concentrations at the two surfaces c_1 at $x = 0$ and c_2 at $x = d$ remain constant (stationary case).

The first integration of Eq. (7.13) then gives

$$\frac{dc}{cx} = \text{constant} \qquad (7.14)$$

A constant concentration gradient exists in the film perpendicular to the film's surface, and consequently there is a constant diffusion flux in the x-axis direction according to Eq. (7.3c) at every location between $x = 0$ and $x = d$. Integrating Eq. (7.14) again leads to

$$\frac{c - c_1}{c_2 - c_1} = \frac{x}{d} \qquad (7.15)$$

and the amount of the flux through the film is

$$J_x = -D \frac{dc}{dx} = D \frac{c_1 - c_2}{d} \qquad (7.16)$$

7.2.2
Nonsteady State

A number of solutions exist by the integration of the diffusion equation (7.12), which are dependent on the so-called initial and boundary conditions of special applications. It is not the goal of this section to describe the complete mathematical solution of these applications or to make a list of the most well-known solutions. It is much more useful for the user to gain insight into how the solutions are arrived at, their simplifications, and the errors stemming from them. The complicated solutions are usually in the form of infinite series from which only the first or first few members are used. In order to understand the literature on the subject, it is necessary to know how the most important solutions are arrived at so that the different assumptions affecting the derivation of the solutions can be critically evaluated.

It is highly recommended to solve, or at least to read, the examples and their solutions given in this chapter. The necessary mathematical data are given in corresponding tables, and only a pocket calculator is needed. One goal followed with these very simple examples is to obtain a feeling for the many possibilities of problem solving with easily manageable mathematical tools. But it reveals also the limitations of such methods due to the enormous variety and complexity of practical problems for which a user-friendly software is needed. In Chapter 9, the examples from this chapter are again solved with such a tool and in addition much more complex case studies will be treated. The mathematical background for these is the same as that shown in Chapters 7 and 8, but the problem solving today is only possible with corresponding programs.

Most solutions of the diffusion equation (7.12) are taken from analogous solutions of the heat conductance equation that has been known for many years:

$$\frac{\partial T}{\partial t} = \kappa \frac{\partial^2 T}{\partial x^2} \tag{7.17}$$

which can be directly applied to diffusion problems. The standard reference work on the mathematics of diffusion is by Crank 1975, from which most of the solutions contained in this chapter have been taken. The solutions themselves have their origins in the older and more comprehensive reference work on heat conductance in solids by Carslaw and Jaeger 1959.

The selection of diffusion equation solutions included here are as follows: diffusion from films or sheets (hollow bodies) into liquids and solids as well as diffusion in the reverse direction; diffusion-controlled evaporation from a surface; influence of barrier layers and diffusion through laminates; influence of swelling and heterogeneity of packaging materials; coupling of diffusion and chemical reactions in filled products as well as permeation through packaging.

7.2.3
Diffusion in a Single-Phase Homogeneous System

The diffusion problem is simplest to solve analytically if the diffusing substance is concentrated at the beginning of the process in an infinitely thin sheet (plane) and then diffuses perpendicular to the plane of this sheet into an infinite liquid media found on both sides of the sheet. The flowing away from or diverging from the source

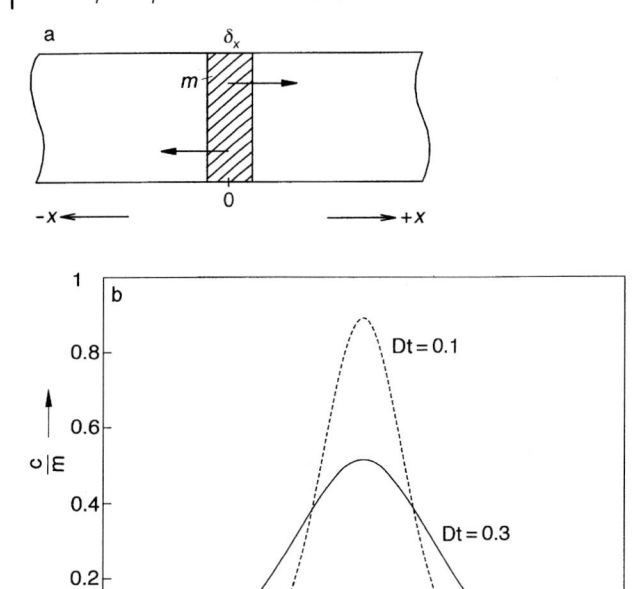

Figure 7.4 (a) Two-sided diffusion from an infinitely thin layer (source); (b) distribution of concentrations for different values of the product Dt.

is, once more, a graphic example for the expression of the diffusion equation in the form represented in Eq. (7.6): $-\partial c/\partial t = \text{div } J$. A model corresponding to this situation can be represented by a long cylindrical shape made from a polymeric material, e.g., polyethylene, with a cross-section of 1 cm². In the middle of the material, there is a very thin layer of material colored with a pigment which acts as a diffusion source (Figure 7.4(a)). The color molecules then diffuse outward toward both ends of the bar along the x-axis of the coordinate system without reaching the ends of the bar during the time interval considered.

At the beginning of diffusion, time $t=0$ and the total amount of color having mass m are located at position $x=0$. Because of the theoretically infinitely thin layer δx of the color source, the initial concentration there is infinite and the concentrations at all other positions of the bar are zero. The solution of the diffusion equation (7.12) is immediately given as

$$c = \frac{A}{t^{1/2}} \exp\left(-\frac{x^2}{4Dt}\right) \tag{7.18}$$

where A is the integration constant whose formation can be easily checked from the partial derivatives $\partial c/\partial t$ and $\partial^2 c/\partial x^2$ from Eq. (7.12). The expression in Eq. (7.18) is symmetric with respect to $x=0$ because of x^2 and goes to zero if x becomes positively or negatively infinite and $t>0$. After substitution

$$\frac{x^2}{4Dt} = q^2 \qquad dx = 2(Dt)^{1/2}dq \qquad (7.19)$$

and because the total amount m is obtained, which means

$$m = \int_{-\infty}^{+\infty} c\, dx \qquad (7.20)$$

one can write

$$m = 2AD^{1/2} \int_{-\infty}^{+\infty} \exp(-q^2)dq = 2A(\pi D)^{1/2} \qquad (7.21)$$

The values for A resulting from Eq. (7.21) are used in Eq. (7.18), and then one obtains the solution

$$c = \frac{m}{2(\pi Dt)^{1/2}} \exp\left(-\frac{x^2}{4Dt}\right) \qquad (7.22)$$

for the spreading of the color by diffusion. The increased spreading with time can be seen in Figure 7.4(b).

In the above case, half of the substance diffuses in the positive direction and the other half in the negative direction of the x-axis.

If an absolute barrier is now assumed to exist at position $x=0$ so that diffusion can occur only in the direction $x>0$, then the half of m diffusing in the $x<0$ is reflected by the barrier and overlaps the other half diffusing in the $x>0$ direction (Figure 7.5(a)). Because the symmetry of the curve of Eq. (7.22) with respect to the source at the position $x=0$, one obtains a solution for diffusion in a half open media

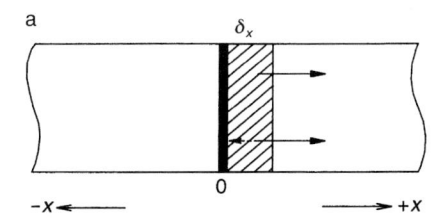

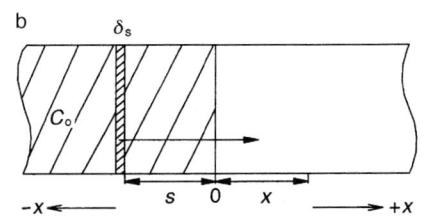

Figure 7.5 (a) Single layer diffusion from a source with a barrier on one side; (b) diffusion from an infinitely thick layer represented as coming from infinitely many sources.

that is double the value of Eq. (7.22):

$$c = \frac{m}{(\pi Dt)^{1/2}} \exp\left(-\frac{x^2}{4Dt}\right) \tag{7.23}$$

The requirement of a barrier layer at $x=0$ is expressed mathematically by the boundary condition of $\partial c/\partial x = 0$ at $x \leq 0$.

The first complication for the application of the diffusion equation (7.12) comes when the complete left half of the plastic bar, $x < 0$, is uniformly and completely colored with coloring agent which can diffuse in the direction of $x > 0$ (Figure 7.5(b)). The concentration of the color is expressed by the finite concentration of c_0. In order to find a solution to the problem, the colored region $x < 0$ is thought of as being divided into an infinite number of layers perpendicular to the x-axis. In doing this, the problem can be related to an infinite number of diffusion sources and the mathematical solution can be arrived at by overlapping many solutions of the form of Eq. (7.18).

Considering the thickness δs of such a source (Figure 7.5(b)), one gets the amount of substance contained in the cross-section of the bar, $c_0\,\delta s$, because it has the unit surface area. One obtains the expression for the concentration c_s of the color originating from this source at a distance s at time t according to Eq. (7.22):

$$c_s = \frac{c_0 \delta s}{2(\pi Dt)^{1/2}} \exp\left(-\frac{s^2}{4Dt}\right) \tag{7.25}$$

The integration of c_s over all layers δs gives with Eq. (7.25) the concentration $c(x, t)$ at any position $x > 0$ at time t:

$$c_s(x, t) = \frac{c_0 \delta s}{2(\pi Dt)^{1/2}} \exp\left\{-\frac{(x+s)^2}{4Dt}\right\}$$

$$c(x, t) = \sum_{s=0}^{\infty} c_s(x, t) = \frac{c_0}{2(\pi Dt)^{1/2}} \int_{s=0}^{\infty} \exp\left\{-\frac{(x+s)^2}{4Dt}\right\} ds$$

$$= \frac{c_0}{2(\pi Dt)^{1/2}} \int_z^{\infty} e^{-\eta^2} 2(Dt)^{1/2} d\eta = \frac{c_0}{\pi^{1/2}} \int_z^{\infty} e^{-\eta^2} d\eta \tag{7.26}$$

with

$$\frac{x+s}{2(Dt)^{1/2}} = \eta \quad ds = 2(Dt)^{1/2} \quad \text{for} \quad s = 0 \ \ \eta = \frac{x}{2(Dt)^{1/2}} = z \tag{7.27}$$

With $c_s(x, t)$, the concentration coming from the source is designated for position x at time t at a distance s from the initial point.

In order to make the right side of Eq. (7.26) easier to use, the following relationship can be considered:

$$\frac{2}{\pi^{1/2}} \int_z^{\infty} e^{-\eta^2} d\eta = \frac{2}{\pi^{1/2}} \int_0^{\infty} e^{-\eta^2} d\eta - \frac{2}{\pi^{1/2}} \int_0^z e^{-\eta^2} d\eta \tag{7.28}$$

$$= 1 - \mathrm{erf}(z) = \mathrm{erfc}(z)$$

Table 7.1 Table of different error function forms.

z	erf z	erfc z	F(z)	z	erf z	erfc z	F(z)
0.00	0.000000	1.000000	0.00000	1.10	0.880205	0.119795	0.59827
0.05	0.056372	0.943628	0.05401	1.20	0.910314	0.089686	0.62146
0.10	0.112463	0.887537	0.10354	1.30	0.934008	0.065992	0.64236
0.15	0.167996	0.832004	0.14908	1.40	0.952285	0.047715	0.66126
0.20	0.222703	0.777297	0.19098	1.50	0.966105	0.033895	0.67841
0.25	0.276326	0.723674	0.22965	1.60	0.976348	0.023652	0.69405
0.30	0.328627	0.671373	0.26540	1.70	0.983790	0.016210	0.70834
0.35	0.379382	0.620618	0.29850	1.80	0.989091	0.010909	0.72144
0.40	0.428392	0.571608	0.32921	1.90	0.992790	0.007210	0.73349
0.45	0.475482	0.524518	0.35775	2.00	0.995322	0.004678	0.74460
0.50	0.520500	0.479500	0.38431	2.10	0.997021	0.002979	0.75488
0.55	0.563323	0.436677	0.40907	2.20	0.998137	0.001865	0.76441
0.60	0.603856	0.396144	0.43220	2.30	0.998857	0.001143	0.77326
0.65	0.642029	0.357971	0.45382	2.40	0.999311	0.000689	0.78150
0.70	0.677801	0.322199	0.47407	2.50	0.999593	0.000407	0.78919
0.75	0.711156	0.288844	0.49306	2.60	0.999764	0.000236	0.79640
0.80	0.742101	0.257899	0.51090	2.70	0.999866	0.000134	0.80310
0.85	0.770668	0.229332	0.52767	2.80	0.999925	0.000075	0.80950
0.90	0.796908	0.203092	0.54347	2.90	0.999941	0.000041	0.81540
0.95	0.820891	0.179109	0.55836	3.00	0.999978	0.000022	0.81540
1.00	0.842701	0.157299	0.57242				

where the error function erf (z) is given by

$$\text{erf}(z) = \frac{2}{\pi^{1/2}} \int_0^z \exp(-\eta^2)\, d\eta \quad \text{erf}(-z) = -\text{erf}(z) \quad \text{erf}(0) = 0 \quad \text{erf}(\infty) = 1 \tag{7.29}$$

for which complete tables are available (Table 7.1). The complement of erf (z) is designated as erfc (z) and is also given in Table 7.1. The solution Eq.(7.26) can now be expressed in a convenient and easily useable form:

$$c(x,t) = \frac{1}{2} c_0 \, \text{erfc}\left\{ \frac{x}{2(Dt)^{1/2}} \right\} \tag{7.30}$$

The shape of the concentration curve is shown in Figure 7.6.

At position $x = 0$, $c = 0.5\, c_0$ for all values of $t > 0$. The amount of substance diffused into the uncolored portion of the bar up to time t (shaded region after $x > 0$) is equal to the amount of substance diffusing out of the colored portion (shaded region $x < 0$).

Example 1

A 10-cm high, cylindrical-shaped wheel of cheese contains a homogeneously dispersed ingredient with a concentration $c_0 = 100$ mg/kg. A second similar wheel of the same type of cheese without this ingredient is laid on top of the first wheel.

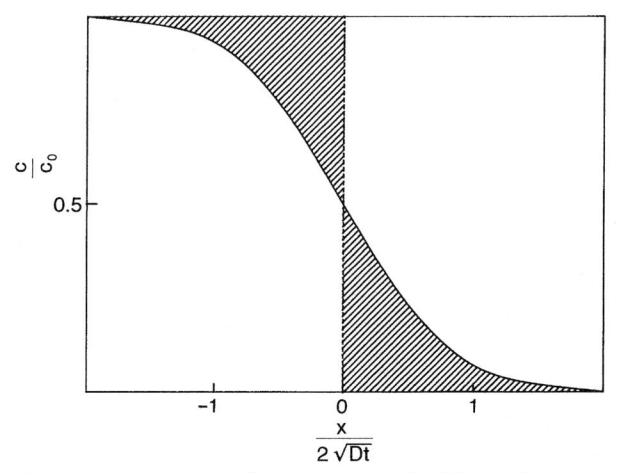

Figure 7.6 Concentration distribution curve for diffusion from an infinitely thick initial layer.

Assuming there is intimate contact between the two wheels of cheese, what is the concentration of this ingredient in the second block of cheese at a depth of 1 mm after 25 h of contact? $D = 3E{-}7 \text{ cm}^2/\text{s}$.

This problem corresponds to the example in Figure 7.5(b). A 10-cm thick wheel of cheese can be considered to be infinitely thick with respect to the diffusion coefficient provided the contact time is not too long. Equation (7.30) can be used to solve the problem.

For $x = 0.1$ cm, $t = 25 \text{ h} \cdot 3600 \text{ s/h} = 90{,}000 \text{ s}$, and $D = 3E - 7 \text{ cm}^2/\text{s}$, one calculates

$$z = \frac{x}{2(Dt)^{1/2}} = \frac{0.1}{2(3E - 7 \times 90000)^{1/2}} = 0.3$$

Looking up the value for erfc z in Table 7.1, erfc(0.3) = 0.671373, and using this value in Eq. (7.30), the concentration of the ingredient at this time and distance can be calculated to be

$$c(x, t) = \frac{1}{2} c_0 \text{ erfc}(z) = 0.5 \times 100 \times 0.671373 = 34 \text{ mg/kg}$$

Example 2

What would the distance from the surface of the second wheel for the 34 mg/kg concentration from Example 1 be after (a) 3 months and (b) after 1 year? Assume that the storage conditions remain constant and the properties of the cheese wheels also remain constant during these times.

Because the z values from Eq. (7.30) will always lead to the same concentration (i.e., 34 mg/kg), one can simply solve z for the distance

After 3 months, $(3\,\text{mo} \times 30\,\text{d/mo} \times 24\,\text{h/d} \times 3600\,\text{s/h} = 77,76,000\,\text{s}$

$$z = \frac{x}{2 \cdot (D \cdot t)^{1/2}} = \frac{x}{2(3E-7 \cdot 7776000\,\text{s})^{1/2}} = 0.3$$

and therefore $x = 0.92\,\text{cm}$.

After 1 year:

$$z = \frac{x}{2(Dt)^{1/2}} = \frac{x}{2(3E-7 \times 31536000\,\text{s})^{1/2}} = 0.3$$

and therefore $x = 1.85\,\text{cm}$.

7.2.3.1 Dimensionless Parameters and the Proportionality of Mass Transfer to the Square Root of Time

In order to compare results of studies that are expressed in different quantities, dimensionless representations are always preferred. Examples of dimensionless quantities are the relative concentration c/c_0 already mentioned above and the parameter appearing in the error function $z = x/2(Dt)^{1/2}$ in Figure 7.6. Systems described with the help of the same model but differing from one another with respect to material constants, e.g., D values, can have the same z and c/c_0 values at different times. As a result, whole series of curves can be represented by a single, easy to read curve.

Since the same z values always lead to the same c/c_0 values, the distance x_c, which is the distance of the diffusion front having concentration c, has traveled from the surface with a constant initial concentration c_0 to time t, the definition of z is given as

$$x_c = 2(Dt)^{1/2}z \tag{7.31}$$

Like in the solution given for the diffusion out of a bar colored on one side bar in Eq (5.30), it can be seen that the same c/c_0 values always result for the same z values. This means that the diffusion front of a given concentration c is proportional to the square root of Dt.

The error function used in solving the above diffusion problem occurs as a consequence of the summation of an infinite number of infinitely thin colored layers, which themselves bring about an exponential distribution of the concentration. Because of the error function's significance for numerous practical cases, this solution will be treated in somewhat more detail.

In the same manner, one obtains a solution to the diffusion equation starting with a colored layer having a finite thickness $2d$ and an initial concentration c_0 in both directions of the unbounded x-axis (Figure 7.7):

$$c = \frac{1}{2}c_0\left[\operatorname{erf}\left(\frac{d-x}{2(Dt)^{1/2}}\right) + \operatorname{erf}\left(\frac{d+x}{2(Dt)^{1/2}}\right)\right] \tag{7.32}$$

In the next example, a short section of length l is cut from the plastic bar (Figure 7.8). The bar is uniformly colored from one end up to a layer thickness of d with a pigment having concentration c_0. The zero position of the x-axis is assigned to the colored end of the bar. With the mathematical boundary conditions $x = l$, $\partial c/\partial x = 0$, one finds that the diffusion of a color molecule in the uncolored section cannot go further than the theoretical barrier existing at $x = l$ and is reflected back. For illustration purposes, the representations of the concentration profiles of this process are shown by straight lines with different slopes in Figure 7.8. The reflected path of the curve from a source

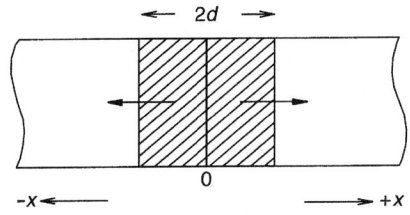

Figure 7.7 Two-sided diffusion from an infinitely thick layer.

δ_s now overlaps the original curve and the concentration at a given location x of the bar is the sum of the two contributions. A further reflection takes place at the other end of the bar at $x = 0$ and then again at $x = l$ and so forth, whereby every reflected curve section overlaps the previous section.

Because the original curve is already represented as the sum of two error functions, the complete system is represented by a series of error functions:

$$c = \frac{1}{2} c_0 \sum_{n=0}^{\infty} \left[\operatorname{erf}\left(\frac{d + 2nl - x}{2(Dt)^{1/2}}\right) + \operatorname{erf}\left(\frac{d - 2nl + x}{2(Dt)^{1/2}}\right) \right]$$
$$+ \frac{1}{2} c_0 \sum_{n=0}^{-\infty} \left[\operatorname{erf}\left(\frac{d + 2nl - x}{2(Dt)^{1/2}}\right) + \operatorname{erf}\left(\frac{d - 2nl + x}{2(Dt)^{1/2}}\right) \right] \quad (7.33)$$

Even though the above method of solution of the diffusion equation (7.12) becomes impractical for complicated cases, it illustrates the appearance of the error function in problems where diffusion from an infinite number of sources occurs and the solution is obtained in the form of an infinite series as a result of the overlapping of diffusion streams. The overlapping diffusion streams are due to an infinite number of repeated reflections at the ends of the diffusion path which are spaced finite distances apart.

As seen in Figure 7.8, the decrease in concentration is shown by sloping lines so that after each reflection the corresponding amount relative to the total concentration c becomes smaller. Due to the exponential character of the solution, the decrease is much more rapid than in the simplified representation shown in the figure and the series converge very rapidly, so that after a few terms the total concentration at a given location and time stays practically constant. There are other different methods for solving the diffusion equation in Eq. (7.12), which are described in mathematics books. Older methods, in particular separation of variables x and t, are worth mentioning. They also produce infinite series in their solutions in the form of the Fourier trigonometric series.

Further, a very elegant analytical method uses the Laplace transforms (Kreyszig 1993). In addition to analytical solutions, the possibility exists to obtain numerous

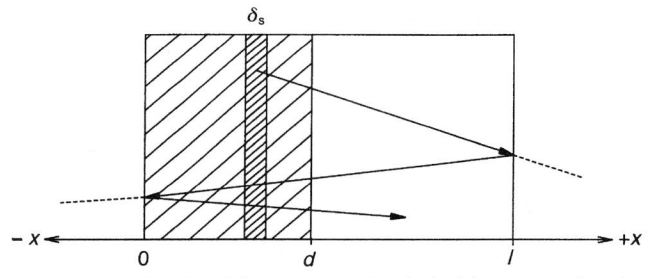

Figure 7.8 Single-sided diffusion from a finitely thick layer into a finite layer of the same material.

7.2 Solutions of the Diffusion Equation

exact solutions using numerical methods with the help of computers. The advantage of numerical methods lies primarily with their application for complicated cases, e.g., for nonconstant diffusion coefficients, for which there are no analytical solutions.

Example 3

A 100-μm thick plastic film contains an initial concentration of 100 mg/kg of some additive. This film is brought in direct contact with another 100-μm thick plastic film of the same material initially containing no additive. Assuming ideal contact between the two films (i.e., no boundary conditions exist to hinder the transfer across the interface). The exterior sides of the films are not permeable (they are in contact with a glass or metal surface). The diffusion coefficient of the additive is $3E{-}7\ \text{cm}^2/\text{s}$ for both the films. What is the concentration on the outside of the second film after 1 min contact time?

This example corresponds to Figure 7.8 in the text. The solution can be obtained using Eq. (7.33). Putting $d = 0.01$ cm, $l = 0.02$ cm, and $x = 0.02$ cm, one gets a constant sum of $c = 13.5$ mg/kg after two steps $n = 0$ and $n = 1$:

$$c = \frac{1}{2} \cdot c_0 \sum_{n=0}^{\infty} \left\{ \text{erf}\left(\frac{d + 2n \cdot l - x}{2(Dt)^{1/2}}\right) + \text{erf}\left(\frac{d - 2n \cdot l + x}{2(Dt)^{1/2}}\right) \right\}$$

$$+ \frac{1}{2} \cdot c_0 \sum_{n=0}^{-\infty} \left\{ \text{erf}\left(\frac{d + 2n \cdot l - x}{2(Dt)^{1/2}}\right) + \text{erf}\left(\frac{d - 2n \cdot l + x}{2(Dt)^{1/2}}\right) \right\}$$

for $n = 0$: $\left(\dfrac{d + 2n \cdot l - x}{2(Dt)^{1/2}}\right) = \left(\dfrac{0.01 + 2 \cdot 0 \cdot 0.02 - 0.02}{2(3E - 7 \cdot 60)^{1/2}}\right) = -1.1785$

for $n = 0$: $\left(\dfrac{d - 2n \cdot l + x}{2(Dt)^{1/2}}\right) = \left(\dfrac{0.01 + 2 \cdot 0 \cdot 0.02 + 0.02}{2(3E - 7 \cdot 60)^{1/2}}\right) = 3.5355$

for $n = 1$: $\left(\dfrac{d + 2n \cdot l - x}{2(Dt)^{1/2}}\right) = \left(\dfrac{0.01 + 2 \cdot 1 \cdot 0.02 - 0.02}{2(3E - 7 \cdot 60)^{1/2}}\right) = 3.5355$

for $n = 1$: $\left(\dfrac{d - 2n \cdot l + x}{2(Dt)^{1/2}}\right) = \left(\dfrac{0.01 - 2 \cdot 1 \cdot 0.02 + 0.02}{2(3E - 7 \cdot 60)^{1/2}}\right) = -1.1785$

for $n = -1$: $\left(\dfrac{d + 2n \cdot l - x}{2(Dt)^{1/2}}\right) = \left(\dfrac{0.01 - 2 \cdot 1 \cdot 0.02 - 0.02}{2(3E - 7 \cdot 60)^{1/2}}\right) = -5.8926$

for $n = -1$: $\left(\dfrac{d - 2n \cdot l + x}{2(Dt)^{1/2}}\right) = \left(\dfrac{0.01 + 2 \cdot 1 \cdot 0.02 + 0.02}{2(3E - 7 \cdot 60)^{1/2}}\right) = 8.2496$

$$c = \frac{1}{2} 100 \left\{ \text{erf}(-1.1785) + \text{erf}(3.5355) + \text{erf}(3.5355) + \text{erf}(-1.1785) \right\}$$

$$+ \frac{1}{2} 100 \left\{ \text{erferf}(-1.1785) + \text{erf}(3.5355) + \text{erf}(-5.8926) + \text{erf}(8.2496) \right\}$$

$$= 50\{-0.9012 + 1 + 1 + -0.9012\} + 50\{-0.9012 + 1 - 1 + 1\}$$
$$= 9.88 + 4.94 = 14.8\ \text{mg/kg}$$

After terms higher than $n=1$, the error function terms start canceling themselves out.

7.2.3.2 Comparison of Different Solutions for the Same Special Cases

Various methods can give different expressions for the solution of the same application. Even though these lead to the same result, the solutions of problems in the form of infinite series can converge at varying rates. Consequently some solutions are favored over others, depending on the parameters under consideration. Finally, the considerations of the homogenous plastic bar model will be used as an example to show the differences between different solutions.

A plastic bar of infinite length (e.g., 1 m or longer) is uniformly colored with an initial color concentration of c_0 (Figure 7.9), except for a thin layer in the middle with thickness $d = 2\,l$ (approximately 1 cm). As a simplified approximation, it is assumed that the concentration of the color at the location $x = \pm l$ remains constant at c_0. The boundary conditions are expressed mathematically as

$$\begin{aligned} c &= c_0 & x &= \pm l & t &> 0 \\ \frac{\partial c}{\partial x} &= 0 & x &= 0 & t &> 0 \end{aligned} \quad (7.34)$$

The condition of $\partial c/\partial x = 0$ at the location $x = 0$ expresses the requirement that no diffusion can take place through the axis of symmetry at $x = 0$. This leads to the same result as single-sided diffusion in a layer having half the thickness.

For the solution of the diffusion equation, Eq. (7.12), two different series expressions can be obtained:

$$\begin{aligned} 1.\; c &= c_0 \sum_{n=0}^{\infty} (-1)^n \operatorname{erfc}\left\{\frac{(2n+1)x}{2(Dt)^{1/2}}\right\} + c_0 \sum_{n=0}^{\infty} (-1)^n \operatorname{erfc}\left\{\frac{(2n+1)+x}{2(Dt)^{1/2}}\right\} + \\ 2.\; c &= c_0 - \frac{4c_0}{\pi} \sum_{n=0}^{\infty} \frac{(-1)^n}{2n+1} \exp\left[-D(2n+1)^2 \pi^n \frac{t}{4l^2}\right] \cos\frac{(2n+1)\pi x}{2l} \end{aligned} \quad (7.35)$$

The first series converges very rapidly for not too large values of Dt/l^2, in other words for relatively short diffusion times. For $Dt/l^2 = 1$, the concentration ratio c/c_0 at location $x = 0$, $c/c_0 = 0.9590 - 0.0678 + 0.0008 = 0.8920$ and for $Dt/l^2 = 0.25$,

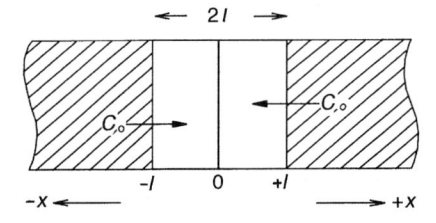

Figure 7.9 Two-sided diffusion in an infinitely thick layer.

$c/c_0 = 0.3146 - 0.0001 = 0.3145$. The trigonometric series in Eq. (5.35) converges rapidly for large t values. For $Dt/l^2 = 1$, it is $c/c_0 = 1 - 0.01080 = 0.8920$ and for $Dt/l^2 = 0.25$, $c/c_0 = 1 - 0.6872 + 0.0017 = 0.3145$.

7.2.4
Diffusion in Multiphase Systems

In this section, the important cases for food packaging are treated. These cases differ from the previous examples in that mass transfer takes place across an interface between two different media with different characteristics, e.g., with different diffusion coefficients. If the value of a quantity is desired, for example the concentration of the substance transported across the interface in one of the two media, then a mass balance must be considered that takes into account the ratio of the contact surface area and the volume of the corresponding medium.

7.2.4.1 Diffusion in Polymer/Liquid Systems

For the sake of conformity, in the following every quantity related to the packaging is designated with the index P; and the quantities related to the food are labeled with the index L.

Figure 7.10(a) shows a model that describes the mass transfer of a component dissolved in the filled product L, e.g., an aroma compound, into the packaging material P.

The model is based on the following assumptions:

1. A component i in the liquid phase with an initial concentration $c_{L,0}$ is sorbed onto the contact surface area A between the liquid and packaging and subsequently diffuses into the matrix of the packaging. In so doing, there is a decrease in concentration in the region of the contact surface which leads to further transport of i from the matrix of the liquid to the contact surface.

2. The mass transfer, controlled mainly by diffusion taking place in the packaging during storage, is several orders of magnitude lower than diffusion in the liquid phase. The difference is even greater when mixing (convection) occurs by shaking, e.g., during transport. It can be assumed that the concentration of component i in L, $c_{L,t}$, is dependent on time t but not on the distance x from the contact surface.

3. A constant distribution of i between L and P takes place that is independent of concentration of i and time. For relatively small concentrations of i (<1%), this approximate assumption is fulfilled and one defines the partition coefficient K as a constant ratio of the concentration i in the packaging material at time t on the contact surface $c_{P,t}$ (d_P) to the concentration of i in the liquid independent of location at the same time, $c_{P,t}$:

$$K = \frac{c_{P,t}(d_P)}{c_{L,t}} = \frac{c_{P,\infty}}{c_{L,\infty}} \qquad (7.36)$$

where $K = K_{P/L}$ is the ratio at $t = \infty$ of the equilibrium concentrations of i in P, $c_{P,\infty}$ to that in L, $c_{L,\infty}$. This concentration ratio is also sometimes referred to as the relative solubility constant, S_r, of i in P (relative to $c_{L,t}$).

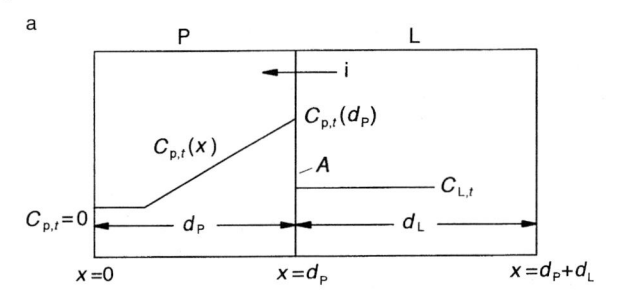

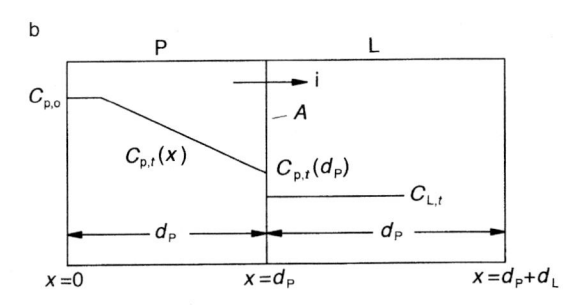

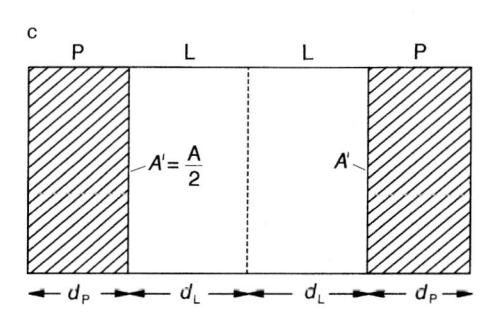

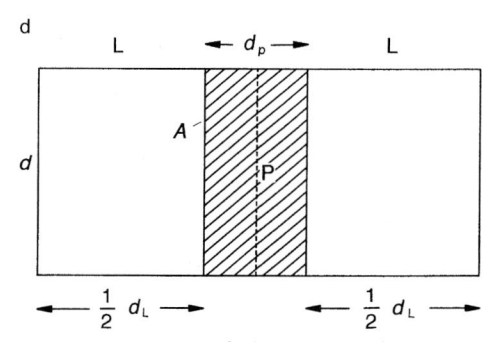

Figure 7.10 Mass transfer between a packaging material and a liquid product. (a) Diffusion out of the package into the liquid; (b) diffusion out of the package into the liquid; (c) cross-section of a representative container; (d) two-sided contact of a package material with a liquid.

4. The second important quantity influencing the mass transport is the diffusion coefficient D_P of i in P. For relatively low concentration ranges assumed for i in L, D_P is assumed to be constant. The diffusion controlled mass transport rate of i in P leads to a decrease in concentration of i with increasing distance from the contact surface (Figure 7.10(a)). Particularly in the initial stages of diffusion, the total amount of substance i transferred into the package can be concentrated in a region near to the contact surface next to L, while the location-dependent concentration of i in P in the matrix of the packaging is equal to zero.

5. The mass transport is assumed to occur in the x direction perpendicular to the contact surface. Even though the geometry of the packaging/product system influences the amount of mass transport occurring (see below), it is of minor significance for most practical cases.

6. All the above assumptions are valid for mass transfer in the reverse direction as well. This means that the migration of component i from the package into the product is also described (Figure 7.10(b)). By considering the corresponding initial conditions, the mathematical solution of the problem results in the same form.

7. The contact between packaging and product shown in Figures 7.10(a) and (b) is single sided. This means that the external surface of the packaging at location $x=0$ is assumed to be impermeable to i. The model also establishes an absolute barrier layer at the location $x=d_P+d_L$ which simplifies the representation of the problem. A representation closer to conditions in practice is shown in Figure 7.10(c) for a plastic container with a wall thickness of d_P. The single difference to Figure 7.10(a) is that the sum of the two contact surfaces A' in Figure 7.10(c) is replaced by $A=2A'$ in Figures 7.10(a) and (b). In the literature, one frequently finds a two-sided contact with the packaging using the same model shown by the representation in Figure 7.10(d). Because the axis of symmetry at $x=0$ serves as a barrier layer in the mathematical boundary conditions, the expression for the solution is not changed when instead of the half layer thickness $d_p/2 = 1$ for two-sided contact of P, the actual layer thickness d_P with single-sided contact is used. This is because in the symmetrical model in Figure 7.10(d), the total layer thickness of the liquid d_L is taken into consideration. The symmetrical model in Figure 7.10(d) also illustrates the common two-sided contact migration measurement practice in which a film or sheet is immersed in a liquid. One obtains the corresponding volumes of the packaging, $V_P = d_P A$, and liquid, $V_L = d_L A$, using the layer thicknesses d_P, d_L, and the contact surface area A. With the corresponding densities of the liquid, ρ_L, and packaging, ρ_P, the mass of liquid, $m_L = \rho_L V_L$ and mass of packaging, $m_P = \rho_P V_P$ can be calculated. In many practical cases, the assumption $\rho_L = \rho_P \cong 1$ can be made for simplification without significant error.

With dimensionless quantities α and T:

$$\alpha = \frac{1}{K}\frac{V_L}{V_P} = \frac{1}{K}\frac{d_L}{d_P} \quad \text{and} \quad T = \frac{Dt}{d_P^2} \qquad (7.37)$$

one obtains for the mass transfer by diffusion of i from a well-mixed liquid (Assumption 2) into a package or the migration in the opposite direction the general expression from Crank:

$$\frac{m_t}{m_\infty} = 1 - \sum_{n=1}^{\infty} \frac{2\alpha(1+\alpha)}{1+\alpha+\alpha^2 q_n^2} \exp(-q_n^2 T) \tag{7.38}$$

Equation (7.38) is a solution of the diffusion equation (7.12) for the models shown in Figure 7.10, where m_t is the mass of i diffusing up to time t from L through the boundary surface A into the package or opposite direction and m_∞ is the amount which has migrated at equilibrium.

The parameters q_n in the series are the positive roots of the trigonometric identity $\tan q_n = -\alpha\, q_n$. Several values of this parameter for various α and n are given in Table 7.2. The values of q_n lie between $n\pi$ (for $\alpha = 0$) and $(n-1/2)\pi$ (for $\alpha = \infty$). For $\alpha \ll 1$, $q_n \cong n\pi/(1+\alpha)$, and for the remaining α values, $q_n \cong [n - \alpha/2(1+\alpha)]\pi$.

The solutions of Eq. (7.38) converge rapidly for long diffusion times, while for short times, e.g., at the beginning of diffusion ($T \cong 0.001$), approximately 50 terms are needed. Even though this equation is no obstacle for today's PCs, it is more convenient for short times to use the form of the solution based on the error function:

$$\frac{m_t}{m_\infty} = (1+\alpha)\left[1 - \exp(z^2)\,\mathrm{erfc}(z)\right] \tag{7.39}$$

with

$$z = \frac{T^{1/2}}{\alpha} = \frac{K}{d_L}(Dt)^{1/2} \tag{7.40}$$

In Table 7.1, the values for the function contained in brackets in Eq. (7.39) are given as

$$F(z) = 1 - \exp(z^2)\,\mathrm{erfc}(z) \tag{7.41}$$

Table 7.2 Roots of $\tan q_n = -\alpha q_n$.

α	q_1	q_2	q_3	q_4	q_5	q_6
∞	1.5708	4.7124	7.8540	10.9956	14.1372	17.2788
9.0000	1.6385	4.7359	7.8681	11.0057	14.1451	17.2852
4.0000	1.7155	4.7648	7.8857	11.0183	14.1549	17.2933
2.3333	1.8040	4.8014	7.9081	11.0344	14.1674	17.3036
1.5000	1.9071	4.8490	7.9378	11.0558	14.1841	17.3173
1.0000	2.0288	4.9132	7.9787	11.0856	14.2075	17.3364
0.6667	2.1746	5.0037	8.0385	11.1296	14.2421	17.3649
0.4286	2.3521	5.1386	8.1334	11.2010	14.2990	17.4119
0.2500	2.5704	5.3540	8.3029	11.3349	14.4080	17.5034
0.1111	2.8363	5.7172	8.6587	11.6532	14.6870	17.7481
0.000	3.1416	6.2832	9.4248	12.5664	15.7080	18.8496

Equation (7.39) is particularly suitable for $T < 1$ and $\alpha < 100$. A convenient rational approximation of the error function formula in Eq. (7.39) is

$$\frac{m_t}{m_\infty} = (1 + \alpha) \left[1 - \sum_{n=1}^{5} a_n r^n \right] \quad (7.42)$$

with $r = 1/(1 \times 0.3275911\, T^{1/2}/\alpha)$, $a_1 = 0.25482592$, $a_2 = -0.28449636$, $a_3 = 1.42141371$, $a_4 = -1.45315207$, and $a_5 = 1.061405429$. For $T > 1$ or $\alpha > 100$, this approximation must not be used (Chang et al. 1988).

The solutions in Eqs. (7.38), (7.39), and (7.42) are valid for material transfer of a component i from food into the package (Figure 7.10(a)), as well as for the migration from packaging into the food (Figure 7.10(b)) under the assumptions of the described model. However, because at the beginning of diffusion in the first case the total amount m_0 of i is in L and in the second case it is in P, the values of m_t and m_∞ relative to m_0 must be different for the two cases.

1. Mass transfer from L into P
   The mass balance for i is given as

   $$V_L c_{L,\infty} + V_P c_{P,\infty} = V_L c_{L,0} = m_{L,0} \quad (7.43)$$

   where $c_{L,0}$ is the initial concentration of i in L. For the amount $m_\infty = m_{P,\infty}$, the amount of substance i in P after reaching equilibrium is obtained from Eq. (7.43) with the definition $K = c_{P,\infty}/c_{L,\infty}$ from Eq. (7.36) when Eq. (7.37) is taken into consideration:

   $$m_{P,\infty} = V_P c_{P,\infty} = \frac{V_L c_{L,0}}{1 + V_L/(V_P K)} = \frac{V_L c_{L,0}}{1 + \alpha} = m_{L,0} \frac{1}{1 + \alpha} \quad (7.44)$$

   The ratio of $m_{P,\infty}$ and $m_{L,0}$ labeled $U_{P,\infty}$ shows the fraction of the total amount of i in the package at equilibrium:

   $$U_{P,\infty} = \frac{m_{P,\infty}}{m_{L,\infty}} = \frac{1}{1 + \alpha} \quad (7.45)$$

   For $\alpha = 1$, then up to half of i would diffuse into the package at equilibrium.

2. Migration from P into L
   The total amount $m_{P,0}$ of i is contained in P at time $t = 0$ and the mass balance is expressed as

   $$V_L c_{L,\infty} + V_P c_{P,\infty} = V_P c_{P,0} = m_{P,0} \quad (7.46)$$

   The amount of substance transferred into the food at equilibrium $m_\infty = m_{L,\infty} = V_L c_{L,\infty}$ is obtained by combining Eqs. (7.36) and (7.37):

   $$m_{L,\infty} = V_P c_{P,\infty} = \frac{V_P c_{P,0}}{1/\alpha + 1} = m_{P,0} \frac{\alpha}{1 + \alpha} \quad (7.47)$$

   and related to $m_{P,0}$ the fraction of the total amount is given by

   $$\frac{m_{L,\infty}}{m_{P,\infty}} = \frac{\alpha}{1 + \alpha} \quad (7.48)$$

The fraction of i diffused from L into P up to time t, from $m_{L,0} = V_L c_{L,0}$ and the fraction migrated from P into L up to time t, from $m_{P,0} = V_P c_{P,0}$ are

$$\frac{m_{P,t}}{m_{L,0}} = \frac{m_{P,t}}{m_{P,\infty}} \frac{1}{1+\alpha} \tag{7.49}$$

and

$$\frac{m_{L,t}}{m_{P,0}} = \frac{m_{L,t}}{m_{L,\infty}} \frac{\alpha}{1+\alpha} \tag{7.50}$$

Example 4

Ten 4-cm diameter, circular 200-μm thick plastic film pieces are mounted on a stainless steel wire and placed in a glass vial containing 100 mL solvent. What percentage of the additives initially contained in the plastic migrate into the liquid over the 24-hour period ($D_P = 2.10E - 10 \text{ cm}^2/\text{s}$)? Note that the plastic additives are readily soluble in the solvent, the solvent has low viscosity, and the solvent does not swell the plastic.

This case corresponds to Figure 7.10(b) with variation in Figure 7.10(d). Because the additives are readily soluble in the solvent, $K_{P/L} \cong 1$ can be assumed in Eq. (7.36). The volume of the plastic is

$$V = 10\pi r^2 h = 10\pi \cdot 2 \text{ cm}^2 \cdot 0.02 \text{ cm} = 2.51 \text{cm}^3$$

Using Eq. (7.37), one gets

$$\alpha = \frac{1}{K_{P/L}} \cdot \frac{V_L}{V_P} = \frac{1}{1} \cdot \frac{100}{2.51} = 39.8$$

Given the two-sided contact of the liquid with the plastic $0.5\, d_P = 0.5 \cdot 0.02 \text{ cm} = 0.01 \text{ cm}$ and thus with Eq. (7.37), one gets for T

$$T = \frac{D_P t}{d_P^2} = \frac{2.1E - 10 \text{ cm}^2/\text{s} \cdot (24 \cdot 60 \cdot 60 \text{ s})}{(0.01 \text{ cm})^2} = 0.181$$

With $\alpha = 39.8$, one uses the values for α equal to infinity (∞) in Table 7.2 for the roots of $\tan q_n \cdot = -\alpha \cdot q_n$. Carrying out calculations with Eq. (7.38) for the fraction of additive migrating at time t to what would migrate at $t = \infty$

$$\frac{m_t}{m_\infty} = 1 - \sum_{n=1}^{\infty} \frac{2\alpha(1+\alpha)}{1+\alpha+\alpha^2 q_n^2} \exp\left(-q_n^2 T\right)$$

$$= 1 - \frac{2 \cdot 39.8(1+39.8)}{1+39.8+39.8^2 1.5708^2} \exp\left(-1.5708^2 \cdot 0.181\right)$$

$$- \frac{2 \cdot 39.8(1+39.8)}{1+39.8+39.8^2 4.7124^2} \exp\left(-4.7124^2 \cdot 0.181\right)$$

$$= 1 - 0.65544 - 0.001657 = 0.473$$

Note that for the summation, the second term is quite small. Because the mass balance for migration out of plastic into a liquid (Eq. (7.47)) shows $m_{L,\infty} = m_{P,0}$

$$m_{L,\infty} = m_{P,0} \frac{\alpha}{1+\alpha} = m_{P,0} \frac{39.8}{1+39.8}$$

$$\therefore m_{L,\infty} \cong m_{P,0}$$

Therefore, the percentage of additive that has migrated from the polymer in 24 h is according to Eq. (7.49) is 46.1%:

$$U_{L,t} = \frac{m_{L,t}}{m_{L,\infty}} \cdot \frac{\alpha}{1+\alpha} = 0.473 \cdot \frac{39.8}{1+39.8} = 0.461$$

Example 5

What percentage of the additives migrate out of the plastic into the liquid in Example 4 when the partition coefficient $K_{P/L} = 133$?

Starting with Eq. (7.38), one first calculates m_t/m_∞:

$$\alpha = \frac{1}{K_{P/L}} \cdot \frac{V_L}{V_P} = \frac{1}{133} \cdot \frac{100}{2.51} = 0.300$$

$$T = \frac{D_P \cdot t}{d_P^2} = \frac{2.1E-10\,\text{cm}^2/\text{s} \cdot (24 \cdot 60 \cdot 60\,\text{s})}{(0.01\,\text{cm})^2} = 0.181$$

$$\frac{m_t}{m_\infty} = 1 - \sum_{n=1}^{\infty} \frac{2\alpha(1+\alpha)}{1+\alpha+\alpha^2 q_n^2} \exp(-q_n^2 T)$$

$$= 1 - \frac{2 \cdot 0.3(1+0.3)}{1+0.3+0.3^2 2.5093^2} \exp(-2.5093^2 \cdot 0.181)$$

$$- \frac{2 \cdot 0.3(1+0.3)}{1+0.3+0.3^2 5.2937^2} \exp(-5.2937^2 \cdot 0.181)$$

$$= 1 - 0.13368 - 0.00128 = 0.865$$

Note that the values for q_n are estimated by linear interpolation of Table 7.2 values.

Now the fraction migrated from the polymer into the liquid is calculated:

$$U_{L,t} = \frac{m_{L,t}}{m_{L,\infty}} \cdot \frac{\alpha}{1+\alpha} = 0.865 \cdot \frac{0.3}{1+0.3} = 0.20$$

The percentage remaining in the polymer is 20%. Compared with Example 4, this result illustrates the effect of the larger partition coefficient where the migrant is more favorably retained in the polymer as opposed to the liquid.

Example 6

Solve Example 4 using Eq. (7.39) and compare the two results.
Starting with $\alpha = 39.8$ and $T = 0.181$ from Example 4 value for z is calculated:

$$z = \frac{T^{1/2}}{\alpha} = \frac{0.181^{1/2}}{39.8} = 0.01069$$

Entering this value for z in Eq. (7.39), one can solve for m_t/m_∞:

$$\frac{m_t}{m_\infty} = (1+\alpha)\left[1 - \exp(z^2)\operatorname{erfc}(z)\right]$$

$$= (1+39.8)[1 - \exp(0.01069^2)\operatorname{erfc}(0.01069)]$$
$$= (1+39.8)[1 - \exp(0.01069^2)\cdot(0.98795)] = 0.487$$

Then the fraction migrated is calculated by using the mass balance equation:

$$U_{L,t} = \frac{m_{L,t}}{m_{L,\infty}} \cdot \frac{\alpha}{1+\alpha} = 0.487 \cdot \frac{39.8}{1+39.8} = 0.475$$

Thus 47.5% of the additive in the polymer migrates in 24 h which is very close and within experimental error to the result in Example 4 of 46.1%.

Note that the values of erfc(0.01069) are estimated from Table 7.1 values by linear interpolation.

Example 7

Edible oil is stored in a plastic bottle with an external diameter of 10 cm and with a wall thickness of 2 mm. What percent of the antioxidant contained in the plastic bottle migrates after (a) 100 days and (b) after 2 years into the oil when the antioxidant has a diffusion coefficient of $D_P = 1E - 11 \text{ cm}^2/\text{s}$?

(a) This example corresponds to the case shown in Figure 7.10(c). Calculating α, T, and z as follows:

$$\alpha = \frac{1}{K_{P/L}} \cdot \frac{V_L}{V_P} = \frac{1}{K_{P/L}} \cdot \frac{d_L}{d_P} = \frac{1}{1} \cdot \frac{4.8}{0.2} = 24$$

$$T = \frac{D_P \cdot t}{d_P^2} = \frac{1E - 11 \text{ cm}^2/\text{s} \cdot (100 \cdot 24 \cdot 60 \cdot 60 \text{ s})}{(0.2 \text{ cm})^2} = 0.00216$$

$$z = \frac{T^{1/2}}{\alpha} = \frac{0.00216^{1/2}}{24} = 0.00194$$

For small times one can use Eq. (7.39), and performing linear interpolation on the z values between 1 and 0.05 in Table 7.1, one gets

$$\frac{m_t}{m_\infty} = (1+\alpha)\left[1 - \exp(z^2)\operatorname{erfc}(z)\right]$$

$$= (1+24)[1 - \exp(0.00194^2)\operatorname{erfc}(0.00194)]$$
$$= (1+24)[1 - \exp(0.00194^2)\cdot(0.997813)] = 0.0546$$

$$U_{L,t} = \frac{m_{L,t}}{m_{L,\infty}} \cdot \frac{\alpha}{1+\alpha} = 0.0546 \cdot \frac{24}{1+24} = 0.0524$$

Thus $100 \cdot m_t/m_\infty = 5.24\%$ migrates.

(b) Using Eq. (7.39), one gets

$$T = \frac{D_P \cdot t}{d_P^2} = \frac{1E - 11 \text{ cm}^2/\text{s} \cdot (2 \cdot 365 \cdot 24 \cdot 60 \cdot 60 \text{ s})}{(0.2 \text{ cm})^2} = 0.01577$$

$$z = \frac{T^{1/2}}{\alpha} = \frac{0.01577^{1/2}}{24} = 0.005232$$

Using the same Eq. (7.39) and performing a linear interpolation on the erfc values for z between 0.05 and 1.0 in Table 7.1, one gets

$$\frac{m_t}{m_\infty} = (1+\alpha)\left[1 - \exp(z^2)\,\text{erfc}(z)\right]$$

$$= (1+24)[1 - \exp(0.00523^2)\,\text{erfc}(0.00523)]$$
$$= (1+24)[1 - \exp(0.00523^2) \cdot (0.994104)] = 0.1467$$

$$U_{L,t} = \frac{m_{L,t}}{m_{L,\infty}} \cdot \frac{\alpha}{1+\alpha} = 0.1467 \cdot \frac{24}{1+24} = 0.141$$

Thus $100 \cdot m_t/m_\infty = 14.1\%$.

Example 8

Plastic film 100-μm thick are placed between 3-mm thick slices of cheese. How many milligrams of plastic additive are found per kg cheese after being in contact for 1 day given the initial concentration of additive is $c_{P,0} = 1$ g/kg and the diffusion coefficient in the plastic is $D_P = 2E - 10 \text{ cm}^2/\text{s}$? The diffusion coefficient in the cheese is $D_L = 1\,E - 7\,\text{cm}^2/\text{s}$ and the partition coefficient between the plastic and cheese is $K_{P/L} = 1$.

This problem corresponds to the example shown in Figure 7.10(c). First, it is necessary to calculate α and T:

$$\alpha = \frac{1}{K_{P/L}} \cdot \frac{V_L}{V_P} = \frac{1}{K_{P/L}} \frac{d_L}{d_P} = \frac{1}{1} \cdot \frac{0.3}{0.01} = 30$$

$$T = \frac{D_P \cdot t}{d_P^2} = \frac{2E - 10 \text{ cm}^2/\text{s} \cdot (24 \cdot 60 \cdot 60 \text{ s})}{(0.01 \text{ cm})^2} = 0.1728$$

Using Eq. (7.38), one calculates m_t/m_∞:

$$\frac{m_t}{m_\infty} = 1 - \sum_{n=1}^{\infty} \frac{2\alpha(1+\alpha)}{1+\alpha+\alpha^2 q_n^2} \exp(-q_n^2 T)$$

$$= 1 - \frac{2 \cdot 30(1+30)}{1+30+30^2 1.5708^2} \exp(-1.5708^2 \cdot 0.1728)$$

$$\quad - \frac{2 \cdot 30(1+30)}{1+30+30^2 4.7124^2} \exp(-4.7124^2 \cdot 0.1728)$$

$$= 1 - 0.53931 - 0.0020026 = 0.45869$$

Given that

$$K_{P,L} = 1 = \frac{c_{P,\infty}}{c_{L,\infty}},$$

$$\therefore c_{L,\infty} = c_{P,\infty}$$

One can use the mass balance equation (7.46) to calculate $c_{L,\infty}$.

Using the mass balance equation (7.50) to calculate the concentration of additive in the cheese, one gets

$$V_L \cdot c_{L,\infty} + V_P \cdot c_{P,\infty} = V_P \cdot c_{P,0} = m_{P,0}$$
$$0.3 \text{ cm}^3 \cdot c_{L,\infty} + 0.02 \text{ cm}^3 \cdot c_{L,\infty} = 0.02 \text{ cm}^3 \cdot 1 \text{ mg/cm}^3$$
$$c_{L,\infty} = 0.0625 \text{ mg/cm}^3 = 62.5 \text{ mg/kg}$$

By definition

$$\frac{c_{L,t}}{c_{L,\infty}} = \frac{m_{L,t}}{m_{L,\infty}}$$

Then solving for $c_{L,t}$, one gets

$$\frac{c_{L,t}}{62.5} = 0.45869$$

$$\therefore c_{L,t} = 28.7 \text{ mg/kg}$$

In order to take into account the influence of the rate of diffusion in the cheese, Eq. (7.58) is used to calculate β.

$$\beta = \frac{1}{K_{P/L}} \cdot \left(\frac{D_L}{D_P}\right)^{1/2} = \frac{1}{1} \cdot \left(\frac{1E-7}{2E-10}\right)^{1/2} = 22.4$$

The effect of β on Eq. (7.59) versus Eq. (7.60) without β is about $\beta/(1+\beta) = 0.957$, 4.3% smaller.

With Eqs. (7.38) and (7.50), taking the mass balance into account, the migrated amount $m_{L,t}$ through the contact surface A during time t can be calculated as follows:

$$\frac{m_{L,t}}{A} = c_{P,0} \rho_P d_P \left(\frac{\alpha}{1+\alpha}\right) \left[1 - \sum_{n=1}^{\infty} \frac{2\alpha(1+\alpha)}{1+\alpha+\alpha^2 q_n^2} \exp\left(-D_P t \frac{q_n^2}{d_P^2}\right)\right] \quad (7.51)$$

The following equation (Eq. (7.52)) represents the simplified form of Eq. (7.51) for $\alpha \gg 1$:

$$\frac{m_{L,t}}{A} = c_{P,0} \rho_P d_P \left[1 - 2 \sum_{n=1}^{\infty} \frac{1}{q_n^2} \exp\left(-D_P t \frac{q_n^2}{d_P^2}\right)\right] \quad (7.52)$$

$$q_n = (2n-1)\pi/2.$$

Equation (7.53) is an alternative migration equation for small t values using the error function

$$\frac{m_{L,t}}{A} = c_{P,0}\rho_P d_P \alpha \left[1 - \exp\left(\frac{D_P t}{d_P^2 \alpha^2}\right) \text{erfc}\left(\frac{\sqrt{D_P t}}{d_P \alpha}\right)\right] \quad (7.53)$$

$m_{L,t}/m_{L,\infty} \leq 0.5$

The following equation (Eq. (7.54)) is a simplified migration equation for $K \leq 1$ and relatively small t values, for which an infinite thickness of P is assumed:

$$\frac{m_{L,t}}{A} = \frac{2}{\sqrt{\pi}} c_{P,0}\rho_P (D_P t)^{1/2} = 1.128 c_{P,0}\rho_P (D_P t)^{1/2} \cong c_{P,0}\rho_P (D_P t)^{1/2} \quad (7.54)$$

The maximum amount of migration derived from the mass balance is

$$\frac{m_{L,\infty}}{A} = c_{P,0}\rho_P d_P \left(\frac{\alpha}{1+\alpha}\right) \quad (7.55)$$

Two typical examples of food packages with the corresponding values of the needed parameters are shown below, together with the results obtained with Eqs. (7.51) to (7.54):

$A = 600\,\text{cm}^2$, $d_P = 0.02\,\text{cm}$, $\rho_P = 1\,\text{g/cm}^3$, $t = 864{,}000\,\text{s}$ (10d), $c_{P,0} = 1000\,\text{mg/kg}$, D_P $1.0\text{E} - 10\,\text{cm}^2/\text{s}$, $K_{P/L} = 1$.

Calculated with equations	$V_L = 1000\,\text{cm}^3$, $\alpha = 83$, $m_{L,t}/A$ (mg/dm^2)	$V_L = 300\,\text{cm}^3$, $\alpha = 25$, $m_{L,t}/A$ (mg/dm^2)
Eq. (7.51)	1.042	1.030
Eq. (7.52)	1.047	1.047
Eq. (7.53)	1.049	1.049
Eq. (7.54)	1.049	1.049

The maximum amounts $m_{L,\infty}/A$ calculated with Eq. (7.55) are 1.98 and 1.92 mg/dm^2, respectively.

Example 9

Solve Example 8 using the approximation equation solution in Eq. (5.57) and compare the two results.

Given that $V_L = Ad_L = 2\,\text{cm}^2 \times 0.3\,\text{cm} = 0.6\,\text{cm}^3$, one can then calculate $c_{L,t}$ using Eq. (7.57):

$$c_{L,t} = \frac{m_{L,t}}{V_L} \cong c_{P,0}\frac{A}{V_L}(D_P t)^{1/2} = 1000\frac{2}{0.6}\left(2 \times 10^{-10} \times 24 \times 60 \times 60\right)^{1/2}$$

$$= 13.9\,\text{mg/kg}$$

This is a difference of 6.4% between the two results which is well within most experimental migration measurement errors.

In order to use the migration equations, especially the generally accepted Eq. (7.51), values for the partition coefficient K of the migrant between P and L and the diffusion coefficient D_P of the migrant in P are needed. For migrants with a high solubility in the foodstuff or simulant, the value $K_{P/L} = 1$ can be used and a worst-case estimation is obtained in this way.

For migrants with a low solubility in the foodstuff or simulant water, $K_{P/L} = 1000$ can be used to obtain a worst-case estimation (see also Chapters 4, 9 and 15).

Currently, there exists only a limited number of reliable diffusion coefficients due to the enormous requirements needed for the experimental determination. However, even for diffusion coefficients, useful estimation procedures exist (see Chapters 6 and 15). The diffusion coefficient at a given temperature T depends on the nature of the polymer, the mass and structure of the solute, and on the activation energy E_a in the diffusion process.

The material transport from a liquid, assumed to be well mixed, into packaging and the migration from packaging into a liquid both vary proportionally to the square root of time and the square root of the diffusion coefficient. While in the beginning phase (approximation equation is only valid for small z values, meaning short times), the mass transfer of i into the package is proportional to $K_{P/L}$, the migration of i from the package is independent of $K_{P/L}$. The partition coefficient plays a deciding role in the sorption (solution) of i in the packaging layer in contact with the liquid. This leads to the total amount of sorbed material being concentrated in a thin layer of packaging material in contact with the liquid, and the transport process in the initial stage is independent of the material thickness. In contrast, the migration process into the liquid takes place independent of $K_{P/L}$. Due to good mixing in the initial stages of migration, the total amount of material i is transported away from the contact layer of liquid into the volume of the liquid so that the concentration in the liquid contact layer goes to zero. The rate of diffusion of i out of the package is the rate determining step and is independent of the layer thickness d_P.

With longer migration times, the partition coefficient also plays a deciding role through the α value because for $\alpha \ll 1$ ($K_{P/L} \gg 1$), $m_{L,\infty}/m_{P,\infty} \to \alpha$ and subsequently only a very small fraction of $m_{P,0}$ migrates into L (Eq. (7.50)) (Figure 7.11).

7.2.4.2 Influence of Diffusion in Food

The diffusion coefficient in the filled product must be taken into account in liquids that are not well mixed and in viscous and solid foods. This is done through the definition of a further dimensionless parameter β:

$$\beta = \frac{1}{K}\left(\frac{D_L}{D_P}\right)^{1/2} \tag{7.56}$$

which, in addition to the parameters $K_{P/L}$ and D_P, contains the diffusion coefficient of i in the food. This dimensionless parameter can be combined with the approximation formula in Eq. (7.54) in the following way:

$$\frac{m_{L,t}}{A} = \frac{2}{\sqrt{\pi}} c_P \rho_P \frac{\beta}{1+\beta} (D_P t)^{1/2} \tag{7.57}$$

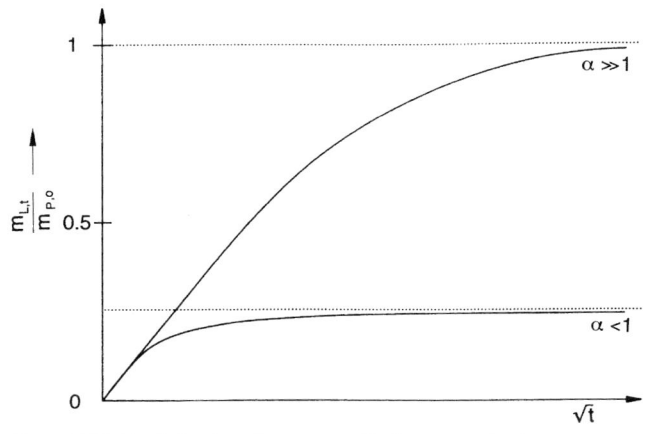

Figure 7.11 The behavior of mass transfer from a packaging material into food for different α values.

From this expression, two limiting cases can be derived as follows:

1. Where $D_L \gg D_P$ and $K_{P/L} \leq 1$, then $\beta/(1+\beta) \to 1$ and Eq. (7.57) goes to Eq. (7.54). This means that for high diffusion rates in the food, the rate of migration is determined by diffusion into packaging. The same result is obtained for $D_L \cong D_P$ and $K_{P/L} \ll 1$ meaning for approximately equal diffusion coefficients of i in L and P that transport through the packaging determines the rate of the whole process if i dissolves much better in L than in P. The flux of n through the unit surface area for equal diffusion coefficients is directly proportional to its concentration.

2. Where $D_L < D_P$ and $K_{P/L} > 1$, then $\beta/(1+\beta) \to \beta$ which in this case gives the following expression instead of Eq. (7.57):

$$\frac{m_{L,t}}{A} = \frac{2}{\sqrt{\pi}} c_P \rho_P (D_L t)^{1/2}$$

Here the migration rate of i in the food is determined by the value of the diffusion coefficient in the food as well as by the partition coefficient. The concentration $c_{L,t}$ of migrants that are poorly dissolved in the food ($K_{P/L} > 1$) increases more slowly than when they are more easily dissolved (Reid et al. 1980).

The exact expression for the differential equation (7.12) that takes into consideration the diffusion in food and finite values for V_P and V_L is extremely complicated. For such problems, numerical solutions of the diffusion equations for multilayers and corresponding computer programs are needed, as will be shown in the next chapters.

7.2.5
Surface Evaporation

A substance diffusing through the packaging toward the external environment will enter the atmosphere in the absence of a barrier layer. Water or aroma loss through permeation and drying of a printed film by evaporation of the residual solvent are common examples of this process. In cases where the substance reaching the surface has a very low vapor pressure, e.g., a plasticizer, the rate of evaporation can be slower than the rate of diffusion through the packaging, thus determining the rate of the entire process. At very low evaporation rates, the entire process can reach a standstill or lead to "sweating" in the case of low solubility of the plasticizer in the plastic.

It is also possible to have the reverse process whereby a substance condenses out of the atmosphere on the package surface, e.g., water, with subsequent diffusion into P. If the packaging surface (or also that of the food) is dry, i.e., the surface has a lower water partial pressure than that in the gas phase G, then water can be absorbed, A.

In order to mathematically describe the evaporation and condensation processes, the simplifying assumption is used that the rate of material transport through the surface is directly proportional to the difference between the concentration $c_{A,P}$ on the package surface at that time and the concentration $c_{A,G}$ in G in equilibrium with the partial pressure of the substance in the atmosphere. Using this assumption, one obtains the boundary condition for the surface at the location $x = 0$:

$$-D_P \frac{\partial c_P}{\partial x} = k \left(c_{A,G} - c_{A,P} \right) \tag{7.58}$$

with the proportionality constant k. If $c_{A,G} > c_{A,P}$, condensation will take place and if $c_{A,G} < c_{A,P}$, then evaporation will take place.

The general solution for this problem in the form of the dimensionless ratio m_t/m_∞ according to Crank is

$$\frac{m_t}{m_\infty} = 1 - \sum_{n=1}^{\infty} \frac{2L^2 \exp(-\beta_n^2 D_P t / d_P^2)}{\beta_n^2 (\beta_n^2 + L^2 + L)} \tag{7.59}$$

with $L = \dfrac{d_P k}{D_P}$ and $\beta = 1$.

Values of the positive roots of the equation $\beta \tan \beta = L$ are given in Table 7.3. Here, m_t is the amount of material taken up by the packaging or evaporated from the surface up to time t, and m_∞ is the corresponding amount at equilibrium. In Figure 7.12, the

Table 7.3 Roots of $\beta \tan \beta = L$.

L	β_1	β_2	β_3	β_4	β_5	β_6
0.00	0.0000	3.1416	6.2832	9.4248	12.5664	15.7080
0.01	0.0998	3.1448	6.2848	9.4258	12.5672	15.7086
0.10	0.3111	3.1731	6.2991	9.4354	12.5743	15.7143
0.20	0.4328	3.2039	6.3148	9.4459	12.5823	15.7207
0.50	0.6533	3.2923	6.3616	9.4775	12.6060	15.7397
1.00	0.8603	3.4256	6.4373	9.5293	12.6453	15.7713
2.00	1.0769	3.6436	6.5783	9.6296	12.7223	15.8336
5.00	1.3138	4.0336	6.9096	9.8928	12.9352	16.0107
10.00	1.4289	4.3058	7.2281	10.2003	13.2142	16.2594
100.00	1.5552	4.6658	7.7764	10.8871	13.9981	17.1093
∞	1.5708	4.7124	7.8540	10.9956	14.1372	17.2788

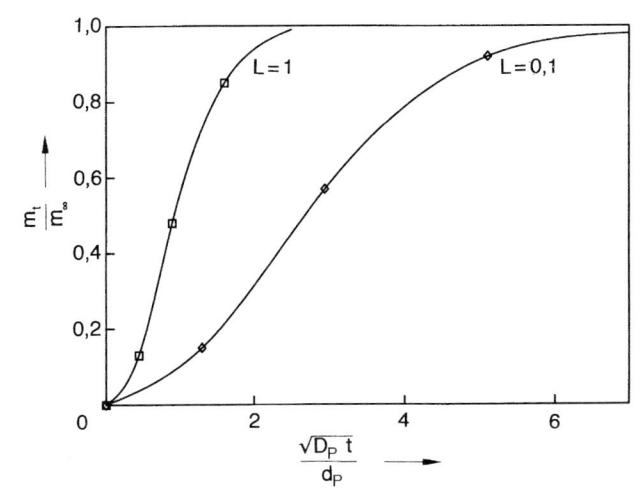

Figure 7.12 Sorption or desorption curves in the valid range of Eq. (7.58) for different L values.

ratio of m_t/m_∞ is given as a function of the dimensionless quantity $(D_P t/d_P^2)^{1/2}$ for various L values. In the absence of evaporation, the curves show a linear increase at the beginning of diffusion (Figure 7.11), while the obvious curving shown in Figure 7.12 for small k values is caused by the slower evaporation process.

7.2.6
Permeation Through Homogeneous Materials

Steady-state permeation which follows Fick's first law has been previously described in Eq. (7.16).

Assuming the concentration of i in P has a constant value $c_{P,1}$ at the surface ($x=0$), a constant value $c_{P,2}$ at the other surface ($x=d_P$), and at the beginning of permeation the concentration inside of P has the value $c_{P,0}$ ($t=0$), and then a nonsteady state of diffusion will take place leading to a change in the concentration $c_{P,t}$ within P. For simplification, one can set $c_{P,0}=0$ and $c_{P,2}=0$. The resulting amount of mass diffusing through the package up to time t is then given as

$$m_t = A d_P c_{P,1} \left[\frac{D_P z}{d_P^2} - \frac{1}{6} - \frac{2}{\pi^2} \sum_{n=1}^{\infty} \left\{ \frac{(-1)^n}{n^2} \exp(-D_P n^2 \pi^2 t/d_P^2) \right\} \right] \quad (7.60)$$

This equation becomes asymptotic to the straight line:

$$m_t = \frac{A d_P c_{P,1}}{d_P} \left(t - \frac{d_P^2}{6 D_P} \right) \quad (7.61)$$

as $t \to \infty$. The intersection of this straight line with the t-axis at location Θ is

$$\Theta = \frac{d_P^2}{6 D_P} \quad (7.62)$$

This is Barrer's equation for determining the diffusion coefficient using permeation measurements (Figure 9.1). The steady-state permeation flux is given by the slope of the straight line (7.61):

$$J = \frac{m_t}{At} = D_P \frac{c_{P,1}}{d_P} \tag{7.63}$$

This expression is identical to Eq. (7.16) for $c_{P,2} = 0$.

7.2.7
Permeation Through a Laminate

Diffusion through a barrier layer is a special case of diffusion through a laminate film composed of several layers with different thicknesses and diffusion coefficients. The mathematical treatment of the case of nonsteady state is complicated and can be solved in a general way only with numerical methods as shown in the following sections of this and the next chapter. The steady-state permeation case allows the overall transport to be simply treated. Let n films with thicknesses $d_{P1}, d_{P2}, \ldots$

$\Delta C = \frac{J d_{P,1}}{D_{P,1}} + \frac{J d_{P,2}}{D_{P,2}} + \cdots + \frac{J d_{P,n}}{D_{P,n}} = (R_1 + R_2 + \cdots + R_n)J$, d_{Pn} with corresponding diffusion coefficients $D_{P1}, D_{P2}, \ldots, D_{Pn}$ be bound together in a laminate. Because in steady state, the flux J of the diffusing substance i is the same through every individual component of the laminate, one obtains an expression for the concentration gradient as follows:

$$\Delta C = \frac{J d_{P,1}}{D_{P,1}} + \frac{J d_{P,2}}{D_{P,2}} + \cdots + \frac{J d_{P,n}}{D_{P,n}} = (R_1 + R_2 + \cdots + R_n)J \tag{7.64}$$

with the resistance $R_1 = d_{p1}/D_{P1}$ etc.

The total resistance related to the diffusion is then the sum of the individual resistances, and the total flux is practically determined by the layer with the smallest diffusion coefficient.

Diffusion through a liquid boundary layer can be treated as a special case of diffusion through a laminate. With large $K_{P/L}$ values, that is low solubility of component i in a liquid food, the material transport through the interface A can also be determined from the contribution of diffusion in L under conditions of thorough mixing. Van der Waals attraction forces between the package surface and the molecules of L in intimate contact with P lead to the formation of a thin but immobile layer in which the diffusion coefficient of i in L, D_L, controls mass transport (the Nernst diffusion layer). An example will be treated in Chapter 9.

7.2.8
Concentration Dependence of the Diffusion Coefficient

At dilute concentration D_P is usually constant. When swelling is caused by either fat, water, essential oils, or other organic components found in the product, then D_P can become concentration dependent in the region of a boundary layer in P. In such

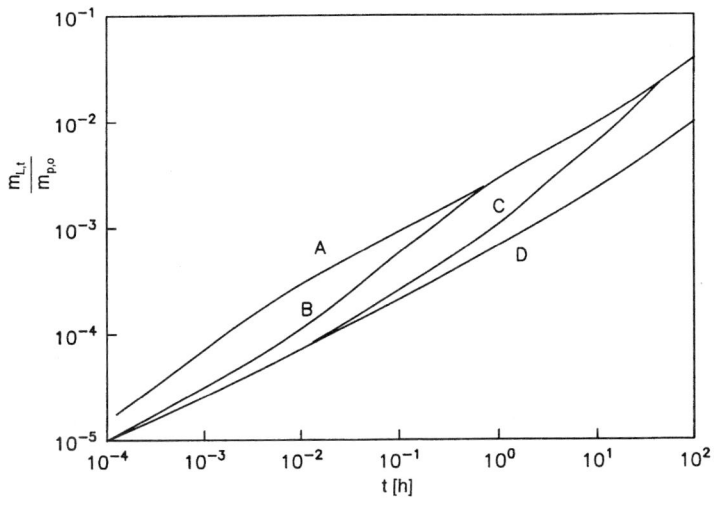

Figure 7.13 Migration from a system with swelling under various conditions (Chang et al. 1988). $D_{P,Q} = 16E-10\,\text{cm}^2/\text{s}$, $D_P = 10E-10\,\text{cm}^2/\text{s}$; $t_0 = 0$; v_Q: $E-5\,\text{cm/s}$ in A, $E-6\,\text{cm/s}$ in B, $E-7\,\text{cm/s}$ in C, and $E-8\,\text{cm/s}$ in D.

cases, the diffusion equation (7.12) is no longer valid and the general form of the diffusion equation (7.11) must be used.

In this case, $D = D(c)$ is a function of the concentration c of the substance causing the swelling. The literature holds numerous recommended solutions for treating such cases, none of which are universally applicable.

A general way for solving problems of this type is to use numerical integration in combination with a representative model suitable for the specific case. In the initial stages when the food or another product is brought in contact with the package ($t = 0$), the migration of the substance i from P into L takes place with a constant D_P because the swelling processes require a certain amount of time before they affect the migration process of i. After this initial contact phase, the swelling front moves into P with a certain speed v_Q (Figure 7.13). In the region $x > x_Q$, the diffusion of i takes place with D_P and in the region $x < x_Q$ with $D_{P,Q} > D_P$. The swelling front x_Q moves into P with the speed v_Q:

$$x_Q = v_Q(t - t_0) \quad (7.65)$$

whereby $t_0 > 0$ signifies the initial contact phase before swelling takes place. The result of such a process can be qualitatively seen in Figure 7.13 (Chang et al. 1988).

7.2.9
Diffusion and Chemical Reaction

When a first-order irreversible chemical reaction (e.g., oxygen absorption and oxidation) takes place simultaneously with diffusion in food for example, then one obtains the following expression from the general mass transfer equation (7.10):

$$\frac{\partial c}{\partial t} = D\frac{\partial^2 c}{\partial x^2} - kc \tag{7.66}$$

where k is the reaction rate constant. If the reaction takes place in a relatively thin layer near the surface or boundary layer of L to P, then one can consider L as a half-open medium (infinitely thick). This leads to a considerable simplification of the mathematical treatment. Furthermore, letting $c_{L,o}$ be a constant surface concentration, one obtains the absorbed amount m_t up to time t:

$$m_t = Ac_{L,0}(D_L/k)^{1/2}\left[\left(kt + \frac{1}{2}\right)\mathrm{erf}\,(kt)^{1/2} + (kt/\pi^{1/2}e^{-kt})\right] \tag{7.67}$$

For large kt values, the $\mathrm{erf}(kt)^{1/2}$ goes to one and

$$m_t \rightarrow Ac_{L,0}(D_L/k)^{1/2}\left(t + \frac{1}{2k}\right) \tag{7.68}$$

that means m_t increases linearly with t. For very small values of kt, one obtains

$$m_t = Ac_{L,0}(D_L/k)^{1/2}\left(1 + \frac{1}{2}kt\right)(D_L t/\pi)^{1/2} \cong Ac_{L,0}(D_L/k)^{1/2}\left(1 + \frac{1}{2}kt\right)(D_L t)^{1/2} \tag{7.69}$$

When $k \rightarrow 0$, only diffusion without reaction takes place:

$$m_t \cong Ac_{L,0}(D_L t)^{1/2} \tag{7.70}$$

Because the diffusion process and the reaction occur in the same medium L, the ratio of A/V_L does not come into consideration.

7.3
Numerical Solutions of the Diffusion Equation

7.3.1
Why Numerical Solutions?

Despite a large number of analytical solutions available for the diffusion equation, their usefulness is restricted to simple geometries and constant diffusion coefficients. The boundary conditions, which can be analytically handled, are equally simple. However, there are many cases of practical interest where the simplifying assumptions introduced when deriving analytical solutions are unacceptable. For example, the diffusion process in polymer systems is sometimes characterized by markedly concentration-dependent diffusion coefficients, which make any analytical result inapplicable. Moreover, when the analytical solutions are generally expressed in the form of infinite series, their numerical evaluation is no trivial task. That is, the simplicity of the adopted models is not necessarily reflected by an equivalent simplicity of evaluation (Koonin and Meredith 1990).

To obtain solutions to the diffusion equation, which more realistically models practical situations (where, for example, the diffusion coefficient or the boundary conditions are nonlinear), one must resort to numerical methods. Basically, these imply restricting the solution of the diffusion problem to a set of grid points, conveniently distributed within the integration domain, and approximating the involved derivatives by discrete schemes. Such an approach leads to a system of linear equations, having as unknowns the solution values at the grid points. The linear system can be solved in principle by any classical method, even though, for the sake of computational efficiency, more specialized methods are recommended. The numerical discretization methods affect the essence of the physical model much less than the analytical approximations do, allowing for much more complex diffusion problems to be treated (Burden and Faires 1985).

7.3.2
Finite-Difference Solution by the Explicit Method

We consider for now the one-dimensional diffusion equation, with constant diffusion coefficient D:

$$\frac{\partial c}{\partial t} = D\frac{\partial^2 c}{\partial x^2} \tag{7.71}$$

Such an equation is useful for describing the time evolution of the concentration profile of some diffusant across a plane sheet of given thickness L and infinite transverse extension. In order to model a particular experimental arrangement, this equation must be solved in conjunction with certain *initial* and *boundary conditions*. We will consider that Eq. (7.71) is subject to the *initial condition*

$$c(x, t_0) = c^0(x) \quad x \in [0, L] \tag{7.72}$$

which means that the concentration profile at the initial moment t_0 is given over the entire sheet thickness. However, the solution of the *initial value* (or *Cauchy*) *problem* defined by Eqs. (7.71) and (7.72) cannot be uniquely determined unless supplementary *boundary conditions* for $t > t_0$ are specified. For simplicity, we will assume that the concentration values at the outer surfaces of the sheet are constant for any $t \geq t_0$

$$c(0, t) = c_0^0 \quad c(L, t) = c_L^0 \tag{7.73}$$

Such boundary conditions, specifying the *values* of the solution, are known as *Dirichlet boundary conditions*. The so-called *Neumann boundary conditions*, which define the *derivative* of the solution on the boundaries, form another important category, considered among others later in this chapter.

The method we use to approximate the solution to the problems (7.71) to (7.73) is based on finite difference schemes for the derivatives involved by the diffusion equation (and, in general, by the boundary conditions, too). The straightforward

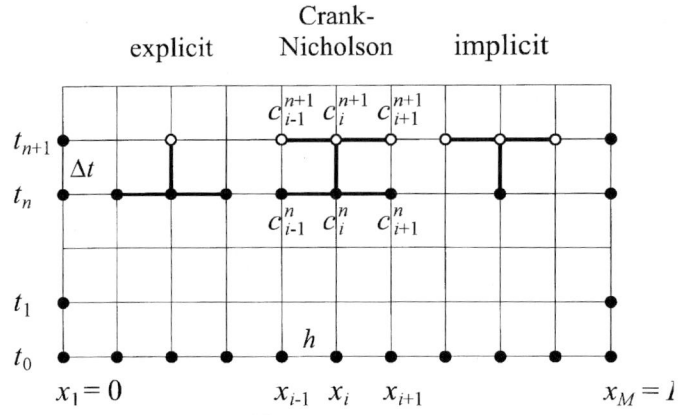

Figure 7.14 Space–time grid for the one-dimensional diffusion equation, evidencing the explicit forward-difference, implicit backward-difference, and Crank–Nicholson discretization schemes.

approach is to choose equally spaced points along the x and t axes, covering the space–time integration domain by a regular rectangular grid (Figure 7.14). Denoting by h and Δt the corresponding mesh constants (with the stipulation that L/h is an integer), the grid points are defined by the discrete coordinates:

$$\begin{aligned} x_i &= (i-1)h & i &= 1, 2, \ldots, M \\ t_n &= n\Delta t & n &= 0, 1, 2, \ldots \end{aligned} \tag{7.74}$$

Here M represents the number of spatial grid points, and the spatial mesh constant is given by

$$h = L/(M-1) \tag{7.75}$$

We use the notation $c_i^n \equiv c(x_i, t_n)$. The time derivative of c at point (x_i, t_n) can be obtained from its Taylor series in t for constant $x = x_i$:

$$c_i^{n+1} = c_i^n + \Delta t \left(\frac{\partial c}{\partial t}\right)_{i,n} + \frac{1}{2}(\Delta t)^2 \left(\frac{\partial^2 c}{\partial t^2}\right)_{i,n} + \cdots \tag{7.76}$$

Taking the linear approximation and expressing the first-order time derivative, one obtains

$$\left(\frac{\partial c}{\partial t}\right)_{i,n} = \frac{c_i^{n+1} - c_i^n}{\Delta t} + O(\Delta t) \tag{7.77}$$

$O(\Delta t)$ signifies that in the above approximation, the leading term that was neglected is of the order Δt (we have divided Eq. (7.76) by Δt to get Eq. (7.77)). This is the so-called *Euler forward-difference* scheme. While it is only first-order accurate in Δt, it has the advantage that it allows for the quantities at time step $n+1$ being calculated only from those known at time step n.

The discrete approximation for the second-order spatial derivative $(\partial^2 c/\partial x^2)_{i,n}$ at $x = x_i$ results in a similar manner, namely by expressing the concentrations at the neighboring grid points x_{i-1} and x_{i+1} from the Taylor series in x at constant $t = t_n$:

$$c_{i+1}^n = c_i^n + h\left(\frac{\partial c}{\partial x}\right)_{i,n} + \frac{1}{2}h^2\left(\frac{\partial^2 c}{\partial x^2}\right)_{i,n} + \cdots$$
$$c_{i-1}^n = c_i^n - h\left(\frac{\partial c}{\partial x}\right)_{i,n} + \frac{1}{2}h^2\left(\frac{\partial^2 c}{\partial x^2}\right)_{i,n} + \cdots$$
(7.78)

On adding we find

$$\left(\frac{\partial^2 c}{\partial x^2}\right)_{i,n} = \frac{c_{i+1}^n - 2c_i^n + c_{i-1}^n}{h^2} + O(h^2) \qquad (7.79)$$

This second-order approximation is a *centered-difference* scheme, since it expresses the spatial derivative at point i by means of data from symmetrically distributed points. All the implied information is known at time step n.

By substituting relations of Eqs. (7.77) and (7.79) in Eq. (7.71), one obtains the following finite-difference approximation to the diffusion equation at point (x_i, t_n):

$$\frac{c_i^{n+1} - c_i^n}{\Delta t} = D\frac{c_{i+1}^n - 2c_i^n + c_{i-1}^n}{h^2} \qquad (7.80)$$

Having in view only the way the time derivative was approximated, this is the *forward-difference representation* of the diffusion equation and it is of the order $O(h^2 + \Delta t)$. Slight rearrangement yields a formula that expresses the time-propagated solution c_i^{n+1} for any interior spatial grid point in terms of the other quantities known at time step n:

$$c_i^{n+1} = \lambda c_{i-1}^n + (1 - 2\lambda)c_i^n + \lambda c_{i+1}^n \qquad i = 2, 3, \ldots, M-1 \qquad (7.81)$$

where

$$\lambda = \frac{D\Delta t}{h^2} \qquad (7.82)$$

The concentration values on the boundaries, c_1^{n+1} and c_M^{n+1}, generally result from the boundary conditions and, within the simple adopted model, are seen to be constant:

$$\begin{aligned} c_1^{n+1} &= c_1^n = c_0^0 \\ c_M^{n+1} &= c_M^n = c_L^0 \end{aligned} \qquad (7.83)$$

Since the solution of Eq. (7.81) propagated at time step t_{n+1} is expressed solely in terms of data from time step t_n, not requiring any previous information, the forward-difference scheme is said to be *explicit*, and its essence can be extracted from Figure 7.14, too.

The explicit nature of the recursive process described by Eqs. (7.81) to (7.83) becomes even more apparent if using matrix notation for the involved linear system

$$\mathbf{c}^{n+1} = \mathbf{B} \cdot \mathbf{c}^n \qquad n = 0, 1, 2, \ldots \qquad (7.84)$$

The components of the column-vector $\mathbf{c}^n$ are the values of the solution from all spatial grid points at time step t_n:

$$\mathbf{c}^n = [c_1^n \ c_2^n \ \cdots \ c_{M-1}^n \ c_M^n]^T \tag{7.85}$$

and the propagation matrix $\mathbf{B}$ has tridiagonal structure, i.e., except for the main diagonal and the neighboring upper and lower codiagonals, all elements are equal to 0:

$$\mathbf{B} = \begin{bmatrix} 1 & 0 & & & 0 \\ \lambda & 1-2\lambda & \lambda & & \\ & \ddots & \ddots & \ddots & \\ & & \lambda & 1-2\lambda & \lambda \\ 0 & & & 0 & 1 \end{bmatrix} \tag{7.86}$$

When solving the one-dimensional diffusion equation (7.85) by the explicit forward-difference formulation described above, one is faced under certain conditions with severe numerical *instability* problems. This means that instead of yielding a relatively smooth spatial profile, the algorithm develops oscillations, which grow exponentially in time, "unweaving" the solution and making it unusable. This critical behavior occurs when the used time step exceeds a certain upper limit for a given spatial mesh constant and is caused by the increasing dominance of round-off errors.

In order to emphasize the critical relationship between the time step and the spatial step, we consider the one-dimensional diffusion equation with constant diffusion coefficient $D = 1$:

$$\frac{\partial c}{\partial t} = \frac{\partial^2 c}{\partial x^2} \quad x \in [0, 1] \quad t > 0 \tag{7.87}$$

subject to the simple boundary conditions

$$c(0, t) = c(1, t) = 0 \quad t > 0 \tag{7.88}$$

and initial condition

$$c(x, 0) = \sin(\pi x) \quad x \in [0, 1] \tag{7.89}$$

It can be easily verified that the analytical solution to this problem is

$$c(x, t) = e^{-\pi^2 t}\sin(\pi x) \tag{7.90}$$

We investigate the behavior of the numerical solution to problems (7.87) to (7.89) at the moments $t = 2.0$, $t = 2.5$, and $t = 3.0$ for two different time steps, by using the constant spatial step $h = 0.05$.

Figure 7.15 shows the spatial concentration profiles obtained by using the time step $\Delta t = 0.00125$, corresponding to $\lambda = 0.5$ (λ is defined in Eq. (7.82)). As one may notice, apart from the inaccuracies caused by the finite spatial step size, the profiles resulted from the numerical solution (depicted with dotted lines) fairly reproduce the analytical results (continuous lines).

Figure 7.16 shows the spatial concentration profiles obtained with the slightly increased time step $\Delta t = 0.0013$, corresponding to $\lambda = 0.52$. Even though the solution

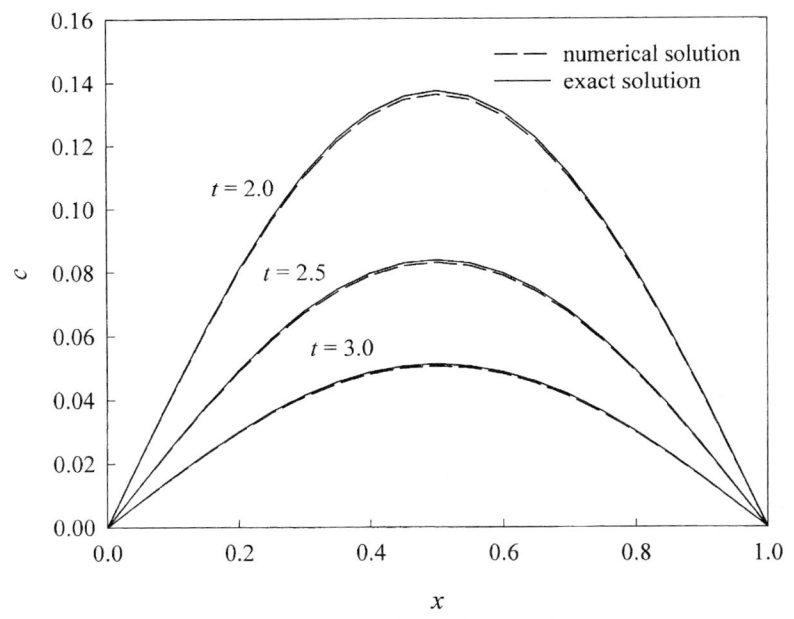

Figure 7.15 Exact and numerical solutions obtained by the explicit method for the Cauchy problem (8.17) to (8.19), by using the spatial step $h = 0.05$ and the time step $\Delta t = 0.00125$.

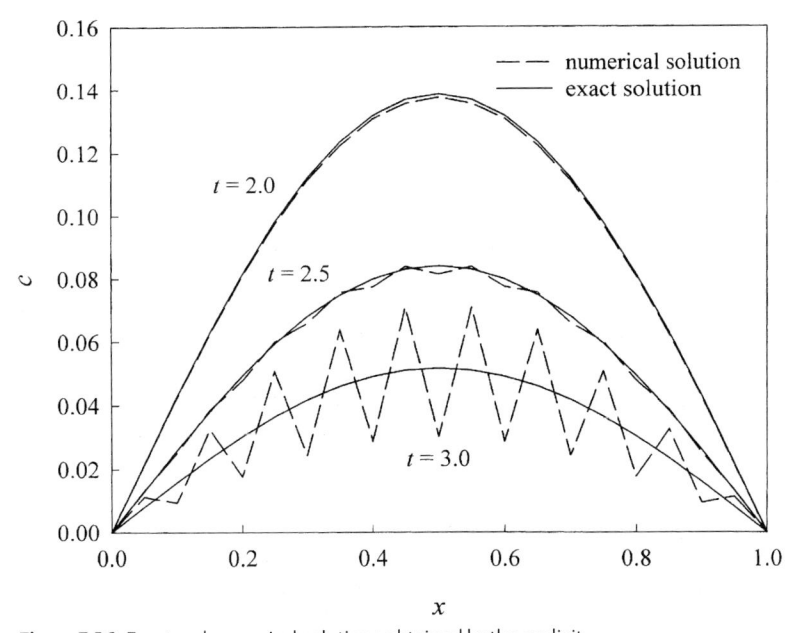

Figure 7.16 Exact and numerical solutions obtained by the explicit method for the Cauchy problem (8.17) to (8.19), by using the spatial step $h = 0.05$ and the time step $\Delta t = 0.0013$.

at $t = 2.0$ can hardly be distinguished from the one obtained with $\Delta t = 0.00125$, it is apparent that at $t = 2.5$ instabilities begin to develop and they dominate the solution entirely at $t = 3.0$. Hence, a seemingly insignificant change in the time step leads to a dramatic qualitative change of the solution. This indicates that the value $\lambda = 1/2$ is critical and it separates two domains of numerical parameters characterized by different behavior of the solution: for $\lambda < 1/2$ the propagation of the solution is stable, while for $\lambda > 1/2$ it turns out to be unstable.

7.3.2.1 von Neumann Stability Analysis

An intuitive way of investigating the stability properties of a finite-difference scheme is the *von Neumann stability analysis*, which we briefly outline as follows. The von Neumann analysis is *local*, being based on the assumption that the coefficients of the difference equation are so slowly varying in space and time as to be considered constant. Under such assumptions, the *eigenmodes* (the independent solutions) of the difference equation may be written in the general form

$$u_i^n = \xi^n \exp[\iota k(i-1)h] \qquad (7.91)$$

ι stands for the imaginary unit (not to be confused with the spatial index i), k is the spatial wave number, which can take any real value, and $\xi = \xi(k)$ is the so-called *amplification factor*, which is a complex function of k. Apart from the spatial details, the essential feature of the eigenmodes is their time dependence through the time step index n, as integer powers of the amplification factor.

The time propagation of the solution is considered to be *stable* if the amplification factor satisfies the condition

$$= \xi(k)| < 1 \qquad (7.92)$$

since no exponentially growing modes of the difference equation can exist under such circumstances.

In order to express the amplification factor for the forward-difference representation of the one-dimensional diffusion equation, one has to replace the general form of Eq. (7.91) of the eigenmodes into the difference equation (7.81):

$$\xi = \lambda \exp(-\iota kh) + (1 - 2\lambda) + \lambda \exp(\iota kh)$$

By combining the exponentials and employing the trigonometric identity $1 - \cos x = 2 \sin^2(x/2)$, one obtains for the amplification factor

$$\xi = 1 - 4\lambda \sin^2(kh/2) \qquad (7.93)$$

Use of the von Neumann stability criterion, Eq. (7.88) leads to the condition

$$0 < \lambda < 1/2 \qquad (7.94)$$

which, taking into account the definition in Eq. (7.82) of λ, becomes

$$\Delta t < \frac{1}{2}\frac{h^2}{D} \qquad (7.95)$$

The significance of this result is that the time step Δt insuring the stability of the algorithm is limited by an upper bound, which is proportional to the diffusion time across a cell of width h. This makes the explicit scheme, characterized by forward-time differencing, *conditionally stable* and proves that the value $\lambda = 1/2$ is indeed critical.

7.3.2.2 The Crank–Nicholson Implicit Method

To obtain an algorithm that is *unconditionally stable*, we consider an implicit discretization scheme that results from using backward finite differences for the time derivative. The corresponding difference equation is most conveniently obtained by approximating the diffusion equation at point (x_i, t_{n+1}):

$$\frac{c_i^{n+1} - c_i^n}{\Delta t} = D \frac{c_{i+1}^{n+1} - 2c_i^{n+1} + c_{i-1}^{n+1}}{h^2} \tag{7.96}$$

Analogous to the forward-difference method previously discussed, it is only first-order accurate in Δt. The only formal difference with respect to the forward-difference equation (8.10) appears to be the fact that the space derivative is evaluated at time t_{n+1}, but not at t_n.

By rearranging the terms in Eq. (8.26), the following system of linear equations results:

$$-\lambda c_{i-1}^{n+1} + (1 + 2\lambda)c_i^{n+1} - \lambda c_{i+1}^{n+1} = c_i^n, \, i = 2, 3, \ldots, M-1 \tag{7.97}$$

where, as before, $\lambda = D\Delta t/h^2$. By contrast with the forward-difference method, the propagated concentrations c_i^{n+1} cannot be explicitly expressed, but result from solving the above set of simultaneous linear equations at each time step. For this reason, the discussed backward-difference scheme is said to be *fully implicit*. The implicit nature of this method can also be observed from Figure 8.1, which indicates the data involved.

By using matrix notation, the linear system in Eq. (7.97) can be written as

$$\mathbf{A} \cdot \mathbf{c}^{n+1} = \mathbf{c}^n \quad n = 0, 1, 2, \ldots \tag{7.98}$$

with matrix $\mathbf{A}$ having tridiagonal structure

$$\mathbf{A} = \begin{bmatrix} 1 & 0 & & & 0 \\ -\lambda & 1+2\lambda & -\lambda & & \\ & \ddots & \ddots & \ddots & \\ & & -\lambda & 1+2\lambda & -\lambda \\ 0 & & & 0 & 1 \end{bmatrix} \tag{7.99}$$

It is apparent from the first and last rows of this matrix that again the simple *Dirichlet boundary conditions* of Eq. (7.73) have been considered. Since $\lambda > 0$, the matrix $\mathbf{A}$ is positive definite and diagonally dominant. For solving system in Eq. (7.98), the very efficient Crout factorization method for linear systems with tridiagonal matrix can be applied (see Press et al. 1986, §2.4).

The implicit backward-difference algorithm does not show the stability problems encountered in the case of the explicit forward-difference method, and this results immediately by analyzing the expression of the amplification factor

$$\xi = \frac{1}{1 + 4\lambda \sin^2(kh/2)} \qquad (7.100)$$

which obviously satisfies the von Neumann stability criterion $|\xi(k)| < 1$ for any step size Δt. Hence, the backward-difference scheme is *unconditionally stable*. Provided the solution of the differential equation satisfies the usual differentiability conditions, the local truncation error of the method is of the order $O(h^2 + \Delta t)$. The weakness of this method is, however, the low order of the truncation error with respect to time, requiring comparatively much smaller time intervals than the spatial step size.

A second-order method with respect to both space and time can be derived by approximating the diffusion equation at time step $t_{n+1/2} \equiv t_n + \Delta t/2$ and employing a centered-difference scheme for the time derivative, too. Considering the Taylor series in t at constant $x = x_i$

$$\begin{aligned}
c_i^{n+1} &= c_i^{n+1/2} + (\Delta t/2)\left(\frac{\partial c}{\partial t}\right)_{i,n+1/2} + \tfrac{1}{2}(\Delta t/2)^2 \left(\frac{\partial^2 c}{\partial t^2}\right)_{i,n+1/2} + \cdots \\
c_i^{n} &= c_i^{n+1/2} - (\Delta t/2)\left(\frac{\partial c}{\partial t}\right)_{i,n+1/2} + \tfrac{1}{2}(\Delta t/2)^2 \left(\frac{\partial^2 c}{\partial t^2}\right)_{i,n+1/2} + \cdots
\end{aligned} \qquad (7.101)$$

On subtracting, we find the time derivative sought

$$\left(\frac{\partial c}{\partial t}\right)_{i,n+1/2} = \frac{c_i^{n+1} - c_i^n}{\Delta t} + O((\Delta t)^2) \qquad (7.102)$$

The spatial derivative at time step $t_{n+1/2}$ can be approximated by taking its average over the time steps t_n and t_{n+1}. Hence, the discretized diffusion equation takes the form

$$\frac{c_i^{n+1} - c_i^n}{\Delta t} = \frac{D}{2}\frac{(c_{i+1}^{n+1} - 2c_i^{n+1} + c_{i-1}^{n+1}) + (c_{i+1}^n - 2c_i^n + c_{i-1}^n)}{h^2} \qquad (7.103)$$

This is the so-called *Crank–Nicholson scheme* and, formally, it could have been obtained by simply averaging the explicit forward-difference and implicit backward-difference schemes. By conveniently grouping the terms, the following system of linear equations results:

$$\begin{aligned}
-\lambda c_{i-1}^{n+1} + (1+2\lambda)c_i^{n+1} - \lambda c_{i+1}^{n+1} &= \lambda c_{i-1}^n + (1-2\lambda)c_i^n + \lambda c_{i+1}^n, \\
i &= 2, 3, \ldots, M-1
\end{aligned} \qquad (7.104)$$

with

$$\lambda = \frac{D}{2}\frac{\Delta t}{h^2} \qquad (7.105)$$

The discretized system in Eq. (7.104) can be represented in the matrix form as

$$\mathbf{A}\cdot\mathbf{c}^{n+1} = \mathbf{B}\cdot\mathbf{c}^n \quad n = 0, 1, 2, \ldots \qquad (7.106)$$

where matrices $\mathbf{A}$ and $\mathbf{B}$ have tridiagonal structure and are given by Eqs. (7.99) and (7.86), respectively. Since $\mathbf{A}$ is positive, definite, and diagonally dominant, it is nonsingular, thus allowing for Crout factorization method for tridiagonal linear systems to be applied to obtain $\mathbf{c}^{n+1}$ from $\mathbf{c}^n$ for any $n = 0, 1, 2, \ldots$. It has to be noted that the evaluation of the right-hand-side members of the system in Eq. (7.106) requires a little more work than in the case of the fully implicit scheme, implying the multiplication of the tridiagonal matrix $\mathbf{B}$ with the "old" solution vector $\mathbf{c}^n$.

The amplification factor resulting from Eq. (7.104) is

$$\xi = \frac{1 - 4\lambda \sin^2(kh/2)}{1 + 4\lambda \sin^2(kh/2)} \qquad (7.107)$$

and, consequently, the Crank–Nicholson method turns out to be *unconditionally stable*. Due to its stability and satisfactory order of convergence $O(h^2 + (\Delta t)^2)$, the Crank–Nicholson scheme is the recommended approach for any simple diffusion equation.

In order to demonstrate the beneficial influence of the higher order accuracy with respect to time of the Crank–Nicholson method, we consider again the initial value problems (7.87) to (7.89), which is solved both by the fully implicit and the Crank–Nicholson schemes. We investigate the numerical solution at time $t = 3.0$ by using the constant spatial step $h = 0.05$. Figure 7.17 shows the spatial concentration profiles obtained by using the time step $\Delta t = 0.025$, which is 20 times larger than the largest value to insure the stability of the explicit method (see Figure 7.15). As one may notice, the profile yielded by the Crank–Nicholson algorithm (depicted with long dash) agrees very well with the analytical result (continuous line) even for the large time step considered. By contrast, the solution obtained by the fully implicit method (short dash) departs quite substantially from the exact solution. Hence, it should be

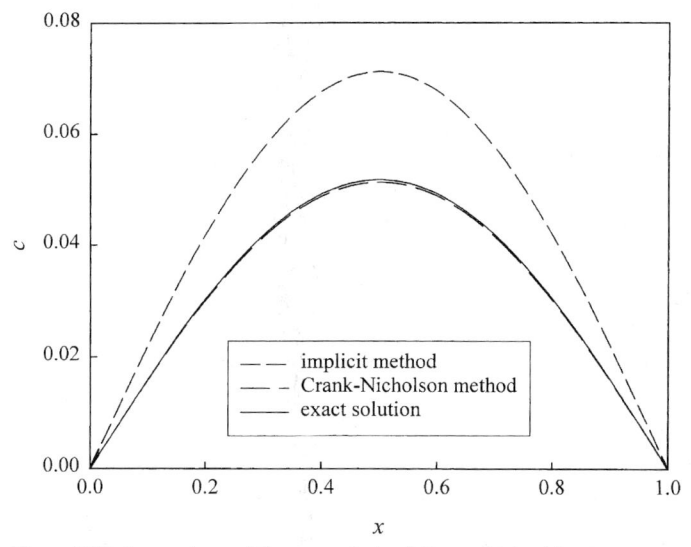

Figure 7.17 Comparison of the numerical solutions obtained by the fully implicit and Crank–Nicholson methods for the Cauchy problem (8.17) to (8.19). The spatial step $h = 0.05$ and the time step $\Delta t = 0.025$ have been used.

obvious that the stability of the fully implicit scheme (absence of oscillations) does not automatically guarantee high accuracy.

7.3.3
Spatially Variable Diffusion Coefficient

If the diffusion coefficient is spatially variable, the one-dimensional diffusion equation has the form

$$\frac{\partial c}{\partial t} = \frac{\partial}{\partial x}\left(D\frac{\partial c}{\partial x}\right) \tag{7.108}$$

The correct way to differentiate this equation relies on the following centered-difference approximation of the spatial derivative at time step t_n:

$$\frac{\partial}{\partial x}\left(D\frac{\partial c}{\partial x}\right)\bigg|_{x_i,t_n} \simeq \frac{1}{h}\left[\left(D\frac{\partial c}{\partial x}\right)\bigg|_{x_{i+1/2},t_n} - \left(D\frac{\partial c}{\partial x}\right)\bigg|_{x_{i-1/2},t_n}\right]$$
$$\simeq \frac{1}{h}\left[D_{i+1/2}\frac{c_{i+1}^n - c_i^n}{h} - D_{i-1/2}\frac{c_i^n - c_{i-1}^n}{h}\right] \tag{7.109}$$

The midpoints $x_{i+1/2} = (x_i + x_{i+1})/2$ and the corresponding values of the diffusion coefficient $D_{i+1/2} = D(x_{i+1/2})$ have been introduced to ensure appropriate centering of the implied derivatives. A convenient approximation for $D_{i+1/2}$ results from considering the average of the values at the neighboring grid points:

$$D_{i+1/2} = \tfrac{1}{2}(D_i + D_{i+1}) \tag{7.110}$$

In the case of unequal spacing of the spatial grid points, Eq. (7.109) can be generalized to give

$$\frac{\partial}{\partial x}\left(D\frac{\partial c}{\partial x}\right)\bigg|_{x_i,t_n} \simeq \frac{2}{x_{i+1} - x_{i-1}}\left[D_{i+1/2}\frac{c_{i+1}^n - c_i^n}{x_{i+1} - x_i} - D_{i-1/2}\frac{c_i^n - c_{i-1}^n}{x_i - x_{i-1}}\right].$$

Application of the Crank–Nicholson method based on the spatial difference scheme in Eq. (7.109) results in the following discretized form of the diffusion equation:

$$\frac{c_i^{n+1} - c_i^n}{\Delta t} = \frac{1}{2h}\left[D_{i+1/2}\frac{c_{i+1}^{n+1} - c_i^{n+1}}{h} - D_{i-1/2}\frac{c_i^{n+1} - c_{i-1}^{n+1}}{h}\right]$$
$$+ \frac{1}{2h}\left[D_{i+1/2}\frac{c_{i+1}^n - c_i^n}{h} - D_{i-1/2}\frac{c_i^n - c_{i-1}^n}{h}\right] \tag{7.111}$$

Rearrangement of terms leads to the system of linear equations

$$-\lambda_{i-1/2}c_{i-1}^{n+1} + (1 + 2\lambda_i)c_i^{n+1} - \lambda_{i+1/2}c_{i+1}^{n+1}$$
$$= \lambda_{i-1/2}c_{i-1}^n + (1 - 2\lambda_i)c_i^n + \lambda_{i+1/2}c_{i+1}^n, \quad i = 2, 3, \ldots, M - 1 \tag{7.112}$$

with

$$\lambda_i = \frac{D_i}{2} \frac{\Delta t}{h^2} \quad (7.113)$$

The system (7.112) should be completed with appropriate equations resulting from the boundary conditions, and it can be solved, in principle, by the same factorization method of Crout for systems with tridiagonal matrix.

When the diffusion coefficient depends not only on the spatial coordinate but also on the local concentration, the discretization of the diffusion equation proceeds in a similar manner, except that one has to solve at each time step not a system of linear equations, but a quite complicated *set of coupled nonlinear equations*.

7.3.4
Boundary Conditions

In the treatment of explicit and implicit difference methods, we have used Dirichlet type boundary conditions, for the sake of simplicity, which specify the values of the solution on the boundaries. A more general type of boundary condition can be defined in the form of a linear combination of the solution and its derivative. Considering in particular the left boundary, such a *mixed boundary condition* can be written as

$$\left[\alpha c(x,t) + \beta \frac{\partial c(x,t)}{\partial x} \right]_{x=0} = \gamma \quad (7.114)$$

For $\beta = 0$ this is a Dirichlet type condition, while for $\alpha = 0$ it is a Neumann type condition. α, β, and γ may eventually be functions of time. A practical example for a mixed boundary condition is the evaporation condition:

$$\left. \frac{\partial c(x,t)}{\partial x} \right|_{x=0} = \alpha [c(0,t) - c_e] \quad (7.115)$$

where c_e is the equilibrium surface concentration.

The simplest finite-difference representation of the mixed boundary condition in Eq. (7.114) may be readily obtained by considering the forward-difference scheme for the spatial derivative, implying the concentrations on the boundary, c_1^n, and at the neighboring interior grid point, c_2^n

$$\alpha c_1^n + \beta \frac{c_2^n - c_1^n}{h} = \gamma \quad (7.116)$$

However, this approximation is of the order $O(h)$, and since the difference equations for all discussed (explicit and implicit) propagation methods are $O(h^2)$, it is definitely not the best choice.

An improved $O(h^2)$ finite-difference representation of the boundary condition (7.104) results by approximating the solution in the vicinity of the boundary by the second-order *Lagrange interpolating polynomial* passing through the points (x_1, c_1^n),

(x_2, c_2^n), and (x_3, c_3^n) (equally spaced grid points are assumed):

$$\tilde{c}(x, t_n) = \frac{1}{2h^2}\left[(x-x_2)(x-x_3)c_1^n - 2(x-x_1)(x-x_3)c_2^n + (x-x_1)(x-x_2)c_3^n\right] \tag{7.117}$$

The interpolation properties of this uniquely defined parabola can easily be verified by direct evaluation for x_1, x_2, and x_3, respectively. The derivative of polynomial (7.117) is given by

$$\frac{\partial \tilde{c}(x, t_n)}{\partial x} = \frac{1}{2h^2}\left[(2x-x_2-x_3)c_1^n - 2(2x-x_1-x_3)c_2^n + (2x-x_1-x_2)c_3^n\right] \tag{7.118}$$

and for $x = x_1$, where we wish to approximate the boundary condition, it takes the particularly simple form

$$\left.\frac{\partial \tilde{c}(x, t_n)}{\partial x}\right|_{x=x_1} = \frac{-3c_1^n + 4c_2^n - c_3^n}{2h} \tag{7.119}$$

By replacing this approximation in the boundary condition (7.114), one is left with the supplementary equation

$$\alpha c_1^n + \beta \frac{-3c_1^n + 4c_2^n - c_3^n}{2h} = \gamma \tag{7.120}$$

Adding this equation to the set of difference equations resulting from the diffusion equation no longer preserves the tridiagonal structure of the system matrix. Indeed, appearing as the first equation of the system, the coefficient of c_3^n does not lie on the principal diagonal or on one of the two neighboring codiagonals. In such cases, one has to resort to general methods for solving the linear system, such as Gaussian elimination or the LU factorization method of Crout.

An absolutely analogous treatment may be applied to the right boundary, for which the boundary condition reads

$$\alpha c_M^n + \beta \frac{c_{M-2}^n - 4c_{M-1}^n + c_M^n}{2h} = \gamma \tag{7.121}$$

A third possibility of approximating the mixed-type boundary condition, widely used, implies considering a fictitious external grid point x_0 and the corresponding concentration c_0^n. The centered-difference approximation of the boundary condition becomes

$$\alpha c_1^n + \beta \frac{c_2^n - c_0^n}{2h} = \gamma \tag{7.122}$$

The unknown (and in principle "useless") value of c_0^n may be eliminated by using the diffusion equation approximated at the grid point x_1 (see Eq. (7.109)):

$$\left.\frac{\partial c}{\partial t}\right|_{x_1, t_n} \simeq \frac{1}{h}\left[D_{3/2}\frac{c_2^n - c_1^n}{h} - D_{1/2}\frac{c_1^n - c_0^n}{h}\right] \tag{7.123}$$

with $D_{3/2} \equiv (D_1 + D_2)/2$ and $D_{1/2} \equiv (D_0 + D_1)/2 = D_1$. On replacing c_0^n from Eq. (7.122) in Eq. (7.123) and by considering the Crank–Nicholson scheme, the following difference equation results:

$$\frac{c_1^{n+1} - c_1^n}{\Delta t} = \frac{1}{2h}\left[(D_{1/2} + D_{3/2})\frac{c_2^{n+1} - c_1^{n+1}}{h} + \frac{2D_{1/2}}{\beta}(\alpha c_1^{n+1} - \gamma)\right] \\ + \frac{1}{2h}\left[(D_{1/2} + D_{3/2})\frac{c_2^n - c_1^n}{h} + \frac{2D_{1/2}}{\beta}(\alpha c_1^n - \gamma)\right]$$ (7.124)

which can be rearranged to become the first equation of the system (7.112). A similar treatment may be applied to approximate the right boundary condition, which then becomes the last equation of the system. In the particular case of Neumann boundary conditions modeling impermeable surfaces ($\alpha = \gamma = 0$), the two equations can be cast in the simple form

$$(1 + 2\lambda_1)c_1^{n+1} - 2\lambda_1 c_2^{n+1} = (1 - 2\lambda_1)c_1^n + 2\lambda_1 c_2^n \\ -2\lambda_M c_{M-1}^{n+1} + (1 + 2\lambda_M)c_M^{n+1} = 2\lambda_M c_{M-1}^n + (1 - 2\lambda_M)c_M^n$$ (7.125)

where λ_i is given by Eq. (7.113). Obviously, these approximations of the boundary conditions preserve the tridiagonal structure of the matrix of the system (7.112).

In certain applications, it is convenient to impose (eventually, instead of one of the local boundary conditions) a *global condition* upon the total amount of diffusant, in the form of an integral over the entire spatial region:

$$\int_0^L c(x,t)dx = Q$$ (7.126)

Such a normalization condition can be readily discretized by considering, for example, the simple trapezoidal rule for performing the numerical quadrature

$$h\left(\tfrac{1}{2}c_1^n + \sum_{i=2}^{M-1} c_i^n + \tfrac{1}{2}c_M^n\right) = Q$$ (7.127)

The quantity Q might be related, for example, to the total amount of diffusant at the initial moment, but it could equally be a function of time.

7.3.5
One-Dimensional Diffusion in Cylindrical and Spherical Geometry

There are certain practical diffusion problems, which can be treated most appropriately in cylindrical or in spherical coordinates. In many cases, choosing the natural coordinate system allows for the coordinates to be separated, and one is left with the simpler problem of dealing with one-dimensional diffusion along the radial coordinate. Basically, the only technical complication, which arises as compared to the one-dimensional diffusion in Cartesian coordinates treated so far, concerns the approximation of the spatial derivative of the concentration involved by the diffusion equation.

Assuming constant diffusion coefficient, the equation describing the radial diffusion in cylindrical coordinates may be written as

$$\frac{\partial c}{\partial t} = D \frac{1}{r}\frac{\partial}{\partial r}\left(r \frac{\partial c}{\partial r}\right) = D\left[\frac{\partial^2 c}{\partial r^2} + \frac{1}{r}\frac{\partial c}{\partial r}\right] \qquad (7.128)$$

In order to approximate its solution, we establish a grid of equally spaced points in the interval $[0, R]$ of the radial coordinates. Denoting the corresponding mesh constant by h, the grid points are defined by the discrete coordinates

$$r_i = (i-1)h \quad i = 1, 2, \ldots, M \qquad (7.129)$$

A second-order representation of the right-hand side of Eq. (7.128) at point (r_i, t_n) is obtained by using centered differences for the spatial derivatives

$$\begin{aligned}\left[\frac{\partial^2 c}{\partial r^2} + \frac{1}{r}\frac{\partial c}{\partial r}\right]_{r_i, t_n} &\cong \frac{c_{i+1}^n - 2c_i^n + c_{i-1}^n}{h^2} + \frac{1}{(i-1)h}\frac{c_{i+1}^n - c_{i-1}^n}{2h} \\ &= \frac{(2i-1)c_{i+1}^n - 4(i-1)c_i^n + (2i-3)c_{i-1}^n}{2(i-1)h}\end{aligned} \qquad (7.130)$$

A special treatment has to be applied to the central grid point $r_1 = 0$. The singularity arising in the second term is overcome by imposing the natural boundary condition that the first-order derivative vanishes at $r_1 = 0$. By introducing a fictitious grid point $r_0 = r_1 - h$, this condition may be approximated by the second-order centered-difference scheme

$$\left.\frac{\partial c}{\partial r}\right|_{r_1, t_n} \cong \frac{c_2^n - c_0^n}{2h} = 0 \qquad (7.131)$$

This relation is practically useful only to eliminate further occurrences of c_0^n. Still, the second term of the spatial derivative, $(1/r)(\partial c/\partial r)$, implies an indeterminate form of the type 0/0. Making use of L'Hospital's rule results in the following representation:

$$\left[\frac{\partial^2 c}{\partial r^2} + \frac{1}{r}\frac{\partial c}{\partial r}\right]_{r_1, t_n} = \left[2\frac{\partial^2 c}{\partial r^2}\right]_{r_1, t_n} \cong 2\frac{c_2^n - 2c_1^n + c_0^n}{h^2} = 4\frac{c_2^n - c_1^n}{h^2} \qquad (7.132)$$

Having established the appropriate finite-difference expressions of the spatial derivatives, the diffusion equation may be approximated as follows:

$$\begin{aligned}\left.\frac{\partial c}{\partial t}\right|_{r_1, t_n} &= 4D_1 \frac{c_2^n - c_1^n}{h^2} \quad i = 1 \\ \left.\frac{\partial c}{\partial t}\right|_{r_i, t_n} &= D_i \frac{(2i-1)c_{i+1}^n - 4(i-1)c_i^n + (2i-3)c_{i-1}^n}{2(i-1)h} \quad i = 2, 3, \ldots, M-1\end{aligned} \qquad (7.133)$$

The equation for the central point ($i = 1$) actually plays the role of "inner" boundary condition. The above system should be completed with one more boundary condition for the outer point $r_M = R$. Irrespective of the type of the used time difference scheme (explicit, fully implicit or Crank–Nicholson), the further treatment of the resulting system of difference equations is absolutely analogous to the one developed for Cartesian coordinates.

The radial diffusion equation in spherical coordinates may be written for constant diffusion coefficient as

$$\frac{\partial c}{\partial t} = D \frac{1}{r^2} \frac{\partial}{\partial r}\left(r^2 \frac{\partial c}{\partial r}\right) = D\left[\frac{\partial^2 c}{\partial r^2} + \frac{2}{r}\frac{\partial c}{\partial r}\right] \tag{7.134}$$

Following the pattern employed in the case of the cylindrical coordinates, one obtains the following finite-difference approximations:

$$\left.\frac{\partial c}{\partial t}\right|_{r_1,t_n} = 6D_1 \frac{c_2^n - c_1^n}{h^2} \quad i = 1$$

$$\left.\frac{\partial c}{\partial t}\right|_{r_i,t_n} = D_i \frac{ic_{i+1}^n - 2(i-1)c_i^n + (i-2)c_{i-1}^n}{(i-1)h} \quad i = 2, 3, \ldots, M-1$$

$$\tag{7.135}$$

The treatment of the various types of outer boundary conditions (for $r_M = R$) and of the complete system of difference equations is again analogous to the case of the Cartesian coordinates.

7.3.6
Multidimensional Diffusion

The extension of the methods described so far to multidimensional diffusion problems is straightforward in principle. However, in such an attempt one is faced with a quite considerable increase in computational effort.

Let us consider for simplicity the two-dimensional diffusion equation as follows:

$$\frac{\partial c}{\partial t} = D\left(\frac{\partial^2 c}{\partial x^2} + \frac{\partial^2 c}{\partial y^2}\right) \tag{7.136}$$

for which we wish to find the solution in the rectangular domain $[0,a] \times [0,b]$. We assume that $c(x, y, t)$ is known over the whole spatial region at $t = t_0$ and it is prescribed on the boundary for any $t > t_0$.

Applying the Crank–Nicholson scheme to Eq. (8.66), relative to a space–time grid characterized by the points

$$x_i = (i-1)h, \quad y_j = (j-1)h \quad t_n = n\Delta t \tag{7.137}$$

results in the following difference equation:

$$c_{i,j}^{n+1} = c_{i,j}^n + \lambda\left(\delta_x^2 c_{i,j}^{n+1} + \delta_y^2 c_{i,j}^{n+1} + \delta_x^2 c_{i,j}^n + \delta_y^2 c_{i,j}^n\right) \tag{7.138}$$

Here, $\lambda = D\Delta t/(2h^2)$ and the second-order, centered-difference operators δ_x^2 and δ_y^2 are defined by

$$\delta_x^2 c_{i,j}^n = c_{i+1,j}^n - 2c_{i,j}^n + c_{i-1,j}^n$$
$$\delta_y^2 c_{i,j}^n = c_{i,j+1}^n - 2c_{i,j}^n + c_{i,j-1}^n \tag{7.139}$$

While in the case of the spatially one-dimensional diffusion, the set of difference equations features a tridiagonal matrix, the system in Eq. (8.68) can be shown to have a *block tridiagonal matrix*, which requires the use of special solving methods.

A very powerful method for the solution of system in Eq. (8.68), widely in use, is the so-called *alternating-direction implicit method*, which is based on the idea of *splitting* each time step into two substeps of size $\Delta t/2$. In each substep, the concentration is kept constant in one of the spatial dimensions, while in the other one it is treated implicitly:

$$c_{i,j}^{n+1/2} = c_{i,j}^n + \lambda \left(\delta_x^2 c_{i,j}^{n+1/2} + \delta_y^2 c_{i,j}^n \right)$$
$$c_{i,j}^{n+1} = c_{i,j}^{n+1/2} + \lambda \left(\delta_x^2 c_{i,j}^{n+1/2} + \delta_y^2 c_{i,j}^{n+1} \right) \qquad (7.140)$$

The first equation treats the x direction implicitly, keeping the y direction unaffected. In the second equation, the role of the two directions is interchanged. Slight rearrangement of the system

$$(1 - \lambda \delta_x^2) c_{i,j}^{n+1/2} = (1 + \lambda \delta_y^2) c_{i,j}^n$$
$$(1 - \lambda \delta_y^2) c_{i,j}^{n+1} = (1 + \lambda \delta_x^2) c_{i,j}^{n+1/2} \qquad (7.141)$$

evidences the great advantage of this method: at each substep, it requires only the solution of a tridiagonal system. Hence, instead of solving at each time step a system with block tridiagonal structure, one has to solve two systems with simple tridiagonal structure.

References

Burden, R.L. and Faires, J.D. (1985) *Numerical Analysis*, Boston: Prindle, Weber & Schmidt.

Carslaw, H.S. and Jaeger, J.C. (1959) *Conduction of Heat in Solids* (1959), Oxford, Clarendon Press.

Chang, S.-S., Guttman, C.M., Sanchez, I.C. and Smith, LE. (1988) In *Food and Packaging Interactions* (ed. J. Hotchkiss) ACS Symposium Series No. 365, Washington.

Crank, J. (1975) *Mathematics of Diffusion*, 2nd ed. Clarendon Press, Oxford University Press, Oxford.

Koonin, S.E. and Meredith, D.C. (1990) *Computational Physics*, Addison-Wesley, Redwood CA.

Kreyszig, E. (1993) *Advanced Engineering Mathematics*, 7th ed. Wiley, New York.

Press, W.H., Teukolsky, S.A., Vetterling, W.T. and Flannery, B.P. (1986) *Numerical Recipes in C*, Cambridge University Press, Cambridge.

Reid, R.C., Sidman, K.R., Schwope, A.D. and Till, D.E. (1980) *Ind. Eng. Chem. Prod. Res. Dev.*, **19**, 580–587.

8
Solution of the Diffusion Equation for Multilayer Packaging
Valer Tosa and Peter Mercea

8.1
Introduction

In industrialized nations the majority of foodstuffs one finds nowadays in commerce are displayed and sold in packaged form. In the past the packaging itself was generally made of monolayer homogenous materials, i.e., different types of polymers, paper/cardboard, glass, ceramics or metals. It is well documented that during the contact between such a packaging and the food a net transfer (migration) of impurities/substances from the packaging into the food may take place. How such a migration process can be described/quantified theoretically was shown in Chapter 7. In the last decades, a steadily increasing number of multilayer (ML) materials are used for food packaging. A ML structure can be designed to add up the properties of the different materials it is made off, i.e., chemical inertness, low permeability for certain substances, physical strength, and/or good optical appearance. Thus, a ML packaging offers a series of advantages over a monolayer one. However, one must account that a migration of impurities/substances from a ML structure into the packaged food takes place, too. Such a migration process may originate from the contact layer (CL) of the ML with the food and/or from other layers, adhesives and/or imprinted surfaces of the ML structure. For quality assurance and/or consumer protection purposes one investigates the migration from ML packaging with experimental tools. Usually, a migration experiment is carried out under "worst case" conditions and the concentration of substances/migrants released from the CL of the ML into a food simulant is monitored. The results obtained are then interpreted in the light of the existing consumer protection laws. Because usually a ML is made of quite different types of materials the unfolding of a migration process from a ML is more complex than that from a monolayer material. Therefore, a consistent interpretation of the obtained experimental migration results is often possible only by using an adequate mathematical model and algorithms for substance migration from the ML structure. In the following sections, a brief presentation of how the migration from a ML can be quantified theoretically will be given.

Plastic Packaging. Second Edition. Edited by O.G. Piringer and A.L. Baner
Copyright © 2008 WILEY-VCH Verlag GmbH & Co. KGaA, Weinheim
ISBN: 978-3-527-31455-3

The diffusion of a substance within a monolayer or ML packaging and transfer (migration) from it into a foodstuff or food simulant is in most cases controlled by diffusion and partitioning processes. In Chapter 7, the mathematical description of transport processes occurring within a series of packaging systems was presented and discussed. The transport equation was presented in its most general form, in which diffusion, convection, and chemical reaction were accounted for. Particular forms for the diffusion equations were presented, together with their analytical solutions for some cases of practical importance for food packaging. However, such (relatively simple) analytical solutions can be derived only for a packaging made of a monolayer homogenous material (polymeric or of other nature) in which the diffusion coefficient of the substance/migrant is constant and the food or food simulant, F, is "well-stirred," i.e., F is a liquid in which at any time, t, there are no migrant concentration gradients. In Section 7.2.9, it was mentioned that for MLs only the steady-state permeation of a substance through the ML structure could be treated theoretically with simple analytical equations. However, the steady-state permeation from a ML into a food accounts only for a relatively limited number of situations occurring in framework of migration processes from MLs into foods. Therefore, one finds out that in the practice of food packaging there are many cases that cannot be quantified theoretically with (relatively) simple analytical equations. This is already the case for the relatively simple system of a monolayer packaging in contact with solid food/medium, F. In such a situation a time-dependent diffusion of the migrating substance into F takes place, i.e., for a certain time there is a migrant concentration gradient in F, $c_{F,t}(x) \neq$ const.

When one considers now the time dependent migration from a ML structure in whatever type of food – "well-stirred" liquid, viscous substance or even a solid one – one cannot quantify these mass transfer processes with analytical algorithms. To obtain solutions of transport/migration equations in such situations one must appeal to numerical methods.

The principles of a numerical method to solve the diffusion equation for a monolayer packaging in contact with a liquid F were presented in Section 7.2. In the followings, this topic will be extended to the one-dimensional (1D) diffusion problem for ML materials in contact with various types of foods. In this respect, a brief presentation of the main numerical approaches developed to solve this mass transfer problem will be made. Then the presentation will be focused on a numerical method developed to solve the transport equations for a ML packaging in contact with any type of homogenous foodstuffs, F. This method is based on a finite difference technique and was developed in 1D for the general case in which the transport processes are controlled not only by the diffusion coefficients (D_i) in the packaging, and respectively, foodstuff (D_F) but also by partition coefficients (K_{ij}) between any two adjacent layers i and j of the packaging as well as between the packaging and foodstuff (K_{pf}).

8.2
Methods for Solving the Diffusion Problem in a Multilayer (ML) Packaging

The diffusion of a substance in a homogenous material (packaging) can be treated theoretically with rather simple tools when the diffusion coefficient, D, in the material is constant and there is a 1D transfer between this material and an adjacent homogenous medium. But if one considers a 2D or even a realistic 3D diffusion in a ML material in which, for example, in each layer, k, the substance has different diffusion coefficients D_k, and initial concentrations, $c_{P,0}(k)$, the theoretical treatment of the process is far more complex. In addition, if one considers the food, F, diffusing into the ML packaging and inducing there spatial and temporal changes of the D_k, then the problem becomes particularly complex.

The numerical treatment of the diffusion equation is, by itself, a large field of research in mathematics. It is beyond the scope of this publication to treat any aspect of this topic in detail. Therefore, in the followings only some general aspects of how one can solve numerically a diffusion equation for a ML packaging in contact with a food will be presented.

The main numerical methods applied to the study of diffusion in ML materials are the finite element method (FEM) (Huebner, 1975), the boundary element method (BEM) (Muzizo and Solaini, 1987), and finite difference (FD) procedures (Clever and Wassel, 1985).

The FEM is a numerical analysis technique developed and used especially by practitioners in solid mechanics and structural engineering. FEM was broadly extended to various other fields, due to its remarkable power and flexibility. The FEM allows a considerable degree of freedom in putting computational elements where one needs them, which is very important when one deals with highly *irregular* geometries. In Figure 8.1, an example on a FD and, respectively, a FEM discretization of a general domain is given.

The next step in a FEM analysis is to assign nodes to each element, define nodal values for the field variables, and then choose the type of interpolating function to represent the variation of the field variable over the element. The field variable can be, for example, pressure, temperature, displacement, stress, impurity concentration, or

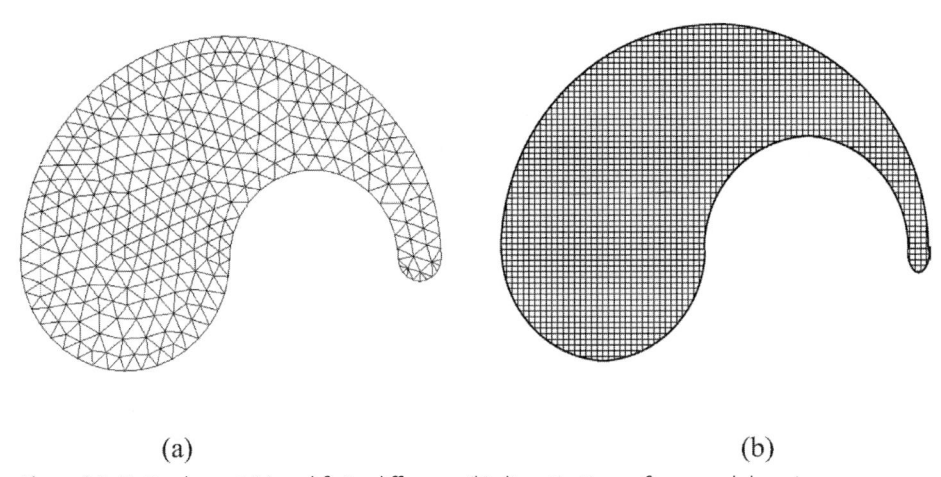

Figure 8.1 Finite element (a) and finite difference (b) discretizations of a general domain.

some other quantity. Certain continuity requirements are imposed at the nodes and along the element boundaries.

The third important step is to find the element properties, in other words to determine the matrix equations expressing the properties of the individual elements. These equations are derived using the mathematical approach to the physical phenomenon one wishes to model.

Further, it is necessary to find the properties of the overall system modeled by the network of elements. To do this one must assemble all the element properties. In simpler words one must gather the matrix equations expressing the behavior of the elements and form the matrix equations expressing the behavior of the entire system. The matrix equations for the system have the same form as the equations for the individual element except that they contain many more terms because they include all nodes. Before the system equations are ready for solution they must be modified to account for the boundary conditions of the problem.

The last step is to solve the system of equations by using standard solution techniques if the equations are linear or alternative approaches if they are nonlinear. The result is the field values in all nodal points of the mesh.

Sometimes additional calculations are needed. For example, if one calculates the distribution of the impurity/substance concentration over a ML structure, one may wish to integrate it to find the total amount of impurity left after migration, thus, to find the impurity which migrated in the food.

Depending on the complexity of the system, and on the desired accuracy to solve the problem, the number of elements might amount to tens of thousands. From the above considerations, it is obvious that FEM is a very powerful method, which is now applied in various practical fields, including diffusion in a ML polymer system with functional barrier (Han *et al.*, 2003; Pettersen *et al.*, 2004; Safa and Abbes, 2002). A dedicated software for such applications was also developed (Roduit *et al.*, 2004) and is commercially available. However, using FEM has also a number of disadvantages: (i) one needs a powerful computer and sophisticated (usually expensive) software, (ii) input/output data might be large and tedious to prepare/interpret, and (iii) the method is susceptible to user-induced modeling errors (poor choice of element type, distorted elements, geometry not adequately modeled). Therefore, a good knowledge of the method basics is desirable, which is not always accessible to a nontrained user.

The BEM is also an important technique in the computational solution of engineering or scientific problems. In BEM only the boundary is discretized. Hence, the mesh generation is done only on the surface, being considerably simpler for this method than for the FEM. Boundary solutions are obtained directly by solving the set of linear equations. However, field valued in the domain can be evaluated only after the boundary solutions have been obtained.

In the FD approach, the starting step is to set a grid of points conveniently distributed within the integration domain, as in Figure 8.1(b). This procedure replaces the derivatives of the differential equations by appropriate discrete formulas, leading to a system of linear equations having as unknowns the solution values at the grid points. The solution of the system can be obtained by standard numerical procedures. When the system matrix has a sparse form (with only a few nonzero elements) one can use more specialized methods that are faster and more accurate.

Very recently (2007) the finite volume method, related to the FD method mentioned above, was applied to solve the problem of migration from a ML to food. The interested reader may access the portal http://h29.univ-reims.fr for more information and for an on-line engine to estimate the migration.

Apart from the purely numerical approaches mentioned above for solving the diffusion equations in ML (or composite) materials, analytical techniques combined with numerical solving are also available. Comprehensive surveys of such methods are presented in Refs. (Ozisik, 1980; Ozisik et al., 1983; de Monte, 2000; Tittle, 1965; Carslaw and Jager, 1959; Siegel, 1999; Haji-Sheikh and Mashena, 1987). The difficulty with these analytical techniques is that when the number of layers increases the implementation of the analytical solutions becomes very complex and cumbersome. This is due to the necessity of solving a Sturm–Liouville eigenvalue problem, which is not of conventional type because there is a discontinuity in the coefficient functions. To remedy this inconvenience, a finite analytic method was developed (Chen and Chen, 1984) and applied (Ramos and Giovannini, 1992) for unidimensional composite layer.

There is no agreed approach to the analysis of transient diffusion in ML media. Consequently, there is no method of any generality for solving diffusion problem, and no established yardstick for assessing a candidate solution method. The chosen solution depends mainly on problem complexity and user capabilities.

8.3
Solving the Diffusion Equation for a Multilayer Packaging in Contact with a Foodstuff

First the FD method developed in Tosa 2004 for solving the diffusion equations for a ML material in contact with a liquid food, F, will be described bellow.

The starting point in this approach is similar to those presented in detail Section 7.3.2 for solving with a FD technique the 1D diffusion equation for the case of a monolayer in contact with a liquid. Explicit, fully implicit and Cranck–Nicolson methods were presented there in conjunction with Dirichlet, Neumann and mixed type boundary conditions. The 1D diffusion of a impurity/substance of concentration c from a homogenous monolayer into a liquid foodstuff is governed by the 1D diffusion equation

$$\frac{\partial c}{\partial t} = D \frac{\partial^2 c}{\partial x^2} \tag{8.1}$$

which can be discretized in a uniform mesh in the x direction, in which concentration c_i is defined at node i of the mesh. If h is the mesh size and ϑ an adjustable parameter, the above equation becomes

$$\frac{c_i^{n+1} - c_i^n}{\delta t} = D \left[\vartheta \frac{c_{i-1}^{n+1} - 2c_i^{n+1} + c_{i+1}^{n+1}}{h^2} + (1-\vartheta) \frac{c_{i-1}^n - 2c_i^n + c_{i+1}^n}{h^2} \right] \tag{8.2}$$

For $\vartheta=0$ and $\vartheta=1$ one obtains the explicit, and respectively, the fully implicit method, whereas for $\vartheta=1/2$ one gets the so-called Crank–Nicolson method which has the advantage of being unconditionally stable and second order accurate in time.

If one rearranges the terms in Eq. (8.2) such as to separate the unknowns c_i^{n+1} in the l.h.s. and the known quantities c_i^n in the r.h.s. of the equations one gets

$$-r\vartheta c_{i-1}^{n+1}+(1+2r\vartheta)c_i^{n+1}-r\vartheta c_{i+1}^{n+1} = r(1-\vartheta)c_{i-1}^n+[1-2r(1-\vartheta)]c_i^n+r(1-\vartheta)c_{i+1}^n \tag{8.3}$$

where $r = D\delta t/h^2$.

Now one solves the diffusion equation for a 1D single domain in which one places a mesh with $M+1$ points, and one writes Eq. (8.3) for the $M-1$ interior points. So results a system of $M-1$ equations with $M+1$ unknowns. To solve this system two more equations are needed and they can be derived from boundary conditions at the left and right side of the domain. The most general form of the boundary condition is

$$P(t)c(x,t) + Q(t)\frac{\partial c(x,t)}{\partial t} = R(x,t) \tag{8.4}$$

They are called boundary conditions of mixed type, and will be written for both node zero and node M. One can easily see that for $P=0$ the Neumann type, while for $Q=0$ the Dirichlet-type boundary conditions, are obtained (see also Section 7.3.2).

To derive the two algebraic equations the concept of fictitious node (Smith 1985) was used. It was supposed that at the left of node zero (see Figure 8.2) one can place an additional node x_{-1} with the corresponding concentration c_{-1}. Equation (8.3) centered to node zero can be written as

$$-r\vartheta c_{-1}^{n+1}+(1+2r\vartheta)c_0^{n+1}-r\vartheta c_1^{n+1} = r(1-\vartheta)c_{-1}^n+[1-2r(1-\vartheta)]c_0^n+r(1-\vartheta)c_1^n \tag{8.5}$$

For the same node it is also necessary to discretize the boundary condition (8.4), at times t and $t+\delta t$

$$Pc_0^n + Q\frac{c_1^n - c_{-1}^n}{2h} = R$$

$$Pc_0^{n+1} + Q\frac{c_1^{n+1} - c_{-1}^{n+1}}{2h} = R \tag{8.6}$$

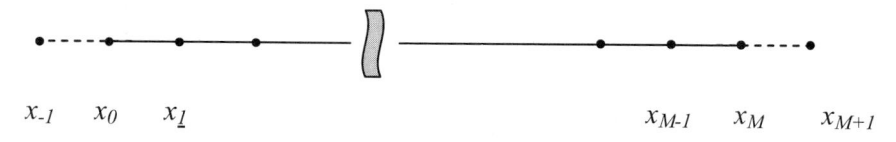

Figure 8.2 The fictitious node method for boundary conditions.

8.3 Solving the Diffusion Equation for a Multilayer Packaging in Contact with a Foodstuff

Eliminating c_{-1}^{n} and c_{-1}^{n+1} from Eqs. (8.5) and (8.6) the desired equation for node zero results

$$\left[1+2r\vartheta\left(1-\frac{P\Delta x}{Q}\right)\right]c_{0}^{n+1}-2r\vartheta c_{1}^{n+1} = \left[1-2r(1-\vartheta)\left(1-\frac{P\Delta x}{Q}\right)\right]c_{0}^{n}$$

$$+2r(1-\vartheta)c_{1}^{n}-\frac{2rR\Delta x}{Q} \tag{8.7}$$

In an analog mode, assuming a fictitious node x_{M+1} (see Figure 8.2) at the right of node M, and writing the diffusion equation for the node M, and the two equations corresponding to the boundary conditions, one gets the corresponding equation for the M node

$$-2r\vartheta\, c_{M-1}^{n+1}+\left[1+2r\vartheta\left(1+\frac{P\Delta x}{Q}\right)\right]c_{M}^{n+1} = 2r(1-\vartheta)c_{M-1}^{n}$$

$$+\left[1-2r(1-\vartheta)\left(1+\frac{P\Delta x}{Q}\right)\right]c_{M}^{n}+\frac{2rR\Delta x}{Q} \tag{8.8}$$

Equations (8.3), (8.7) and (8.8) form a system of $M+1$ equations with $M+1$ unknowns which can be solved numerically. In matricial notation the system can be written as: $\mathbf{A}\mathbf{c}^{n}=\mathbf{B}\cdot\mathbf{c}^{n+1}$, $n=0, 1, 2, \ldots$, where n denotes the index of the time step. The matrices $\mathbf{A}$ and $\mathbf{B}$ of this system are both tridiagonal and thus the system can be solved by noniterative methods which are described in standard textbooks (for example, see Press et al. 1986).

The fictitious point method described above can be used now to derive the equations for the interface conditions in the case of a ML structure. To illustrate the method developed let us consider the simplest case of a two-layer material (packaging).

Sketched in Figure 8.3 is a system made of two homogenous layers, with densities ρ_k, diffusion coefficients D_k, and thicknesses L_k, $k=1, 2$, in contact with a foodstuff of density ρ_F. The diffusion in the two layers is governed by the partial differential equations

$$\frac{\partial c_k}{\partial t} = D_k \frac{\partial^2 c_k}{\partial x^2} \quad k=1,2 \tag{8.9}$$

where k index identify the layer. For the beginning one can assume in the food-stuff, F, a very high diffusion coefficient $D_F \gg D_1$ and D_2, or a well-stired medium, which means that the substance concentration in the food can be considered uniform at any given time. However, this FD method is valid also for applications where the two-layer packaging is in contact with a homogenous foodsuff in which the magnitude of D_F is such that a time dependent diffusion of the substance in F takes place. For the time being we will focus our discussion on the two-layer system.

Initial conditions are written as $c(x, 0) = c_{k0}$, meaning a constant initial concentration of substance (impurity), usually a nonzero value in one layer and zero

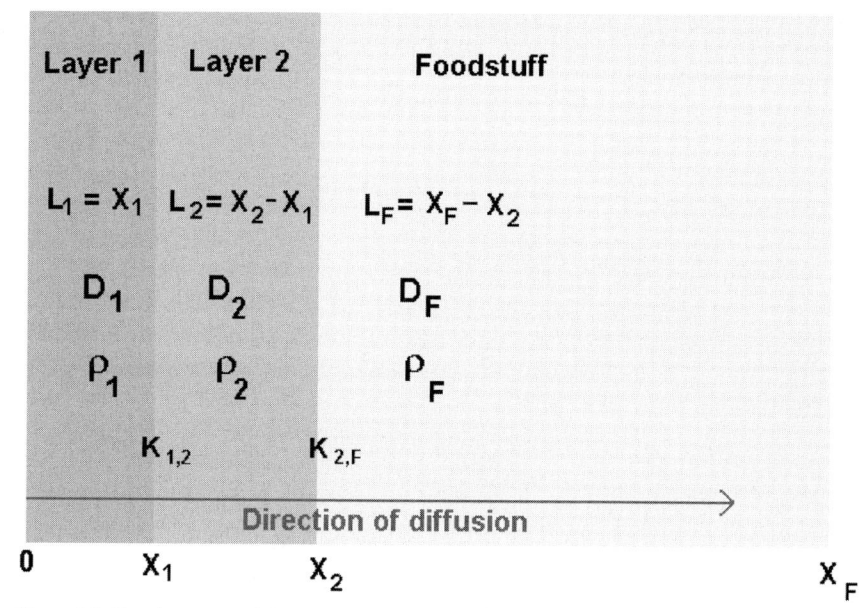

Figure 8.3 Two-layer material system in contact with foodstuff.

concentration in the other layer/s and, respectively, in the foodstuff. However, the method described here is not limited to a constant concentration but can employ a general function $c_0(x)$, specified by values at the mesh points.

A first boundary condition state that there are no flux of matter through the outer wall of the system (for $x = 0$):

$$D_1 \frac{\partial c_1}{\partial x}\bigg|_{x=0} = 0 \qquad (8.10)$$

and two other interface conditions impose the flux continuity (no accumulation of matter) at interlayer contact

$$D_1 \frac{\partial c_1}{\partial x}\bigg|_{x=X_1} = D_2 \frac{\partial c_2}{\partial x}\bigg|_{x=X_1} \qquad (8.11)$$

$$D_2 \frac{\partial c_2}{\partial x}\bigg|_{x=X_2} = D_F \frac{\partial c_F}{\partial x}\bigg|_{x=X_2} \qquad (8.12)$$

In addition, at any given time t, the concentrations at these boundaries obey the conditions

$$c_1 = k_{12} c_2 \quad \text{for} \quad x = X_1 \qquad (8.13)$$
$$c_2 = k_{2Fc F} \quad \text{for} \quad x = X_2 \qquad (8.14)$$

which are imposed by the partition coefficients at the interface at $x = X_1$ between the homogenous layers and at $x = X_2$ between the contact layer and the foodstuff F.

Let us denote by A and B the two domains of the two layers of the packaging material. The spatial meshes built to solve the diffusion equations in the two domains

8.3 Solving the Diffusion Equation for a Multilayer Packaging in Contact with a Foodstuff

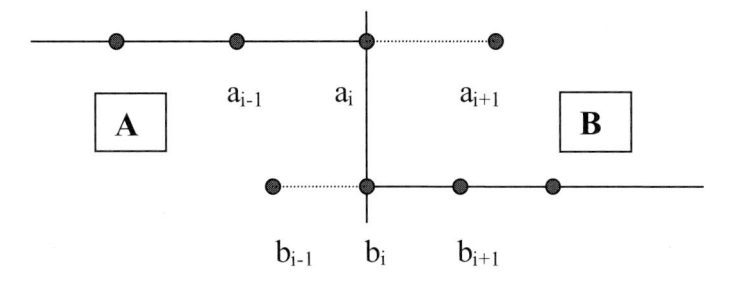

Figure 8.4 The fictitious node method applied to the interface between two different polymer layers. Here, a_i and b_i denote concentrations at node i, in layers A and B, respectively.

may have different step sizes. In Figure 8.4, the meshes at the boundary between two domains are presented. To simplify the notation, the concentrations are denoted by a_i in the A domain extending from the left to the boundary and by b_i in the B domain extending from the boundary to the right. The step sizes will be denoted by δx_a and δx_b, respectively. Two fictitious points, a_{i+1} and b_{i-1}, are also introduced, in the same manner as for mixed type boundary conditions. As one can note, at the interface between two layers, there are two different values of the concentrations a_i and b_i for the same nodal point i. This describes, in fact, the jump in substance concentration at the interface due to different solubility of the substance in domains A and B, respectively. The boundary conditions of the type given by Eq. (8.13), and Eq. (8.14) impose this discontinuity. In fact, the physical reality is a bit more complicated than this description, The jump in substance concentration at the interface takes place in a very thin but spatially finite region of both layers. More elaborated numerical techniques, like FEM, can describe more accurately this jump by imposing an appropriately fine discretization mesh over this region. However, the FD method presented above is simple enough to show the results of the physical process of diffusant partitioning at any interface of a ML structure. Moreover, from the reliability testes made it was found that, if in the boundary regions of an interface the discretization mesh is fine enough, this FD method produces correct and reliable data (see below Section 8.4).

As for the case of external points of a single domain one can apply the fictitious point method for the interface nodal points a_i and b_i. For this purpose one writes the algebraic equations derived from diffusion equation for the interior points a_i and b_i in the two domains. With $r_a = D_a \Delta t/(\delta x_a)^2$, $r_b = D_b \Delta t/(\delta x_b)^2$, and $\bar{\vartheta} = 1 - \vartheta$ these equations are

$$-r_a \vartheta a_{i-1}^{n+1} + (1+2r_a\vartheta) a_i^{n+1} - r_a \vartheta a_{i+1}^{n+1} = r_a \bar{\vartheta} a_{i-1}^n + [1-2r_a \bar{\vartheta}] a_i^n + r_a \bar{\vartheta} a_{i+1}^n \tag{8.15}$$

$$-r_a \vartheta b_{i-1}^{n+1} + (1+2r_a\vartheta) b_i^{n+1} - r_a \vartheta b_{i+1}^{n+1} = r_a \bar{\vartheta} b_{i-1}^n + [1-2r_a \bar{\vartheta}] b_i^n + r_a \bar{\vartheta} b_{i+1}^n \tag{8.16}$$

Because two unknown fictitious concentrations were introduced into the system, one needs two additional equations to solve the system. These equations are provided from writing the boundary conditions as given by Eqs. 8.11–8.14. By using an expression for the flux of matter through the boundary given in Tosa 2004, and writing the boundary conditions for time t and $t+\delta t$, after some algebraic manipulations, one gets eventually

$$-2r_a\vartheta a_{i-1}^{n+1}+(1+2r_a\vartheta-2p_ar_a\vartheta)a_i^{n+1}+2p_ar_a\vartheta kb_i^{n+1}$$
$$= 2r_a\bar{\vartheta} a_{i-1}^n[1-2r_a\bar{\vartheta}(1-p_a)]a_i^n-2p_ar_a\bar{\vartheta} kb_i^n \qquad (8.17)$$

for a_i concentration, and

$$2p_br_b\vartheta a_i^{n+1}+[1+2r_b\vartheta(1-p_bk)]b_i^{n+1}-2r_b\vartheta b_{i+1}^{n+1}$$
$$= -2p_br_b\bar{\vartheta} a_i^n+[1-2r_b\bar{\vartheta}(1-p_bk)]b_i^n+2r_b\bar{\vartheta} b_{i+1}^n \qquad (8.18)$$

for b_i concentration. In matricial form, the resulting system can be written as

$$\mathbf{A}\mathbf{c}^{n+1} = \mathbf{B}\cdot\mathbf{c}^n \qquad n=0,1,2,\ldots \qquad (8.19)$$

where $\mathbf{c}^n$ and $\mathbf{c}^{n+1}$ are arrays containing concentrations $c = [a_0, a_1, \ldots, a_{M_a}, b_0, b_1, \ldots, b_{M_b}]^T$ at time t and $t+\delta t$, respectively, and the matrix $\mathbf{A}$ has the form

$$\begin{bmatrix} 1+2r_a\vartheta p_-^a & -2r_a\vartheta & & & & & & \uparrow \\ -r_a\vartheta & 1+2r_a\vartheta & -r_a\vartheta & & & & & \text{layer} \\ & \cdots & & \cdots & & \cdots & & a \\ & & -2r_a\vartheta & 1+2r_a\vartheta(1-p_a) & 2p_ar_a\vartheta & & & \downarrow \\ & & \uparrow & 2p_br_b\vartheta & 1+2r_b\vartheta(1-kp_b) & -2r_b\vartheta & & \\ & & \text{layer} & & \cdots & \cdots & \cdots & \\ & & b & & & -r_b\vartheta & 1+2r_b\vartheta & -r_b\vartheta \\ & & \downarrow & & & & -2r_b\vartheta & 1+2r_b\vartheta p_+^b \end{bmatrix}$$

$$(8.20)$$

with $p_\pm^{a,b}=1\pm P\delta x_{a,b}/Q$. Matrix $\mathbf{B}$ has a similar form but q is replaced by $q-1$.

One can easily observe that Eq. (8.20) preserves the tridiagonal form of the system matrix, which is nonzero elements on the matrix diagonal and two parallel lines with it, and zero elements otherwise. From the analytical point of view the extension of the method to three and more layers is quite straightforward.

8.4
Development of a User-Friendly Software for the Estimation of Migration from Multilayer Packaging

Based on the above-described scheme numerical algorithms were developed to calculate the spatiotemporal evolution of substance (migrant) concentration in a

8.4 Development of a User-Friendly Software for the Estimation of Migration from Multilayer Packaging

packaging material consisting of up to ten homogenous layers, all densities ρ_k, are constant (Tosa, 2004; Brandsch et al., 2005). The usefullness of these algorithms was increased by extending their solutions to problems where the ML structure is in contact with solid or viscous food in which a time dependent diffusion and not an instantaneous mixing takes place. This allows one to use the algorithms to estimate migrant concentration profiles in solid and/or highly viscous foods, too.

As compared to those given in Eq. (8.2) a modification was introduced in the numerical algorithms of the FD method. Instead of discretizing each layer i with a uniform spatial mesh of size h_i a nonuniform mesh was deployed accross each layer of the system. This mesh shows a higher density of nodes in the vicinity of the interfaces. So the higher rate of concentration variation in the boundary regions of each layer is "smoother" approximated by linear equations of the numerical procedure. As a result more accurate solutions of the diffusion equations are obtained.

The numerical algorithms of this FD method were then implemented into a computer program which can be run on a regular PC (Tosa, 2003). A major concern with this computer program was to check if it produces correct results. For this a series of test were designed/conducted and the results obtained will be presented below.

In a first test a ML system made of n homogenous layers, $2 < n < 10$, with different thicknesses, $L_i \neq L_j$ for any i and j, was considered. It was assumed that all n layers are homogenous and exhibit the same density, ρ_i. Further, it was assumed in all layers the mobility of the migrant and its initial concentration is the same, i.e., all D_i, and $c_{i,0}$ are the same. The partition coefficients at the interfaces between the layers were taken unitary, all $K_{ij} = 1$. The ML was assumed to be in contact with a noncontaminated, $c_{F,0} = 0$, liquid food of finite volume, V_F, and density, ρ_F. The partition coefficient K_{jF} between the contact layer j and F was also taken unitary $K_{jF} = 1$. The migration from this ML into F was then computed with the numerical program varying the L_i, D_i, $c_{i,0}$ and, respectively, V_F in a broad range of values.

Looking to the ML structure defined above one finds out that, in fact, it is a homogenous monolayer "sliced" in n layers (thickness $L = \sum L_i$ density ρ_i, with constant D_i and $c_{i,0}$) in contact with a liquid of volume, V_F, and density, ρ_F. We already know, see Chapter 7, for example, that for such a system the diffusion/migration equation admits a (relatively simple) analytical solution, see Section 7.2.3, and migration values can be computed, for example, with a user-friendly software (Mercea et al. 2005) (see also Chapter 9). Thus, all runs performed with the numerical program (Brandsch et al., 2005) to estimate the migrant concentration in the liquid food, $c_{F,t}$ (numeric), were computed also with the analytical solutions of the migration equation (Mercea et al., 2005). The results obtained were then compared through the ratio: $\Omega = [c_{F,t} \text{(numeric)} - c_{F,t} \text{(analytic)}]/c_{F,t} \text{(analytic)}$. It was found that in all cases investigated $\Omega < 0.01\%$. This means, in fact, that in all cases tested the results obtained with numerical algorithms deviate less than that 0.01% from those calculated with the analytical equation.

The numerical program was then tested for a ML structure made of different types of materials with the same density ρ_i and different D_i in each layer n. For the beginning the partition coefficients at the interfaces between the layers were assumed to be unitary; all $K_{ij} = 1$ and $K_{jF} = 1$. In such a system there are different rates of diffusion in each layer but there is no concentration discontinuity (jump) at the interface between two layers. Let us assume now that initially, for $t = 0$, the migrant is present only in one of the layers of the ML structure, $c_{i,0} \neq 0$ but all other $c_{j,0} = 0$. For $t > 0$ the diffusion of the migrant within the ML takes place and as soon as it reaches the CL–F interface a migrant transfer into F occurs, too. Such a process determines, during the time needed by the ML-food system to reach equilibrium, concentration gradients in each of the n layers and in F, $c_{i,t}(x)$ and $c_{F,t}(x) \neq$ const. By running the numerical program under these assumptions it was checked if the computed concentrations $c_{i,t}(x)$ and $c_{F,t}(x)$ as well as their first derivatives, $dc_{i,t}(x)/dx$ and $dc_{F,t}(x)/dx$ are continuous or not in each layer of the ML-food system and, respectively, at their interfaces.

The results obtained for an ML with $n = 5$, $c_{3,0} \neq 0$ and D_i of different values in each layer are shown in Figure 8.5.

From this figure one can see that the migrant concentrations are continous at the interfaces (see, for example, the enlarged section between layers 4 and 5) while in each layer and F there are different concentration profiles (gradients). This type of test was performed for L_i, D, c and, respectively, V_F varying again in broad ranges of values. The result obtained showed that the continuity of the concentrations at the

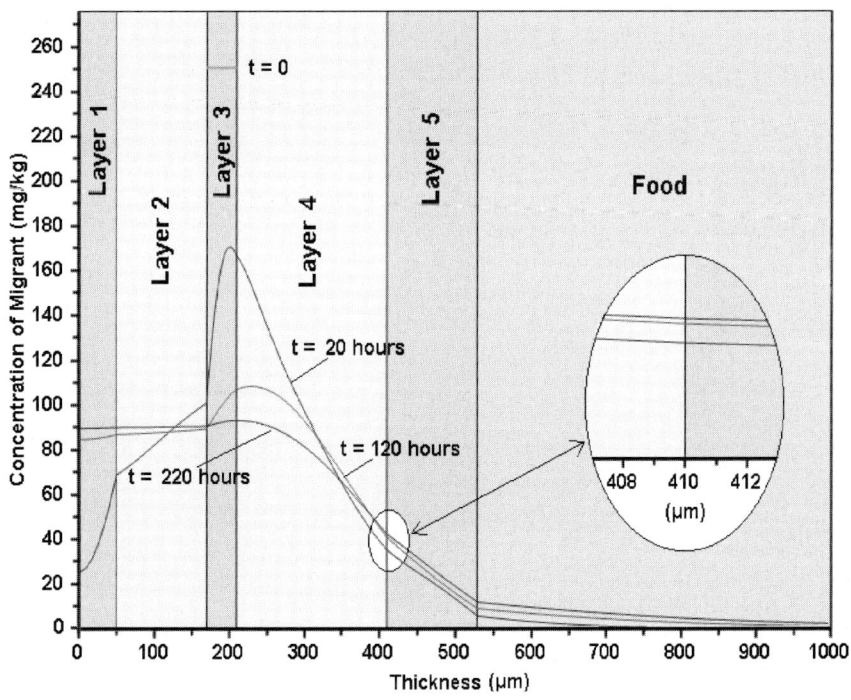

Figure 8.5 Calculated spatial profiles of the concentration for a five-layer system in contact with a highly viscous food.

8.4 Development of a User-Friendly Software for the Estimation of Migration from Multilayer Packaging

interfaces is fulfilled with less than 0.02% relative error for a discretization mesh number of 200 or larger.

For the type of migration problems illustrated in Figure 8.5 the time dependent diffusion/migration equations cannot be solved with analytical methods, see Section 8.2. Thus, a comparison between results obtained with the numerical program and an analytical equation is not possible. However, one can compare the results given by the numerical algorithms for very long diffusion times, $c_{F,t}$ (numeric) for $t \to \infty$, i.e., when the ML-food system reaches equilibrium, with a simple mass balance calculation for the given ML-food system, $c_{F,t} = c_{i,0} L_i / (L_F + \Sigma L_i)$. The results of these tests showed a relative difference of less than 0.001% between the two $c_{F,t}$'s. This is a proof of the fact that the numerical algorithms correctly calculate for such ML-food systems the equilibrium concentrations.

In reality, when a ML packaging is made of several different types of materials, not only the D_i and ρ_i are different in each layer n but also the partition coefficients at the interfaces between any layers i and j are usually nonunitary, $K_{ij} \neq 1$. Thus, the accuracy of the numerical computer program was tested for such a situation, too. During a time dependent diffusion process it was checked, how the ratio of the concentrations, c_i/c_j, calculated with the numerical program at the $n-1$ interfaces of the multilayer agree with the corresponding K_{ij} defined as input values. The quantity $|K_{ij} - c_i/c_j|/K_{ij}$ represents a measure of the computation accuracy. A five-layer ML-food system as presented in Figure 8.5 was used but this time at each interface i,j it was assumed that $K_{ij} \neq 1$. The computation of the migration process was done warying the number of discretization mesh points betwenn 50 and 400. In Figure 8.6, the $|K_{ij} - c_i/c|/K_{ij}$ result obtained at one of the interfaces of the ML is shown.

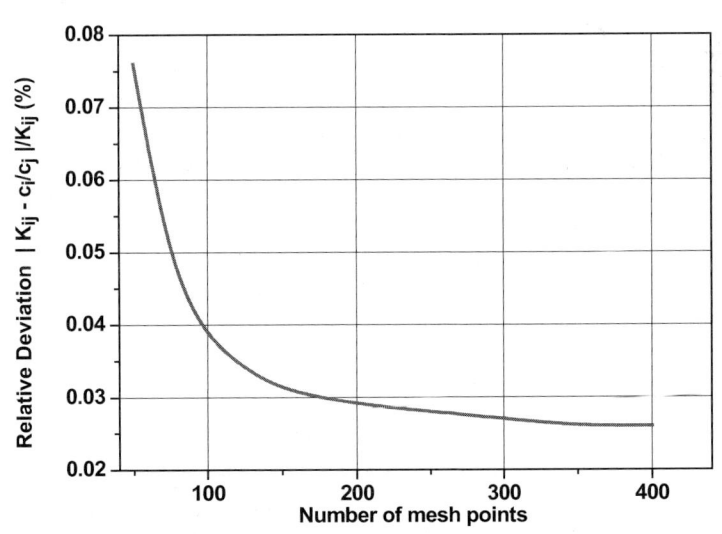

Figure 8.6 Relative deviation between the calculated ratio c_i/c_j and the partitioning K_{ij}, as a function of the number of discretization mesh points in each layer.

From Figure 8.6, one can see that the number of mesh points used in the discretization of each layer of the system plays an important role in the accuracy of the calculations. Below 200 points a rapid decrease of the accurcy is observed while doubling the number of mesh points to 400 brings only a marginal increase of the accuracy. Test were performed for $5 < n < 10$ and L_i, D_i, c_i and, respectively, each K_{ij} varying in a broad range. The result obtained showed patterns similar to those given in Figure 8.6. A practical conclusion was drawn from these tests. For ML systems of relevance in food packaging the numerical computer program developed in Tosa 2003 reproduces, for a number of mesh points exceeding 200, with an accuracy of 0.05% or better the partitioning coefficients K_{ij}. Doubling of the number of mesh points from 200 to 400 points increases the accuracy only slightly while the computation time increases four-fold.

For the ML-food systems analyzed so far, in this chapter, a basic assumption was that during the migration process there is neither loss of migrant from the system nor degradation/transformation of the migrant whithin the system. The quantity of migrant, $q_{i,t}$, found, at any diffusion/migration time t, in layer i can be calculated by integrating the concentration profile $c_{i,t}(x)$ over the thickness L_i, of the layer

$$A \int c_{i,t}(x)dx = V_i \frac{q_{i,t}}{M_i} = \frac{q_{i,t}}{\rho_i}, \quad \text{thus,} \quad q_{i,t} = A\rho_i \int c_{i,t}(x)dx = A\rho_i I_{i,t} \quad (8.21)$$

where V_i, M_i, and ρ_i are, respectively, the volume, mass and density of layer i and A its cross-section. Similarly, the quantity of the migrant species found in the food of volume V_F, mass M_F and density ρ_F is given by

$$q_{F,t} = A\rho_F \int c_{F,t}(x)dx = A\rho_F I_{F,t} \quad (8.22)$$

Thus, at any migration time t the total quantity of the migrant in the ML-food system will be

$$q_t = q_{F,t} + \sum_{i=1}^{n} q_{i,t} \quad (8.23)$$

In a closed ML-food system is which mass conservation is assumed q_t should be equal with the initial amount of migrant, q_0, in the ML structure. To test the "mass conservation" capability of the numerical program presented and discussed in Section 8.3 a series of tests were designed and conducted.

For a ML structure as given in Figure 8.5, the simulations were performed by discretizing each layer of the ML-food system with a mesh of 200 nodes and warying the migration time in the range of $0.1\,s < t < 30h$. A sample of the results obtained are given in Figure 8.7 From here one can see that the computed q_t deviate more from q_0 at the beginning of the migration process. At $t = 0$ the migrant concentration profile in the ML is step like, i.e., in each layer the c_i are constant over the entire thickness of the layer. When the diffusion/migration in the ML-food system starts, $t > 0$, near its interfaces the migrant concentration profiles exhibit very steep variations in space. This introduces errors in calculating the integrals I_i and I_F in Eqs. (8.21) and (8.22) which determines the somewhat higher errors in calculating q_t. With the progress of the migration process the concentration profiles

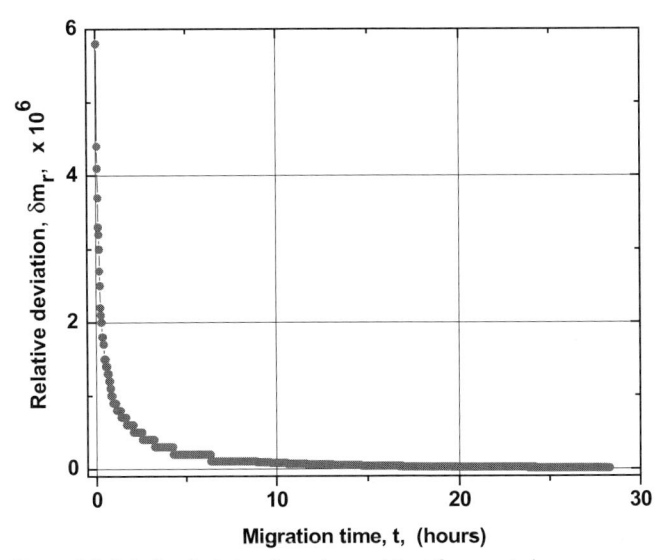

Figure 8.7 Relative deviation $\delta q_r = (q_t - q_0)/q_0$ of computed q_t from the initial amount of migrant in the ML, q_0, as function of migration time.

near the interfaces become smoother which leads to an increased accuracy of the integrations in Eqs. (8.21) and (8.22) and consequently to smaller deviations $\delta q = q_t - q_0$.

The tests were performed for different ML structures, $5 < n < 10$ and for layer parameters L_i, D_i, c_i, K_{ij} and K_{jF} varying in broad ranges. The results obtained showed patterns similar to that given in Figure 8.7 and the relative deviation δq_r never exceeded 0.1%. These tests also showed that the "mass conservation" accuracy of numerical calculations depends on how the migrant is initially distributed in the ML, δq_r is smaller when initially the migrant is distributed in all layers.

Concluding these tests one can state that, for applications from the field of food packaging in ML materials, the computer program developed on the foundation of the numerical FD method presented in Section 8.3 delivers accurate results. The level of computational errors detected so far are orders of magnitude lower than the usual accuracy of the migration experiments in this field. This allowed the development of a user-friendly software for migration estimation from ML packaging into foods and food simulants (see Section 9.2 in Chapter 9).

References

Brandsch, R., Mercea, P., Piringer, O. and Tosa, V. (2005) *Studia Physica*, **50**, 157.

Carslaw, H.S. and Jager, J.C. (1959) *Conduction of Heat in Solids*, 2nd edn, Oxford University Press, London.

Chen, C.J. and Chen, H.C. (1984) Finite analytic numerical method for unsteady two-dimensional Navier–Stokes equations. *J. Comput. Phys.*, **53**, 209.

Clever, R.M. and Wassel, A.T. (1985) Three-dimensional transient heat conduction in a multilayer medium. *J. Spacecraft Rockets*, **22**, 211.

de Monte, F. (2000) Transient heat conduction in one-dimensional composite slab. a 'natural' analytic approach. *Int. J. Heat Mass Transfer*, **43**, 3607.

Haji-Sheikh, A. and Mashena, M. (1987) Integral solution of diffusion equation. Part I. General solution. *J. Heat Tranfer (Trans ASME)*, **109**, 551.

Han, J.-K., Selke, S.E., Downes, T.W. and Harte, B.R. (2003) *Packag. Technol. Sci.*, **16**, 107.

Huebner, K.H. (1975) *The Finite Element Method for Engineers*, Wiley, New York. see also Tamma, K.K. and Yurko, A.A. (1989) Finite element thermal modeling/analysis formulation for layered composite materials. *Numer. Heat Transfer, Part B*, **15**, 73.

Ozisik, M.N. (1980) *Heat Conduction*, Wiley, New York.

Ozisik, M.N., Michailov, M.D. and Vulkanov, N.L. (1983) Diffusion in composite layers with automatic solution of the eigenvalue problem. *Int. J. Heat Mass Transfer*, **26**, 1131.

Mercea, P., Piringer, O. and Petrescu, L. (2005) *Migratest Lite-Instruction Manual and Software*, Fabes GmbH, München.

Mercea, P., Tosa, V., Petrescu, L. and Piringer, O. (2007) *Migratest EXP Software*, Fabes GmbH, Munich, 2007.

Muzizo, A. and Solaini, G. (1987) Boundary integral equation analysis of three-dimensional conduction in composite media. *Numer. Heat Transfer*, **15**, 239.

Pettersen, M.K., Gällstedt, M. and Eie, T. (2004) *Packag. Technol. Sci.*, **17**, 43.

Press, W.H., Teukolsky, S.A., Vetterling, W.T. and Flannery, B.P. (1986) *Numerical Recipes*, Cambridge University Press, Cambridge.

Ramos, F.M. and Giovannini, A. (1992) Finite analytic numerical method for transient heat diffusion in layered composite materials. *Numer. Heat Transfer, Part B*, **22**, 305.

Roduit, B., Borgeat, Ch., Cavin, S., Fragnière, C. and Dudler, V. (2004) *ILSI 3rd International Symposium on Food Packaging*, Barcelona/Spain, October 2004, www.akts.com.

Safa, L. and Abbes, B. (2002) *Packag. Technol. Sci.*, **15**, 53.

Siegel, R. (1999) Transient thermal analysis of parallel translucent layers by using Green's functions. *J. Thermophys. Heat Transfer (AIAA)*, **13**, 10.

Smith, G.D. (1985) *Numerical Solutions of Partial Differential Equations. Finite Difference Methods*, 3rd edn, Oxford University Press, Oxford.

Tittle, C.W. (1965) Boundary value problem in composite media: quasi-orthogonal functions. *J. Appl. Phys.*, **36**, 1486.

Tosa, V. (2003) *Proceedings of Isotopic and Molecular Processes*, Cluj-Napoca, Romania.

Tosa, V. (2004) *Solution of the Diffusion Equation for Multilayer Packaging*, FABES GmbH-Report, Munich.

9
User-Friendly Software for Migration Estimations

Peter Mercea, Liviu Petrescu, Otto Piringer and Valer Tosa

9.1
Introduction

As shown in Chapters 6–8, the mass transfer (migration) from plastic materials into a contact medium (for example, foods or pharmaceuticals) is a predictable physical process, which can be quantified with appropriate mathematical algorithms. In principle for this the mass transport equation, see Section 7.2, must be solved for the system formed by the plastic material and contact medium. The very solution of this equation depends on the initial and boundary conditions in the system. For the calculation of the net mass transfer from the plastic to the medium a series of parameters, i.e. diffusion and partitioning coefficients, geometrical dimensions, and initial migrant concentrations, must be known. In the United States, the calculation of potential migration levels is used since more than a decade as an additional tool in support of regulatory decisions.

Aware of these facts the Directorate General for Research of the European Commissions from Brussels launched a few years ago a project for: "Evaluation of Migration Models to be used under Directive 90/128/EEC." The final report of this project (Hinrichs and Piringer, 2002) summarizes the area where, based on the scientific evidence nowadays available, migration modeling can be used as an alternative method for compliance purposes. As a follow up of these results, the European Union has passed in 2002, the theoretical migration estimation (ME) as a conformity and quality assurance instrument for food contact materials. In the EU "Plastics Directive" 2002/72/EC, Article 8, Section 4 the following is stated:

> "The verification of compliance with the specific migration limits provided for in paragraph 1 may be ensured by the determination of the quantity of a substance in the finished material or article, provided that a relationship between that quantity and the value of the specific migration of the substance has been established either by an adequate experimentation or by the application of generally recognized diffusion models based on scientific evidence. To demonstrate the noncompliance of

Plastic Packaging. Second Edition. Edited by O.G. Piringer and A.L. Baner
Copyright © 2008 WILEY-VCH Verlag GmbH & Co. KGaA, Weinheim
ISBN: 978-3-527-31455-3

a material or article, confirmation of the estimated migration value by experimental testing is obligatory."

Details on which type of plastic material and foods are covered already by this paragraph of law as well as the basic mathematical algorithms to be used for the migration estimation are summarized in a "Practical Guide/Note for Guidance" for "Estimation of migration by generally recognized diffusion models" (Chapters 13 and 15).

This document states a series of conditions that should be considered when one intends to determine the compliance of a certain plastic food contact material by estimating theoretically the migration.

Theoretical specific migration estimations can be accepted on a case-by-case basis using scientific evidence. A reliable model for many situations occurring in practice is based on the following general requirements:

- In most cases from practice a plastic food contact material or article can be regarded as a single layer polymer film/sheet (P) of finite and constant thickness (d_P) being in contact with a liquid food (F) of finite volume (V_F).
- It is assumed that during the manufacturing process of P the migrant is distributed homogenously in P (initial concentration $c_{P,0}$) and that the amount of migrant does not influence the matrix of P.
- It is assumed that there is no boundary resistance for the transfer of the migrant between P and F.
- It is assumed that the interaction between P and F is negligible and no swelling of P by uptake of F occurs during the migration process.
- The migrant is homogenously distributed (well mixed) in F and the sum total amount of migrant in P and F is constant during the migration process.

For a one-dimensional P–F system which obeys the above conditions the diffusion/migration follows generally accepted physical laws of diffusion and Eq. (7.12) describing this process can be solved analytically, see Chapter 7 and Crank . The analytical solution of Eq. (7.12) is in this case that given in Eq. (7.51), which can be used to calculate the migration level in a given plastic-food/simulant system. The input parameters needed to calculate with Eqs. (7.51) and (7.37) a time-dependent migration level, $c_{F,t}$ (in mg/kg or weight ppm) in a packaged good F are:

A	the contact area of P in with F (dm^2)
$c_{P,0}$	the initial concentration of migrant in P (mg/kg)
ρ_P	the density of P (g/cm^3)
ρ_F	the density of F (g/cm^3)
t	the migration time, (s)
d_P	the thickness of P, (cm)
V_F	the volume of F (cm^3)
D_P	the diffusion coefficient of migrant in P (cm^2/s)
$K_{P/F}$	the partition coefficient of the migrant between P and F.

Excepting two of these parameters, all are given either by the geometry of the plastic sample and food (thickness, volume, contact area, etc.) or by their basic physicochemical properties (density, composition, etc.). The remaining two parameters are the diffusion coefficient of the migrant in the plastic, D_P, and the partition coefficient of the migrant between the plastic and the food (simulant) $K_{P,F}$. Both of them play a crucial role in determining the level of migration into a real application. The problem with these two coefficients is that usually their exact value is not known *a priori* but must be determined either by an experimental or theoretical method. For compliance purposes it is useful to estimate these values theoretically in a "worst case" scenario with respect to migration. This is of primary interest from a regulatory standpoint. In the framework of the EU documents mentioned above this requirement was met by assuming that

- For cases where the migrant solves well in the food or food simulant the partitioning coefficient is $K_{P,F} = 1$. For all other cases, that means where the migrant does not solve well in food the or food simulant, one takes $K_{P,F} = 1000$.
- The "real" diffusion coefficient D_P of the migrant in the matrix of the polymer is replaced with a polymer specific "upper bound" diffusion coefficient, D_P^* which can be calculated with Eq. (15.3), known also as the FABES Formula. Using this D_P^* in Eq. (7.38) leads to "worst case" migration estimations which can be regarded as an additional consumer protection safety margin.

In Eq. (15.3), D_P^* depends, in fact, only on three parameters. Two of them, the molecular weight of the migrant, M_r, and the absolute temperature, T, respectively, are not linked to the polymer. The third one, A_P, is primarily linked to the properties of the polymer reflecting its "conductance" toward the diffusion of migrants. From Eq. (15.3), one can see that A_P depends not only on the nature of the polymer but also on the temperature at which the migration takes place. As a general rule A_P decreases with the increasing stiffness of the polymer matrix. Consequently for polymers such as polyethylene A_P is larger than for stiff chain polymers such as polyesters and polystyrenes. For polyethylene and polypropylene polymers, the diffusion coefficient databases given in Appendix I were used in an EU project (Reid et al., 1980) to establish "upper bound" A_P values for each polymer. With these values "upper bound" D_P^*s are obtained with Eq. (15.3) and implicitly "worst case" migration levels estimated with Eq. (7.38). A similar procedure was used for a series of other polymers used in packaging applications, the results being summarized in Table 9.1.

For a plastic packaging made from one of the polymers given in Table 9.1, if one knows the M_r of the migrant and T as well as all input parameters listed above one can quite easily calculate, even with a scientific pocket calculator, with Eq. (7.38) a "worst case" migration scenario and compare the obtained result with the specific migration limit (SML) of the given migrant. To make such a procedure more user-friendly the *FABES GmbH* research company released a couple of years ago the dedicated software MIGRATEST©Lite (Mercea and Piringer 2004).

Table 9.1 Values for the A'_p and τ parameters for calculation of "upper bound" diffusion coefficients with the FABES formula.

Polymer	A'_p	τ (K)	Temperature of use (°C)
LDPE/LLDPE	11.5	0	<80/<100
HDPE	14.5	1577	<120
PP (homo and random)	13.1	1577	<120
PP (rubber)	11.5	0	<100
PS	0	0	<70
HIPS	1.0	0	<70
PET	6	1577	<175
PBT	6.5	1577	<175
PEN	5.0	1577	<175
PA	2.0	0	<100

9.2
MIGRATEST©Lite – A User-Friendly Software for Migration Estimations

The aim in developing MIGRATEST©Lite was to provide to a large spectrum of potential users from industry and research and development as well as from the enforcement laboratories a user-friendly tool for a quick and easy estimation of migration of substances from plastic (polymeric) films into foods and food simulants. A special emphasis was to conceive the software in such a manner to include the actual aspects and data from the EU documents related to migration regulation and especially to those mentioned in summary above. The next four sections are dedicated to offer a brief presentation of MIGRATEST©Lite, along with some representative case examples.

9.2.1
Basic Features of MIGRATEST©Lite and Input Data Menus

This software was conceived to be installed on IBM compatible personal computers and was programmed to work with a Microsoft® Windows® operating system. The hardware requirements to install and work with this software are met nowadays by all commercial PCs. The installation of MIGRATEST©Lite is straightforward from a CD-ROM disk as a self-installing executable program. Once the installation procedure and registration of the software is completed the program can be started from the Programs menu.

A *control panel*, Figure 9.1, is used as main dialogue interface of the program. From here one can input or import all parameters needed for a migration estimation and activate the migration calculation procedures. Once these calculations are done the various pull-down menus of the program can be used to edit a Migration Estimation Report.

To produce a new set of input data in MIGRATEST©Lite one uses the five pop-up windows which can be accesses, for example, either from the pull-down menus of the

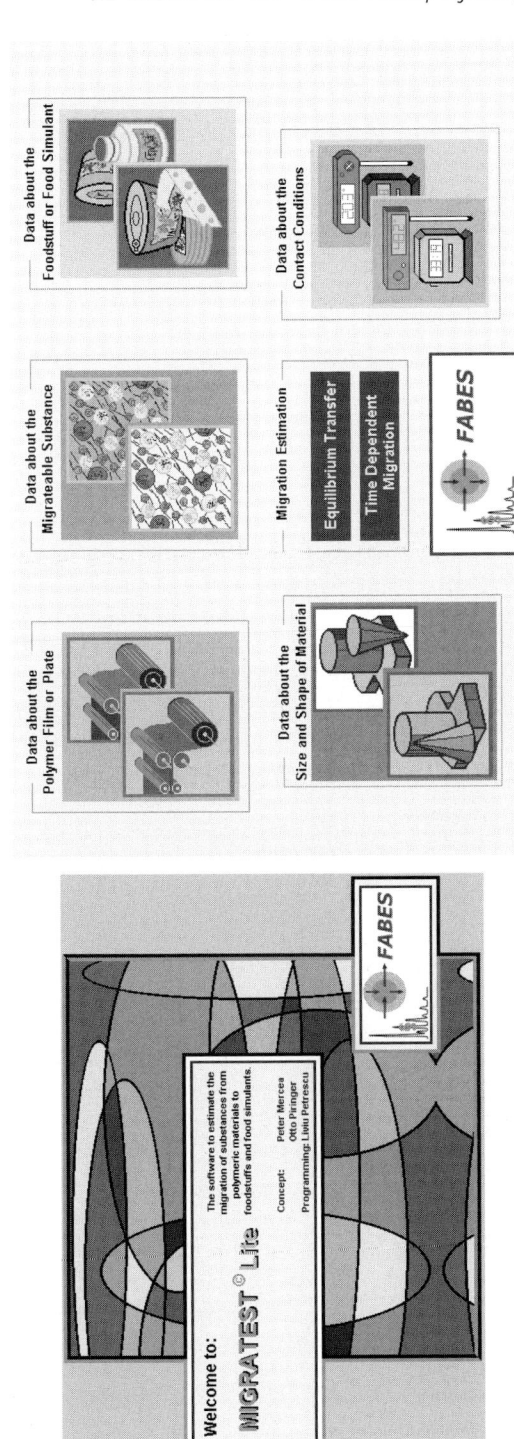

9.2 MIGRATEST©Lite – A User-Friendly Software for Migration Estimations

Figure 9.1 The welcome window and main control panel of the software MIGRATEST©Lite.

program or from the command-buttons placed on the tool bar. A brief description of the features of these pop-up windows is given in the followings.

To input the data characterizing the plastic material one uses the pop-up window shown in Figure 9.2. Here, one specifies the nature of the polymeric sample, its constant thickness, d_P, and density, ρ_P.

A special feature of MIGRATEST$^{©}$Lite is the possibility to chose the plastic material from two predefined lists that, in fact, contain the information given above in Table 9.1.

The pop-up windows shown in Figures 9.3 are for entering the input data for the migrant.

Here one specifies the nature of the migrant, its molecular weight, M_r, and initial concentration, $c_{P,0}$, in the matrix of the plastic material. Another important feature implemented in this pop-up window is the calculation of the diffusion coefficient, D_P.

As one can see from Figure 9.3 the user of MIGRATEST$^{©}$Lite can select the type of the migrant from two predefined data banks, namely a list of "Additives & Monomers which can be considered for migration modeling" and the "Synoptic Document of the European Commission," respectively. The first data bank comprises more than 300 substances, which were listed in the Practical Guide mentioned above while the second data bank gathers more than 3400 substances. Search functions help the user to find a substance in these data banks.

The calculation of D_P is done in MIGRATEST$^{©}$Lite in the framework of the pop-up windows shown in Figure 9.4.

There are two important features here. First, one can specify the so-called temperature pattern for the migration process by choosing one of the three options below:

- during the whole migration process the contact temperature is constant
  $T_1 =$ constant,
- during the migration process there are two constant temperature regimes
  $T_1 =$ const and $T_2 =$ const, and
- the migration takes place at three constant temperatures
  $T_1 =$ const, $T_2 =$ const, and $T_3 =$ const.

The second feature is the calculation of D_P with the FABES Formula given in Eq. (15.3). When one uses this menu in conjunction with a polymer selected from one of the two predefined lists mentioned above the software automatically imports into the FABES formula the corresponding parameters listed in Table 9.1.

In order to perform a migration estimation with MIGRATEST$^{©}$Lite input data about the *food* or *food simulant* which is in contact with the polymeric packaging are needed. To specify these inputs one uses the pop-up windows presented in Figure 9.5.

Here, one can select/specify the type of food/simulant in which the substance migrates by selecting data from:

- the list of food simulants used in the EU (Directive 97/48/EC) or
- the list of real foods given in the EU document 85/572/EEC.

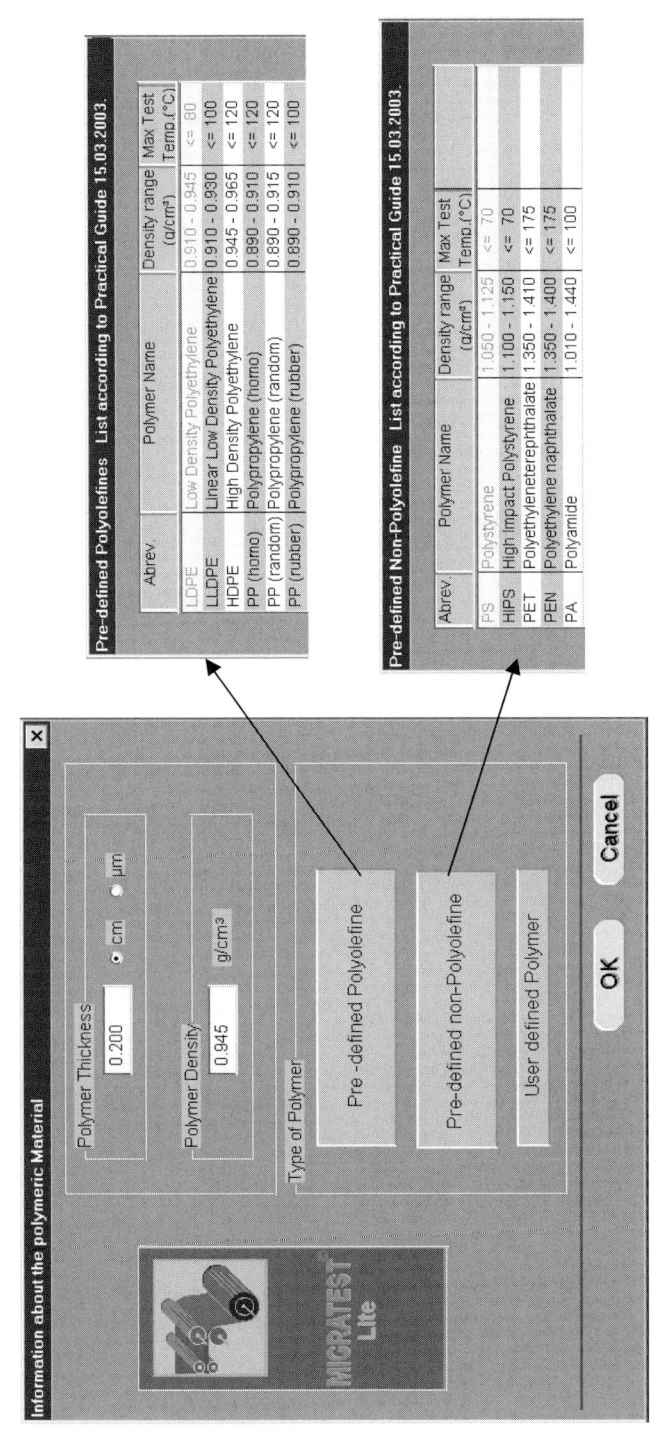

9.2 MIGRATEST© Lite – A User-Friendly Software for Migration Estimations | 269

Pre-defined Polyolefines List according to Practical Guide 15.03.2003.

Abrev.	Polymer Name	Density range (g/cm³)	Max Test Temp.(°C)
LDPE	Low Density Polyethylene	0.910 - 0.945	<= 80
LLDPE	Linear Low Density Polyethylene	0.910 - 0.930	<= 100
HDPE	High Density Polyethylene	0.945 - 0.965	<= 120
PP (homo)	Polypropylene (homo)	0.890 - 0.910	<= 120
PP (random)	Polypropylene (random)	0.890 - 0.915	<= 120
PP (rubber)	Polypropylene (rubber)	0.890 - 0.910	<= 100

Pre-defined Non-Polyolefine List according to Practical Guide 15.03.2003.

Abrev.	Polymer Name	Density range (g/cm³)	Max Test Temp.(°C)
PS	Polystyrene	1.050 - 1.125	<= 70
HIPS	High Impact Polystyrene	1.100 - 1.150	<= 70
PET	Polyethyleneterephthalate	1.350 - 1.410	<= 175
PEN	Polyethylene naphthalate	1.350 - 1.400	<= 175
PA	Polyamide	1.010 - 1.440	<= 100

Figure 9.2 Pop-up window to input data about the polymer.

270 | *9 User-Friendly Software for Migration Estimations*

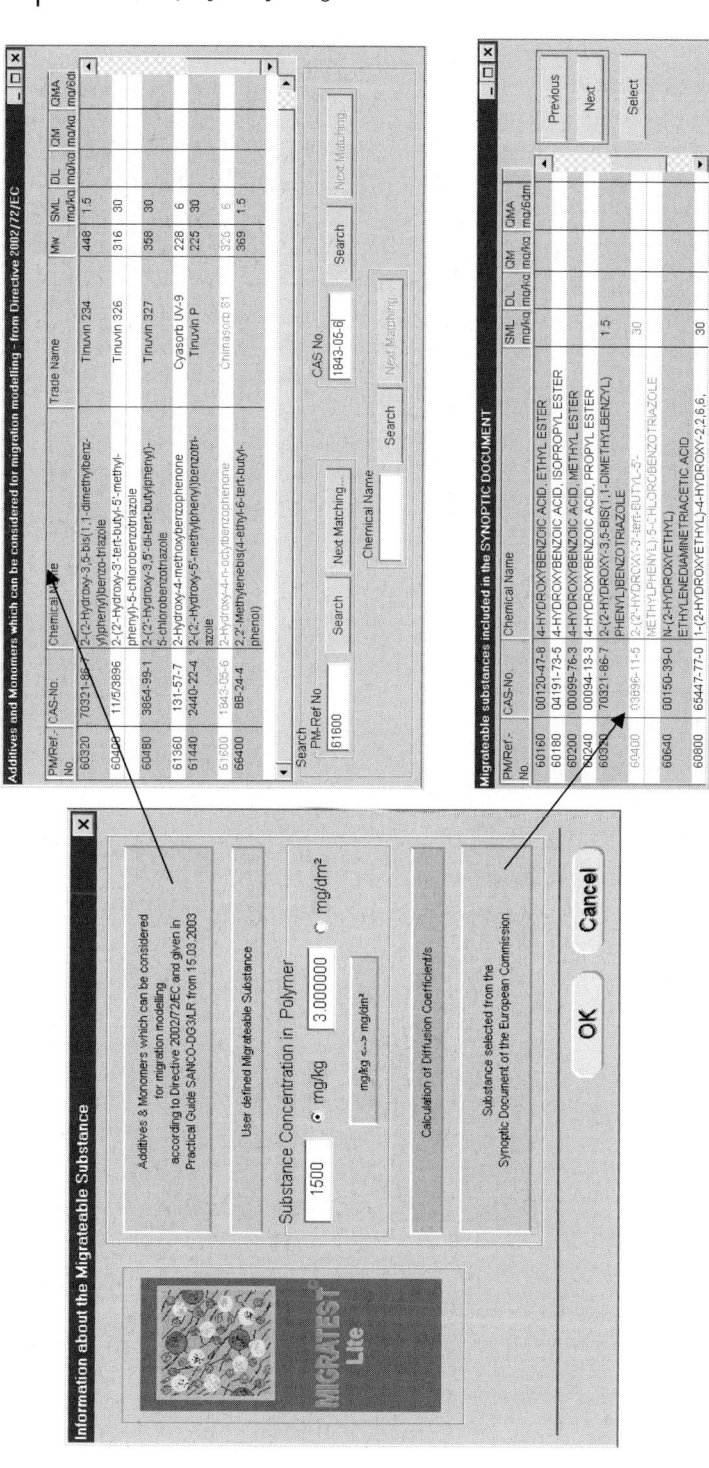

Figure 9.3 Pop-up window to input data about the migrant.

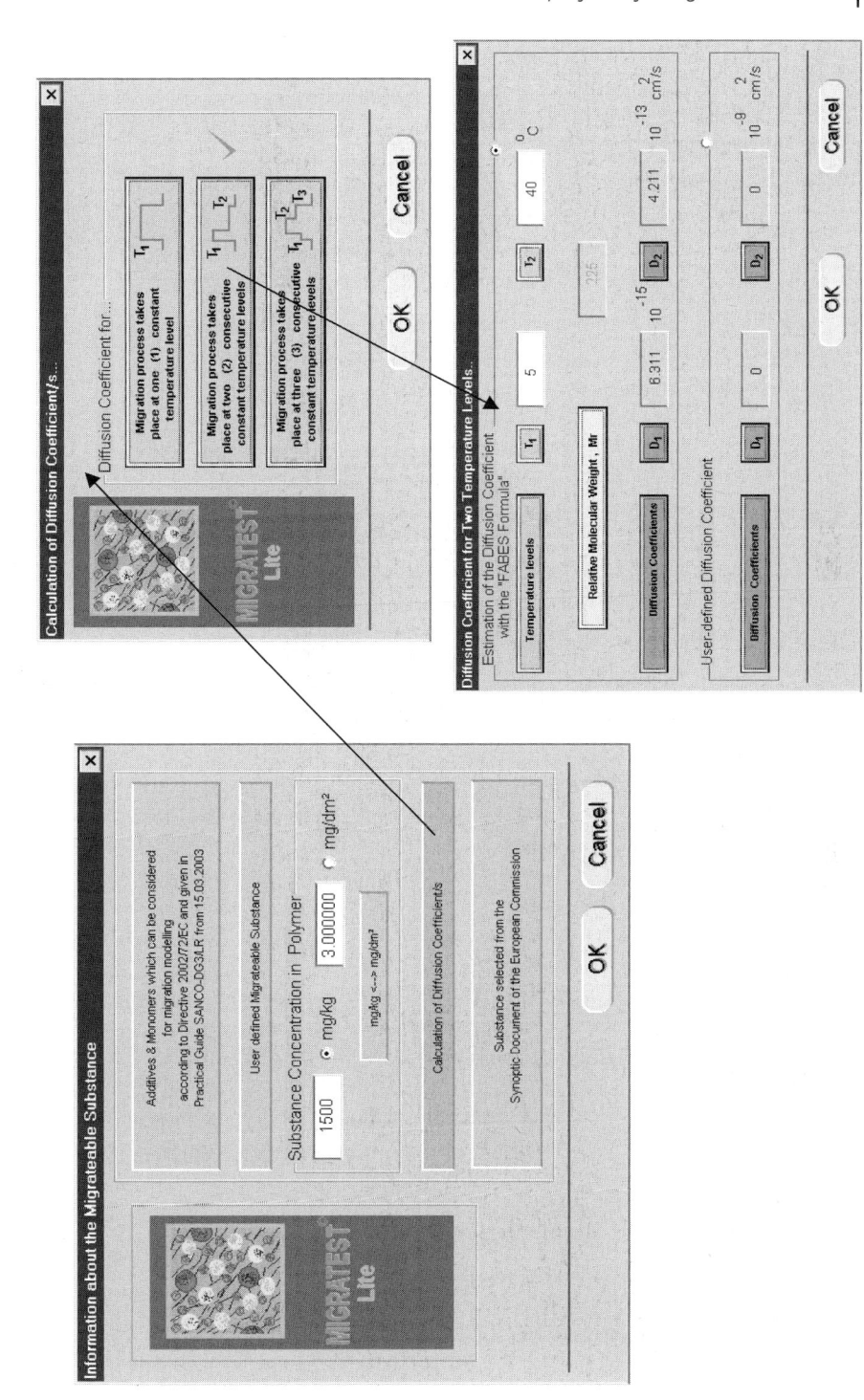

9.2 MIGRATEST©Lite – A User-Friendly Software for Migration Estimations | 271

Figure 9.4 Pop-up windows for the calculation of the migrant diffusion coefficient D_P.

272 | *9 User-Friendly Software for Migration Estimations*

Information about the Foodstuff or Food Simulant

- Food Simulants
- Real Foodstuff from the EEC List 85/572
- User defined Foodstuff

OK | Cancel

Food Simulants according to EU Directive 97/48/EC...

Group	Test Substance
	Food Simulants from EU Directive 97/48/EC
A	Distilled Water
B	3% Acetic Acid in Water
C	10% Ethanol in Water
D	Olive Oil

OK | Cancel

Real Foodstuff from the EEC List 85/572

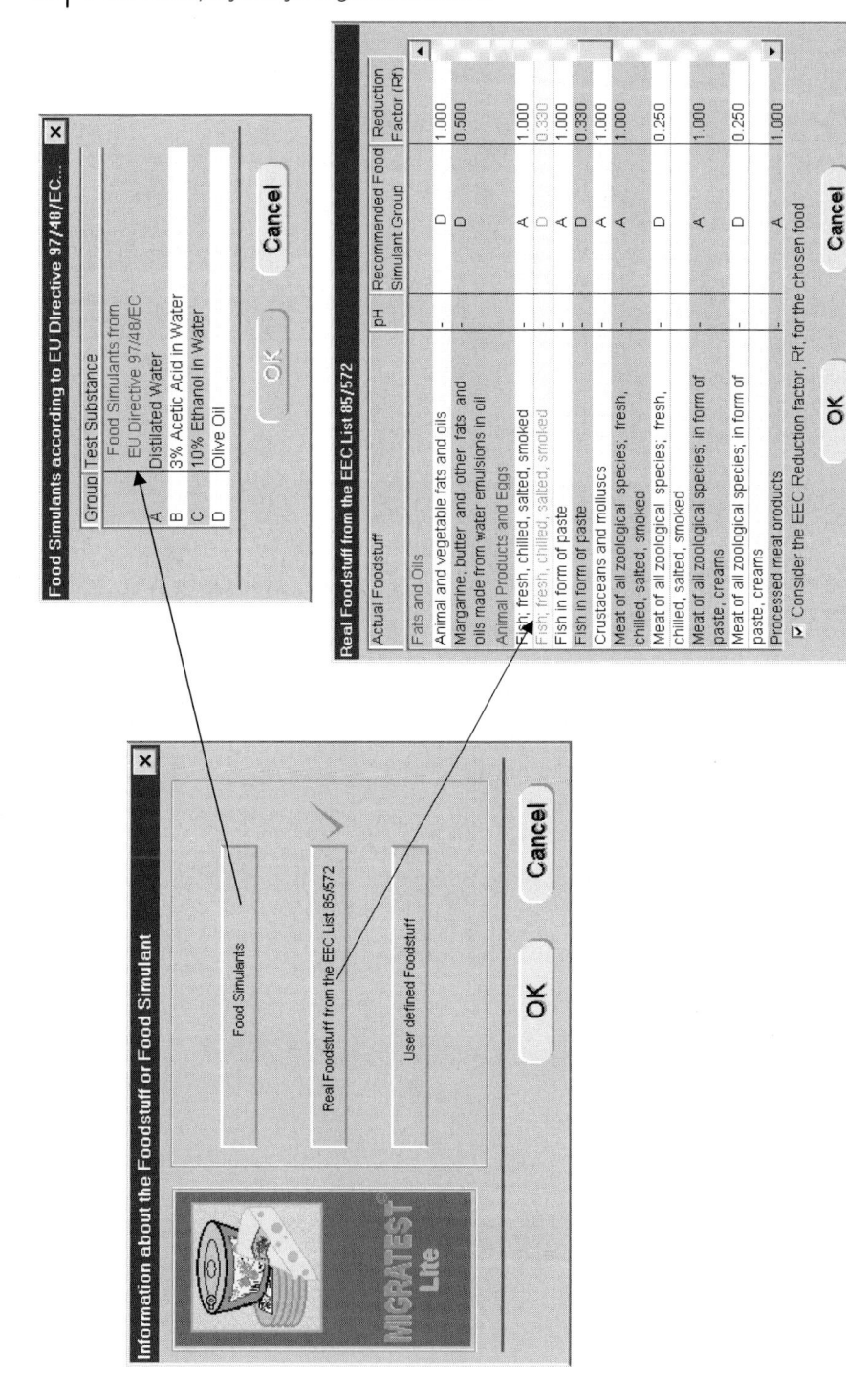

Actual Foodstuff	pH	Recommended Food Simulant Group	Reduction Factor (Rf)
Fats and Oils			
Animal and vegetable fats and oils	-	D	1.000
Margarine, butter and other fats and oils made from water emulsions in oil	-	D	0.500
Animal Products and Eggs			
Fish, fresh, chilled, salted, smoked	-	A	1.000
Fish, fresh, chilled, salted, smoked	-	D	0.330
Fish in form of paste	-	A	1.000
Fish in form of paste	-	D	0.330
Crustaceans and molluscs	-	A	1.000
Meat of all zoological species, fresh, chilled, salted, smoked	-	A	1.000
Meat of all zoological species, fresh, chilled, salted, smoked	-	D	0.250
Meat of all zoological species, in form of paste, creams	-	A	1.000
Meat of all zoological species, in form of paste, creams	-	D	0.250
Processed meat products	-	A	1.000

☑ Consider the EEC Reduction factor, Rf, for the chosen food

OK | Cancel

Figure 9.5 Pop-up windows to input data about the food or food simulant.

The convention resulting from the EU Directives determines that in MIGRATEST©Lite the density, ρ_F, of these food simulants and real foods is considered to be $1\,\text{g/cm}^3$.

Besides the options to select an input data about the contact medium from one of the two lists mentioned above the user of MIGRATEST©Lite can also define a medium of her/his choice. For this one uses the option "User defined Foodstuff" from Figure 9.5.

An important parameter for the estimation of migration with Eq. (7.51) is the ratio between the area of the plastic packaging, A, and the volume, V_F, of the packaged good/food. To input this information for the algorithms of MIGRATEST©Lite one uses the pop-up windows shown in Figure 9.6.

Here one has the options:

- to chose the conventional EU packing – with an area-to-volume ratio of $A/V_F = 600\,\text{cm}^2/1000\,\text{cm}^3$,
- to chose the conventional FDA packing – with an area-to-volume ratio of $A/V_F = 645.16\,\text{cm}^2/1000\,\text{cm}^3$, or
- to specify the dimensions of one from the nine predefined geometrical packaging shapes (rectangular, cylindric, conic, etc.) or to specify the dimensions of one of the three predefined types of sample plates.

As mentioned in Chapter 7 the migration from a plastic material into a contact medium strongly depends on the contact temperature, T, and its duration, t, as well as on the partitioning of the migrant between the plastic and contact medium, see for example, Eq. (7.51). In MIGRATEST©Lite one can input these data in the pop-up window shown in Figure 9.7.

The test temperatures T and times t as recommended by the EU Directive 97/48/EC for migration testing are summarized in two small databases from which the user can select the test conditions for its migration estimation run. On the other hand, in MIGRATEST©Lite there is also an option to input user-defined contact conditions.

For the partitioning coefficient, $K_{P,F}$, one has in principle the following possibilities (Hinrichs and Piringer 2004):

- to use the convention established in the EU Practical Guide, Ref. , and chose $K_{P,F} = 1$ or $K_{P,F} = 1000$ for migrants which are highly soluble or, respectively, less soluble in the contact medium or
- to define a real value of $K_{P,F}$ (known from experiment and/or literature).

Important notice! It is known that $K_{P,F}$ may depend noticeably on test temperature/s T. This is a rather complex dependence that is influenced by a series of parameters linked to the nature of the migrant and food/simulant. Because of the complexity of the problem it was beyond the scope of MIGRATEST©Lite to implement a "temperature sensitive" $K_{P,F}$. It means that if the user specifies for a certain T a $K_{P,F}$ but afterward changes T the program will not recalculate $K_{P,F}$.

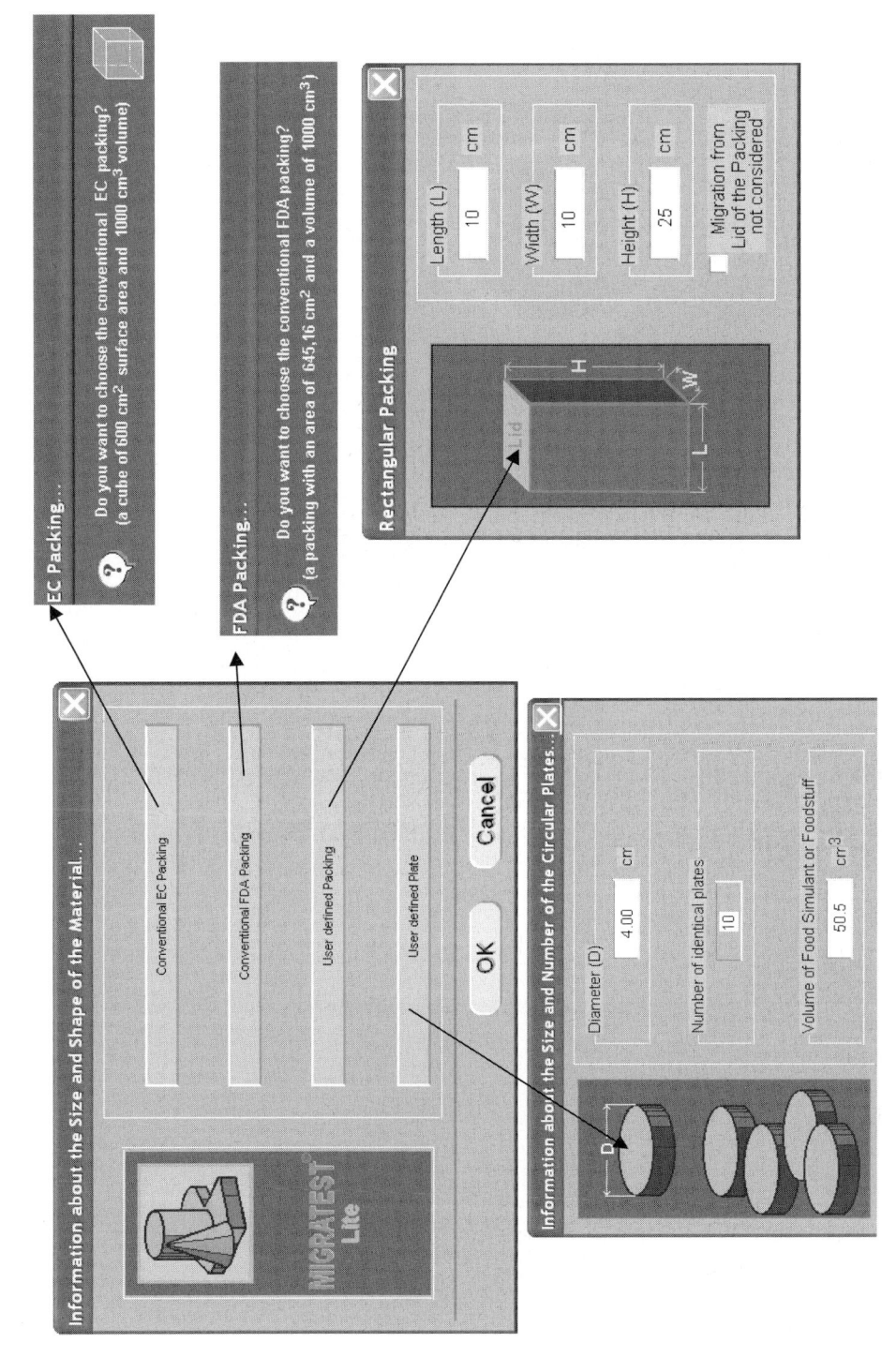

Figure 9.6 Pop-up windows to input data about the size and shape of the plastic packaging.

9.2 MIGRATEST© Lite – A User-Friendly Software for Migration Estimations | 275

Test Temperatures T1 and T2 ...

Contact Temperature (°C)	Test Temperature T (°C)
T* ≤ 5.0	5
5.0 < T* ≤ 20	20
20 < T* ≤ 40	40
40 < T* ≤ 70	70
70 < T* ≤ 100	100
100 < T* ≤ 121	121
121 < T* ≤ 130	130
130 < T* ≤ 150	150
150 < T*	175

Select T1   Select T2

Test Duration for Migration at T1 and T2 ...

Actual Contact Time t* (hours)	Test Time t (hours)
t* ≤ 0.5	0.5
0.5 < t* ≤ 1.0	1.0
1.0 < t* ≤ 2.0	2.0
2.0 < t* ≤ 4.0	4.0
4.0 < t* ≤ 24	24.0
t* > 24	240.0

Select t1   Select t2

Solubility or Migrant in Foodstuff or Simulant

First Temperature Level
○ Insoluble   ● Soluble   Partitioning Coefficient: 5000

Second Temperature Level
○ Insoluble   ● Soluble   Partitioning Coefficient: 1000

Contact Conditions and Solubility for two Temperature Levels

- Contact Temperatures according to Directive 97/48/EC
- Contact Durations according to Directive 97/48/EC
- User defined Contact Data
- Solubility of Migrant in Foodstuff or Food Simulant

OK   Cancel

MIGRATEST® Lite

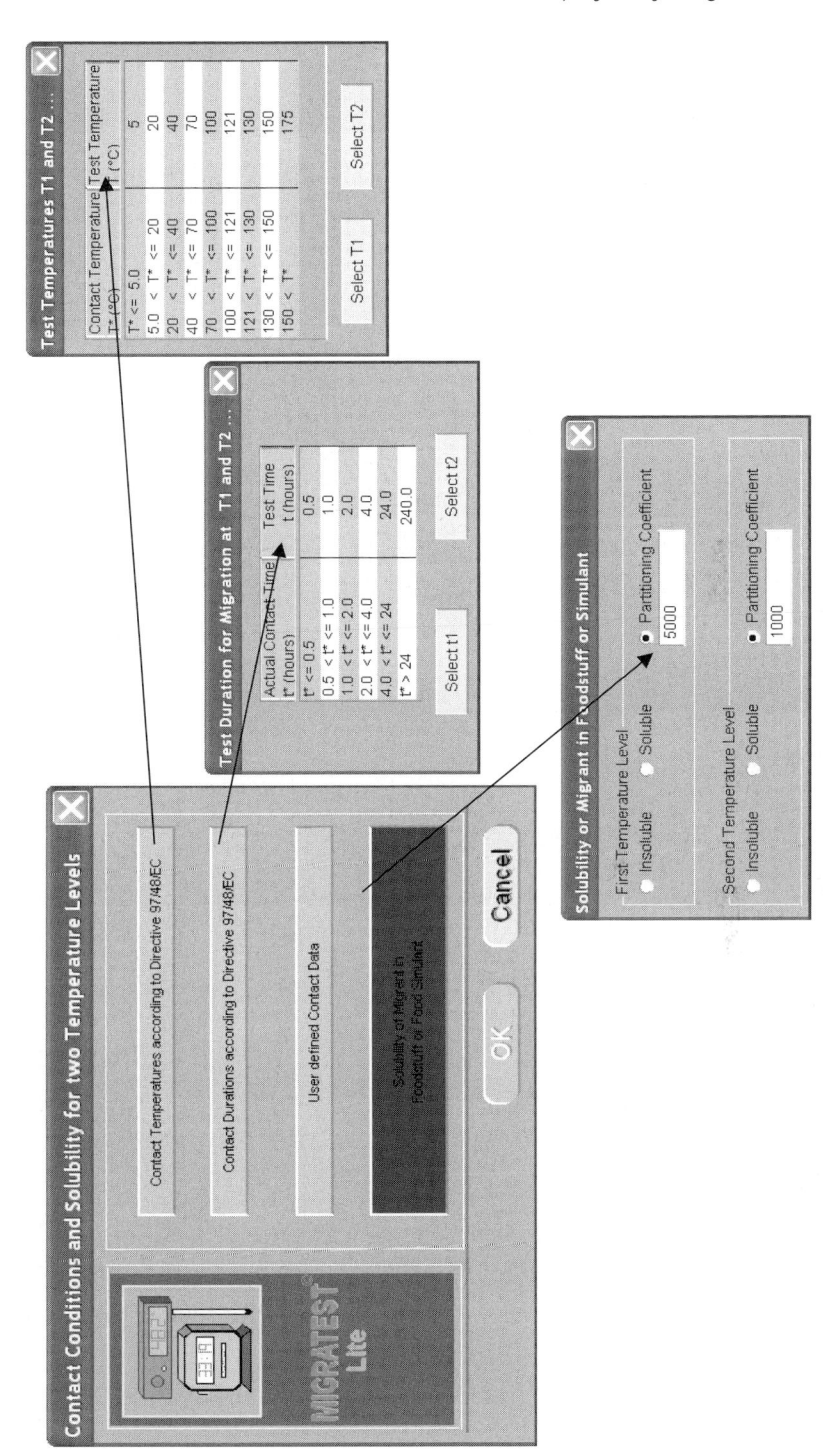

Figure 9.7 Pop-up windows to input data about the contact conditions between plastic material and the food/simulant.

In the software MIGRATEST©Lite a series of simple, warning and error messages help and guide the user through the program. However, a detailed description of these features is beyond the scope of this chapter.

9.2.2
Estimation of Migration with MIGRATEST©Lite

Once all data about the polymer-migrant and food/simulant system and, respectively, contact conditions have been defined/specified one can activate the menus of the main program to perform several types of migration calculations. The procedures to obtain these results are straightforward and user-friendly. Basically, one can run estimations for two different tasks.

First, there are so-called straight "migration estimations" (MEs). Here, one estimates the amount substance migrated from the material (polymeric packaging or sample plate) into a foodstuff or a food simulant.

Then there is a "maximum initial concentrations" (MIC) estimation mode. Here, one can calculate the initial concentration of the migrateable substance in the finished polymeric article for which the migration process would lead to substance concentrations in the foodstuff or food simulant equal to the specific migration limit (SML) of the substance (migrant).

In MIGRATEST©Lite one can compute these migration results for two types of migration process configurations, namely: one sided migrations (OSM) and two sided migrations (TSM), respectively.

From a physical–chemical point of view one can calculate with this software results for two types of migration process assumptions.

First, there is the equilibrium transfer (ET) migration assumption. Here, it is considered that the transfer of the migrateable substance from the polymeric sample into the foodstuff/simulant has already reached a thermodynamic equilibrium level. The ET results are computed using classical mass-balance equations.

Second, there is the time-dependent migration (TDM) assumption which is quantified with Eq. (7.51).

As one can see from Figure 9.8 the program calculates for an ET or TDM three types of migration results:

- the amount of substance, m_F, – in miligrams (mg) – migrated from the polymeric sample into the food/simulant,
- the mass concentration, C_F – in milligrams of migrant *per* kilogram of food (mg/kg) – of the migrant in the food/simulant, and
- the amount of migrated substance per unit contact area, m_F/A – in (mg/dm^2).

For maximum initial concentration (MIC) calculations the software delivers two types of results:

- the maximum initial concentration of substance in the finished polymeric sample, $C_{P,0}^*$ – milligrams of migrant *per* kilogram of polymer (mg/kg) – for which the

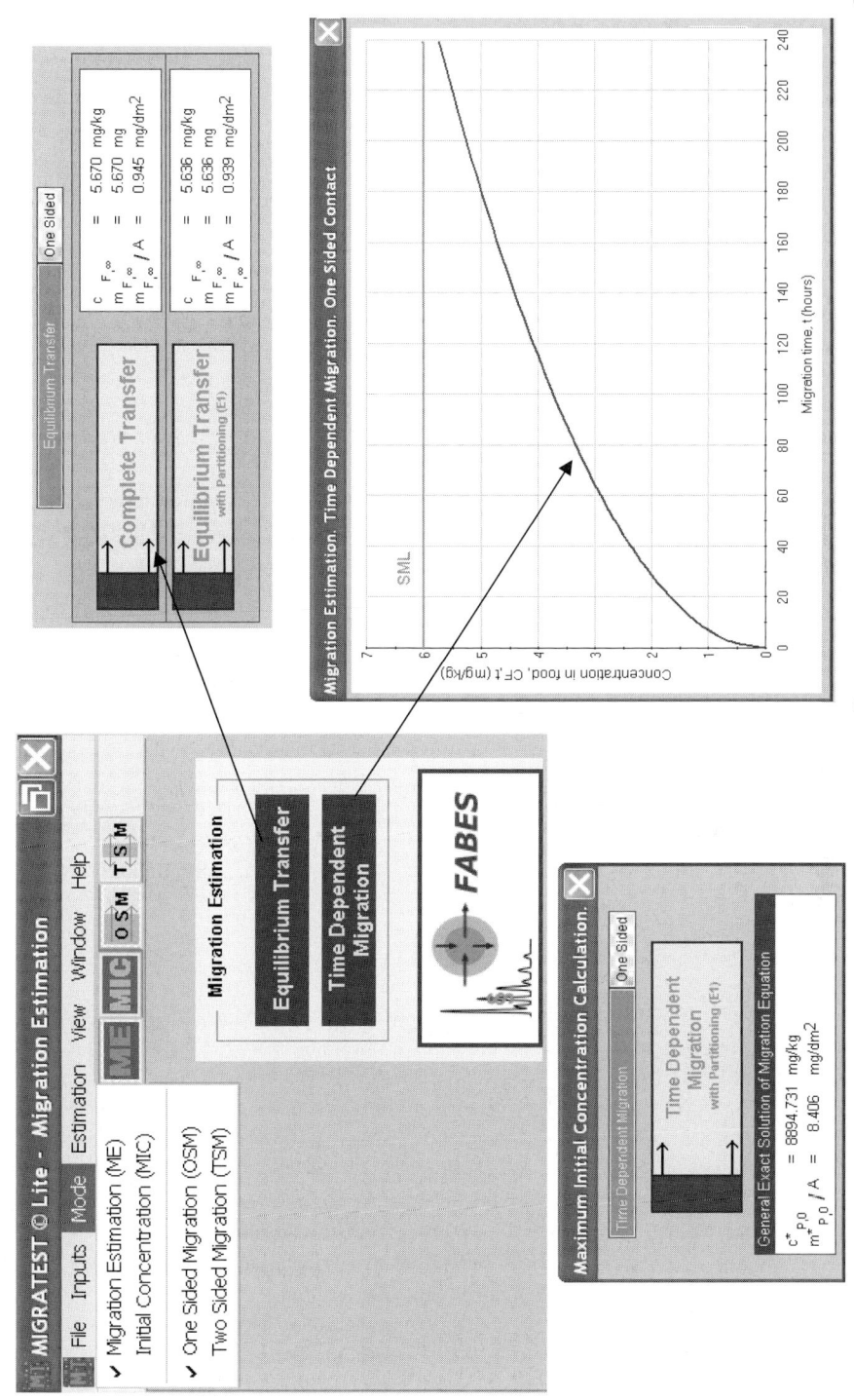

Figure 9.8 Migration estimation menus and estimation options of MIGRATEST© Lite.

migration process would lead to a migrant concentration in the food/simulant equal to SML, and
- the maximum initial amount of migrant in the polymer *per* unit area, $m^*_{P,0}/A$, in (mg/dm²).

9.2.3
Output Information Delivered by MIGRATEST©Lite

The software produces a series of output files which can help the user to analyze and annotate the performed migration estimation and/or archive it for further use.

The input data and information about the type of analytical algorithms used to calculate the migration are summarized in a so-called Input Data Bulletin (IDB) (see Figure 9.9). This bulletin can be previewed on the monitor and/or can be printed.

Then the obtained migration results and graphs are summarized in an Estimation Results Bulletin (ERB).

Both the IDB and/or the ERB can be printed as a hard copy. MIGRATEST©Lite offers the user the possibility to add and/or annotate an ERB with its own comments. This is additional to the "Conclusions" which the program draws automatically by comparing the estimated results with the Specific Migration Limits (SML) from the EU documents.

Another important feature included in the program is the possibility to export independent text files (format *.txt) containing the results for the TDM estimations and used by the program to draw the time-dependent graphs. Such a file can then be used with a program like Microsoft Excel or Origin to draw more complex graphical presentations.

Finally the user has the possibility to save on disk both the input data as well as the estimations results. The program produces specific files with the *.MTS name extension.

9.2.4
Case Examples Computed with MIGRATEST©Lite

In Chapter 7, it was shown how one can solve a series of migration case examples (CEs) by using simple calculation tools and data tables available in the literature. In the following we will show how these CEs can be solved with MIGRATEST©Lite and the results produced by the program in this respect.

The first CE is Example 7.4 from Chapter 7. To solve this CE the software needs as an additional input data only the initial concentration, $c_{P,0} = 1000$ ppm, of the migrant in the circular polymer plates. After specifying all input data – a procedure for which a few minutes are needed – the TDM migration estimation can be started by activating the corresponding command button from the main Control Panel of the software – see Figure 9.1. To plot the time dependence of the relative amount of migrant which left the polymeric circular plates one must use the export feature of the program and generate an export data file *.txt. By importing this file into a graphical program – for example, Origin from Origin Lab – one obtains the graph shown in Figure 9.10.

9.2 MIGRATEST©Lite – A User-Friendly Software for Migration Estimations | 279

MIGRATEST© Lite Report

Estimation performed by:	Peter Mercea	Date: 6/12/2007
Company/Institution:	FABES GmbH	File: Test-007

ESTIMATION RESULTS Bulletin

Type of Estimation: Time Dependent Migration (TDM)
One Sided Migration (OSM)

General Exact Solution - No Consumption Factor was taken into account

$c\ F_t\ =\ 7.083\ mg/kg$
$m\ F_t\ =\ 7.083\ mg$
$m\ F_t\ /\ A\ =\ 1.180\ mg/dm^2$

[Graph: Concentration in food, CF_t (mg/kg) vs. Migration time, t (hours), from 0 to 240 hours]

MIGRATEST© Lite Report

Estimation performed by:	Peter Mercea	Date: 6/12/2007
Company/Institution:	FABES GmbH	File: Test-007

INPUT DATA Bulletin

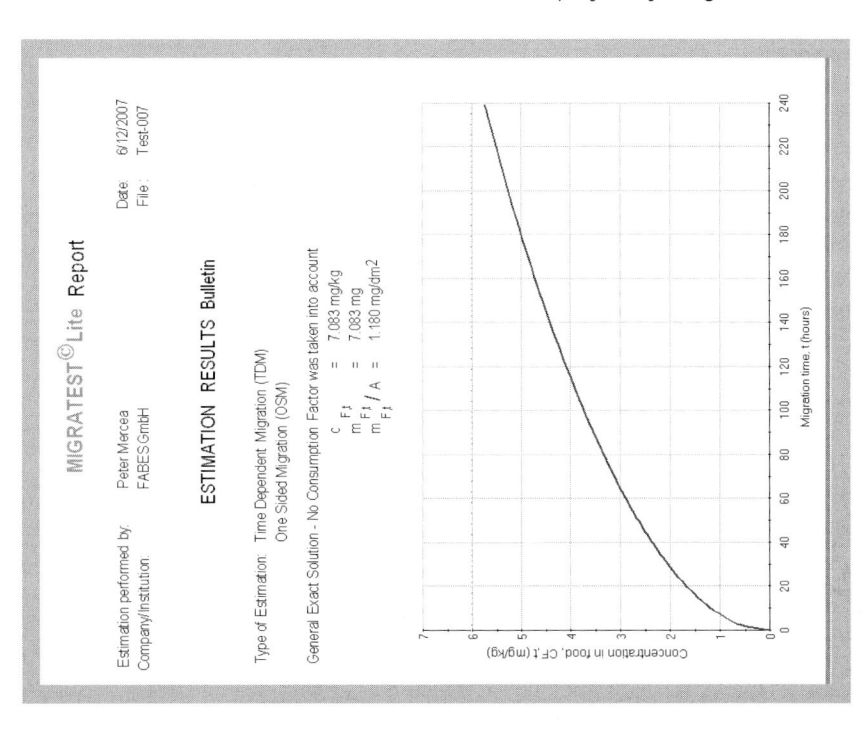

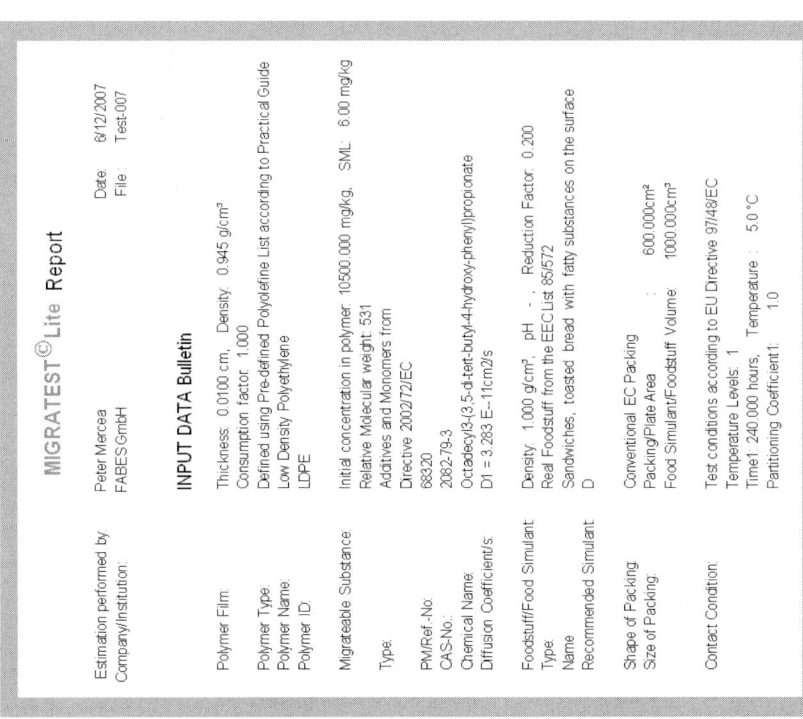

Polymer Film: Thickness: 0.0100 cm, Density: 0.945 g/cm³
Consumption factor: 1.000
Polymer Type: Defined using Pre-defined Polyolefine List according to Practical Guide
Polymer Name: Low Density Polyethylene
Polymer ID: LDPE

Migrateable Substance: Initial concentration in polymer: 10500.000 mg/kg, SML: 6.00 mg/kg
Relative Molecular weight: 531
Type: Additives and Monomers from
Directive 2002/72/EC
PM/Ref.-No: 68320
CAS-No: 2082-79-3
Chemical Name: Octadecyl-3-(3,5-di-tert-butyl-4-hydroxy-phenyl)propionate
Diffusion Coefficient/s: $D1 = 3.283\ E-11\ cm^2/s$

Foodstuff/Food Simulant
Type: Density: 1.000 g/cm³, pH : - , Reduction Factor: 0.200
Name: Real Foodstuff from the EEC List 85/572
Recommended Simulant: Sandwiches, toasted bread with fatty substances on the surface
D

Shape of Packing: Conventional EC Packing
Size of Packing: Packing/Plate Area : 600.000 cm²
Food Simulant/Foodstuff Volume: 1000.000 cm³

Contact Condition: Test conditions according to EU Directive 97/48/EC
Temperature Levels: 1
Time1: 240.000 hours, Temperature : 5.0 °C
Partitioning Coefficient1: 1.0

Figure 9.9 a) MIGRATEST©Lite reports.

a) $K_{P,F} = 1.0$

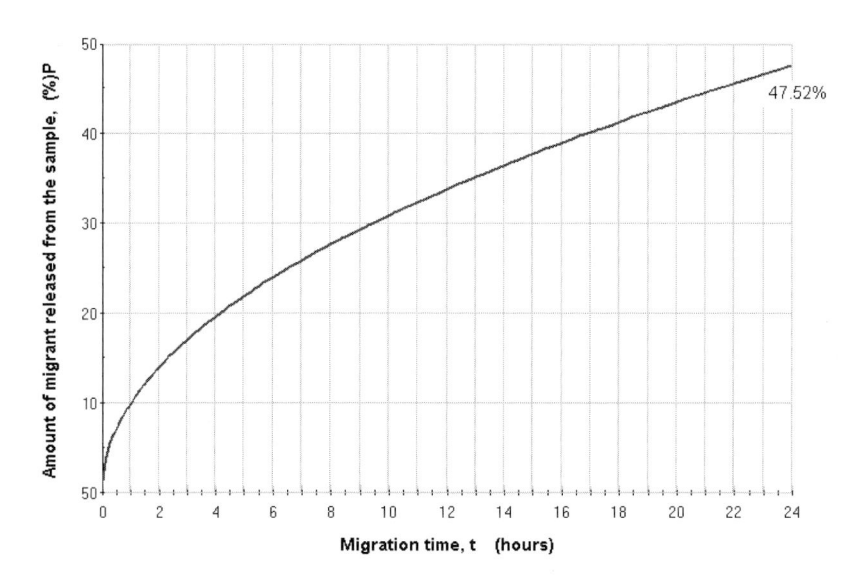

b) $K_{P,F} = 133.0$

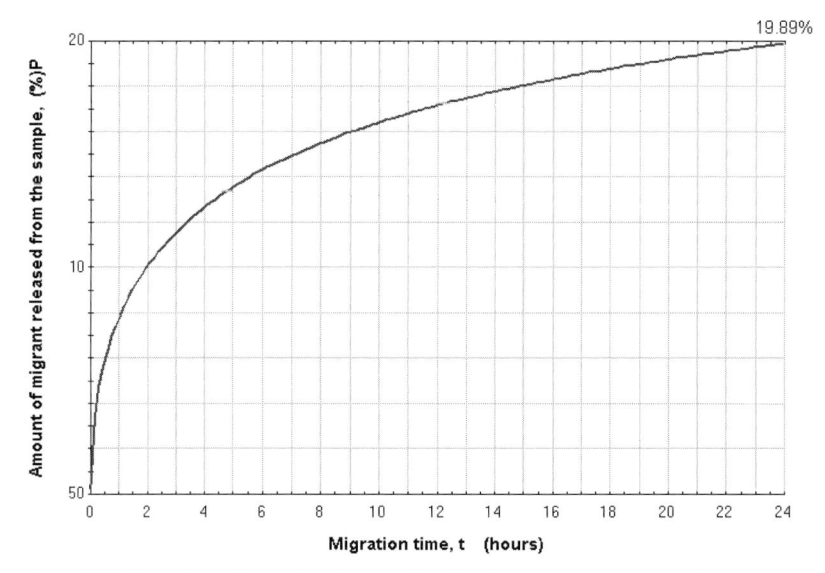

Figure 9.10 Time-dependent migration calculated with MIGRATEST©Lite for Examples 7.4 and 7.5.

From this figure one can see that the result computed with MIGRATEST©Lite is 47.5% which is slightly higher than that given in Chapter 7, 46.1%. This difference is caused by the fact that in Chapter 7 the solutions for the equation $\tan q_n = -\alpha q_n$ were taken only as approximations while in the software this equation is fully solved with a numerical procedure.

The next CE is Example 7.5 from Chapter 7. This is, in fact, the above CE in which the partitioning coefficient, $K_{P,F}$, is taken 133 instead of unity. To implement this new situation in MIGRATEST©Lite requires only to switch in Example 7.4 to the pop-up window "Data about the Contact Conditions" and input 133 in the box corresponding to the partitioning coefficient. With a few more mouse clicks one obtains the TDM calculations and a new graph from which one can see that in this case the relative amount.

The last CE is Example 7.6, in which the migration process from Example 7.4, was calculated with Eq. (7.39) the so-called error function solution of the differential equation. This solution is implemented in the software MIGRATEST©Lite too – in the pull-down menu "Estimations → Approximations for the Time Dependent Migration." This makes it very easy and user-friendly to compute the migration with this formula too. The result obtained is 48.02%, which is again slightly higher than the result reported in Chapter 7 – 47.5%. The difference between the results is caused by the fact that in Chapter 7 the value of the error function was estimated by linear interpolation from Table 7.1 while in MIGRATEST©Lite exact values for this function are used.

9.2.5
Migration Estimations with the MIGRATEST©EXP Software

In Chapters 7 and 8, it was mentioned that in practice there are many situations in which the migration process cannot be quantified with analytical formulae (either the one given in Eq. (7.51) nor other analytical expressions). This is the situation, for example, when migration takes place from a monolayer plastic into a solid food and/or when the packaging is made of a multilayer structure. The same is true when a mono or multilayer plastic swells during the contact with a food, which determines a local change of the diffusion coefficient. Although in principle in these cases too the time-dependent migration can de described with Fick's differential equation, its solution can be obtained only be numerical procedures (Mercea and Piringer XX). A brief presentation of how one can settle such a problem was given in the preceding chapter. However, the development and use of such a computational technique is quite demanding and certainly beyond the possibilities of a scientific pocket calculator. Aware of this situation we developed MIGRATEST© *EXP*, a user-friendly software to address migration estimations where a numerical solving of the differential migration equation is needed.

In the followings a short presentation of the main features of this computational tool will be discussed. Its capabilities will be demonstrated by solving some CEs taken from the practice.

The software MIGRATEST© *EXP* was conceived to be installed on IBM compatible personal computers and was programmed to work with a Microsoft® Windows® XP or Vista operating system. The hardware requirements to install and work with this software are met nowadays by most of the commercial PCs. However, due to the high-

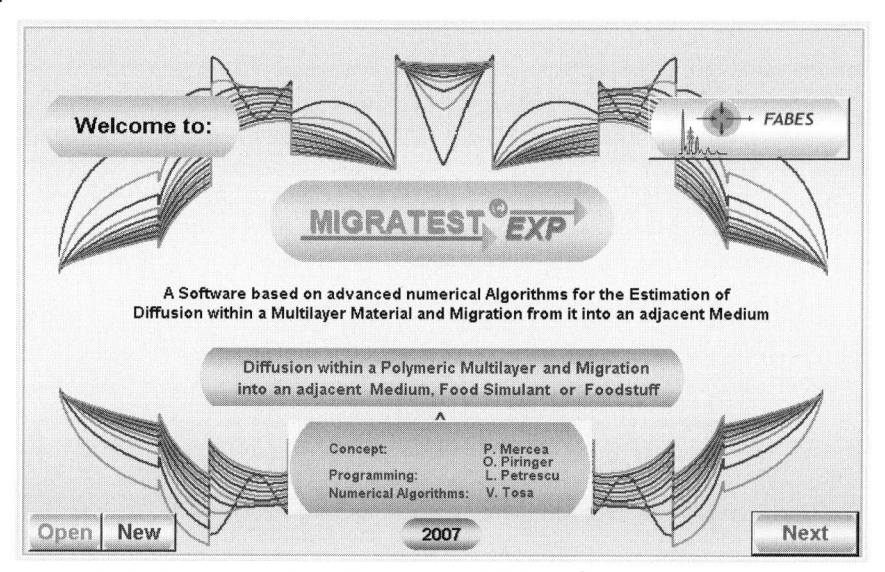

Figure 9.11 The Welcome window of the software MIGRATEST© EXP.

computational work asked from the PC during the running of the numerical algorithms it is recommended to use a powerful PC. The installation of the program is straightforward from a CD-ROM disk as a self-installing executable program. Once the installation procedure and registration of the software is completed the program can be started from the Programs menu. At the beginning the program starts with a welcome window – Figure 9.11 – where the user may chose to open an already saved migration estimation file or start to work on a new case.

When starting a new migration estimation routine with MIGRATEST© EXP the user is asked first to define its multilayer system and the diffusion time, t, and temperature, T, patterns. The program allows the definition of a multilayer with up to five layers and a diffusion pattern with up to three consecutive levels of constant T, see Figure 9.12(a)).

The next step in the data input procedure is to define the nature and geometrical/physical parameters of each layer and the nature of the migrant and its initial concentrations in each of the layers. For this, one uses the input fields shown in Figure 9.12(b)). The information about the type of polymer and migrant can be selected – as in MIGRATEST© Lite – from predefined data banks. The diffusion and partitioning coefficients in the multilayer system are defined in the framework of two special pop-up windows (see Figure 9.12).

From Figure 9.13(a) one can see that the diffusion coefficients can be calculated either with the FABES formula, Eq. (15.3) or with an Arrhenius type of equation $D_P = D_0 e^{-E/RT}$. Once all necessary input data were specified one can start the numerical computation of the diffusion of the migrant in the multilayer structure during its manufacturing and storage using the pop-up window shown in Figure 9.14. Here, the user can fine tune the precision of the computations and its duration by selecting for the discretization of the multilayer an appropriate number of mesh

a)

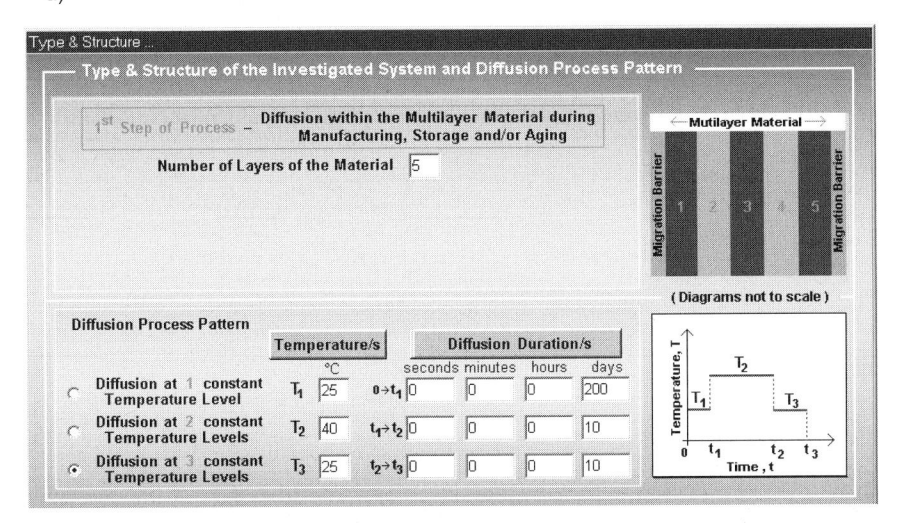

[Screenshot: Type & Structure window showing "Type & Structure of the Investigated System and Diffusion Process Pattern". 1st Step of Process – Diffusion within the Multilayer Material during Manufacturing, Storage and/or Aging. Number of Layers of the Material: 5. Diffusion Process Pattern options: Diffusion at 1 constant Temperature Level (T_1 25, $0 \to t_1$ 0 seconds, 0 minutes, 0 hours, 200 days); Diffusion at 2 constant Temperature Levels (T_2 40, $t_1 \to t_2$ 0, 0, 0, 10); Diffusion at 3 constant Temperature Levels (selected) (T_3 25, $t_2 \to t_3$ 0, 0, 0, 10). Diagram shows Temperature vs Time with T_2 highest between t_1 and t_2, T_1 and T_3 lower. (Diagrams not to scale)]

b)

[Screenshot: Type & Composition window — "Type & Composition of System and Calculation of Diffusion Coefficient/s"]

Type of Material		Layer 1	Layer 2	Layer 3	Layer 4	Layer 5
ID of Material		LDPE	PG 1	PP (ran)	PG 1	HDPE
Density	g/cm³	0.928	1.2	0.903	1.2	0.95
Thickness	cm	0.025	0.0005	0.005	0.0005	0.03

Migrant & its Partitioning

Molec. weight 531

Concentration of Migrant	mg/kg	0	2000	0	2000	0
	mg/dm²	0	0.12	0	0.12	0

Calculation of Diffusion Coefficient/s

			D_1	D_2	D_3	D_4	D_5
T_1 25	°C	cm²/s	4.08 e-10	4.54 e-12	1.02 e-11	4.54 e-12	4.14 e-11
T_2 40	°C	cm²/s	2.19 e-9	2.43 e-11	7.05 e-11	2.43 e-11	2.86 e-10
T_3 25	°C	cm²/s	4.08 e-10	4.54 e-12	1.02 e-11	4.54 e-12	4.14 e-11

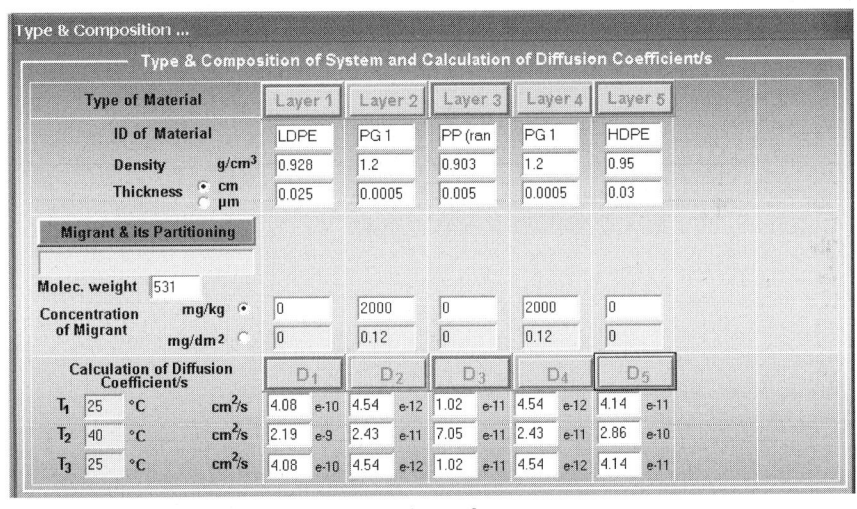

Figure 9.12 The main data input pop-up windows of MIGRATEST© EXP to be used to define the parameters of the migration process.

points. As a rule of the thumb a doubling of the number of mesh points leads to a fourfold increase of the numerical computation time. The progress of computation is shown by a time gauge. The program also displays a message box where the user is informed if the computation is expected to last longer than several minutes. As soon as the numerical computation is finished one can display the obtained results.

a)

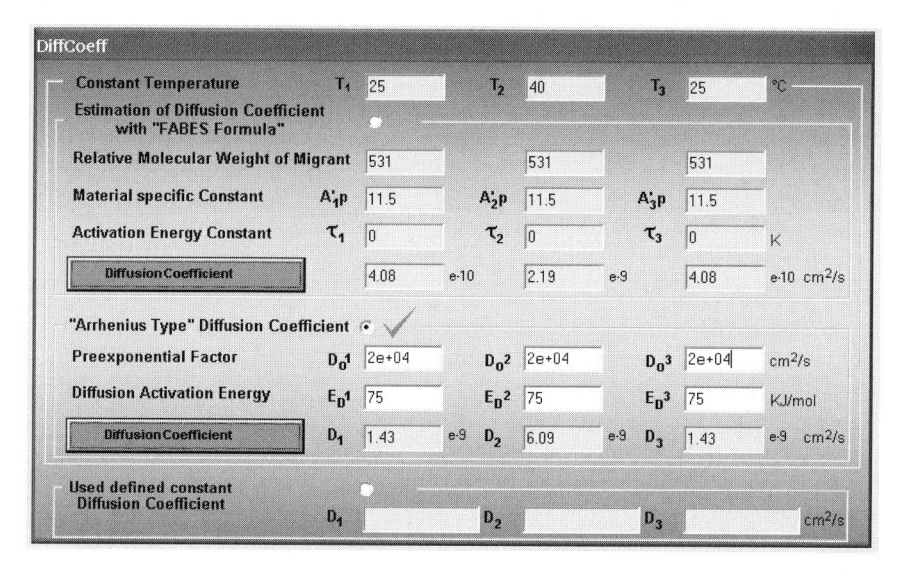

b)

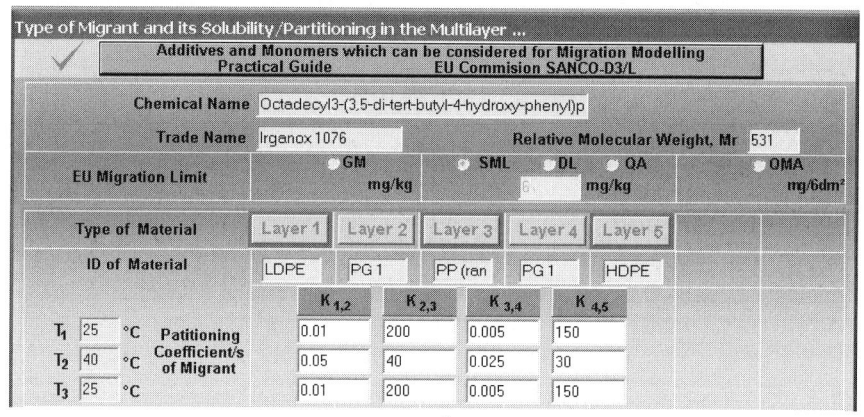

Figure 9.13 The pop-up windows of MIGRATEST© EXP for defining/specifying the diffusion and partitioning coefficients.

There are two types of graphs shown by the program:

- migrant concentration profiles in the multilayer, and
- the time-dependent evolution of the mean migrant concentration in each layer of the system.

These results can be printed on paper, exported to another application (Microsoft® Excel®, for example) or saved as a file on a hard-drive.

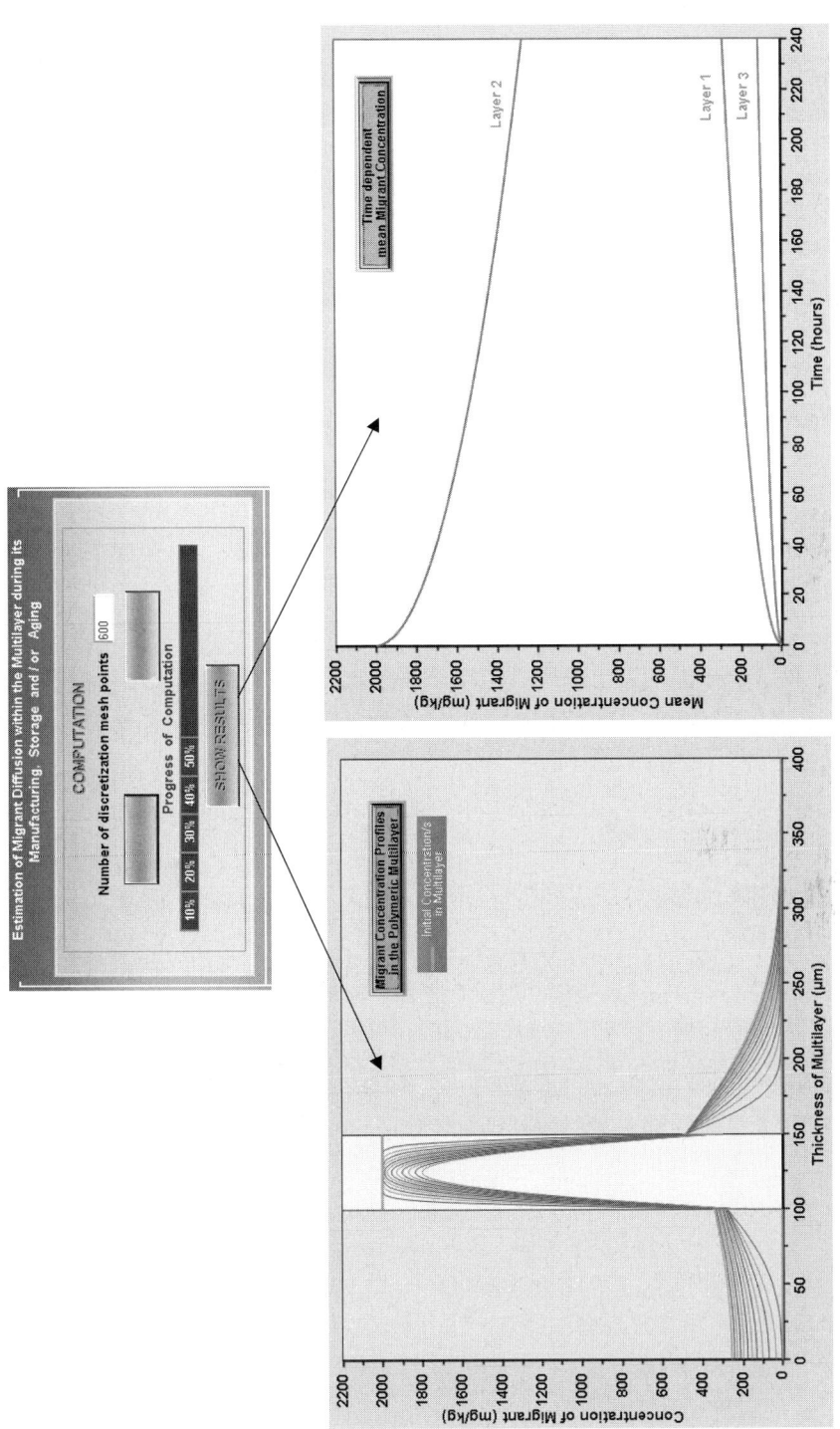

9.2 MIGRATEST© Lite – A User-Friendly Software for Migration Estimations | 285

Figure 9.14 Computation of diffusion and display of results.

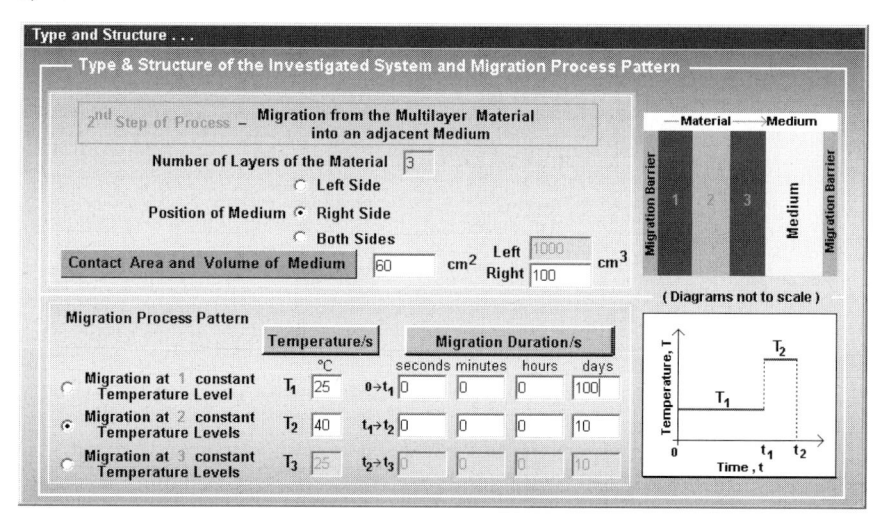

a)

b)

Figure 9.15 The main data input pop-up windows of MIGRATEST© *EXP* to be used to define the parameters of the migration process.

Once the process of migrant diffusion within the multilayer structure during the manufacturing and storage is computed MIGRATEST© *EXP* allows one to estimate the migration from this structure into a contact medium (food, pharmaceutical environment, etc.) For this one uses the pop-up windows and input fields shown in Figure 9.15(a) and (b). The program allows simulations of migration into all types of homogenous contact media in which the migrant exhibits, at a given temperature, a constant diffusion coefficient D_F. The calculation of this diffusion coefficient as well as the specification of the partitioning coefficient at the multilayer – medium interface can be performed in a similar manner as shown in Figure 9.13. The computation of the migration process is started from a pop-up window similar to that shown in Figure 9.14. Here, one can again fine tune the precision and duration of the numerical computation by adjusting the number of mesh points used for the discretization of the multilayer – medium system.

The results obtained are plotted as migrant concentration profiles in the multilayer – contact medium system or as the time-dependent evolution of the mean migrant concentration in each of the layers, see Figure 9.16. A special feature of MIGRATEST©EXP is the possibility to zoom in into a region of the OX-axis. It means that one can enlarge a certain portion of the graphs in order, for example, to better discriminate/show the obtained results – see bottom graphs in Figure 9.16.

Using the various menus of the software one can print on paper, export to another application or save as a file the migration estimation results obtained as well as all input data used to perform the simulations.

9.2.6
Case Examples Computed with MIGRATEST©EXP

In Section 9.2.5, it was already mentioned that one of the simplest situations from practice where the migration process can be quantified with numerical algorithms is the transfer of a substance from a monolayer material/plastic into a solid homogenous contact medium (food or nonfood) in which there is a time-dependent diffusion. It means that the mobility of the migrant in the matrix of the contact medium, D_F, is not large enough to ensure an instant well-mixing of the migrant in the entire mass of the contact medium.. In the food packaging practice this is the case with a large number of foods, for example, dairy products, meat, salamis, or chocolate spreads. Such situations were recently investigated experimentally in the framework of the "Food-Migrosure" EU-Project (Chapter 11), and the results obtained fitted with theoretical computations generated with MIGRATEST©EXP. In Figure 9.17 the results obtained for several real foods are shown and one can see that both D_F and the partitioning coefficient, $K_{P/F}$, at the polymer–food interface depend on the nature of the contact medium.

The software MIGRATEST©EXP can be used to obtain quick computations along with graphical presentation for some of the case examples presented in Chapter 7. Using the input data specified for Examples 7.1 and 7.2 one obtains with this software the result shown in Figure 9.18. From Figure 9.18(a), one can see that at 25 h the migrant concentration at 1 mm deepness into the second wheel of cheese is 33.6 mg/kg while the more approximate calculation procedure used in Chapter 7 led to only a slightly higher value, 34 mg/kg.

The same good agreement is obtained between the migrant concentration profiles calculated with MIGRATEST©EXP and the results given in Chapter 7 for Example 7.2, see Figure 9.18(b) and Section 7.2.3.

With the numerical software MIGRATEST©EXP one can also compute quickly and precisely Example 7.3 which, in fact, is very similar to the examples discussed

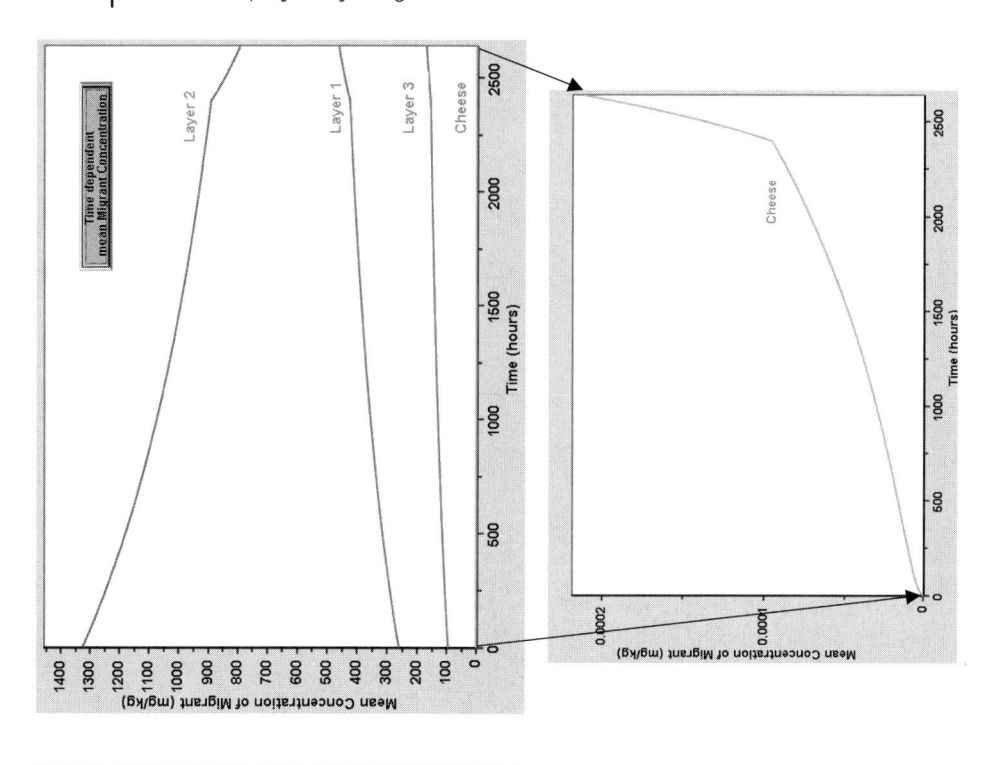

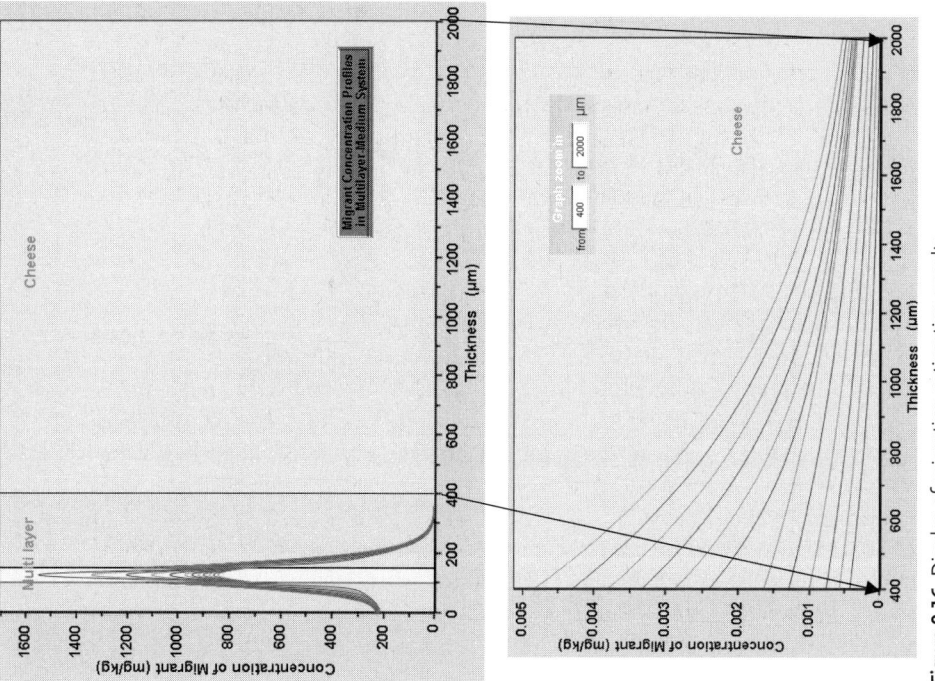

Figure 9.16 Display of migration estimation results.

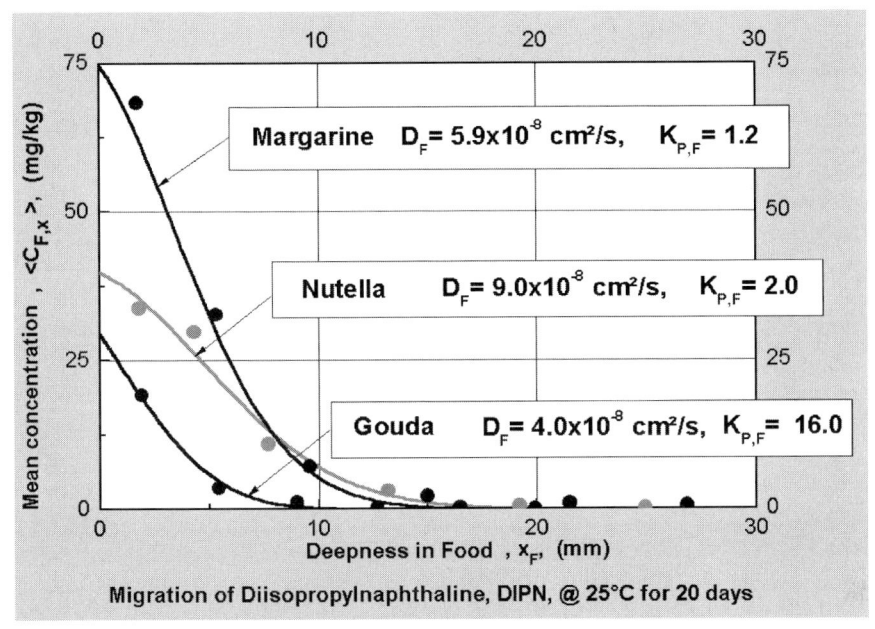

Figure 9.17 Migration of diisopropylnaphthaline, DIPN, from an LDPE packaging film into real foods.

above. The result computed by the software for the migrant concentration at the outer boundary of the second plastic film at migration time, $t = 1$ min, is 9.85 mg/kg. This value is somewhat lower than the 14.8 mg/kg calculated by using Eq. (7.33). This discrepancy results from the assumptions made in the development of Eq. (7.33) that, in fact, gives only an approximate of the concentration profiles in the second film.

One of the simplest migration applications which can be estimated with the MIGRATEST© EXP software is the migration of a residual substance X from a three-layered structure in which a recycled polymer is coextruded between two fresh (virgin) polymers of the same make. Let us assume that the total thickness of such a polyethylene therephthalate, PET, multilayer is 600 μm and that its three layers are equally thick. The residual substance/migrant from the core layer has a molecular weight of 220.4 g/mol and its diffusion coefficient in PET at room temperature is about 7×10^{-16} cm^2/s and 5×10^{-8} cm^2/s at the melting point of this polymer. Further, let us assume that after coextrusion the multilayer cools down to room temperature in about 15 s and afterward the bottle produced is stored at room temperature for a week before being filled with a soft drink. What we would like to estimate is the degree of contamination of the virgin PET layers with the residual substance during the manufacturing and storage of the bottle and then the time needed by this migrant to diffuse through the food contact layer and start contaminating the soft drink. The results computed with MIGRATEST© EXP are shown in Figure 9.19.

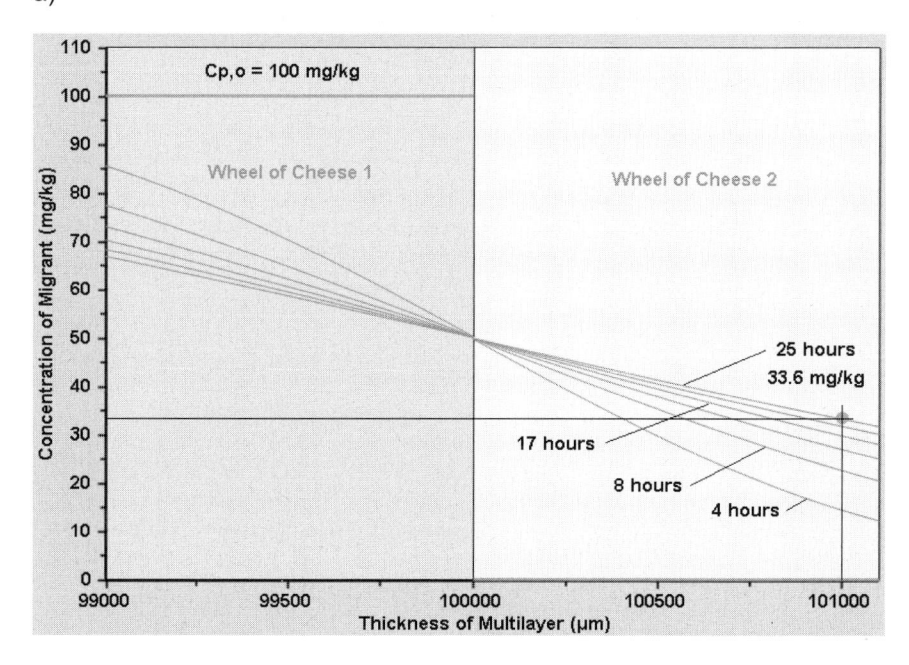

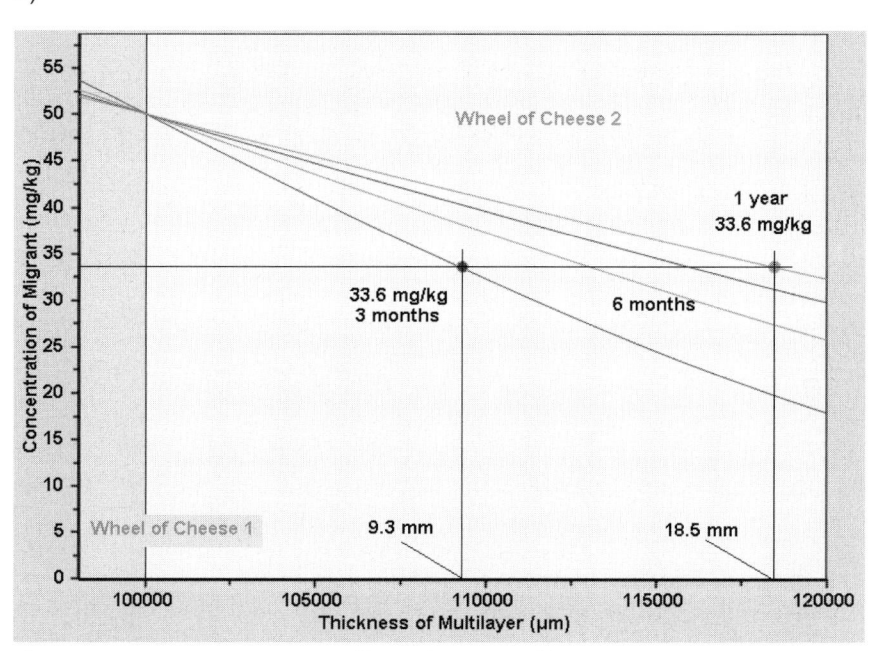

Figure 9.18 Migrant concentration profiles in the vicinity of the interface between the two wheels of cheese as discussed in Examples 7.1 and 7.2.

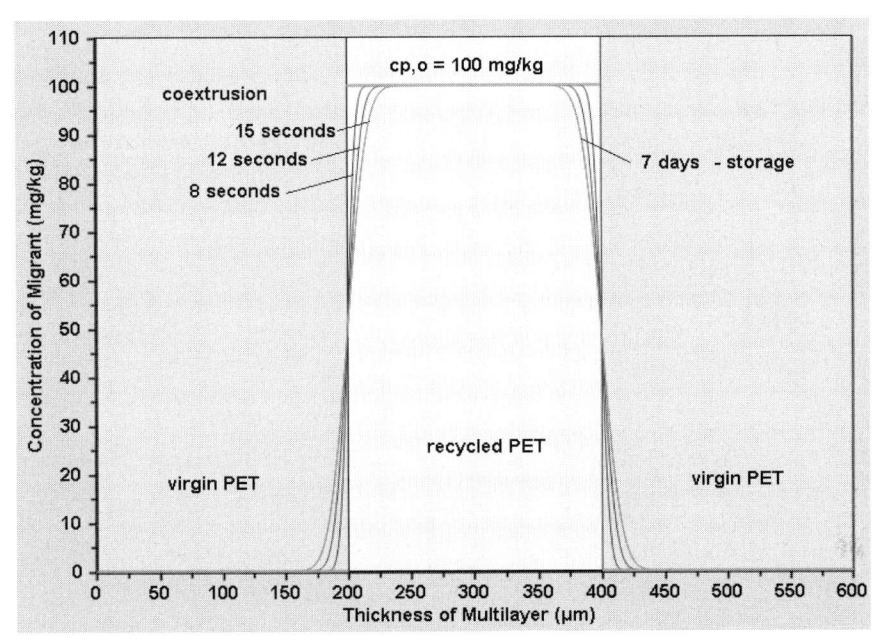

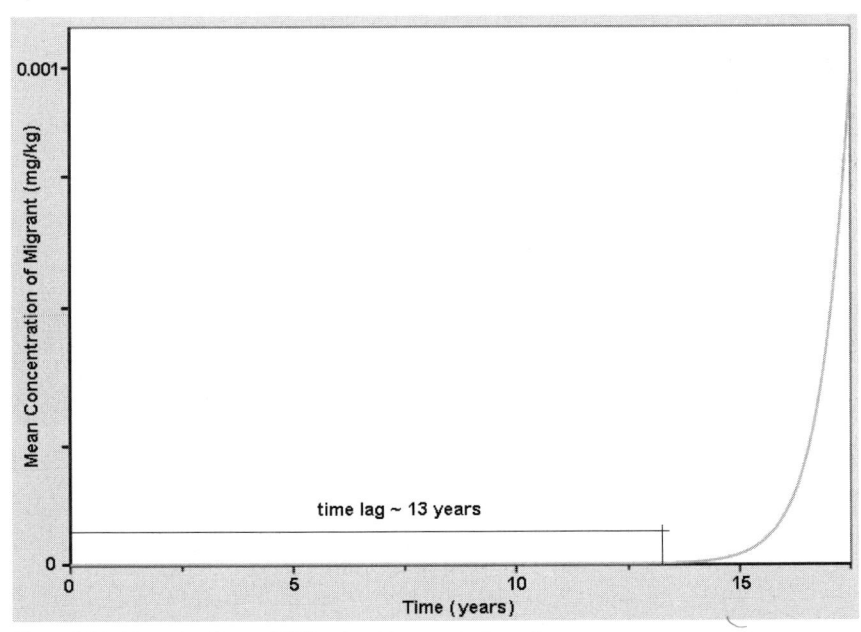

Figure 9.19 Migration from a PET bottle with a recycled core layer.

From Figure 9.19(a) one can see that during the high temperature coextrusion of the PET bottle a certain degree of migration of the residual substance from the recycled layer into the adjacent ones occurs. However, because the PET bottle reaches room temperature already after 15 s this contamination is limited to about 30 μm into the virgin PET layers. During the storage of the bottle for a week at room temperature before its filling with the soft drink there is virtually no more diffusion of the residual substance through the virgin PET layers. Thus, at the time point of the bottling of the soft drink into the PET bottle its inner surface is free of migrant. Because at room temperature the diffusion coefficients of the migrant in PET is very low the time needed by the migrant to reach the inner surface of the bottle and thus, start to contaminate the soft drink is about 13 years, see Figure 9.19(b). The length of this contamination "time lag" strongly depends on the temperature. If, for example, the PET bottle is stored for a week at 60 °C (for example, in a car parked in the sun) the contamination "time lag" not only decreases about one order of magnitude but the migrant concentration in the soft drink is also much higher. This is due to the fact that during the "hot week" the diffusion coefficient of the migrant in the PET layers increased more than two orders of magnitude!

Contamination of the food contact layer often occurs in practice when multilayered packaging materials are piled-up or wound-up in a reel during the time from their manufacturing to use in the food processing plant. This process can also be simulated with the software MIGRATEST© EXP.

The migration of a photo initiator, PHI, from the imprinted outer surface of polypropylene, PP, yoghurt cups was investigated experimentally. After being manufactured in a food packaging plant some of the yoghurt cups were left "free standing" while the other were "piled-up." Both types of cups were stored at room temperature for 30 days and then conditioned at 40 °C for 10 days. Afterward cutouts were taken from each type of PP cup and the migration of the PHI in 95% ethanol was determined for 1 day at 70 °C. For the "free standing" PP cups the mean concentration of PHI in the food simulant was about 6 μg/kg while in the "piled-up" a six times higher mean migrant concentration was found. To fit these experimental results with MIGRATEST© EXP the following assumptions were made, see Figure 9.20(a). During the manufacturing, storage, and conditioning of the PP cups the PHI diffuses from the printing ink layer into the matrix of the cup. This determines a contamination of the yoghurt cup. However, qualitatively and quantitatively this process differs for the two types of PP cups considered. For the "free-standing" cups the diffusion is one-directional from the printing ink layer toward the inner surface of the PP layer (what will become later the food contact layer), see the left graph in Figure 9.20(a). The inner surface of a piled-up PP cup is, in fact, in contact with the printing ink layer of the next PP cup. If the contact between the piled-up cups is good enough, a migration of PHI from a printing ink layer into the PP layer of the preceding PP cup is possible, too. This, invisible "set-off" process was simulated with MIGRATEST© EXP and the result obtained is shown in the right graph of Figure 9.20(a).

To fit the experimental results reported above it was necessary to assume that the rate of PHI diffusion into the PP cup due to the invisible "set-off" is somewhat lower than that due to diffusion from the imprinted surface of the PP cup. But this

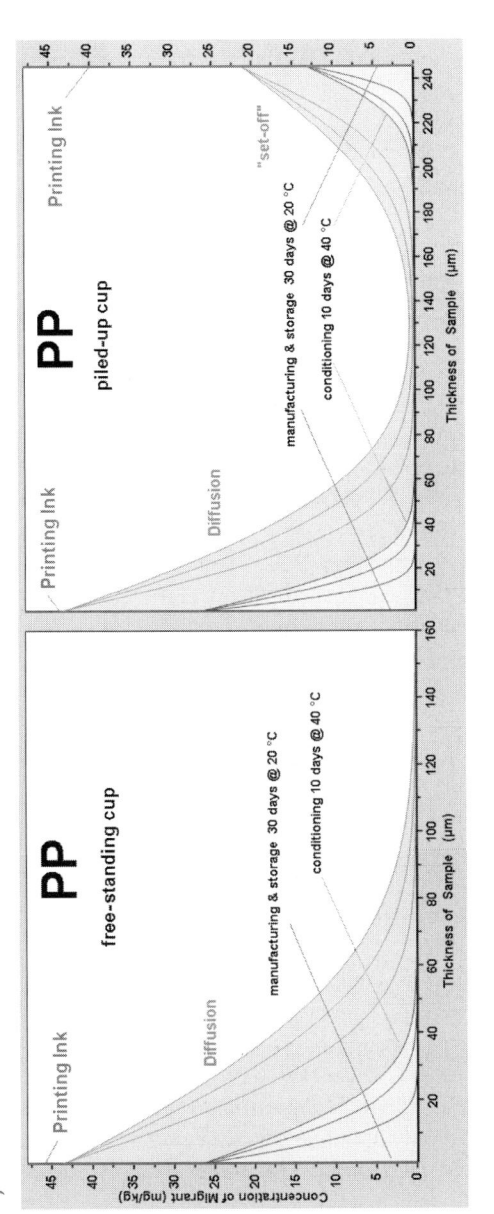

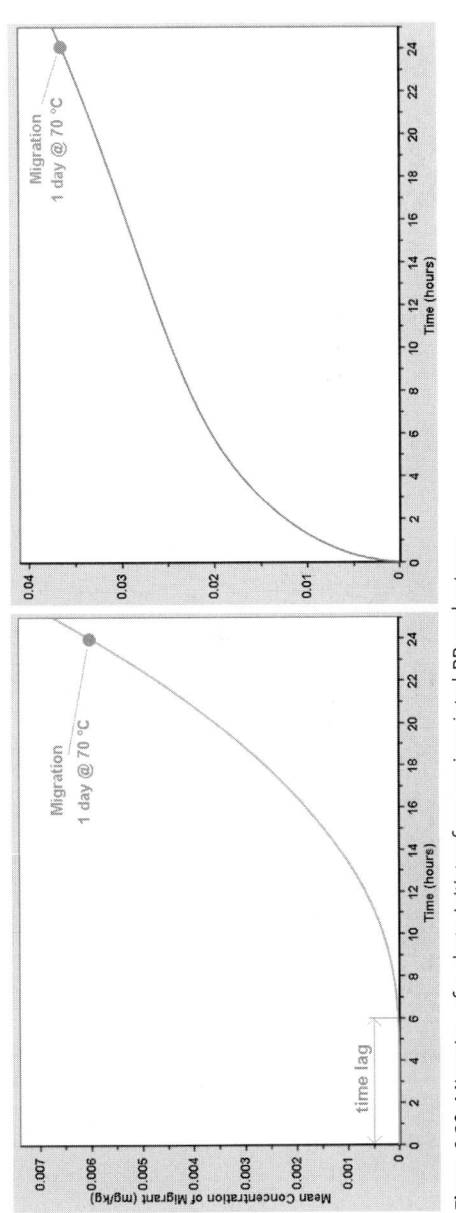

9.2 MIGRATEST© Lite – A User-Friendly Software for Migration Estimations

Figure 9.20 Migration of a photo initiator from an imprinted PP yoghurt cup.

assumption is reasonable if one takes into account that the ink imprinted on a PP cup is in much intimate contact with the polymer than the printing ink of the next cup from the pile-up. With this assumption, a good agreement is obtained between the experimental results and the estimated ones, Figure 9.20(b). The other important conclusion from the simulations performed with MIGRATEST© EXP is the fact that in case of an invisible "set-off" a certain degree of contamination of the packaged food with migrants from an outer layer is possible, even if the multilayer contains a perfect barrier (an aluminum layer, for example). Moreover, this contamination starts as soon as the food in brought into contact with the inner surface of the packaging (see right graph in Figure 9.20(b)) and not only after a certain "time-lag" as in the case of a "free-standing" PP cup, left graph in Figure 9.20(b).

The migration processes from a plastic material into certain types of contact media are often influenced by a boundary resistance at their interface. Such a situation can be simulated with MIGRATEST© EXP by assuming that the boundary resistance occurs, which occurs in a very thin layer at the interface, is caused by a very low partitioning (solubility) of the migrant in this layer. In Figure 9.21, such a scenario was simulated for the case examples discussed in Reid et al. 1980 and which were derived from the problem of plasticizer migration from PVC into aqueous solutions. In such cases, the amount of migration might be very sensitive to type and degree of agitation of the aqueous solution. Let us assume that a 200-µm thick plastified PVC film is in contact with a finite amount of an aqueous solvent, $d_F = 2000$ µm. Further, in the absence of agitation, let us assume that at the interface between the two media a thin boundary layer, about 5 µm, is formed. This layer hinders to a certain degree the migration of the plasticizer because of its very low solubility in the layer.

The migration of the plasticizer with $c_{P,0} = 100$ mg/kg from the PVC sample in the aqueous solution was numerically simulated both with and without taking into account a 5-µm thick boundary layer between them. The results obtained show that assuming a low solubility of the migrant in the boundary layer, $K_{1,2} \sim 10^4$, changes not only the level of migration in F but also the beginning of the migration process the shape of the time-dependent migration curve. In the absence of a boundary resistance the time-dependent migration curve is, up to about 60% from its equilibrium level, linear with the square root of the migration time, $t^{1/2}$. With boundary resistance the same curve is linear with t. From Figure 9.21(a) one can see that as the effective boundary layer thickness is increased, the quantity of the plasticizer migrated into the aqueous solution decreases. The other effect of this boundary layer is a very accentuated slow down of the contamination of the aqueous solution with the plasticizer.

In conclusion one can state that with the development of the user-friendly software MIGRATEST©Lite and MIGRATEST© EXP the interested user has the necessary tools to estimate/simulate a very broad range of migration problems occurring in practice. The examples presented above are only a very few types of problems which can be addressed with these software. Their capabilities are much wider and with a little practice an user from industry, R&D or an enforcement agency, will certainly be able to estimate/simulate most of the specific migration problems encountered in his/her work.

9.2 MIGRATEST©Lite – A User-Friendly Software for Migration Estimations | 295

a)

Migration without boundary resistance
$D_P = 10^{-11}$ cm²/s
$D_F = 10^{-6}$ cm²/s
$K_{P,F} = 1.0$

$\delta_F = 2.5$ μm
$\delta_F = 5$ μm
$\delta_F = 7.5$ μm
$\delta_F = 10$ μm
$\delta_F = 0$

Migration with boundary resistance
$D_P = 10^{-11}$ cm²/s
$D_F = 10^{-6}$ cm²/s
$K_{P,\delta F} = 10^4$

b)

Migration without boundary resistance

$\delta_F = 5$ μm

Migration with boundary resistance
$D_P = 10^{-11}$ cm²/s
$D_F = 10^{-6}$ cm²/s
$K_{P,\delta F} = 10^4$

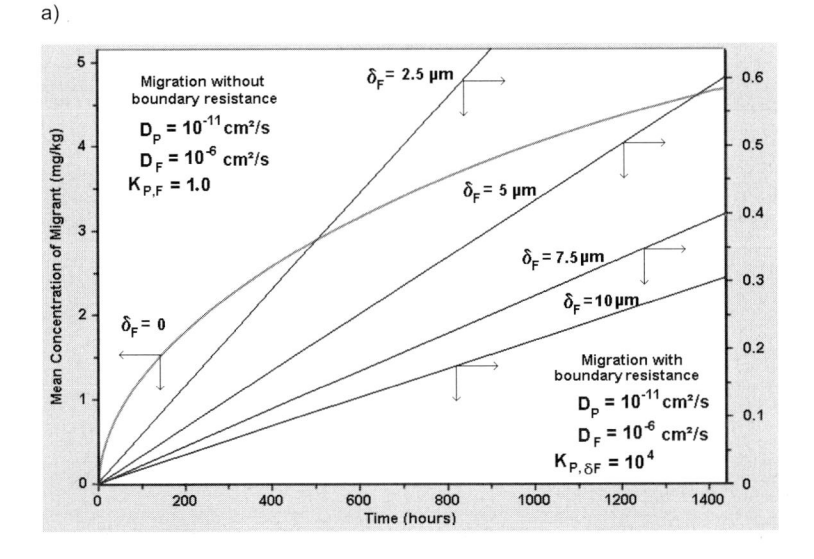

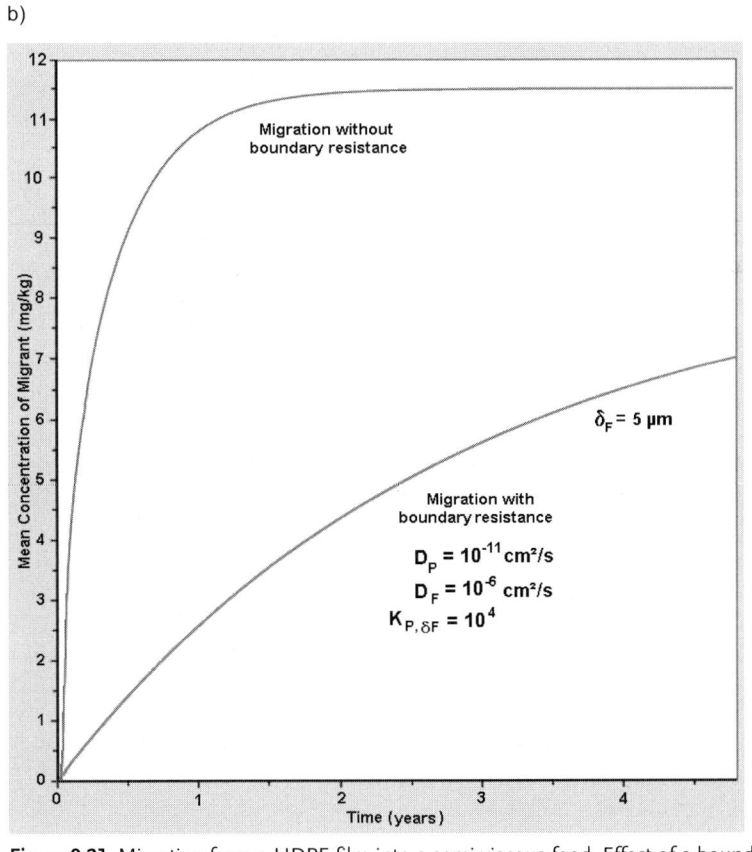

Figure 9.21 Migration from a HDPE film into a semi viscous food. Effect of a boundary layer.

References

Crank, J. (1975) *The Mathematics of Diffusion*, 2nd edn., Claredon Press, Oxford.

Hinrichs, K. and Piringer, O. (eds) (2002) Evaluation of migration models to be used under Directive 90/128/EEC; Final Report Contract SMT4-CT98-7513. EUR 20604 EN; Brussels, European Commission, Directorate General for Research.

Mercea P., Piringer O. (2004), *MIGRATEST© Lite – Instruction Manual*, 3rd revised edition, FABES Gmbh, Munich/Germany.

Reid, R.C., Sidman, K.R., Schwope, A.D., and Till, D.E. (1980) Loss of adjuvants from polymer films to foods or food simulants. Effect of the external phase. *Ind. Eng. Chem. Prod. Res. Dev.* **19**, 580–587.

Website References

Multitemp and Multiwise (INRA, Reims, France): http://www.inra-fra/internet/Produits/securite-emballage/pagefr.htm#4%20multi

EXDiF (office Fédéral de la Santé Publique, Bern, Switzerland): http://www.bag.admin.ch/verbrau/gebrauch/info//f/exdif.htm

SML (AKTS AG, Switzerland): http://www.akts.com

10
Permeation of Gases and Condensable Substances Through Monolayer and Multilayer Structures

Horst-Christian Langowski

10.1
Introduction: Barrier Function of Polymer-Based Packaging

Sensitive foods and encapsulated technical products are generally sensitive to their surroundings, in particular to oxygen and water vapor. Transfer of other substances both into and out of products can also adversely affect the quality of those products, for example loss of flavors from foods and contamination of encapsulated products by foreign outside substances. In practice this means that the materials used to package or encapsulate these products must possess barrier functions against water vapor, oxygen, and many other substances. The magnitude of these barriers must be selected depending on the sensitivity of the products. Figure 10.1 schematically shows the current barrier property requirements that packaging and encapsulating materials have to fulfill for a variety of products. Plastics are often used for packaging/encapsulating due to their flexibility, transparency, low weight, and ease of processing. However, the permeabilities of favorably priced commodity polymers (for food packaging) and also more expensive specialty polymers (for encapsulation of technical devices) to water vapor, oxygen, and other substances are far too high for most applications. Figure 10.2 shows the oxygen and water vapor permeabilities of commodity polymers that are currently used for food packaging applications. Figure 10.3 shows the same parameters for technical polymers used for encapsulation, in polymeric coatings and in adhesives. In both these figures the permeability values are normalized to a thickness of the polymer of 100 µm.

It was against this background that work began in the 1960s to improve the barrier properties of plastic materials. In addition to the straightforward method to combine the less expensive, more permeable commodity polymers with special barrier polymers or with metallic foils in the form of multilayers, other hybrid structures were developed. First successful attempts were made via vacuum coating of polymeric surfaces with thin layers of aluminum, subsequently also with transparent oxides, both types of layers in a thickness range between typically 10 and 100 nm.

Plastic Packaging. Second Edition. Edited by O.G. Piringer and A.L. Baner
Copyright © 2008 WILEY-VCH Verlag GmbH & Co. KGaA, Weinheim
ISBN: 978-3-527-31455-3

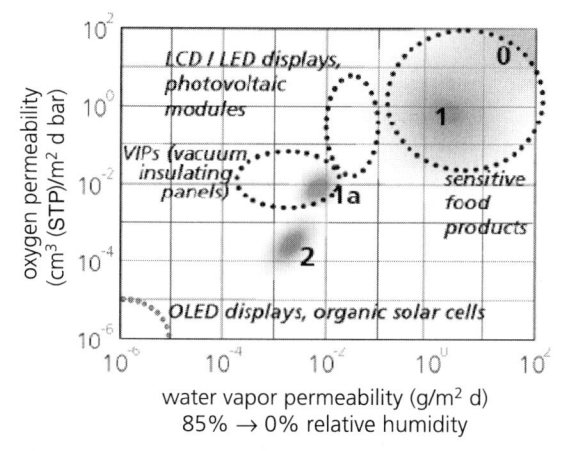

Figure 10.1 Barrier properties as required for different product sectors (dotted circles) and performance of the following encapsulation/packaging materials (shaded areas): (0) polymeric substrates alone, (1) polymeric substrates with one inorganic barrier layer, industrial standard, (1a) polymeric substrates with one inorganic barrier layer, special coating processes, (2) systems with two pairs of inorganic/polymeric layers. Reference temperature: 23°C. After (Langowski, 2003b).

A general remark is to be given already here concerning the usage of the words "film" and "layer": In different scientific and technological areas, the word "film" is used for two different types of structures:

1. For flat, large-area polymeric materials, in thickness ranges from less than one up to several hundreds or even thousands of micrometers.
2. For thin plane structures deposited on top of different types of substrates, in thickness ranges from less than one nanometer up to thousands of nanometers (see, e.g., the name of the related scientific journal *Thin Solid Films*).

To facilitate the discrimination between these two different types of structures, the word "film" will be used in this chapter only for the former one, e.g., for extruded or coextruded large area plastic films. Films of this kind are usually produced as independent structures without the need for a supporting substrate.

For the latter kind of structures, the word "layer" will be used in this chapter instead of "film." "Layer" will also be used for the single elements of multilayer polymer films and containers, where different polymer layers are connected to form one film or one container wall. According to this definition, the major difference between a film and a layer is that the latter does not exist alone, but always in combination with a substrate or together with other layers.

In the simplest case the barrier systems in current use consist of a plastic substrate with a thin barrier layer applied on top of this. However, as this layer can be very easily damaged, such systems are only used for plastic containers (mostly PET bottles). These bottles usually have an interior barrier layer coating which is hence protected

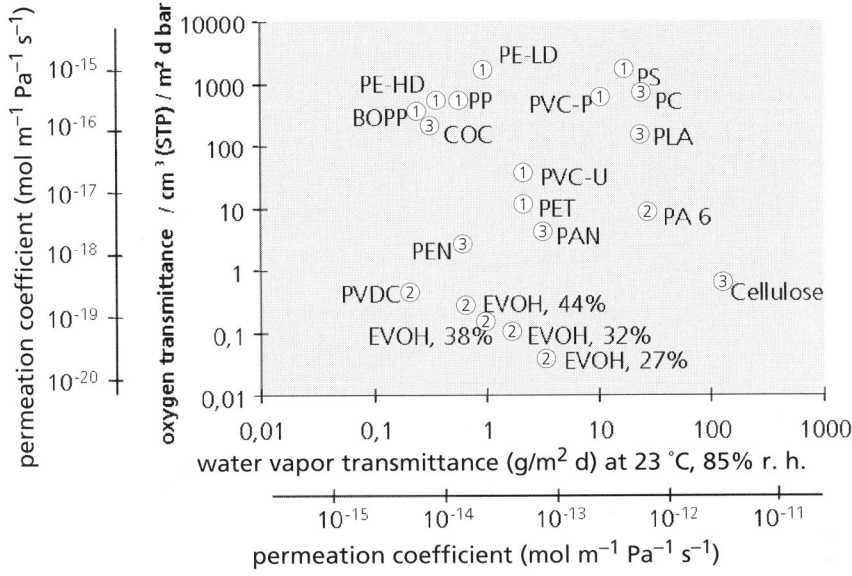

Figure 10.2 Transmittance (i.e., permeability normalized to 100 μm material thickness) for oxygen and water vapor, for typical packaging polymers, at 23°C. Additional scales are shown for permeation coefficients in SI units. ① : Commodity thermoplastics, ② : frequently used barrier polymers, ③ : specialty polymers.

PE-LD	Low-density polyethylene
PE-HD	High-density polyethylene
PP	Polypropylene
BOPP	Biaxially oriented PP
COC	Cycloolefin copolymer
PEN	Polyethylene naphtalate
PET	Polyethylene terephthalate
PAN	Polyacrylonitrile
PA 6	Polyamide 6
PVC-U	Polyvinyl chloride, nonplasticized
PVC-P	Polyvinyl chloride, plasticized
PLA	Polylactic acid
PS	Polystyrene
PC	Polycarbonate
PVDC	Polyvinylidene chloride
EVOH	Ethylene-vinylalcohol copolymer (percentage: fraction of ethylene)
Cellulose	regenerated cellulose hydrate (former name: Cellophane)

from the outside world by the container wall. For film systems, a further polymer film always has to be applied to the barrier layer. In some cases this takes the form of a lacquer layer but usually it is an adhesive which then allows a further sealable film to be laminated on top of this.

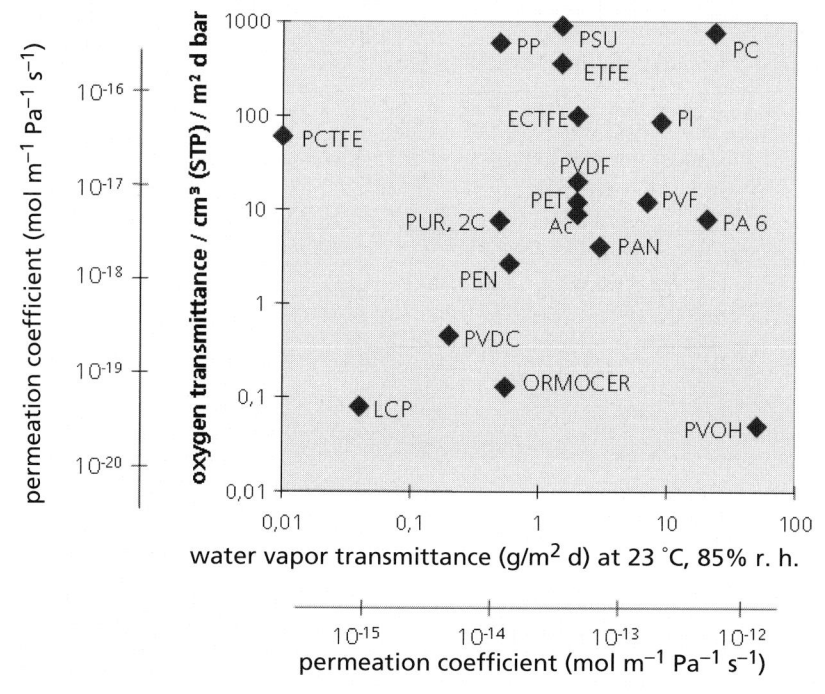

Figure 10.3 Permeability for oxygen and water vapor, normalized to 100-μm thickness (transmittance) for technical polymers, at 23 °C. In addition: scale for permeation coefficients, in SI units.

PSU	Polysulfone
PC	Polycarbonate
PP	Polypropylene
ETFE	Ethylene-tetrafluoroethylene-copolymer
ECTFE	Ethylene-chlorotrifluoroethylene copolymer
PI	Polyimide
PVDF	Polyvinylidene fluoride
PET	Polyethylene terephthalate
PVF	Polyvinyl fluoride
PA 6	Polyamide 6
PAN	Polyacrylonitrile
PEN	Polyethylene naphtalate
PVDC	Polyvinylidene chloride
PCTFE	Polychlorotrifluoroethylene
PVOH	Polyvinyl alcohol
LCP	Liquid crystal polymer
Ac	Acrylate coating
PUR, 2 C	2-Component polyurethane adhesive
ORMOCER	Inorganic–organic hybrid polymer

10.1 Introduction: Barrier Function of Polymer-Based Packaging

To improve the barrier properties of polymers with the help of inorganic materials, an alternative method exists to coating: Inorganic particles can be included in the polymeric matrix. Also in the 1960s, theoretical work showed the options of such a strategy, especially for platelet-shaped particles with a high ratio of length vs. thickness, the so-called aspect ratio, and a high degree of mutual alignment. Nearly 30 years later, polymers filled with nanoparticles were developed which today, in some cases, either serve as an alternative to polymeric-inorganic layer structures or they are simply used for a lower barrier improvement of the bulk polymer, for less stringent barrier requirements.

Today, the following four groups of systems with barrier functions made on the basis of inorganic layers or particles can be found on the packaging market:

1. Plastic film substrates coated with thin inorganic barrier layers by vacuum deposition methods. In packaging applications, these structures are always used with additional polymeric layers or polymeric films on top of the inorganic layer. It can be estimated that by far more than 15 billions of square meters made by this method are used worldwide for packaging laminate films. The majority of these films is made via vacuum coating of films from PET and BOPP with aluminum, a smaller fraction (about 1 billion of square meters, calculated from Smith, 2006) uses transparent oxides (silicon oxide, aluminum oxide) as inorganic barrier materials.

2. Plastic containers with different types of barrier layers. The basic material for them is PET. Overall, an equivalent of about 0.5 billion of square meters is produced on the basis of different techniques. The majority of high-barrier plastic containers are made on the basis of polymeric multilayers, often with additional chemical oxygen absorption functions (so-called scavenger functions) integrated into the polymeric material. To improve the barrier function further, nanoscale platelet-shaped silicate mineral particles are incorporated into the polymeric material. A smaller fraction of plastic high-barrier containers are coated on their interior with thin transparent barrier layers. Chemically, a fraction of these layers is similar to transparent layers from silicon oxide, which are deposited on polymeric film substrates. Another fraction consists of amorphous carbon.

3. A third group of polymeric hybrid barrier materials is found in the market in much smaller quantities than the two groups above. This group consists of nanoparticles embedded in a polymeric matrix as the specific barrier layer. It contains two sub groups:
   - One sub group is mainly represented by polyamides (PA 6 or the aromatic polyamide MXD 6) and polyethylene terephthalate (PET), all filled with exfoliated layer silicate structures in amounts of some per cent by weight.
   - In the other sub group, the barrier is created by a lacquer layer, in which the polymeric binder is again filled with exfoliated layer silicate particles.

4. Finally, a specific class of flexible materials is being developed for technical applications in which the highest barrier properties are required to protect optoelectronic devices, photovoltaic modules, thermal insulators, or thin film batteries (Gross, 2006) predominantly against water vapor and oxygen. They consist of multilayer structures

of alternating thin layers from inorganic and polymeric materials, deposited on a flexible polymeric substrate.

10.2
Permeation Through Polymeric Materials

Although the principles of permeation of substances through polymers have been described in detail in the literature (Comyn, 1985) and relevant data is available for a variety of polymers and substances (Pauly, 1999), the situation is very different for heterogeneous structures incorporating inorganic particles and thin inorganic layers. The key difference between polymers and inorganic materials is that the mechanism of solution-diffusion, which accurately describes most polymer/substance combinations, seldom applies to different kinds of inorganic materials. Already the pioneers in the area of substance transport through solids, polymers, and glasses (Barrer, 1941; Perkins, 1971) realized that in the bulk volume of intact solids and glasses, solution and diffusion of molecules only occurs to a very limited extent and that substance transport through these materials is largely dictated by defects.

With regard to the mechanisms of substance transport through systems comprising polymeric substrates and inorganic barrier layers, in some cases with additional polymeric layers on top, the following distinctions can be made:

In *polymeric materials*, permeating substances are able to be dissolved in a quasi-homogenous way, meaning that the permeation process can also be considered to be *homogenous*. Even when such polymer materials consist of several phases (for example partially crystalline polymers), the size of the structures is relatively small compared to the size of the bulk sample and the phases are statistically distributed, meaning that it is not necessary to consider substance transport processes to be localized. This is valid for single layer and multilayer *polymer substrates*, which are employed in thicknesses ranging from just a few microns up to several hundred microns, and also *polymer layers* formed from lacquers or adhesives, usually having a thickness of a few microns or less. The characteristic parameters for substance transport in many polymers are available in the literature (Comyn, 1985; Pauly, 1999).

For *inorganic layers*, which are applied to polymer substrates in thicknesses ranging from about 10 nm to several hundred nanometers, substance transport is mainly localized *at defects*. The substance flow through the intact layer material is negligibly small compared to substance transport through the defects. The diameters of the defects range from less than a few microns down to some tenths of a nanometer. Decisive are the size distribution and number of defects. These factors depend on the substrate material and the quality of the surface of the substrate, the coating process, and also the layer material itself. This will become evident below. The fact that substance transport through inorganic layers is localized at defects necessitates that the permeation process be considered differently to permeation in homogenous polymers.

Thin layers containing many pores in the nanometer and subnanometer range represent a limiting case in which a clear separation between homogenous and localized transport mechanisms is no more possible. In this case the sizes of the defect structures are small compared to the size of the layer and the pores are numerous and statistically distributed. This can once again be considered a quasi-homogenous material, which does however also contain larger defects. In practice, however, these pores are very difficult to characterize and their existence mainly has to be derived indirectly from experimental findings.

Finally, for particles which are included in a polymeric matrix, other aspects have to be regarded. In spite of the fact that they are usually statistically distributed inside the polymeric matrix, their shape and their orientation plays a decisive role on the final permeability of the resulting hybrid material.

10.2.1
Substance Transport Through Monolayer Polymer Films

The theoretical basis for the permeation of substances through polymers has already been laid down in Chapter 7. As a vast number of polymers have been studied thoroughly over many years, a lot of information can be found in the relevant literature (see for example Comyn, 1989 and Pauly 1999).

In addition to the considerations for diffusion processes laid down in Chapter 7, let us remember the elementary steps of the permeation of a substance through a polymeric sample (see especially Piringer 2000) :

- Adsorption of the substance at the surface of the polymeric substrate, in higher net amounts at the side where the concentration of the substance in the surrounding environment is higher.

- Solution of the substance, first in the near-surface region, later – as a follow-up of the diffusion process – in the whole bulk of the polymeric substrate

- Diffusion inside the polymer substrate bulk, under the effect of concentration gradients, as treated in Chapter 7.

- Desorption from the polymeric substrate, in higher net amounts from the side where the concentration of the substance in the surrounding environment is lower.

In the discussion which follows, only the case where the medium surrounding the polymer substrate is gaseous is considered. Then the permeating substance is characterized by its partial pressure, in the case of water vapor also via the relative humidity which at a given temperature is proportional to the partial pressure.

The *sorption coefficient/solubility coefficient S* determines the concentration c of a substance in the polymer material under equilibrium conditions at partial pressure p. Thus, it links the equations from Chapter 7 with the practical conditions regarded in this chapter. If the solubility is constant, Eq. (10.1) holds, usually named Henry's law.

$$c = Sp \qquad (10.1)$$

At the start of a permeation process, concentration gradients are time-dependent. The resulting flow of substance is described by Fick's first and second laws of diffusion, see Eqs. (7.3c) and (7.4b), respectively.

A flow equilibrium (stationary state) and a concentration profile inside the sample that is independent of the time are only established after the so-called time lag, θ. (Barrer, 1941).

For thin polymer films exposed to oxygen and water vapor, this equilibrium is attained over a period ranging from a few hours to several days. For more complex layer structures, such as those described below, and for thicker films and permeating molecules of larger size, the establishment of the stationary state may take months or even years. A description of these induction processes as a function of time is available in a one-dimensional form for simpler laminate structures of polymer materials (see for example Barrie, 1963; Crank, 1992). With regard to the combinations of polymeric and inorganic layers under consideration here, only first approaches can be found in the literature (Graff, 2004).

The discussion which follows considers a flat sample in stationary state in which a stationary flow regime has become established and the flow of substance is constant as a function of time. It is also assumed that the diffusion coefficient D and solubility coefficient S are both constant for a given combination of substance and polymer.

Under these conditions, the concentration of the substance in the surface zone of the polymer is $c_1 = Sp_1$ on one side of the sample and $c_2 = S p_2$ on the other side, p_1 and p_2 being the related partial pressures of the substance in the gaseous space in contact with the polymer.

Substitution into Eq. (7.16) gives

$$J = -D(c_2 - c_1)/d = DS(p_1 - p_2)/d \equiv P(p_1 - p_2)/d \tag{10.2}$$

The so-defined *permeation coefficient* P is the product of diffusion coefficient and solubility coefficient:

$$P = DS \tag{10.3}$$

The actual flow rate F through a sample of thickness d and area A is given by

$$F = JA = PA(p_1 - p_2)/d = QA(p_1 - p_2) \tag{10.4}$$

where Q is now the permeability, which is a function of the permeation coefficient and material thickness given by

$$Q = P/d \tag{10.5}$$

The temperature dependence of diffusion coefficient, sorption coefficient, and permeation coefficient can be described by Arrhenius-type equations

$$D = D_0 \exp(-E_D/RT) \quad S = S_0 \exp(-E_S/RT) \quad P = P_0 \exp(-E_P/RT) \tag{10.6}$$

where T is the absolute temperature, E_D, E_S, and E_P are activation energies (specific for a material and respective coefficients) and R is the gas constant.

In practice, the permeability of a 100-μm-thick sample of material, Q_{100}, often named *transmittance*, which is directly proportional to P, is usually taken instead of the permeation coefficient.

The time behavior of the transported substance in the initial phase – before the steady state of flux has been established – can be described for a sheet of a homogeneous material materials having thickness d and the diffusion coefficient D in one-dimensional form via Eq. (7.60) giving the value for the time lag θ via: $\theta = d^2/6D$.

10.2.2
Substance Transport Through Multilayer Polymer Films (Laminates)

For the permeability of laminates, namely layer systems comprising different polymer materials, the so-called ideal laminate theory is often applied. According to this, different polymers respectively having permeation coefficients P_i and thicknesses d_i give rise to the familiar relationship for the overall permeability of the layer system Q_{ovl}:

$$Q_{ovl}^{-1} = \sum_i (d_i/P_i) = Q_1^{-1} + Q_2^{-1} + Q_3^{-1} + \ldots \quad (10.7)$$

which is analogous to Eq. (7.64).

It should be noted, however, that this relationship no longer applies as soon as a laminate contains even a single inorganic layer. As will be shown later, approximations can be derived for certain layer systems and these permit the use of Eq. (10.7) again under certain limited conditions.

The time behavior of the transported substance in the initial phase, however – before the steady state of flux has been established – can no more be calculated by Eq. (7.60), because layers of different thickness and different values for the coefficients of diffusion and solution, D and S, are involved. If the material consists of several different polymer layers, the exact time behavior becomes extremely complicated. However, a method to calculate the time lag for laminates consisting of up to three layers has been derived (Barrie, 1963). In the case of a three-layer system consisting of

- layer (1) of the thickness d_1, with diffusion coefficient D_1 and solubility coefficient S_1
- layer (2), with the related data d_2, D_2, and S_2, respectively,
- and layer (3), with the related data d_3, D_3, and S_3,

the following equation for the resulting time lag θ_{123} can be applied:

$$\theta_{123} = \frac{\frac{d_1^2}{D_1}\left(\frac{d_1}{6D_1S_1} + \frac{d_2}{2D_2S_2} + \frac{d_3}{2D_3S_3}\right) + \frac{d_2^2}{D_2}\left(\frac{d_1}{2D_1S_1} + \frac{d_2}{6D_2S_2} + \frac{d_3}{2D_3S_3}\right) + \frac{d_3^2}{D_3}\left(\frac{d_1}{2D_1S_1} + \frac{d_2}{2D_2S_2} + \frac{d_3}{6D_3S_3}\right) + \frac{S_2 d_1 d_2 d_3}{D_1 S_1 D_3 S_3}}{\frac{d_1}{D_1S_1} + \frac{d_2}{D_2S_2} + \frac{d_3}{D_3S_3}}$$

(10.8)

It should be noted that here – in contrary to the permeability of such a system under steady-state conditions as given by Eq. (10.7) – the time lag may be affected by a change in the sequence of the layers, e.g., by a change of layers (1) and (2).

For the simpler case of a two-layer system, consisting of

- layer (1) of the thickness d_1, with diffusion coefficient D_1, and solubility coefficient S_1
- and layer (2), with the related data d_2, D_2, and S_2, respectively,

the following expression is obtained, which is now independent of the sequence of the layers:

$$\theta_{12} = \frac{\frac{d_1^2}{D_1}\left(\frac{d_1}{6D_1 S_1} + \frac{d_2}{2D_2 S_2}\right) + \frac{d_2^2}{D_2}\left(\frac{d_1}{2D_1 S_1} + \frac{d_2}{6D_2 S_2}\right)}{\frac{d_1}{D_1 S_1} + \frac{d_2}{D_2 S_2}} \tag{10.8a}$$

This expression allows us to compare different cases of two-layer laminates:

To improve the barrier properties of polymer films, they are often coated with other polymers of lower permeation coefficient. Let us assume that a film of thickness d_1, the permeation coefficient $P_1 = D_1 S_1$ and the permeability $Q_1 = P_1/d_1$ has been coated by a layer of another polymer that is thinner by a factor of 10, i.e., of a thickness $d_2 = 0.1\,d_1$. Let us assume further that the thinner layer has a permeation coefficient P_2 that is 10 times smaller than that of the first layer, i.e., $P_2 = 0.1\,P_1$. In this case, both the layers would have exactly the same permeability and – according to Eq. (10.7) – the total permeability of the resulting laminate, Q_{ovl}, would be just half of that of the initial film before coating:

$$Q_{ovl}^{-1} = Q_1^{-1} + Q_2^{-1} = 2Q_1^{-1} \quad \text{or} \quad Q_{ovl} = 0.5 Q_1$$

To obtain a second polymer of a permeation coefficient lower by a factor of 10 compared to the first one, however, different selections are possible:

> One could, for instance, select a polymer that has – compared to the initial film – a lower solubility, but the same diffusion coefficient for the permeating substance, i.e., $D_2 = D_1$ and $S_2 = 0.1\,S_1$.

> Another option would be to use instead a material with the same solubility S for the permeating substance, but a diffusion coefficient that is lower by a factor of 10, i.e., $D_2 = 0.1\,D_1$ and $S_2 = S_1$.

> There are also barrier polymers characterized by a very low diffusion coefficient, but higher solubility. Such a polymer, still with the same permeation coefficient, can be simulated via $D_2 = 0.01 D_1$ and $S_2 = 10 S_1$.

All these cases would lead to the same final permeability of the laminate, but the related time lag would show substantial differences, as can be calculated easily with the help of Eq. (10.8a). In relation to the time lag for the permeation of the substance through the initial monolayer film, θ_1, with $\theta_1 = d_1^2/6D_1$, the time lag for the coated two-layer films, θ_{12}, would be

- for case 1, with $D_2 = D_1$ and $S_2 = 0.1 S_1$: $\theta_{12} = 2.02\,\theta_1$
- for case 2, with $D_2 = 0.1 D_1$ and $S_2 = S_1$: $\theta_{12} = 2.2\,\theta_1$
- and for case 3, with $D_2 = 0.01 D_1$ and $S_2 = 10 S_1$: $\theta_{12} = 4\,\theta_1$

The intuitive assumption that it is possible to obtain the resulting time lag in a laminate by simply adding the time lag values for the individual layers is, however, wrong. It does only work in this example because the permeability values of the individual layers are equal. This fact can be verified by calculating case 4 with $D_2 = 0.1 D_1$, $S_2 = 0.1 S_1$, $P_2 = 0.01 P_1$ and $Q_2 = 0.1 Q_1$, leading to $Q_{ovl} = 0.09\, Q_1$, and $\theta_{12} = 2.94\, \theta_1$.

These examples show us that, in order to obtain a long lag time of a laminate, the best technical solutions are achieved by coating with polymers of low diffusion coefficient for the permeating substance.

10.2.3
Units for Different Parameters

A somewhat problematic issue regarding the technical literature concerns the use of different units for the various parameters:

Quantities of substances should ideally be given in SI units by their mass or their number of molecules (e.g., gram or mol). In daily practice, however, quantities of substances that are gaseous under practical conditions, for example oxygen, are usually indicated in cm^3 (STP), namely gas volume units under standard conditions (standard temperature and pressure (STP) = 1013 hPa, 0 °C).

For water and other condensable substances, however, their mass is usually given. If the amount of substances that are liquid under normal conditions is again given in cm^3 (STP), this means that their mass has been transformed into gas volume under standard conditions by assuming a behavior like an ideal gas.

For partial pressures of gases, figures are usually given in Pascal or bars, whereas for water vapor, however, partial pressures are mainly given via their equivalents in terms of relative humidity.

As an example, a transmittance for oxygen as usually given in the technical literature as $1\,cm^3$ (STP) $\times\, 100\,\mu m/(m^2\, d\, bar)$ is equivalent to a permeation coefficient given in SI units of $5.16 \times 10^{-19}\, mol\, m^{-1}\, Pa^{-1}\, s^{-1}$.

For water vapor, a similar comparison can be made: A transmittance of $1\,g \times 100\,\mu m/m^2\, d$ measured at a difference of relative humidity from one side of the sample to the other of 85% to 0%, at 23 °C is equivalent to a permeation coefficient of $2.69 \times 10^{-14}\, mol\, m^{-1}\, Pa^{-1}\, s^{-1}$.

In Figs. 10.2 and 10.3, both technical units and SI units have been indicated. Later on, however, only units as used in the technical literature will be applied.

10.3
Substance Transport Through Single and Multilayer Polymer Substrates Combined with One Inorganic Barrier Layer

10.3.1
Numerical Modeling

The permeation of substances through a laminate structure comprising a polymer substrate combined with a thin inorganic layer which contains defects, was first described nearly 50 years ago (Prins,1958). Since then a great deal of experimental and theoretical work has been undertaken.

Most of the theoretical studies have been based on a model which considers a thin inorganic layer – which is assumed to be coherent at a thickness of at least 10 nm – that has been applied to a polymer substrate. The model assumes that substance transport through this layer occurs in the regions where the layer is interrupted by defects and that the rest of the layer can be deemed to be impermeable. The permeation properties of the polymer substrate are described by the permeation coefficient (P) which, as described in the previous section, depends on the nature of the polymer substrate, the temperature, and the permeating substance. Depending on the theoretical approach adopted, the defects are considered to be either square or circular holes.

After passage of the time lag, there is a time-independent three-dimensional concentration profile of the permeating substance in the polymeric substrate. Figure 10.4 (Hanika, 2004) shows this using the example of a $(2\,\mu m)^2$ defect in a metal layer on a plastic substrate made of polyethylene terephthalate (PET). It can be seen that the concentration gradient is greatest in the direct vicinity of the defect and is very small at considerable distances away from the defect. From this concentration profile and the known parameters of the polymer substrate, it is possible to calculate the permeability of an individual defect in the metal layer when in combination with the polymer substrate. As already demonstrated in Chapter 7, such geometry usually can no more be calculated analytically and requires numerical treatment.

For circular defects of radius r_0 in an inorganic layer, evenly distributed, with n defects per unit area, on a polymeric substrate of thickness d, permeation coefficient P and permeability Q_0, Prins and Hermans (Prins, 1958) found approximate solutions for the three-dimensional diffusion equations (7.2) and (7.3). They applied different approximations to obtain the resulting permeability of the coated substrate, Q_b, for different ratios of substrate thickness d and defect radius r_0 (see the original paper for details).

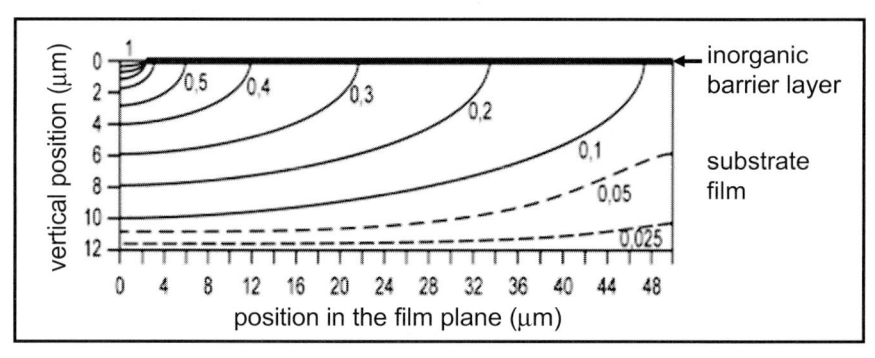

Figure 10.4 Concentration profile of oxygen dissolved in a 12-μm-thick PET substrate film below a defect (upper left) of a size of $(2\,\mu m)^2$ in an inorganic barrier layer. Lines of equal relative concentrations are shown (figures). The highest concentration ($c = 1$) in the polymer is to be found at the defect in the barrier layer, the lowest concentration ($c = 0$) at the lower edge of the substrate film (From Hanika, 2004).

In the first limiting case, the size of the defects is large compared with the thickness of the substrate. This means either a standard film combined with an inorganic coating that contains large defects or a standard inorganic coating on an extremely thin substrate film. In such a case the permeability Q_b of the coated substrate becomes

$$Q_b \approx P d^{-1} n \pi r_0^2 = Q_0 n \pi r_0^2 \tag{10.9a}$$

This equation simply means that exactly that fraction of the substrate that is not covered by the inorganic barrier layer, i.e., $n\pi r_0^2$, contributes to the permeability.

In another limiting case, the size of the defects is very small compared with the thickness of the substrate, like a standard inorganic coating on a thick substrate film. Here, Q_b becomes

$$Q_b \approx 3.7 P n r_0 \tag{10.9b}$$

Note that in this case, the permeability of the coated substrate is apparently independent of its thickness.

For the intermediate case, provided that $d/r_0 > 0.3$, the resulting permeability becomes

$$Q_b \approx P d^{-1} n \pi r_0^2 (1 + 1.18 \, d r_0^{-1}) \tag{10.9c}$$

In (Hanika 2004), a model with square defects was applied to investigate systematically the influence of different size and shape of the defects and of the substrate thickness, by means of numerical solutions of the three-dimensional diffusion equation. Figure 10.5 exemplifies how the permeability of the substrate-layer system changes when the substrate thickness is varied. It can be seen – in concert with Eq. (10.9b) – that above a *critical substrate thickness* d_k (see Figure 10.5) the permeability of the substrate-layer system remains virtually constant.

Other findings of the numerical simulation work of (Hanika, 2004) are as follows:

- Neighboring defects can be considered not to influence each other when the distance between them is greater than the *critical defect–defect distance* L_k.
- Rectangular defects have the same permeability as square defects of the same area provided their length/width ratio does not exceed 3. Cracks having a length/width ratio of greater than 10 have much higher permeabilities than square or circular defects of the same area.

These results give rise to the following conclusions and simplifications:

- If the thickness of the substrate is greater than the *critical thickness*, its effect can be neglected. This leads to the simplifications described in the next section (barrier improvement factor).
- The permeabilities of individual defects of different sizes can be calculated individually and then summed, if the distance between them is greater than the *critical defect–defect distance*.

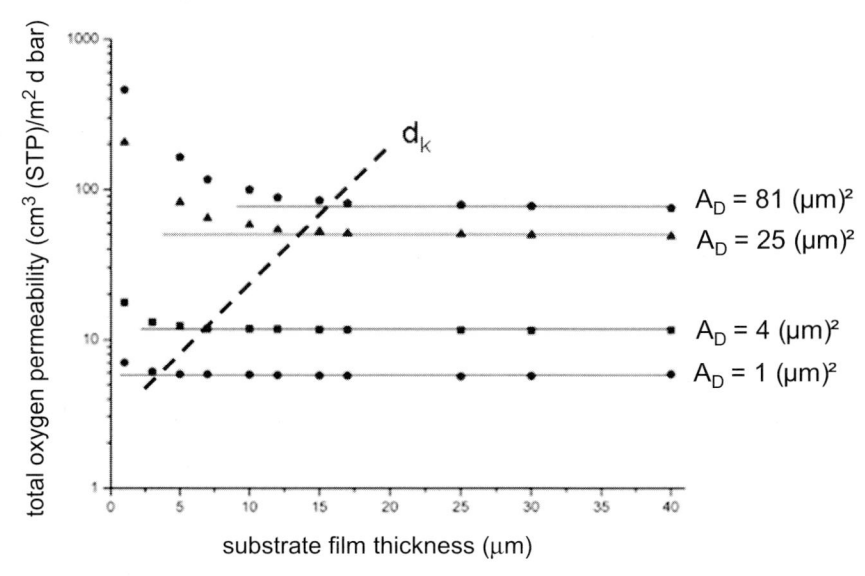

Figure 10.5 Influence of substrate film thickness on the oxygen permeability of a PET film coated with an inorganic layer. Curves have been calculated for 10^4 defects per cm^2 and for four different defect sizes A_D, ranging from $(1\,\mu m)^2$ up to $(9\,\mu m)^2$ (taken from Hanika 2004). Also shown is the critical substrate thickness d_k, above which the total permeability stays independent on substrate thickness.

- Defects of different shapes can be considered to be squares or circles of the same area, provided their length-to-width ratio does not exceed 3.

In this model the critical substrate thickness is solely dependent on the defect size, whilst the critical defect–defect distance is also dependent on the thickness of the substrate. Here the critical substrate thickness d_k is 2 to 2.5 times the defect diameter and the critical defect–defect distance L_k is either 15 times the defect diameter or 3 times the substrate thickness depending on which number is the larger (Langowski 2003a; Hanika, 2002a, 2004; Langowski, 2002b, 2003, 2003b).

To combine the results of different numerical simulations, Hanika 2004 derived a heuristic analytical formula from his numerical results, allowing to approximate the permeability Q of a polymeric substrate of thickness d with permeation coefficient P, coated with an inorganic layer with evenly distributed defects of equal area A_D which are separated by an average distance L. (Note that the number of defects per unit area, n, is then correlated to L via $L = n^{-1/2}$)

$$Q = Pd^{-1}A_D L^{-2}\left[1 - \exp(-0.507 d^{-1}\sqrt{A_D}) + 0.01 L^{-2} A_D\right]^{-1} \tag{10.10}$$

In reality, defects in barrier layers do not show a uniform size, but a size distribution. In this case, the permeability can be calculated for every size class of defect area $A_{D,i}$ and average defect distance L_i valid for this size class according to Eq. (10.10a) and the resulting permeability can be summed over the single classes:

$$\frac{Q}{Q_0} = \sum_i \left(\frac{\frac{A_{D,i}}{L_i^2}}{1-\exp\left(-0.507\frac{\sqrt{A_{D,i}}}{d}\right)+0.01\frac{A_{D,i}}{L_i^2}} \right) \quad (10.10a)$$

(Here, Q_0 is the permeability of the noncoated substrate film, with $Q_0 = Pd^{-1}$)

10.3.2
Simplification: Barrier Improvement Factor

In practice, large efforts are needed to obtain information about the number of defects and their size distribution and this is particularly so when the inorganic layers are transparent. The permeabilities of the noncoated substrate, Q_0, and the coated substrate, Q_b, however, are standard parameters for describing the quality of the manufacturing process. The measured values are often used to derive a barrier improvement factor (BIF) which is defined by

$$\text{BIF} = Q_0/Q_b \quad (10.11)$$

The barrier improvement factor depends on the value of Q_0 and is therefore still dependent on the substrate thickness. This means that it is not easy to compare coatings on substrates of different thickness. However, utilizing the aforementioned fact that the permeability of a substrate/inorganic layer system, Q_b, is independent of the substrate thickness over a large range, this opens up another opportunity for simplification. If the standard material-specific permeability for a material thickness of 100 µm (transmittance, Q_{100}) is taken (which is proportional to the permeation coefficient, see above), the following relationship approximately describes a substrate having a thickness above the critical thickness in combination with an inorganic layer having a certain numbers of defects:

$$\text{BIF}_{100} \approx Q_{100}/Q_b \quad \text{or} \quad Q_b \approx Q_{100}/\text{BIF}_{100} \quad (10.12)$$

In reality, the newly defined parameter BIF_{100} corresponds to the barrier improvement factor that would be measured after coating of a polymer substrate of 100-µm-thickness compared to the noncoated substrate. Instead of the BIF according to Eq. (10.11), it allows to compare the permeability of different polymer substrates of varying thickness onto which inorganic layers having the same defect structure have been applied (see Langowski 2002b).

One important factor is to be mentioned here: The permeability of a substrate coated with an inorganic barrier layer cannot be completely independent of the thickness of the substrate. At very large values of the substrate thickness, the permeability of the substrate alone will reach the value of the permeability of the coated substrate as obtained from Eq. (10.9b) or Eq. (10.12). Starting at this point, an increase in substrate thickness will again lead to a reduction of the permeability of the coated substrate. This fact has been included by Hanika in the calculations leading to Eqs. (10.10) and (10.10a).

Figure 10.6 shows the resulting relations between substrate thickness and permeability for coated and noncoated substrate films. The values have been calculated from Eq. (10.10) for the average defect size (equivalent to defect areas below 0.5 $(\mu m)^2$) and density on industrially coated films for the permeation of oxygen through the following film configurations:

- a BOPP substrate film of varying thickness, coated with an inorganic layer of constant size and density of defects, in comparison to the substrate film alone;
- a PET substrate film, coated and noncoated, for the same situation.

It can be clearly seen that the permeability remains again virtually constant over a large range of the substrate thickness (see also Figure 10.5) up to some hundreds of micrometers and that the BIF at a substrate thickness of 100 μm, BIF_{100}, assumes values that are larger than 10. In such a case, the application of the concept shown above toward different film configurations as demonstrated in the following is justified.

However, as soon as barrier coatings of poorer quality are concerned, the barrier improvement will be much less and the point where the substrate thickness becomes important for the overall permeability will be at a lower substrate thickness. In such a case, the value formally calculated for BIF_{100} may even be lower than one. In such a case, the concept is no more applicable.

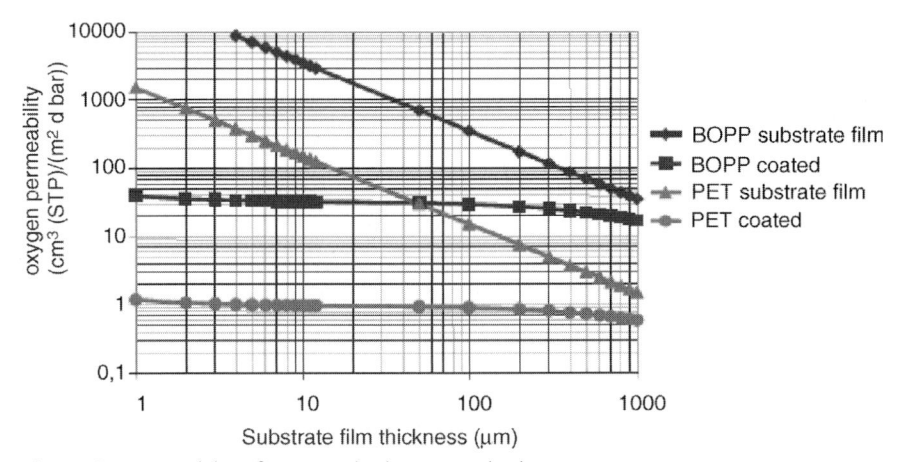

Figure 10.6 Permeability of noncoated substrates and substrates coated with a barrier layer of realistic defect structure, calculated according to Eq. (10.10), shown in dependence on substrate thickness (Source: Fraunhofer IVV, internal results).

10.3.3
Multilayer Polymer Substrates Combined with One Inorganic Layer

The above approach allows a simplified consideration of multilayer substrate films whose top layer has been coated with an inorganic barrier layer. Such layer systems are widely used in practice and are manufactured by coextruding or lacquering substrates with a thin (few μm thick) special outer polymer layer in preparation for a subsequent vacuum coating step (see for example Hanika 2003; Davis 1998,1999). Such systems are often found to have surprisingly good barrier properties after the inorganic coating process. This can be explained as follows (see Figure 10.7): First, we will only consider the outer polymer layer directly adjacent to the inorganic layer. If this polymer layer is of greater thickness than the *critical thickness* (see above) resulting from the defect size distribution in the inorganic layer, then the permeability of the system is determined solely by this polymer layer and the inorganic layer in accordance with Eq. (10.12). In other words: This usually relatively thin outer polymer layer adjacent to the inorganic coating can be considered to be equivalent to a much thicker, say 100-μm-thick polymer substrate which is then coated with an inorganic layer. If the polymer material for this outer layer has been appropriately chosen and is a material with a low permeation coefficient, then the polymer outer layer/inorganic layer system dictates the overall permeation properties of the whole layer system. Conversely, an unsuitable outer polymer layer having a high permeation coefficient (or a boundary layer formed as a result of unsuitable pretreatment) may result in no noteworthy barrier improvement compared to the noncoated substrate despite the presence of an inorganic barrier layer.

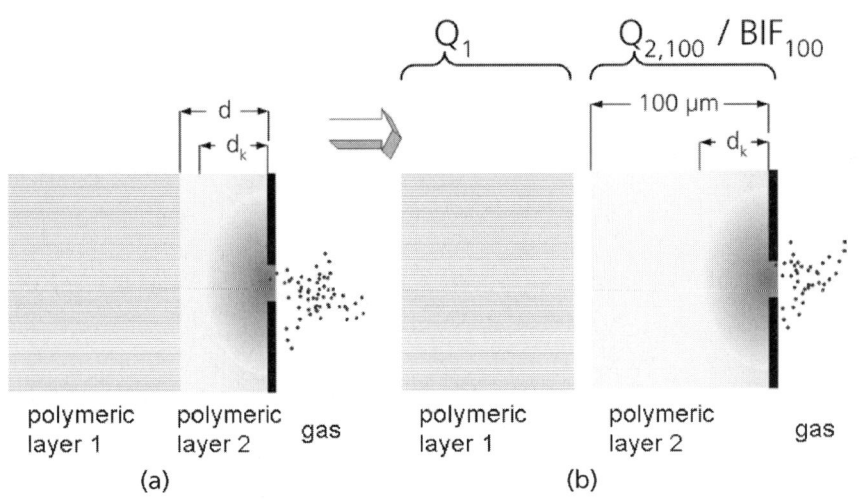

Figure 10.7 An inorganic barrier layer containing defects, coated on a double layered substrate onto a substrate layer with thickness d (no. 2). (a) Real configuration, (b) equivalent representation consisting of layer no. 1 and a 100-μm-thick layer no. 2 coated with the same inorganic layer. From (Langowski, 2005), with kind permission of Wiley-VCH.

This simplification allows Eq. (10.7) to be used again to describe the combination of the multilayer polymeric substrate in combination with the inorganic coating:

When $Q_{1,100}$ is the normalized permeability of the outer polymer layer and Q_2 is the nonnormalized permeability of the other polymer layers, the whole layer structure can be approximated as a two-layer laminate in the following way: The first layer of the laminate consists of a combination of the inorganic layer and adjacent partial-layer of the polymeric substrate having together the permeability Q_b. In accordance with Eq. (10.12), the permeability Q_b is approximately given by

$$Q_b \approx Q_{1,100}/\mathrm{BIF}_{100}$$

The second layer of the laminate consists of the rest of the polymeric substrate – but without the polymeric layer adjacent to the inorganic layer –, having the permeability Q_2. The overall permeability from Eq. (10.7) then becomes

$$Q_{ovl}^{-1} = Q_b^{-1} + Q_2^{-1}$$

If the value of $Q_{1,100}$ is less than the value of $Q_{2,100}$, then, virtually independent of the actual thickness of the outer polymer layer, the even lower value of Q_b relative to $Q_{1,100}$ dictates the overall permeability of the whole layer system.

10.3.4
Polymer Substrates Combined with an Inorganic Barrier Layer and Other Polymer Layers on Top of the Inorganic Layer

In most cases, the thin inorganic layers on polymer substrates are covered with other polymer layers. These are mostly adhesives for laminating on other films, or special lacquers.

Numerical simulations have also been carried out on such systems (Hanika, 2004; Müller, 2002). Figure 10.8, for example, shows the normalized oxygen concentration in a structure comprising an inorganic layer embedded between two different polymer materials whose permeation coefficients differ by more than a factor of 100. It can be seen that the concentration gradient in the polymer layer of higher permeability (PE-LD) is much lower than that in the polymer layer of lower permeability (PET) and that again the highest concentration gradient is to be found directly below the defect in the inorganic layer. In this case, the PET – inorganic layer portion of the structure dictates the overall permeability.

Similar calculations were carried out on three-layer systems and four-layer systems (Hanika, 2004). The results are not shown here but they do allow the following general conclusions to be drawn:

- When an inorganic layer lies between two polymer layers having very different permeation coefficients, the combination of the polymer material with the lower permeability and the inorganic layer dictate the overall permeability.

- When the two polymers have the same or very similar permeation coefficients, the permeability is halved compared to a system comprising a single polymer layer and the inorganic layer.

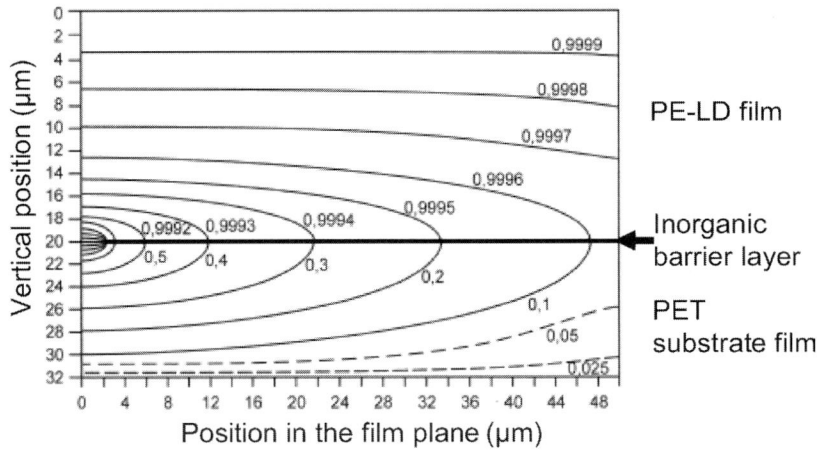

10.3 Substance Transport Through Single and Multilayer Polymer Substrates

Figure 10.8 Normalized concentration profile of oxygen in a three-layer structure, consisting of a highly permeable polymer (PE-LD, upper zone), an inorganic layer (center) with defects (center, left side) and a PET substrate film (lower zone). The highest oxygen concentration ($c = 1$) is at the upper edge of the laminate, the lowest ($c = 0$) at the lower edge (modified, from Hanika, 2004).

In practice the defect size distribution is usually unknown. It can, however, usually be assumed that the thicknesses of the polymer substrates and layers are above the critical layer thickness (see above). Equation (10.7) can therefore be extended to such a three-layer system. This now comprises (see Figure 10.9):

- The first polymer layer (the substrate film) having a permeability normalized to 100 μm thickness of $Q_{1,100}$.
- An inorganic layer whose defect structure is characterized by the barrier improvement factor BIF_{100}.
- The second polymer layer (e.g., an adhesive) having a permeability normalized to 100 μm thickness of $Q_{2,100}$.

In this case the inorganic layer, characterized by its barrier improvement factor, influences *both polymer layers*. The overall permeability of the three-layer system, Q_{ovl}, is approximately given by (see Langowski, 2002b)

$$Q_{\text{ovl}}^{-1} \approx (BIF_{100}/Q_{1,100}) + (BIF_{100}/Q_{2,100}) = BIF_{100}(Q_{1,100}^{-1} + Q_{2,100}^{-1}) \quad (10.13)$$

If, for example, there is a third polymer layer of permeability Q_3 adjacent to the second polymer layer, then Eq. (10.12) becomes

$$Q_{\text{ovl}}^{-1} \approx BIF_{100}(Q_{1,100}^{-1} + Q_{2,100}^{-1}) + Q_3^{-1} \quad (10.13a)$$

This approach allows the overall permeability of many combinations of inorganic and polymer layers to be approximately calculated, provided the normalized barrier improvement factors, BIF_{100}, have been determined from measurements on the

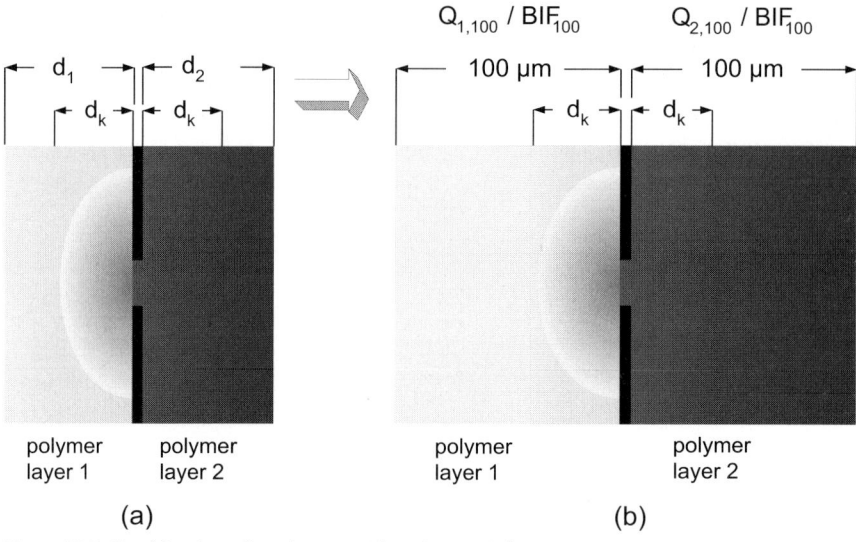

Figure 10.9 Combination of a substrate with an inorganic barrier layer, covered by a second polymeric layer (a) and the equivalent representation (b), the latter consisting of two 100-μm-thick polymer films with the inorganic layer in between. From (Langowski, 2005), with kind permission of Wiley-VCH.

individually coated polymer substrates and provided that the permeation coefficients and thicknesses of the polymer layers are known.

The limitations of this approach are once again to be repeated: As soon as no substantial barrier improvement is to be observed or calculated for a 100-μm-thick substrate film, thus leading to a value of BIF_{100} lower than one, the approach is no more applicable. In reality (see Table 10.1 in Section 10.6) the validity can be verified for many types of industrial and experimental coatings, where values for BIF_{100} between 1.9 and as high as 480 are to be observed for the permeation of oxygen through different coatings on standard 12-μm-thick PET film. Already for the permeation of water vapor through the same systems, however, lower values for BIF_{100}, namely between about 1 and 57 are to be seen.

10.3.5
Temperature Behavior of the Structures Shown Above

Equations (10.9) up to (10.13) show one important property: The permeability of a film coated with an inorganic layer or of even more complex structures can be separated into a term that is only dependent on the properties of the polymers involved and a geometry factor that depends on the defect structure of the inorganic layer. Thus, if the defect model is applicable, the permeability of different types of laminates would only depend on the temperature characteristics of the polymers in contact with the inorganic layers. Usually, this can be described in terms of the Arrhenius relation given in Eq. (10.6). Thus, the temperature dependence of the permeability of a coated film should follow the same function as the permeation coefficient of the polymer, namely $P = P_0 \exp(-E_P/RT)$. If different polymers are involved, as in the case of the sequence substrate film – inorganic layer – polymer topcoat, the temperature dependence of both polymers has to be considered in the same way by taking into account the activation energies for their permeation coefficients, E_p.

What is usually neglected in this consideration is the aspect that defect sizes in inorganic layers may also show a variation with temperature, e.g., in the case of stress cracks in the layer on a substrate that expands upon heating. For such a case, quantitative approaches are still missing.

10.3.6
Substance Transport Through Thin Polymer Layers Having Inorganic Layers on Both Sides

If a laminate system is made from two substrates vacuum coated with inorganic layers and then by laminating them with their coated sides together, practice shows that such a system has a considerably lower permeability than the permeability that is estimated by using, for example, Eqs. (10.7) and (10.12) for the resulting system consisting of: substrate film – inorganic barrier layer – adhesive layer – inorganic barrier layer – substrate film. The reason for this is that the thickness of the adhesive layer, which is now embedded between two inorganic layers, is far below the critical thickness (see above). This effect is utilized in a variety of technical applications, in particular for the manufacture of flexible materials with extremely low permeabilities ("flexible ultrabarriers") (Langowski, 2003b). Currently the most important of these processes is to apply alternating polymer layers and thin inorganic layers in subsequent steps onto a polymer substrate (Shaw, 1993, 1994; Affinito, 1996, 1997, 1999; Yializis 2000).

In order to understand these effects, numerical and analytical calculations have been carried out on relevant structures (Roberts, 2002; Hanika, 2004; Henry, 2004; Schaepkens, 2004; Miesbauer, 2007). The base structure for the modeling here was a free-standing polymer layer covered with inorganic layers on both sides which, as before, have a statistical distribution of defects. In reality, such a structure is obtained when a polymer substrate is first coated with an inorganic barrier layer, then with a polymer layer, for instance a lacquer, followed by deposition of another inorganic barrier layer, and if the presence of the substrate is neglected in a first approximation. For a good barrier effect, the position of defects in the inorganic layer on one side must however be independent from the other side, namely no defect must be replicated through the polymer layer. If this condition is fulfilled, the base structure can be simplified (see Figure 10.10). As a simplification, it is assumed here that the defects in both the layers are present in the form of a square grid of side length a and that the defect-grids in both layers are positioned relative to each other such that a defect in one layer, when projected onto the other layer, lies exactly in the center of four defects.

The distance from a defect in the top layer to a defect in the lower layer is the shortest effective path via which permeating molecules can pass through the

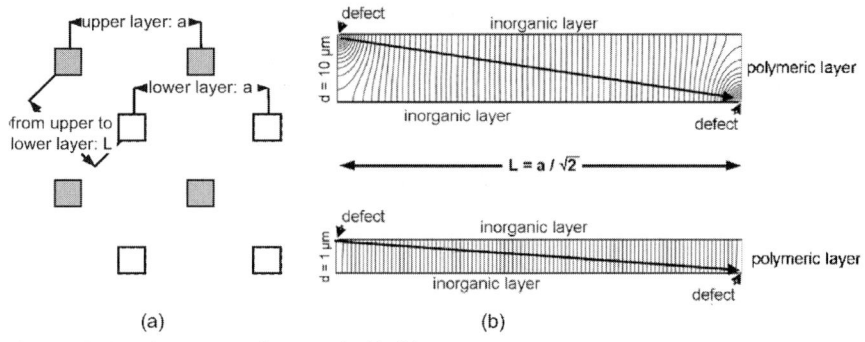

Figure 10.10 Polymeric interlayer, embedded between two inorganic barrier layers containing defects. (a) Simplified representation of defects via two staggered quadratic lattices in upper and lower inorganic layers, top view. (b) Cross-sectional view along the shortest path for permeating molecules, including the related concentration profiles, for polymeric interlayers of different thickness. After (Hanika, 2003).

intermediate polymer layer; this distance is L, where $L = a/\sqrt{2}$, if the usually small thickness of the intermediate polymer layer is neglected.

Exemplary results of numerical modeling of the permeation through the structures shown in Figure 10.10 are given in Figure 10.11. There, three different distances (distance a in Figure 10.10) between the defects within one inorganic layer have been used as a basis. Also shown are – as a reference – the permeability figures for substrate films of identical permeation coefficient as the thin polymer layers, covered by one inorganic layer of identical defect distance. They give the following trends for the barrier properties of this kind of layer structures:

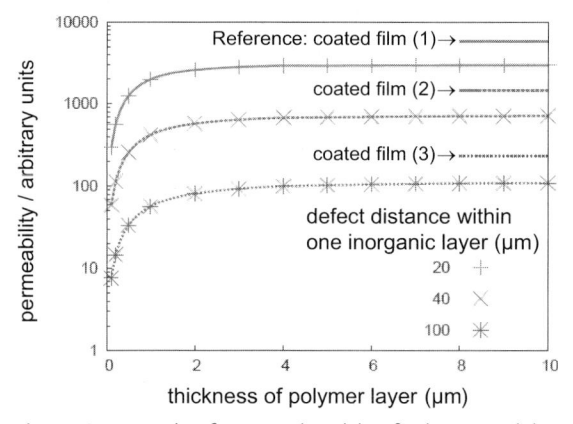

Figure 10.11 Results of numerical modeling for the permeability of a polymer layer of varying thickness embedded between two thin inorganic layers as shown in the right part of Figure 10.10. Reference figures are given for substrate films which consist of the same polymer, but have only been coated on one side (Miesbauer, 2007).

10.3 Substance Transport Through Single and Multilayer Polymer Substrates

When the intermediate layer is sufficiently thick, the permeability of the three-layer structure is just half of that of the corresponding two-layer structure that has been taken as reference. This is the result that would be obtained by simply applying the ideal laminate theory.

When the intermediate polymeric layer is made as thin as possible, the overall barrier properties of the layer system can be reduced by up to a factor of ten with respect to the permeability calculated for the thicker three-layer structures. However, the intermediate polymer layer must have a certain minimum layer thickness in order to avoid that defects in the first inorganic layer be replicated through the polymer layer in a way that defects are formed at the same positions in the second inorganic layer. This requirement means there is a limit on the achievable barrier effect, which is to be expected at a thickness of the polymeric interlayer of some hundreds of nanometers and possibly a barrier improvement by a factor of five in relation to thicker structures.

A combination of a substrate coated by an inorganic layer, a subsequent thin intermediate polymer layer, a second inorganic layer, and a subsequent polymer coating may be (in a thought experiment) separated in the middle of the inorganic layers, giving the following segments as to be seen in Figure 10.12: A substrate coated with a barrier layer, a triple layer as described just above and another substrate coated with a barrier layer. If the permeability Q^* is assigned to the triple layer, following the

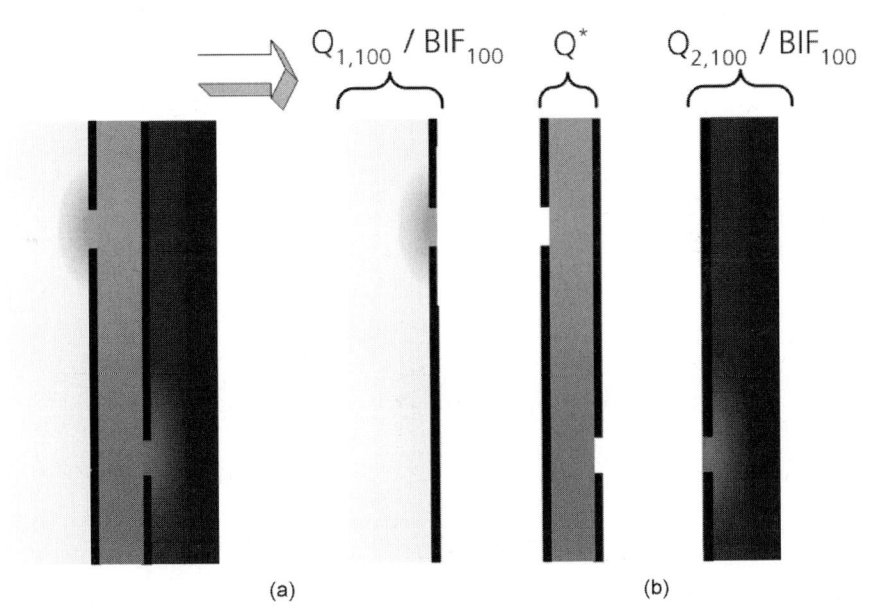

Figure 10.12 Combination of a substrate with an inorganic barrier layer, covered by a polymeric layer, another inorganic barrier layer and a second polymeric layer (a) and the equivalent representation (b) See further explanations in the text. From (Langowski, 2005), with kind permission of Wiley-VCH.

considerations represented by Figure 10.11, a common barrier improvement factor, BIF_{100}, is given to the inorganic layers and the external polymeric layers have the transmittance figures $Q_{1,100}$ and $Q_{2,100}$, respectively, then their total permeability is given by

$$Q_{\mathrm{ovl}}^{-1} \approx (\mathrm{BIF}_{100}/Q_{1,100}) + Q^{*-1} + (\mathrm{BIF}_{100}/Q_{2,100}) \tag{10.14}$$

10.5
Substance Transport Through Polymers Filled with Particles

Filling of polymers with particles to improve their barrier properties is a method that has been used for decades. Already 40 years ago, Nielsen developed a theory which is able to describe all changes in permeability observed so far (Nielsen 1967).

For barrier improvements, dense, solid and nonporous filler particles should be selected. If the polymer is filled by a volume fraction Φ of these particles, the permeation process only takes place in the remaining volume fraction of the polymer, $1 - \Phi$. This effect gives the first contribution to a reduction in permeability. A second contribution results from an increase of the effective path length for the permeating molecules, see Figure 10.13.

This increase in effective path length is described by the so-called *tortuosity factor* τ, defined by the average path length for a molecule diffusing through the filled polymer, divided by the macroscopic thickness of the polymer sample. As a result, the ratio of the permeation coefficient of an unfilled polymer, P_0, and the permeation coefficient P_1 of a sample of a filled polymer becomes

$$P_1/P_0 = (1-\Phi)/\tau \tag{10.15}$$

If, as depicted in Figure 10.13, the filler particles have a rectangular cross section of length L and thickness w, their *aspect ratio* α is given by

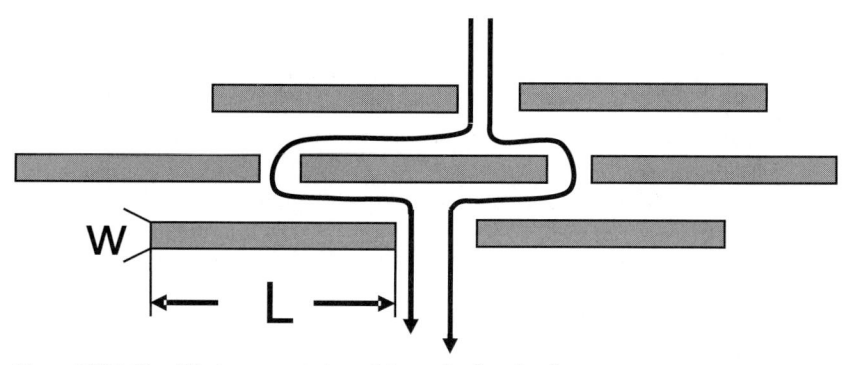

Figure 10.13 Simplified representation of the path of molecules through a polymer filled with aligned rectangular platelets of thickness w and length L.

$$\alpha = L/w \tag{10.16}$$

In the optimum case for barrier improvement, many thin platelet-shaped particles should be uniformly distributed in the polymer and their orientation should be parallel to the plane of the film. In this case, the tortuosity factor can be derived from geometric principles and the resulting effective permeation coefficient P_1 of the filled polymer becomes (Nielsen 1967)

$$P_1 = \frac{P_0(1-\Phi)}{1+\frac{\alpha\Phi}{2}} \tag{10.17}$$

This expression allows an estimation of the minimum permeability that can be obtained by filling of polymers with particles of rectangular cross section that are aligned parallel to the surface. It is obvious that any deviation from such an ideal arrangement will lead to a lower reduction in permeability relative to the unfilled polymer.

In addition to the considerations leading to Eq. (10.25), Nielsen also regarded situations which would offer additional pathways to the permeating molecules.

A practically relevant assumption is that the permeating molecules would not only be dissolved inside the polymer, but also exist in a liquid phase located in an interfacial region between filler particles and the polymer matrix, e.g., in cases of capillary condensation. This would lead to an increased effective permeation coefficient for types of molecules which are able to condense, the most important of which is water.

Often, filler particles will be porous instead of solid. In such a case, an increase in permeability up to values which are higher than for the unfilled polymer is to be expected. As a consequence, deviations from Eq. (10.20) toward higher effective permeation coefficients will often be observed in practice and numerous possible origins for such deviations exist, for which it will be difficult to identify the right one.

As all effects can be described by the volume fraction Φ and the aspect ratio α alone, the actual size of the particles does not have a direct influence on the barrier improvement.

10.6
Experimental Findings: Polymer Films and One Inorganic Barrier Layer

For most vacuum coated polymeric substrates, the barrier improvement obtained by coating is taken as the characterizing factor. In the beginning, differences observed between different coating materials or different coating methods were interpreted in terms of a possible permeability of the bulk material of the permeating layers (Krug, 1990). At present, the sum of results shows that it is more likely that the tendency of the layer materials to form stress failures and the ability of the coating process to cover particles and other surface irregularities are the factors that make the difference between methods and materials for barrier coating. Table 10.1 gives an overview on oxygen and water vapor barrier properties of PET film coated with different materials by different methods. PET substrate film in a thickness of 12 μm has become a

Table 10.1 Comparison of different inorganic layers on PET substrates (film substrates unless otherwise indicated). Typical figures for oxygen and water vapor permeability of coated substrates obtained on experimental and industrial samples.

Substrate/ layer structure	Inorganic layer, deposition method	References and typical figures of oxygen permeability cm³(STP)/m² d bar at 23 °C	References and typical figures of water vapor permeability g/m² d at 23 °C, 85% → 0% r. h.
PET substrate, 12 μm Polymeric substrate/ thin inorganic single or multiple layer	Without coating	80–100(Q_{100} between 9.6 and 12) (Krug et al. 1993)	14–16(Q_{100} between 1.7 and 1.9) (Hanika, 2004)
	Al, boat evaporation	0.5 (Utz, 1992, Moosheimer et al. 2001; 0.2–0.8 Langowski, 1999a)	0.2–0.5
	SiO_x, electron beam evaporation	<2 (Hartwig, 1992; Krug et al. 1993; Lohwasser et al. 1995)	<2 (Lohwasser et al. 1995)
	AlO_x, boat evaporation, reactive	1.6 (Kelly, 1993)	≈0.4 (Hartwig et al. 1992; Krug et al. 1993)
	AlO_x, electron beam evaporation	<4 (Hoffmann et al. 1994)	0.1 (Langowski, 1999a)
			≈0.4 (Kelly, 1993)
			<2 (Hoffmann et al. 1994)
	SiO_x–Al_2O_3 double layer electron beam evaporation	2.2 (Schiller et al. 1994)	0.4 (Schiller et al. 1994)
		5 (Misiano et al. 1993)	≈0.9 (Misiano et al. 1993)
		<2 (Ohya et al. 2000)	<0.5 (Ohya et al. 2000)
	Al_2O_3 + SiO_2 double layer electron beam evaporation	<1.0 (Yamada et al. 2000)	≈0.5 (Yamada et al. 2000)
		≈2.3 (Phillips et al. 1993)	≈0.5 (Phillips et al. 1993)
	SiO_x, PECVD → film substrates	<2 (Casey et al. 1999)	≈0.25 (Casey et al. 1999)
		1 (Finson and Felts, 1994)	(Finson and Felts, 1994)
	→containers, inside coated	<0.1–0.36 (Berger, 2006, Lüpke, 2005)	
	AlO_x, sputtering	0.03 (Langowski, 2002a)	0.03 (Langowski, 2002a)
	AlO_xN_y, sputtering	1 (Erlat et al. 2001)	≈0.07 (Erlat et al. 2001)
	Al, sputtering	0.02 (Benmalek and Dunlop, 1995)	<0.05 (Benmalek and Dunlop, 1995)

10.6.1
Structures and Defects in Inorganic Barrier Layers on Polymer Substrates

As shown above, defects in thin inorganic barrier layers play a decisive role for the barrier properties of the final laminates. According to their size, these defects may be separated into two groups:

- Macroscopic defects, ranging from some nanometers to several micrometers. These defects can be identified via atomic force microscopy (AFM), scanning electron microscopy, and optical microscopy (the latter method only for metallic layers)
- Microscopic defects (pores) from some nanometers down to the sub-nm-range. These structures are small enough to allow condensation of water and other substances. As they cannot be identified sufficiently well by microscopic methods, their existence has only been shown by indirect methods so far.

The following parameters are responsible for the formation of macroscopic defects:

- The properties of the surfaces of the polymeric substrate: Often, so-called antiblock articles are added to the substrate polymer to avoid strong adhesive forces between the adjacent layers in the stack of one film roll. Typical particle sizes are some micrometers, typical concentrations some 1000 ppm (w/w). This means that about 10^5 particles per cm^2 are to be found on a typical film surface, leading to average particle distances of about 35 μm, in concert with practical findings (see Figure 10.14). Defects in the inorganic barrier layer will be generated around these

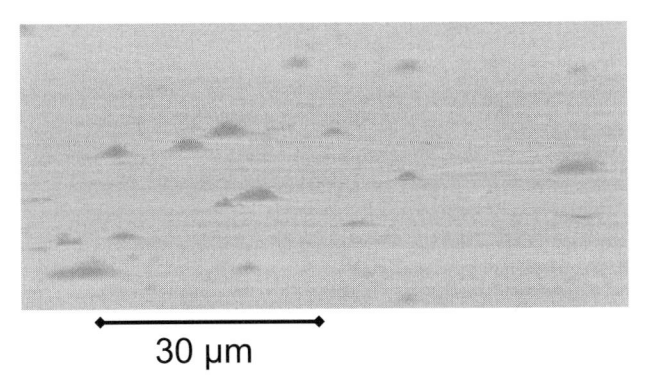

Figure 10.14 Surface of a standard film from biaxially oriented polypropylene (BOPP) with protruding antiblock particles. Image made in a scanning electron microscope at high tilt angle. (The scale shown by the micrometer bar is only valid in horizontal direction.) Source: Fraunhofer IVV.

particles. Moreover, there will also be a contact between the antiblock particles from the rear side of the substrate film with the freshly deposited inorganic barrier layer, upon winding the coated film onto a roll in the vacuum web coater. Such a contact will lead to damage of the inorganic barrier layer, as to be seen in Figure 10.15.

A specific technical measure against these effects is to produce specific multilayer substrate films in which only the rear film surface carries those antiblock particles (Fayet, 2005). This method, however, does not avoid the contact between coated side and rear side of the films. For this purpose, size and shape of the antiblock particles in the rear side can be additionally optimized, but systematic investigations have not been published so far.

- Another possible source of defects in thin barrier layers is to be expected in the surface topography of the substrate itself. Typical film surfaces show average roughness figures r_a below some nanometers. So far, no clear correlation between substrate surface topography and resulting barrier properties could be found.

- The simplest possible source for defects is represented by particulate contaminations on the substrate surface. Due to the many steps of polymer conversion, there is a substantial probability for such contamination, particularly because polymer films are usually manufactured outside clean room conditions.

The dependence of the permeability of a polymer substrate coated by an inorganic barrier layer is shown in Figure 10.16 for aluminum on PET (Hanika, 2004). This figure shows that – for a lower thickness of the coating – external defects cannot be the sole responsible factor for the limited barrier properties of thin inorganic layers on polymer substrates. The permeability of the coated PET substrate film shows a clear decrease with increasing Al layer thickness up to about 50 nm. A further increase in layer thickness does not lead to substantial improvements in barrier properties. Similar findings are reported for silicon oxide layers (da Silva Sobrinho, 1999, 2000). The usual interpretation for this behavior is the following: In the early stage of the layer growth, the layer is still not coherent, giving way for an additional permeation until a continuous layer has been formed. In the end, this layer – together with its remaining defect structure – determines the best attainable barrier in combination with the substrate. A further increase in layer thickness gives only a slight barrier improvement, possibly due to a further reduction of the effective size of the remaining defects.

10.6.2
Comparison of Model Calculations and Experimental Results for Combinations of Polymer Films and One Inorganic Barrier Layer

A test of the validity of the defect model – and the determination of its limits – are extremely important for further improvements in barrier properties of flexible materials, because the actual permeation mechanisms are decisive for possible technical measures.

If the defect model applies, the following observations should be made when different types of coated substrate films are investigated:

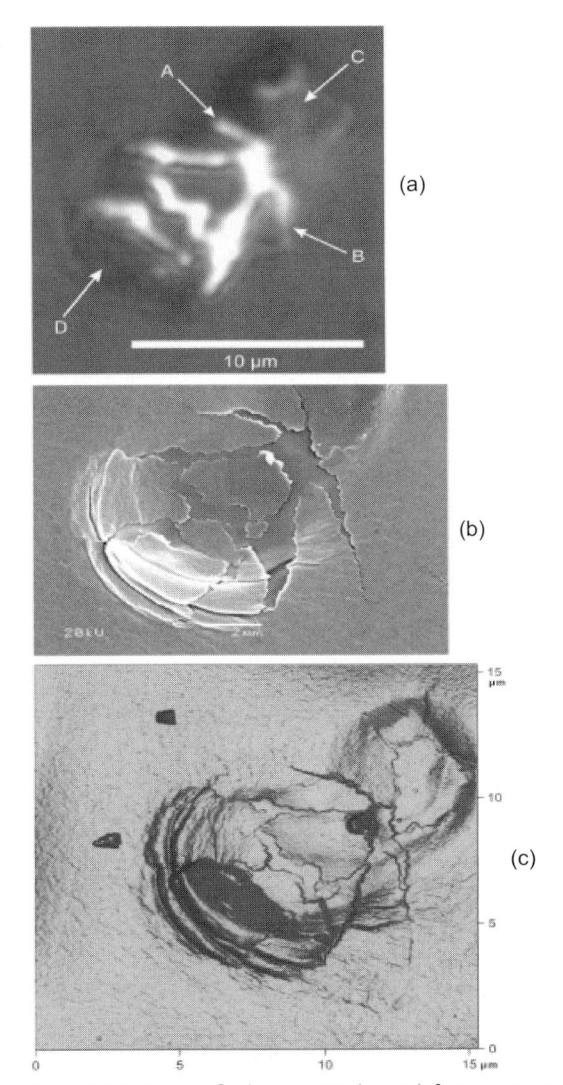

Figure 10.15 Image of a characteristic larger defect – as created by the contact with a spherical antiblock particle – in an aluminum layer on a BOPP substrate, recorded via different methods: (a) optical microscopy, numerical aperture 0.4, (b) scanning electron microscopy, (c) atomic force microscopy for illustration of the capability of different methods to analyze surface topography of layers deposited on polymer substrates (from Hanika, 2004).

- There should be a clear correlation between number and size of the defects in the inorganic layer and the permeability of the coated film
- The ratio of the permeabilities of the noncoated vs. the coated film should be the same for different permeating substances as the deposited inorganic barrier layer

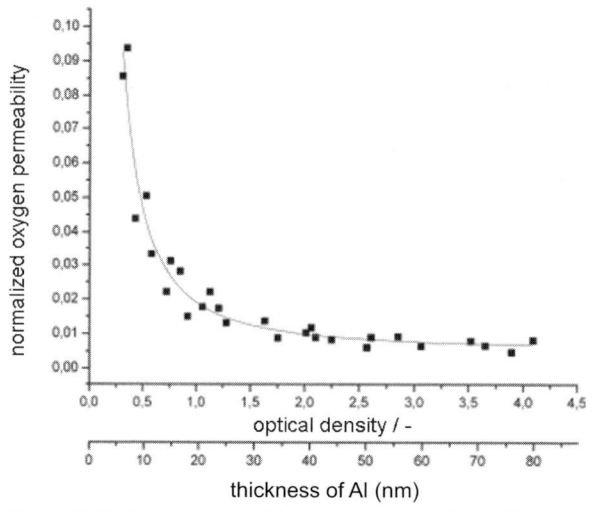

Figure 10.16 Oxygen permeability of a 12-μm-thick PET film coated with Al layers of varying thickness, which is proportional to their optical density. The oxygen permeability has been normalized to the permeability of the uncoated substrate film (from Hanika, 2004).

would reduce the permeability of the polymer substrate by the afore-described barrier improvement factor for all permeating substances to the same extent.
- There should be only a small effect of the thickness of the deposited inorganic layer, provided a minimum thickness has been achieved in coating.
- Also, the dependence of the permeability on the temperature should show no difference between the polymer substrate alone and the coated polymer substrate.

An overview on experiments of this type is presented in Table 10.2.

The experimental findings that have been obtained by different authors can be summarized according to the indicated criteria in the following way:

- Number and size of defects: It is extremely difficult to obtain this figure in a way that would be representative for larger areas, especially on nonmetallic inorganic layers. Therefore, in spite of many attempts, there are not too many coherent results available. Quantitative validation of the defect model was obtained for the permeation of oxygen through aluminum-coated polymer substrates by characterizing the defect size distribution in the aluminum layers and calculating the expected permeabilities (see Figure 10.17). There was good agreement between the experimental and calculated data for oxygen (Hanika 2004; Hanika et al. 2003). Water vapor permeabilities, however, were found to be considerably higher than the values calculated from the defect size distributions.
- Type of permeating substance: For most systems it is found that the permeabilities of the coated and noncoated substrates for permanent gases and CO_2 do not depend on the nature of the permeating molecules. Water vapor is however

virtually always found to be an exception: Coated substrates are observed to have considerably higher permeabilities than one would expect from the permeability measurements with permanent gases, see also Figure 10.18. For some layer systems helium behaves similarly (Bichler, 1996). Also for flavor substances (see Chapter 9), barrier improvements obtained by coating are much less than for oxygen and even lower than for water vapor. However, for selected aluminum oxide layers applied by reactive sputtering, the improvements for oxygen and water vapor permeabilities were found to be in the same proportions (Langowski, 2003b), as to be seen in Figure 10.18.

- For the influence of the thickness of the inorganic layer, similar behavior as shown in Figure 10.17 is reported. The major differences observed here are to be found in the layer material and the deposition process. Apparently, coating processes that supply impinging atoms or molecules of higher energy (like sputtering or plasma-enhanced chemical vapor deposition) give layers of a lower defect density, at lower thickness.
- In many cases, the temperature behavior of the permeability follows the same characteristic for coated and noncoated films. Deviations toward higher apparent activation energies are explained by (Henry, 2000; Erlat, 2001, 2002) via additional permeation mechanisms in the bulk of the inorganic layer

10.6.3
Apparent Additional Transport Mechanisms for Water Vapor

The decisive problem is therefore to explain the additional transport mechanisms for water (possibly also flavor substances) and to subsequently derive counteracting measures, which is important for the practical use of these layer systems. The most probable additional transport mechanism for water molecules as discussed in the literature (Langowski, 1992; Roberts, 2002; Henry, 2001; Affinito, 2004) relates to porous structures in the inorganic layers. So far, there has only been an indirect access to size and abundance of pores in barrier layers on polymer substrates. However, a simplified modeling of water transport through filled pores/capillaries can be made under the following assumptions:

The water transport occurs along the fraction of capillaries that are penetrating the layer possessing a radius of some tenths of a nanometer, as suggested by (Roberts, 2002; Affinito, 2004). These can be described in terms of an overall dimensionless volume fraction η. The interior of these pores will be completely filled with water of the density ρ of about $\rho = 1 \, \text{g cm}^{-3}$ above a relative humidity of a few percent at room temperature (Schubert 1982).

Furthermore, the transport process is supposed to be determined solely by self-diffusion of the water molecules in the filled pores. Here, a self-diffusion coefficient of $D = 2 \times 10^{-5} \, \text{cm}^2/\text{s}$ may be assumed (Lechner, 1992).

This model would be equivalent to a polymeric structure in which water is dissolved homogeneously, with the effective volume density $\eta \, \rho$, which is equivalent to an effective solubility coefficient of the same magnitude. It can be shown that for values of η of less than 10^{-8}, which is far below the measurement accuracy of

Table 10.2 Overview of available results for transport of substances through systems of polymeric substrates and one inorganic barrier layer, in dependence on layer properties, permeating substance and temperature.

Layer sequence	Experimental results						Theoretical studies: permeability in dependence on...
	Material of inorganic layer, production via	Barrier improvement only	Experimental investigations: permeability in dependence on...				
			Number and size of defects	Type of permeating substance	Thickness of inorganic layer	Temperature	Number and size of defects	Temperature

Layer sequence	Material of inorganic layer, production via	Barrier improvement only	Number and size of defects	Type of permeating substance	Thickness of inorganic layer	Temperature	Number and size of defects	Temperature
Different polymeric substrates/one thin inorganic barrier layer	Al, boat evaporation		(Utz, 1995; Hanika, 2004; Hanika et al., 2003a, 2002b; Utz, 1994)	(Utz, 1994; Moosheimer et al. 2000)	(Krug et al. 1993; Utz, 1993; Hartwig et al. 1992; Hanika, 2004; Langowski, 2002b; Utz, 1994; Moosheimer and Langowski 1999)	(Hanika 2004; Utz, 1994)	(Langowski, 2003a; Hanika et al. 2002a; Decker and Henry, 2002; Prins and Hermans 1958; Hanika 2004; Langowski, 2003b; Hanika et al. 2003a, 2003b; Fayet et al. 2005; da Silva Sobrinho et al. 2000; Hanika et al. 2002b; Utz, 1994; Moosheimer and Langowski, 1999; Müller and Weiser, 2002; Rossi and Nulman 1993; Beu and Mercea, 1990; Jamieson and Windel, 1983; Czeremuszkin, 2000, 2001, 1999; Deak and Jackson, 1993; Czeremuszkin et al. 2003a; Roberts et al. 2002; Chatham, 1996; Rossi and Nulman, 1993; da Silva Sobrinho et al. 1998)	(da Silva Sobrinho et al. 2000; Roberts et al. 2002; Tropsha et al. 1992; Henry et al., 2003, 2001)

10.6 Experimental Findings: Polymer Films and One Inorganic Barrier Layer

Process					
AlOx, boat evaporation, reactive	(Kelly, 1993, 1994)				
SiOx, SiNx, TiOx, plasma-enhanced chemical vapor deposition (PECVD)	(Walter et al., 2003; Weikart and Smith 1995; Anderle et al., 1997; Wood and Chatham, 1993)	(Fayet et al., 2005; da Silva Sobrinho et al. 1999; Czeremuszkin et al. 2003a; da Silva Sobrinho et al. 1998)	(Bichler et al. 1996; Langowski, 1999b; Moosheimer et al. 2000)	(da Silva, Sobrinho et al. 1998; Smith et al. 2002; Casey et al. 2002; Denmler 1999; Felts et al. 2001; Nelson and Chatham, 1991; da Silva Sobrinho et al. 2000)	(da Silva Sobrinho et al. 2000; Tropsha and Harvey 1992)
ITO, AlOxNy, SiOx, AlOx, sputtering		(Erlat et al. 2001; Henry et al. 2000)		(Erlat et al. 2001; 2002; Henry et al. 2000; Schiller et al. 2001)	(Erlat et al. 2001; Henry, et al. 2001; Erlat, et al. 2002; Henry et al. 2000)
SiOx, AlOx, electron beam evaporation		(Kessler 1994)	(Bichler et al 1996; Roberts et al 2002; Langowski 1999b; Henry et al 1998, 2002)	(Krug et al 1993; Hartwig et al 1992; Schiller et al 1994; Yamada et al 2000; Roberts et al 2002; Kessler, 1994; Henry et al 2002)	(Kessler 1994; Henry et al 2001, 2002, 1998; Nörenberg et al. 2000)

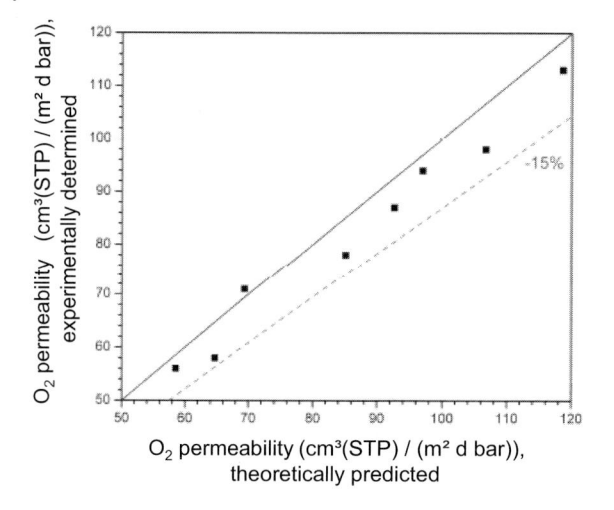

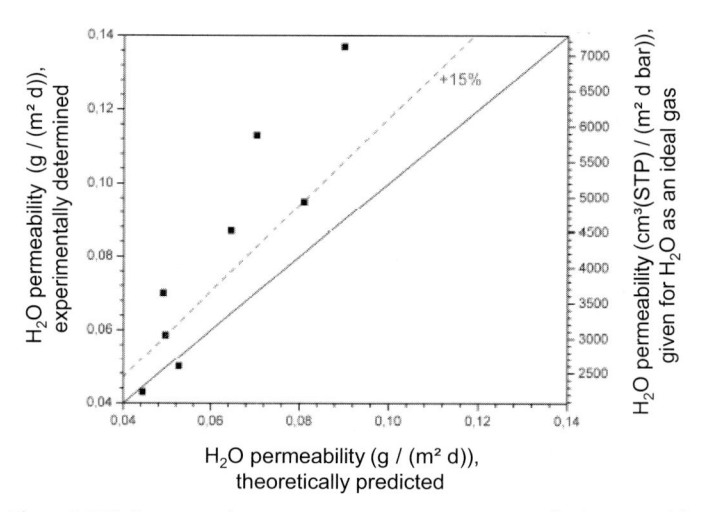

Figure 10.17 Oxygen and water vapor permeability for BOPP substrates coated with aluminum barrier layers. Abscissae: values predicted from defect distributions as determined via optical microscopy. Ordinates: experimentally determined figures. The bisecting line would be valid for total agreement between theory and experiment, the dashed line shows a deviation of 15%, a typical range for experimental errors in this area (from Hanika 2004).

currently available characterization methods for such layers, a substantial increase in water vapor permeability would be detected, namely about $40 \, g/m^2 \, d$ at a layer thickness of 40 nm.

This estimation, although a rough one, shows that virtually any deviation toward higher values of water vapor permeability through inorganic barrier layers on polymeric substrates can be explained by transport through pores. These pores

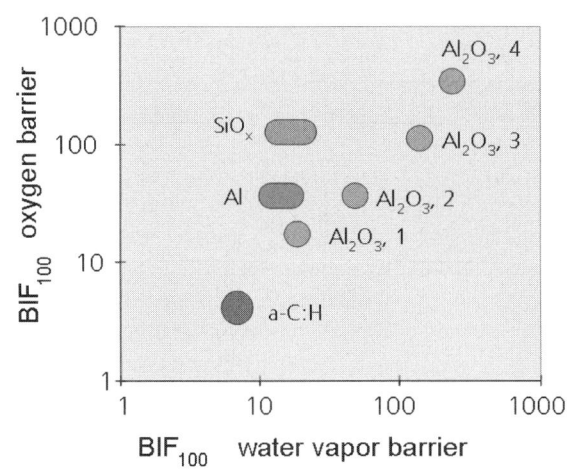

Figure 10.18 Barrier improvement factors BIF_{100} (normalized to 100-μm substrate film thickness), for different layer materials, coating methods and substrates, as shown below. Values for BIF_{100} extend to 370 (for oxygen) and to 230 (for water vapor) (from Langowski, 2003b).

Layer material	Method of coating, substrates
Al	Boat evaporation, onto standard industrial film substrates (BOPP, PET)
SiO_x	Different methods from laboratory to industrial scale (electron beam coating, plasma assisted, plasma assisted chemical vapor deposition, reactive sputtering) onto standard PET and PEN substrate films
a-C:H	Experimental coating with amorphous carbon via plasma assisted chemical vapor deposition, on standard PET film
Al_2O_3, 1	Reactively sputtered aluminum oxide, on standard PET film, Al_2O_3 thickness: 50 nm
Al_2O_3, 2, 3, 4	Reactively sputtered aluminum oxide, on special PET film, Al_2O_3 thickness values: (2): 20 nm, (3): 50 nm, (4): 200 nm

need only to be present in volume concentrations far below present characterization methods, irrespective of details of their possible complex shape (Langowski, 2005).

Using the values for the diffusion coefficient and solubility coefficient for oxygen in water (Lechner, 1992) in a similar estimation, it can be readily shown that water-filled capillaries are of no significance at all for oxygen transport compared to the existing oxygen transport through larger defects.

For the real existence of such pores, some indication is given by BET adsorption studies on aluminum layers detached from polymer substrates (Hanika, 2004). The results show that these aluminum layers have a specific surface area for the adsorption of molecular nitrogen that exactly corresponds to the geometric surface area. However, when experiments were carried out on the same samples with water vapor roughly twice the surface area was measured, due to pore structures in the sub-nanometer region (Hanika, 2004).

10.6.4
Properties of Systems with at least One Inorganic Layer Embedded Between to Polymer Layers or Films

In daily practice, the configuration of a polymer substrate, coated by an inorganic barrier layer and, further on, by additional polymeric layers are the most abundant examples, e.g., in packaging laminates made from aluminum metallized PET or BOPP film. In the literature, however, much less systematic investigations are to be found on this or similar structures, most probably due to two reasons:

- The inorganic layer is not as easily accessible as before for different methods of characterization.
- The complexity of different steps of the conversion process induces an additional variability of the results, impeding the establishment of a correlation between the different parameters.

An overview of the related literature is given in Table 10.3.

In spite of the probably existing additional path for water vapor permeation, the results do not show contradictions to the validity of the defect model.

10.7
Experimental Findings: Combinations of Polymer Films and More Than One Inorganic Barrier Layer

In the course of the rising need for flexible ultrabarrier materials, the number of related publications shows a significant increase over the last years. Table 10.4, therefore, shows just a selection of results which allow for a comparison between experimental findings and theoretical modeling.

So far, most of the results can be explained by applying the ideal laminate theory (Eq. (10.7)) in a way that the multilayer systems are regarded as if each pair of polymeric and inorganic layer were one polymeric layer with lower permeability (see especially Henry, 2006). The following example may explain the procedure and also the limitations:

In (Henry, 2006), the following values are reported (water vapor permeability in $g/(m^2\ d)$ at 30°C and a difference in relative humidity across the sample of 100% to 0%):

Permeability of the 12-µm-thick PET substrate film: $44.1\ g/(m^2\ d)$, of the PET substrate film coated with an acrylate layer of about 0.6 µm thickness: $31.2\ g/(m^2\ d)$. Using Eq. (10.7) in the form

$$Q_{ovl}^{-1} = Q_1^{-1} + Q_2^{-1} \quad \text{or} \quad Q_2^{-1} = Q_1^{-1} - Q_{ovl}^{-1}$$

a water vapor permeability for the acrylate layer alone of about 107 g/(m² d) can be derived. Taking into account the thickness of substrate and layer, the values of Q_{100} are: 0.64 g/(m² d) for the acrylate layer material, compared to 5.3 g/(m² d) for PET. This means that the permeation coefficient of the acrylate layer is lower than that of the PET substrate by a factor of about 8.3.

As the first step to produce ultrabarrier multilayer stacks, PET was coated with an Al layer, followed by coating the Al layer with the same acrylate layer material as above. For the water vapor permeability of the combination PET/Al inorganic coating/acrylate, a value of 3.0 g/(m² d) was obtained. This is a surprisingly low improvement, as – if the combination Al inorganic coating/acrylate layer would be considered alone – a value for BIF_{100} of less than 1 would be obtained for the effect of the Al on the barrier properties of the acrylate coating. Therefore, the validity of the model shown in Section 10.3 has to be already questioned here for this case.

After coating the side of the acrylate coating with another layer of Al, leading to the structure PET/Al inorganic coating/acrylate/Al inorganic coating, a value for the water vapor permeability of 1.2 g/(m² d) was measured. If once again the PET substrate would be neglected and only the structure Al inorganic coating/acrylate/Al inorganic coating would be considered, a water vapor permeability of about 1.5 g/(m² d) is to be expected form the ideal laminate theory in the following way: If we (once again in a thought experiment) split the acrylate layer in the middle, we now would have two acrylate layers covered by inorganic barrier layers which reduce their permeability by a given barrier improvement factor, thus leading to half of the permeability of a combination from the Al layer and an acrylate layer alone. As, from Figure 10.12 and Eq. (10.14), we would expect much lower values for the permeability, the comparison to the actually measured value of 1.2 g/(m² d) shows that the ideal laminate theory can be applied to this system without assuming additional barrier effects like those described in Chapter 4. This finding is in line with the fact that already the initial barrier improvement due to the Al layer was relatively low in the first combination PET/Al inorganic coating/acrylate.

This fact shows that to obtain ultrabarrier properties via multilayer systems, it is not useful to combine just many layer pairs of higher permeability. Moreover, it is essential that good inorganic barrier coatings with a low density of defects and, therefore, a high barrier improvement are used. As Figure 10.12 shows, in such a case configurations of the type inorganic layer – thin polymeric layer – inorganic layer may already be as efficient as a multiple layer stack.

10.8
Experimental Findings: Polymers Filled with Platelet-Shaped Particles

Since the early 1990s, different polymers have been modified via incorporation of platelet-shaped inorganic particles. In most reported cases, particles of a size in the nanometer scale in one dimension have been used. The main reason for this is that the optical properties of the initial polymer are least disturbed and that the mechanical properties may even be improved by this method. In most

Table 10.3 Overview of available results for transport of substances through systems of polymeric substrates, one inorganic barrier layer and one additional polymeric layer, ir dependence on layer properties, permeating substance and temperature.

Layer sequence	Experimental results		Experimental investigations: permeability in dependence on				Theoretical studies: permeability in dependence on...
	Material of inorganic layer, production via	Barrier improvement only	Number and size of defects	Thickness of inorganic layer	Type of permeating substance	Temperature	Number and size of defects
Polymeric substrate/ thin inorganic barrier layer/polymeric cover layer	Al, boat evaporation	(Shaw and Langlois, 1994; Langowski, 1999a; Bichler et al 1998; Henry et al 2005; Moosheimer et al 2000)	(Hanika et al 2002a; Langowski 2002b; Utz, 1995)	(Hanika et al 2002a; Langowski 2002b)	(Moosheimer et al., 2000)		(Langowski, 2003a; Langowski, 2002a, 2002b; Müller and Weisser, 2002)
	SiO$_x$, electron beam evaporation	(Amberg-Schwab et al 1998; Moosheimer and Langowski, 1999; Moosheimer et al 2000; Langowski, 1999a)			(Moosheimer et al 2000)		
	AlO$_x$, sputtering						(Henry et al 2001, 2002)

Polymeric substrate/ polymeric intermediate layer/thin inorganic barrier layer	Al, boat evaporation	(Shaw and Langlois, 1994; Hanika et al 2002a; Davis and Peiffer, 1998, 1999; Moosheimer and Langowski, 1999; Henry et al 2005; Bichler et al. 1998; Seserko et al 1998)
	SiO_x, electron beam evaporation	(Davis and Peiffer, 1998, 1999; Moosheimer and Langowski, 1999; Seserko et al 1998)
	AlO_x, electron beam evaporation	(Davis and Peiffer, 1998; Moosheimer and Langowski, 1999; Seserko et al 1998)
	AlO_xN_y, sputtering	(Erlat et al 2002; Henry et al 2004)
	AlO_x, sputtering	(Henry et al 2004)
	TiO_xN_x, PECVD	(Henry et al 2004)
	SiN_x, PECVD	(Schaepkens et al 2004)
	SiN_x, ETP	(Creatore et al 2005)

Table 10.4 Overview of selected results for transport of substances through systems of polymeric substrates, at least two inorganic barrier layers and at least one intermediate polymeric layer, in dependence on layer properties.

Layer sequence	Experimental results		Theoretical studies
	Material of inorganic layer, production via	Barrier improvement only	
Polymeric substrate/ polymer layer (optional)/thin inorganic layer/thin polymer layer/thin inorganic layer/ further pairs of layers(optional)	Al, thermal evaporation	(Shaw and Langlois 1994; Henry et al. 2005; Davis and Peiffer 1999)	(Hanika et al. 2003b; Erlat et al. 2004; Czeremuszkin et al. 2003b; Schaepkens et al. 2004): for one polymeric layer between two inorganic layers
	Al, electron beam evaporation	(Henry et al. 2006)	(Langowski, 2005): for further sequences of layer pairs
	AlO_x, electron beam evaporation	(Affinito et al. 1996)	(Graff et al. 2004): for up to five pairs of layers
	AlO_x, sputtering	(Affinito et al. 1996): up to seven pairs of layers	(Affinito and Hilliard 2004): modeling of an inorganic porous matrix, filled with polymeric material
	SiN_x, PECVD	(Schaepkens et al. 2004) up to two pairs of layers	(Henry et al. 2006): for two pairs of layers
	SIN_x, ETP	(Creatore et al. 2005): up to two pairs of layers	

cases, layered silicate minerals like montmorillonite are applied because of their ability to exfoliate and to form very thin particles with a thickness down to one nanometer, at lateral dimensions up to several microns, i.e., thousands of nanometers. Also for this technique, a barrier improvement factor can be used to indicate the additional effect of the nanoparticles. In contrast to Eqs. (10.11) and (10.12), this barrier improvement factor is independent of the thickness of the sample, because here, the filler particles are dispersed homogeneously inside the polymer, which gives a situation different from an inorganic layer on the surface of a polymer substrate.

At present, materials on the basis of PA 6, the aromatic polyamide MXD-6 and PET are to be found on the market. It must be stated that permeability data given on

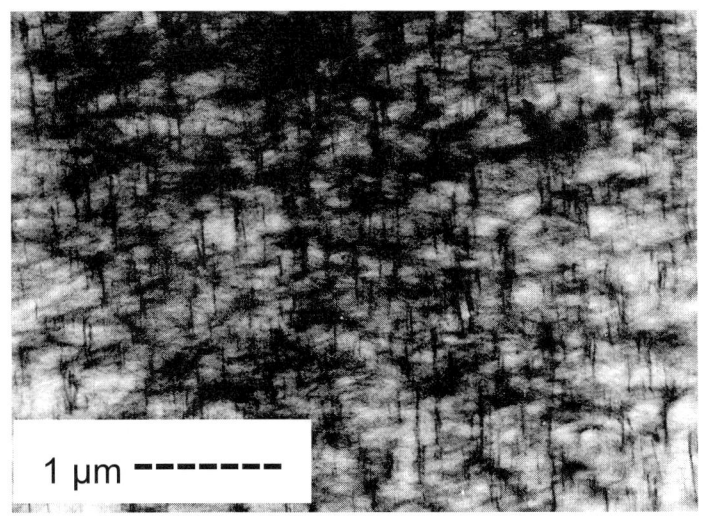

Figure 10.19 Cross section of a PA 6 film with nanoscale layer silicate filler particles after extrusion, imaged via transmission electron microscopy (reproduction with kind permission of Lanxess).

supplier data sheets guarantee much lower improvement factors (usually factor 2) than the experimental values shown in the table below.

An example of a commercial product on the basis of PA 6 is shown in Figure 10.19. Apparently, the requirements for a high aspect ratio and a good parallel alignment of the platelets have been met in this case.

A selection of experimental results for different types of polymers, obtained in research scale, is given in Table 10.5. Unfortunately, filling grades by volume – as needed for a calculation of the expected permeability according to Eq. (10.17) – are not often available. Therefore, as an approximation, filling grades given by weight have been set to be equal to the values by volume, for the following reasons: Usually, the layer silicates undergo a surface modification prior to their incorporation into the polymer. In addition, their content in the polymeric matrix is determined after the experiment by incineration and weighing of the residual ashes. So, in spite of the fact that the density of the untreated minerals is usually more than twice that of the polymers, the treated ones will have an effective density similar to the polymer matrix due to the high organic content of the treated and exfoliated layers.

Experimental and estimated theoretical figures in Table 10.5 show a generally good agreement, given the large uncertainties in the determination of the filling grade and especially the aspect ratios. As a general trend for experimental results and the industrial practice, practical improvement figures are often lower than theoretically expected. As to be seen from Eq. (10.17), the crucial issue is to isolate (exfoliate) the single layers from the minerals and thus to achieve a high aspect ratio. Unwanted agglomerates of the layers do not only impair barrier properties by reduced aspect ratios and probably poorer alignment, but also by their known tendency to

Table 10.5 Experimental results for polymers filled with nanoparticles on the basis of exfoliated, surface-modified layer silicate minerals. Reductions of the permeability relative to the unfilled polymer are given as barrier improvement factor (BIF). Where available, aspect ratios and filling grades are indicated, the latter in percent by volume (v/v) or by weight (w/w).

System/authors	BIF for O_2	BIF for water vapor	Filling grade[a] (%) Aspect ratio α	Theoretical value for BIF (Eq. (10.17))
PA 6/synthetic mica (Yasue 2000)	6	3.5	4, w/w	
PET/Na-montmorillonite (Frisk 1999)	6	–	1, v/v, $\alpha = 500$	3,5
	11	–	1, v/v, $\alpha = 1000$	6.1
	18	–	3, v/v, $\alpha = 500$	8.8
	>45	–	$\alpha = 1500$	40.5
EVOH/modified kaolinite (Lagaron, 2005)	4.6	1.3	5, $\alpha = 80$	3.2
PLA/modified kaolinite (Lagaron, 2005)	1.9	1	4, $\alpha = 80$	2.7
PLA/synthetic fluorine mica, organically modified (Sinha 2003)	5.7	–	10 w/w, $\alpha = 275$	16
PI/modified montmorillonite (Yano 2000)	2.5	2.2	2 w/w, $\alpha = 200$	3.1
	13.9	7.2	8 w/w, $\alpha = 200$	9.8

[a] Filling grades in % w/w are often determined from the ash content of the polymer after incineration and thus should be higher in reality.

incorporate water vapor between the single layers. This may explain why improvements for the water vapor barrier are once again lower than for the oxygen barrier.

10.9
Experimental Findings: Permeation of Flavors Through Mono- and Multilayer Films and Combinations with Inorganic Barrier Layers

With the following experimental results, the options to improve the barrier properties of films for flavor substances via different coating methods will be demonstrated. All have been obtained on a common substrate film, namely biaxially oriented polypropylene, BOPP. In all cases, this substrate material was a three-layer system, with a central layer of the polypropylene homopolymer and 1-µm-thick outer layers of a propylene–ethylene copolymer, to improve heat sealing properties. Due to the chemical similarity of those two polymers, they may be regarded as identical.

A specific "cocktail" from flavor substances was applied as described in (Franz 1995). This cocktail consists of the following substances, in the sequence of their molecular weight: cis-3-hexenol, isoamyl acetate, D-limonene, menthol, citronellol, diphenyl oxide, and linalyl acetate. To ensure that all substances in the measurement cell have the same vapor pressure, they are not used in their pure form, but are applied diluted in adapted concentrations in polyethylene glycol (PEG 400). This procedure gives a common vapor pressure of all substances of 0.1 Pa (10^{-6} bar).

As a first example, measurement curves obtained on a 16-μm-thick BOPP film are shown in Figure 10.20. Obviously, the steady-state equilibrium is reached for all substances within about one week, which demonstrates that BOPP is not a good barrier for flavor substances. From these figures, permeation coefficients P, diffusion coefficients D and solubility coefficients S can be derived from the slope of the curves in their linear part, giving P with help of the vapor pressure of the substances and the thickness of the film, via Eq. (10.4). Extrapolation of the linear part of the curves toward zero permeated amount gives the time lag and the diffusion coefficient, via Eq. (7.60). The solubility coefficient finally is derived from P and D via Eq. (10.3).

Related values are given in Table 10.6, for two different commercial BOPP samples as obtained from different suppliers. Apparently, sample 2 shows generally lower values for the coefficients of diffusion and permeation, probably due to a higher degree of orientation which will lead to a denser molecular structure. What is also obvious is the high solubility for all flavor substances in the two BOPP samples.

One frequently applied method for improvement of the limited flavor barrier properties of BOPP is to coat it on both sides with a 1-μm-thick layer of an acrylate lacquer. Related measurements have been published in (Franz, 1995), obtained on another BOPP sample. Overall, the permeation rates for different substances could be reduced by factor 3.8 (for citronellol) up to factor 35 (for diphenyl oxide). As an example, limonene was taken to calculate the complete figures for P, D, and S, both for the substrate film as well as for the polymer of the acrylate coating, see Table 10.7.

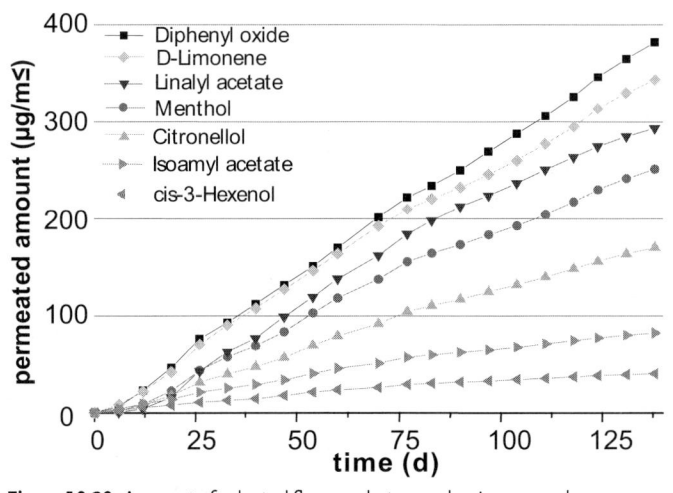

Figure 10.20 Amount of selected flavor substances having passed a 16-μm-thick BOPP film, in dependence on time. Source: Fraunhofer IVV, see also (Moosheimer, et al. 2000).

Table 10.6 Diffusion coefficients, D, solubility coefficients, S, and permeability coefficients, P, measured for two different commercial BOPP samples (source: Fraunhofer IVV).

Substance	Molecular weight	D (cm^2 s^{-1}) Sample 1	D (cm^2 s^{-1}) Sample 2	S (g cm^{-3} bar^{-1}) Sample 1	S (g cm^{-3} bar^{-1}) Sample 2	P (g cm^{-1} s^{-1} bar^{-1}) Sample 1	P (g cm^{-1} s^{-1} bar^{-1}) Sample 2
Cis-3-Hexenol	100	2.9×10^{-12}	1.2×10^{-12}	16.9	9.8	4.9×10^{-11}	1.2×10^{-11}
Isoamyl acetate	130	2.9×10^{-12}	1×10^{-12}	17.5	18.7	5.1×10^{-11}	1.9×10^{-11}
D-Limonene	136	2.2×10^{-12}	1.1×10^{-12}	88.6	59.6	2×10^{-10}	6.6×10^{-11}
Menthol	156	1.8×10^{-12}	3.4×10^{-13}	46.2	104	8.3×10^{-11}	3.6×10^{-11}
Citronellol	156	1.8×10^{-12}	6.8×10^{-13}	203	185	3.7×10^{-10}	1.3×10^{-10}
Diphenyl oxide	170	2×10^{-12}	8.5×10^{-13}	540	653	1.1×10^{-9}	5.6×10^{-10}
Linalyl acetate	196	1.4×10^{-12}	3.4×10^{-13}	27.9	148	3.9×10^{-11}	5×10^{-11}

In this representation, the difference between the permeation characteristics between the acrylate coating and the PET can be seen, in relation to the exemplary calculations shown in Section 6.2.2. In spite of the fact that both materials have a similar permeation coefficient, the acrylate has a diffusion coefficient that is lower by a factor of about 30 in relation to PET, whereas the solubility of limonene in PET is more than a factor of 10 lower than in the acrylate. Thus, according to Eq. (10.8), the acrylate coating will lead to a much longer time lag compared to a possible additional PET layer on BOPP whereas the PET will lead to a lower absorption of the flavor in the polymer, a phenomenon known by practitioners as "flavor scalping."

The effect of applying an inorganic barrier coating on top of the BOPP film is shown in Figure 10.21. Here, layers of aluminum and silicon oxide have been deposited on the same type of substrate film in industrial scale (Moosheimer, 2000). Obviously, the barrier improvements – which range from virtually no improvement for diphenyl oxide up to an improvement by factor 5 for cis-3-hexenol – are much lower than those usually observed for oxygen and even for water vapor. Although the mechanism behind this deviation is still open to discussion, it is very probable that the origin is the same as has been discussed in Section 6.3 for the higher permeation rate of water vapor through inorganic barrier layers, namely effects due to condensation of the substances within porous structures or along interfaces.

Table 10.7 Coefficients for diffusion, solubility, and permeability, for the permeation of limonene through BOPP, acrylate coating, and PET as reference (Franz 1995).

Material	D (cm^2 s^{-1})	S (g cm^{-3} bar^{-1})	P (g cm^{-1} s^{-1} bar^{-1})
BOPP film	2.1×10^{-12}	42	1.56×10^{-10}
Acrylate coating	2×10^{-15}	14	3×10^{-14}
PET (reference)	6×10^{-14}	0.13	1×10^{-14}

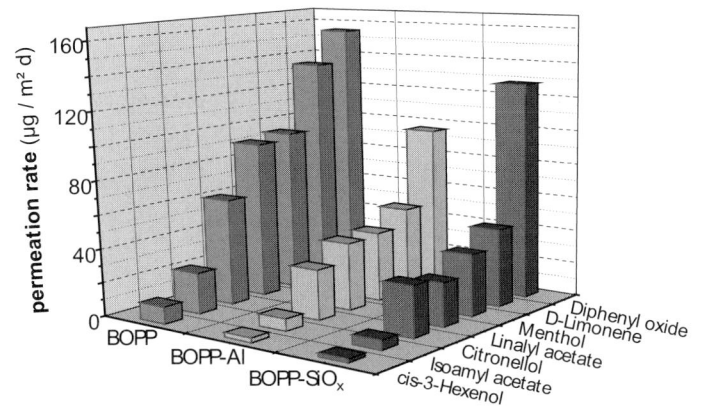

Figure 10.21 Permeation rates measured for the noncoated 16-μm-thick BOPP film (see Figure 10.20) without coating and after coating with two different inorganic barrier layers from aluminum (Al) and silicon oxide (SiO$_x$). Source: Fraunhofer IVV, see also (Moosheimer et al. 2000).

In Figure 10.22, the effects of a further lamination process are demonstrated: Both types of inorganic coated BOPP films were laminated to the same type of noncoated BOPP substrate film via a two-component solventless polyurethane adhesive, on the side of the inorganic coating. For comparison, a laminate from two identical bare BOPP films was made with the help of the same adhesive.

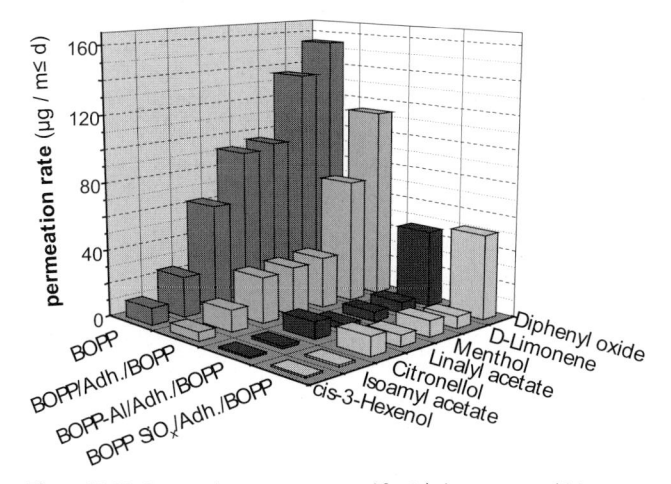

Figure 10.22 Permeation rates measured for (a) the noncoated 16-μm-thick BOPP film (see Figs. 10.20 and 10.21) without coating, (b) after lamination of two BOPP films together with a two-component polyurethane adhesive ("Adh."), (as a reference for iii and iv) and after laminating a BOPP film onto the BOPP substrate films after coating with Al (c) and SiO$_x$ (d), onto the inorganic coating. Source: Fraunhofer IVV, see also (Moosheimer et al. 2000).

The latter results in a barrier improvement slightly above a factor of 2 in relation to the initial BOPP film, a trivial result because the film thickness has been doubled by this procedure. For the laminates that incorporate an inorganic barrier layer, improvements are observed in relation to a bare single BOPP film which range from about factor 3 (for diphenyl oxide) up to factor 23 (for Isoamyl acetate). So, the improvement effect of the combined processes of vacuum coating and lamination does overall not supersede that of the acrylate coating. In addition, one should remember that the thickness of the laminated films is nearly two times that of the acrylate coated films.

10.10
Conclusions

There are different methods to improve the barrier properties of polymer materials which make use of different permeation mechanisms: Coating or lamination with high barrier polymers, coating with inorganic barrier layers, filling of the basic polymers with platelet-shaped particles and aligning them in the conversion process.

For gases such as oxygen, the highest efficiency in barrier improvement can be obtained via combinations of polymer substrates, inorganic layers, and additional polymeric layers, thus embedding the inorganic barrier. For this kind of structures and permeants, the permeabilities can be described using a defect model. In order to calculate the permeability of more complex layer systems from alternating inorganic and polymeric layers, simple calculation methods can be derived from the results of numerical simulations.

Thin polymer layers embedded between two inorganic layers are the most effective structural elements for manufacturing flexible materials having extremely high barrier properties for future applications.

Improvement effects obtained by filling polymers with platelet-shaped particles, preferably nanoparticles, and by aligning these particles in the course of the extrusion process still lead to lower improvement effects. Probably, there is still large room for technical improvement.

An additional transport mechanism is evident for the permeation of water vapor, possibly also flavor substances through layer structures. Most probably this involves capillary condensation. In order to manufacture layer systems having very high barrier properties, the porosity of materials must hence be restricted to an absolute minimum. Moreover, there is a strong need for a scientifically sound clarification of the exact transport mechanism because only that will allow for a targeted improvement strategy

References

Affinito, J.D., Eufinger, S., Gross, M.E., Graff G.L. and Martin, P.M. (1997) *Thin Solid Films*, **308**, 19–25.

Affinito, J.D., Graff, G.L., Shi, M.-K., Gross, M.E., Mounier, P.A. and Hall M.G. (1999) *42nd Annual Technical Conference*

Proceedings, Society of Vacuum Coaters, Albuquerque, pp. 102–107.

Affinito, J.D., Gross, M.E., Coronado, C.A., Graff, G.L., Greenwell E.N. and Martin, P.M. (1996) *39th Annual Technical Conference Proceedings, Society of Vacuum Coaters, Albuquerque*, pp. 392–397.

Affinito, J. and Hilliard, D. (2004) *47th Annual Technical Conference Proceedings, Society of Vacuum Coaters, Albuquerque*, pp. 563–593.

Amberg-Schwab, S., Hoffmann, M., Bader, H. and Gessler, M. (1998) *J. Sol-Gel Sci. Tech.*, 141–146.

Anderle, F., Henrich, J., Dicken, W., Grünwald, H., Liehr, M., Nauenburg, K. and Beckmann, R. (1997) DE Patent 19722205.

Barrer, R.M. (1941) *Diffusion in and Through Solids*, University Press, Cambridge, UK.

Barrie, J.A., Levine, J.D., Michaels, A.S. and Wong, P. (1963) *Faraday Soc.*, **59**, 869–878.

Benmalek, M. and Dunlop H.M. (1995) *Surface and Coatings Technology*, **76–77**, 821–826.

Berger, F. (2006) *Pack Report*, **3**, 41–43.

Beu, T.A. and Mercea, P. (1990) *Mat. Chem. Phys.*, **26**, 309–322.

Bichler, C., Bischoff, M., Langowski, H.-C. and Moosheimer, U. (1996) *Verpackungs-Rundschau, Interpack 96 Special*, **47**, E36–E39.

Bichler, Ch., Mayer, K., Langowski, H.-C. and Moosheimer, U. (1998) *41st Annual Technical Conference Proceedings, Society of Vacuum Coaters, Albuquerque*, pp. 349–354.

Casey, F., Smith A.W. and Ellis, G. (1999) *42nd Annual Technical Conference Proceedings, Society of Vacuum Coaters, Albuquerque*, pp. 415–419.

Chatham, H. (1996) *Surf. Coatings Technol.*, **78**, 1–9.

Comyn, J. (ed.)(1985) *Polymer Permeability*, Chapman & Hall, London.

Crank, J. (1992) *The Mathematics of Diffusion*, 2nd edition, Oxford Science Publications, Oxford, Great Britain.

Creatore, M., Klaasse Bos, C.G. and Hamelinck, A.C.M. (2005) *48th Annual Technical Conference Proceedings, Society of Vacuum Coaters, Albuquerque*, pp. 163–168.

Czeremuszkin, G., Latrèche, M., da Silva Sobrinho, A.S. and Wertheimer, M.R. (1999) *42nd Annual Technical Conference Proceedings, Society of Vacuum Coaters, Albuquerque*, pp. 176–180.

Czeremuszkin, G., Latrèche, M. and Wertheimer, M.R. (2000) *43rd Annual Technical Conference Proceedings, Society of Vacuum Coaters, Albuquerque*, pp. 408–413.

Czeremuszkin, G., Latrèche, M. and Wertheimer, M.R. (2001) *44th Annual Technical Conference Proceedings, Society of Vacuum Coaters, Albuquerque*, pp. 178–183.

Czeremuszkin, G., Latrèche, M. and Wertheimer, R. (2003a) *46th Annual Technical Conference Proceedings, Society of Vacuum Coaters, Albuquerque*, pp. 586–591.

Czeremuszkin, G., Latrèche, M. and Wertheimer, M.R. (2003) PCT Application WO 03/005461 A1.

da Silva Sobrinho, A.S., Latrèche, M., Czeremuszkin, G., Klemberg-Sapieha, J.E. and Wertheimer, M.R. (1998) *J. Vac. Sci. Technol. A*, **16**, 3190–3199.

da Silva Sobrinho, A.S., Czeremuszkin, G., Latrèche, M., Dennler, G. and Wertheimer, M.R. (1999) *Surf. Coatings Technol.*, **116–119**, 1204–1210.

da Silva Sobrinho, A.S., Czeremuszkin, G., Latrèche, M. and Wertheimer, M.R. (2000) *J. Vac. Sci. Technol. A*, **18**, 149–157.

Davis, R. and Peiffer, H. (1998) *41st Annual Technical Conference Proceedings, Society of Vacuum Coaters, Albuquerque*, pp. 505–506.

Davis, R. and Peiffer, H. (1999) *42nd Annual Technical Conference Proceedings, Society of Vacuum Coaters, Albuquerque*, pp. 511–512.

Deak, G.I. and Jackson, S.C. (1993) *36th Annual Technical Conference Proceedings, Society of Vacuum Coaters, Albuquerque*, pp. 318–323.

Decker, W. and Henry, B. (2002) *45th Annual Technical Conference Proceedings, Society of Vacuum Coaters, Albuquerque*, pp. 492–502.

Dennler, G., Wertheimer, M.R., Houdayer, A. and Ségui, Y. (2001) *44th Annual Technical*

Conference Proceedings, Society of Vacuum Coaters, Albuquerque, pp. 476–481.

Erlat, A.G., Henry, B.M., Grovenor, C.R.M. and Tsukahara, Y. (2002) 45th Annual Technical Conference Proceedings, Society of Vacuum Coaters, Albuquerque, pp. 503–508.

Erlat, A.G., Henry, B.M., Ingram, J.J., Mountan, D.B., McGuigan, A., Howson, R.P., Grovenor, C.R.M., Briggs G.A.D. and Tsukahara, Y. (2001) Thin Solid Films, 388, 78–86.

Erlat, A.G., Schaepkens, M., Kim, T.W., Heller, C.M., Yan, M. and McConnelee, P. (2004) 47th Annual Technical Conference Proceedings, Society of Vacuum Coaters, Albuquerque, pp. 654–659.

Fayet, P., Jaccoud, B., Davis, R. and Klein, D. (2005) 48th Annual Technical Conference Proceedings, Society of Vacuum Coaters, Albuquerque, pp. 237–240.

Felts, J.T. (1991) 34th Annual Technical Conference Proceedings, Society of Vacuum Coaters, Albuquerque, pp. 99–104.

Finson, E. and Felts, J. (1994) 37th Annual Technical Conference Proceedings, Society of Vacuum Coaters, Albuquerque, pp. 139–143.

Franz, R. (1995) In: Foods and Packaging Materials – Chemical Interactions (eds C. Ackermann, M. Jägerstadt and T. Ohlsson), The Royal Society of Chemistry, Cambridge, Great Britain, pp. 45–50.

Frisk, P. and Laurent, J. (1999) U.S. Patent 5,876,812.

Graff, G.L., Williford, R.E. and Burrows, P.E. (2004) *J. Appl. Phys.*, **96** (4), 1840–1849.

Gross, M.E., Olsen, L.C., Bennett, W.D., Bonham, C.C., Martin, P.M., Graff, G.L. and Kundu, S. (2006) 49th Annual Technical Conference Proceedings, Society of Vacuum Coaters, Albuquerque, pp. 139–142.

Hanika, M. (2004) Ph.D. Thesis, TU München, Munich, Germany.

Hanika, M., Langowski, H.-C. and Moosheimer, U. (2002a) 45th Annual Technical Conference Proceedings, Society of Vacuum Coaters, Albuquerque, pp. 523–528.

Hanika, M., Langowski, H.-C., Moosheimer, U. and Peukert, W. (2002) *CIT*, **74** (7), 984.

Hanika, M., Langowski, H.-C., Moosheimer, U. and Peukert, W. (2003) *Chem. Eng. Technol.*, **26** (5), 605–614.

Hanika, M., Langowski, H.-C. and Peukert, W. (2003) 46th Annual Technical Conference Proceedings, Society of Vacuum Coaters, Albuquerque, pp. 592–599.

Hartwig E., Meinel, J., Krug, T. and Steiniger, G. (1992) 35th Annual Technical Conference Proceedings, Society of Vacuum Coaters, Albuquerque, pp. 121–127.

Henry, B.M., Assender, H.E., Erlat, A.G., Grovenor, C.R.M., Briggs, G.A.D., Miyamoto, T. and Tsukahara, Y. (2004) 47th Annual Technical Conference Proceedings, Society of Vacuum Coaters, Albuquerque, pp. 609–614.

Henry, B.M., Erlat, A.G., Grovenor, C.R.M., Briggs, G.A.D., Miyamoto, T. and Tsukahara, Y. (2002) 45th Annual Technical Conference Proceedings, Society of Vacuum Coaters, Albuquerque, pp. 514–518.

Henry, B.M., Erlat, A.G., Grovenor, C.R.M., Briggs, G.A.D., Miyamoto, T. and Tsukahara, Y. (2003) 46th Annual Technical Conference Proceedings, Society of Vacuum Coaters, Albuquerque, pp. 600–605.

Henry, B.M., Erlat, A.G., Grovenor, C.R.M., Briggs, G.A.D., Tsukahara, Y., Miyamoto, T., Noguchi, N. and Niijima, T. (2000) 43rd Annual Technical Conference Proceedings, Society of Vacuum Coaters, Albuquerque, pp. 373–378.

Henry, B.M., Erlat, A.G., Grovenor, C.R.M., Deng, C.-S., Briggs, G.A.D., Miyamoto, T., Noguchi, N., Niijima, T. and Tsukahara, Y. (2001) 44th Annual Technical Conference Proceedings, Society of Vacuum Coaters, Albuquerque, pp. 469–475.

Henry, B.M., Howells, D., Topping, J.A., Assender, H.E. and Grovenor, C.R.M. (2006) 49th Annual Technical Conference Proceedings, Society of Vacuum Coaters, Albuquerque, pp. 654–657.

Henry, B.M., Roberts, A.P., Grovenor, C.R.M., Sutton, A.P., Briggs, G.A.D., Tsukahara, Y., Yanaka, M., Miyamoto, T. and Chater, R.J.

(1998) *41st Annual Technical Conference Proceedings, Society of Vacuum Coaters*, Albuquerque, pp. 434–439.

Henry, B.M., Topping, J.A., Assender, H.E., Grovenor, C.R.M. and Marras, L. (2005) *48th Annual Technical Conference Proceedings, Society of Vacuum Coaters*, Albuquerque, pp. 644–648.

Hoffmann, G., Ludwig, R., Meinel, J. and Steiniger, G. (1994) *37th Annual Technical Conference Proceedings, Society of Vacuum Coaters*, Albuquerque, pp. 155–160.

Jamieson, E.H.H. and Windle, A.H. (1983) *J. Mat. Sci.*, **18**, 64–80.

Kelly, R.S.A. (1993) *36th Annual Technical Conference Proceedings, Society of Vacuum Coaters*, Albuquerque, pp. 312–316.

Kelly, R.S.A. (1994) *37th Annual Technical Conference Proceedings, Society of Vacuum Coaters*, Albuquerque, pp. 144–148.

Kessler, K. (1994) Ph.D. thesis, ETH Zürich, Zürich, Switzerland.

Krug, T.G. (1990) *33rd Annual Technical Conference Proceedings, Society of Vacuum Coaters*, Albuquerque, p.163.

Krug, T., Ludwig, R. and Steiniger, G. (1993) 36th Annual Technical Conference Proceedings, Society of Vacuum Coaters, Albuquerque, pp. 302–305.

Lagaron, J.M., Cabedo, L., Cava, D., Feijoo, J.L., Gavara, R. and Gimenez, E. (2005) *Food Additives Contam.*, **22** (10), 994–998.

Langowski, H.-C. (1999) in Wertschöpfung bei der Folienextrusion, Tagung Bad Homburg, 28. und 29. September 1999. VDI - Gesellschaft Kunststofftechnik, VDI-Verlag: Düsseldorf, Germany, pp. 49–77.

Langowski, H.-C. and Utz, H. (1992) *ZFL*, **43** (9), 520–526.

Langowski, H.-C. (2003a) *46th Annual Technical Conference Proceedings, Society of Vacuum Coaters*, Albuquerque, pp. 559–565.

Langowski, H.-C. (2003b) *Galvanotechnik*, **94** (11), 2800–2807.

Langowski, H.-C. (2005) *Vakuum in Forschung und Praxis*, **17** (1), 6–13.

Langowski, H.-C., Melzer, A. and Schubert, D. (2002) *45th Annual Technical Conference Proceedings, Society of Vacuum Coaters*, Albuquerque, pp. 471–475.

Lechner, M.D (ed.),(1992) *D'Ans-Lax, Taschenbuch für Chemiker und Physiker*, 4th edition Springer, Berlin, Germany, Vol. 1, p. 627 and p. 699.

Lohwasser, W., Frei, O., Severus H. and Wisard, A. (1995) *38th Annual Technical Conference Proceedings, Society of Vacuum Coaters*, Albuquerque, pp. 40–46.

Lüpke, E. February(2005) Proceedings OTTI-Profiforum Kunststoff-Verpackungen, Freising, pp. 133–153.

Miesbauer, O. and Langowski, H.-C. (2007) to be published in Vakuum in Forschung und Praxis.

Misiano, C., Simonetti, E., Cerolini, P., Staffetti, S., Taglioni, G., Pasqui, A. and Fusi, F:D.O.B. (1993) *36th Annual Technical Conference Proceedings, Society of Vacuum Coaters*, Albuquerque, pp. 307–311.

Moosheimer, U. and Langowski, H.-C. (1999) *42nd Annual Technical Conference Proceedings, Society of Vacuum Coaters*, Albuquerque, pp. 408–415.

Moosheimer, U., Langowski, H.-C. and Melzer, A. (2000) *43rd Annual Technical Conference Proceedings, Society of Vacuum Coaters*, Albuquerque, pp. 385–390.

Moosheimer, U., Melzer, A. and Langowski, H.-C. (2001) *44th Annual Technical Conference Proceedings, Society of Vacuum Coaters*, Albuquerque, pp. 458–463.

Müller, K. and Weisser, H. (2002) *Packag. Technol. & Sci.*, **15**, 29–36.

Nelson, R.J. and Chatham, H. (1991) *34th Annual Technical Conference Proceedings, Society of Vacuum Coaters*, Albuquerque, pp. 113–117.

Nielsen, L.E. (1967) *J. Macromol. Sci. Chem.*, **A1** (5), 929–942.

Nörenberg, H., Deng, C., Kosmella, H.-J., Henry, B.M., Miyamota, T., Tsukahara, Y., Smith, G.D.W. and Briggs, G.A.D. (2000) *43rd Annual Technical Conference Proceedings, Society of Vacuum Coaters*, Albuquerque, pp. 347–351.

Ohya, T., Suzuki, S., Yoshimoto, M., Iseki, K., Yokoyama S. and Ishihara, H. (2000) *43rd*

Annual Technical Conference Proceedings, Society of Vacuum Coaters, Albuquerque, pp. 368–372.

Pauly, S. (1999) Permeability and diffusion data, In: *Polymer Handbook* (eds J. Brandrup, E.H Immergut and E.A. Grulke), 4th edition Wiley, New York, p. VI/543.

Perkins W.G. and Begeal, D.R. (1971) *J. Chem. Phys.*, **54**, 1683–1694.

Phillips, R.W., Markantes, T. and LeGallee, C. (1993) *36th Annual Technical Conference Proceedings, Society of Vacuum Coaters, Albuquerque*, pp. 293–301.

Piringer, O.G. (2000) in : *plastic packaging Materials for Food-Barrier Function, Mass Transport, Quality Assurance and Legislation* (eds. O.G. Piringer and A.L. Baner), Wiley-VCH Weinheim.

Prins, W. and Hermans, J. (1958) *J. Phys. Chem.*, **63** (5), 716–719.

Roberts, A.P., Henry, B.M., Sutton, A.P., Grovenor, C.R.M., Briggs, G.A.D., Miyamoto, T., Kano, M., Tsukahara, Y. and Yanaka, M. (2002) *J. Membrane Sci.*, **208**, 75–88.

Rossi, G. and Nulman, M. (1993) *J. Appl. Phys.*, **74** (9), 5471–5475.

Schaepkens, M., Kim, T.W., Erlat, A.G., Yan, M., Flanagan, K.W., Heller, C.M. and McConnellee, P.A. (2004) *J. Vac. Sci. Technol.*, **A22**, 1716–1722.

Schiller, S., Neumann, M., Morgner, H. and Schiller, N. (1994) *37th Annual Technical Conference Proceedings, Society of Vacuum Coaters, Albuquerque*, pp. 203–210.

Schiller, N., Straach, S., Fahland, M. and Charton, C. (2001) *44th Annual Technical Conference Proceedings, Society of Vacuum Coaters, Albuquerque*, pp. 184–188.

Schubert, H. (1982) *Kapillarität in Porösen Feststoffsystemen*, Springer, Berlin, Germany.

Seserko, P., Löbig, G., Ludwig, R., Kukla, R., Hoffmann, G. and Steiniger, G. (1998) *41st Annual Technical Conference Proceedings, Society of Vacuum Coaters, Albuquerque*.

Shaw, D.G. and Langlois, M.C. (1993) *36th Annual Technical Conference Proceedings, Society of Vacuum Coaters, Albuquerque*, pp. 348–351.

Shaw, D.G. and Langlois, M.C. (1994) *37th Annual Technical Conference Proceedings, Society of Vacuum Coaters, Albuquerque*, pp. 240–244.

Sinha Ray, S., Yamada, K., Okamoto, M., Ogami, A. and Ueda, K. (2003) *Chem. Mater.*, **15**, 1456–1465.

Sinha Ray S. and Bousmina, M. (2007) in: *Polymeric Nanostructures and Their Applications. Vol. 2 Applications* (ed. H.S. Nalwa), American Scientific Publishers, Los Angeles, CA, pp. 2–97.

Smith, A.W. and Copeland, N. (2006) *49th Annual Technical Conference Proceedings, Society of Vacuum Coaters, Albuquerque*, pp. 636–641.

Smith, A.W., Copeland, N., Gerrerd, D. and Nicholas, D. (2002) *45th Annual Technical Conference Proceedings, Society of Vacuum Coaters, Albuquerque*, pp. 525–529.

Tropsha, Y.G. and Harvey, N.G. (1997) *J. Phys. Chem. B.*, **101**, 2259–2264.

Utz, H. (1992) *Verpackungs-Rundschau*, **43** (3), 17–24.

Utz, H. (1993) *Verpackungs-Rundschau*, **44** , (8) 51–58.

Utz, H. (1995) Ph.D. Thesis TU München, Munich, Germany.

Walter, M., Heming, M., Spallek, M. and Zschaschler, G. (1994) DE Patent 44 38 359.

Weikart, C. and Smith, T. (2003) *46th Annual Technical Conference Proceedings, Society of Vacuum Coaters, Albuquerque*, pp. 486–490.

Wood, L. and Chatham, H. (1992) *35th Annual Technical Conference Proceedings, Society of Vacuum Coaters, Albuquerque*, pp. 59–62.

Yamada, Y., Iseki, K., Okuyama, T., Ohya T. and Morihara Y. (1995) *38th Annual Technical Conference Proceedings, Society of Vacuum Coaters, Albuquerque*, pp. 28–31.

Yano, K., Usuki, A., Okada, A., Karauchi, T. and Kamigaito, O. (1993) *J. Polym. Sci. A, Polymer Chemistry*, **31**, 2493–2498.

Yasue, K., Katahira, S., Yoshikawa, M. and Fujimoto, K. (2000) *In: Polymer-Clay Nanocomposites; Wiley Series in Polymer Science* (eds T.J. Pinnavaia and GW. Beall), Wiley, New York, pp. 111–125.

Yializis, A., Mikhael, M.G. and Ellwanger, R.E. (2000) *43rd Annual Technical Conference Proceedings, Society of Vacuum Coaters, Albuquerque,* pp. 404–407.

11
Migration of Plastic Constituents

Roland Franz and Angela Störmer

Packaging plastics contain substances of low or medium molecular weight such as residual monomers and other starting substances as well as additives such as antioxidants or lubricants. These substances may migrate into the packed good after filling and form undesirable levels of concentrations in the product until consumption. Therefore, the purpose of migration testing is to ensure product safety and protect the consumer, by controlling the levels of constituents from the packaging in the product. Applicants and users of migration tests in the first instance are from industrial laboratories, such as from polymer or plastic additives producers. But manufacturers of packaging materials such as bottle and film producers as well as converters also apply or commission migration testing to demonstrate compliance with food law, for quality assurance purposes, or for submission of new product petitions. Finally and, in principal with highest responsibility, the end users of product packaging, the food or pharmaceutical industry themselves, need migration tests. In the last five years, increasing awareness of this issue has grown and spread in the food industry, this also as a consequence of crises in context with the detection of unwanted compounds in foods and originating from packaging materials. As another group of laboratories involved in migration testing, independent certifying laboratories as well as public or governmental control and surveillance laboratories must be mentioned. The importance of realistic migration data was recognized for estimation of exposure of the consumer concerning packaging constituents for which concentration data of migrants in foods are one major key parameter needed besides food consumption and packaging usage data (Foodmigrosure, 2006; Franz, 2005; ILSI Europe, 2007).

11.1
Definitions and Theory

11.1.1
Migration, Extraction, and Adsorption

The term "migration" is used to describe the process of mass transfer from a packaging material into its contents, in particular when these are liquid or semiliquid

(Chapters 1 and 7). During the migration process a packaging material, in general, withstands the product contact conditions and does not alter its mechanical or diffusion properties too significantly.

The term "extraction" is often used in the same sense as "migration" which may lead to confusion in the context with solvent extraction of packaging materials. Therefore, the term "extraction" is defined in the following as a very intensive interaction process with a packaging material where a solvent penetrates the packaging material and generates swelling, thus altering its mechanical and diffusion properties significantly and even extremely.

The term "adsorption" is used to describe a particular case of migration where the content of the packaging material is a so-called dry product and where migrants are transferred through the gas phase onto the product and adsorbed on its surface.

11.1.2
Functional Barrier

The term "functional barrier" is used to describe the effect of a particular multilayer packaging structure where the migrant from a layer behind which is not in contact with the product (food), is hindered or delayed in it mass transfer process to enter the food. Functional barriers may act through their advantageous diffusion properties (Chapters 7 and 9), thus causing time delay effects in migration or, alternatively, may reduce migrants concentration in food just based on their favorable solubility properties for the migrant and thus limiting the maximum possible concentration of the migrant in the product (Chapter 4) by partitioning effects. It needs to distinguish between absolute (so-called zero migration cases) and transmissive functional barriers.

11.1.3
Legal Migration Limits and Exposure

The European food packaging regulatory systems applies so-called specific migration limits (SMLs) as well as an overall or global migration limit (OML). SML values are laid down in the positive list system of European Directives such as 2002/72/EC and amendments and have been derived from toxicological data in connection with some conservative conventional assumptions. The migration limit, expressed in milligrams per kilogram food (simulant), is calculated from the toxicological magnitudes acceptable or tolerable daily intake (in milligram substance per kilogram body weight per day) by multiplication with the factor 60 which (in kilograms) is the conventionally assumed body weight of an average European person. It is further assumed that this average person eats 1 kg food per day which is packed by 6 dm^2 plastic material and that the substance is present in the food at the SML concentration (Ashby *et al.*, 1997). It should be kept in mind that SMLs established in this way do contain inbuilt safety margins in the range of 100–1000 relative to so-called No adverse effect levels (Katan, 1996). SML values range typically from 10 to 20 µg/kg (ppb) level for specific substances of toxicological concern such as vinyl chloride (VC), butadiene, acrylonitrile, and others up to the lower milligram per kilogram (ppm) level for less toxic principles such as ethylenediamine (12 ppm), 1-octene (15 ppm) or monoethylene glycols (MEG)/diethylene glycols (DEG) (30 ppm). The OML limit has conventionally been set at 60 mg of the sum of all migrating substances per kilogram of food or at 10 mg/dm^2 food contact area of the package with the intention to impose on food contact materials (FCMs) and articles the requirement that they should be as much as possible inert and not release any unnecessary amounts of their constituents into food. In principle, the overall migration (OM) is understood to be the sum of all specific migrations (SMs) but due to volatility of potential migrants and the conventional character of the European OM tests it is strongly recommendable in general not to establish correlations between OM and SMs, although in selected cases the link may be justified. So, opposite to SM, OM has no toxicological dimension. Therefore, it is astonishing how much time, costs, and efforts both the legislator and the industry have dedicated to the development of standardized, conventional OM test procedures.

Since toxicological considerations for SML values are linked to the question of exposure: "how much of a substance would be consumed with the food by the consumer," measured SM values in food simulants should either be linked with those in foods using correlation factors or at best should simulate the situation with foods very closely. To estimate then from food SM values exposure scenarios, more information such as representative food consumption behavior of the consumer as well as information on the frequency of use of certain FCMs for the particular food of interest are required. Establishing exposure estimation models has been a challenge for numerous European projects and activities. And in fact, probabilistic exposure modeling on the basis of food consumption and packaging usage data has made enormous progress as can be recognized from numerous, very recent scientific publication (Duffy et al., 2006a, 2006b, 2007; Gibney and van der Voet, 2003; Gilsenan et al., 2003; Oldring, 2006; Holmes et al., 2005; Castle 2003; Vitrac and Leblanc, 2007). In a streamlining activity, very recently a guidance document for exposure estimation has been composed and cross-checked for applicability by experts in this area (ILSI Europe, 2007). One of the driving forces in Europe to endeavor exposure-oriented migration evaluation was, besides recognition of the weaknesses of the European system, clearly the US FDA's migration-to-exposure calculation principle (US FDA, 2002, 2006) which is a deterministic exposure model and has the advantage of being conveniently applicable but which, on the other hand, has also a significant degree of convention and appears not to be applicable as it is for Europe (Chapter 12).

The European harmonization process to regulate FCMs and articles is largely finalized on plastics but has so far not covered other, very challenging areas such as paper and board, coatings, adhesives, printing inks. Due to the fact that in these areas thousands of substances can be used, either intentionally on purpose or as non-intentionally added substances (NIAS) such as technical impurities or chemical reaction products formed during manufacture and converting where only a small fraction may be toxicologically evaluated, further regulatory means than a positive list system with innumerable SMLs are needed. A possible option which – in analogy to US FDA's threshold-of-regulation policy (US FDA, 1995) – has currently been adopted in Europe in context with the functional barrier concept is a so-called threshold-of-no-concern concentration for migrating substances with no explicit regulatory restriction and under the assumption that these substances would not be reproduction toxic (CMR) carcinogenic, mutagenic, substances according to European Commission (2007). The benchmark value for such a threshold is established at the 10 µg/kg (ppb) limit which is a generally recognized as analytical detection limit and has been set as SML for highly toxic migrants such as VC or butadiene. Finally, a further benefit would be to establish also a so-called zero-migration policy in Europe in cases where, because of the particular food packaging application, a migrant would have no possibility at all to be transferred into the food.

11.1.4
Parameters Determining Migration

Besides the geometric dimensions of a given product/packaging system which influence migration, mass transfer from a plastic into product consists essentially of kinetic (diffusion of the migrant both in the polymer and in the product) and thermodynamic (solubility of migrant in the product and equilibrium partitioning between packaging and product) factors. Once a molecule has moved to and arrived at the interface between plastic and product, the extent of mass transfer into the product depends on its viscosity and, very crucially, on its properties to dissolve a migrant. This solubilization property, for example of a food, is largely a function of the food's fat content; however, other food parameters such as the emulsifying character or terpene oil content have been recognized as further relevant ones. Since most plastic constituents do have lipophilic rather than hydrophilic character, migration increases generally strongly in fatty foods and especially so in the case where the fat or oil represents the outer phase of the food matrix. With polar plastics such as polyamides, a reversed situation exists for the particular polar monomeric or oligomeric migrants where aqueous foods increase the extent of migration.

Migration as a result of diffusion in plastics and transfer from there into a product is linked predominantly to the volatility or molecular weight of the migrating (organic) substances and to the basic diffusivity of a polymer type (Chapters 6 and 15). For a given plastic, this means that the mobility of a migrant decreases with increasing molecular weight. For a given molecule, this means that its mobility in different polymers is dependent on the elasticity or thermoplastic properties of the polymer. In those cases where the migration process is still in its very initial phase and therefore fully diffusion controlled, the effect of the partition coefficient K is nearly negligible with the practical consequence that the extent of migration in this situation is largely independent of the product or a simulant. In those cases where the migration process has reached an equilibrium, extent of migration is controlled by the partition coefficient K of a substance between the two phases polymer and product. K is defined as the concentration ratio at equilibrium of a migrant in the polymer, $C_{P,e}$, divided by the one in the product or a liquid simulant, $C_{L,e}$, at equilibrium (Chapter 4). The K value depends mainly on the polarity of the substance and the polarities of these two phases. Applying the rule "like dissolves like," when considering a plastic-packaging material for instance for contact with an aqueous product, one would as a consequence select a nonpolar polymer (e.g. polyolefin) that will not be attacked by water. In such a system strongly polar substances prefer the product phase and nonpolar substances prefer the packaging material. However, when filling a fatty product into the same hydrophobic polymer type, a completely different partitioning behavior from polymer into foodstuff takes place. K values range over several orders of magnitude depending on the polarities of polymer involved, the food simulant, and the nature of the migrant.

Concerning the migrating molecule itself, as a rule, it should always be noted that a migrant cannot exceed its solubility. This is, in particular, important for aqueous simulants where many lipophilic substances are fully migration controlled by partitioning effects between the simulant and the food contact layer. Based on toxicological considerations, the relevant molecular weight range for migrants is considered to end at a 1000 Da. Larger molecules are considered from a biophysical point of view not to be physiologically absorbed in the gastrointestinal tract.

In migration testing mass transport in the opposite direction may occur when using organic solvents as product simulants. This process can be described as sorption into the polymer and can lead to partial or even complete swelling of the polymer structure with consequences for the practical migration test.

For a better understanding of the influence of a simulant on migration the following relationship is discussed: The average migration distance made by the diffusing additive i in the polymer during time t should be designated x_i and its diffusion coefficient be D_i. Analogously, x_j should be considered to be the distance which the migration front of the simulant has moved in the polymer during the same time t and D_j be the diffusion coefficient of the simulant in the polymer. From the theory of diffusion (Chapter 7), the following ratio for x_i/x_j can be derived:

$$\frac{x_i}{x_j} = \left(\frac{D_i}{D_j}\right)^{1/2} \qquad (11.1)$$

Now, in the case of $x_i \ll x_j$, the polymer additive i will be "overrun" by the simulant before it can migrate out of the polymer. Depending on the solubility of the simulant in the polymer (K_j value), the migration behavior of i will then be influenced in an accelerated way. In the case, however, that $x_i \gg x_j$, the migration of i from the plastic is not influenced by the migration of the simulant.

From the above discussion the following conclusions can be drawn with respect to the selection of a product simulant:

1. When significant interactions occur between fat or oil and a plastic (Piringer 1990), two borderline cases must be distinguished:
   Case (1a): When $D_i \gg D_j$, then according to equation 11.1 the previously mentioned case occurs where $x_i \gg x_j$ and the migration from the packaging is not affected by the uptake of fat by the plastic. The migration of substances whose

molecular weights are much less than that of triglycerides remains unaffected by the uptake of fat because, as a rule, the diffusion coefficient decreases with increasing molecular weight. The molecular weights of the triglycerides in question range roughly from 600 to 1000 Da. Substances with molecular weights up to and including this range can be considered to meet this assumption.

Case (1b): When $D_i \ll D_j$, then according to equation 11.1 migration in the polymer is controlled by the fat penetration meaning the sorbed fat leads to higher migration values (Vom Bruck *et al.* 1986).

2. For volatile food simulants with low molecular weights, for example heptane, isooctane, ethanol, the relationship $D_i \ll D_j$ is valid for practically all polymer constituents. However, here again, two borderline cases must be differentiated as well:

    Case (2a): If the partition coefficient of the simulant $K_j = 1$ (i.e., the simulant is well dissolved by the polymer), the amount of sorbed simulant *j* is so large that it causes severe swelling of the polymer. As a consequence, total extraction of the migrant *i* from the polymer takes place provided that the solubility of migrant in the simulant allows that. This is, for example, the case where isooctane comes into contact with nonpolar polyolefins.

    Case (2b): For $K_j \ll 1$, the sorption of *j* is so low that the D_i value and consequently the migration of migrant *i* is not significantly influenced. This is, for example, the case where polar ethanol comes in contact with nonpolar polyolefins at normal test temperatures (up to 40 °C). At higher temperatures the solubility of ethanol (as well as that of fat) increases in polyolefins. However, by diluting the ethanol with water, the undesired increase in migration due to higher temperatures can be avoided and a situation with a diffusion control in the polymer according to case (2b) can be achieved again (Piringer, 1993).

If one wishes to use a volatile simulant as a lower molecular weight alternative to the natural edible oils or synthetic triglycerides, then one must determine according to the above discussion whether the simulant belongs to case (2a) or (2b). In the case of (2a), a time factor (in an accelerated sense) must be considered because of the extraction effect caused by swelling. This means a series of studies must be carried out to determine at which contact time does the same mass transfer take place compared to oil or actual food. Moderate temperature increase can also be used at the same time as a further variable. In the case of (2b), the time factor does not need to be considered as long as the test temperature is the same or very similar for the simulant as it is for the oil.

11.2
Indirect Migration Assessment

Migration assessment can be achieved principally on two different routes. Besides the direct determination via a migration experiment as described in Section 11.3, indirect access to migration values is the method of choice in many cases and should always be considered at first. Also, indirect assessment can be achieved by numerous pathways on the basis of what is discussed in the following. As a general principle and which must be understood from a legislation point of view, indirect migration assessment allows only to demonstrate compliance of an FCM but not noncompliance. In cases of doubt, compliance can finally be demonstrated by direct migration measurement when the experiment has been carried out appropriately.

The question by which approach migration should be controlled best cannot be answered by recommending only one and the same test principle for all cases. Nowadays a number of options, including mathematical modeling of diffusion processes, are available and all of them can be justified for application under certain circumstances. It is recommendable and largely already common practice in migration evaluations to start with a total mass transfer scenario as the worst-case assumption. Initial considerations to any compliance testing should always start with the question: "Can a given migration limit be exceeded or even reached at all?" For this the initial concentration of a migrant in the packaging must be known or should be reasonably judged. When this simple calculation leads to exceeding a migration limit, then it is advisable and necessary to achieve further refinements to the total mass transfer approach. This can be achieved by consideration of the migrant's solubility and taking into account partition coefficient K effects as well as the role of the diffusion coefficient D as important material-specific parameters for migration. Today, for these latter considerations, migration modeling tools have been developed in the recent years and for which convenient softwares from the market or freewares from the internet can be applied (Chapter 9).

In any case it is wise not to start in all cases of compliance testing immediately with a SM test but to consider the material-related, geometric, and time–temperature application parameters of the individual food/packaging system to be evaluated. More specifically, one should consider the migration potential of a given migrant in a plastic before initiating a long-term and time-consuming migration test. The second consideration should then address the question: "What is the most economic way to get a realistic migration value or to demonstrate compliance of an FCM?"

11.2.1
Worst-Case (Total Mass Transfer) Assumption

The most simplistic and direct approach is to assume total mass transfer into a product or simulant and compare this worst-case value with a given SML restriction. This requires either exact knowledge of the initial concentration of a migrant in a polymer ($C_{P,0}$), or the compositional analysis in the plastic. Vice versa, a given SML restriction can be transformed into a corresponding maximum concentration in the plastic, again under the assumption of total mass transfer, and compared to the experimentally determined $C_{P,0}$ (Chapters 7 and 15).

Conventionally, maximum $C_{P,0}$/SML ratios (ppm in plastic/ppm in food) of 100 have been assumed many years ago and even implemented in the legislation – as an example the European VC restrictions of QM = 1 mg/kg and SML = not detectable at

0.01 mg/kg may be cited. However, under total mass transfer assumptions, the thickness of the respective polymer layer must be taken into account. It is obvious that for very thin film thicknesses a measured $C_{P,0}$ value may exceed a given QM value without posing any health risk to a consumer. In other words, the $C_{P,0}$/SML ratio is a function of the film thickness and for very thin films may end up at ratio orders of magnitude higher than the conventional one of 100. In Figure 11.1, this relationship is depicted for VC monomer: the graph shows a correlation between thickness of polyvinyl chloride (PVC) material and the VC concentration in PVC which corresponds to the SML (0.01 mg/kg) when total mass transfer would occur. From the graph it can be seen that at a thickness of 120 µm, the QM restriction coincides with the SML restriction. For smaller thickness the corresponding concentration increases, for instance, up to factors 6 or 24 (times QM) for thicknesses of 20 or 5 µm. On the other hand, for higher thicknesses, the QM restriction would be too relaxed when assuming total mass transfer. The rigid PVC inherent low diffusivity, however, ensures that also for thick PVC materials VC migration will hardly ever reach or exceed the SML value. Today in a time of using migration modeling, it gets very evident that the VC QM restriction is too restrictive and, as a consequence, generates

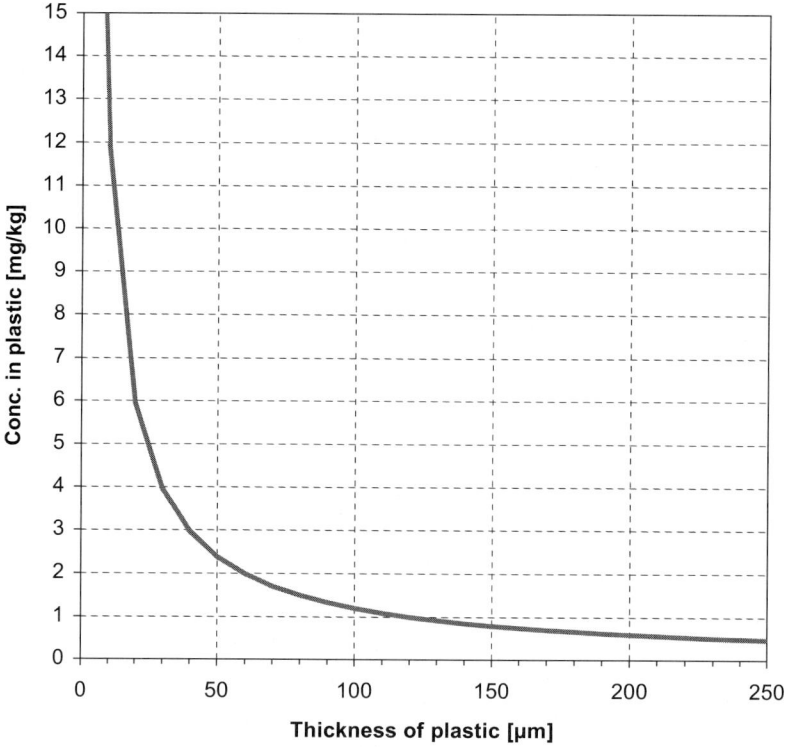

Figure 11.1 Correlation between thickness-dependent concentration of vinyl chloride in PVC material (density = 1.4 g/cm³) and SML = 0.01 mg/kg food under total mass transfer assumption at a surface-to-volume ratio of 6 dm²/kg food.

a juridical conflict with the VC SML restriction. Which of both restrictions is legally binding, if the QM is clearly exceeded, but the SML would never be reached?

With the exception of this particular case where both a QM and a SML exist, in all other cases under the premise of total mass transfer where a indirect migration assessment demonstrates the impossibility of exceeding a given legal SML restriction criterion, full compliance testing has been achieved and no further migration assessment or testing is necessary.

11.2.2
General Considerations: Taking Solubility and/or Low Diffusivity of Certain Plastics into Account

In those cases where the above assessment leads to an uncertain conformity status, other considerations can be taken into account.

One important consideration should always be directed to the solubility of a migrant. For example, *n*-octadecyl-3-(4-hydroxy-3,5-di-*tert*-butylphenyl)propionate (Irganox 1076) is a very widespread antioxidant used for the production of plastics and has an SML restriction of 6 mg/kg food. The solubility of this additive in water is in the ppt (ng/L) range and can be expected to be also extremely low (ppb to ppm) in highly aqueous systems such as the official EU simulants 3% acetic acid (B) and 10% ethanol (C) and even at higher ethanol contents up to 50% (Chapter 4). With these figures any attempt to measure migration into food simulants A (water), B, and C appears to be not only unreasonable but also a waste of resources. Considering then the log $P_{O/W}$ value ($=13$) as a rough measure for the expected order of magnitude for the partition coefficient $K_{P/F}$ between a lipophilic polymer such as a polyolefin and aqueous simulants of this additive, any attempt to determine migration into aqueous food simulants, including 50% ethanol, can at maximum be a challenge for the analyst in achieving the necessary analytical sensitivity but can never end in a noncompliance result. Indeed, a $\log(K_{P/L})$ value of 15 was calculated for LDPE/water system by using the so-called vapor pressure index system method (Chapter 4).

As another practical example, bis(2-ethylhexyl)adipate (DEHA), a common plasticizer with an SML restriction of 18 mg/kg food can be mentioned. Its solubility in water is around 1 mg/L and log $P_{O/W} = 6.1$. Assuming for instance a concentration of DEHA in the plastic (practical example: plasticized PVC) of 15% (150,000 ppm) and setting the log $P_{O/W}$ equal to the partition coefficient $K_{P/L}$, then at equilibrium a maximum concentration of 0.15 ppm could be established in water and for highly aqueous systems such as 10% ethanol or 3% acetic acid. Again, any attempt to measure migration in these simulants is totally superfluous (see also Chapter 15).

Another important consideration as to whether a migration limit can be exceeded or not should be directed to the basic diffusivity of a given plastic. Polyethylene terephthalate (PET), an important food contact plastic, can be mentioned as a relevant case example.

It is well known that PET has a very favorable kinetic migration in the sense that it exhibits a very low migration velocity of its constituents. In addition, for manufacturing of PET bottles usually very low amounts of additives are added. Therefore, due to the high inertness of PET and the low content of migratable substances, consequently a very low migration can be expected which is in agreement with the observation made during routine analysis of PET FCMs that always very little migration was found. To substantiate this thesis better and for validation purposes, a systematical migration study program on PET bottles was carried out by Störmer et al. (2004). In this study, sampling was carried out such that it was representative for the German PET bottle market. Measured parameters were OM as well as specifically the PET monomers monoethylene glycol MEG, diethylene glycol DEG, and terephthalic acid as well as antimony. The test conditions applied include aqueous, alcoholic, as well as fatty simulants. In addition, possibilities to accelerate and simplify migration testing of PET bottles were explored.

Concerning OM results it was found that all investigated PET bottles had an OM value (contact conditions 10 days at 40 °C) below 0.5 mg/dm^2 in all three food simulants: 3% acetic acid, 10% ethanol, and the alternative fat simulant 95% ethanol. This is not only far below the limit of 10 mg/dm^2 as laid down in the EU Plastics Directive 2002/72/EC (European Commission, 2002) but also demonstrates that the measured values are smaller than the analytical uncertainty of the method as indicated in the EN 1186 standards (CEN, 2002). For regulatory evaluation the uncertainty which is based on ring trials carried out within CEN Working Group 1 of TC194/SC1 is set to be 2 mg/dm^2 according to Annex I of the EU Plastics Directive (European Commission, 2002). Even more severe extraction tests with 95% ethanol under conditions of 24 h at 50 °C and 2 h at 60 °C gave comparably low OM results.

Concerning SM results it was found that in most of the investigated PET bottles, the SM of the PET monomers MEG and DEG, respectively, was below the analytical detection limit of ∼0.5 mg/dm^2 for MEG and DEG. In some migration solutions, signals were found close to this detection limit. Terephthalic acid was in all cases below the detection limit of ∼0.1 mg/dm^2. Consequently, for all PET bottles the SM values were extremely far below the SMLs of 30 mg/kg food (simulant) or 5 mg/dm^2 contact area for the sum of the glycols and 7.5 mg/kg or 1.25 mg/dm^2 for terephthalic acid. Interestingly, it was checked whether the OM test could be used to control these relatively high SMLs by spiking of migration solutions with MEG and DEG at a level of 30 mg/kg and determining then gravimetrically the recovery of glycols using the OM determination procedure according to EN 1186. The recovery was found to be between 1% and 4% which can be explained by the water vapor volatility of the glycols and demonstrates the nonsuitability of the OM test approach for SM control in this particular case. In addition, SM of antimony was determined on seven randomly chosen PET bottles using aliquots of the migration solutions of 3% acetic acid and 95% ethanol. In these tests, measured antimony concentrations ranged between 0.16 and 0.57 µg/dm^2 in 3% acetic acid and were below the detection limit of 0.3 µg/dm^2 in 95% ethanol. Compared to the EU legal specific antimony migration limit of 40 µg/kg food (simulant) or 6.6 µg/dm^2 contact area, again a high safety margin was observed.

From the results and on the basis of migration theory it was concluded that against the food regulatory requirements as set by the OML and the SMLs for terephthalic acid and MEG/DEG, PET bottles obviously can never offend simply due to the PET inherent physical diffusion properties. Therefore, conventional migration testing turns out to be superfluous in this case. Only significant amounts of substances when located on the surface (which was not the case in any of the investigated bottles in this study) could influence the OM. Such an eventual situation, however, could be covered by a very rapid test, for example, 2 h at 60 °C contact with 95% ethanol or using even a shorter surface solvent washing or rinsing test.

11.2.3
Migration Assessment of Mono- and Multilayers by Application of Complex Mathematical Models

Mathematical models to describe migration processes FCMs are presented and discussed in Chapters 7–9. Over the last 15 years such models have been developed starting from monolayer plastics being in contact with liquid food simulants where the migration rate determining step was allocated to the plastic. Kinetic and thermodynamic effects in the food simulant or the food had been ignored or conventionally controlled by placing appropriate physicochemical parameters. It is important to note that EU legislation has incorporated migration modeling in 2002 to check compliance in case of certain plastics when in contact with food simulants (Directive 2002/72/EG, Article 8, European Commission, 2002), which was based on the outcome of an important European project (Piringer and Hinrichs, 2001; Begley *et al.*, 2005). Since this time migration modeling has further evolved in two dimensions: first, multilayer models have been established (Piringer *et al.*, 1998; Brandsch *et al.*, 2002; Reynier *et al.*, 2002) since most food contact articles consist of more than one layer and second, it was found that when attempting to model migration into foodstuffs, then partitioning effects between the food contact layer and the food need more thorough consideration which was the reason that research has focused on this issue in the last 5 years. One important research project dealing with this issue was the "Foodmigrosure" EU project, which is presented and discussed in more detail in Section 11.7.

11.2.4
Multilayers

Whereas EU legislation has made enormous progress and can be considered to be nearly completed for monolayer materials, there is, on the other hand, not yet a satisfactory regulatory approach available for multilayer plastics. Here, in principle, there are two options: (1) each plastic layer must comply with current positive lists in EU legislation or (2) only the food contact layer must comply and act as a barrier against substances from the layers behind. Typically, plastics multilayers exhibit a migration behavior which is different from monolayer plastics and is characterized by a delayed migration profile. As a consequence, the question arises whether currently applied migration test conditions are still applicable for multi layers. From experience, research data (AiF project, 2004; Brandsch *et al.*, 2005), and migration

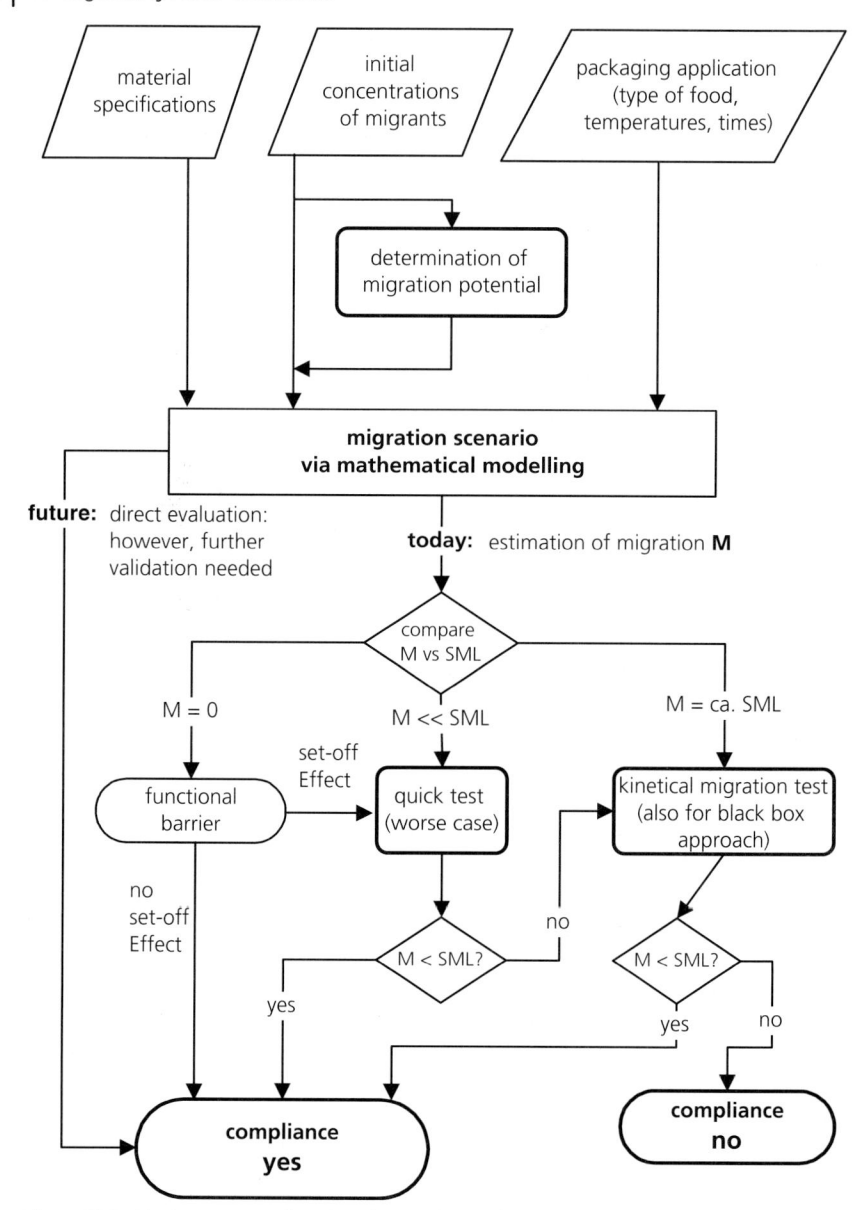

Figure 11.2 Modern concept for migration testing and assessment of multilayer packaging structures.

theory, it appears that migration test conditions should be adapted to the requirements coming from multilayer migration characteristics. In Figure 11.2, a recommended decision tree flow scheme for indirect or direct migration testing of multilayer plastics including migration assessment steps is given.

11.3
Migration Experiment

Until today, the control of transfer from plastics-packaging materials into foods has mainly been based on the measurement of the substance(s) in the food or simulant after certain specified, and in most cases standardized, contact conditions. Each migration test can be subdivided in two phases: (1) the preanalytical migration exposure phase (migration experiment) and (2) the pure analytical phase, where the OM value or the specific migrant must be determined in the respective food or simulant as precisely and reproducibly as possible. The various possibilities to perform the migration exposure phase are described and discussed in this chapter, the analytical phase in Section 11.4.

Furthermore, the migration exposure phase can be distinguished in two principal types:

- *conventional direct migration measurements* where a sample is placed in contact with a food or simulant in a manner representing the contact conditions of actual conditions in use.

- *alternative semidirect migration test approaches* where a sample is kept in contact with an appropriate simulant in such a manner that a strong interaction between simulant and plastic takes place ("more severe test conditions") and, although shorter contact times are then applicable, at least equal or exaggerated extents of migration are obtained.

11.3.1
Direct Migration Measurement in Conventional and Alternative Simulants

The principle of direct migration measurement is to measure either directly in foodstuffs or, more commonly, using agreed and authorized food simulants as laid down in EU Directive 97/48/EC (see Section 11.3.3.1 and Chapter 13). When measuring directly in food, the food regulatory evaluation is restricted to the investigated food. The food simulants shall mimic as closely as possible a given food packaging application, but the measured values may be transferred for the evaluation of different food types. The second advantage of food simulants is the simpler matrix which makes analysis easier. OM determination is only possible in simulants because of the nonspecific nature of the test. When the SMLs are in the lower ppb range, the necessary analytical sensitivity can often only be reached when using simulants. Some substances react with food ingredients and are no more accessible for extractive methods then. Those substances are therefore only measurable in suitable simulants.

When using olive oil or another oil as fat simulant, similar analytical problems occur as above described for the foods because of complex matrix oil. The OM test with oil may cause severe analytical problems or cannot be carried out because of too strong interactions of the oil with the material (e.g., at higher temperatures) or interferences as discussed in detail in Section 11.4.1. When the use of oil as fat simulant is technically not feasible, the substitute fat simulants 95% ethanol, isooctane, and/or modified polyphenylene oxide (MPPO, Tenax®) for high-temperature applications of 100 °C and higher are used according to Table 4 in Directive 97/48/EC. MPPO in general and 95% ethanol in case of nonpolar polymers as polyolefins show no or only little interaction with the polymer. Therefore, these tests can also be considered as direct migration tests. The isooctane acts as accelerated extractor and is discussed in Section 11.3.2.

The advantage of direct migration test is that the results can be directly and *definitely* compared with legally prescribed migration limits, thus allowing immediately a statement of conformity or disapproval of the test sample. The disadvantage of direct measurement has recognized more and more during recent years: analysis of migrants in complex food simulants, such as oils and fats, is often very time consuming and costly and at the same time relatively poor in terms of analytical sensitivity and precision. The second disadvantage of direct migration tests is the long contact time of 10 days for materials and articles intended for long-term storage.

The correlation of migration testing in simulants and migration into real foods is discussed in Section 11.7.

11.3.2
Accelerated Migration Tests: Alternative Migration Tests

The principle of accelerated migration tests is to apply more severe test conditions by using volatile solvents with strong interactions toward the plastic, to enhance the migration rate from the plastic. Thus, the extraction test is based on an accelerated mass transport mechanism where the diffusion coefficients of migrants are increased by several orders of magnitude compared to the original migration test. As a rule, extraction tests are designed such that they make use of the following principle:

$$\text{Polar polymer} + \text{polar migrant} + \text{polar solvent}$$
$$= \text{worse case} =$$
$$\text{nonpolar polymer} + \text{nonpolar migrant} + \text{nonpolar solvent}$$

Following this principle, semidirect and generally quick extraction tests have been established with the aim of determining the migration potential for assessment of the worse case migration. These tests do not need to be exhaustive as an extraction for determination of the content in the material ($C_{P,0}$) would be but at least as severe as or more severe than the conventional migration test using olive oil. This type of test can be considered to be semidirect because it produces an extraction value which can be directly compared to a legal restriction. But at the same time this value is an exaggerated one and does not always correspond to the real (lower) migration value. To illustrate the principle of accelerated migration tests, the migration or extraction kinetics of the monomer laurolactam out of a Nylon 12 film (polylaurolactam) at 40 °C into isooctane and into 95% ethanol is shown in Figure 11.3. The polar polymer Nylon 12 is strongly swollen by 95%

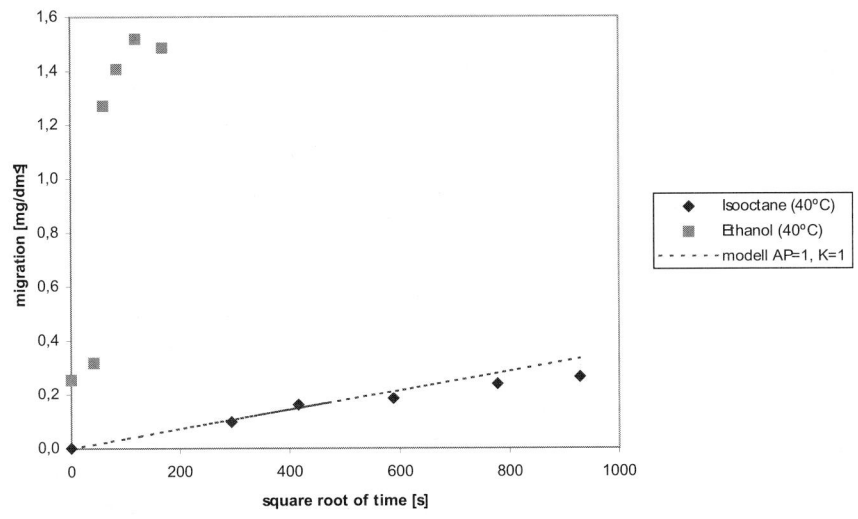

Figure 11.3 Migration of laurolactam from PA12 film into isooctane and 95% ethanol at 40°C.

ethanol and rapidly extracted whereas the nonpolar isooctane does not interact with the polymer and laurolactam shows a slow migration.

The legal basis for applying such alternative tests instead of migration test in oil is given in EU Directive 97/48/EC Chapter IV. Using such alternative tests, conformity of a material with the migration limits can be checked. Exceeding the limit cannot be proven using a worse case test but only using a conventional direct migration test.

An example for such a rapid extraction test as alternative to OM into fatty food simulants is published as CEN standard EN 1186-15 (CEN 2002; Berghammer *et al.* 1994). The method uses nonpolar isooctane and/or polar ethanol as extraction solvents depending on the polarity of the FCM. According to results obtained by this method on a large number of samples and additionally taking physicochemical considerations into account, the obtained extraction efficiency was generally found to be equivalent to or higher than OM results obtained under the following test conditions: 10 days at 40 °C, 2 h at 70 °C, 1 h at 100 °C, 30 min at 121 °C, and 30 min at 130 °C, as specified in Council Directive 82/711/EEC and its amendment 97/48/EC (European Commission, 1997). Ensuring as complete as possible an extraction of the potential migrants requires a strong interaction of the extraction solvent with the sample, for example, by swelling. For this purpose, isooctane is used for plastic materials and articles containing nonpolar food contact layers, such as polyolefins. For test samples with polar food contact plastics such as polyamide or PET, 95% ethanol (v/v in water) is used. In case of polystyrenes, plasticized PVC and other polymers especially with medium polarity where the identification of the suitable extraction medium is not obvious, two parallel extraction tests are conducted using both the proposed extraction solvents. The higher value obtained is taken as the relevant result. In the case of asymmetrical structures such as plastic laminates and coextruded plastics, the nature of the food contact layer determines the selection of

Table 11.1 Use of extraction solvents and test conditions in relation to polymer types.

Polymer type of the food contact layer	Extraction solvent	Extraction conditions
Polyolefines	iso-octane	24 hours at 40 °C
Polyamides	95% ethanol	24 hours at 40 °C
Polystyrene	iso-octane and 95% ethanol	24 hours at 40 °C
Polyethylene terephthalate	95% ethanol	24 hours at 50 °C
Polyvinyl chloride (plasticised)	iso-octane and 95% ethanol	24 hours at 40 °C
Polyvinyl chloride (rigid)	95% ethanol	24 hours at 50 °C
In case of doubt or unknown	iso-octane and 95% ethanol	24 hours at 50 °C

the extraction solvent(s). Table 11.1 gives an overview of the allocation of extraction solvents and test conditions to polymer types.

The rapid extraction test was primarily developed for flexible packagings less than 300 μm in thickness. However, if this extraction test is applied to materials with higher thickness and the result does not exceed the OML, the material can be considered to be in compliance. If the test result exceeds the allowed OML, regardless of the film thickness of the test material, then this worse case test may be repeated by a test which represents more realistic the intended usage conditions. Instead of total immersion, the single-sided mode (using a test cell) may be applied or an alternative direct test (see Section 11.3.1) or the conventional fat test which is in the end the decisive test.

In any case the rapid extraction test was designed to demonstrate compliance in the case of extraction values lower than the OML. Because of its nature as worse case condition, the test cannot disapprove a material whose extraction value exceeds the limit.

Conditions differing from those described above are of course possible as much quicker tests. But in all cases, a reliable relationship between the short test and the full migration test must be established. In addition, it is also of practical and economic interest to design these tests so that they can be applied as broadly as possible, that is, in most laboratories without too high an investment. Another quick extraction test has been proposed especially tailored for rigid PVC material (Tice and Cooper 1997). Treating the samples with methanol for 2 h under reflux conditions provided values which were considerably higher than those achieved under conventional olive oil conditions but still remained far below the OML, thus demonstrating fully legal conformity of the test materials.

The substitute fat test using isooctane, for example polyolefines, as defined by Table 4 of EU Directive 97/48/EC is a further example of a semidirect, accelerated migration test. A substitute test is applied in cases where technical or analytical difficulties are connected with the regular fatty food simulants (see also Section 11.3.1). For instance, if a 10 days/40 °C olive oil test on polyolefins is analytically impossible, it can be replaced by a 2 days/20 °C extraction with isooctane. In this case, a suitable time point has been chosen on the kinetic curve of an extraction process where an empirically satisfying agreement has been found between isooctane extractions and fat migration tests into olive oil (De Kruijf and Rijk 1988). Numerous examples of measured OM values have been collected and the interested reader can find a published data compilation summarized for different polymer types (Van Battum 1996). In case of polar polymers which do not or only slightly get swollen in contact with isooctane, the substitute conditions might be too week. Stoffers *et al.* (2003) compared migration of laurolactam from PA12 films at 2 h/100 °C into olive oil with the substitute condition 1.5 h/60 °C into isooctane and found clearly lower migration into isooctane. To obtain comparable values longer contact time would be necessary. In this case, conditions for an accelerated test are underestimating; on the other hand when comparing migration kinetics at the same temperature (40 °C, 60 °C, 80 °C), isooctane showed a somewhat higher diffusion coefficient, that is also in contact with the polar nylon polymer PA12 there is a slight interaction of isooctane with the polymer (Stoffers *et al.*, 2003).

For high-temperature applications again, a semidirect test strategy is applied for the substitute tests using isooctane and also 95% ethanol, empirically based on corresponding comparative test results. For instance, a high-temperature fat test under test conditions of 2 h/150 °C can be replaced by a 3 h/60 °C isooctane extraction. However, it is important to note that nearly all of these comparative long-term/high-temperature migration versus short-term/low-temperature extraction measurements have focused more or less on just the OM and do not include sufficiently SMs (De Kruijf and Rijk, 1994). As a consequence, further research work is necessary to correlate substitute test conditions for SM purposes where the chemical and thermal stability of migrants as well as the possible formation of breakdown products and solubility questions related to individual migrants must be taken into account.

The quick extraction tests in general produce higher migration values than the corresponding conventional test. Therefore, they might be unfavorable when the aim of the test is to correlate such a value with the reputation of a test sample in the competition of the lowest migration value. However, when the testing costs can be decreased in this way by 50–70% and conformity can still be shown, somewhat higher results by the alternative tests seem to be only a question of getting accustomed to extraction values.

11.3.3
Choice of Appropriate Test Conditions

11.3.3.1 Food Simulants
The choice of food simulants depends on the intended use of the material or article. The official food simulants as laid down in Directive 97/48/EC (European Commission 1997) are water (simulant A), 3% acetic acid in water (simulant B), 10% ethanol in water (simulant C), and olive oil (simulant D). Olive oil may be replaced by another triglyceride oil, for example sunflower oil, HB 307 (synthetic triglyceride), Miglyol (colorless synthetic triglyceride), which might help solving matrix interference problems. In the case that the use of oil is technically not feasible, for example, because of interferences, Directive 97/48/EG gives in Chapter III the conditions for

the substitute simulants 95% ethanol, isooctane, and MPPO (trade name Tenax®) for high-temperature applications (>100 °C). A new simulant 50% ethanol is introduced in the first amendment of Directive 85/572/EEC (Directive 2007/19/EC: European Commission 2007) for milk and milk products. Tenax® is discussed to be introduced as official simulant for dry foods at ambient temperature conditions.

In the ideal case the applied simulants shall simulate the interaction with the real food in a similar or slightly exaggerated manner. Water shall represent neutral aqueous foods, 3% acetic acid acidic aqueous foods, 10% ethanol alcoholic foods like wine and beer, and simulant D fatty foods. As real foods show in most cases a combination of these properties, the migration test are carried out using different simulants and the highest obtained value is compared to the legal migration limit. More generally related to the physical properties of the foods, these combinations are described in Chapter I of Directive 97/48/EC, for specific food groups in Directive 85/572/EEC; additional examples are given in Chapter IV of the Note for Guidance (EFSA 2006). These allocations of simulants to foods are in most cases conventions and do not represent the ideal case of similar interactions. This issue is discussed in detail in Section 11.7. When a material or article is intended to be used for all type of food, tests with the simulants B, C, and D are carried out. Water is omitted according to 97/48/EEC Chapter I 2.1 because it will give in any case the same or lower results than 3% acetic acid or 10% ethanol.

With the same argumentation it is possible to test only with the most severe simulant and to omit the others if experience exists which it is. For example, if SM of a lipophilic additive shall be tested, it would be sufficient to use only a fat simulant for compliance testing because migration into an aqueous simulant would be distinctly lower.

Alternative migration (no interaction with the polymer) and extraction (accelerated migration by swelling) simulants save costs and time. The use is possible and recommended if it is known by experience, scientific evidence, or comparative experimentation that the test is at least as severe as the fat test (Directive 97/48/EC Chapter IV: European Commission, 1997).

The use of substitute tests is justified, when the migration test carried out with each of the possible simulants D is found to be inapplicable due to technical reasons connected with the migration test, for example interferences, incomplete extraction of oil, absence of stability of the weight of the plastics, excessive absorption of fatty food simulant, and reaction of components with the fat (EN 1186-1, CEN 2002). When substitute tests shall be performed, all substitute simulants are used and the higher value is taken for compliance evaluation. By derogation from this rule, a test may be omitted if it is generally recognized by scientific evidence that it is not appropriate for that sample. This might be the case when the test is exaggerated strong for that type of material or not enough strong. When migration properties of the material are known, again it is possible to choose only the strongest substitute or that which would be theoretically most similar to the interaction with oil.

For high-temperature applications, we use only Tenax® as simulant. Al Nafouri and Franz (1999) compared migration from PP samples at 121 °C into oil, Tenax®, isooctane, and 95% ethanol. Oil, Tenax®, and isooctane resulted in comparable OM values except for that material with high amount of volatiles which got lost during the gravimetric determination of isooctane solution. The simulant 95% ethanol (4.5 h/60 °C) gave in all cases lower values, which is caused by the less strong interaction with the polymer. SM of volatile migrants was highest onto Tenax®; semivolatiles like the oligomers were found in comparable amounts onto Tenax® and in isooctane whereas nonvolatiles like the additives Irganox 1076 or Irgafos 168 were not found onto Tenax. This might be caused by the totally different mechanism of substance transfer onto Tenax compared to the fluid simulants. While in fluid simulants, the substances on the polymer surface are directly solved in the simulant, using Tenax the substances are transferred in the gas phase and then adsorbed by the simulant surface (Chapter 7). This is the mechanism which also occurs in case of dry foods. Therefore, Tenax also seems to be a suitable simulant for dry foods at ambient temperatures (20 °C, 40 °C), which is already shown for various types of dry foods (see also Section 11.7).

Substituting the 10 days/40 °C test, in case of polyolefins with high thickness, the substitute test with isooctane (2 days/20 °C) might result in exaggerated values whereas 95% ethanol, which does not interact with the polymer, gives more realistic fat migration values at 10 days/40 °C. In case of medium polar polymers like polystyrene, both isooctane and 95% ethanol strongly interact with the polymer and cause higher migration as fat would do. US FDA (2002) recommends 50% ethanol as fat simulant for polystyrene as well as for the polar polymers PET and rigid PVC.

11.3.3.2 Time–Temperature Conditions

Time–temperature conditions are chosen according to the intended use of the final FCM or article. The conditions are classified in categories which are laid down in Table 3 of Directive 97/48/EC (European Commission 1997) (Chapter 13 of this book). When a material or article shall be used for various time–temperature conditions, for example storage at ambient and refrigerated temperature, the more severe is applied for the test and the evaluation is valid for this and all less severe conditions. In case a heating step (e.g., hot fill or sterilization) is applied before long-term storage, the test conditions will be combined (e.g., 30 min/121 °C followed by 10 days/40 °C). In case of hot fill defined as maximum filling temperature of 85 °C which cools down below 70 °C within 15 min, and subsequent long-term storage, the hot fill test (2 h/70 °C or 30 min/100 °C) can be omitted according to Chapter II 4.3 of Directive 97/48/EC because the 10 days/40 °C test is conventionally considered as more severe. For an FDA petition or food contact notification which requires similar test conditions, the combined test is necessary also in this case (condition C "hot fill above 66 °C": 2 h/70 °C + 238 h/40 °C or 30 min/100 °C + 239.5 h/40 °C) (US FDA 2002, Appendix II). According to EU Directive 97/48/EC, Chapter II 2, test conditions of 4 h/100 °C and 2 h/175 °C for aqueous simulants and the fat simulant, respectively, are conventionally considered as the most severe tests which could be applied at all. With other words, at these testing conditions maximum migration is deemed to be reached. The beneficial practical meaning is that independent of the real packaging fill and storage conditions which may even include longer contact times, it is not necessary to test longer and to omit the otherwise lengthy test conditions of 10 days at

40 °C usually applicable for long-term storage. In case the intended high-temperature application (for instance, baking conditions) would exceed the maximum regulated temperature of 175 °C, we, nevertheless, recommend to test at these higher temperatures because beginning degradation of the material may occur and initiate very rapidly increased migration.

In the EU, for compliance testing, it is today normally sufficient to test migration at one time point. For approval of a new substance, it is required by FDA and recommendable also for EU to perform a kinetic test. FDA requires four time points which are defined depending on the intended usage conditions (US FDA 2002, Appendix II). When knowing the kinetic behavior of a migrant in a given packaging application, this better enables a realistic estimate of migration at other contact times than the measured ones and allows better plausibility checking of the results by mathematical modeling.

A situation which is not yet regulated in the EU is how to test multilayer packaging structures for migration of substances originating from the layers behind the food contact layer. This is the more remarkable because most packaging applications consist of more than one layer and this is even more sensitive if one considers printing ink and adhesive form separate layers with a certain migratory potential. To find an approach to this issue considerable work has been done in the last 15 years in the area of migration research on multilayer packaging (Franz et al., 1996, 1997; Piringer et al., 1998; Simal-Gandara et al., 2000a; Feigenbaum et al., 2005; Dole et al., 2006). All these research activities came to the conclusion that multilayer migration testing must consider migration kinetics (either by direct measurements or indirectly through diffusion/migration acceleration factors or modeling) and cannot be based reliably on an one time point measurement according to the traditional or conventional EU migration test principle (Chapter 9). In these studies, diffusion behavior of various packaging materials has been investigated at different temperatures to derive activation energies for mass transport processes in polymers. With known values for the activation energy in a particular case, it is possible to calculate acceleration factors and to design timely shortened tests which are deemed to provide the same migration result as the realistic longer term packaging application. However, for simplicity reasons and for better reproducibility, a less sophisticated but scientifically justifiable approach can be thought of: this approach considers known relationships between diffusion coefficients D_i at different temperatures T_i and migration times t_i and includes the so-called *factor 10 rule*. In the above mentioned scientific research work, it was found that the relevant activation energies for migrants in polymers range typically from 80 to 100 kJ/mol. As a consequence, for the most EU relevant migration test temperatures $T_1 = 20\,°C$ (normally, in fact room temperature that is $\pm 23\,°C$) and $T_2 = 40\,°C$, the corresponding diffusion coefficients D_1 and D_2 (in a given polymer and for a given migrant), respectively, are roughly correlated by a *factor 10*:

$$\frac{D_{1(T=20°C)}}{D_{2(T=40°C)}} = \frac{1}{10} \tag{11.2}$$

From the known proportionality of migration with $(Dt)^{1/2}$ (Chapters 7 and 9) the same migration is obtained when

$$\frac{D_1}{D_2} = \frac{t_2}{t_1} \tag{11.3}$$

From this the following correlations which can be useful for a simple multilayer test concept can be made when assuming infinite thickness (upto 60% of maximum migration).

1. The migration time needed at 20 °C to reach the same value as after 10 days at 40 °C is *10 times* longer.
2. The migration value reached after a given time is at 20 °C *10 times* lower compared to 40 °C.
3. In the linear phase of a migration curve, the migration rate at 40 °C is *10 times* higher than at 20 °C.
4. The time lag observed from a 40 °C migration test is *10 times* shorter compared to 20 °C.

These relationships are also applicable to the other most important EU official migration test temperatures, in particular when comparing 40 °C with 60 °C as well as the room temperature with the 0–5 °C range. It should be noted that this rule has a certain degree of convention but it is fully supported by most recent scientific research work in this area. In addition, from activation energy considerations and because the most packaging multilayers are not infinitely thick, the extrapolations from the 40 °C to lower temperatures contain a safety margin.

11.3.3.3 Surface-to-Volume Ratio

The conventional surface-to-volume ratio in the EU is 6 dm^2/kg, which is derived from a cube with 1 dm edge length (density is calculated as 1 kg/L). In the experiment, usually a somewhat higher surface-to-volume ratio of 1 dm^2/100 ml is used (Annex I 2002/72/EG, EN 1186-1). Fillable articles are usually tested at the real surface-to-volume ratio. In the United States, the conventional surface-to-volume ratio is 10 g/in^2 equal to 6.45 dm^2/kg (US FDA, 2002).

Especially at SM testing when applying a higher surface-to-volume ratio, a concentration step might be avoidable to reach the necessary sensitivity. In case of good solubility of the migrant in the simulant, the surface-to-volume ratio can be increased without influence on the area-related migration value. The same is possible if migration is mainly determined by the diffusion in the polymer (e.g., from PET or rigid PVC). If the partition coefficient is on the side of the polymer, that is if the substance is good soluble in the polymer but poor in the simulant, then the concentration in the simulant remains more or less constant when the surface-to-volume ratio is varied. In this case a simple proportional calculation to other surface-to-volume ratios is not allowed but the partition coefficient and the ratio of simulant volume to polymer volume in the experiment need to be considered. For example, the plasticizer diethylhexyl adipate will be nearly totally extracted by a fat simulant from a plasticized PVC independent from the simulant volume but migration into water needs to be measured at the real surface-to-volume ratio. By using a software for migration modeling as described in Chapter 9, the influence of the surface-to-volume ratio is automatically considered.

Theoretically, these considerations at the two extreme cases can be derived from equation. which is transformation of the equation defining the partitioning coefficient $K_{P/L}$.

$$\frac{m_{L,e}}{V_L} = \frac{m_{P,0}}{V_L + K_{P/L} \cdot V_P} \qquad (11.4)$$

In case the partition equilibrium is on food simulant side, $K_{P/L}$ is small and the product of partitioning coefficient $K_{P/L}$ and polymer volume V_P is negligibly small compared to the simulant volume V_L. Then the absolute migrating mass $m_{L,e}$ is independent from the simulant volume. So concentration in the simulant can be increased by reduction of the simulant volume at constant polymer volume without influencing the area-related migration test result.

In case the partition equilibrium is totally on the polymer side ($K_{P/L} \gg 1000$), the simulant volume V_L is negligibly small compared to the product of partitioning coefficient $K_{P/L}$ and polymer volume V_P. Then at equilibrium the concentration in the simulant is independent from the simulant volume but equals the quotient of the initial concentration in the polymer and $K_{P/L}$.

11.3.3.4 Migration Contact

Migration contact can be carried out by total immersion or single-sided test by using a cell, a pouch, or filling of a container as described in EN 1186 series (OM) (CEN 2002) and EN 13130-1 (SM) (CEN, 2004). In the following text the advantages and disadvantages of these methods are discussed.

The single-sided test is the most realistic way to test packaging materials which are with only one side in contact with the food and especially recommended for asymmetric multilayer constructions. Containers can be filled with the real filling volume in case of small and medium volumes. Sealable films can be sealed to pouches with the contact side as outer side and totally immersed. If this is not possible, pouches with the contact side inside can be filled with simulant. As the presence of a small air bubble normally cannot be prevented which might additionally grow during storage at elevated temperature, then especially the latter pouch method has the disadvantage of a higher uncertainty in the contact area. The best choice is the migration cell. The first cell was developed by Tice (Pira International). This was a relatively large metal cell with 1.25 dm^2 contact area (EN 1186-1 Cell A). We have long-term experience with glass cells, where the film is fixed between the flat flange of a beaker and the lid. The simulant is filled in through an opening in the lid. Such a cell is shown in Figure 11.4 in an improved version by FABES GmbH and glass blower Gassner, Munich. The glass cells are available in various diameters. The diameter is chosen depending on the needed simulant volume and the available sample area, for most cases the cell with 0.4 dm^2 contact area filled with 40-ml simulant is appropriate. Larger contact areas increase the risk of leakage; the same is for samples with buckles or creases.

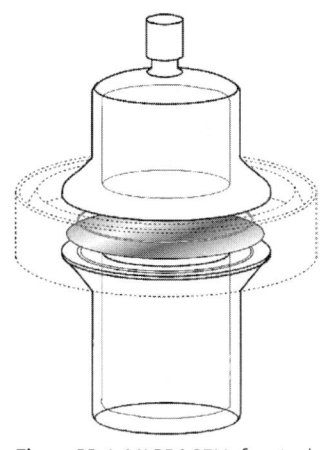

Figure 11.4 MIGRACELL for single-sided migration testing.

Total immersion is the easiest way to testing migration. Inner and outer surface of the material are in contact with the simulant; migration occurs from both sides. Thin materials will approach partitioning equilibrium or will be nearly totally extracted during migration contact. At total immersion the equilibrium will be reached faster compared to a single-sided contact but the total migrating mass is the same. Therefore, for thin materials, only one-sided surface area is used for migration calculation. When migration is mainly determined by the diffusion in the material, doubling the contact surface doubles the migration velocity and therefore the migrating mass. For these samples, the total surface area (inner and outer surface) is used for migration calculation. Conventionally and independent from the polymers and their diffusion properties, the material thickness is used to distinguish these two cases. Migration of samples with a thickness of 500 μm and higher is calculated using the total surface area, samples with a thickness less than 500 μm using the one-sided surface area. These considerations are only valid for monomaterials or symmetric multilayer constructions. Knowing the diffusion behavior of a material, this conventional rule at which minimum thickness migration will be mainly determined by diffusion in the material can be adapted to the material properties. For example, migration from segments of PET bottle walls can be calculated on the basis of scientific evidence using the total surface area even they have a thickness of only 300 μm or less. The edges are usually neglected. But when the surface area of the edges is greater than 10% of the measured area, the edges are included in the calculation (EN 1186-1, CEN, 2002). Examples for such geometry are injection molded test sticks of 120 mm × 10 mm × 2 mm size. Especially cut edges might have a different morphology and migration behavior than the smooth surface of a material and be stronger attacked by the simulant. For example in our experience, lacquer layers, which had been stable against the simulant in a single-sided test, had been partly destroyed at the edges in the total immersion test resulting in a considerably higher migration value.

If the migration value in the total immersion test is higher than the allowed limit, the experiment may be repeated using the more realistic single-sided contact.

11.4
Analysis of Migration Solutions

11.4.1
Overall Migration

Determination of OM is described in EN 1186 series of CEN (European Committee for Standardization) standards (CEN, 2002) and in numerous papers and books (Ashby *et al.*, 1997; Katan 1996b; De Kruijf and Rijk, 1988; Tice, 1997). The method will not be taken up in detail here but only the principle and the possible problems shall be discussed. In general, the gravimetric methods loose volatile migrants totally or at least partially. For covering volatile migrants no methods are developed up to now.

11.4.1.1 Aqueous and Alternative Volatile Simulants
The method is described in EN 1186 parts 3, 5, 7, and 9 for aqueous simulants and parts 14 and 15 for isooctane and 95% ethanol as substitute or alternative fatty food simulants, respectively. The test principle is such that the migration is determined as the mass of nonvolatile residue after evaporation of the simulant. The residue is dried to weight constancy at 105 °C. Therefore, the method covers only such substances which are not volatile at these conditions. The result is expressed in milligrams per square decimeter surface area of the test specimen or in milligrams per kilogram of filling according to Article 2 of the Plastics Directive 2002/72/EG (European Commission, 2002). The measured value is compared to the OML given by the Plastics Directive taking the analytical tolerance of this method into account. On the basis of ring trials at CEN, the analytical tolerance of these methods is set 2 mg/dm^2 or 12 mg/kg according to Annex I of the Plastics Directive 2002/72/EC (European Commission, 2002).

11.4.1.2 Olive Oil
The method is described in parts 2, 4, 6, 8, and 13 (Method A) of EN 1186 (CEN 2002). Because of its nonvolatility a simple gravimetric determination of olive oil is not possible. Therefore, a complicated indirect method is applied. The mass difference of the sample before and after migration contact is determined. As the sample after contact still contains residues of the olive oil which is sticking on the surface or migrated into the polymer, the sample is extracted with pentane and the oil is quantified after derivatizing to fatty acid methyl esters. This value is used for correction of the mass difference. In other words (EN 1186-13A, CEN, 2002): Migration into the olive oil is calculated by subtracting the mass of olive oil retained by the test specimen from the mass of the test specimen after removal from the olive oil, then subtracting this mass from the initial mass of the specimen.

In this procedure inheres a high analytical imprecision up to such analytical interferences which makes the application of the method in many cases not feasible.

- Subtracting high values obtained by weighing the sample in order to determine a much smaller OM value causes inherent imprecision. At too high sample masses differences of some milligrams are too small related to the accuracy of the balance.
- The sample weight of polar plastics depends highly on their water content and therefore on the environmental humidity. Exact moisture conditioning is necessary in these cases to assure comparable conditions at weighing before and after migration contact.
- The sample takes oil up during migration contact. The oil migrates into the plastic, especially in increased amounts at higher temperatures. If back extraction is not complete, the experiment ends up with too low or negative migration values.
- Analytical determination of the absorbed oil amount may be disturbed due to many possible gas chromatography (GC)–flame ionization detection (FID) interferences, for example the use of fatty acid esters as additives. According to the Plastics Directive, some 40 or 50 interfering compounds are in the positive list.
- At high-temperature testing (>100 °C), the increased oil absorption often causes strong problems. Temperature differences of 1 °C may cause a sharp rise of oil uptake which might be a source of irreproducible results. Such nonlinear increase of oil uptake at increasing temperatures had been shown, for example, for polypropylene between 120 and 130 °C (own results, unpublished data). High oil uptake plasticizes the material and changes completely the properties. Such effects often occur only with pure fats and oils but not with partially fatty foods and are not compensated by reduction factors. The obtained migration value might therefore be not realistic for real foods.

Even when the performance of an oil test looks good and these problems do not appear in an distinctly remarkable way, all these points add to a high analytical uncertainty and imprecision. The Plastics Directive 2002/72/EG sets 3 mg/dm^2 or 20 mg/kg food simulant as value for the analytical uncertainty (at an OML of 10 mg/dm^2 or 60 mg/kg which means a 30% analytical uncertainty). The results of proficiency test schemes (e.g., FAPAS) often show a much higher scattering of the laboratories results.

The many working steps needed to come to the result cause a high personal workload and make the test extremely time and cost extensive. At high-temperature tests, handling of the hot oil requires high care.

These extreme disadvantages of the oil test had been the driving force to develop more rapid, easier, more precise, and more cost-efficient alternative tests.

11.4.1.3 Modified Polyphenylene Oxide (Tenax®)

The OM method using MPPO is described in EN 1186-13 Method B for high-temperature applications (>100 °C) (CEN 2002). MPPO is a porous polymer with a high adsorption capacity. It has a high molecular weight (500,000–1,000,000 g/mol), a very high-temperature stability ($T_{max} = 350\,°C$), a high surface area, and a low specific mass (0.23 g/cm^3). The surface of the article to be tested is covered with MPPO (4 g/dm^2) and held at the selected time–temperature test conditions. The heating takes place in a conventional oven even if the samples are intended for use in a microwave oven. After contact, the adsorbent is extracted using diethyl ether. Finally, the extract is evaporated to dryness using a nitrogen stream and the residue remaining is determined gravimetrically. According to the standard EN 1186-13B, the maximum temperature applicable is 175 °C. Higher temperatures can be applied but as oxygen is present, MPPO starts to get oxidized and cannot be fully reconditioned to a white powder.

11.4.2
Specific Migration

Only for a little part of the approved substances with restriction validated or even published methods are available. Sources for such methods are presented in the following sections. Development and validation of new methods are discussed in Section 11.5.

11.4.2.1 Vinyl Chloride EU Directives

As a consequence of EU Directive 78/142/EEC (European Commission 1978) that introduced a limitation of VC monomer both as residual amount in final articles (QM: 1 mg/kg) intended to come into contact with foodstuffs and in migration to food (SML: not detectable; limit of detection (LOD): 0.01 mg/kg), the corresponding necessary analytical methods were developed between several European expert laboratories and laid down as agreed methods in EU Directives 80/766/EEC and 81/432/EEC, respectively (European Commission 1981). This piece of the EU harmonization process was too time- and work-consuming to continue in this way. The VC Directives, therefore, remain a unique feature in EU food packaging legislation which was found to be impractical for generalization.

11.4.2.2 EN 13130 Series

Validation and standardization of analytical methods is a recognized basic task of the European Committee for Standardization (CEN). Within the CEN organization, a working group, CEN TC 194/SC1/WG2, has produced fully validated methods for 15 plastic monomers that have been published as European norms (EN) within the EN 13130 series (CEN, 2004). While the Part 1 of this multipart standard gives general guidance to the SM test methodology prior to analysis of the specific migrant, the remaining seven parts are pure analytical methods for the determination of monomers in food simulants or plastics. Table 11.2 gives an overview of the methods published as standard in the EN 13130 series.

During the period 1993–1996, a European project was conducted within the Standards, Measurements, and Testing program of DG XII (Research). The scope of this BCR (S,M&T) project "Monomers" was to fill the tremendous gap in analytical methods by development and prevalidation of methods of analysis for 36 monomers) selected from the "Plastics Directive" positive lists. The project was carried out by an European consortium of 13 laboratories from 9 different member states, under the coordination of the "Fraunhofer-Institute of Process Engineering and Packaging" (Fraunhofer IVV) Freising, Germany, and the main partner "TNO-Nutrition and

Table 11.2 Overview of standard methods in EN 13130 series (2004).

No.	Title	Restriction (mg/kg)
Part 1	Guide to the test methods for specific migration of substances from plastics into food and food simulants and the determination of substances in plastics and the selection of conditions of exposure to food simulants	–
Part 2	Determination of terephthalic acid in food simulants	SML: 7.5
Part 3	Determination of acrylonitrile in food and food simulants	SML: not detectable, LOD: 0.02
Part 4	Determination of 1,3-butadiene in plastics	QM: 1
Part 5	Determination of vinylidene chloride in food simulants	SML: not detectable, LOD: 0.05
Part 6	Determination of vinylidene chloride in plastics	QM: 5
Part 7	Determination of MEG and DEG in food simulants	SML (T): 30
Part 8	Determination of isocyanates in plastics: • 2,6-toluene diisocyanate • diphenylmethane-4,4'-diisocyanate • 2,4-toluene diisocyanate • hexamethylene diisocyanate • cyclohexyl isocyanate • 1,5-naphthalene diisocyanate • diphenylmethane-2,4'-diisocyanate • 2,4-toluene diisocyanate dimer • phenyl isocyanate	QM (T): 1 (expressed as NCO)

Food Research" Zeist, The Netherlands. From the 36 target monomers (see Table 11.3 and 11.4), the project has elaborated 33 prevalidated methods of analysis for the determination of the SM of a selection of monomers listed with a restriction in at that time actual Plastics Directive 90/128/EEC and its amendment 92/39/EEC (Franz and Rijk 1997). As these methods are up to now not fully validated but only in two or three laboratories, they have not been published as standard methods but in 2005 as technical specifications within the EN 13130 series (CEN/TS 13130-9 to 13130-28, CEN, 2005a).

11.4.2.3 Further Standard Methods

Further standard methods exist for the epoxy derivatives bisphenol A-diglycidyl ether (BADGE) and bishphenol F-diglycidyl ether (BFDGE) and their hydroxy and chlorinated derivatives in food simulants (EN 15136, CEN, 2006a). The method was developed for a sum restriction of BADGE, hydroxy, and chlorinated derivatives of 1 mg/kg. BADGE and derivatives are determined before and after total hydrolysis to BADGE·$2H_2O$. In the meantime, the restriction was changed to 9 mg/kg for BADGE and hydroxy derivatives and 1 mg/kg for the sum of chlorinated derivatives. The method is still applicable for low levels. At higher migration it has to be modified to specific quantification of all derivatives before hydrolysis and, for example, using the hydrolysis only for check on matrix interferences.

Table 11.3 Overview of prevalidated CEN/TS technical specifications in EN 13130 series (CEN 2005a).

No.	Title	Restriction (mg/kg)
Part 9	Determination of vinyl acetate in food simulants	SML: 12
Part 10	Determination of acrylamide in food simulants	SML: not detectable, LOD: 0.01
Part 11	Determination of aminoundecanoic acid in food simulants	SML: 0.01
Part 12	Determination of 1,3-benzenedimethanamine in food simulants	SML: 0.05
Part 13	Determination of 2,2-bis(4-hydroxyphenyl)propane in food simulants	SML: 0.6 (1997: restriction is 3)
Part 14	Determination of 3,3-bis(3-methyl-4-hydroxyphenyl)-2-indolinone in food simulants	SML: 1.8
Part 15	Determination of 1,3-butadiene in food simulants	SML: not detectable, LOD: 0.02
Part 16	Determination of caprolactam and caprolactam, sodium salt in food simulants	SML: 15
Part 17	Determination of carbonyl chloride in plastics	QM: 1
Part 18	Determination of 1,2-dihydroxybenzene, 1,3-dihydroxybenzene, 1,4-dihydroxybenzene, 4,4′-dihydroxybenzophenone, and 4,4′-dihydroxybiphenyl in food simulants	SML: 6, 2.4, 0.6, 6, and 6, respectively
Part 19	Determination of dimethylaminoethanol in food simulants	SML: 18
Part 20	Determination of epichlorohydrin in plastics	QM: 1
Part 21	Determination of ethylenediamine and hexamethylenediamine in food simulants	SML: 12 and 2.4, respectively
Part 22	Determination of ethylene oxide and propylene oxide in plastics	QM: 1 each
Part 23	Determination of formaldehyde and hexamethylenetetramine in food simulants	SML: 15 each
Part 24	Determination of maleic acid and maleic anhydride in food simulants	SML (T): 30
Part 25	Determination of 4-methylpentene in food simulants	SML: 0.02
Part 26	Determination of 1-octene and tetryhydrofuran in food simulants	SML: 15 and 0.6, respectively
Part 28	Determination of 2,4,6-triamino-1,3,5-triazine in food simulants	SML: 30
Part 29	Determination of 1,1,1-trimethylolpropane in food simulants	SML: 6

11.4.2.4 Methods of Analysis in Petitions to the European Commission

As another source of analytical methods for monomers and additives, the numerous technical dossiers submitted to the Scientific Committee of Food or to the European Food Safety Authority (EFSA) through many European companies should be mentioned. According to the Commission's request to the petitioners, these methods

Table 11.4 Further methods in BCR project "Monomers," not published in EN 13130 series.

Title	Restriction (mg/kg)
Determination of ethyleneimine in food simulants	SML: 0.01
Determination of methylacrylonitrile in food simulants	SML: 0.02
Determination of 1,3-phenylenediamine in plastics	QM: 1
Determination of trialkyl(C5–C15)acetic acid, 2,3-epoxypropylester in food simulants	SML: 6

should have been written in a CEN standard format and meet current analytical requirements. Normally, however, these methods were established in assessing front line human exposure under the envisaged contact application, and were not always suitable for general control purposes. Nevertheless, there seems to be a large potential for technically suitable methods to be further evaluated and processed to a generally agreed level of validation on the Europe-wide scale. References to those methods can be found on the website of the Joint Research Centre of the European Commission in Ispra: http://crl-fcm.jrc.it.

11.4.2.5 Methods in Foods (Foodmigrosure Project)

Between March 2003 and September 2006 the Foodmigrosure EU project, contract number QLK-CT2002-2390, was carried out with the intention to develop a migration model tool to allow prediction of mass transfer of constituents from plastic FCMs into foodstuffs (see Section 11.7). In the course of the project analytical determination methods for model migrants in a series of foodstuffs had to be developed as a basis for the systematic migration core studies of the project on which the model should be built on. As model migrants 18 different organic chemical compounds were selected according to the criteria: representativeness with regard to physical and chemical properties such as chemical structure, molecular weight (104–532 Da), polarity or solubility in foods, volatility but also intended use as monomer/other starting substance/additive or unwanted occurrence as contaminant in FCMs. Part of these model migrants were selected since they had been incorporated in plastic films produced as candidate certified reference materials within the European project "Specific Migration" under contract number G6RD-CT-2000-0411 so that extremely useful migrant release systems were already available at the start of the project (Stoffers et al. 2004). Furthermore, 25 representative foods and food categories were selected as matrices into which the mass transport behavior of the selected model migrants was studied. For this, the elaboration of analytical methods was an inevitable requirement. The analytical project work started with a "triangular" approach to select those three foods on each corner of the triangle representing the extremes of different food components (fat, protein, and carbohydrate) which could influence the sampling and work-up procedure prior to the analytical determination. From this, minced turkey or chicken breast (<5% fat), orange juice (containing pulp, not filtered), and Gouda or Emmental cheese (~30% fat content in fresh matter) were selected that were thought to represent major food categories, both with respect to consumption and toward the physicochemical properties influencing the migration process from FCMs. In an EU report (Paseiro et al., 2006a), the collection of analytical determination methods for all 18 model migrants and at least for the 3 "triangle corner foods" are compiled in a harmonized written format. These methods have been successfully applied to all of the 25 foods selected for the project, some of them after minor modifications to meet the specific requirements of a particular foodstuff. In a second EU report, existing literature and strategies for development of methods in food are presented (Paseiro et al., 2006b, see also Section 11.5.2).

11.5
Development of Methods, Validation, and Verification

11.5.1
Establishing (Juristically) Valid Performance of Methods

Ideally and strictly speaking also legally prescribed, the positive list system in Directive 2002/72/EC and its follow-ups would formally only be enforceable on the basis of fully validated analytical methods for SM determinations. Furthermore, EU Regulation No. 882/2004 European Commission 2004b states that analytical procedures for compliance testing with food laws are to be carried out on the basis of validated methods. However, since full collaborative trials according to ISO 5725 (ISO 1994) are very time consuming and expensive and because of the large number of SML values to be validated, it is immediately quite obvious that achieving this ideal situation is an economical impossibility. In addition, the time frame to fulfill such a task would exceed the dimensions of any real-life requirement. Furthermore, many of the positively listed plastic constituents obviously have such a low commercial relevance that the question of absurdity would also be raised in these cases. As a consequence, there is clearly a need for pragmatic solutions to this problem. Since provision of "fully validated" methods turns out to be impossible, certain minimum requirements to method validation should be agreed upon at a European level to produce so-called generally agreed or accepted methods. Possible ways out of the situation are in-house validation procedures carried out by one laboratory, which, however, have to fulfill generally agreed requirements for single laboratory validation. This strategy may be assisted by the definition of minimum requirements for test method precision based on the so-called Horwitz trumpet (Horwitz, 1988, 1995) which links repeatability to concentration. As an economic alternative to ISO 5725 and obeying full validation ring trials, small collaborative trials with two or three laboratories can also be considered.

It is generally recognized and accepted that analytical methods must be suitable for the intended use. Method validation is known as the process used to confirm that a procedure is fit for a particular analytical purpose. This process, an essential part of analytical quality assurance, can be described as the set of tests used to establish and document performance characteristics of a method. The performance characteristics of a method are experimentally derived values for the fundamental parameters of importance in assessing the suitability of the method (Horwitz 1988, 1995; Thompson and Wood 1993, 1995; Eurachem 1998; FAO, 1998; US EPA, 1995; US FDA, 1993; Thompson et al., 2002). These parameters include:

Applicability:	Includes matrix, analyte, and species being measured, concentration ranges and the purpose for which it is suited, limitations of the method.
Selectivity:	The ability to discriminate between the target analyte and other substances in the test sample.
Calibration:	The calibration curve is a graphic representation of the detection system's response as a function of the quantity of analyte.
Accuracy:	The closeness of agreement between a test result and the accepted reference or true value.
Precision:	The closeness of agreement between independent test results obtained under stipulated conditions.
Range:	The interval of concentration within which the analytical procedure demonstrates a suitable level of precision and accuracy.
Limit of quantification:	The lowest amount or concentration of analyte in a sample which can be quantitatively determined with an acceptable level of precision and accuracy.
Limit of detection:	The smallest amount or concentration of analyte in a sample that can be reliably distinguished, with stated significance, from the background or blank level.
Sensitivity:	A measure of the magnitude of the response caused by a certain amount of analyte.
Ruggedness:	The resistance to change of an analytical method when minor deviations are made in the experimental conditions of the procedure.
Practicability:	The ease of operation, in terms of sample throughput and costs, to achieve the required performance criteria and thereby meet the specified purpose.

A fully validation of a method includes a collaborative trial for which internationally accepted protocols have been established (Horwitz, 1988, 1995; ISO, 1994). These protocols require a minimum number of laboratories and test materials to be included. However, only for a very limited number of substances such as fully validated methods exist (e.g., EN 13130 part 2–8) as well as for the OM methods (EN 1186). Methods developed in the European project "Development of Methods of Analysis for Monomers" (SM&T project MAT1-CT92-0006), which are now published as CEN Technical Specifications (CEN/TS 13130 part 9–28, CEN, 2005a), have been only validated by two laboratories up to now. For many other substances, methods are only validated in a limited way within one laboratory.

The background to whether acceptable validation of analytical methods for migration testing could be achieved faster and at less cost than by fully validation is discussed within the CEN report "Validation and interpretation of analytical methods, migration testing and analytical data for materials and articles in contact with food. Part 1. General considerations" (CEN, 2006b). Practical guidelines for single laboratory validation of analytical methods are given by IUPAC (Thompson *et al.*, 2002) and by Eurachem (1998). Single laboratory validation is used to show the suitability of a method before entering the (expensive) ring trial or in cases where conducting a ring trial is not practicable. A next higher level of validation is including a second laboratory to confirm the performance obtained. A (limited) set of single laboratory validation data should be collected by each laboratory even when using fully validated standard methods to show its ability to carry out the methods with a comparable performance.

The validation data are the basis for the estimation of the analytical uncertainty of the method.

11.5.2
A Practical Guide for Developing and Prevalidation of Analytical Methods

In the following, a practical guide for a step-by-step procedure is presented to establish a validated method of analysis for determination of both a specific migrant in a food simulant and the residual concentration in a plastic. This procedure was first developed and then applied in an European project (Franz and Rijk, 1997) and found to be very practical. It should be considered as a recommendation based on the great practical experience of the analysts involved.

The development procedure consists of the following eight steps:

1. Scope of the method

    Basically, two types of method must be taken into account:
    - Analysis of a specific migrant in a food simulant (SML methods)
    - Analysis of a specific migrant in a polymer (QM methods)

    Generally, the method to be developed should allow quantitative analysis of the analyte at the required restriction limit in all the official food simulants, including substitutes or alternatives, and/or in the polymer, respectively. That means that for very low SML values which are assumed to be in the range of the detection limit, the aim should be to obtain a detection limit equal to or even lower than the restriction criterion. For other, higher SML and QM values, the aim should be to obtain a detection limit at least ten times below the legal or self-defined restriction criterion. It should also be kept in mind that the method description should provide the relevant intralaboratory precision data (at the required SML/QM value) according to ISO 5725 (ISO, 1994).

    The most suitable analytical methodology should be selected based on the required performance characteristics. A sound literature search is always of great help with respect to known methods for the respective analyte and matrices. In most cases the search results will not directly provide the method wanted but will allow the most likely successful analytical approach to be set up. In this context, preconsiderations should address the most appropriate sample work-up procedure as well as the suitable analytical separation and detection system. The question of direct analysis of the analyte or a derivate formed after chemical reaction should be clarified. Finally, some thoughts should already be given to the question of chemical stability of the analyte in the given matrices under the applied conditions.

Paseiro *et al.* reviewed within the EU project Foodmigrosure (QLK1-CT2002-2390) existing literature to a range of substances and developed strategies for methods determining these substances in plastic materials and food starting from the physical chemical properties (Sendón García *et al.*, 2006; Sanches-Silva *et al.*, 2006; Paseiro *et al.*, 2006b).

2. Setting up the chromatographic and detection system

    It exceeds the scope of this chapter to give more background and details on analytical chemistry. The corresponding scientific knowledge and technical information have been described elsewhere (for instance, Schomburg 1987; Lee *et al.* 1984; Chapman, 1986; and many other lecture books).

    Having rationalized the most suitable analytical principle as a result from step 1, it is necessary to demonstrate the adequate specificity and sensitivity of the analytical system. This aim can be achieved by carrying out an initial feasibility study where the following points need in-depth consideration:
    - availability and purity of reference standards;
    - purity requirements for chemicals, reagents, and solvents;
    - safety considerations;
    - selection of sampling and chromatographic instrument;
    - choice of separation column;
    - suitable detection system;
    - optimization of instrument parameters;
    - appropriate internal standard;
    - solvent to be used for preparation of stock and standard solutions.

    The feasibility exercise should include preparation of a concentrated (stock) solution as well as diluted standard solutions of various concentrations and establishing a first calibration curve. From the data obtained, preliminary conclusions should be drawn with respect to the approximate precision, its working range, and LOD. Finally, the results should provide sufficient evidence with respect to the workability of the intended analytical approach. If the method appears inappropriate, it must be optimized by methodological improvements, instrument changes, or application of a completely different analytical technique. If no satisfactory improvements can be achieved, a possible way out of the problem may be through compromising the acceptance limits.

3. Preparation and measurement of calibration samples

    When the initial study has been successfully completed, the performance characteristics should be investigated. As a first step on this way, calibration samples should be prepared in order to prove the calibration with respect to fulfilling general acceptance limits for linearity and repeatability performance.

Starting from two independent stock solutions, two sets of analyte calibration solutions should be prepared. The solutions should preferably consist of the same medium (i.e., either food simulant for SML methods or swelling/extraction solvent for QM methods) to be used for the final determination of the SM or the residual amount in the plastic. Since the method's performance characteristics are to be established in relation to the intended use, it is not necessary to check the method's linearity over the full range of the equipment. Therefore, at least five concentration levels are required spanning the given restriction criterion value (or expected concentration value) from $0.1x$ value to $2.0x$ value, provided this is within the LOD. Solutions without any analyte (blanks) should be analyzed as well. In the case of standard addition procedures, five levels should also be analyzed spanning the QM restriction value by standard additions ranging from $0.5x$ value to $5.0x$ value.

All calibration and blank samples should be measured in triplicate (three injections of one sample) and the calibration graph should be constructed by plotting the detection signal obtained for the analyte (preferably peak area rather than peak height) relative to that of the internal standard versus analyte concentration. With respect to the correlation coefficient obtained (usually "R") from the – in most cases – linear regression line, a minimum value of $R = 0.9996$ should be defined as a general acceptance limit. Deviation from this minimum requirement to linearity should only occur in exceptional cases. But the correlation factor gives only limited information on the linearity (Analytical Methods Committee, 1988; Van Loco et al., 2002). Additionally a visual check of the calibration graph is necessary. A graph plotting the residues between measured and calculated values at the different calibration levels might also be useful. Objective information on linearity is given by statistical tests (Lack-of-fit test: Massart et al. 1997; Mandel's fitting test: Mandel 1964). Full statistical evaluation of the calibration graph provides useful data about the method's performance characteristics over the applied calibration range such as the standard error of the procedure, s_x, or the standard error of estimate, s_y. On the basis of 95% probability level, the corresponding confidence bounds should be calculated. The two independently prepared sets of calibration samples should coincide with the upper and lower confidence bounds as another general acceptance limit with respect to repeatability performance. From the data of the calibration samples the detection limit can be obtained (see Section 11.5.4).

4. Within laboratory (repeatability conditions) precision according to ISO 5725

The precision of an analytical method is the degree to which individual determinations of a series of standards agree. Since in general only one laboratory is involved in the development of the method, the precision, as determined by one laboratory by one operator over a relatively short time, is defined as repeatability "r" (ISO, 1994; compare also Section 11.5.3.

For determining "r," the following procedure is recommended:

- *SML methods:* for conventional or alternative food simulants, at least six samples should be prepared, having the same concentration *at* the restriction criterion (SML value). All the samples should be measured by at least double injections and the detector signals obtained should be evaluated using the calibration graph established as described under step 3 above.

- *QM methods:* for the analysis of polymer matrices, 12 samples should be prepared for headspace sampling technique or 6 samples for liquid injection. In each case, the series of samples should be prepared in the polymer/swelling solvent system with all samples using the same concentration *at* the restriction criterion (QM value). Headspace samples are measured only once and liquid injection samples in duplicate. If possible, analyte-free polymer should be used here. Again the spiked concentrations should be verified by standard addition calibration procedure carried out as described above under step 3. When conducting an additional series of measurements using only the swelling solvent as the matrix without polymer and comparing results to those obtained above, the influence of the polymer matrix on the detection of analyte can be investigated.

From the results obtained the repeatability standard deviation "S_r" as well as the repeatability limit "r" can be calculated on a 95% probability level according to equation 11.5 (see also Chapter 11.5.6)

$$r = 2.8 S_r \qquad (11.5)$$

In addition, the results can also be used to calculate the mean recovery percentage as (the ratio of measured concentration/nominal concentration) ×100 and its standard deviation in the case of direct analyte determinations without any sample work-up. In cases where a sample work-up procedure such as extraction or chemical derivation has been applied, the mean recovery can be determined by comparing the detector response for the analyte signal after sample work-up with the response obtained from a standard dissolved in pure solvent.

5. Development of an appropriate confirmation procedure

Whenever a measured value exceeds a certain threshold (an internally defined limit or a legal restriction criterion) then a confirmation procedure is recommended or even necessary. The purpose of confirmation analysis is to prove or disapprove the measurement result obtained by the usual analytical method. Generally, the difference from the confirmation procedure compared to the usual test method should be due to only either the use of a completely different separation column (with completely different retention behavior) in the same detection system or the use of an alternative detection method with sufficient sensitivity. For the latter case and especially for GC methods, the preferred procedure should be to apply analyte selective mass spectroscopy (MS) detection. In some cases, derivatization of the analyte followed by MS detection can also be the method of choice. In the case of HPLC methods, different polarity of another column in connection with full exploitation of modern UV diode array detection systems may be useful to selectively allow confirmation of the analyte. Using LC-MS or LC-MS-MS highly selective detection of analytes is possible. It is extremely important to make sure that the confirmation procedure works at the restriction criterion level or other self-defined concentration limit!

6. Stability check on stock and standard solutions

    Stability tests are understood to be time-dependent measurements of a stock and a standard solution at different temperature conditions, for instance at ambient temperature (approximately −22 °C), normal refrigerator conditions (2–8 °C), and at deep freezing temperatures (approximately −20 °C). Stability tests should always be carried out with the exclusion of light. Under these storage conditions, stock and standard solutions should be monitored for constancy of initial analyte concentration. This can be achieved by comparison against freshly prepared solutions. Storage time should be extended to at least three months or until a decrease of 50% or more has been observed. Sampling frequency depends on the decrease rate of the solutions. It is wise to commence stability checks early enough when starting method development work. The aim here is to find out the optimum storage conditions and maximum practical storage time. Internal standards, if applied, should also be investigated.

7. Workability of the test method under practical conditions

    After successful completion of all the development steps described above, the analyst still cannot be sure that the developed method will work under realistic conditions. The workability of the method therefore has to be proved. There are two major reasons why this workability test has to be carried out: First, it should be demonstrated that the method is not affected by interferences migrating from the polymer matrix. Second, it needs to be clarified whether the analyte is stable under the contact conditions applied during the migration exposure, to avoid false-negative migration results. Therefore, a suitable plastic material containing a high residual level of the analyte under investigation should be available for the following experiments:

    - *SML methods:* The selected polymer sample should be brought into contact with the food simulants under the relevant time–temperature conditions. In general, a migration test applying the total immersion principle using olive oil and 10% ethanol at test conditions of 10 days at 40 °C is sufficient. The determination should be performed in triplicate with double injections for analysis of the food simulants. In cases where the analyte level in the migration solutions is found to be below the detection limit, the migration solutions should be fortified with the migrant at the restriction criterion level or some other concentration of concern and measured again. In parallel, to check for migrant stability in the migration solutions, the relevant food simulants should be fortified at the level of concern to ensure that it is sufficiently higher than the LOD. If the test level concerned is in the range of the LOD, then the threefold concentration should be applied. The food simulants spiked in this way should be stored under appropriate time–temperature conditions and recovery of the analyte determined by cross-checking against freshly prepared solutions.

    - *QM methods:* In case of headspace analysis, triplicate determination of the concentration of the analyte in the selected polymer sample should be performed by the standard addition procedure using the polymer/swelling solvent system. The comparison to a calibration curve of the analyte in the pure swelling solvent allows significant polymer matrix effects to be recognized. When extracting semivolatiles or nonvolatiles from a polymer, using the standard addition approach in most cases a homogeneous distribution of added substance levels in the polymer sample is not feasible. Recovery cannot be determined this way. Completeness of extraction can only be checked by analysis of subsequent extractions. An impact of interferences from the matrix to the result is tested in these cases by standard addition to the ready prepared extract (after removing the polymer sample). In all cases, the stability of the analyte in the swelling or extracting solvent should be studied by fortification at the QM concentration or other relevant level and determination of recovery under the applied swelling and polymer extraction conditions.

8. Method description and reporting

Once the method has been established and validated, it should be described in full detail such that it can be carried out by any other analyst. Besides the numerous experimental details relating to the chemicals, solvents, and solutions used and the chromatographic parameters, important observations such as the findings about the stability of standard solutions should be laid down appropriately in the method description as notes or remarks. But potential health risks to the analytical operator should also be addressed, for instance in a warning note at the beginning of the method description. The following structure of a method description, which was agreed upon as a CEN standard format, is a recommended example.

Foreword:	Optional paragraph explaining about the background or history of the method.
1. Introduction:	This chapter gives a rationale why it was necessary to establish this method.
2. Scope:	In this section, the range of applications for the method should be indicated.
3. Principle:	This paragraph summarizes the applied analytical principle, including sample preparation techniques.
4. Reagents:	It is necessary to describe in full detail the origin and purity of chemicals and solvents, the preparation of stock and standard solutions or other solutions, such as the mobile phase in the case of HPLC analysis. In conjunction with a given set of analytical parameters, the chromatogram obtained or at least an indication of retention times obtained for the analyte and the internal standard should be presented.
5. Apparatus:	This chapter should describe the complete set of instrumental and other analytical parameters as well as special laboratory equipment and analytical accessories such as size and type of sample vials, pipettes, and syringes, standard laboratory glassware and equipment accepted.

(continued)

6. Samples:	In this section, the preparation of test samples, blanks, and calibration samples has to be described, together with an indication of the minimum number of samples needed. If necessary, precautions should be mentioned, for instance to avoid cross-contamination of samples in the case of volatiles or to minimize chemical degradation in the case of unstable analytes.
7. Procedure:	Here it is necessary to provide details as to how the analytical measurement of test, blank, and calibration samples is executed and how the obtained data are evaluated. The measured concentration of the analyte obtained in this way may need further transformation into a different dimension and this should also be addressed in this section.
8. Confirmation:	When a certain critical concentration value has been measured and found excessive, then it may be recommendable or even necessary to confirm the result or the identity of the quantified analyte by means of another analytical technique, for instance, by specific detection using MS. This confirmation procedure should be clearly presented in this paragraph.
9. Precision data:	This chapter should give an insight into the validation procedure applied and report the most important performance characteristics: • the achieved LOD or LOD range; • the achieved repeatability criteria, that is the r values in the different food simulants or in the polymer matrix and the concentration range where they have been determined; and • if available the determined reproducibility, that is the R value and the critical difference, that is the CrD_{95} value, as obtained in the most usual situation, that is one laboratory carries out n measurements Chapter 11.
10. Test report:	The test report should contain all necessary documentation such as • date of analysis and reporting; • clear identification of the test laboratory and the responsible analyst; • analyte and method of test, including references; • sample details like origin and specification, type of food/simulant/material/article, reception date, and storage conditions; • results expressed in milligram analyte per kilogram food simulant or plastic material, • details of confirmation procedure, if any; and • reasons for modifications introduced into the test method, if any.

This format is obligatory for method descriptions in EU petitions for approval of new substances for FCMs and also described in the Note for Guidance (EFSA, 2006). Practical examples can be found in the EN 13130 series (CEN, 2004) and the BCR project "Monomers" report (Franz and Rijk, 1997).

11.5.3
Validation Requirements for EU Food Contact Petitions and US FDA Food Contact Notifications

The requirements to the analysis methods to be provided in EU food contact petitions are described in the Note for Guidance (EFSA, 2006), that for FDA food contact notifications or petitions in FDA chemistry recommendations (US FDA, 2002). Data on reproducibility, recovery, detection limit, and stability of the analyte at migration test conditions are requested. An appropriate way to determine reproducibility is described in Section 11.5.2, step 4. At EU level, recovery is mainly focused on sample preparation steps and may be determined using spiked simulant blanks. For checking if other substances out of the plastic sample interfere with the analyte, recovery should be determined by spiking real migration solutions at one level at least. FDA requires in any case spiking of the migration solution with the longest contact time at three levels (half, full, double expected concentration), each in triplicate, instead of spiking of blanks. At nondetects, the detection limit shall be verified by spiking at the detection limit and at least one further level. The ranges of acceptable recoveries are given in Table 11.5. Stability testing, which is explicitly required for EU petitions, is described in Section 11.5.2, step 7.

The method description is part of the EU petition and shall follow the CEN format as described in Section 11.5.2, step 8.

The specific requirements of EU and FDA can easily be combined so that the same set of measurements and reports may be used for petitions/notifications to both authorities.

11.5.4
Determination of the Detection Limit

There are two principle ways to determine the detection limit (LOD): directly by repeated analysis of blanks and statistically from the calibration line. If the analyzed

Table 11.5 Acceptable recoveries and relative standard deviations of fortification experiments according to FDA chemistry recommendations (US FDA 2002).

Levels in food or food simulants[a] (mg/kg)	Acceptable average recovery (%)	Acceptable relative standard deviation (%)
<0.1	60–110	<20
>0.1	80–100	<10

[a] If 0.001 mg of a substance is extracted from a square inch of packaging material into 10 g of food or food simulant, the estimated concentration is 0.1 mg/kg.

concentration is near or below the detection limit or if the requirement is non-detectable, LOD should be determined precisely. If the working range of the method that covers the migration limit or the found concentrations is far above the analytical observable detection limit, a rough estimate would be sufficient.

The detection limit can be obtained from the threefold or fivefold signal to noise ratio in case that there is no peak in the blank at the retention time of the analyte. A more precise method would be to fortify six blanks with the lowest acceptable concentration and to determine the standard deviation. If is a signal found at the respective retention time, the mean area and the standard deviation of this signal in the blank (without fortification) is determined. The Eurachem Guide on method validation (Eurachem, 1998) proposes to use 10 independent samples. The detection limit is obtained from the mean concentration in the blank (if any) plus the k-fold standard deviation. Usually factor $k = 3$ is used. Huber (2003) describes the procedure to calculate the detection limits and to derive these factors k statistically from the probability of a false positive result, that is detection of a nonexisting analyte. Factor $k = 3$ equals to a 99% confidence interval (1% probability of a false positive detection) at 9 or 10 independent single determinations. At a confidence interval of 95% which is the most common used one, at six independent single determinations $k = 2.2$. In case of means of double injections, at five samples $k < 2$.

According to DIN 32645 (DIN, 1994), detection limit is defined as substance concentration at which, at a given α, a substance is detected with a probability of 50% (β error, probability of false negatives) (Figure 11.5). A prerequisite for the suitability of this tool is the homogeneity of the variances or standard deviations over the range of the calibration line (the variances of the first and the last calibration point should be similar). This is normally only the case between the lowest detectable concentration and its tenfold value. Otherwise using this formula is statistically not correct and the result only a rough estimate. In any case, the result of the calibration line method

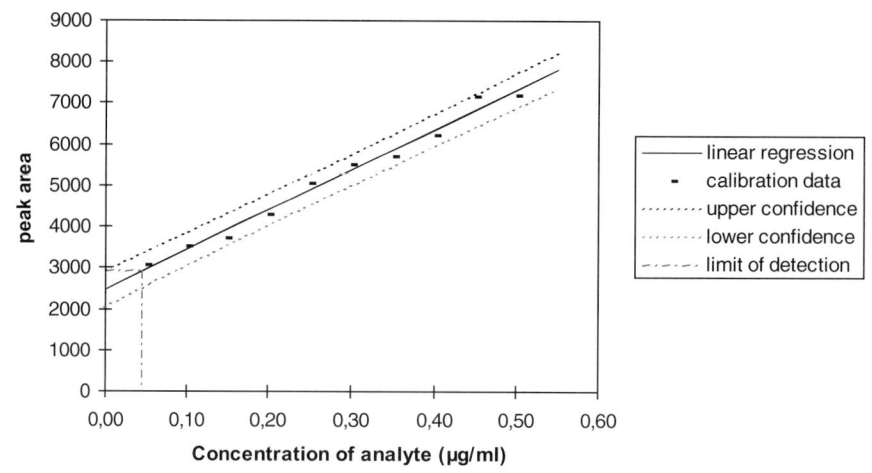

Figure 11.5 Calibration line, detection limit, and 95% confidence interval according to DIN 32645. Example data from DIN 32645.

should be compared to the signal to noise ratio. If the latter gives a higher detection limit, this value or determination via the standard deviation of the blank should be used.

The detection limit via the calibration line method is calculated according to equation 11.6

$$x_{LOD} = s_{x0} t_{n-2,\alpha} \sqrt{\frac{1+n}{n} + \frac{x^2}{\sum(x_i - x)^2}} \tag{11.6}$$

with x_{LOD} the detection limit, s_{x0} the standard error of the procedure, $t_{n-2,\alpha}$ the Student's t factor (single sided) for $n-2$ degrees of freedom and a type 1 error probability α (normally 5% corresponding to a 95% confidence interval), n the number of calibration points (single determinations), x_i the calibration level at i, and $\bar{x}$ the mean of all calibration levels.

11.5.5
Analytical Uncertainty

Estimation of the analytical uncertainty is an important tool to show confidence in the analytical results and for evaluating results at the specification or legal limit. An authority may only object to a material if the measurement value is significantly higher than the legal limit, that is higher than the legal limit plus the analytical uncertainty. Accredited laboratories need to have procedures to estimate the analytical uncertainty according to EN ISO/IEC 17025 (EN ISO/IEC, CEN 2005b). The principal procedures and definitions are laid down in the ISO Guide of Uncertainty of measurements (GUM) (ISO, 1993).

Applying fully validated methods (e.g., standard methods), then data on intralaboratory repeatability and interlaboratory reproducibility are already available from which the analytical uncertainty can be obtained as described in Section 11.5.6. In this case, a laboratory has only to check if its proficiency fits to the proficiency data given in the standard (detection limit, linearity, repeatability). Furthermore, it has to be checked if all analytical steps have been tested in the validation ring trial or if the laboratory applies additional steps which are not described in the standard, for which then the uncertainty needs to be estimated.

For many analytes fully validated standard methods do not exist. In this case, the laboratory has to estimate the uncertainty itself using the in-house validation data, precision data from laboratory equipment (e.g., scale, pipettes) as well as expert judgment. More specifically, estimation of analytical uncertainty of chemical analytical methods with examples is described in the Eurachem/CITAC Guide (Eurachem, 2000). In the first step, the measurand has to be clearly specified. Then in the second step, the uncertainty sources need to be identified and best to be collected in a source-and-effect diagram. In a third step, it should be simplified by grouping sources covered by existing data. The uncertainty of grouped components should be quantified, then of the remaining components (negligible ones may be omitted) and all uncertainties shall be converted into standard deviations. In a forth step, the single uncertainties shall be combined in principle according to the Gaussian error propagation function. The equations and simplifications in special cases are described in the Eurachem Guide. For obtaining the uncertainty in terms of a 95% confidence interval, the standard deviation has to be multiplied with a factor k which is normally 2. Only in case of only a few replicate measurement data (<6 degrees of freedom), the Student's t factor is used as k (Eurachem, 2000).

All the standard methods (except for OM) and usually also the in-house methods start with the measurand in the ready-prepared migration solution, that is the uncertainty impact of the migration contact is not considered. To estimate this uncertainty contribution is difficult. The properties of the material, those of the migrating substance of interest, especially its partitioning coefficient between the material and the simulant, the time, the temperature, and the volume play a role, but at the various points on the kinetic migration curve in different extent. In the beginning of the migration curve, when migration is proportional to the square root of time, deviations of time and temperature have a large impact on uncertainty whereas in partitioning equilibrium migration does not increase anymore with the contact time. Deviations in the volume and losses during migration contact have a higher impact to the uncertainty in case when a high partitioning coefficient (low solubility in the simulant) restricts migration or when the ratio between simulant volume and packaging volume is small. This reflection considers already that in the end of the migration experiment the volume has to be defined or the simulant to be filled to a defined volume. Otherwise the volume error might be unacceptably high. In EU project G6RD-CT2000-00411 "Specific Migration" a feasibility study was performed on the production and certification of reference materials for SM (Stoffers et al. 2004, 2005). Using those future certified reference materials, the laboratories shall be enabled to test their proficiency in carrying out the migration contact. Six material–migrant combinations had been shown to be suitable (LDPE//Irganox 1076/Irgafos 168, LDPE//1,4-diphenylbutadiene, HDPE//Chimasorb 81/Uvitex OB, PPhomo//Irganox 1076/Irgafos 168, HIPS with 1% mineral oil//styrene, PA 6// caprolactam). The certification exercise was only performed in 4 partner laboratories. The first step is done, but for using these materials for estimating the uncertainty of performing the migration contact, a new and larger scale of these materials has to be produced and the certification parameters need to be tested in a ring trial with a larger number of laboratories.

11.5.6
Use of the Precision Data from Fully Validated Methods

A relevant juristically valid statement about the precision of a method can only be made after defining the performance characteristics obtained from interlaboratory trial study (round robin, collaborative trial), as for instance described in ISO 5725 (ISO, 1994). This study is used to determine the statistical key data about the precision of a method.

ISO uses two terms, "trueness" and "precision," to describe the accuracy of a measured value. "Trueness" refers to the closeness of agreement between the average value of a large number of test results and the true or accepted reference value. "Precision" refers to the closeness of agreement of test results, or in other words, the variability between repeated tests. The standard deviation of the measured value obtained by repeated determinations under the same conditions is used as a measure of the precision of the measurement procedure, whereas the repeatability standard deviation "s_r" describes the intralaboratory precision and the reproducibility "s_R" the variation between different laboratories. In other words, if the test result as an average of several individual measurements is obtained with the same method from an identical test sample, in the same laboratory, by the same analyst, with the same instrumentation, over a short period of time, then the study takes place under "repeatability" conditions. On the other hand, "reproducibility" conditions occur when the measurements take place following the same procedure and using identical samples but in different laboratories using different analysts with different instrumentation.

The repeatability limit, "r," is the within-laboratory precision and describes the maximum expected value of the difference between two individual test results obtained under repeatability conditions at a defined significance which is in most cases a probability level of 95%. Similarly, the reproducibility limit, "R," describes the analogous between-laboratory precision. An important assumption for the use of r and R in practice is that they have been determined in an interlaboratory test in which the participating laboratories represent those potential candidate appliers of the particular analytical procedure. For the determination of r and R, the method of analysis must be described very clearly and in detail to eliminate as many differences between laboratories as possible. Particular precautions are necessary with regard to the homogeneity and stability of the sample to be studied in the interlaboratory test. Clearly the sample must withstand transport conditions and arrive unaltered at the participating laboratories.

The statistical model for estimating the precision of the analytical method assumes that every individual measurement result y is the sum of three components:

$$y = m + B + e \tag{11.7}$$

Here m represents the average of all values for the material studied (the characteristic level), B the scattering between the laboratories, and e the random deviation in results occurring in every measurement. The characteristic level m must not necessarily agree completely with the true value. There may be a difference $(m - m_y)$ from the true value due to a systematic error in the measurement procedure (bias). For contributions B and e, it is assumed that they approximately follow the normal distribution. Then the variance of B, var(B), is the variance between laboratories (σ_L^2). This includes the scattering between different analysts and different instruments. The variance of e, var(e), is referred to as the internal variance of a laboratory (σ_W^2). The average of all the internal variances of the participating laboratories in an interlaboratory test is expressed as the repeatability variance σ_r^2. While r depends only on the repeatability variance, R is determined by the sum of the repeatability variances and the variance between all laboratories.

The standard deviations of repeatability and reproducibility are given by σ_r and $\sigma_R^2 = (\sigma_L^2 + \sigma_r^2)^{1/2}$ and it follows:

$$r = t\sqrt{2}\sigma_r \quad \text{and} \quad R = t\sqrt{2}\sigma_R \tag{11.8}$$

The factor $\sqrt{2}$ is based on the fact that r and R are related to the difference between two measurement results. For distributions which are approximately normal and in the case of not too small a number of measurements, the factor t is the two-sided Student's t factor and does not vary much from 2 at a probability level of 95%. As approximate value one can use the factor 2.8 for $t \cdot 2^{1/2}$. In practice, the true repeatability and reproducibility standard deviations are not known. They are replaced with estimated values s_r and s_R from the interlaboratory study with a limited number of participating laboratories and measurement results.

Repeatability r and reproducibility R are then expressed as:

$$r = 2.8 s_r \quad \text{and} \quad R = 2.8 s_R \tag{11.9}$$

The precision of a standard measurement method is expressed using the values of r and R. It can be verbally described as:

Repeatability: The difference between two individual measurement results, which an analyst obtained on the identical sample material with the same instrument within the shortest time span possible, will on average not exceed the repeatability limit r more than once in 20 cases, provided the measurement procedure has been correctly carried out.

Reproducibility: The difference between two individual measurement results, reported by two laboratories for identical sample material, will on average not exceed the reproducibility limit R more than one time in 20 cases provided the measurement procedure has been correctly carried out.

When precision shall be expressed as a probability level of 99%, the values for r and R must be multiplied by factor 1.25.

The precision data r and R can be used for the decision if two measurement results are significantly different or if a result is significantly different from a given value (e.g., the SML or a certified value of a reference material). Two values are significantly different when their difference is higher than critical difference CrD_{95} obtained from the precision data.

In the case of comparison of data from the same sample, if the difference of two averages or of an individual value and a given value exceeds the corresponding critical difference, then this deviation should be considered suspect. There could be a specific reason why the critical difference is exceeded and this should be rationalized. In particular, if the given or reference value is a true or correct value, then the suspected difference can point to a bias in the measured result.

In the case of comparing measured data with an SML, the critical difference evaluation system allows the decision whether a legal restriction criterion has been exceeded or not.

The calculation of the critical difference is shown in the following four cases: (1) comparison of two measurement rows in one laboratory, (2) comparison of results of two laboratories, (3) comparison of a measurement result of one laboratory with a given value, and (4) comparison of a group of laboratories with a given value.

1. In one laboratory, two groups of measurements are carried out.

   In one laboratory, two groups of measurements are carried out under repeatability conditions whereby the first group of n_1 measurements gives an average value of y_1 and the second group of n_2 measurements gives an average value of y_2. With r being the repeatability limit (for two individual measurement results), the critical difference $CrD_{95}(y_1 - y_2)$ is then:

$$CrD_{95}(|\bar{y}_1 - \bar{y}_2|) = r\left(\frac{1}{2n_1} + \frac{1}{2n_2}\right)^{1/2} \qquad (11.10)$$

   In the case of $n_1 = n_2 = 1$, then by definition one obtains r as the critical difference.

2. Two laboratories conduct more than one measurement each.

   One laboratory carries out n_1 measurements with an average of y_1 while a second laboratory obtains an average of y_2 for n_2 measurements. The critical difference between the two is then:

$$CrD_{95}(|\bar{y}_1 - \bar{y}_2|) = \left[R^2 - r^2\left(1 - \frac{1}{2n_1} - \frac{1}{2n_2}\right)\right]^{1/2} \qquad (11.11)$$

   By definition, for the special case where $n_1 = n_2 = 1$ the formula simplifies to R and for $n_1 = n_2 = 2$, one obtains

$$CrD_{95}(|\bar{y}_1 - \bar{y}_2|) = \left(R^2 - \frac{r^2}{2}\right)^{1/2} \qquad (11.12)$$

3. The mean value from one laboratory is compared with a given value.

   One laboratory has carried n measurements under repeatability conditions and has obtained an average value y which is compared with a given value m_0 (e.g., SML). Then one obtains the critical difference as

$$CrD_{95}(|\bar{y}_1 - m_0|) = \frac{1}{2^{1/2}}\left[R^2 - r^2\left(\frac{n-1}{n}\right)\right]^{1/2} \qquad (11.13)$$

4. The mean value of several laboratories is compared with a given value.

   A number of p laboratories have carried out n_i measurements and obtained the average values y_i ($i = 1, 2, \ldots, p$). The overall mean value over y_i, $\bar{\bar{y}}$, is compared with a given value m_0. One obtains the following expression for the critical difference:

$$CrD_{95}(|\bar{\bar{y}} - m_0|) = \frac{1}{(2p)^{1/2}}\left[R^2 - r^2\left(1 - \frac{1}{p}\sum_i \frac{1}{n_i}\right)\right]^{1/2}, \quad \bar{\bar{y}} = \frac{1}{p}\sum_i \bar{y}_i$$

$$(11.14)$$

11.6
Sources of Errors

During method development and validation, a number of practical difficulties may occur and need control. An already well-known major phenomenon that can cause problems to the analyst is, for instance, insufficient or even zero recovery of analytes from the migration test solution. Possible reasons for that may be as follows:

1. the chemical instability of analytes under migration test conditions due to oxidation, chemical binding to food simulant, (acid catalyzed) hydrolysis, or ethanolysis or
2. volatilization during migration exposure and sample preparation (Rijk 1993).

To illustrate and put into practice what has been said so far, several examples of methods of analysis are presented in the following text, together with some specific difficulties and problems related to SM determination methods.

11.6.1
Highly Volatile Migrants

Butadiene, $CH_2=CH-CH=CH_2$ [CAS No. 106-99-0; PM/Ref. No. 13630], is commonly copolymerized with styrene and butadiene to make ABS or BS food contact plastics or rubbery impact modifiers, for example, for high impact polystyrene. Butadiene is a suspected carcinogen with extreme volatility (bp $-4.5\,°C$) and low water solubility. This makes it very difficult to handle migration and calibration samples especially when the matrix is of highly aqueous character such as the aqueous food simulants.

The method developed in the BCR project (Franz and Rijk, 1997; CEN/TS 13130-15) to determine butadiene in all of the official food simulants and probably also in real foodstuffs was prevalidated by a collaborative trial with three laboratories. It was found appropriate in principle for the quantitative determination of butadiene at a range of 0.01–0.1 mg/kg in food simulants. Indeed the LOD was found to be in the range 4–9 µg/kg, thus being even in the worst case significantly lower than originally presumed when establishing the Plastics Directive limit of 0.02 mg/kg. The level of butadiene in a food or food simulant is determined by headspace GC with automated sample injection and by FID. Quantification is achieved using an internal standard (*n*-pentane) with calibration against relevant food simulant samples fortified with known amounts of butadiene.

During the method development and validation work in the project, severe problems had been observed with respect to volatilization of butadiene. Therefore, it is important and crucial to take the following into account when planning and designing a migration test: From migration experiments carried out at 10 days for 40 °C it was recognized that irreproducible considerable loss (up to 90%) can result from volatilization of 1,3-butadiene when using aqueous food simulants. Just opening and closing vials containing calibration solutions caused significant headspace losses of the volatile analyte, which is due to very unfavorable partitioning from the aqueous phase to the headspace. On the other hand, olive oil samples were found to provide satisfactory recoveries, due to the much better solubility of butadiene in this nonpolar matrix. As a consequence, migration exposure of plastic materials to an aqueous food simulant in a test cell or glass container combined with sampling steps to prepare food simulant aliquots for analysis even when using gas tight containers will most likely lead to irreproducible results due to uncontrollable loss of analyte.

As a solution for this kind of problem, the idea of two restriction types for butadiene was born – having an SML (not detectable at detection limit 0.02 mg/kg) and alternatively a restriction for residual butadiene in the finished material (QM 1 mg/kg). Indeed, compliance testing with respect to the QM limit of butadiene in plastic according the EN13130-4 standard method, which also originates from the German official analytical methods according to §35 LMBG (now § 64 LFGB), is in all cases highly recommendable since this method is much easier and straightforward and therefore, much less error prone.

11.6.2
Reaction with Food/Simulant Constituents

The two homologous aliphatic diamines ethylenediamine (SML = 12 mg/kg) and hexamethylenediamine (SML = 2.4 mg/kg) are commonly used as bifunctional monomers for polycondensation reactions. Hexamethylenediamine or 1,6-diaminohexane, $C_6H_{16}N_2$ [CAS No. 124-09-4, PM Ref. No. 1840], which is most well known as a polyamide (Nylon 66) monomer, is also copolymerized with sebacic acid to form Nylon 6/10, or with isophthalic acid. Besides that, it is applied as a curing agent for epoxy resins. Practical packaging applications are vacuum and modified atmosphere packs, boil-in-packs for packaging meat, fish, coffee, and snack foods. In the field of rigid containers, monolayer or multilayer bottles for refilling with soft drinks and water are on the market. Ethylenediamine or 1,2-diaminoethane, $C_2H_8N_2$ [CAS No. 107-15-3, PM Ref. No. 16960] is also used to make some nylons and thermosetting resins. It finds application as a reactive hardener in epoxy resins and in stabilizing rubber latexes. Examples of practical applications are adhesives, moisture barrier coatings for paper, cellophane or others, and corrosion inhibitor for aluminum alloys.

In the BCR project, a group method was developed for both diamines HMDA and EDA in the same way (Franz and Rijk, 1997; Demertzis et al., 1995, CEN/TS 13130-21). During the project work a remarkable observation was made: Stability tests in olive oil as a food simulant carried out under test conditions 10 days/20 °C and 10 days/40 °C indicated that both diamines could no longer be recovered, whereas in aqueous food simulants, nearly 100% recovery was obtained under the same test conditions. To investigate the mechanism of diamine disappearance a model experiment was carried out. A 1:1 mixture by mass of olive oil and diamines was stored for 10 days at 40 °C. Then the mixture was analyzed by supercritical fluid chromatography (SFC) using FID detection and compared with the original olive oil SFC pattern. The result is depicted in Figure 11.6. It can be recognized that the

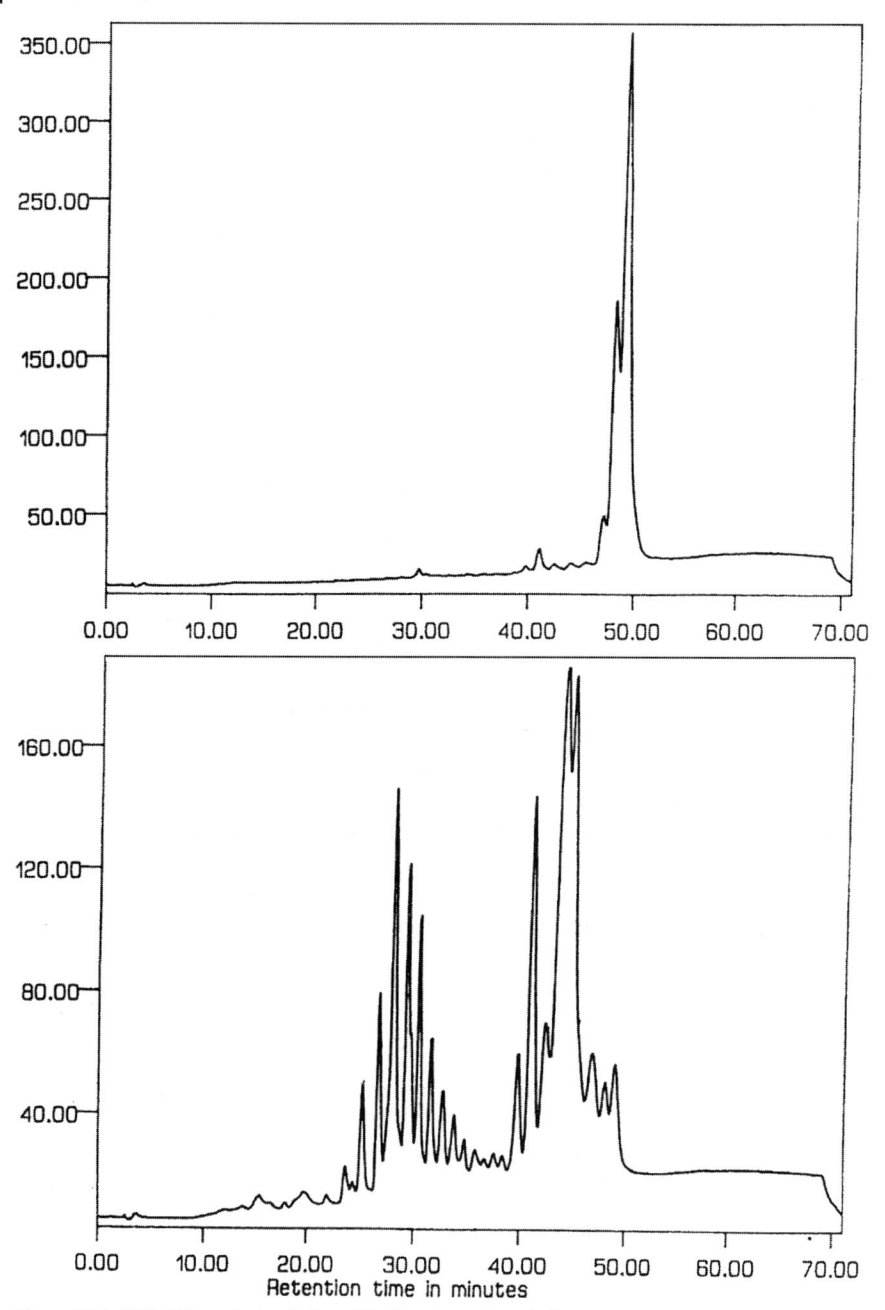

Figure 11.6 SFC–FID analysis of olive oil before (upper) and after reaction (lower) with an EDA/HMDA mixture.

original olive oil triglyceride peaks are nearly completely transformed into a series of different SFC peaks with lower molecular weights. The only reasonable explanation is that the triglycerides react with the diamines to form transamidation products. This was confirmed by LC-MS analysis that demonstrated that the products formed contain the moiety of the diamines.

An important conclusion from these findings was that even though this analytical method works in principle with olive oil as a food simulant, the migration test using olive oil or another fat simulant can provide false-negative results. Therefore, the method should only be applied in the case of short exposure periods with olive oil. The example shows the importance of a recovery check with spiked simulant applying the same time–temperature migration test conditions (compare to Section 11.5.2). As a consequence of these findings, the scope of the analytical method was extended from the determination of the diamine monomers in the aqueous food simulants and in olive oil to the substitute food simulants 95% (v/v) ethanol and isooctane. 95% ethanol should be preferred because the solubility of EDA and HMDA in isooctane is lower and therefore the partitioning coefficient between packaging and simulant is higher than in olive oil. This will cause an underestimation of migration when using isooctane as fat simulant.

Another example for reaction of migrants with the simulants during migration contact is the epoxy compound BADGE (2,2-bis(4-hydroxyphenyl)propane-bis-(2,3-epoxypropyl) ether or bisphenol A diglycidyl ether, CAS No. 1675-54-3, Ref. No. 13510). The epoxy group hydrolyzes during migration contact in aqueous simulants (Table 11.6) (Paseiro Losada *et al.* 1993; Philo *et al.* 1997). This is considered in the SML which is defined as sum of BADGE and its hydrolysis products BADGE·H_2O and BADGE·$2H_2O$ (SML(T) = 9 mg/kg in EU Regulation 1895/2005 (European Commission 2005). With chlorine-containing compounds, for example, in PVC, BADGE reacts to chlorine derivatives already in the material. Because of their different toxicological properties, the chlorine derivative have a separate SML(T) of 1 mg/kg food. At long-term storage of food or olive oil over several months, BADGE disappears slowly by binding on protein components (Petersen, 2003; Cortizas Castro, 1999). But at 10 days/40 °C contact in olive oil no significant losses are observable.

11.6.3
Migrants in Reactive Processes (e.g., Primary Aromatic Amines from Adhesives)

In reactive processes, for example, during curing of polyurethane adhesives, the concentration of the relevant monomeric migrants is decreasing with the curing time

Table 11.6 Hydrolysis of BADGE in aqueous food simulants at 40 °C

Food simulant	Half-life time (days)
Distilled water	1.1
3% (w/v) acetic acid	0.15
15% (v/v) ethanol	1.4

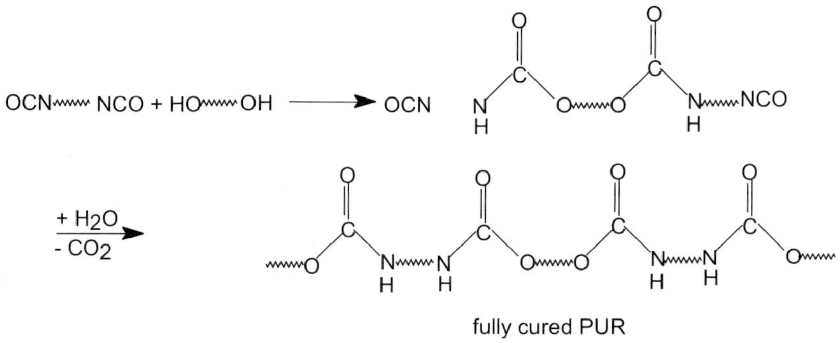

Figure 11.7 Reaction of isocyanates and alcohols to polyurethanes.

as they react to polymeric chains. At the end of the curing process all monomer should have been reacted. Therefore, migration is strongly dependent on the sampling time point and, furthermore, in a not fully cured material migration and hardening might be interfering and opposite processes.

Bifunctional isocyanates react with bifunctional alcohols to polyurethane chains (Figure 11.7). Not fully reacted free isocyanate might migrate to the packaging surface and form amines in contact with moisture of the food (Figure 11.8). Primary aromatic amines are considered to be carcinogens. Aliphatic isocyanates and the corresponding amines are of toxicological relevance as well.

Because of this high reactivity of the isocyanates there is not an SML but a residual content limit QM(T) for the sum of all isocyanates which is 1 mg/kg in the finished article calculated as NCO. Migration of primary aromatic amines is regulated as general specification in Annex V of the Plastics Directive 2002/72/EC (version of EU Directive 2007/19/EC) that plastic articles and materials shall not release them in a detectable quantity (detection limit 10 μg/kg food or food simulant).

The QM(T) limitation for NCO was originally directed to polyurethane plastics with a homogeneous distribution of isocyanates and not for a two or more layer structure combined through thin polyurethane adhesive layers. Since isocyanates are not stable in food simulants, a QM instead of an SML was established to allow in principle enforcement of an isocyanate restriction. However, in practice, isocyanates-based PUR layers are part of multilayer structures and the QM is then related to the whole laminate. Depending on the thicknesses of the individual layers of a laminate, a certain relatively high initial isocyanate concentration in the PUR layer can lead to completely different mass transfers into the food. Especially in cases where an

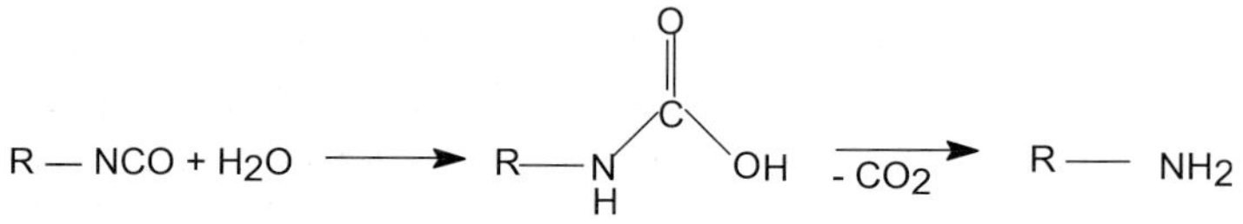

Figure 11.8 Reaction of isocyanates and water to amines.

isocyanate layer is separated from the food only by a thin barrier, the QM of 1 mg/kg NCO in the finished product does not assure that detectable amine migration will not occur. On the other hand, if the QM is related only to the PUR layer(s), it is an unrealistic (also from toxicological sight) and technically not feasible requirement for these thin layers.

Consequently, it appears that the determination of NCO in laminates is of limited use when accounting the chemical reactivity of the NCO group. Furthermore, determination of isocyanates according to EN 13130-8 is complicated and error prone. It needs an experienced laboratory but even then, matrix interferences make analysis in several cases difficult or impossible. In principle, migration testing for primary aromatic amines should be more appropriate and replace the determination of aromatic isocyanates in plastic.

The migration test shall represent the situation in the roll at the sampling time point. Access of air and humidity to the film as well as light and elevated temperatures cause accelerated decrease of isocyanates. Therefore, samples are transported best as small rolls. Because of the possible influence of humidity present in the outer layers of a roll, the samples must be taken from inner roll layers. Based on experience, 5–10 layers are recommendably removed prior to sampling. This ensures that the specimen taken was not in contact with the atmosphere before. The sampling distance from the cutting edge of the roll should be more than 10 cm or better from the middle part of the roll. Samples should be measured immediately. For transport to the test laboratory, samples should be protected from humidity, elevated temperatures, and light. If the samples are not analyzed immediately after sampling, it should be noted that the curing process and chemical degradation of isocyanates as well as amines are ongoing during sample storage and may occur at a higher rate compared to the roll. Then the results of the taken samples will not represent the roll. Migration/extraction solutions in 3% acetic acid can be stored refrigerated for a short period. The isocyanates status might be conserved by deep freezing the film when protected from air and humidity, for example, by wrapping in aluminum foil.

Since aromatic amines are formed from isocyanates by hydrolysis and due to their weekly basic character, 3% acetic acid is generally considered to be the most severe food simulant for aromatic amines. Isocyanates as well as the amines can chemically react with olive oil and other fats. With ethanol, isocyanates react to ethyl urethanes. Therefore, in 95% ethanol as alternative fat simulant as well as 10% aqueous ethanol, migrating isocyanates are not completely subsumed by measurement of aromatic amines. At higher temperature applications, for example sterilization, 3% acetic acid migrating into the laminate might cause hydrolysis of the polyurethane. In case of nonacidic fillings, water would be the more appropriate choice then.

Laminates are often used for dry foods (snack products, coffee, instant coffee, coffee whitener etc.). Directives 97/48/EC and 85/572/EEC do not require a migration test for dry foods although migration into or onto dry food is principally possible. In comparison to that, the test with 3% acetic acid, however, is likely to give too exaggerated values relative to the intended application. The alternative use of Tenax

(MPPO) as simulant for dry foods appears to be a suitable option but is not yet sufficiently validated for isocyanate and amine migration.

Conventionally, migration of primary aromatic amines is tested at 2 h/70 °C contact. It has been shown in many cases that this condition is similar to 10 days/40 °C. Having thick or high-barrier layers between the polyurethane layer and the food this short-time contact might be less severe. On the other hand, in case of not fully cured adhesives, the 40 °C storage condition causes accelerated reaction of isocyanates to polyurethanes, whereas at 70 °C migration is more rapid then curing reaction.

Primary aromatic amines may be measured by the photometrical sum method (ASU, 1995) or by a specific method. For the photometric method, the primary aromatic amines are derivatized to an azo dye and quantified as aniline hydrochloride equivalents. The method does not consider differences in reactivities compared to the aniline hydrochloride standard and not shifting of the wavelength of absorption maximum. Because of this, the value can underestimate real migration. Therefore, it was agreed within CEN TC 194/SC1/TG9 that this method is suitable as a first screening method and shows conformity if migration is below the detection limit of 2 ppb of this method. At higher migration only specific methods are able to show conformity with the specification of nondetectable at 10 ppb.

11.7
Migration into Food Simulants in Comparison to Foods

Basically, the European food packaging compliance testing concept has been a migration modeling system as long as it exists. It applies the use of food simulants for migration testing and correlates the obtained migration data with foodstuffs to conclude whether the so derived concentrations in foodstuffs will be acceptable or not. In fact, it is an experimental modeling system that makes use of only four food simulants with the preassumption that they serve as model contact media for all types of foods. To take account of the generally assumed higher extraction efficiency of the fat simulant D (olive oil) a reduction factor concept is legally applied (EU Directive 85/572/EEC: European Commission 1985) to adjust the experimental results obtained from food simulant D to migration levels in foods. Although this is a very crude and largely conventional approximation, in principle, it is a usable concept for pure compliance testing. However, it needs to emphasize that the applied concept and the correlations between simulant and foods have been established ~30 years ago. And, since scientific knowledge in the area of migration research has almost exponentially evolved, major deficiencies linked to this European concept were getting more and more evident. Again and again discrepancies and miscorrelations were found when comparing the actual migration in foodstuffs with that obtained from the fat simulant D according to the correlation as set by the EU legislation. An important study that had a completely different aim, namely to develop a fatty contact test method in support of EU Directive 85/572/EEC (Castle et al., 2000), revealed with its systematic research program clearly the weakness of the fat simulant D reduction factor (DRF)

Migration into olive oil

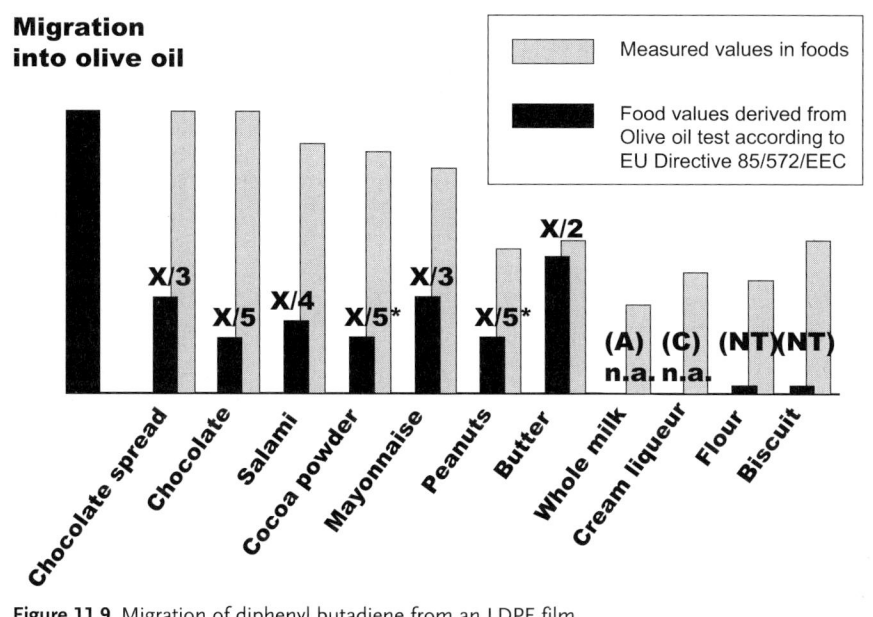

Figure 11.9 Migration of diphenyl butadiene from an LDPE film into olive oil (standard EU test at 10 days/40 °C) in comparison to various foods (at realistic contact conditions). According to EU Directive 85/572/EEC: X/n (2, 3, 4, or 5) are the reduction factors for fatty foods, simulants (A) and (C) were/are applicable for whole milk and cream liqueur, and flour and biscuit as dry foods are not to be tested (NT).

concept. Figure 11.9 shows a comparison of migration data obtained for olive oil (fat simulant D) and various foods. The data demonstrate in that particular case that in foods such as chocolate spreads the same migration values are obtained as for olive oil, and other foods such as mayonnaise or cocoa powder would also be largely underestimated. Even dry foods such as flour or biscuit for which the Directive does not require migration testing at all cause very significant migration. Finally, the comparison indicated that migration into milk products is likely to be underestimated when using aqueous simulants A and C for simulation. This observation had been made long before (O'Neill et al., 1994). In this study, the migration behavior of styrene monomer from PS into different milk products and water ethanol mixtures was investigated and the authors concluded that 50% ethanol in water would simulate milks appropriately. The scientific reason for that was found in the gas/liquid partition coefficients which were also determined for styrene in the investigated foods and simulants. These coefficients were largely higher in pure water compared to the milks and were found more or less identical for 50% ethanol and for regular milk containing 3.5% fat. It should be noted that today this miscorrelation has been corrected in the meantime by the European legislator who prescribes now 50% ethanol as the simulant to be used for milk products (Directive, 2007/19/EC: European Commission, 2007).

From what we know today, one important reason for these observed miscorrelations is that the legally applied correlation concept according to EU Directive 85/572/EEC ignores the migrant releasing matrices, the packaging materials themselves. The correlation depends severely on the type of plastic materials (criterion: its basic diffusivity) and for a given plastic material, in particular for polyolefins (which have a high diffusivity), on the thickness. Moreover, the molecular weight of the migrant plays also a role as well as the test temperature because these parameters are linked with the diffusion coefficient. For better illustration of this discussion, migration modeling according to the recognized state of the art according to the Piringer model (Baner et al., 1998; Piringer and Hinrichs, 2001) was applied to demonstrate how the migration behavior under standard conditions of 10 days at 40 °C for any migrant up to a molecular weight of 1000 g/mol from two different plastics would be in dependency of the film thickness and the fat concentration (in %) in a food packed with 6 dm^2 packaging film/kg food (Franz 2005). In this thought experiment, the percentage of fat is expressed by the actual volume of the fat fraction in the food and by assuming that only the pure fat phase is in contact with the packaging film. Consequently, the 100% fat food (pure oil) is expressed by having 1000 ml olive oil in contact per 6 dm^2, whereas the 20% and 5% fat-containing foods are expressed by having 200 ml and 50 ml olive oil in contact per 6 dm^2, respectively. For the migrant, lipophilic character is assumed so that it would dissolve only in the fat phase and not in the aqueous phase of the food. Figures 11.10 and 11.11 show the situation for

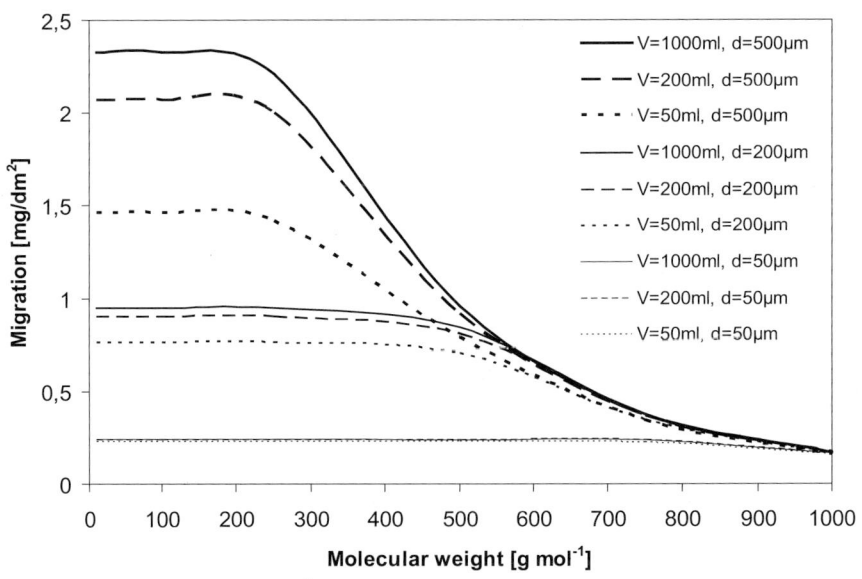

Figure 11.10 Migration in mg/dm^2 in dependency of the molecular weight up to 1000 g/mol from an HDPE film ($C_{P,0}$ = 500 ppm, 10 days/40 °C, contact area 6 dm^2/kg food) as a function of HDPE film thickness d and fat content in food (expressed as volume V of fat fraction in food, that is when V = 1000 ml or 50 ml, $\Rightarrow$ food has 100% or 5% fat).

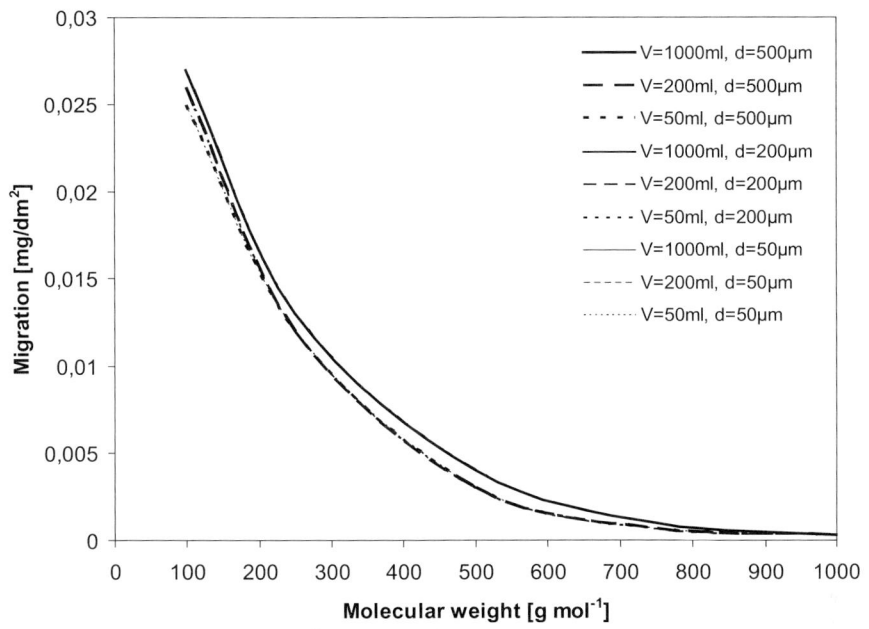

Figure 11.11 Migration in mg/dm² in dependency of the molecular weight up to 1000 g/mol from a PS film ($C_{P,0}$ = 500 ppm, 10 days/40 °C, contact area 6 dm²/kg food) as a function of PS film thickness d and fat content in food (expressed as volume V of fat fraction in food, that is when V = 1000 ml or 50 ml, ⇒ food has 100% or 5% fat).

HDPE and PS as two representatives for higher and lower diffusivity, respectively. Both plastics are modeled each at three different thicknesses (50, 200, and 500 µm). The initial concentration $C_{P,0}$ of migrant is assumed to be 500 ppm. It can be seen that for the thin film with 50 µm thickness the migration is independent of the fat content of a food since always almost complete extraction of the migrants takes place. For the 200 µm thick film the pure fat and the 20% fat food show again almost the same migration behavior and only the 5% fat food exhibits somewhat lower migration. Only for the thick HDPE film there is a considerable influence of the fat content in the food on the migration behavior. This is however only for the smaller molecules up to 450 g/mol. With bigger molecules again similar migration behavior can be found. In case of the PS example, migration is more or less the same in all cases because it is completely controlled by the diffusion in the polymer. It is also important to note that the extent of migration is one to two orders of magnitude lower than the lowest migration curve observed for the HDPE film. This confirms the complete diffusion control due to the much lower diffusivity of PS compared to HDPE. From the PS migration curves it gets evident that simulant DRFs of up to 5 are obviously not applicable anymore where extreme low diffusion occurs. On the other hand, from the HDPE figures it gets evident that DRFs are not applicable to thin films and to situations where almost exhaustive migration takes place. This has, in fact, also been corrected by the European legislator (European Commission, 2007) with the clause that DRFs are not applicable when the migration exceeds 80% of the maximum possible migration.

These examples demonstrate that the current experimental EU compliance test approach is problematic because food simulant-based migration test results do not always assure compliance with the legal requirements which, in the final instance, are applicable to foodstuffs. In other words: in numerous cases, compliance shown with the food simulant test can be contradicted by a migration value determined in the food itself. Consequently, modeling migration into foods could assist or even replace decisions taken on food simulant migration data. One practical case that made this very evident was the so-called isopropyl thioxanthone (ITX) crisis which came up by the end of 2005. ITX was used as a photoinitiator in food packaging inks and was transferred from the printing ink layer from the outside of the packaging material to the food contact side either by setoff (on the roll or in stacked beakers before filling) or permeation through the packaging structure. Migration data, which were made publicly available (EFSA, 2005), demonstrated that ITX migration highly depended on the nature of the foods. For instance, for clear fruit juices ITX migration was not detectable ($<5\,\mu g/L$) but for cloudy drinks migration values ranged up ~ 300 ppb ($\mu g/L$). Similar high values were also obtained for milk and soy drinks. It should be noted that the food simulants A, B, and C would also not give any detectable migration above $5\,\mu g/L$, which is simply due to the low solubility of ITX in these aqueous matrices. Very recently a study has been undertaken to investigate the reasons for the higher migration in cloudy fruit juices using as the case example orange juice in contact with an LDPE film containing diphenyl butadiene, an optical brightener. After removal of solids by centrifugation and filtration from the orange juice matrix, diphenyl butadiene migration into the remaining clear filtrate was almost not detectable anymore compared to a high migration value obtained from the orange juice (Sanches-Silva et al., 2007).

As the weaknesses of EU Directive 85/572/EEC correlations have been recognized since some years, an attempt was made with the Foodmigrosure EU project (Foodmigrosure project (finished, 2006), Franz, 2005) to generate a better scientific basis for understanding the migration into foods as a basis for any necessary changes of the legislation. This project was generated in context with European Union's consumer protection and health care policies according to which one important aspect is the exposure of consumers to undesirable chemicals in the diet. FCM are one potential contamination source and therefore of particular interest for food exposure assessment. On the other hand, scientific investigations concerning the migration potential and behavior of food-packaging materials have demonstrated that diffusion in and migration from FCM are foreseeable physical and, in principle, mathematically describable processes. Because of this situation and the current state-of-the-art in migration science, the Foodmigrosure research project (European Union Contract No. "QLK1-CT2002-2390") was initiated within the fifth Framework Programme of the European Commission from March 2003 until September 2006. Project aim was to extend currently existing migration models (which have been demonstrated to be applicable for less complex matrices such as food simulants) to foodstuffs themselves and to establish a novel and economic tool for estimation of consumer exposure to chemicals migrating from food contact plastic materials. The more scientific aim was to increase knowledge of the mechanisms of diffusion of organic compounds in foodstuffs and provide data on the partitioning effects between FCMs and foods. Today the latter aspect is increasingly regarded as a fundamental influence parameter for migration into foods. As a result, the Foodmigrosure project has provided besides a new compilation of analytical standard methods for specific compliance testing in foods an extensive set of kinetical migration data as well as concentration profiles in all food categories for model migrants released from a plastic material after defined time–temperature conditions. From these data sets physicochemical parameters (diffusion coefficients in foods and partition coefficients between plastics and foods) were derived as basis for a migration model applicable to foodstuffs. Important conclusions had to be drawn from the project findings with respect to European legislation on migration testing:

1. EU assumptions for correlations between the fat test with olive oil and fatty foods are very problematic and are rather too relaxed than realistic, if possible at all.
2. The aqueous simulants A and C do largely *not* reflect what occurs in aqueous foods, which is due to their too low solubility for many or even most migrants.

As a consequence from these findings, compliance testing needs a more thorough consideration what measured into-food simulant migration values would mean in terms of actually occurring migration in foods (Foodmigrosure, 2006). Obviously, EU Directive 85/572/EEC needs a thorough revision concerning more realistic correlations between simulants and foods. This is even more important when considering that exposure-oriented migration is today one of the modern key aspects in ensuring compliance and safety of food-packaging materials.

Indeed, the final intention of into-food migration determination or evaluation is to have information about the possible concentration(s) of packaging constituents to which a consumer typically would be exposed to. Besides information of concentration(s) in food, further information data on plastics-packaging usage factors and food consumption data are needed (which can be obtained from other sources). But also without these additional information into-food migration modeling can be used as an, admittedly, exaggerating instrument for orientation purposes. For instance, it can serve to identify unproblematic exposure levels that can be ruled from further considerations or it can help to identify critical exposure scenarios where more refinement and accurate studies are necessary.

As a useful example the potential migration from PET bottles is considered in the following condition. This packaging material is typically used for carbonated soft drinks, isotonic beverages, ice teas, mineral water, and other. It has a relatively high market share that makes it interesting for exposure estimation. PET exhibits a very low-diffusion behavior and can lead to relatively small migration values only. In particular, the following migration scenario is modeled. A PET bottle with 1 L volume and 8 dm^2 inner contact surface is assumed to be representative for the exposure estimation. The average contact temperature with food should be room temperature (23 °C). The diffusion coefficient D_P for a given migrant with molecular weight up to 1000 g/mol is the corresponding one as set in the recognized migration model for food simulants (Piringer and Hinrichs 2001). Since at this moment, no validated migration model for foodstuffs is finally established, as a worst-case assumption the solubility for any migrant is assumed to be high which means $K_{P/F} = 1$. In practice, this means a migration scenario is modeled in this way which represents the situation of a PET bottle filled with pure fat or simulant D, olive oil. This modeled scenario, which is shown in Figure 11.12, can then be considered as a worse case for any food item potentially filled into the bottle. In Figure 11.12, the migration rates are expressed or lined out as functions of the molecular weight of a potential migrant and as storage time-related curves with contact times between 14 days and 1 year. To explain the meaning of the figure, for instance, a substance with molecular weight = 100 g/mol and present in PET material at initial concentration $C_{P,0} = 10$ ppm migrates into food at a level of 20 ppb after 1 year. Consequently, when assuming a standard 60 kg body weight person, the maximum possible exposure for the consumer from this type of bottle would be 0.02 mg per person and day or 0.33 µg per kg body weight per day. In this way, the migration behavior for any chemical substance of interest of concern from a PET bottle when filled with any foodstuff can be derived regardless to the type and nature of the particular foodstuff of interest or concern. The only information which is needed is the concentration of the migrant in

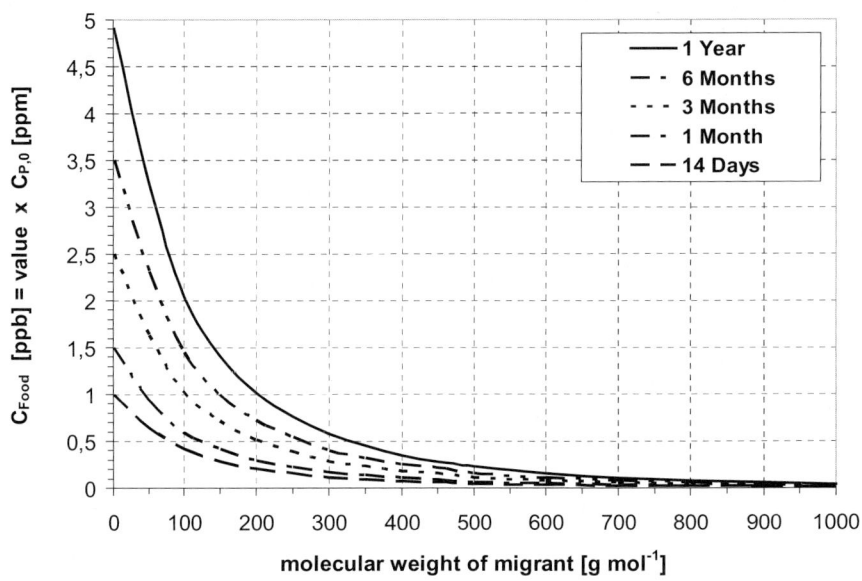

Figure 11.12 Modeling migration (worst-case exposure) at room temperature (23°C) into food packed in a 1 L volume PET bottle (8 dm² contact surface) as a function of molecular weight, residual content $C_{P,0}$ of a substance in the bottle wall, and storage time until consumption and under the assumption of high migrant solubility in the food. [For instance: a substance with molecular weight = 100 and present in PET with $C_{P,0} = 10$ ppm migrates into food at a level of 20 ppb after 1 year.]

the PET material. Again for worst-case considerations appropriate assumptions can be made from the generally available technical informations and specifications of PET bottle materials. Moreover, the modeled curves can also be used to predict the behavior of any PET packaging type independent of the material thickness because for thin films the migration rates would decrease and therefore the modeled curves would overestimate the situation. Also for other packaging surface-to-volume ratios the curves would either give directly relatively precise information or could be, of course, modeled for any ratio. In conclusion, the modeled curves allow very quick answers for any migrant of interest present in PET-packed foodstuff. Moreover, it should be noted that similar curves as in Figure 11.12 would be obtained for other low-diffusivity plastics such as rigid PVC, polycarbonate, polymethyl methacrylates, general purpose polystyrene, and other. As a consequence for all these materials, worst-case modeling in analogy to Figure 11.12 could largely simplify the discussion on possible exposures from these plastics (Chapters 6, 7 and 9).

11.8 Consideration of Non Intentionally Added Substances (NIAS) and Other not Regulated Migrants

For many years, NIAS, for example reaction by-products in the raw materials, degradation products, had been considered only when filing a petition for approval of a new substance but normally not for regular compliance testing. They have no SMLs but they are included in the general requirement of "not to endanger human health" of Article 3 EU Regulation No. 1935/2004. In the recent years, these substances come more and more in the focus of both, the legislator and the industry. Now in the fourth amendment to the Plastics Directive, 2002/72/EC, assessment of such contaminants and reaction products is required from industry in accordance with internationally recognized scientific principles (Directive, 2007/19/EC, "whereas (13)": European Commission, 2007). However, for this requirement no specific guidance documents do exist. But on the other hand, EC Directive 2007/19/EC outlines a possibility for evaluation of migration of nonauthorized substances by using the functional barrier concept. In multilayer structures nonauthorized substances may be used in non-food-contact layers if migration (including setoff) is below 0.01 mg/kg (EU Directive 2007/19/EC Article 7a, for definition compare also Section 11.1.2). Not allowed to be used are substances which are known or suspicious to be carcinogenic, mutagenic, or reproduction toxic. In United States, substances may be excluded from authorization when dietary exposure is below 0.5 mg/kg daily diet (21 CFR 170.39, US Government 2004 and US FDA, 1995). The migration value is multiplied with the packaging material specific consumption factor to obtain the dietary exposure (US FDA, 2002). The consumption factor is the statistical ratio of the use of the target material to the whole of all FCMs.

Intentionally added substances migrating from layers which are not specifically regulated like printing inks or adhesives are evaluated in a similar way. The ITX crisis in 2005/2006 (see above in Section 11.7) made it obvious that migration and setoff of printing ink components from the outside of a packaging may not be neglected. After transfer to the food contact side of the packaging, these substances are rapidly transferred into the food after package fill. Even in materials with very low diffusivity like polystyrene or PET setoff occurs. Printing ink components can plasticize the interface layer by setoff and cause a more rapid migration into food than expected from the diffusivity properties of the material. The Council of Europe recommends that migration of not evaluated substances should be below 10 µg/kg food (Council of Europe, 2006). The European Printing Ink Association proposes a similar procedure in future (EUPIA, 2007). Substances with not known toxicological properties should not migrate at levels higher than 10 µg/kg. For substances migrating up to 50 µg/kg absence of a mutagenic potential should be shown.

Under the NIAS also substances originating from use or misuse by the consumer in repeated use packagings (e.g., bottles) or recycling materials have to be subsumed. For recycling processes especially for PET bottle to bottle recycling, guidelines have been developed with respect to cleaning efficiency to such post consumer contaminants and safety evaluation (Franz et al., 2004; BfR, 2000; AFSSA, 2006) and an EU Regulation is in preparation. In cleaning/filling lines for repeated used bottles efficient detection systems are necessary to sort out bottles that are not sufficiently cleaned during the washing process.

From an analytical view, common to all these cases is that the identity of the potentially migrating substances is a priori not exactly known. Analytical screening assays are necessary. Identification can be obtained by mass spectrometric detection. To have higher concentrations in the solutions, identifications should be carried out best on the level of direct extracts from the materials. Semiquantitative estimates of concentrations in extracts and migration solutions are possible by GC–FID. From theory, the FID signal correlates to the mass of carbon atoms per time unit and therefore the signal should be more or less independent from the chemical structure and can be quantified via the signal of a universal internal standard. The use of *tert*-butyl-hydroxyanisol or other chemicals such as toluene as internal standards is recommendable. For nonvolatiles that are only measurable by HPLC such a generally applicable detection method with sufficient sensitivity for semiquantitative determination based on universal calibration is not yet available.

The evaluation should focus on additional substances apart from the typical oligomers. Oligomers from the approved polymers are considered as less reactive and less toxicological relevant. The reactive groups from the monomers disappear during polymerization reaction and are only present at the ends of the chains. In a multilayer material, peaks from approved polymer layers can be distinguished, for example, from peaks originating from adhesive or printing ink layers by comparing the chromatograms of the finished materials with those from the pure polymer films. Even knowing that every screening method will not detect all possible substances, this is the best available way at present to tackle this issue of unknown components in the materials, also known as the so-called forests of (GC) peaks.

References

AFSSA (Agence Française de Securité Sanitaire des Aliments). (2006) Opinion of the french food safety agency (AFSSA) on the assessment of health risks associated with the use of materials made from recycled poly (ethylene terephthalate) intended for or placed in contact with foodstuffs and drinking water. AFSSA, Mandate no 2001-SA-0315.

AiF Project (No. 13040 N). (2004) Final report "Development of a cost efficient concept for testing and evaluation of mass transport between thin packaging layers and foodstuffs for quality assurance and technological optimisation of plastics based multilayer packaging applications." Supported by the German Arbeitsgemeinschaft industrieller Forschungsvereinigungen (AiF) "Otto von Guericke" e.V.

Al Nafouri, A.J. and Franz, R. (1999) A study on the equivalence of olive oil and the EU official substitute test media for migration testing at high temperatures. *Food Addit. Contam*, **16**, 419–431.

Analytical Methods Committee. (1988) Uses (proper and improper) of correlation coefficients, Analyst 113, 1469–1471.

Ashby, R., Cooper, I., Harvey, S. and Tice, P. (1997) Food Packaging Migration and Legislation. Pira International. *Leatherhead*.

ASU (Amtliche Sammlung von Untersuchungsverfahren nach) § 64 LFGB. (1995) *Untersuchung von Lebensmitteln, Bestimmung von primären aromatischen Aminen in wässrigen Prüflebensmitteln*. L 00.00-6. Beuth-Verlag, Berlin.

Baner, L., Brandsch, J., Franz, R. and Piringer, O. (1996) The application of a predictive migration model for evaluating the compliance of plastic materials with European food regulations. *Food Addit. Contam*, **13** (5), 587–601.

Begley, T., Castle, L., Feigenbaum, A., Franz, R., Hinrichs, K., Lickly, T., Mercea, P., Milana, M., O'Brien, A., Rebre, S., Rijk, R. and Piringer, O. (2005) Evaluation of migration models that might be used in support of regulations for food contact plastics. *Food Addit. Contam*, **22** (1), 73–90.

Begley, T.H. and Hollifield, H.C. (1995) Food Packaging Made from Recycled Polymers: Functional Barrier Considerations. in *Plastic, Rubber and Paper Recycling: A pragmatic approach. ACS Symposium Series 609* (eds C.P. Rader, S.D. Baldwin, D.D. Cornell, G.D. Sadler and R.E. Stockel) American Chemical Society, Washington DC. pp. 445–457.

Berghammer, A., Bücherl, T. and Malter, C. (1994) Rapid extraction method for the determination of potential migrants from flexible packaging coated materials. *Verpack.-Rundsch*, **45** (7), 41–45.

BfR (Bundesinstitut für Risikobewertung). (2000) Use of mechanical recycled plastic made from polyethylene terephthalate (PET) for the manufacture of articles coming in contact with food, available from. http://www.bfr.bund.de.

Brandsch, J., Mercea, P., Rüter, M., Tosa, V. and Piringer, O. (2002) Migration modelling as a tool for quality assurance of food packaging. *Food Addit. Contam*, **19** (Suppl.), 29–41.

Brandsch, R., Gruner, A., Palzer, G., Piringer, O. and Franz, R. (2005) Migrationsuntersuchungen zur Prüfung und Bewertung von Stoffübergängen aus flexiblen Mehrschichtkunststoffverpackungen in Lebensm. *Dtsch. Lebensm.-Rundsch*, **101** (12), 549–559.

Bush, J. Gilbert, J. and Goenaga, X. (1994) *Spectra for Identification of Monomers in Food Packaging*, Kluwer Academic Publishers, Dordrecht, Boston, London.

BgVV. (1995) Statement of the German Commission on Plastics of the BgVV: Bundesgesundheitsblatt, 38, 73–74.

Castle, L., Honeybone, C.A., Read, W.A. and Boenke, A. (2000) *BCR Information Chemical Analysis 'Establishment of a Migration Test Method for Fatty Contact'*, EU Report 19376 EN.

Castle, L. (2003) Approaches to assess risk and assign priorities to chemicals used to make food contact materials. Final report for FSA project A03023 (available from library&info@foodstandards.gsi.gov.uk).

CEN. (2002) European Standard Series EN 1186, Materials and articles in contact with foodstuffs – plastics, Part 1–15. European Committee for Standardization, Brussels.

CEN. (2004) European Standard Series EN 13130, Materials and articles in contact with foodstuffs – plastics, Part 1–8. European Committee for Standardization, Brussels.

CEN. (2005a) Technical Specification CEN/TS 13130, Materials and articles in contact with foodstuffs – plastics, Part 9–28. European Committee for Standardization, Brussels.

CEN. (2005b) EN ISO/IEC 17025, General requirements for the competence of testing and calibration laboratories (ISO/IEC 17025:2005/Cor.1:2006). European Committee for Standardization, Brussels.

CEN. (2006a) EN 15130, Materials and articles in contact with foodstuffs – certain epoxy derivatives subject to limitation. Determination of BADGE, BFDGE and their hydroxy and chlorinated derivatives in food simulants. European Committee for Standardization, Brussels.

CEN. (2006b) CEN/TR 15356-1, Validation and interpretation of analytical methods, migration testing and analytical data for materials and articles in contact with food. Part 1. General considerations. European Committee for Standardization, Brussels.

Chapman, J.R. (1986) *Practical organic mass spectrometry*, John Wiley & Sons, New York.

Cortizas Castro, M.D. (1999) Ph.D. thesis. Universidad de Santiago de Compostela, Faculdad de Farmacia.

Council of Europe. (2006) Policy statement concerning packaging inks applied to the non-food contact surface of food packaging. Includes Resolution ResAP (2005)2 on packaging inks applied to the non-food contact surface of food packaging. Available at: http://www.coe.int/Soc-Sp.

De Kruijf, N. and Rijk, M.A.H. (1988) Iso-octane as fatty food simulant: Possibilities and limitations. *Food Addit. Contam*, **5**, 467–483.

De Kruijf, N. and Rijk, M.A.H. (1994) Test methods to simulate high-temperature exposure. *Food Addit. Contam*, **11**, 197–220.

Demertzis, P.G., Simal-Gándara, J. and Franz, R. (1995) A convenient group method for the gas chromatographic determination of aliphatic diamines in the four official EC food simulants. *Dtsch Lebensm.-Rdsch*, **91** (2), 35–38.

DIN. (1994) *Deutsche Norm DIN 32645: Chemische Analytik: Nachweis-, Erfassungs- und Bestimmungsgrenze: Ermittlung unter Wiederholbedingungen; Begriffe, Verfahren, Auswertung.* Deutsches Institut für Normung e.V, Berlin.

Dole, P., Voulzatis, Y., Vitrac, O., Reynier, A., Hankemeier, T., Aucejo, S. and Feigenbaum, A. (2006) Modelling of migration from multi-layers and functional barriers: Estimation of parameters. *Food Addit. Contam*, **23** (10), 1038–1052.

Duffy, E., Hearty, A.P., Flynn, A., McCarthy, S. and Gibney, M.J. (2006b) Estimation of exposure to food packaging materials. 2: Patterns of intakes of packaged foods in Irish children aged 5–12 years. *Food Addit. Contam*, **23** (7), 715–725.

Duffy, E., Hearty, A.P., Gilsenan, M.B. and Gibney, M.J. (2006a) Estimation of exposure to food packaging materials. 1: Development of a food packaging database. *Food Addit. Contam*, **23** (6), 623–633.

Duffy, E., Hearty, A.P., McCarthy, S. and Gibney, M.J. (2007) Estimation of exposure to food packaging materials. 3: Development of consumption factors and food-type distribution factors from data collected on Irish children. *Food Addit. Contam*, **24** (1), 63–74.

EFSA (European Food Safety Authority). (2005) Opinion of the Scientific Panel on Food Additives, Flavourings, Processing Aids and Materials in contact with food on a request from the Commission related to 2-isopropyl thioxanthone (ITX) and 2-ethylhexyl-4-dimethylaminobenzoate (EHDAB) in food contact materials

(Question numbers EFSA-Q-2005-240 & EFSA-Q-2005-241). *EFSA J.*, **293**, 1–15.

EFSA (European Food Safety Authority). (2006) Note for guidance for petitioners presenting an application for the safety assessment of a substance to be used in food contact materials prior to its authorisation. Available at: www.efsa.europa.eu/en/science/afc/afc_guidance/722.html.

EUPIA (European Printing Ink Association). (2007) Guideline on printing inks applied to the non-food contact surface of food packaging materials and articles. Available at:www.eupia.org.

Eurachem. (1998) The fitness for purpose of analytical methods – a laboratory guide to method validation and related topics. Available at: www.eurachem.org.

Eurachem. (2000) in *Eurachem/CITAC Guide Quantifying Uncertainty in Analytical Measurement*, 2nd edition (eds S.L.R. Ellison, M. Rosslein and A. Wiliams), Available at: www.eurachem.org.

European Commission. (1978). Council Directive of 30 January 1978 on the approximation of the laws of the Member States relating to materials and articles which contain vinyl chloride monomer and are intended to come into contact with foodstuffs (78/142/EEC). *Official J. Eur. Communities*, **L44**, 15–17. Corrigendum L163, 24.

European Commission. (1980) Commission Directive 80/766/EEC of 8 July 1980 laying down the Community method of analysis for the official control of the vinyl chloride monomer level in materials and articles which are intended to come into contact with foodstuffs. *Official J. Eur. Communities*, **L213**, 42–46.

European Commission. (1981). Commission Directive 81/432/EEC of 29 April 1981 laying down the Community method of analysis for the official control of vinyl chloride released by materials and articles into foodstuffs. *Official J. Eur. Communities*, **L167**, 6–11.

European Commission. (1985). Council Directive of 19. December 1985 laying down the list of simulants to be used for testing migration of constituents of plastic materials and articles intended to come into contact with foodstuffs (85/572/EEC). *Official J. Eur. Communities*, **L372**, 14–21.

European Commission. (1997) Commission Directive 97/48/EC of 29 July 1997 amending for the second time Council Directive 82/711 EEC laying down the basic rules necessary for testing migration of the constituents of plastic materials and articles intended to come into contact with foodstuffs. *Official J. Eur. Communities*, **L222**, 10–15.

European Commission. (2002) Commission Directive 2002/72/EC of 6 August 2002 relating to plastic materials and articles intended to come into contact with foodstuffs. *Official J. Eur. Communities*, **L220**, 18–58.

European Commission. (2004a) Regulation (EC) No 1935/2004 of the European Parliament and of the Council of 27 October 2004 on materials and articles intended to come into contact with food and repealing Directives 80/590/EEC and 89/109/EEC. *Official J. Eur. Communities*, **L338**, 4–17.

European Commission. (2004b) No 882/2004 of the European Parliament and of the Council of 29 April 2004 on official controls performed to ensure the verification of compliance with feed and food law, animal health and animal welfare rules. *Official J. Eur. Communities*, **L191**, 1–51.

European Commission. (2005) Commission Regulation (EC) No 1895/2005 of 18 November 2005 on the restriction of use of certain epoxy derivatives in materials and articles intended to come into contact with food. *Official J. Eur. Communities*, **L302**, 28–32.

European Commission. (2007) Commission Directive 2007/19/EC amending Directive 2002/72/EC. *Official J. Eur. Communities*, **L91**, 17–36.

FAO. (1998) Validation of analytical methods for food control. Report of a Joint FAO/IAEA Expert Consultation, December 1997, FAO

Food and Nutrition Paper No. 68, FAO, Rome.

Feigenbaum, A., Dole, P., Aucejo, S., Dainelli, D., De la Cruz Garcia, C., Hankemeier, T., N'Gono, Y., Papaspyrides, D., Paseiro, P., Pastorell, S., Pavlidou, S., Pennarum, PY., Saillard, P., Vidal, L., Vitrac, O. and Voulzatis, Y. (2005) Functional barriers: Properties and evaluation. *Food Addit. Contam*, **22** (10), 956–967.

Feron, V.J., Jetten, J., de Kruijf, N. and van den Berg, F. (1994) Polyethylene terephthalate bottles (PRBs): A health and safety assessment. *Food Addit. Contam*, **11** (5), 571–594.

Franz, R. (1995) Permation of flavour compounds across conventional as well as biodegradeable polymer films, in *Foods and Packaging Materials – Chemical Interactions* (eds Jägerstad Ackermann and Jägerstad Ohlsson), The Royal Society of Chemistry, Cambridge, pp. 45–50.

Franz, R. (2005) Migration modelling from food contact plastics into foodstuffs as a new tool for consumer exposure estimation. *Food Addit. Contam*, **22** (10), 920–937.

Franz, R., Bayer, F. and Welle, F. (2004) Guidance and criteria for safe recycling of post consumer polyethylene terephthalate (PET) into new food packaging applications. EU report EUR21155.

Franz, R., Huber, M. and Piringer, O.G. (1997) Presentation and experimental verification of a physico-mathematical model describing the migration across functional barrier layers into foodstuffs. *Food Addit. Contam*, **14** (6–7), 627–640.

Franz, R., Huber, M., Piringer, O.G., Damant, A.P., Jickells, S.M. and Castle, L. (1996) Study of functional barrier properties of multilayer recycled poly(ethylene terephthalate) bottles for soft drinks. *J. Agric. Food Chem*, **44**, 892–897.

Franz, R. and Rijk, R. (1997) Development of methods of analysis for monomers and other starting substances with SML and/or QM limits in Directives 90/128/EEC and 92/39/EEC. European Commission BCR information: Chemical analysis, EU report 17610 EN, ECSC-EC-EAEC. *Brussels, Luxembourg*.

Foodmigrosure. (2006) Proceedings of the Foodmigrosure Closing Conference on 26–27 September 2006 in Baveno, Italy, available from http://crl-fcm.jrc.it. Foodmigrosure EU Project Contract No. QLK1-CT2002-2390, www.foodmigrosure.org.

Gibney, M.J. and van der Voet, H. (2003) Introduction to the Monte Carlo project and the approach to the validation of probabilistic models of dietary exposure to selected food chemicals. *Food Addit. Contam*, **20** (Suppl. 1), S1–S7.

Gilsenan, M.B., Thompson, R.L., Lambe, J. and Gibney, M.J. (2003) Validation analysis of probabilistic models of dietary exposure to food additives. *Food Addit. Contam*, **20** (Suppl. 1), S61–S72.

Holmes, M.J., Hart, A., Northing, P., Oldring, P.K.T., Castle, L., Stott, D., Smith, G. and Wardman, O. (2005) Dietary exposure to chemical migrants from food contact materials: A probabilistic approach. *Food Addit. Contam*, **22**, 907–919.

Horwitz, W. (1988) Protocol for the design, conduct and interpretation of method performance studies. *Pure Appl. Chem*, **60**, 855–864.

Horwitz, W. (1995) Protocol for the design, conduct and interpretation of method performance studies. *Pure Appl. Chem*, **67**, 331–343.

Huber, W. (2003) Basic calculations about the limit of detection and its optimal determination. *Accred. Qual. Assur*, **8**, 213–217.

ILSI Europe. (2007) Guidance for Exposure Assessment of Substances Migrating from Food Packaging Materials. Monography prepared under the responsibility of the ILSI Europe Packaging Material Task Force, Brussels, Belgium, available at http://europe.ilsi.org/activities/taskforces/foodchain/PackagingMaterials.htm.

ISO. (1993) Guide to the Expression of Uncertainty in Measurement, Geneva.

ISO. (1994) International Standard ISO 5725: Accuracy (trueness and precision) of Measurement Method and Results, 12.

Katan, L.L. (1996a) Effects of migration, in *Migration from Food Contact Materials* (ed. L.L. Katan), Blackie Academic, London & Professional, (Chapter 2).

Katan, L.L. (ed.) (1996b) *Migration from Food Contact Materials*, Blackie Academic & Professional, London.

Lee, M.L., Yang, F.J. and Bartle, K.D. (1984) Open tubular column gas chromatography, *Theory and Practice*, John Wiley & Sons, New York.

Mandel, J. (1964) *The statistical analysis of experimental data*, Wiley, New York.

Massart, D.L., Vandeginste, B.G.M., Buydens, L.M.C., De Jong, S., Lewi, P.J. and Smeyers-Verbeke, J. (1997) *Handbook of Chemometrics and Qualimetrics. Part A*, Elsevier, Amsterdam.

Oldring, P.K.T. (2006) Exposure estimation – the missing element for assessing the safety of migrants from food packaging materials, in *Chemical Migration and Food Contact Materials* (eds K.A. Barnes, R. Sinclair and D. Watson), Woodhead Publishing, pp. 122–157.

O'Neill, E.T., Tuohy, J.J. and Franz, R. (1994) Comparison of milk and water/ethanol mixtures with respect to mono styrene migration from a polystyrene packaging material. *Int. Dairy J*, 4, 271–283.

Paseiro, P., Simoneau, C. and Franz, R. (2006a) Compilation of analytical methods for model migrants in foodstuffs: Collection of method descriptions. European Commission – Joint Research Centre, European Report EUR 22232 EN, © European Communities 2006 (printed in Italy).

Paseiro, P., Simoneau, C. and Franz, R. (2006b) Compilation of analytical methods for model migrants in foodstuffs: Review of analytical methodologies. European Commission – Joint Research Centre, European Report EUR 22552 EN, © European Communities 2006 (printed in Italy).

Paseiro Losada, P., Simal Lozano, J., Paz Abuín, S., López Mahía, P. and Simal-Gándara, J. (1993) Kinetics of the hydrolysis of bisphenol A diglycidyl ether (BADGE) in water-based food simulants. *Fresenius J. Anal. Chem*, 345, 527–532.

Petersen, H. (2003) Ph.D. thesis. Universität Hamburg, Fachbereich Chemie, Deutschland.

Philo, M.R., Damant, A.P. and Castle, L. (1997) Reactions of epoxide monomers in food simulants used to test plastics for migration. *Food Addit. Contam*, 14 (1), 75–82.

Philo, M.R., Jickells, S.M., Damant, A.P. and Castle, L. (1994) Stability of plastics monomers in food-simulating liquids under European Union migration test conditions. *J. Agric. Food Chem*, 42, 1497–1501.

Piringer, O. (1990) Ethanol und Ethanol/Wasser-Gemische als Prüflebensmittel für die Migration aus Kunststoffen. *Dtsch. Lebensm.-Rundsch*, 86 (2), 35–39.

Piringer, O., Franz, R., Huber, M., Begley, T.H. and McNeal, T.P. (1998) Migration from food packaging containing a functional barrier: Mathematical and experimental evaluation. *J. Agric. Food Chem*, 46, 1532–1538.

Piringer, O. and Hinrichs, K. (2001) Final report of the EU project contract SMT-CT98-7513, "Evaluation of Migration Models," DG Research, Brussels.

Piringer O.G. (ed.) (1993) *Verpackungen für Lebensmittel – Eignung, Wechselwirkungen, Sicherheit*. VCH Verlagsgesellschaft, Weinheim, New York, Basel, Cambridge. (Chapter 4).

Reynier, A. Dole, P. and Feigenbaum, A. (2002) Integrated approach of migration prediction using numerical modelling associated to experimental determination of key parameters. *Food Addit. Contam*, 19 (Suppl.), 42–55.

Rijk, R. (1993) Migration cells to test volatile substances. Proceedings of International Conference on "Materials for Food Packaging," March 10 and 11, PACKFORSK Gothenburg, Sweden.

Sanches-Silva, A., Cruz Freire, J.M., Franz, R. and PaseiroLosada, P. (2007) Time-temperature study of the kinetics of migration of diphenyl butadiene from polyethylene films into aqueous foodstuffs. Food Research International, in press.

Sanches-Silva, A., Sendón García, R., Cooper, I., Franz, R. and PaseiroLosada, P. (2006) Compilation of analytical methods and guidelines for the determination of selected model migrants form plastic packaging. *Trends Food Sci. Technol*, **17**, 535–546.

Schomburg, G. (1987) *Gaschromatographie – Grundlagen, Praxis, Kapillartechnik. 2. Auflage*, VCH Verlagsgesellschaft mbH, Weinheim.

Sendón García, R., Sanches Silva, A., Cooper, I., Franz, R. and Paseiro Losada, P. (2006) Revision of analytical strategier to evacuate different migrants from food packaging materials. *Trends Food Sci. Technol*, **17**, 354–366.

Simal-Gandara, J., Sarria-Vidal, M., Koorevaar, A. and Rijk, R. (2000b) Tests of potential functional barriers for laminated multilayer food packages. Part I: Low molecular weight permeants. *Food Addit. Contam*, **17**, 703–711.

Simal-Gandara, J. Sarria-Vidal, M. and Rijk, R. (2000a) Tests of potential functional barriers for laminated multilayer food packages. Part II: Medium molecular weight permeants. *Food Addit. Contam*, **17**, 815–819.

Stoffers, N.H., Brandsch, R., Bradley, E.L., Cooper, I., Dekker, M., Störmer, A. and Franz, R. (2005) Feasibility study for the development of certified reference materials for specific migration testing. Part 2: Estimation of diffusion parameters and comparison of experimental and predicted data. *Food Addit. Contam*, **22**, 173–184.

Stoffers, N.H., Dekker, M., Linssen, J.P.H., Störmer, A., Franz, R. and Van Boekel, M.A.J.S. (2003) Alternative fatty food simulants and diffusion kinetics of nylon 12 food packaging. *Food Addit. Contam*, **20**, 949–959.

Stoffers, N.H., Störmer, A., Bradley, E.L., Brandsch, R., Cooper, I., Linssen, J.P.H. and Franz, R. (2004) Feasibility study for the development of certified reference materials for specific migration testing. Part 1: Initial migrant concentration and specific migration. *Food Addit. Contam*, **21**, 1203–1216.

Störmer, A., Franz, R. and Welle, F. (2004) New concepts for food law compliance testing of polyethylene terephthalate bottles. *Dtsch. Lebensm.-Rundsch*, **100**, 44–47.

Tice, P. (1997) European Committee for Standardization (CEN) – Progress with standard test methods, in *Food Packaging Migration and Legislation* (eds R. Ashby, I. Cooper, S. Harvey and P., Tice) Pira International, (Appendix to Chapter 2, p. 37).

Tice, P. and Cooper, I. (1997) Rationalizing the testing of food contact plastics, in *Food Packaging Migration and Legislation* (eds. R. Ashby, I. Cooper, S. Harvey and Tice P.) Pira International, Leatherhead, (Chapter 5, p. 155).

Thompson, M., Ellison, LR. and Wood, R. (2002) Harmonized guidelines for single laboratory validation of methods of analysis. *Pure Appl. Chem.*, **74**, 835–855.

Thompson, M. and Wood, R. (1993) The International harmonised protocol for the proficiency testing of (chemical) analytical laboratories. *Pure Appl. Chem*, **65**, 2123–2144.

Thompson, M. and Wood, R. (1995) Harmonised guidelines for internal quality control in analytical chemistry laboratories. *Pure Appl. Chem*, **67**, 49–56.

US EPA. (1995) Guidance for methods development and methods validation for the Resource Conservation and Recovery Act (RCRA) Program, Washington, USA, 1995. Available at: www.epa.gov/epaoswer/hazwaste/test/pdfs/methdev.pdf.

US FDA (Food and Drug Administration). (1993) *Technical Review Guide: Validation of chromatographic methods*, Center for Drug Evaluation and Research (CDER), Rockville, Maryland, USA.

US FDA (Food and Drug Administration). (1995) Food and Drug Administration: Food Additives: Threshold of regulation for substances used in food contact articles. *Federal Register*, **60**, 36582–36596.

US. FDA (Food, Drug Administration). (2002) Guidance for Industry: Preparation of food contact notifications and food additive petitions for food contact substances: Chemistry recommendations. Final Guidance. Center for Food Safety and applied Nutrition. Office of Food Additive Safety (available at: http://www.cfsan.fda.gov/~dms/opa2pmnc.html).

US FDA (Food and Drug Administration). (2006) Guidance for Industry: Estimating Dietary Intake of Substances in Food. Center for Food Safety and applied Nutrition. Office of Food Additive Safety (available at: http://www.cfsan.fda.gov/guidance.html).

US Government. (2004) Code of Federal Regulations 21 Part 170.39 Threshold of regulation for substances used in food-contact articles. http://www.access.gpo.gov/nara/cfr/waisidx_04/21cfr170_04.html.

Van Battum D. February (1996) Alternative Fatty Food Simulants – A fact finding exercise. EU Commission Research Report N33, Revision 1.

Van Loco, J., Elskens, M., Croux, C. and Bernaert, H. (2002) Linearity of calibration curves: Use and misuse of the correlation coefficient. *Accred. Qual. Assur,* **7**, 281–285.

Vitrac, O. and Leblanc J-Ch. (2007) Consumer exposure to substances in plastic packaging I. Assessment of the contribution of styrene from yogurt pots. *Food Addit. Contam,* **24** (2), 194–215.

Vom Bruck, C.G., Bieber, W.-D. and Figge, K. (1986) Interactions between food and packaging materials and its consequences on migration, in *Food Packaging and Preservation. Theory and Practice* (ed. M. Mathlouti), Elsevier Applied Science, London, pp. 39–66.

12
US FDA Food Contact Materials Regulations

Allan Bailey, Layla Batarseh, Timothy Begley and Michelle Twaroski

12.1
Introduction

In the United States (US), regulatory processes have changed significantly in the past decade with regard to components of food contact articles or materials, often referred to as indirect food additives, allowing industry a variety of options for obtaining authorization for their safe use. A thorough understanding of the US regulatory processes for these substances, presently referred to as food contact substances[1] (FCS), allows industry to determine the most appropriate regulatory option based on the intended use. In this chapter, we discuss the US Food and Drug Administration's (FDA) regulatory authority and premarket safety evaluation of FCSs.

12.2
Regulatory Authority

Two acts are pertinent to any discussion regarding the regulation of food contact materials in the US. These are the 1958 Food Additives Amendment to the Federal Food, Drug, and Cosmetic Act (FFDCA) and the 1969 National Environmental Policy Act (NEPA). A brief discussion of the authority granted to the FDA under each follows.

12.2.1
Federal Food, Drug and Cosmetic Act (FFDCA)

The US Congress granted authority to the FDA to regulate food additives in the 1958 Food Additives Amendment to the FFDCA. A food additive is defined as "... any substance the intended use of which results or may reasonably be expected to result,

---

1) Section 409(h)(6) of the Act defines an FCS as "any substance intended for use as a component of materials used in the manufacturing, packing, packaging, transporting or holding of food if such use is not intended to have a technical effect in such food."

Plastic Packaging. Second Edition. Edited by O.G. Piringer and A.L. Baner
Copyright © 2008 WILEY-VCH Verlag GmbH & Co. KGaA, Weinheim
ISBN: 978-3-527-31455-3

directly or indirectly, in its becoming a component or otherwise affecting the characteristics of any food."[2] The food additive definition contains certain exclusions, such as those for color additives, substances whose use is generally recognized as safe, and substances approved for their intended use prior to September 6, 1958.

As detailed in the FFDCA, a food additive shall be deemed unsafe unless it conforms to an exemption (for investigational use), a regulation listing or, as explained below, an effective food contact notification (FCN).[3] Moreover, in the US, food additives require premarket evaluation before introduction into interstate commerce. Such an evaluation can lead to an approval through a food additive petition resulting in the publication of a regulation authorizing its intended use, a Threshold of Regulation (TOR) exemption resulting in a website listing, or an effective FCN resulting in a website listing for an individual notifier. The food additive petition ("petition") process is codified in Title 21, Part 171, of the Code of Federal Regulations (denoted as 21 CFR 171.1 through 171.130), while the TOR exemption and the FCN processes are codified in 21 CFR 170.39 and 21 CFR 170.100, respectively. As TOR exemptions and FCNs are appropriate for the majority of exposures encountered in food contact applications (i.e., exposures of 0.5 ppb or $\leq$1000 ppb, respectively), a more detailed discussion of these processes is provided below. A comparison of the current regulatory options available for an FCS is shown in Table 12.1. Of particular note is the applicability of an FCN with regard to both the range of allowable exposures and substances.

In 1995, the FDA established the TOR exemption process[4] (US Food and Drug Administration 1995). To obtain an exemption, two criteria must be met and confirmed by FDA; first, the estimated daily intake from the proposed use of the substance must be less than or equal to 1.5 μg per person per day (equivalent to a dietary concentration of less than or equal to 0.5 μg per kg of food consumed)[5] and, second, the substance must not be known to be a carcinogen in man or animals. FDA may decline to grant an exemption if the substance is a potent toxin or if there is a reason, based on the chemical structure of the substance, to suspect that the substance is a carcinogen. In addition, the substance must not contain a carcinogenic impurity with a tumor dose of 50 (TD_{50})[6] value of less than 6.25 mg per kg body weight per day. Alternatively, if the substance is currently regulated for direct addition to food and the estimated daily intake from the proposed use is less than or equal to 1% of the acceptable daily intake value, a TOR exemption from the need for a regulation may also be granted.

In 1997, the FDA Modernization Act amended Section 409 of the FFDCA to establish a new process, referred to as the FCN process, whereby food additives that are FCSs can be deemed safe for their intended use (US Food and Drug Administration 2002). The new FCN process[7] is intended as the primary method of

2) FFDCA, Section 201(s).
3) FFDCA, Section 409(a).
4) Codified in 21 CFR 170.39.
5) For the purpose of this chapter, estimated daily intake and dietary concentration are collectively referred to as exposure.
6) For the purpose of this chapter, the TD_{50} is the feeding dose that causes cancer in 50% of the test animals when corrected for tumors found in control animals.
7) Described in 21 CFR 170.100 through 170.106.

Table 12.1 General and specific listing regulations for food ingredients and packaging: Title 21 of the Code of Federal Regulations.

Regulation	Description
25	Environmental impact considerations
170	General information on food additives and notification process
171	General information on food additive petitions
172	Direct food additive regulations
173	Secondary direct food additive regulations
174–178	Indirect food additive regulations
175	Adhesives and Coatings
176	Paper and paperboard
177	Polymers
178	Adjuvants and production aids
179	Irradiation of foods
180	Substances permitted on an interim basis
181	Prior-sanctioned food ingredients
182	Generally recognized as safe (GRAS) substances
184	Direct substances affirmed GRAS
186	Indirect substances affirmed GRAS
189	Prohibited substances

authorizing new uses of an FCS; however, the petition process is still available. As codified in 21 CFR 170.100(c), a petition must be submitted to FDA for an FCS (unless FDA agrees to accept an FCN) in the cases where the cumulative estimated daily intake (which takes into account the estimated daily intakes for all authorized uses) is greater than 3 mg per person per day (or 0.6 mg per person per day for biocidal compounds) or where there exists a bioassay[8] on the FCS that FDA has not reviewed and which is not clearly negative for carcinogenic effects. FDA has generally agreed to accept an FCN rather than a petition if the sponsor submits a draft FCN and consults with FDA (termed "prenotification consultation" or PNC) prior to submission of an FCN, such that there is adequate time for a thorough review of the safety information.

In regard to the available options, industry/submitters should examine the comparisons outlined in Table 12.2. In contrast to the TOR exemption and petition processes, the FCN process results in an authorization for only the notifier and manufacturer/supplier listed in the FCN as opposed to the "generic" listing in the CFR that results from a petition approval. A TOR exemption is effective for any manufacturer/supplier of the FCS; however, unlike a petition, a CFR listing is not generated as a result of the safety review. Instead, FDA maintains a list of effective FCNs and TOR exemptions on its website (a list of current website addresses is contained in Table 12.3). Furthermore, unless the FDA objects, an FCN becomes

[8] Bioassay herein describes a chronic feeding study for the assessment of the potential of a chemical to produce carcinogenic effects.

Table 12.2 Regulatory options relevant to food contact materials with regard to the food additive petition (petition), food contact notification (FCN), and threshold of regulation (TOR) exemption processes.

Factor	Petition process	FCN process	TOR process
Requirements	Data requirements specified in 21 CFR 171.1; environmental requirements are specified in 21 CFR 25.15 and 40	Data requirements specified in 21 CFR 170.101; environmental requirements same as petitions	Data requirements specified in 21 CFR 170.39; environmental requirements same as petitions
Review period	180 Day review period after filing that can be "reset" with a new filing date as a result of substantive amendment	120 Day review period after acceptance that cannot be reset	Review period is variable, averages 45 days
Review outcome	Federal Register (FR) publications; regulation listing in 21 CFR 170–199	Notification letters; listing on CFSAN website	Letter to submitter; listing on CFSAN website
Legality	Not legal until a regulation publishes	Legal if the Agency has no objections after the 120 day review period	Not legal until letter is received
Ownership	Generic to all manufacturers	Exclusive to the supplier or manufacturer named in the FCN	Same as petitions
Confidentiality	Disclosure of submission and data during review; automatic disclosure of environmental assessments at the time of filing	Disclosure after 120 day review period, including disclosure of environmental assessments	Same as petitions; automatic disclosure of environmental assessments at the time of receipt of submission
Qualifying exposure	None	Cumulative dietary concentration of less than 1000 µg per kg food consumed	Dietary concentration of less than or equal to 0.5 µg per kg consumed

effective 120 days after the acceptance date and the FCS may be legally marketed for the proposed use. In contrast, TOR exemption requests and petitions do not have a statutory time frame default approval like FCNs and, unlike petitions, FCN submissions are confidential until the 120 day effective date or the date the Agency objects to the submission. Alternatively, if a notifier decides to withdraw an FCN during the review process, the submission of the FCN and its contents remain confidential.

Table 12.3 Useful links on the "Food Ingredients and Packaging: Food Contact Substance Notification Program" section of the CFSAN Website (http://www.cfsan.fda.gov/~lrd/foodadd.html).

Contents	Location (URL)
Inventory of effective FCNs	http://www.cfsan.fda.gov/~dms/opa-fcn.html
Inventory of environmental records for effective FCNs	http://www.cfsan.fda.gov/~rdb/opa-envt.html
Administrative guidance	http://www.cfsan.fda.gov/~dms/opa2pmna.html
Chemistry guidance	http://www.cfsan.fda.gov/~dms/opa2pmnc.html
Toxicology guidance	http://www.cfsan.fda.gov/~dms/opa2pmnt.html
Environmental guidance	http://www.cfsan.fda.gov/~dms/opa2eg.html
FDA Form 3480	http://www.cfsan.fda.gov/~dms/pmnforms.html
CEDI/ADI database	http://www.cfsan.fda.gov/~dms/opa-edi.html
Threshold of Regulation (TOR) guidance	http://www.cfsan.fda.gov/~dms/torguid.html
Inventory of TOR exemption requests	http://www.cfsan.fda.gov/~dms/opa-torx.html
CFR listing through GPO	http://www.access.gpo.gov/nara/cfr/waisidx_01/21cfrv3_01.html
List of indirect additives used in packaging	http://www.cfsan.fda.gov/~dms/opa-indt.html
Redbook 2000	http://www.cfsan.fda.gov/~redbook/red-toca.html
Toxicology templates	http://www.cfsan.fda.gov/~dms/opatoxtm.html

Under all three processes, the submitter bears the burden of demonstrating that the intended use of the FCS is safe. In reviewing the submitter's determination of safety, FDA uses the "reasonable certainty of no harm" safety standard.[9] Thus, the data and information in all three processes are generally comparable for a given level of consumer exposure and, internally, FDA's safety review and standard are equivalent for the three options.

12.2.2
National Environmental Policy Act (NEPA)

The FDA has environmental responsibilities under NEPA of 1969. NEPA requires Federal agencies to take, to the fullest extent possible, environmental considerations into account in the planning and making of their major (where the responsible official fails to act and that failure is reviewable by courts; i.e., it is subject to judicial review) and final (meets the finality test under the Administrative Procedure Act) Agency decisions. NEPA is divided into two sections. Title I contains the broad statement of national environmental policy and the "action-forcing" components of the Statute. Title II establishes the Council on Environmental Quality (CEQ) which is the regulatory body charged with overseeing NEPA implementation. In accordance with Title II, the regulations of CEQ require Federal agencies to develop their own

---

[9] Codified in 21 CFR 170.3(i).

regulations to comply with the procedures and achieve the goals of NEPA.[10] FDA's NEPA-implementing procedures are set forth in 21 CFR Part 25 to supplement the CEQ regulations.

NEPA is a declaration of the Nation's environmental policy and goals. It supplements FDA's authority under the FFDCA and other public health statutes but it does not supersede these statutes. It does not require substantive FDA decisions to favor environmental protection over other considerations mandated by other statutes the FDA administers, such as human health. NEPA is a full disclosure statute that requires public involvement and it is a broad statute that considers all aspects of the human environment. In addition, NEPA applies abroad and requires Federal agencies to identify those actions that may have trans-boundary environmental effects. The FDA considers authorizing the uses of FCSs under all three regulatory options (FAP, TOR, and FCN) to be both major and final Agency actions and, thus, all are subject to NEPA considerations.

12.3
Premarket Safety Assessment

12.3.1
Introduction

FDA's safety assessment relies on evaluating probable consumer exposure to an FCS, including all constituents or impurities, as a result of the proposed use and other authorized uses, and ensuring that such exposures are supported by the available toxicological information. It is important to understand that the safety assessment focuses on those substances that would be expected to become components of food as a result of the proposed use of the FCS. A general discussion of the recommended chemistry, toxicology, and environmental information for a submission relating to an FCS follows. Detailed technical guidance documents can be found on the Agency's website (www.cfsan.fda.gov/~dms/opa-guid.html). An industry/stakeholder considering an FCN submission must address all three of these areas in order for their submission to be complete.

12.3.2
Chemistry Information

The recommended chemistry information includes the discussion and submission of supporting data on the identity, manufacture, stability, technical effect, and proposed use of the FCS, all of which are used to identify and estimate consumer exposure to the various substances originating from the proposed use of the FCS. Exposure estimates usually involve combining migrant concentrations in food with information on the use of articles that contain the FCS. The following discussion

---

10) Codified in 40 CFR 1500–1508.

will focus on what may be considered the most pivotal aspect of the safety assessment, i.e., estimating consumer exposure.

12.3.2.1 Migrant Levels in Food

In general, migrant levels in food are determined by one or more of the following methods: (1) accelerated migration studies conducted with food-simulating liquids under the most severe anticipated conditions of use; (2) the assumption of 100% migration to food using actual use or residue levels; or (3) mathematical modeling of mass transfer from polymers to food based on a thermal processing-extended storage scenario using actual use or residue levels.

There are two categories for using an FCS: single-use and repeat-use applications. Single-use describes an article that will be used one time, such as a plastic bottle for holding beverages, whereas repeat-use articles will be used over an extended period and will contact food repeatedly, such as an o-ring, a conveyor belt used in a food processing plant, or a food tray. The protocols for migration studies are generally intended to model the thermal treatment and extended storage (or contact) of the article containing the FCS. The intended use conditions of the FCS (i.e., use level and article(s), food types, time/temperatures conditions, single or repeat use) are crucial in determining test parameters (i.e., test sample(s), food simulants, and accelerated time/temperature protocols). In other words, the intended use of articles containing the FCS determines the appropriate food types and classifications (Table 12.4) as well as the accelerated time/temperature conditions (Table 12.5) that should be used to determine migrant concentrations in food. It is important to note that these migration testing protocols for a new FCS or use bear no relation to the "end-test" specifications listed in the CFR for compliance evaluation.

In some cases where the use level of the FCS or residue level of a migrant is low, it may be possible to dispense with migration studies altogether by assuming 100% migration of the substance to food. Although actual migration studies might result in a lower estimate of migration to food, hence a lower exposure, such studies would be unnecessary if the resulting exposure is sufficiently low or otherwise supported by the available toxicological information. Alternatively, mathematical modeling of migration based on the principles of mass transfer from polymers to foods, an approach also recognized by the European Food Safety Authority, can be used along with a consideration of the intended conditions of use (Begley *et al.*, 2005). Migration modeling is discussed in Section II, Part D.5, of FDA's chemistry guidance document and elsewhere in this text.

12.3.2.2 Packaging Information

Information on single-use food contact articles that contain the FCS is captured by packaging factors, which include both consumption factors (CF) and food-type distribution factors (f_T). The CF describes the fraction of the daily diet expected to contact specific packaging materials and represents the ratio of the weight of all food contacting a specific packaging material to the weight of all packaged food. CF values for select packaging categories (e.g., polymer and paper), specific food-contact polymers (e.g., low-density polyethylene), and applications (e.g., microwave susceptors) are summarized in the chemistry guidance document and presented here in Table 12.6.

Table 12.4 Classification of food types and recommended food simulants for food contact articles[a].

Type	Description	Classification
I	Nonacid, aqueous products; may contain salt, sugar or both (pH > 5), e.g., raspberries, maple syrup, consomme, ripe olives	Aqueous
II	Acid, aqueous products; may contain salt, sugar or both, and including oil-in-water emulsions of low- or high-fat content, e.g., vinegar, mayonnaise, orange juice, cream dressing	Acidic
III	Aqueous, acid or nonacid products containing free oil or fat; may contain salt, and including water-in-oil emulsions of low- or high-fat content, e.g., crab, lobster, ground beef, bacon, chicken, oleomargarine	Fatty
IV	Dairy products and modifications	
	A. Water-in-oil emulsions, high- or low-fat, e.g., cheddar cheese, Swiss cheese, butter	Fatty
	B. Oil-in-water emulsions, high- or low-fat, e.g., milk, ice cream, cottage cheese, sweet cream (40%)	Aqueous
V	Low-moisture fats and oils, e.g., Lard, peanut oil	Fatty
VI	Beverages	
	A. Containing up to 8% alcohol, e.g., beer	Low-alcohol
	B. Nonalcoholic, e.g., soda	Aqueous
	C. Containing more than 8% alcohol, e.g., distilled spirits, vodka	High-alcohol
VII	Bakery products (other than those under types VIII or IX)	
	A. Moist bakery products with surface containing free fat or oil	Fatty
	B. Moist bakery products with surface containing no free fat or oil	Aqueous
VIII	Dry solids with the surface containing no free fat or oil, e.g., macaroni, shredded wheat, corn meal, coffee	Dry
IX	Dry solids with the surface containing free fat or oil, e.g., potato chips, French fried potatoes, broiled meat and fish, fried chicken, pop corn	Fatty

[a]Recommended food simulants (a more detail discussion is in the chemistry guidance document): aqueous/acidic/low-alcohol foods – 10% ethanol; high-alcohol foods – 50% ethanol; fatty foods – aqueous ethanol solutions or food oil.

The f_T describes the fraction of all food contacting each material that is aqueous (aq), acidic (ac), alcoholic (al), and fatty (fat). The f_T accounts for the variable nature of foods contacting each packaging material. Their use in calculating exposure is critical, as migration is dependant on several factors, including the nature of the food matrix, i.e., the food type. Although migration might be highest in fatty foods for a particular FCS, use of the article containing the FCS might also be extremely limited in its application to fatty foods. Applying the f_T to the migrant levels in food simulants allows for a "weighted average" migration to food, denoted as $\langle M \rangle$, to be used in estimating exposure.

The expression relating migrant levels in food and packaging factors to estimate exposure (expressed as dietary concentration, DC, in units of mass of migrant per mass food consumed) for single-use articles is

$$DC = CF\langle M \rangle = CF\sum (M_{FSL})(f_T)$$

Table 12.5 Time-temperature conditions of use for single- and repeat-use.

Designation	Name or description	Protocol
Single use		
A	High temperature, heat sterilized or retorted above 100°C	121°C/2 h, 40°C/238 h
B	Boiling water sterilized	100°C/2 h, 40°C/238 h
C	Hot-filled or pasteurized above 66°C	100°C/0.5 h, 40°C/239.5 h or 66°C/2 h, 40°C/238 h
D	Hot-filled or pasteurized below 66°C	66°C/0.5 h, 40°C/239.5 h
E	Room temperature filled and stored (no thermal treatment in the container)	40°C/240 h
F	Refrigerated storage (no thermal treatment in the container)	20°C/240 h
G	Frozen storage (no thermal treatment in the container)	20°C/120 h
H	Frozen or refrigerated storage; ready-prepared food intended to be reheated in container at time of use	100°C/2 h
Repeat use		
	Repeated use in contact with food	Highest intended use temperature for the longest time
High temperature use		
	Dual-ovenable trays	Highest intended conventional oven temperature for the longest time
	Microwavable containers	See guidance document
	Microwave heat-susceptor applications	See guidance document

where M_{FSL} is the migration level in each food simulating liquid and f_T is the factor for the simulant intended to model each type of food. The expression FDA uses in relating DC and estimated daily intake, EDI, assuming a consumption of 3 kg (1.5 kg liquid and 1.5 kg solid) food per person per day is

$$\text{EDI} = \text{DC} \times 3\,\text{kg food per person per day}$$

Exposure estimates derived from packaging factors are "averages" across the US population and may be thought of as "per capita" estimates. FDA believes that this "per capita"-based approach to estimate exposure to components of food contact articles is appropriate because consumer selection of food is not generally dependent on the type of packaging; rather, it is dependent on the eating habits and spending preferences of the consumer. In fact, one criticism of FDA's approach to estimate exposure for components of packaging materials is the assumption that a food(s) eaten by a given consumer will have been packaged with the same material 100% of the time. For example, if a notifier proposes use of an antioxidant in high-density polyethylene, the consumer is assumed to ingest the selected food(s) only if it is packaged in the high-density polyethylene containing the antioxidant, even though the food(s) may be packaged in other materials as well.

Table 12.6 Select packaging factors for several packaging categories (a complete list may be found in Appendix IV of the chemistry guidance document).

Category	CF	Packaging factors			
		f(aqueous)	f(acid)	f(alcohol)	f(fat)
Metal – uncoated	0.03	0.54	0.25	0.01	0.20
Metal – polymer coated	0.17	0.16	0.35	0.40	0.09
Paper – uncoated and clay coated	0.1	0.57	0.01	0.01	0.41
Paper – polymer coated	0.2	0.55	0.04	0.01	0.40
Low-density polyethylene (LDPE)	0.12	0.67	0.01	0.01	0.31
Linear LDPE (LLDPE)	0.06	"	"	"	"
High density polyethylene (HDPE)	0.13	"	"	"	"
Polypropylene (PP)	0.04	"	"	"	"
Polyethylene terephthalate (PET)	0.16	0.01	0.97	0.01	0.01
Nylons	0.02	0.10	0.10	0.05	0.75
Adhesives (consistent with Section 21 CFR 175.105)	0.14				
Retort pouch	0.0004				
Microwave susceptor	0.001				

FDA modifies these packaging factors as new information on the use of packaging materials in the marketplace becomes available. In fact, the more information a submitter can provide on the specific scenarios of use, such as subdividing packaging or resin categories with marketing information, the more accurately FDA can estimate exposure. As summarized recently by Heckman 2005, this use of packaging factors for determining consumer exposure contrasts with the approach utilized by other regulatory bodies, such as the EU.

For repeat-use food-contact articles that contain the FCS, FDA takes into account information such as the weight of food contacting a known area of a representative repeat-use article, contact time of food with the repeat-use article, and the average lifetime of the article. This allows the calculation of a representative food mass-to-surface area ratio for the use and the extrapolation of the representative migrant levels in food over the entire service lifetime of the article. In general, exposure estimates for repeat-use applications are lower than those for single-use, primarily because of the large food mass-to-surface area ratio used in the calculations. In fact, submissions for only repeat use are generally ideal candidates for TOR exemption requests.

12.3.3
Toxicology Information

FDA's toxicological assessment is based on a tiered approach and is consistent with the general principle that increased exposure leads to increased potential health risks; however, the inherent toxicity of a structural/functional class of chemicals is also considered, as demonstrated by the separate requirements for biocides. The following section discusses FDA's approach to the evaluation of toxicological information in the overall safety assessment of an FCS.

12.3.3.1 Safety Assessment

Tiered Testing Recommendations and Noncarcinogenic Assessments FDA does not ordinarily suggest conducting toxicity tests at DCs less than or equal to 0.5 µg per kg food consumed; rather, a literature search and a structural comparison to known carcinogens are requested. This approach and the safety information are the minimum regardless of submission type. It is noteworthy that the tiered approach expands the test requirements with increasing exposure (Table 12.7). Accordingly, the information needed to support safety at these low doses is required for any level of exposure. The use of structural comparisons or structure–activity relationship (SAR) analysis has long been important in prioritizing toxicity concerns. With the development of the TOR and FCN processes, an FDA review group now reviews every submission explicitly from this perspective.

At the next tier of exposure, between 0.5 and 50 µg per kg food consumed, FDA recommends assessing the genetic toxicity of the substance using *in vitro* assays. These assays are less expensive and time consuming than a bioassay and they provide information on the ability of the substance to cause genetic damage, a key event in the development of cancer (Pitot III and Dragan 2001). As many of the chemicals involved in the production of FCSs are likely to be of wide interest, based on other uses and/or subject to another regulatory authority, these studies are often available in the open literature. These assays have varying specificities and sensitivities, as recently evaluated by Kirkland *et al.* 2005, and measure different endpoints. Therefore, when differences in individual test results arise, the collective data may be difficult to interpret with regard to the overall carcinogenic potential of the substance. Accordingly, by examining multiple endpoints of genetic toxicity, a broad assessment of the substance's ability to cause genetic damage is obtainable. Pairing these data with SAR analysis allows for a more comprehensive review of a chemical's potential to be mutagenic and/or carcinogenic.

At exposures of greater than 50 µg per kg food consumed, in addition to the data requested at the lower tiers, FDA recommends the completion of an *in vivo* test for chromosomal damage using rodent hematopoietic cells (i.e., mouse micronucleus assay) and two subchronic studies, one rodent and one nonrodent. Additional studies may be suggested based on the results of the subchronic studies and/or SAR analysis. FDA uses this information to determine an acceptable daily intake (ADI) value for the substance from the most sensitive species (hence the need for both rodent and nonrodent data). In determining an ADI value, FDA uses the no observed effect level (NOEL) or the no observed adverse effect level (NOAEL, if the use of a NOAEL is appropriate based on the endpoint examined), and uncertainty (or safety) factors. Table 12.8 details the determination of an ADI value and the applicable uncertainty factors used by the FDA. Once an ADI value is determined, this value is compared to the cumulative estimated daily intake (CEDI). If the ADI value is greater than the

Table 12.7 Toxicology testing recommendations for food contact substances based on dietary concentration (DC) and corresponding estimated daily intake (EDI) values[a].

Exposure	Recommendation
DC of $\leq 0.5\,\mu g\,kg^{-1}$ (EDI of $\leq 1.5\,\mu g$ per person per day)	• No toxicity testing recommended • Available information on potential mutagenicity and carcinogenicity, published and unpublished, should be submitted and discussed • Structural similarity of the substance to known carcinogens or genotoxic chemicals should be discussed, if appropriate
CDC of $> 0.5\,\mu g\,kg^{-1}$ but $\leq 50\,\mu g\,kg^{-1}$ (CEDI of $>1.5\,\mu g$ per person per day but $\leq 150\,\mu g$ per person per day)	Recommendations for DC of $\leq 0.5\,\mu g\,kg^{-1}$ and: • Genetic toxicity tests on the substance • An *in vitro* cytogenetic test in mammalian cells or an *in vitro* mouse lymphoma tk± assay • A test for gene mutations in bacteria
CDC of $>50\,\mu g\,kg^{-1}$ but $<1000\,\mu g\,kg^{-1}$ (CEDI of $>150\,\mu g$ per person per day but $<3\,mg$ per person per day)	Recommendations for DC of $\leq 0.5\,\mu g\,kg^{-1}$ and: • Genetic toxicity tests on the substance 1. A test for gene mutations in bacteria 2. An *in vitro* cytogenetic test in mammalian cells or *in vitro* mouse lymphoma tk± assay 3. An *in vivo* test for chromosomal damage using rodent hematopoietic cells • Potential toxicity of the substance should be evaluated by two subchronic (90 day) oral toxicity tests, one in a rodent species and the other in a nonrodent species • Results from these studies or other available information may trigger the need for longer term (1-year or 2 year) or specialized (e.g., reproductive/developmental toxicity, neurotoxicity, etc.) tests
CDC of $\geq 1000\,\mu g\,kg^{-1}$ (EDI of $\geq 3\,mg$ per person per day)	Recommend Food Additive Petition containing the data listed above for lower exposures and: • Two-year carcinogenicity bioassays in two rodent species (one study should include *in utero* phase) • A two-generation reproductive study in rats with a teratology phase • Other specialized studies, as appropriate

[a]Note that the cumulative exposures are based on nonbiocidal chemicals; biocidal tiers are $1/5$th the cumulative dietary concentration (CDC) and cumulative estimated daily intake (CEDI) values expressed. DC and CDC values are in units of mass of migrant per mass food consumed. EDI and CEDI values are in units of mass of migrant per person per day. Abbreviations are as follows: µg: microgram; kg: kilogram; mg: milligram; $\leq$: less than or equal to; $<$: less than;' and $>$: greater than.

CEDI, the substance is considered safe for the proposed use. If the ADI value is less than the CEDI, the substance is not considered safe for its proposed use and more information, either toxicological or exposure, may be used to establish safety. In general, ADI values are not calculated for substances with positive genotoxic or neoplastic findings.

Table 12.8 Examples of FDA's approach to toxicity safety assessment depending on endpoint of concern.

Endpoint	Example of assessment approach
Nonneoplastic endpoint	Derivation of an acceptable daily intake (ADI) value using the no observed effect level (NOEL): ADI (mg kg^{-1} bw per day) = NOEL/(UF1 × UF2 × ...); (ADI)* 60 kg per person = mg per person per day Uncertainty factors (UF): • 10 extrapolation from animals to humans • 10 intraspecies variability • 10 for less than chronic exposure Typical FDA uncertainty factors: • 100–200 for chronic toxicity studies • 1000–2000 for subchronic toxicity studies • 1000 where reproductive and developmental effects are severe and/or irreversible • 100 where reproductive and developmental effects are not severe or are reversible
Neoplastic endpoints	
Food additive	Delaney clause applies – assessment is qualitative (positive or negative)
Constituent – genotoxic	Multiply unit cancer risk by EDI to obtain worst case upper bound lifetime cancer risk. • Unit cancer risk (UCR) is defined as the sum of the slopes of lines drawn from the lowest apparent effective dose of the chemical through zero for each tumor site • Tumors arising from multiple sites are assumed to be independent and are added to obtain the overall UCR • Lowest dose at which significant neoplastic findings are reported is used to calculate UCR
Constituent – nongenotoxic	• Evaluate using threshold approach applying applicable uncertainty factors detailed above

At exposures of greater than 1000 µg per kg food consumed, FDA recommends the completion of two 2-year carcinogenic bioassays in rodents (one with an *in utero* phase) to determine the carcinogenic potential of the substance, a reproductive/developmental study, and other specialized studies as warranted by the results of shorter term studies, structure–activity relationship analysis, exposure or other available information. Again, substances with DCs greater than 1000 µg per kg food consumed are not acceptable for submission of an FCN without previous agreement with FDA.

Carcinogenic Assessments With regard to a food additive, Section 409(c)(3)(a) of the FFDCA states that "... no additive shall be deemed safe if it is found to induce cancer when ingested by man or animal...." This is the so called food additive or anti-cancer Delaney Clause. The Delaney Clause applies to the additive itself, not to constituents of the additive or impurities that may be present [(Scott v. FDA, 728 F. 2d 322 (US 6th Cir. 1984)]. Accordingly, individual constituents of the food additive may be evaluated under the general safety standard using applicable risk assessment procedures. This regulatory interpretation is often referred to as the Constituents Policy (US Food and Drug Administration 1982). Although the level of exposure from an FCS may not necessarily require performing a bioassay, if such studies are available, these data are considered pivotal and must be submitted regardless of the level of exposure expected.

In the safety assessment of the FCS, neoplastic data on constituents of an additive are reviewed to determine whether or not the constituent is a carcinogen. If there are no positive carcinogenic findings, this resolves any cancer issues and other toxicity data are considered in the safety assessment of the constituent. On the other hand, if the constituent is determined to be a carcinogen, risk assessment procedures are used to determine whether or not the proposed use would present more than a negligible risk. The unit cancer risk derived from the linear, low-dose extrapolation is used in combination with the exposure estimate to determine the upper-bound lifetime cancer risk. Details of this approach for constituents are provided in Table 12.8 and have been elaborated upon elsewhere (Lorentzen 1984; FDA 2006). It is important to note that this approach is applicable only to constituents of food additives and not to food additives themselves, which are subject to the Delaney prohibition.

Use of Structure Activity Relationship (SAR) Analysis As mentioned above, the structure of a chemical may be used to predict its toxicological profile based on analogies to chemicals that have been tested for the endpoint of concern. This is the basis of SAR analysis. FDA has historically considered this predictive approach valuable is in the safety assessment as exemplified in Redbook 2000 and the analysis performed in establishing the TOR exemption policy by Rulis 1989. Moreover, 21 CFR 170.39(a)(1) specifies that the substance that is the subject of the TOR exemption request will be granted an exemption only if it has not been shown to be a carcinogen in humans or animals, and *there is no reason, based on the chemical structure of the substance, to suspect that the substance is a carcinogen.* This allows FDA to consider the carcinogenic potential of a substance whose structure suggests that it may be a potent carcinogen even if the use of the substance would result in a DC less than or equal to 0.5 ppb.

At the FDA, a team with expertise in SAR analysis and analysis software routinely apply SAR analysis in the evaluation of an FCS. Details regarding SAR analysis and its incorporation in the safety review were recently elaborated upon by Bailey et al. 2005. Briefly, SAR analysis is performed on all migrating substances using a two-phase approach, qualitative SAR, and, if necessary, quantitative SAR. Qualitative SAR analysis involves using experts systems and software programs that compare the queried structure to tested substances using substructure, similarity, and biophores to produce a concern level. Quantitative SAR, on the other hand, involves examining the data of the analogs, identified during the qualitative run or elsewhere, in order to obtain potency values that can be used in a predictive analysis. The results of these analyses are combined with the toxicological data provided or identified for the compound in question to perform a weight of evidence safety assessment. For example, SAR analysis, when paired with mixed results of multiple genetic toxicity assays, can be useful in determining if the genotoxic substance is related to known carcinogens and what level of risk it presents given the exposure. Moreover, using this tool, FDA can more effectively determine if expensive, long term studies, such as bioassays, are necessary to ensure safety given the exposure and what is known about compounds in a given class of chemicals.

12.3.3.2 General Considerations

There are several important considerations to keep in mind with regard to toxicity data. First, although FDA has established a tiered approach to testing, these are not rigid criteria but points that trigger examination of additional endpoints. As the safety standard is "a reasonable certainty of no harm," it is important to consider a comprehensive approach to the data and how the structure of the substance relates to its potential toxicity. As an example, though toxicological testing is not ordinarily requested for exposures less than or equal to 0.5 μg per kg food consumed, it may be requested for a substance that contains biophores indicating mutagenic activity. The same can be said for reproductive SAR analysis at exposures of less than or equal to 50 μg per kg food consumed. In addition, notifiers are required to submit and discuss all available, relevant data regardless of the level of exposure. For this reason, the FDA receives numerous studies that would not be required based on the estimated exposure. Some of these studies, such as 28-day studies, are not considered appropriate for setting ADI values; rather, they are used to assess toxicity which may be the basis for requesting longer terms studies to ensure safety.

Submitters should thoroughly search all publicly available databases, including those of international and foreign regulatory agencies, for applicable information with regard to their safety analysis. A thorough review of available information often avoids questions and requests for additional information during FDA's initial review. Additionally, correspondence between FDA and submitters regarding available, previously determined safety analyses conducted by FDA is often time saving and can assist in the preparation of a safety assessment. This is important if exposure is expected to a substance that was evaluated prior to the inception of the FCN process, because the substance may not have been evaluated for mutagenicity when it was evaluated for general toxicity. In such cases, the submitter proposing a new use may be requested to submit genetic toxicity studies to supplement the available data.

Over the last several years, the FDA has developed several toxicology assessment tools to aid industry in the submission process. The *Redbook 2000 Toxicological Principles for the Safety Assessment of Food Ingredients* (US Food and Drug Administration 2004) details testing guidelines for commonly requested toxicology tests. Additionally, the FDA has developed toxicology templates to aid in the review and evaluation of commonly requested tests. Lastly, FDA-developed ADI and unit cancer risk values are available either on FDA's website or through correspondence with the Agency.

12.3.4
Environmental Information

FDA's assessment of the environmental impact of FCSs is the same under the FCN, TOR, and petition review processes; however, there is a major difference in the availability of the environmental record. NEPA is a full disclosure statute and Section 1506.6(a) (Public Involvement) of the Council on Environmental Quality regulations implementing NEPA states that Federal agencies shall make diligent efforts to involve the public in preparing and implementing their NEPA procedures. As a result, the environmental record for a petition and a TOR exemption request are made publicly available at FDA's Dockets at the time of acceptance. However, the FFDCA requires that the FCN remain confidential during the 120-day review period.[11] Because NEPA does not supersede the main statute under which the Agency functions, the environmental record for an FCN is not made publicly available until FFDCA permits its availability; specifically, after an FCN becomes effective.

The Agency will not accept an FCN, a TOR exemption request, or a petition for review if the environmental component is missing or deficient (21 CFR 25.15); thus, every one of these submissions must contain either an environmental assessment or a claim of categorical exclusion from the need to prepare an environmental assessment (EA).

12.3.4.1 Claim of categorical exclusion

A claim of categorical exclusion applies to Agency actions that do not individually or cumulatively affect the quality of the human environment and, therefore, do not require the preparation of either an EA or an environmental impact statement (EIS). An adequate claim of categorical exclusion[12] must include a citation of the 21 CFR 25.32 subsection under which the exclusion is warranted (Table 12.9), a statement of compliance with the categorical exclusion criteria, and a statement that, to the submitter's knowledge, there are no extraordinary circumstances that will require the submission of an EA.

Tiered Testing Recommendations: In accordance with 40 *CFR* 1508.4 and 21 *CFR* 25.21, FDA will require at least an EA for any normally excluded action if extraordinary circumstances indicate that the proposed action may have a significant environmental affect. An extraordinary circumstance may be shown by data available to either the agency or industry sponsor and may be based on production, use, or disposal of a substance. Data available to the agency include public information, information in the submission, and information the agency has received in other submissions for the same or similar substances. Examples of extraordinary

---

11) FFDCA section 409(h).   **12)** Definition codified in 21 CFR 25.15.

Table 12.9 Claims of categorical exclusions applicable to food-contact materials: Title 21, Parts 25, Section 32 of the Code of Federal Regulations (denoted as 21 CFR Section 25.32).

21 CFR 5.32 subsection	Category of substances
i	Substance is present in the finished food-packaging material at not greater than 5% by weight (wt.%) and is expected to remain with finished food-packaging material through use by consumers or when the substance is a component of a coating of a finished food-packaging material
j	Substance is to be used as a component of a food-contact surface of permanent or semipermanent equipment or of another article intended for repeated use
q	Substance that is registered by the US Environmental Protection Agency (US EPA) under the Federal Insecticide, Fungicide and Rodenticide Act (FIFRA) for the same use requested in the FCN
r	Substance that occurs naturally in the environment when the action does not alter significantly the concentration or distribution of the substance, its metabolites, or degradation products in the environment

circumstances that may apply to CFSAN actions are listed at the Agency's website[13]. The CEQ regulations have defined "significantly" to aid in determining if an action may affect significantly the quality of the human environment. This definition should be considered when evaluating whether extraordinary circumstances exist that may warrant the submission of at least an EA.

Significance: Section 1508.27 of CEQ regulations states that "Significantly" as used in NEPA requires considerations of both context and intensity:

1. *Context:* This means that the significance of an action must be analyzed in several contexts such as society as a whole (human, national), the affected region, the affected interests, and the locality. Significance varies with the setting of the proposed action. Both short- and long-term effects are relevant.

2. *Intensity:* This refers to the severity of impact. Responsible officials must bear in mind that more than one agency may make decisions about partial aspects of a major action. Things to consider when assessing intensity are listed under Section 1508.27(b) of CEQ regulations.

The Council on Environmental Quality's view is that the information submitted in a request for categorical exclusion is usually sufficient to determine that the exclusion is applicable to the requested action. Therefore, FDA has formulated its categorical exclusions to include specific criteria so that, in most instances, a categorical exclusion can be determined simply by citing the exclusion listed in 25 CFR 25.32 or confirmed by review of other information contained in the submission,

[13] Accessible at http://www.cfsan.fda.gov/~dms/opa2eg.html#partiii

available in the Agency's files, or published in the open literature. However, in limited instances, it may be necessary to submit additional information to the Agency to establish that the criteria for a categorical exclusion have been met, particularly for exclusions claimed under 25 CFR Sections 25.32(i), 25.32(q), and 25.32(r).

Inadequacies in a claim of categorical exclusion: A careful evaluation of the general environmental guidance available at the Agency's website will aid submitters to prepare an adequate claim of categorical exclusion.[14] An inadequate claim may delay the review process of a submission. Examples of inadequacies in a claim of categorical exclusion are citation of the wrong exclusion, claim lacks explicit statement that the exclusion complies with the applicable criteria and that there are no extraordinary circumstances that would require preparation of an EA, exclusion applicable to only some, not all, of the uses requested in the FCN, the use requested in the claim is different from the use requested in other sections of the FCN, and there is insufficient evidence in the notification for the agency to determine if the exclusion criteria are met.

12.3.4.2 Environmental Assessment (EA)

Any industry-initiated action that is subject to NEPA must contain an adequate environmental assessment (EA), as defined in 21 CFR 25.40, if it does not qualify for a claim categorical exclusion or when extraordinary circumstances exist that would make an exclusion unwarranted, as determined by the submitter or by the Agency. The majority of proposed uses of FCSs qualify for categorical exclusion under 21 CFR 25.32 (i), (j), (q), or (r); those that do not qualify for categorical exclusion require at least an EA. There are three proposed uses of FCSs requiring at least an EA: (1) substances used in the production and processing of food and are not intended to remain with food, (2) processing aids used in the production of food-packaging material and are not intended to remain as components of finished packaging, and (3) components of food-packaging material present at greater than 5% by-weight (wt.%) of the finished packaging. For example, if the FCS is for use as a nonchemically bound adjuvant in all polymers, the proposed use would qualify for categorical exclusion under 21 CFR 25.32(i) if it is present at not-greater-than 5 wt.% of the finished food-packaging material and almost all of the market volume will remain with the packaging. However, if the FCS is a chemically-bound component of a polymer (such as a monomer) used to make finished food-packaging material, then the submission must contain an EA, even if the FCS is present at not-greater-than 5 wt.% of a polymer, if the polymer is used at levels greater than 5 wt.% of the finished food-packaging material. In these cases, the substance that is the subject of FDA's review under NEPA regulations is the material that will be introduced into the environment; specifically, the food additive or polymer containing the FCS. In general, for substances incorporated in polymers, as blends or covalently bound, FDA believes that little or no introduction of these substances into the environment will result from their use because they are almost completely incorporated into the polymeric food-packaging material and essentially all of them are expected to remain

---

14) Accessible at http://www.cfsan.fda.gov/~dms/opa2eg.html#partii

with the packaging throughout use of the product. Consequently, the EA for polymeric substances or new uses of polymeric substances should focus on the impact of the proposed use on energy and resources.

An adequate environmental assessment is one that addresses the relevant environmental issues and contains sufficient information to enable the Agency to determine whether the proposed action may significantly affect the quality of the human environment. It must contain a brief discussion of: (1) the need for the proposed action, (2) introductions, fate, and effects of the substances in the environment, (3) alternatives to the proposed action, and (4) of the environmental impact of the proposed use as a result of use and disposal of the substance; except for these requirements, the Agency's NEPA-implementing regulations do not specify what data the EA must include to demonstrate that the proposed use is not expected to cause a significant impact on the environment . However, the agency is often unable to determine what impact the proposed use may have on the environment without additional information about the amounts of substances introduced into the environment, the fate of these substances in the environment, and the effects of such amounts on organisms in the environment. It is the submitter's responsibility to demonstrate no significant impact on the environment as a result of the proposed use of the FCS. However, the agency published guidance documents on its website to provide submitters with additional recommendation for information to be included in an EA to help the agency determine the potential of the proposed use to lead to significant environmental impacts as a result of use and disposal of the article.

The environmental fate and effects data included in an EA could be either actual data or obtained by using prediction models such as the US Environmental Protection Agency's ECOSAR and EPIWIN (EPA, 2007). FDA will accept data obtained from testing either under its own testing methods or those of any other entity such as the US EPA, the Organization for Economic Co-operation and Development, or other international regulatory bodies. Regardless of the method used to determine no significant impact on the environment, the submitter should consider the physical/chemical properties of the FCS (water solubility, dissociation constants in water, n-octanol/water partition coefficient (K_{ow}), and vapor pressure or Henry's Law constant), environmental depletion mechanisms (adsorption coefficient (K_{oc}), aerobic and anaerobic biodegradation, hydrolysis, and photolysis), and effects data (aquatic toxicity mostly).

12.3.4.3 Polymeric Food Packaging Materials

The Agency has the responsibility to consider the impact of its actions on the use of natural resources and energy as required under NEPA, Section 102 (b)(6), which states that one goal of NEPA is to, ". . . enhance the quality of renewable resources and approach the maximum attainable recycling of depletable resources. " As a result, the EA for a new polymeric food-additive (if it makes greater than 5 wt.% of the finished food packaging) or a new use of a polymeric material should focus on the identity of the polymer and on how it may impact the environment as a result of disposal by affecting solid waste management practices such as recycling (resources and energy use), landfilling (groundwater contamination), and incineration (acid gas emissions). Specifically, an EA should discuss the environmental impacts of the proposed use as a result of the disposal of the polymeric packaging. In general, unlike for human safety determinations, FDA does not review the environmental impact of contaminants present in polymeric materials.

The EA should identify fully the new polymeric material and provide relevant physical, thermal, and mechanical properties such as melting point, glass transition point, intrinsic viscosity, melt flow, and crystallinity. This information is especially important to assess how the new polymer will affect use of energy and resources if it will replace recycled material. Additional information needed to assess such impact is: (1) technical effect of the proposed substance, (2) estimated yearly market volume (confidential), (3) types of packaging, e.g., films, multi-component containers, bottles or other rigid containers, (4) size of containers (mass and volume), (5) intended food applications (specific food and beverage products), (6) method of disposal of the proposed packaging, and (7) currently used packaging materials that the proposed additive may compete with and replace.

Questions Relevant to Impact on Recycling When assessing the environmental impact of a new polymeric substances or a new use of a polymeric substance, the following specific questions should be considered: (1) Is the polymer intended to replace a similar substance already in use? Will it not change the potential uses or disposal pattern of the replaced packaging? (2) Will the polymer replace packaging that is recycled but it will not be recycled itself? If so, to what extent will it reduce the quantity of an existing material that is recycled, and how will it be distinguished from other recyclable packaging? (3) Will the polymer replace packaging that is recycled and will it be recycled itself? If so, then: (4) Will it fit into established source separation, collection, reprocessing systems or will new systems have to be developed? (5) Will the material be processed separately or only with commingled materials? (6) Will the package be distinctly marked to encourage recycling and minimize contamination of existing recycling systems? (7) What markets exist for the recycled material?, and (8) Does the FCS, if it is a modifier to the base polymer, open new markets for the modified polymer?

12.3.4.4 Inadequacies in EAs

A careful evaluation of the general environmental guidance available at the Agency's website at http://www.cfsan.fda.gov/~dms/opa2eg.html will aid submitters to prepare an adequate EA. An inadequate EA may delay the review process of a submission. Examples of inadequacies in an EA are: (1) the use requested in the EA is different than the use requested in other sections of the FCN, (2) lacks an explicit statement that there are no extraordinary circumstances that apply to the site of production of the substance, (3) not independent of the submission, (4) does not include estimates of environmental introduction concentrations, (5) does not include physical, thermal, and mechanical properties for polymeric substance, (6) lacks a discussion of potential impact on solid waste management strategies, and (7) contains confidential information.

Incorporation by reference: An EA may incorporate material by reference when the effect will be to cut down on bulk without impeding agency and public review of the action. The incorporated material shall be cited in the EA and its content summarized in the EA to the extent possible. No material may be incorporated by reference unless it is reasonably available for inspection by potentially interested persons. Material based on proprietary data which is itself not available for review and comment must not be incorporated by reference.

Cumulative impact: The agency has the obligation to consider the cumulative impact of its actions. Section 1508.7 of CEQ regulations defines cumulative impact as, "The impact on the environment which results from the incremental impact of the action when added to other past, present, and reasonably foreseeable future actions regardless of what agency (Federal or non-Federal) or person undertakes such other actions. Cumulative impacts can result from individually minor but collectively significant actions taking place over a period of time."

12.4
Final Thoughts

Exposure estimates, developed from migrant levels in food and information on food contact uses, determine the amount of toxicological information needed to support the proposed use of an FCS. Limitations on use conditions may be imposed by the submitter, possibly at the suggestion of FDA, to reduce exposure to either the FCS or to a constituent due to safety concerns identified during the review. For example, if the available toxicological information is limited and, thus, supports only a low exposure in accord with the tiered testing scheme, the submission may propose a narrow or limited use. As additional chemistry or toxicological data become available, additional submissions for expanded use of the FCS may be considered. As another example, consider two monomers used in the manufacture of polymeric food contact articles, one of which is commonly used as a "base" monomer and the other used as a minor monomer. Through its evaluation of numerous submissions that the FDA has processed, FDA has determined that some polymeric articles, manufactured from certain monomers, are used only in such niche or specialty applications that they are not expected to see wide use in food packaging. As such, the migrants of these articles usually require minimal data to ensure safety. This approach allows the FDA not only to fine tune the safety assessment for the particular FCS and use in question, but also to reassess the safety assessment of FCSs and their constituents on a continuing basis. If concerns are raised in the reassessment, they may result in postmarket action on any permitted uses of the FCS.

In applying the hazard information gained in toxicology testing to a safety evaluation, it is essential to assess the hazard in the context of consumer exposure. For instance, if the substance has a large toxicological database of information, the safety assessment may be made using the data deemed most relevant to the exposure level being evaluated. As an example, consider the proposed use of a substance with a dietary concentration of 1 µg per kg food consumed and supporting data consisting of genetic toxicity tests and a subchronic study. At 1 μg per kg food consumed, the genetic toxicity studies would be thoroughly reviewed because it can provide information relevant at such a low level while the subchronic study would be subjected to a preliminary review. If the preliminary review of the subchronic study indicates a safety concern based on the margin of exposure, the study should be thoroughly reviewed to establish an acceptable daily intake value. An acceptable daily intake value need not be calculated if the exposure and preliminary review do not warrant an extensive evaluation of the additional data to ensure safety.

12.5
Conclusions

Though the last decade has witnessed a change in the US with regard to regulatory processes for components of food contact articles, the safety standard has remained unchanged since the Food Additives Amendment of 1958. FDA's approach to the safety assessment of these substances is exposure driven, in that it is specific to the intended use and the resultant dietary exposure, which determines the amount of toxicological data consistent with the tiered requirements. Structure activity relationship analysis or the pairing of structure–activity relationship analysis with short-term genetic toxicology data can be used to determine the carcinogenic potential of a substance in lieu of available data. Potentially carcinogenic constituents with bioassay or analog data are evaluated using quantitative risk assessment principles and, when data are available and exposure warrants review, acceptable daily intake values are established for comparison to cumulative exposure values.

Currently, FDA is exploring the development of refined tiers for multiple endpoints of genetic and reproductive toxicity, the safety evaluation of low-molecular weight oligomeric fractions of polymeric substances, and the use of market share in the refinement of packaging factors. FDA's goal has, and will continue to be, the use of all available information and the identification of the accompanying uncertainties in the safety analysis to develop better guidance and thorough, efficient safety evaluations.

References

Bailey A.B., Chanderbhan R., Collazo-Braier N., Cheeseman M.A. and Twaroski M.L. (2005) The use of structure–activity relationship analysis in the food contact notification program. *Regul. Toxicol. Pharmacol.*, **42**, 225–235.

Begley T., Castle L., Feigenbaum A., Franz R., Hinrichs K., Lickly T., Mercea P., Milana M., O'Brien A., Rebre S., Rijk R. and Piringer O. (2005) Evaluation of migration models that might be used in support of regulations for food contact plastics. *Food Add. Contamin.*, **22** (1), 73–90.

Heckman J. (2005) Food Packaging Regulation in the United States and the European Union. *Regul. Toxicol. Pharmacol.*, **42**, 96–122.

Kirkland D., Aardema M., Henderson L. and Muller L. (2005) Evaluation of the ability of a battery of three *in vitro* genotoxicity tests to discriminate rodent carcinogens and non-carcinogens I. Sensitivity, specificity and relative predictivity. *Mutat. Res.*, **584** (1–2), 1–256.

Lorentzen R. (1984) FDA procedures for carcinogenic risk assessment. *Food Technol.*, **38** (10), 108–111.

Pitot III, H.C. and Dragan, Y.P. (2001) Chemical Carcinogenesis, in *Klaassen, Casarett and Doull's Toxicology: The Basic Science of Poisons* McGraw-Hill, New York, pp. 241–319.

Rulis, A. (1989) Establishing a threshold of regulation, *Risk Assessment in Setting National Priorities* (eds J. Bonin and D. Stevenson), Plenum, New York, pp. 271–278.

US Environmental Protection Agency (2007) Pollution Prevention (P2) Framework – EPIWIN Suite and SMILES, accessible online at http://www.epa.gov/oppt/exposure/pubs/episuitedl.htm.

US Food and Drug Administration (1982) Policy for regulating carcinogenic chemicals in food and color additives: Advance notice of proposed rulemaking. *Federal Register*, **47** (64), 14464–14470.

US Food and Drug Administration (1995) Food additives: Threshold of regulation for substances used in food-contact articles (Final Rule). *Federal Register*, **60** (136), 36582–36596.

US Food and Drug Administration (2002) Food Contact Substance Notification System (Final Rule). *Federal Register*, **67** (98), 35724–35731.

US Food and Drug Administration (2004) Redbook 2000. Toxicological Principles for the Safety Assessment of Food Ingredients (website referenced in Table 12.3)

13
Community Legislation on Materials and Articles Intended to Come into Contact with Foodstuffs
Luigi Rossi

13.1
Introduction

At the end of the 1950s, following the first legal provisions adopted by the US Food and Drug Administration (FDA) in 1958 in the sector of materials intended to come into contact with foodstuffs (FCM), the German and Italian authorities began issuing the first regulations in the field of migration. These rules, which in Germany were adopted in the form of Recommendation, were designed to avoid excessive release into foodstuffs of the substances contained in the materials, especially in plastics, and above all to rule out the possibility of a health hazard to the consumer as a result of the toxicity of some of the substances used to manufacture these materials. One needs only think of the migration of certain monomers regarded as carcinogenic (e.g., vinyl chloride, acrylonitrile), the release of certain highly toxic metals such as lead and cadmium from ceramic surfaces, the presence of nitrosamines in certain types of rubber used for teats and soothers, the finding of semicarbazide, phthalates and ITX in baby food, etc.

Subsequently, France, the Netherlands, Belgium, and Spain also issued similar laws, although each provision differed from the other not so much in the objectives behind them (consumer protection) as in the ways of achieving those objectives.

In the European Union (EU) the differences in the provisions adopted at national level may create problems for packaging companies, which were forced to adjust production to the country of destination and apply for authorization to use a new material. This led to a growing need to approximate ("harmonize") various laws and thereby remove legal barriers to Community trade in packaged food which, with the abolition of customs duty and the new systems of sales (supermarkets) and lifestyles (prepackaged food), had developed enormously.

In addition to this need there was the growing awareness on the part of the press and public opinion of everything to do with health protection and the ever more pressing demand for regulation from the professional associations, which wanted to have legal safeguards and to be able to advertise their products as safe by the law. This explains the proliferation of national and community rules and the need to harmonize the national rules.

In 1972 the European Commission (EC) drew up a broad program of action designed to harmonize all existing national laws in the field of materials intended to come into contact with food (plastics, paper, ceramics, rubber, etc.). In practice, rather than harmonizing laws and standards which were often too different to be reconciled, this meant drawing up a Directive (=Community legal act requiring a national implementing law) or a Regulation (directly applicable at national level) to replace national laws. In this chapter we will limit to describe the main aspects of current Community legislation, i.e., the legislation adopted at the Community level and therefore valid in EU.

13.2
Community Legislation

It is not useful to list the EU Directives/Regulations as they are increasing continuously and, then, the list will be very soon not updated. Therefore, it is better to provide you an EC website where it is possible to find the references of all adopted legislative and their link with the texts itself (http://ec.europa.eu/food/food/chemicalsafety/foodcontact/legisl_list_en.htm).

The directives adopted can be divided into three categories:

- directives applicable to all materials and articles,
- directives applicable to one category of materials and articles,
- directives relating to individual or groups of substances.

13.2.1
Directives/Regulations Applicable to all Materials and Articles

13.2.1.1 Framework Directives/Regulation
The Commission initially drew up a framework Directive, 76/893/EEC, which establishes two general principles:

- The principle of the "inertness" of the material and the "purity" of the foodstuffs, whereby the materials and articles must not transfer to foodstuffs any of their constituents in quantities which could "endanger human health and bring about an unacceptable change in the composition of the foodstuffs or a deterioration in the organoleptic characteristics thereof." It should be pointed out that this rule applies not only to packaging but also to all articles whose surface can come into contact with food at any stage of production, storage, transport, and consumption. For practical reasons only water supply plants and antiques are excluded from the field of application.

Figure 13.1 Symbol for food safe material.

- The principle of "positive labeling," whereby materials and articles intended to come into contact with foodstuffs must be accompanied by the words "for food contact" (or an appropriate symbol) or, where there are restrictions on their use, with some indication of the limitations in question so that consumers or users are informed of the potential use and limitations of the materials and articles they buy. Only for materials and articles, which by their nature are clearly intended to come into contact with foodstuffs, do Member States have the option of not imposing such labeling at the retail stage.

In 1980 the framework Directive was completed by an implementing Directive 80/590/EEC, laying down the symbols to affix to materials and articles (see Figure. 13.1).

In 1989 the framework Directive 76/893/EEC was replaced by Directive 89/109/EEC. In confirmation of the principles set out above, the new Directive laid down the sectors in which the Commission is asked to establish specific Community rules (see Section 13.2.2) and the criteria and procedures to be followed in the drafting the specific directives. These requirements can be summarized as follows:

- The Commission must, as far as possible, satisfy rigorous health criteria and thus consult the Scientific Committee for Food (SCF) on any regulation with implications as regards health.
- Specific directives and amendments to existing directives will be adopted by the Regulatory Committee procedure, in this instance the Standing Committee on Foodstuffs (CPDA), which is composed of the representatives of the Member States.

In 2004 Directives 89/109/EEC and 80/590/EEC were repealed and replaced by the new Framework Regulation 1935/2004/EC. This new text is much more extended and introduced additional rules following the new EU policy on a "better regulation" and increasing transparency. In this new Regulation the field of application was enlarged to include also the materials and articles which "can reasonably be expected to be brought into contact with food or to transfer their constituents to food under normal or foreseeable conditions of use." Moreover, the group of materials set out in

Directive 89/109/EC was extended to include also other materials indicated in italics in the following list:

1. *Active and intelligent materials and articles*
2. *Adhesives*
3. Ceramics
4. Cork
5. Rubbers
6. Glass
7. *Ion-exchange resins*
8. Metals and alloys
9. Paper and board
10. Plastics
11. *Printing inks*
12. Regenerated cellulose
13. *Silicones*
14. Textiles
15. Varnishes and coatings
16. Waxes
17. Wood

It was also clarified that the new Framework Regulation applies also to active and intelligent materials, as the interpretation of the previous framework Directive to these new applications was unclear. Because the active emitter materials may bring about changes in the composition or organoleptic characteristics of food, a new article (Article 4) was inserted to permit these changes, provided that the changes comply with the Community or, in absence, with national provisions applicable to food. Adequate labeling is provided for the active and intelligent materials to avoid the ingestion by the consumer of nonedible parts of the content inside the packaging and to indicate to the enforcement authorities that the materials or articles are active and/or intelligent and that, therefore, special rules apply.

Moreover, other rules were introduced in the new Regulation. Particularly important was the addition of the so-called supporting documentation to the declaration of compliance, already existing in the previous texts. Article 16 states that the specific measures shall require that materials and articles be accompanied by a written declaration stating that they comply with the rules applicable to them and that "appropriate documentation shall be available to demonstrate such compliance." That documentation shall be made available to the competent authorities on demand. This article should be read together with Article 16 on traceability which requires "business operators to have in place systems and procedures to allow identification of the businesses from which and to which materials or articles and, where appropriate, substances or products covered by this Regulation and its implementing measures used in their manufacture are supplied." In practice, it is requested that in any stage of the manufacture of a material and article a declaration of compliance should be delivered to the next business operator and that documentation should be available to demonstrate this compliance such as good quality of the substance used, with respect to any existing restrictions and rule. These new provisions as indicated in whereas (18) will "facilitate control, the recall of defective products, consumer information and the attribution of responsibility."

The Framework Regulation established that a Community Reference Laboratory (CRL) and national reference laboratories should be selected in accordance with the Regulation (EC) No 882/2004 on control. The Commission Regulation (EC) No 776/2006 of 23 May 2006 selected as CRL the Joint Research Centre of the European Commission (Ispra – Italy) (JRC).

Finally, the new Framework Regulation describes in any detail the procedure for an authorization of a new substance or the modification of his legal status. Adequate deadlines (6 months, or under certain conditions 12 months) have been imposed in the evaluation of substances to the new European Food Safety Authority (EFSA), which replaced the SCF. The technical dossier containing the relevant information for the risk assessment of the substances (chemical and physical properties, migration data, toxicology testing) should be available to the public unless some data (to be justified) are considered confidential by the petitioner, and in the case of conflicts, by the EC (Articles 8–12 and 20).

It was also clarified that the data included in the above-mentioned technical dossier "may be used for the benefit of another applicant, provided that the Authority considered that the substance is the same as the one for which the original application was submitted, including the degree of purity and the nature of impurities, and that the other applicant has agreed with the original applicant that such information may be used" (Article 21).

It should be noted that the new Regulation provides that the new rules for materials indicated with the terms "measures" in the text may be adopted in the form of Regulations, Directives and Decisions and not only in the form of Directive as set out before. Because normally the European texts are quite precise and do not leave rooms to Member States to modify the rules, it is expected that in the future the preferred form to adopt new rules will be taken with Regulations and Decisions and no longer Directives.

13.2.1.2 Regulation on Good Manufacturing Practice

A new Regulation (EC) 2023/2006 applicable to all materials was adopted on 22 December 2006 and applicable as from 1 August 2008 on Good Manufacturing Practice (GMP). It stated that the business operators shall:

- Establish and ensure adherence to a *quality assurance system* which should verify that "the materials and articles are in conformity with the rules applicable to them and the quality standards necessary for their intended use".

- Establish and maintain an effective *quality control system*, i.e., "the systematic application of measures established within the quality assurance system that ensure compliance of starting materials and intermediate and finished materials and articles with the specification determined in the quality assurance system".

- Ensure for printing inks applied to the nonfood contact side of a material or article that substances are not transferred into food by set-off or transfer through the substrate. Moreover, the printed surface shall not come into direct contact with food.
- Establish and maintain appropriate documentation in paper or electronic format with respect to specifications, manufacturing formulae and processing which are relevant to compliance and safety of the finished material or article.

13.2.2
Directives Applicable to One Category of Materials and Articles

Having defined the general framework, the Commission began to study three of the principal materials to be dealt with at the Community level, these being regenerated cellulose film, ceramics and plastics. One of the reasons for this choice had to do with the possibility of using the rules for these three sectors as models for other, similar, ones (see Table 13.1). The main results obtained are described below.

13.2.2.1 Directive on Regenerated Cellulose Film

The first specific Directive was adopted in 1983, this being Directive 83/229/EEC on regenerated cellulose film (RCF). Based on a number of existing national

Table 13.1 Harmonization plan.

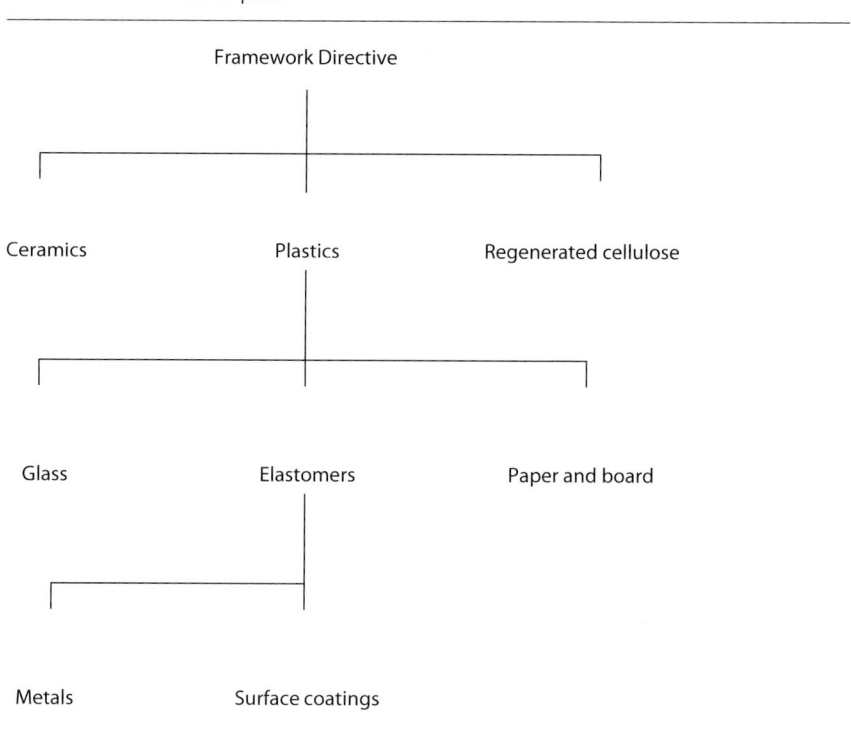

```
                        Framework Directive
                               |
        ┌──────────────────────┼──────────────────────┐
     Ceramics               Plastics          Regenerated cellulose
                               |
        ┌──────────────────────┼──────────────────────┐
      Glass                Elastomers           Paper and board
                               |
        ┌──────────────────────┤
     Metals              Surface coatings
```

regulations, the Community Directive lays down rules which differ from those laid down for plastics (see below). The technical impossibility of applying migration tests to these materials based on simulating liquids, limited use of these materials in food packaging (only for solid or semisolid foodstuffs) because of its technological properties, and the possibility of using a limited number of substances in the manufacture of the finished material were the reasons for treating RCFs differently from plastics.

For the preparation of the Directive the Commission collected the documentation and evaluated about 150 substances or groups of substances considered necessary to manufacture the finished article by the CIPCEL (European professional association in the sector) (CEC, Reports of the Scientific Committee for Food, 6th Series, 1978). The Directive established

- a list of authorized substances and
- restrictions on the composition of the material.

Only for monoethylene glycol (MEG) and diethylene glycol (MEG), which under certain circumstances can be transferred in unacceptably high quantities, have migration limits in food been provided for in Directive 86/388/EEC. The positive list has been amended on two occasions (Directives 86/388/EEC and 92/15/EEC) and the Commission took the opportunity of the need to amend certain rules to codify all the directives adopted (Directive 93/10/EEC).

In 2004 a new amendment (Directive 2004/14/EC) to the codified Directive was adopted. The main raison of this new amendment was the request of CIPCEL to authorize a new type of regenerated cellulose film with a coating consisting of plastics, which is compostable and biodegradable in accordance with the requirement of packaging waste Directives. Therefore, the limited number of polymeric coatings authorized not fully compostable and biodegradable have been suppressed and replaced by all the polymers authorized by plastic Directives.

13.2.2.2 Directive on Ceramics

In 1984 Directive 84/500/EEC on ceramic articles was approved. It lays down the specific migration limits for the two elements, according to their intended uses, along with the essentials of the method for checking those limits (see Table 13.2). The analysis method, on the other hand, is contained in a CEN standard (EN 1388-1:1995).

At national level, the checking of compliance with the rules increased during the last time and showed that particularly some imported ceramics did not fulfill with the EU requirements. A more severe amendment was, therefore, adopted in 2005 (Directive 2005/31/EC). It extends to ceramics the obligation established in the Framework Regulation 1935/2004 that the manufacturer or the seller established within the Community shall issue a declaration of compliance containing, between the others, the identity and the address of the manufacturer and the confirmation that the ceramic article meets relevant EU requirements. This declaration shall be accompanied by the so-called supporting documentation containing the results of the analysis carried out, the test conditions and the name and the address of the laboratory

Table 13.2 Specific migration limits for lead and cadmium. The test is carried out in total darkness at 22 °C for 24 h using 4% acetic acid (v/v) as the simulating liquid.

Category	Type of article	Lead	Cadmium
Category 1	Nonfillable articles and fillable articles of internal depth not exceeding 25 mm	$0.8\,mg/dm^2$	$0.07\,mg/dm^2$
Category 2	All other fillable articles	4.0 mg/l	0.3 mg/l
Category 3	Cooking utensils; packaging with a capacity of more than 3 liters	1.5 mg/l	0.1 mg/l

that performed the testing. Moreover, the amendment, taking into account the technological progress, establishes a set of performance criteria that the analytical method must comply without any indication of the method to be chosen.

13.2.2.3 Directive on Plastics Materials

Finally, the Commission began to draw up rules for the most complex and important area of packaging, that of plastic materials. With the wide divergence of national regulations, their poor scientific basis and the need imposed by the legislative procedure to consult a whole series of committees and to come to an agreement of all unanimously (until the enforcement of 89/109/EEC) and then, after the adoption of the Single Act which revised the decision-making procedures, by a qualified majority, the Commission has been obliged to take a very cautious, step-by-step approach toward harmonization. EU legislation (Directive 2002/72/EC and its 4th amendments, Directive 82/711/EEC and its two amendments and Directive 85/572/EEC and its amendment, included in the 4th Amendment) has now established:

- a list of authorized substances at the Community level,
- restricted use of certain substances,
- a system of checking migration.

Other very important rules have been introduced with the adoption of the so-called 4th Amendment (Directive 2007/19/EC of 30 March 2007), e.g.,

- the Functional Barrier
- the Fat (exposure) Reduction Factors
- the Declaration of Compliance
- specific rules for infants and young children
- special restrictions for certain phthalates now authorized at EU level
- new simulant for milk and milk products

13.2.2.4 Field of Application

The current legislation applies (a) to materials and articles made up of one or more layers exclusively of plastics material and (b) to "Plastic layers or plastic coatings, forming gaskets in lids that together are composed of two or more layers of different types of materials are also included" (see later the 4th Amendment).

The surface coatings which are applied to materials other than plastics are therefore outside its scope as well as "(a) the materials and articles which are supplied as antiques; (b) covering or coating materials, such as the materials covering cheese rinds, prepared meat products or fruits, which form part of the food and may be consumed together with this food; (c) fixed public or private water supply equipment."

13.2.2.5 EU List of Authorized Substances

This list, contained in the codified Directive 2002/72/EC and its amendments 2004/19/EC and 2005/79/EC, concerns mainly monomers and other starting substances as well as the majority of the additives, including few substances which constitute the means of polymerization (PPA = Polymerization Production Aids) if they may be used also as additives. Aids to polymerization and colors are excluded from the Community list as the Commission has not yet established a policy in these problematic issues.

The list of monomers and starting substances related to the type of plastics regulated is complete. It should be stressed that the provisional list contained in Section B of Directive 2002 is no longer valid as the period of its validity is exceeded as from 1 January 2005.

The list of additives is not yet complete. However Directive 2004/19/EC stated that:

- "the Commission shall establish, by 31 December 2007 at the latest, the date when that list shall become a positive list," i.e. a list authorized additives, to the exclusion of all others;

- "the Commission shall establish, by 31 December 2007 at the latest, a provisional list of additives which may continue to be used after 31 December 2007 subject to national law until the Authority has evaluated them."

Only substances having the requested technical dossier in accordance with SCF/EFSA guidelines and permitted in one or more of the Member States no later than 31 December 2006 are included in this so-called provisional list.

The inclusion in this provisional list ensures their legal existence until a decision is taken about their introduction into the Community positive list. This means that these substances, if they are not yet evaluated by EFSA, may be marketed in accordance with national rules even when a positive list is enforced provided the data requested but EFSA are available and accepted for an evaluation. While all additives other than those described before will be forbidden as from the date when the incomplete list of additives becomes a positive list.

All of the more important information on the evaluation of substances can be found in the document entitled "Synoptic" inside the SANCO website (http://ec.europa.eu/comm/food/food/chemicalsafety/foodcontact/index_en.htm) and for the more recent evaluations in EFSA website (http://www.efsa.europa.eu/science/afc/afc_opinions/catindex_en.html) as the synoptic document may not be updated on time.

13.2.2.6 Restricted Use of Authorized Substances (OML, SML, QM, and QMA)
The legislation in force provides for two main types of restriction.

The first, applying to all substances, provides that they may not be released alone or together with others in quantities greater than 60 mg/kg or 10 mg/dm^2 per material or article. This is an overall migration limit (OML) which is designed on the one hand to impart a certain inertia to the material intended to come into contact with the food so as to guarantee its purity and on the other hand to avoid setting a special migration limit for each substance.

The second one, which applies to isolated substances, provides that they must not migrate in quantities higher than a certain value fixed according to the toxicological characteristics of each of them. This is what is called the specific migration limit (SML), the value of which is generally established according to the acceptable daily intake (ADI) or the tolerable daily intake (TDI) laid down by the SCF/EFSA for the substance and the quantity of food containing the substance released by the plastics material.

It is a diffused mistake to consider that the European risk assessment is based on the comparison of TDI with the migration level. In fact, there is no sense to compare a hazard expressed in mg/kg body weight with a migration level expressed in mg/kg of food. The mistake is due to the fact that the EC, pending the establishment of precise rules to estimate the exposure of the consumer (mg (of substance)/person/day), conventionally assumed that a person of 60 kg ingest daily 1 kg of food (solid or liquid) containing the substance at the maximum permitted level "Conventional Exposure." Therefore, there is a numeric coincidence between the specific migration (mg/kg of food) and the exposure person (related to 1 kg of person) to be ingested by a person in one day.

The rules applied by the EU in establishing the SML are as follows:

- For substances classified by the SCF/EFSA in category 1 (having an ADI) or 2 (having a TDI), the SML is obtained by multiplying the ADI and the TDI by 60 as it is assumed that a person weighs 60 kg.

- For substances classified in category 3 the SML is equal to the restriction, if any, fixed by SCF/EFSA.

- For substances classified by the SCF/EFSA in category 4A ("Substances for which it was not possible to establish an ADI or a TDI but which could be used if the substance which migrates in the food or in food simulants were not detectable by an acknowledged sensitive method") an SML equal to a detection limit conventionally fixed at a level of 0.01 mg/kg of food has been set in the Community directives, sometimes together with an analytical tolerance of 100%.

However, when, for example, the substances decomposes in food simulants or the analytical methods for the determination in food simulants are not available, the restrictions expressed as SML are often replaced by residue limits of the substances in the finished product (QM = "Quantité maximale" in French Language) or by a QMA ("Quantité maximale" referred to an "area" assumed equal to 6 dm^2). For monomers it was assumed that only 1% of the residual monomers may migrate and, therefore, the detection limit of 0.01 mg/kg was transformed in a QM equal to 1 mg/kg.

In the past for only few substances a restriction on use was established. In the 4th Amendment to reduce as much as possible the exposure to phthalates released by the food contact materials as their other sources, besides the SMLs, very severe limitations on their use were set out (see later).

13.2.2.7 Authorization of New Substances

For new substances to be included in the Community lists the industry has to submit a special request accompanied by a technical file containing the set of data which will enable the SCF/EFSA to evaluate the risk associated with the use of that substance. These data, as shown below, were necessary for the toxicological evaluation of a substance and were laid down by the SCF and enforced also by EFSA:

- identity
- physical, chemical and other properties
- use
- migration data
- toxicological data.

With regard to the most important data, i.e., the toxicity data, in principle the SCF/EFSA requires a long-term study plus data on mutagenesis, reproduction, metabolism, etc. The full set of essential toxicological tests is shown below. These studies should be carried out according to prevailing EU or OECD guidelines, including "Good Laboratory Practice":

- 3 mutagenicity studies *in vitro*:
  - A test for induction of gene mutations in bacteria
  - A test for induction of gene mutations in mammalian cells *in vitro* (preferably the mouse lymphoma to assay)
  - A test for induction of chromosomal aberrations in mammalian cells *in vitro*
- 90-day oral toxicity studies, normally in two species;
- Studies on absorption, distribution, metabolism, and excretion;
- Studies on reproduction in one species, and developmental toxicity, normally in two species;
- Studies on long-term toxicity/carcinogenicity, normally in two species.

However in the introduction of SCF/EFSA Guidelines, it is clearly noted that "these guidelines should not be applied or interpreted too rigidly. For example, since the petitioner has knowledge of the identity, use of and potential exposure to the substance requested, and of the database available for it, the petitioner may deviate from the guidelines, provided valid, scientific reasons are given in the application. On the other hand, the petitioner should provide all available data, which are relevant for

Table 13.3 Set of toxicological tests in relation to migration/exposure.

Conditions: migration data (mg/kg)	Toxicological tests required by the SCF/EFSA	Usual decision of the SCF/EFSA
0–0.05	3 mutagenesis tests	- If results are positive: use prohibited or other tests are required - If results are negative: $R^* = 0.05$ mg/kg
0.05–5	3 mutagenesis tests 90-day oral administration test Bioaccumulation test	Depends on toxicological results
5–60	Full set of essential toxicological tests, unless there are good reasons for dispensing with them	Depends on toxicological results

R^* = restriction recommended by SCF/EFSA which may be expressed in a specific migration limit or a value expressed in mg/kg p.c. or in some other way.

the evaluation by the SCF. In all cases the SCF may request additional data, if the data submitted are equivocal or warrant further investigation."

The set of data to be supplied depends on the scale of the migration (see Table 13.3). However, "In determining the appropriate extent of the data set required the migration values should not be regarded as absolute limits but as indicative values."

To assist industry in the procedure for authorization of a new substance, SCF/EFSA and the Commission have drawn up a document entitled "Note for Guidance" which contains the SCF/EFSA Guidelines as well as all the information necessary for preparing the technical file and supplies explanations, suggestions and model letters for the transmission of documents. This document is available in the EFSA website (http://www.efsa.europa.eu/science/afc/afc_opinions/catindex_en.html).

13.2.2.8 Directives on the System of Checking Migration

In 1982 the first directive in the sector, Council Directive 82/711/EEC, laying down a precise reference framework for the system of checking specific and/or overall migration, was adopted. It establishes what simulating liquids (i.e., liquids which can simulate the extractive capacity of foodstuffs), contact times, and temperatures are to be used in migration tests performed under standardized conditions. This reference framework, which may seem unduly rigid given the innumerable conditions of contact in reality, was made flexible by the inclusion of a clause which permits Member States to depart from the standard conditions where these prove to be inadequate in the case in question either for technical reasons or because they are too different from the real conditions.

Moreover, the first amendment to it, Council Directive 93/8/EEC, made the standard conditions for migration tests more flexible by allowing a greater number

Table 13.4 Conditions for migration tests.

Conditions of contact in actual use	Test conditions
Contact time	Test time
$t \leq 0.5\,h$	$0.5\,h$
$0.5\,h < t \leq 1\,h$	$1\,h$
$1.0\,h < t \leq 2\,h$	$2\,h$
$2\,h < t \leq 24\,h$	$24\,h$
$T > 24\,h$	10 days
Contact temperature	**Test temperature**
$T \leq 5\,°C$	$5\,°C$
$5\,°C < T \leq 20\,°C$	$20\,°C$
$20\,°C < T \leq 40\,°C$	$40\,°C$
$40\,°C < T \leq 70\,°C$	$70\,°C$
$70\,°C < T \leq 100\,°C$	$100\,°C$ or reflux temperature
$100\,°C < T \leq 121\,°C$	$121\,°C^a$
$121\,°C < T \leq 130\,°C$	$130\,°C^a$
$130\,°C < T \leq 50\,°C$	$150\,°C^b$
$T > 150\,°C$	$175\,°C^b$

^{a}Use simulant C at reflux temperature.
^{b}Use simulant D at 150 °C or 175 °C in addition to simulants A, B, and C used as appropriate at 100 °C or at reflux temperature.

of possible combinations of times and temperatures and the use of other simulants for the "fat test" in cases where it is not possible to use those previously provided for Table 13.4 shows the conditions which now apply.

A second amendment, Directive 97/48/EEC, laid down the conditions of the use of volatile solvents, e.g., isooctane and ethanol, as test liquids in the "fat test" (see Table 13.5). It provided that these solvents may replace olive oil or the other fat simulants (HB 307, sunflower oil, etc.) if the fat test is not applicable in its basic version for technical reasons or in routine checking.

It is possible to carry out alternative tests using isooctane or ethanol or other solvents if the following conditions are satisfied:

- The values obtained in a "comparative test" are higher than or equal to those obtained in the test carried out with olive oil or other fat simulators.
- The detection limits are not exceeded.

By way of derogation from condition (a), the comparative test may be dispensed with if there is conclusive evidence based on experimental scientific results that the values obtained in the alternative test are equal to or higher than those obtained in the migration test.

Directive 85/572/EEC is a complement of Directive 82/711/EEC and it lays down the simulating liquids to be used for materials or articles intended to come into contact with only one food product or with one specific group of food products.

Table 13.5 Conventional conditions for substitution tests.

Test conditions with simulator D	Test conditions with isooctane	Test conditions with ethanol at 95%	Test conditions with Tenax (modified polyphenylene oxide = MPPO)
10 d – 5 °C	0.5 d – 5 °C	10 d – 5 °C	–
10 d – 20 °C	1 d – 20 °C	10 d – 20 °C	–
10 d – 40 °C	2 d – 20 °C	10 d – 40 °C	–
2 h – 70 °C	0.5 h – 40 °C	2.0 h – 60 °C	–
0.5 h – 100 °C	0.5 h – 60 °C^a	2.5 h – 60 °C	0.5 h – 100 °C
1 h – 100 °C	1.0 h – 60 °C^a	3.0 h – 60 °C^a	1 h –100 °C
2 h – 100 °C	1.5 h – 60 °C^a	3.5 h – 60 °C^a	2 h – 100 °C
0.5 h – 121 °C	1.5 h – 60 °C^a	3.5 h – 60 °C^a	0.5 h – 121 °C
1 h – 121 °C	2.0 h – 60 °C^a	4.0 h – 60 °C^a	1 h – 121 °C
2 h – 121 °C	2.5 h – 60 °C^a	4.5 h – 60 °C^a	2 h – 121 °C
0.5 h – 130 °C	2.0 h – 60 °C^a	4.0 h – 60 °C^a	0.5 h – 130 °C
1 h – 130 °C	2.5 h – 60 °C^a	4.5 h – 60 °C^a	1 h – 130 °C
2 h – 150 °C	3.0 h – 60 °C^a	5.0 h – 60 °C^a	2 h – 150 °C
2 h – 175 °C	4.0 h – 60 °C^a	6.0 h – 60 °C^a	2 h – 175 °C

^{a} Volatile test media are used up to a maximum temperature of 60 °C. It is a precondition for substitution testing that the material or article should withstand the test conditions applied with simulator D. Immerse a test specimen in olive oil in the appropriate conditions. If the physical properties are changed (e.g., melting or deformation), the material is considered to be unsuitable for use at that temperature. If the physical properties are not changed, carry out substitution tests using new specimens.

Because in principle the fatty food simulants have a greater extractive capacity of the fatty food, the result of migration testing is divided by a factor 2 up to 5 called "Simulant D Reductions Factor" (D-RF). The majority of D-RFs have been attributed as follows: if the fatty food is solid a value 5 is allocated; if the fatty food is semisolid a D-RF 3 is allocated; and, finally, if the fatty food is liquid no D-RF is attributed. Exceptionally values 2 and 4 are also attributed. Table 13.6 gives examples of the application of the D-RFs.

It should be noted that the use of fatty simulant (olive oil, HB 307) for checking the compliance with the SML is not yet in application. The 4th Amendment extended up to 1 April 2008, the date where the "fat test" using the simulant shall be applied. This date coincides with the data of application of the correction of migration. However the respect of the SML is ensured by analyzing the migration of the substance in fatty foods itself.

13.2.2.9 Functional Barrier

Directive 2007/19/EC defines "Plastic Functional Barrier (FB)" as a "barrier consisting of one or more layers of plastics which ensures that the finished material or article complies with Article 3 of Regulation (EC) No 1935/2004 and plastics Directives." Behind a plastic functional barrier, nonauthorized substances may be used, provided they are not proved or suspect "carcinogenic," "mutagenic," or

Table 13.6 Some examples taken from the directive laying down the list of simulants to be used in the migration tests. Only the simulant indicated by an "X" may be used. When "X" is followed by an oblique stroke and "3" or "5," the result of the migration tests should be divided by the number indicated, known as the "reduction factor." This figure is conventionally used to take account of the greater extractive capacity of the simulant for fatty foods compared with other types of foods.

Description of foodstuffs	Water	3% acetic acid	15% ethanol	Olive oil
Nonalcoholic beverages, etc.	X	X	–	–
Chocolate, chocolate-coated products, etc.	–	–	–	X/5
Fresh, chilled, salted, or smoked fish	X	–	–	X/3
Animal and vegetable fats and oils, etc.	–	–	–	X
Vinegar	–	X	–	–
Fried potatoes, fritters, and the like	–	–	–	X/5

"toxic to reproduction" and their migration remains below 0.01 mg/kg in food or a simulant. This value was fixed taking into account the difficulties of this type of analysis affected by a large analytical tolerance as well as the minimum tolerated level of detection for the presence of pesticides in baby foods for infants and young children (Directive 2006/125/EC of 5 December 2006, O:J. 339/16 of 5 December 2006).

The introduction of this concept is very useful for business operators and enforcement national authorities not only because it permits the use of substances nonauthorized but also because it indirectly recognizes that at this level of detection the risk for the majority of migrants is considered to be negligible by the law. Therefore, this value is an important reference value to ensure the compliance with article 3 of Framework Regulation of substances not intentionally added substances (NIAS), such as impurities not covered by toxicity testing, reactions products between added substances, oligomers and in general not listed substances.

13.2.2.10 Fat (Consumption) Reduction Factors

The EU Regulation is not based on the evaluation of the real risk assessment of the consumer. In fact the exposure is not estimated on the basis of the measured concentration of the substance migrated in food or food simulants and on the quantity of food ingested. The exposure at EU level is assumed conventionally equal to 1 kg of food or simulant containing the migrant at maximum quantity permitted. This assumption in the case of the ingestion of fatty food is clearly unrealistic, as a person according to the nutritional statistics cannot ingest daily more than 200 g of fat.

The 4th Amendment introduces the so-called Fat (consumption) Reduction Factor (FRF) to take into account the limitation on fat ingestion. Therefore the value obtained in the migration testing with fatty food or fatty simulant shall be divided, under certain specified conditions, by the FRF. This factor is equal to 1 up to 5 and its value depends on the percentage of fat in fatty food. It should be noted that this Directive states that the correction is permitted only under specified conditions and that it can be applied to only certain listed lipophilic substances which likely do not migrate in the aqueous simulant. The result of the multiplication of the DRF with the FRF cannot be greater than 5.

13.2.2.11 Declaration of Compliance

The Framework Regulation in its article 16 set out that a declaration of compliance and supporting documentation should be available. In analogy with the ceramic Directive the plastic Regulation through the 4th Amendment establishes the content of the declaration of compliance.

The written declaration of compliance shall contain the following information:

1. The identity and address of the business operator which manufactures or imports the plastic materials or articles or the substances intended for the manufacturing of these materials and articles.

2. The identity of the materials, the articles, or the substances intended for the manufacturing of these materials and articles.

3. The date of the declaration.

4. The confirmation that the plastic materials or articles meet relevant requirements laid down in this Directive and Regulation (EC) No 1935/2004.

5. Adequate information relative to the substances used for which restrictions and/or specifications are in place under this Directive to allow the downstream business operators to ensure compliance with those restrictions.

6. Adequate information relative to the substances which are subject to a restriction in food, obtained by experimental data or theoretical calculation about the level of their specific migration and, where appropriate, purity criteria in accordance with Directives 95/31/EC16, 95/45/EC17, and 96/77/EC18 to enable the user of these materials or articles to comply with the relevant Community provisions or, in their absence, with national provisions applicable to food.

7. Specifications on the use of the material or article, such as
   - type or types of food intended to be put in contact with;
   - time and temperature of treatment and storage in contact with the food;
   - ratio of food contact surface area to volume used to establish the compliance of the material or article.

8. When a plastic functional barrier is used in a plastic multilayer, the confirmation that the material or article complies with the requirements of Article 7a, paragraphs 2 to 4 of this Directive.

The written declaration shall permit an easy identification of the materials, articles, or substances for which it is issued and shall be renewed when substantial changes in the production bring about changes in the migration or when new scientific data are available.

13.2.2.12 Specific Rules for Infants and Young Children

The 4th Amendment introduces for first time specific rules to improve the protection of infants and young children, since infants ingest more food in proportion to their body weight than adults.

- When a plastic is in contact with baby foods, the OML equal to 60 mg/kg of food or the SML shall always be expressed in mg/kg. For the articles less than 500 ml the expression of the migration in mg/kg represents a more severe restriction.
- The correction by the FRF is not applicable for baby foods.
- For ESBO the SML is lowered to 30 mg/kg.
- For some phthalates used as plasticizers the contact with nonfatty food was forbidden.
- Finally, instead of 4 April 2009, for ESBO, phthalates and other plasticizers used in PVC gaskets in lids the noncompliance with to the new fixed restrictions is anticipated admitted until 30 June 2008.

13.2.2.13 Special Restrictions for Certain Phthalates now Authorized at EU Level

As known phthalates are ubiquitous contaminants derived from many sources. Moreover some of them were suspected to be estrogenic and carcinogens. Even if EFSA allocated full Tolerable Daily Intake (TDI), European Commission in 4th Amendment for some phthalates used in food contact materials established strict limitations on use. The use as plasticizers was largely restricted to those applications which do not have a significant contribution to the total exposure and as regards phthalic acid, bis(2-ethylhexyl) ester (DEHP), and phthalic acid, dibutyl ester (DBP), the use of fatty food application was prohibited and only 50% of the TDI was allocated to the remaining applications as the exposure to them from food consumption is in the range of the TDI (see Table 13.7). No concern exists for the use of the phthalates as technical support agent. Therefore the restriction on use was limited to the maximum quantity in the finished articles (Table 13.8). Finally the 4th Amendment shortened the deadline for the exhaustion of the stock for the articles which are not compliant with these restrictions. The deadline is fixed to 30 June 2008.

Table 13.7 Restrictions on use for phthalates as plasticizers in FCM.

Phthalates[a]	Restrictions on use
DEHP	Only for repeated use articles for nonfatty food contact
DBP	Only for repeated use articles for nonfatty food contact
BBP	Only for single use for nonfatty and food repeated use articles
DIDP	Only for single use for nonfatty and food repeated use articles
DINP	Only for single use for nonfatty and food repeated use articles

[a]DEHP = phthalic acid, bis(2-ethylhexyl) ester; BBP = phthalic acid, benzyl butyl ester; DBP = phthalic acid, dibutyl ester; DIDP = phthalic acid, diisodecyl ester; DINP = phthalic acid, diisononyl ester.

Table 13.8 Restrictions for phthalates as technical support agent in FCM.

Phthalate[a]	% in final product	TDI	% of TDI	SML (mg/kg)
DEHP	0.1	0.05	50	1.5
DBP	0.05 only for PO[b]	0.01	50	0.3
BBP	0.1	0.5	100	30
DIDP	0.1	0.15	100	9
DINP	0.1	0.15	100	

[a]DEHP = phthalic acid, bis(2-ethylhexyl) ester; BBP = phthalic acid, benzyl butyl ester; DBP = phthalic acid, dibutyl ester; DIDP = phthalic acid, diisodecyl ester; DINP = phthalic acid, diisononyl ester.
[b]PO = polyolefines.

13.2.2.14 Simulant for Milk and Milk Products

In September 2005 isopropylisothioxanthone (ITX), a substance originating from the printing ink, applied to the outer surface of the packaging was found in certain foods containing a certain percentage of fat but considered nonfatty food in Directive 85/572/EEC. The substance was transferred from the outer to the inner food contact surface during the storage of the unfilled packaging material through direct contact between the two surfaces. The analysis was carried out by using simulant provided for in the mentioned Directive, i.e., water not detected ITX in those foods. Therefore the 4th Amendment contains an amendment of Directive 85/572/EEC. This amendment provides for milk products the replacement of the water by 50% ethanol considered more appropriate than olive oil or the other fat simulants.

13.2.2.15 Other Complementary Community Initiatives

To help the supervisory authorities and industry to verify the conformity of products with the Directives/Regulations, JRC, acting now as CRL, established the following:

- Plastics reference materials with a certified overall value for migration into the four simulants.

- A bank of standard samples of substances contained in the Community lists, accompanied by corresponding spectral and physical data, to facilitate the identification and quantitative determination of these substances.

- Standard methods for evaluating overall migration into the four simulants and certain specific migrations or contents in the finished products. For this the Commission gave CEN a special mandate. Other specific evaluation methods have been developed as part of the activities of the CRL (JRC).

These activities and information can be found in JRC website related to food contact materials (http://crl-fcm.jrc.it/).

13.2.3
Directives Concerning Individual or Groups of Substances

While legislating on a broad scale, i.e., in relation to various sectors of production, the Commission has also been obliged to lay down rules for individual substances which have been the cause of considerable public concern. This applies to the individual measures related to vinyl chloride monomer used in PVC, monoethylene glycol (MEG), and diethylene glycol (DEG) used in regenerated cellulose film, N-nitrosamines, and N-nitrosatable substances from rubber teats and soothers, certain epoxy derivatives in plastics, the use of azodicarbonamide in plastics, and some plasticizers in gaskets in lids.

13.2.3.1 Directives on Vinyl Chloride

A directive was adopted in 1978 regarding exclusively materials and articles containing free vinyl chloride monomer (VCM) (Directive 78/142/EEC). It lays down the maximum quantity of free monomer permitted in the finished article as 1 mg/kg and states that such materials and articles must not release to the foodstuffs with which they are in contact any amount of vinyl chloride detectable by a method of analysis with a detection limit of 0.01 mg/kg. In 1980 and 1981 two further directives were adopted which lay down the method of analysis for vinyl chloride in the finished article and in foodstuffs, respectively (Directives 80/766/EEC and 81/432/EEC).

13.2.3.2 Directive on MEG and DEG in Regenerated Cellulose Film

In 1985, following application of the safeguard clause by the Federal Republic of Germany, which had observed excessive migration of MEG and DEG from regenerated cellulose film under certain circumstances, the Commission proposed a directive, promptly adopted by the Council (Directive 86/388/EEC), amending the conditions of the use of these substances and establishing a migration limit in foodstuffs of 50 mg/kg for both of them, which was reduced to 30 in Directive 93/10/EEC.

13.2.3.3 Directive on Nitrosamines in Rubber Teats and Soothers

On 15 March 1993, the Commission adopted a Directive concerning the migration of N-nitrosamines and N-nitrosatable substances from elastomer or rubber teats and soothers (Directive 93/11/EEC). This Directive stipulates that these articles must not release any N-nitrosamine and N-nitrosatable substance detectable by a validated method able to detect 0.01 mg/kg of total N-nitrosamines and 0.1 mg/kg of total N-nitrosatable substances. It also specifies the method to be used although a detailed description of the analytical procedure is left to the CEN's TC252/WG5.

13.2.3.4 Regulation on the Restriction of Use of Certain Epoxy Derivatives

Regulation 1895/2005, which replaced Directive 2002/16/EC, prohibits the use of bisphenol-F diglycidyl ether (BFDGF) and novolac glycidyl ethers (NOGE) and some of their derivatives as from 1 January 2005 and permits the use of bisphenol-A diglycidyl ether (BADGE) and some of its derivatives provided the migration does not exceed 9 mg/kg of food or food simulants. The availability of the new data on BADGE suppressed the concern about its carcinogenicity and genotoxicity.

13.2.3.5 Directive on the Suspension of the Use of Azodicarbonamide as Blowing Agent in Plastics

Directive 2004/1/EC (6 January 2004) set out the suspension of the use of azodicarbonamide as blowing agent due to its decomposition into semicarbazide (SEM) during high temperature processing. The SEM at that time was evaluated by EFSA as having a weak genotoxicity *in vitro*. The 4th Amendment establishes definitively the prohibition to the general use of azodicarbonamide. As indicated in the whereas of this last Directive, the Commission considered appropriate this ban as the Authority in its opinion of 21 June 2005 concluded that the carcinogenicity of semicarbazide is not of concern for human health at the concentrations encountered in food, if the source of semicarbazide related to azodicarbonamide is eliminated.

13.2.3.6 Regulation on Some Plasticizers in Gaskets in Lids

Recent data from Member States and Switzerland showed concentrations of ESBO in fatty food greater than OML reaching up to 1150 mg/kg. Plastic or coated gaskets in lids were the source of ESBO migration. With such high values, the TDI of ESBO may be exceeded for consumers. Other plasticizers, which may be used as substitutes of ESBO as they have a higher TDI or migrate to a lesser extent, may also be released at unacceptable levels.

The 4th Amendment clarified that these articles are covered by plastic rules and established for them shorter deadlines for articles not compliant with the fixed restrictions when they are in contact with baby food (4 June 2008 instead of 4 June 2009). However, it is unclear which rules should be applied to these articles and the cited plasticizers as only on 4 April 2008 harmonized levels shall be applied and only on 4 June 2008 the noncompliant articles shall be forbidden. In fact, the national authorities may consider these articles either covered or excluded from the national rules on plastics and coatings.

The Regulation 372/2007 clarifies the legal situation until 30 June 2008 by establishing a total SML for ESBO and its mentioned substitutes up to 300 mg/kg or 50 mg/dm^2 in accordance with the usual rule of article capacity. This value ensures that the TDI is not exceeded and forbids as from 1 July 2008 the articles exceeding the cited provisional values.

13.3
National Law and European Mutual Recognition

The national laws or the German recommendations related to any type of materials are not yet completely harmonized at EU level. For instance as regards the plastics the multilayers composed of materials other than the sole plastics or some categories of substances used in the manufacture of such as adhesives, colorants, solvents, aids to polymerization, and inks are not yet covered by the current incomplete plastic

Regulation. Also many other materials listed in the Framework Regulation such as paper, rubber, wood, etc. are not yet subject to specific EU measures and, therefore, existing national laws apply provided they are in compliance with the requirement of Framework Regulation.

In these cases a possible option to avoid technical obstacles to the trade in EU is given by the application of the so-called principle of mutual recognition set out in Articles 28 and 30 of the Treaty. It states that any product lawfully produced and marketed in one Member State must be admitted to the market of any other Member State. However, the authorities of the importing country can block sale of the product if they can demonstrate legitimate concerns related the protection of public health or environment and the protection of the consumer. Moreover, the measure taken by the importing authorities must proportionate to the objective pursued and it must be the most suitable means of achieving that objective.

Little recourse has so far been had to this principle as the national laws are all motivated by the protection of health. Moreover, the procedure under Article 30 provides for appeal to the Court of Justice in cases of dispute, which is very time-consuming, and the validity of the judgment is limited to the case in dispute. Finally, the implementation of the "principle of mutual recognition" is hampered by several other problems:

- The lack of awareness of enterprises and national authorities about the existence of the mutual recognition principle.
- The legal uncertainty about the scope of the principle and the burden of proof.
- The risk for enterprises that their products will not get access to the market of the Member State of destination.
- The absence of regular dialogues between competent authorities in different Member States.

On 14 February 2007 the Commission has transmitted for adoption to European Parliament and Council of Ministry a draft of Regulation laying down procedures relating to the application of certain national technical rules to products lawfully marketed in another Member State (COM(2007) 36 final).

Its objective is to define the rights and obligations of, on the one hand, national authorities and, on the other, enterprises wishing to sell in a Member State products lawfully marketed in another Member State, when the competent authorities intend to take restrictive measures about the product in accordance with national technical rules. In particular, the proposal concentrates on the burden of proof by setting out the procedural requirements for denying mutual recognition. Moreover, the proposal aims at reducing the risk for enterprises that their products will not get access to the market of the Member State of destination and at enhancing regular dialogues between competent authorities byestablishing one or several "product contact points" in each Member State. Their main task will consist of providing information on technical rules on products to enterprises and to competent authorities in other Member States, as well as providing the contact details of the latter. That will allow public authorities to identify their colleagues in other Member States so that they can easily obtain information from, and start a dialogue with, the competent authorities in other Member States.

In conclusion, a Member State that would restrict a material will have not only to point out incompliance with its applicable legislation (e.g., positive list) but also to provide a scientific justification for the decision. In other words the national authorities cannot rely only on positive lists but have the burden of proving a real risk of the product to human health.

13.3.1
Future Commission Plans

For the time being the Commission will probably seek to add to the present Community legislation other provisions or measures such as follows:

- codification or recast of all plastic Directives in one Regulation to improve the consistency of the texts and to permit the amendments through Regulations;
- establishment of the provisional list of additives for which an accepted technical dossier was sent on 31 December 2006 at latest;
- completion of the list of substances with the insertion of the new evaluated substances and establishment of the date when the list of additives becomes a positive list with the exclusion of all the other substances;
- laying down specific rules for recycled materials and active and intelligent packaging;
- extension of the scope of the future codified/recast Regulation to multilayers composed of two or more materials.

13.4
National Legislations and Council of Europe Resolutions

A full analysis of current legislation in the European Community should also take into account the national rules and recommendations in force in certain countries on articles or materials not yet harmonized by Community specific measures as well as the Council of Europe Resolutions and Guidelines. However this chapter does not examine them.

To consult national legislations useful references can be found in SANCO website (http://ec.europa.eu/food/food/chemicalsafety/foodcontact/index_en.htm) and in internet at www.foodcontactmaterials.com.

To consult Council of Europe (CoE) Resolutions and Guidelines see the CoE website (http://www.coe.int/T/E/Social_Cohesion/soc-sp/Public_Health/Food_contact).

13.5
Conclusions

Harmonization of legislation at the Community level is moving forward slowly but surely. Progress is often slow because the technical and scientific data necessary for selecting the most appropriate measures are not available. Also, the Member States and the professional organizations require problems to be solved at the Community level which have never been solved at the national level, so the Commission has to provide for expensive and complex prenormative research or standardization programs which will make it easier to secure agreement on legislative proposals or reduce negative preconceptions. However, over the last few years the Member States have shown a firmer resolve to speed up progress and the Commission is currently preparing a series of texts (see before) which should make it possible for legislation on plastics to be fully harmonized at the Community level.

14
Packaging Related Off-Flavors in Foods
Albert Baner, Francois Chastellain and André Mandanis

14.1
Introduction

The manufacturing complexity of plastics combined with multiple converting operations is often required to transform them into useful packaging materials, which almost guarantees the presence of potential off-flavors in such materials. Whether or not these potential off-flavors can be detected in the packed product depends on a variety of factors such as the chemical nature of the off-flavor substance, its concentration, converting processes, age of the material, how easily the off-flavor can transfer to the product, the nature of the packed product, the sensitivity of the consumer, etc.

Off-flavors represent one of the major quality issues in the food industry and can result in significant economic damage to a company. Even if they may not represent any health risk, they can seriously damage the quality image of a brand, and the confidence that the consumer has in it. By law, for example, in the EU (89/109/EEC) as well in the US (Federal Food, Drug and Cosmetics Act Section 402(a) (3)), packaging materials or substances transferring from the package to the foodstuff are not allowed to impart unacceptable changes to the organoleptic characteristics of the packed foodstuffs.

In this text, the term off-flavors follows the commonly used definition to include those off-flavors due to internal food reactions (oxidation, Maillard reactions, metabolism of micro-organisms, etc.), or to external sources of contamination (taints due to packaging materials, promotional items, storage, transport, etc.). Strictly defined an off-flavor is an atypical flavor usually associated with deterioration of the food itself and taint is a taste or odor, foreign to the product (ISO, 1992).

The present document focuses on the off-flavors which are associated with packaging materials. Because of the number of raw materials, additives, adhesives, inks, solvents and other chemicals used in the food packaging industry, and the number of suppliers/converters implicated in the manufacture of finished printed materials, many different sources of contamination are possible.

Plastic Packaging. Second Edition. Edited by O.G. Piringer and A.L. Baner
Copyright © 2008 WILEY-VCH Verlag GmbH & Co. KGaA, Weinheim
ISBN: 978-3-527-31455-3

The origin of the problems can be divided into three main categories:

- migration of odorous substances from package to food and to package headspace;
- inadequate protection of food from environmental influences;
- reaction of substances in packaging material with each other or with food components.

Sensory evaluation remains the most widely used method to assess the sensory quality of packaging materials. It represents the starting point in off-flavor investigations to confirm the existence of a complaint or problem. Sensory analysis is also a starting point for subsequent analytical work to identify the cause of the off-flavor contamination and for taking corrective action. Since many different types of off-flavor contamination exist; specific and accurate descriptors are needed to characterize the problem with precision at an early stage of the off-flavor investigation.

Many studies have been published in the scientific literature on specific odor and taint issues. There are also a number of excellent general packaging related off-flavor reviews, which describe the main off-flavors related to packaging materials (Whitfield, 1983; Raamshaw 1985; Harvey, 1963; Risch, 1988; Nijssen, 1991; Henshall 1991; Thompson et al., 1994; Tice, 1996; Lord, 2003). However, none of them provides a comprehensive list of descriptors, which could be used as a basic vocabulary for descriptive sensory analyses of packaging off-flavors.

A glossary of off-flavor descriptors and definitions is presented in Appendix. This glossary aims at collecting the most relevant descriptors, which have been published in the literature over the years (more than 300 referenced documents so far). The descriptors are presented with their definitions in alphabetical order, and classified into different packaging material categories. For each type of taint, explanations are given on the origin of the off-flavor, as well as some recommendations for avoiding these problems in practice. When descriptors are associated with chemical substances, the references to the odor-active compounds and respective precursors are given, together with some typical sensory threshold values. Where available, the descriptor has been translated to various languages: F for French, D for German, and E for Spanish. The symbol → refers to another related descriptor in the glossary with more information. This glossary has been found to be most effective when combined with an odor test training kit consisting of different odor solutions representing 12 typical off-flavors in the form of alcoholic solutions (Huber et al., 2002).

Table 14.1 lists the most common sources of off-flavors and their associated descriptors.

14.2
Sensory Evaluation

Sensory evaluation of packaging materials is generally carried out using olfactory (i.e., Sniff test) or taint transfer (i.e., taste) type tests. Both methods involve comparing a reference material versus a test material using a trained sensory panel. Tice (1996) and Huber et al. (2002) discuss ways of training sensory panels for

Table 14.1 Index of common off-flavors and their descriptors.

Source of taint	Associated descriptors (main descriptor in bold character)
Polyethylene (LDPE, HDPE), polypropylene (PP, OPP)	Candle, musty, paraffin, *PE-odor*, pencil, rancid, soapy, wax
Polyethylene terephthalate (PET)	*Acetaldehyde-like*, coconut, floral, fruity
Polystyrene (PS)	Acrid, burnt plastic, pungent, *styrene-like*
Polyvinyl chloride (PVC)	Linoleum flooring, shower curtain, *vinyl-like*
Polyamide (PA)	Bitter taste
Paper and board materials	Almond, *cardboard*, green, grassy, metallic, mushroom, musty, pine, rancid, sewer-like
Aluminum and tinplate materials	Bitter taste, salty taste, *metallic*, mineral oil
Jute sacks	Bitumen, *jute sack*, mineral oil, naphthalene, tar
Gravure/flexography inks, overlacquers, adhesives	Fruity, glue, *solvent*, sweetish, sharp, toluene-like
Offset/lithography inks	Linseed oil, green, grassy, mineral oil, *offset*, painting, petroleum, rancid, varnish
Serigraphy inks	*Camphorated*
Cold seal coating	Amine, *cold seal*, fishy, latex, rubber
Chlorophenols	Antiseptic, *disinfectant*, hospital, medicinal, pesticide, phenolic
Chloroanisoles	Corked wine, mouldy, *musty*

packaging off-flavors. The underlying physical chemistry of the methods relies on establishing equilibrium between volatiles in the packaging material and the headspace (i.e., olfactory) above it or in a food product/stimulant (i.e., taste).

Standard olfactory methods are all relatively similar and include the German DIN Standard 10955, 1993, BSI (British Standards Institution) Standard BS 3755, European Norm EN 1230-1:2001, and ASTM standards E462-84 and E619-84. The basic principle of the method involves preparing test and reference material samples by taking a standard surface area of each package material, placing them in separate inert sealed containers, warming the containers containing the samples over a period of time to decrease the time to reach equilibrium of the off-flavors between the materials and the headspace of the container. Afterward the containers are returned to room temperature and the intensity and nature of the odors in both container headspaces are sniffed by a trained sensory panel and compared. The difference in odor nature and intensity of the test material versus the control material is scored using a category scale that for example starts with a score of $0 =$ no perceptible difference between samples, $1 =$ odor just perceptible but still difficult to define, $2 =$ moderate odor, $3 =$ moderately strong odor and $4 =$ strong off-odor. The mean, mode, and median of the test scores as well as descriptors of the odors are reported. Interpretation of the significance of the scores depends on the goals of the test and end use application.

The taint transfer test is much like that of the olfactory test except that a small dish containing a food sample is added to the sealed container and a means of controlling relative humidity is incorporated into the container. In this case the taste of the food sample held in the container with the control material is compared to the taste of the food sample stored in the container containing the test material. Once again a category scale similar to the olfactory test is used to score the test. A version of this test has been widely used for confectionery and chocolate but other foods and food simulants can also be used (e.g., butter, mineral oil, confectioners sugar, water). Examples of standard taint transfer tests are ASTM 462-84, EN 1230-2:2001, and BSI BS 3755.

From previous experience it has been found that olfactory testing can be very useful for quality assurance release of incoming lots of production packaging materials or for new packaging material development and qualification. Additionally, the taint test has been used extensively for off-flavor sensitive foods in the chocolate and confectionery industries. However, it is very rare that a material passes the olfactory test and fails a subsequent taint transfer test (Huber et al., 2002). It is important to differentiate between odors that can be detected in the headspace above a package and the ability to taste them in the packed food product. The ability to taste an off-odor in a product depends on its threshold taste in the food. For chocolate products residual solvents coming from a packaging wrapper can be easily detected in the package headspace via sniffing but those same solvent levels will not be detected via a taint taste test in the chocolate even though they have been absorbed by the food.

14.3
Identification of Off-Flavor Compounds

Once an off-flavor problem has been confirmed using sensory analysis it is necessary to either confirm the presence of the suspected substance or to identify the cause of the off-flavor using instrumental analysis. Only by identification of the offending substance(s) can the off-flavor issue be considered, solved, and steps taken to avoid its reoccurrence. The identification of off-flavor compounds in packaging and food has been covered extensively by Saxby (1996) and Baigre (2003) among others.

The technique of choice for identification of off-flavors is GC-MS (gas chromatography-mass spectrometry) coupled with various sample preparation techniques. The identification of odor problems is made much easier than taste problems by using a sniffing or odor port fitted at the end of the analytic column before the MS. The sniffing port allows simultaneous sensory detection via sniffing along with instrumental detection via MS. Positive identification of off-tastes requires spiking or fortifying the food product with the suspected off-flavor substance followed by sensory analysis to confirm their effect. Lord (2003) outlines a decision tree schematic for identification of off-flavors, which incorporates different sensory and instrumental techniques depending on whether there is an off-taste or off-odor.

The most widely used sample preparation techniques are dynamic headspace, static headspace, solid phase microextraction (SPME), and steam distillation. Dynamic headspace also known as purge and trap is probably the most commonly used technique for off-odors. Successful application of the technique requires that the off-flavor is volatile and the technique offers the greatest sensitivity because the off-flavors are thermally desorbed from the sample matrix and then trapped and concentrated on some kind of absorbent matrix (e.g., Tenax TA) for off-odors. Static headspace involves heating the sample matrix in a sealed glass vial thus establishing partitioning between the headspace and sample. The headspace is then sampled. This method is not as sensitive as dynamic headspace but is faster and can be used quantitatively or as a screening method. SPME uses a coated stationary adsorbent phase on a glass fiber to partition substances present in the static headspace above a sample. The sensitivity is not as high as dynamic headspace but it does offer the ability to selectively adsorb different classes of compounds from the headspace. Steam distillation is most suitable for food samples where very difficult and complex matrices are involved. Steam distillation is a solvent codistillation technique where the sample is boiled with water. The steam drives the volatiles from the sample and the volatiles are then condensed into a water immiscible solvent gradually concentrating them. Examples of this technique are the Likens–Nickerson steam distillation extraction apparatus coupled with the Kurderna–Danish evaporation apparatus.

14.4
Physical Chemical Parameters Determining Off-Flavors

Whether or not a certain substance leads to a perceptible quality change for a given application depends on numerous parameters. Compared to toxicologically relevant substances, generally there is no valid limit value, which can be assigned to a substance for all cases.

The influence of a sensory active component from packaging to product is largely determined by the following parameters (Granzer et al., 1986):

- concentration of component in packaging material;
- solubility of component in packaging material (partition gas phase/packaging material);
- solubility of component in food (partition gas phase/food);
- sensory threshold level of component;
- type and intensity of food aroma;
- diffusion rate of component in packaging material;
- diffusion rate of component in food;
- time and temperature of storage;
- ratio of amount of packaging material to amount of food.

Knowledge of all these parameters makes it possible for a case-by-case determination of the limits for avoiding a reduction in quality. The lowest concentration of a substance in air is sufficient to give a perceptible odor is defined as the absolute threshold level and is designated as *OTa* in **Table 14.2**.

The values listed in Table 14.2 are averages of several measurements and published values. The published threshold levels can vary over three orders of magnitude for a

Table 14.2 Absolute odor threshold (OT_a) concentrations of different chemical compounds.

OT_a (mg/m³)	Compound
10^3	Heptane, Octane, Nonane
10^2	Ethanol, Acetone
10^1	Isopropanol (50), Ethylacetate (50), n-Butanol (30), Methyl-ethyl-ketone (30), Ethylglykol (20), Toluene (20), Ethyl-glykolacetate (10), Isopropylacetate (10), Methacrylic Acid
10^0	Butylacetate, Vinylacetate, Acetic Acid, Acrylic Acid, 2-Ethylhexylmethacrylate, 2-Ethyl-hexylacrylate, Methylmethacrylate
10^{-1}	Styrene, Mesithyloxide, n-Butylmethacrylate, Ethylmethacrylate, Methylacrylate
10^{-2}	n-Butylacrylate, Eugenol, Butyric Acid, Chlorophenol
10^{-3}	Ethylacrylate, 2-Nonenal, Ethylmercaptane
10^{-4}	1-Octene-3-one (Mushroom-ketone), Pentylmercaptane
10^{-6}	Vanillin

given compound. This widely scattered range is partly due to the imprecise definition of the perceptible sensory concentration as either a stimulation- or recognition-threshold and partly due to different methods for determining OT_a values and to the sensitivity of the test participants. Some of the variation is also due to the previous use of the olfactometer to determine absolute threshold levels. With the olfactometer it is not possible to separate the substance being studied from traces of other odor active contaminants which can lead to completely wrong OT_a values. The use of GC to separate odor substances from contaminants eliminates the largest source of error in these measurements.

A criterion selection for solvents used in packaging manufacture, e.g., for printing inks, is to select solvents with OT_a values as high as possible with a minimum between 10 and 100 (mg/m³).

The partition coefficient K is an important physical parameter in the sensory influence of an aroma compound on food (Chapters 4 and 9). Partition coefficients include those of a substance between the gas headspace (atmosphere) and packaging material (P), $K_{G/P} = c_G/c_P$, and between the gas (G) and food (L), $K_{G/L} = c_G/c_L$, as well as the partition coefficient between the packaging material and food, $K_{P/L} = K_{G/L}/K_{G/P} = c_P/c_L = S_r$, where the corresponding concentrations in the packaging material, food, and gas are c_P, c_L, and c_G. In Table 14.3 the $K_{G/L}$ values and diffusion coefficients in food, D_L, are given for several solvents in a selection of liquid, fatty, and solid foods at 23 °C. The partition and diffusion coefficients differ by four orders of magnitude in comparison to the differences between the relative molecular masses M_r and boiling points T_B of the pure solvents. The K values of a strongly polar solvent, e.g., ethylene glycol, can vary over three orders of magnitude depending on the polarity of the food (which is related to the food's water content), while a medium polarity solvent, e.g., ethyl acetate, has a much smaller range.

The $K_{G/L}$ values in aqueous systems for the solvents listed in Table 14.3 can be used as approximate values for other solvents with similar structures. The aromatic

Table 14.3 Partition ($K_{G/L}$) and diffusion (D_L) coefficients of several solvents in a selection of liquid, fatty and solid foods at 23 °C (M_r = relative molecular mass, T_B = boiling point).

Solvent	Food	M_r	T_B (°C)	$K_{G/L} \times 10^3$	$D_L \times 10^6$ (cm²/s)
Ethyl acetate	Cocoa butter	88	77	1.5	1.3[a]
	Soft cheese			4.0	0.3
	Butter cookies			15	3.0
	Water			5.3	–
Methyl ethyl ketone	Cocoa butter	72	80	1.3	1.5[a]
	Soft cheese			1.9	0.5
	Butter cookies			12	3.0
	Jam			7.7	–
	Water			1.7	–
Ethanol	Cocoa butter	46	78	7.7	0.9[a]
	Soft cheese			0.59	1.2
	Butter cookies			9.1	3.1
	Jam			0.91	–
	Water			0.29	–
Ethylene glycol	Cocoa butter	90	135	0.23	–
	Soft cheese			0.02	0.5
	Butter cookies			2.2	–
	Jam			0.53	–
	Water			0.006	–

[a] At 0 °C.

hydrocarbon, toluene, is an exception where its partition coefficient in the air/water system has a value of $K_{G/L} = 0.5$.

Compared to the range of $K_{G/L}$ values observed, the D_L values can hardly be differentiated from one another. The measured diffusion values in food having an order of magnitude of 1×10^{-6} (cm²/s) lay between those for liquids (10^{-5}) and those for plastics ($<10^{-7}$). These values are in agreement with the firmness of the fatty food studied. The small variation of the diffusion coefficient allows the values in Table 14.3 to be used for other solvents as estimates. The D_L values investigated allow a simple estimation of the rate of penetration of the solvent into the fatty food with the help of the formula (Chapter 7):

$$d_{L,t} = (2 D_L t)^{1/2} \tag{14.1}$$

where $d_{L,t}$ is the average penetration distance of the solvent into the food up to time t. A diffusion coefficient $D_L = 1.2 \times 10^{-6}$ (cm²/s) for ethanol in soft cheese corresponds to a penetration of approximately 0.5 (cm/day) or 8.7 (cm/year) at 23 °C.

When determining odor or taste threshold levels for a substance in a food or other testing medium, attention should be made to ensure that during the "taste test," the compound studied can be detected in the gas headspace above the food. One defines the relative threshold level of a substance over a food to be the lowest concentration of the substance in food leading to a perceptible odor in the gas headspace over the food at equilibrium. The relative threshold level is designated by OT_r and has the relationship:

$$OT_r = \frac{OT_a}{1000 \cdot \rho_L \cdot K_{G/L}} \text{ (mg/kg)} \qquad (14.2)$$

where the density of the food is designated by ρ_L.

The relative threshold values of solvents in several foods are given in Table 14.4. The values were determined by placing a dilution series of solvent in weighed amounts of food in sealable glass containers and equilibrating overnight at 23 °C. Each test series was composed of a minimum of eight dilution levels (Rüter 1992). A scattering of the threshold levels over an order of magnitude is due to the different sensory sensitivity of individual testers. This relatively narrow range allows the definition of values for the establishment of characteristic odor threshold numbers. However, for sensory evaluation, the lowest value of the most sensitive tester must be given consideration since complaints often originate from such sensitive consumers.

Sensory evaluation differentiates between the stimulation threshold (a just detectable level where a perceptible but not yet definable deviation of the sample from the standard is observed) and the recognition threshold, a level where the odor is identifiable or creates odor problems (a no longer tolerable quality deterioration caused by a definite off-odor and/or taste). The difference between a perceptible and identifiable level is usually only one to two dilution series of a geometric dilution series apart. Therefore, only undifferentiated odor and taste thresholds are given in Table 14.4, because of the very different sensitivities of individual testers. The perceptible (stimulation) levels of a less sensitive tester can overlap with the identifiable (recognition) level of another more sensitive tester.

Ethyl acetate, one of the most commonly used solvents for printing food contact materials, has a potential to cause many sensory problems with its low odor threshold of 10 (mg/kg). Assuming a complete transfer of ethyl acetate from packaging step to the product formation, the threshold levels found in Table 14.4 are reached with a package surface area to product mass of >1 (m^2/kg) based on the basis of 10 (mg) ethyl acetate per m^2 content in the packaging material. This could be the case for small packages or for foods with a low-fill weight (e.g., potato chips). With the present state-of-the-art technology, the residual amounts of ethyl acetate are usually under 10 (mg/dm^2) and can be monitored analytically without difficulty.

The relative threshold levels of acrylates and methyl acrylates in test foods are contained in Table 14.5. The threshold levels pass through a minimum for the ethyl esters. The values of the acrylates lay approximately an order of magnitude lower than the methyl acrylates. The influence of the partition coefficient $K_{G/L}$ can be easily seen when comparing the threshold levels of 2-ethyl-hexylacrylate and acrylic acid. Even though the relative threshold level of acrylic acid is only three times higher than that of the ester, it is found that the relative threshold level of acrylic acid in water is 100 times higher than the ester. This is the consequence of the good aqueous solubility of the polar acrylic acid and the small $K_{G/L}$ values. In sunflower oil, the $K_{G/L}$ value of the nonpolar ester is much smaller than that of the large acrylic acid value, although the relative threshold levels of the two compounds are practically identical.

Table 14.4 Relative odor and taste thresholds of several solvents in different food, OT_r (mg/kg).

Solvent	Potato chips		Coffee		Jelly-Bears		Chocolate	
	Odor	Taste	Odor	Taste	Odor	Taste	Odor	Taste
Cyclohexane	50–100	100–1000	500–1000	200–500	1000–3000	1000–2000	100–500	1000–2000
Toluene	20–500	20–100	100–500	10–20	20–50	500–1000	20–25	20–25
Acetone	20–1000	100–2000	100–2000	1000–5000	100–500	1000–2000	20–100	50–100
Methyl ethyl ketone	50–200	50–500	100–500	200–300	50–100	500–1000	50–100	500–1000
Ethyl acetate	10–50	10–100	100–1000	10–50	20–50	500–2000	20–100	50–100
Isopropyl acetate	10–50	50–200	50–100	10	50–500	500–1000	20–50	50–100
Ethanol	200–1000	200–2000	500–2000	5000–20000	2000	1000–2000	500–2000	1000–2000
Isopropanol	100–1000	1000–3000	500–1000	500	100–500	2000	500–1000	1000–2000
1-Ethoxy-2-propanol	100–1000	500–2000	500–1000	200–600	500	1000–2000	500–1000	500–1000

Table 14.5 Relative odor thresholds of acrylates and methacrylates in test foods, OT_r (mg/kg).

Compound	Water	Sunflower-Oil	10 v-% Ethanol	3 v-% Acetic Acid
Methylacrylate	0.005–0.01	0.005–0.2	0.005–0.2	0.01–0.1
Ethylacrylate	0.0001–0.002	0.001–0.05	0.001–0.01	0.0005–0.0005
n-Butylacrylate	0.002–0.02	0.1–1	0.005–0.1	0.005–0.2
2-Ethyl-hexyacrylate	0.005–0.2	0.2–4	0.01–0.2	0.01–0.1
Acrylic Acid	0.5–10	0.5–10	0.05–2	0.05–1
Methylmethacrylate	0.05–0.5	0.2–10	0.05–1	0.05–0.5
Ethylmethacrylate	0.002–0.05	0.05–1	0.02–0.2	0.01–0.1
n-Butylmethacrylate	0.01–0.1	0.1–4	0.05–0.5	0.05–0.5
2-Ethyl-hexylmethacrylate	0.02–0.5	0.5–10	0.05–0.5	0.05–0.5
Methacrylic Acid	2–100	2–100	–	–

The relatively small values for acrylic acid in the presence of ethanol as well as acetic acid can be caused by a partial ester formation and a small amount of dissociation along with a high partial pressure over the solution.

14.5
Derivation of Threshold Concentrations of Sensory-Active Compounds

Fully converted and printed packaging materials (e.g., laminate plastics) often possess characteristic odors related to the converting processes and materials used. Even though the transfer of odor substances to the food is the most important aspect from a food regulatory view; the package end user will often evaluate the incoming package material itself for odors. When the package material is found to have an odor, the question for the packager becomes how many of these sensory-active substances will be transferred to the packed food?

It is simple to analytically determine the mass of the odor compound per unit mass of packaging material, c, in (mg/kg) (ppm) or the mass of odor compound per unit surface area of packaging c'_P in (mg/m^2). It should be mentioned that in the case of packaging materials (e.g., laminate films) containing a functional aroma barrier between the food and print layer that odor transfer to the food can still take place from the outer packaging material layers via the phenomenon of set-off. Set-off is when the inner and outer layers of material are in direct contact with one another while still in roll stock form. Even though in the case of semipermeable packaging materials, a fraction of the odor compound content is lost into the atmosphere during storage, it is assumed in the derivation of a first upper allowable limit approximation that a complete transfer of the odor compound into the food occurs.

The maximum upper limit value of the odor concentration in the food by complete transfer from the package is

$$c_L < c'_P \frac{A}{m_L} = c_P \frac{m_P}{m_L} \qquad (14.3)$$

where m_L and m_P are the mass of the food and packaging material and A is the inner surface area of the packaging material. Setting $c'_P A/m_L$ equal to the relative threshold level OT_r from Eq. (14.2) one obtains the maximum allowable amount of an odor substance in a packaging material $c'_{P,max}$ before that odor can be detected sensorially:

$$c'_{P,max} = OT_r \frac{m_L}{A} = \frac{OT_a}{1000\, \rho K_{G/L}} \frac{m_L}{A} \quad (\text{mg/m}^2) \tag{14.4}$$

The threshold level of a substance can be decreased by the presence of less sensorially active substances. For example, in a mixture of ethanol, ethyl acetate, ethylene glycol, monoethyl ether, and toluene, the odor threshold level of ethyl acetate was reduced by half and in the case of cookies a decrease by a factor of five was observed. A reason for this may be the adsorption process taking place in the solid food. Compared to the solution processes in the food matrix, the influence of other components on the ethyl acetate partition coefficient during adsorption on the surface is likely to be larger. The repulsion of ethyl acetate from the food's surface increases its partial pressure over the food.

In the previous discussion of the odor limit value concentration, the influence of the solubility of the odor compound in the packaging material has been ignored. When one takes into consideration the $K_{P/L}$ value at equilibrium then instead of Eq. (14.5), one gets the following equation:

$$\begin{aligned} c'_{P,max} &= OT_r \frac{m_L}{A} \left(1 + \frac{m_P}{m_L} K_{P/L}\right) \\ &= \frac{OT_a}{1000\, \rho K_{G/L}} \frac{m_L}{A} \left(1 + \frac{m_P}{m_L} K_{P/L}\right) \end{aligned} \quad (\text{mg/dm}^2) \tag{14.5}$$

Even though the above equation assumes that no diffusion of the aroma substances into the atmosphere takes place, it is a realistic approximation. The relative solubility $K_{P/L}$ of the odor substance in the packaging material can play an important role in critical cases (high A/m_L values) where the ratio m_P/m_F assumes a maximum value for a certain packaging material. If for example, polyolefin packaging materials are used for aqueous foods then $K_{P/L} > 1$, particularly in the case of weakly polar odor compounds, e.g., toluene. The threshold level concentration of the odor compound can also be greatly increased by its high solubility in the packaging. The threshold levels calculated according to Eq. (14.5) do not represent established regulatory levels. However, when these levels are exceeded, their negative influence on the food and subsequent conflicts with food law (e.g., Article 2 of the Directive 89/109/EEC in the EU) cannot be ruled out.

In the above discussion it is assumed that during storage, partition equilibrium of the odor substance between the packaging and food is established. However, this is not always the case. Given the time $t_{1/2}$, which is the time required for half of the solvent contained in the packaging material to be transferred to the food, then one gets

$$t_{1/2} = \frac{\pi}{16} \frac{d_P^2}{D_P} \tag{14.6}$$

Table 14.6 $t_{1/2}$ values for different D_P values.

d_P (μm)	D_P (cm²/s)					
	10^{-8}	10^{-9}	10^{-10}	10^{-11}	10^{-12}	10^{-13}
	$t_{1/2}$					
	(h)	(h)	(h)	(days)	(days)	(days)
10	0.005	0.05	0.54	0.23	2.2	23
50	0.14	1.4	14	5.7	57	570
100	0.54	5.4	54	23	223	2250
200	2.2	22	222	91	910	9100

where d_P is the thickness of the packaging layer and D_P is the diffusion coefficient of the odor compound in the packaging material. Equation (14.6) assumes single-sided migration from a material-layer-material. The $t_{1/2}$ values for different D_P values are found in Table 14.6. It can be seen from this table that in packaging materials with $D_P < 10^{-11}$ (cm²/s) (e.g., the polyolefins), the residual solvent can be transferred to the product in a relatively short time.

As shown previously, diffusion in a solid food (e.g., a soft cheese) occurs very slowly and as a consequence the equilibrium state is not reached during the food's shelf-life. For time $t < t_{1/2}$ one calculates the penetration depth $d_{L,t}$ of the odor compound in the food with Eq. (14.1).

The diffusion coefficient of the odor compound in food is given by D_L. The average concentration $c_{L,t}$ of the odor compound in the outer layer of the food having a thickness of $d_{L,t}$ can be estimated by the equation (Chapter 7):

$$c_{L,t} \cong c_P \left(\frac{2}{\pi} \frac{D_P}{D_L}\right)^{1/2} \quad (14.7)$$

The longer the storage, eventually, the concentration in the outer layer decreases until it reaches the equilibrium concentration. The duration of this decrease depends on the D_L value and this can decrease rapidly with decreasing temperature. As a result it is possible to get a concentration of odor compounds in a thin external layer on frozen foods. Upon rapid thawing there may exist a high concentration of odor compounds in the outer food surface layer so that it is possible to experience a perceptible sensory effect (i.e., the odor concentration exceeds the threshold concentration $c'_{L,max}$).

For example, take a 50-μm thick film with a density of $\rho_P = 1$ (g/cm³), a residual solvent concentration $c_p = 100$ (mg/dm²), $D_P = 1 \times 10^{-8}$ (cm²/s), a package area to food mass ratio of 6 (dm²/kg) and $D_L = 1 \times 10^{-6}$ (cm²/s), the initial concentration in the food at $t_{1/2}$ is calculated to be $c_{L,t} = 160$ (mg/kg) in the food layer thickness of 313 (μm) in contact with the packaging material using Eq. (14.7). This assumes that transfer occurs only into the food from the packaging. This concentration falls to 10 (mg/kg) after 1.3 days or after 13 days when stored at a 25 °C lower temperature assuming that the diffusion will decrease by a factor of 10 at the lower storage temperature.

Appendix: Off-Flavor Descriptors and Definitions

Descriptor	Descriptor definition
Acetaldehyde F: acétaldehyde D: Acetaldehyde E: acetaldehido	*Fruity and floral sensation as demonstrated by acetaldehyde. Tainted mineral water is often described like "coconut" by consumers.* The problem of *acetaldehyde* causing off-flavors to mineral water and other soft drinks packed in polyethylene terephthalate (PET) bottles is well documented (Dong *et al.*, 1980; Steiner *et al.*, 1991; Nijseen *et al.*, 1996). Acetaldehyde is a degradation product of PET, which is mainly formed during the processing of the polymer at high temperature. It is possible to monitor the concentration of acetaldehyde in the PET bottles, but it is not possible to completely remove this odor-active compound from the finished material. The migration of acetaldehyde from PET bottles into water at room temperature is relatively slow, and it can take several weeks until a deviation in flavor can be observed. A recently published study revealed that acetaldehyde can be detected in water already at a concentration of 4 (μg/l) (Haack 2000). → coconut, floral, fruity
Acrid F: âcre D: stechend E: agrio	→ styrene
Almond F: amande D: Mandel E: almendra	*Odor reminiscent of bitter almonds as illustrated by almond liqueur "Amaretto".* In paper and board materials: Associated with *heptanal* and *benzaldehyde* in wood pulp (Tice 1996; Söderhjelm and Eskelinen 1991). → cardboard *In foods:* A case was reported in the literature for cheese. The "almond-like" off-flavor was associated

with *2-methyl-2-pentenal*. The taint was traced to the disinfectant used to clean up the plant equipment (Nijssen and Jetten 1987).
in plastic materials degradation

Amine

F: amine
D: Amin
E: amina

→ cold seal

Antiseptic

F: antiseptique
D: antiseptisch
E: antiséptico

→ disinfectant

Bitter

F: amer
D: bitter
E: amargo

One of the four basic tastes, primarily perceived at the back of the tongue, as illustrated by an aqueous solution of quinine sulfate.

In polyamide materials:
It was reported that *6-caprolactam*, a monomer for polyamide (nylon 6), could develop a disagreeable bitter taste (Stepek *et al.*, 1984).

In aluminum and tinplate materials:
→ metallic

Bitumen

F: bitume
D: Bitumen
E: betún

→ jute sack

Burnt plastic

F: plastique brûlé
D: verbranntes Plastik
E: plástico quemado

→ styrene-like

Camphorated

F: camphré
D: Kampfer
E: alcanforado

Typical odor of camphor-based ointments used in medicine as pain relief.

Acetylacetone (=*2,4-pentanedione*) is known to develop a "camphorated" off-flavor. This substance can be released from *titanium acetylacetone*, an adhesion promoter of gravure inks (Kay 1987).

"Camphorated" off-flavors can also be associated with *isophorone* (=*3,5,5-trimethyl-2-cyclo hexen-1-one*). This substance is used as solvent in the formulation of serigraphy (=screen printing) inks. Even though this printing process is not commonly used for food packaging materials, it is sometimes applied to PVC stickers which are added to the finished food product as promotional items (Leach and Pierce 1993).

Candle	→ PE-odor
F: bougie	
D: Kerze	
E: vela	
Cardboard	*General term common to most paper/board materials. To be avoided and to be replaced by more specific descriptors:*
F: carton	→ almond, green, grassy, metallic, mushroom, musty, pine, rancid, sewer-like
D: Karton	
E: cartón	

Many articles have been published in the literature on odor and taint problems related to paper/board materials (Sjostrom 1950; Bidie 1982; Letourneur 1991; Tice and Offen 1994; Lindell 1997).

Most of the problems are traced to the *fatty acids* and other components from the resin, which remain in the wood pulp after processing, and become oxidized into odor-active compounds such as *aldehydes, ketones,* and *alcohols*. Each of them has particular sensorial attributes like "almond", "metallic" or "mushroom." For the characterization of pulp, paper and board materials, a list of 146 descriptors is available (Söderhjelm and Pärssinen 1985).

Microbiological activity in the water cycle of the production site may form short chain fatty acids that may also contribute to off-flavor (Ziegleder et al., 1995)

Apart from the volatile compounds released from the raw material (i.e., wood pulp), additives, adhesives, inks, and varnishes also contribute to the overall "cardboard" off-flavor.

Catty

F: urine de chat
D: Katzenurin
E: orina de gato

Nasty smell reminiscent of cat's urine as demonstrated by some mercaptan-derived compounds.

Mercaptan-derived compounds can be formed by reaction of some food constituents with residual components from the packaging material. For example, *4-methyl-4-mercaptopentan-2-one* was formed by reaction of *hydrogen sulfide* with *mesityl oxide* in "cook-in-the-bag" ham packed in polyamide-ethylene ionomer laminate. Hydrogen sulfide was released from ham, whereas the printing solvent *diacetone alcohol* was transformed into mesityl oxide. Comparable reactions are assumed to have occurred in other food and packaging systems (Franz et al., 1990).

Chemical

General term common to most chemical products. To be avoided and to be replaced by more specific descriptors:

F: chimique
D: chemisch
E: químico

→ acetaldehyde, naphthalene, styrene, toluene, etc.

Coconut

→ acetaldehyde

F: noix de coco
D: Kokosnuss
E: coco

Cold seal

F: scellage à froid
D: Kaltsiegellack
E: sellado frio

Cold seal, is widely used for high speed sealing of flexible materials, especially for confectionery products and ice cream bars (e.g., flow wrapping). Standard cold seal formulation consists of aqueous *natural rubber* emulsion. The wet emulsion usually contains some *ammonia* to keep the rubber in liquid form, and there usually is a typical "rubber" or "latex" smell that remains after drying. It is well known that cold seal has a limited shelf-life. When the emulsion or the cold seal coating on the film becomes too old (>6 months), it can develop unpleasant "amine" or "fishy" off-odors (Maarse et al., 1987).

→ amine, fishy, latex, rubber

Corked wine

F: goût de bouchon
D: Korkgeschmack
E: vino con corcho

Cork taints in alcoholic drinks (wine, cognac, etc.) are due to treatment of cork with fungicides, or to contamination of damp warehouses by fungi (Henshall, 1991).

→ musty

Disinfectant

F: désinfectant
D: Desinfektionsmittel
E: desinfectante

Typical odor of some industrial agrochemical products containing phenolic compounds such as herbicides and pesticides.

→ antiseptic, hospital, medicinal, pesticide, phenolic

Numerous cases of contamination have been reported in the literature with "disinfectant" or "medicinal" descriptors (Whitfield, 1983; Raamshaw 1985; Maarse et al., 1987; Saxby 1985). The problem is generally traced to the presence of *chlorophenols* and other *halogenated phenols* in very small amount. Chlorophenols have very low sensory thresholds and can impart off-flavors to foods at concentrations as low as 1 (ppb) (Dietz and Traud 1978).

Chlorophenols have been used for many years as intermediates in the manufacture of agrochemical products and wood preservatives such as fungicides, biocides, and herbicides.

In 1991, legal recommendations have been given to reduce the production of *pentachlorophenol* and other derivative products for toxicological reasons (EU, 1991). Even if chlorophenols are now less frequently used, they remain a potential source of contamination for packaging items and food products which are likely to come into contact with treated wooden pallets or shipping floors and containers. Among numerous reported cases, several taint incidents have been specifically traced to *6-chloro-o-cresol*. This powerful odorous compound is used in the formulation of some disinfectants, which occasionally find their way into areas of food production and distribution (Saxby, 1985).

Airborne contamination may occur when nonhermetically packed foods are stored near to agrochemical factories. An example is reported in the literature for biscuits. It was established that *6-chloro-o-cresol* was released in the air by a factory, carried away by the wind on a distance of several miles, and then finally absorbed by the product at a concentration of 0.0001 (ppm). This concentration was high enough to impart the distinct "medicinal" off-flavor to the food (Raamshaw 1985; Goldenberg and Matheson 1975).

Chlorophenols can also be produced by chemical reaction of *phenol* derivatives with *chlorine*. This reaction occasionally occurs with wood, paper and board materials, since phenolic compounds can naturally be present in these materials. Sources of chlorine include chlorinated water, bleaching, and cleaning processes. As an example, the formation of chlorophenols was demonstrated in samples of fiberboard cartons, pinewood, and hardwood pallets, which were treated with a number of commercial cleaning agents (Tindale and Whitfield 1989).

Fat → mineral oil

F: gras
D: fettig
E: grasiento

Fishy → cold seal

F: Poisson
D: Fisch
E: pescado

Floral → acetaldehyde

F: fleuri
D: blumig
E: floral

Fruity *In PET materials:*

F: fruité
D: fruchtig → acetaldehyde
E: fruta

In printed materials:

Ethyl acetate is known to produce a "fruity" off-flavor. It is one of the most common solvents used in the manufacture of inks and laminating adhesives.

→ solvent
Glue → toluene

F: colle → fruity (ethyl acetate)
D: Klebstoff
E: cola

Green, grassy *Odor suggesting the bitterness and astringency of freshly cut grass.*

F: vert, herbe Associated with *aldehydes*, such as *hexanal*,
D: grün, Gras *2-hexenal*, and *2,4-hexadienal*, released from
E: verde, hierba paper and board materials (Söderhjelm and Eskelinen 1991; Letourneur 1991).

→ cardboard

Hospital → disinfectant

F: hôpital
D: Krankenhaus
E: hospital

Jute sack

F: sac de jute
D: Jutesack
E: saco de yute

Odor reminiscent of bitumen and tar as demonstrated by 1,4-dimethylnaphthalene.

The odor of jute sacks is traced to batching oils used to treat the raw fiber. Batching oil formulations for nonfood grade sacks contain 20–25% mineral oil of very poorly refined quality. The oil fraction is often characterized by a large amount of *aromatic hydrocarbons* (15–30%) (Seifert et al., 1975; Grob et al., 1991). Hydrocarbon residues and other oil components can migrate into the raw food ingredients during transport and storage, and eventually contaminate the finished food products. Since 1998 sacks for transport of raw food materials vegetable oil based batching oils have to be used (IOCCC, 1999).

→ bitumen, tar, mineral oil

Latex

F: latex
D: Latex
E: latex

→ cold seal

Linseed oil

F: huile de lin
D: Leinöl
E: aceite de lino

Odor suggesting the rancidity of oil paintings as illustrated by oxidized linseed oil.

Linseed oil is used in offset inks as "solvent". During the oxidation-drying process, the vegetable oils may form malodorous components such as *ketones, aldehydes,* and *carboxylic acids.* Fresh linseed oil is relatively neutral.

→ offset, green, grassy, rancid, varnish

Medicinal

F: médicinal
D: Medizin
E: medicinal

→ disinfectant

Metallic

Harsh and sharp sensation as illustrated by an aqueous solution of 4-(methylthio)butanol.

F: métallique
D: metallisch
E: metálico

In aluminum and tinplate materials:

The taint is related to the corrosive effect that some foods have on aluminum and tinplate materials. Corrosive reactions may occur in uncoated cans as well as in defect lacquered cans resulting in the formation of *metallic salts*. For example, particular care should be taken with 3-piece cans, since the surface area around the seam may go rusty more easily. In the majority of cases, the taste turns to "bitter", but sometimes a mixture of "bitter" and "salty: is observed. Among numerous tested products, milk and beer show the most sensitive behavior (Czukor 1984; Hollaender and Sedlmayr 1989).

→ bitter taste, salty taste

In paper and board materials:

Some substances released from paper and board materials, such as *1-pentene-3-ol* and *1-penten-3-one*, may give rise to "metallic" off-flavors (Letourneur 1991).

→ cardboard

In foods:

"Metallic" off-flavors are not always due to packaging materials. In milk products, internal reactions can produce *oct-1-en-3-one* and *octa-1, cis-5-dien-3-one*. These two substances are known to impart "metallic" off-flavors to foods at very low concentrations (Swoboda and Peers, 1977).

Mineral oil

Odor common to poorly refined mineral oils as illustrated by motor oils.

F: huile minérale
D: Mineralöl
E: aceite mineral

In printed materials:

In offset (or lithography) inks, mineral oils are used to control ink evaporation and to obtain the press-stability needed in the printing machine.

Most of converters use mineral oils of very highly refined quality which does not smell. However, mineral oil of less good quality is sometimes used and can release a typical "motor oil" off-flavor. The characteristic and the intensity of the smell depend on the level of *aromatic hydrocarbons* in the distillate (Doruk and Eichner 1975).

In cans:

Lubricant residues from the can-making presses are considered as one of the major source of taint in canned foods (Hardwick 1978).

→ offset, petroleum, motor oil

Moldy

→ musty

F: moisi
D: Schimmel
E: moho

Mushroom

Associated with mushrooms, yeasts and molds as demonstrated by octenol.

F: champignon
D: Pilz
E: champiñón

Some substances resulting from the auto-oxidation of residual resinous substances in paper and board materials, especially *heptenone* and *octenol,* may give rise to "mushroom-like" off-flavors (Söderhjelm and Eskelinen 1991).

→ cardboard

Musty

Smelling or tasting like old, stale or moldy as illustrated by corked wine.

F: moisi
D: Schimmel
E: moho

Most reported cases are traced to *chloroanisoles*. These substances are produced by microbial methylation of chlorophenols under humid and warm conditions. They are well-known to impart "musty" and "moldy" off-flavors to foods at extremely low concentrations. Typical reported odor threshold values in aqueous solutions are: 4×10^{-4} ppm for *2,4-dichloroanisole,* 3×10^{-8} ppm for *2,4,6-trichloroanisole* and 4×10^{-6} ppm for *2,3,4,6-tetrachloroanisole* (Griffiths 1974).

A case is reported in the literature for cocoa powder filled in multiwalled paper sacks. The product was shipped from Asia to Australia in metal containers. Upon arrival, a strong "musty" off-flavor was detected in the product. It was established that the cocoa powder was first contaminated by *chlorophenols* released from the sacks. Chloroanisoles were then formed from chlorophenols during transport by boat under humid conditions (Whitfield *et al.*, 1984).

A similar "musty" off-flavor was found in several brands of beer packed in aluminum cans. It was found that the internal can lacquer was spoiled with *2,4,6-trichloroanisole*, while the empty cans were being transported in shipping containers to the brewery (Lambert *et al.*, 1993).

→ corked wine, mouldy

In paper and board materials:

Some substances released from paper and board materials such as *heptanal, 2-heptanone, octanol,* and *2-octenal* may give rise to "musty" off-flavors (Letourneur, 1991).

→ cardboard

In polyethylene materials:

→ PE-odor

In printed materials:

4,4,6-trimethyl-1,3-dioxane was identified as being responsible for a "musty" off-odor in a printed plastic film. It was tentatively attributed to the reaction between 2- *methyl-2,4-pentanediol* — coming from an additive used to aid ink adhesion — and *formaldehyde* from an unknown source (*McGorrin et al.*, 1987).

Naphthalene

F: naphtalène
D: Naphthalin
E: naftalina

Strong aromatic odor associated with mothballs.

In jute sacks:

Poorly refined mineral oils used as "batching oil" can transmit *naphthalene,* as well as naphthalene derivatives with a distinct off-flavor (e.g., *dimethyl naphthalene*), to raw food ingredients (Seifert et al., 1975).

→ jute sack

In foods:

Indirect contamination may occur when nonhermetically packed foods are transported or stored near to *naphthalene*-derived products, such as mothballs and fire lighters (British Standard, 1971).

Offset

F: off-test
D: Offset
E: offset

Offset (or lithography) inks consist of pigments dispersed in resin solutions, themselves dissolved in a vehicle composed of vegetable oils and mineral oils.

Other sources of contamination involve impurities of distillates, waxes, and dryers. External parameters such as board pH, temperature, and humidity may also influence the smell of the finished printed material (Doruk and Eichner 1975).

-› linseed oil, mineral oil

Painting

F: peinture
D: Lackfarbe
E: pintura

→ linseed oil

Paraffin

F: parafine
D: Paraffin
E: parafina

→ PE-odor

Pencil

F: crayon
D: Bleistift
E: lápiz

A distinct "pencil-like" off-flavor was observed in milk packed in polyethylene containing the antioxidant *4-4'-thio-bis-(3-methyl-6-(2-methylpropyl)-m-phenol (Santonox R)*. The off-flavor was associated with the migration of antioxidant (vom Bruck and Hammerschmidt 1977).

→ PE-odor

PE-odor

F: odeur de PE
D: PE-Geruch
E: olor de PE

Odor suggesting the rancidity of burnt wax and oxidized paraffin as illustrated by old candles.

The typical odor associated with polyethylene and polypropylene materials is mainly due to the thermal degradation and oxidation of the polymer. Consequently, problems generally occur with overheated materials (high temperature of extrusion or high temperature of sealing), and old materials (oxidized materials). The quality and the quantity of additives used in the process (e.g., slip agents) may also influence the characteristics and the intensity of the smell released from the material.

Numerous volatile organic compounds are formed during thermal oxidation of polyethylene and polypropylene. Substances with a high sensory impact are sometimes present in minute quantities, and are consequently not always detectable by common analytical techniques. Among others, *2,2,4,6,6,-pentamethyl heptane* seems to be related to the degradation of polyethylene (vom Bruck and Hammerschmidt 1977). This compound itself does not smell (60). Saturated and unsaturated carbonyl *ketones* and *aldehydes* were identified as odor-active compounds (Bravo and Hotchkiss 1992).

The odor of polyethylene was also reported as "musty", "soapy," and "rancid". The off-flavor was traced to oxidation products of alkanes and alkenes, such as *1-hepten-3-one* and *2-nonenal*. However, 1-hepten-3-one has also been identified as a reaction product of *2-ethyl-1-hexanol*, which is often used in packaging materials, and 2-nonenal may also be formed from different compounds (Piringer 1981, Koszinowski and Piringer 1983).

→ candle, musty, paraffin, pencil, rancid, soapy, wax

Pesticide

F: pesticide
D: Pflanzenschutzmittel

→ disinfectant

E: pesticida

Petroleum → mineral oil

F: pétrole
D: Petroleum
E: petróleo

Phenolic → disinfectant

F: phénolique
D: Phenolgeruch
E: fenólico

Pine Odor reminiscent of pine tree needles as demonstrated by turpentine.

F: pin
D: Kiefer
E: pino

Associated with *terpenes* in wood and resin materials (e.g., *terpineol, borneol, verbenone, bisabolol, caryophyllene, thujopsene,* and *nerolidol*) (Ziegleder 1998).

→ cardboard

Plastic General term common to most plastic materials. To be avoided and to be replaced by more specific descriptors:

F: plastique
D: Plastikgeruch
E: plástico

For PE/PP materials:

→ PE-odor, candle, musty, paraffin, pencil, rancid, soapy, wax

For PET materials:

→ acetaldehyde-like, fruity, floral, coconut

For PS materials:

→ styrene-like, acrid, pungent, burnt plastic

For PVC materials:

→ vinyl-like

Rancid Old and stale sensation as illustrated by oxidized fat.

F: rance
D: ranzig
E: rancio

In paper and board materials:

Attributed to *aldehydes* such as *octanal, nonenal, nonadienal, decanal,* and *decenal,* resulting from the oxidation of residual resinous substances (Söderhjelm and Eskelinen 1991). Also traced to *butyric acid,* a microbiological metabolic product from the circuit water of paper mill (Ziegleder *et al.,* 1995).

→ cardboard

In polyethylene materials:

→ PE-odor

Rubber

→ cold seal

F: gomme
D: Gummi
E: caucho

Salty taste

One of the four basic tastes as illustrated by sodium chloride.

F: salé
D: salzig
E: salado

Salty tastes can be produced by aluminum and tinplate materials.

→ metallic

Sewer

Nasty smell reminiscent of waste matters released from domestic establishments.

F: égout
D: Abwasser
E: alcantarilla

In paper and board materials, "sewer-like" off-flavors are traced to short chain fatty acids, especially *valeric acid, 2-methyl propionic,* and *3-methyl butyric acid.* These odorous substances are formed through microbiological reactions in the circuit water of paper mills and are then absorbed on the finished paperboard material (Ziegleder *et al.,* 1995).

→ cardboard

Soapy

→ PE-odor

F: savon
D: seifig
E: jabonoso

Solvent

General term common to most organic solvents. To be avoided and to be replaced by more specific descriptors:

F: solvant
D: Lösungsmittel
E: disolvente

→ fruity, glue, sweetish, toluene

There are several sources of organic solvents in the food packaging industry. Apart from gravure and flexographic inks, overlacquers, and laminating adhesives represent another important risk of contamination. Problems arise when these inks, adhesives or any other solvent-based coating are not properly dried. Residual solvents remain on the surface of the printed material or trapped between two plastic layers. They can concentrate in the package headspace and migrate into the food product. Solvents are generally detected by nose, when opening a package for the first time, but they can also modify the sensory properties of foods more drastically.

In order to avoid serious odor and taint problems, the level of residual solvents must be carefully monitored in the finished printed material.

Stale

→ rancid

F: renfermé
D: alt
E: añejo

Styrene

Acrid and pungent sensation as illustrated by burnt polystyrene items (trays, foam cups, yogurt cups, etc.).

F: styène
D: Styrol
E: estireno

Many cases of foods contaminated by polystyrene materials are reported in the literature (Heydanek 1978; Passy 1983; vom Bruck and Hammerschmidt, 1976; Linssen *et al.*, 1991; Haack *et al.*, 1996). The off-flavor is generally traced to the presence of residual *styrene monomer* and residual *ethyl benzene* in the finished material (Durst and Laperle 1990).

Polyester materials may also be a source of taint. For example, shipping containers are sometimes made of polyester reinforced with fiberglass containing several percents of styrene monomer. This can occasionally lead to contamination in nonhermetically packed foods during transport (Brun, 1978; Ilsley, 1980)

→ acrid, pungent, burnt plastic

Sweetish

F: douceâtre
D: süsslich
E: dulzón

Soft and mild flavor common to ether-alcohols.

In printed materials:

Ether alcohols, such as *1-methoxy-2-propanol* and *1-ethoxy-2-propanol,* are used as solvents in the manufacture of gravure and flexographic inks.

→ solvent

Tar

F: goudron
D: Teer
E: bituminado

→ jute sack

Toluene

F: toluène
D: Toluol
E: Toluol

Strong aromatic sensation reminiscent of some commercial glues.

In spite of its intrinsic toxicity, *toluene* is still widely used in some emerging countries. In order to discourage its use in ink composition or/and as cleaning agent for printing equipment some food companies have banned it from their lists of permitted solvents.

→ solvent

Varnish

F: vernis
D: Firnis
E: barniz

→ linseed oil

Vinyl

F: vinyl
D: PVC
E: vinil

Odor common to vinyl-based materials as illustrated by vinyl shower curtains or linoleum flooring.

The odor of PVC (polyvinyl chloride) is attributed to volatile organic compounds (*alcohols, aldehydes, ketones, esters, acids,* etc.), which are formed during the heat processing of the material. Some of these substances have been traced to the degradation of the plasticizers *bis-(diethylhexyl) phthalate* and *bis-(diethylhexyl) adipate,* from the antioxidant *tris-nonylphenyl phosphite,* and from the polymer itself. To be noted that PVC of food grade quality should not smell. (Kim et al., 1987).

Wax

F: cire
D: Wachs
E: cera

→ PE-odor

References

Bidie, R.W. (1982) *Odor Problems in Connection with Paper and Boards,* Cham-Tenero Paper Mills Report, Cham, Switzerland, pp. 1–36.

Bravo, A., Hotchkiss, J.H. and Acree, T.E. (1992) Identification of odor-active compounds resulting from thermal oxidation of polyethylene. *J. Agric. Food Chem.,* **40**, 1881–1885.

British Standard, 1971, Guidance on avoiding odor from packaging materials used for foodstuffs, British Standards PD, 6459.

Brun, S. (1978) Ne stockez pas le vin dans n'importe quelle cuve, Emballages, Novembre, pp. 121–123.

Czukor, B., Böröcz-Szabó, M. and Sallay-Horváth, Zs. (1984) Application of aluminum packaging materials in the food industry, and the effect of aluminum contamination on sensory properties of liquid foods, in: *Euro Food Pack Conference,* 19–21 September, pp. 321–324.

Dietz, F. and Traud, J. (1978) Geruchs- und Geschmacks-Schewellen-Konzentrationen von Phenolkorpen, Gas-Wasserfach, Wasser-Abwasser, **119**, 318–325.

Dong, M., DiEdwardo A.H. and Zitomer, F. May (1980) Determination of residual acetaldehyde in polyethylene terephthalate bottles, preforms, and resins by automated headspace gas chromatography. *J. Chromatogr. Sci.,* **18**, 242–246.

Doruk, M. and Eichner, K. (1975) Über die Geruchsentwicklung bei offsetbedruckten Papieren und Kartons und Möglichkeiten ihrer Verringerung. *Allg. Papier-Runschau,* **29**, 829–830.

Durst, G.L. and Laperle, E.A. (1990) Styrene monomer migration as monitored by purge and trap gas chromatography and sensory analysis for polystyrene containers. *J. Food Sci.,* **55** (2), 522–524.

Franz, R., Kluge, S., Lindner, A. and Piringer, O. (1990) Cause of catty odor formation in packaged food. *Packag. Technol. Sci.*, **3**, 89–95.

Goldenberg, N.H.R. and Matheson (1975) Off-flavors in foods, a summary of experience: 1948–1974. *Chem. Ind. (London)*, 551–557.

Granzer, R., Koszinowski, J., Robinson-Mand, L. and Piringer, O. (1986) *Verpack.-Rundsch.* 37, Techn.-Wissensch. Beilage, pp. 53–58.

Griffiths, N.M. (1974) Sensory properties of the chloroanisoles. *Chemical Senses and Flavor*, **1**, 187–195.

Grob, K., Lanfranchi, M., Egli, J. and Artho, A. (1991) Determination of food contamination by mineral oil from jute sacks using coupled LC-GC. *J. Assoc. Off. Anal. Chem.*, **74** (3), 506–512.

Haack, G. (2000) Sensory thresholds and analytical determination of acetaldehyde in mineral water for the quality control of PET bottles. *Poster Presented at the 2nd International Symposium on Food Packaging Ensuring the Safety and Quality of Foods*, Vienna, November 2000.

Haack, G., Rüter, M. and Piringer, O.G. (1996) Quality decline of packaged food due to undesired flavors: overview and test model, in: *Packaging Yearbook*, edited by National Food Processors Association, Washington, pp. 68–84.

Hardwick, W.A. (1978) Two-piece cans: some flavor problems caused by manufacturing materials or practices. *MBAA Technical Quarterly*, **15** (1), 23–25.

Harvey, H.G. (1963) Survey of odor in packaging of foods, The Institute of Packaging. Edinburgh, pp. 1–39.

Henshall, J.D. (1991) Unwanted flavors in foods, in: *Food Contaminants: Sources and Surveillance*, C.S. Creaser, R. Purchase (eds.) pp. 191–200.

Heydanek, M.G. June (1978) How to spot a stinker – Predicting flavor effects of packaging materials. *Food Product Dev.*, 17–18.

Hollaender, J. and Sedlmayr, M. (1989) Interactions between food and tin plate cans – Shelf life and can quality, *Zeitschrift für Lebensm. Technol. und Verfahrenstechnik*, **40**, (10), 606–614.

Huber, M., Ruiz, J. and Chastellain, F. (2002) *Food Addit. Contam.*, **19** (supplement), 221–228.

Ilsley, D.A. (1980) Containers de fret isolés: Interdépendance produits alimentaires et qualité des PRV, Caoutchoucs et Plastiques, (600), Avril, pp. 63–73.

IOCCC, June(1999) 136-46.Specification for jute and sisal sacks foreseen for the transport of foodstuff, Revised.

ISO (1992) Glossary of Terns Relating to Sensory Analysis, ISO Standard, 5492.

Kay, P. (1987) Avoiding odour and discoloration. *Ink Print*, **5** (1), 30.

Kim, H., Gilbert, S.G. and Hartman, T.G. (1987) Characterization of undesirable volatile compounds in PVC films, in: Frontiers of Flavors, *Proceedings of the 5th International Flavor Conference*, Chalkidiki, Greece, 1–3 July, 249–257.

Koszinowski, J. and Piringer, O. (1983) Die Bedeutung von Oxidationsprodukten ungesättigter Kohlenwasserstoffe für die sensorischen Eigenschaften von. Lebensmittelverpackungen, *Deutsche Lebensm.-Rundschau*, **79** (6), 179–183.

Kringlebotn, I. (1971) Übergang eines "Bleistiftgeschmacks" aus Polyäthylenbeuteln in Frischmilch, *Verpackungs-Rundschau*, **2**, p. 17.

Lambert, D.E., Shaw, K.J. and Whitfield, F.B. (1993) Lacquered aluminium cans as an indirect source of 2,4,6-trichloroanisole. *Chem. Ind.*, (6), 461–462.

Leach, R.H. and Pierce, R.J. (1993) *The Printing Ink Manual*, ed. by Blueprint, London.

Letourneur, F. (1991) Développement de methodes sensorielles et analytiques pour la caractérisation de cartons imprimés destinés à l'emballage de denrées alimentaires, *Thèse présentée à l'Université de Paris*, 28 Mai.

Lindell, H. (1997) Sensory and instrumental characterization of food packaging board, in: *Proceedings of the 1st Task Force Meeting on Paper and Board*, ISPRA, 28–29 October.

Linssen, J.P.H., Janssens, J.L.G.M., Reitsma, J.C.E. and Roozen, J.P. (1991) Sensory analysis of polystyrene packaging material taint in cocoa powder for drinks and chocolate flakes. *Food Addit. Contam.*, **8** (1), 1–7.

Linssen, J.P.H., Janssens, J.L.G.M., Roozen, J.C.E. and Posthumus, M.A. (1993) Combined gas chromatography and sniffing port analysis of volatile compounds of mineral water packed in polyethylene laminated packages. *Food Chem.*, **46**, 367–371.

Lord, T. (2003) Packaging materials as a source of taints, In *Taints and Off-Flavors in Food*, (ed. B. Baigre), CRC Press, Boca Raton, Cambridge, Woodhead Publishing, pp. 64–111.

Maarse, H., Nijssen, L.M. and Angelino, S.A.G.F. (1987) Halogenated phenols and chloroanisoles: occurrence, formation and prevention, Paper presented at the *2nd Wartburg Aroma Symposium*. Eisenach, Germany.

McGorrin, R.J., Pofahl, T.R. and Croasmum, W.R. (1987) Identification of the musty component from an off-odor packaging film. *Anal. Chem.*, **59** (18), 1109–1112.

Nijssen, B. (1991) Off-flavors, in: *Volatiles Compounds in Foods and Beverages*, H. Maarse (ed.), pp. 689–735.

Nijssen, B., Kamperman, T. and Jetten, J. (1996) Acetaldehyde in mineral water stored in polyethylene terephthalate (PET) bottles: odor threshold and quantification. *Packag. Technol. Sci.*, **9** (4), 175–186.

Nijssen, L.M. and Jetten, J. (1987) An unexpected off-flavor in small blocks of packaged cheese, in: *Flavor Science and Technology*, (eds. M. Martens, G.A. Dalen, H. Russwurm)., 127–132.

Passy, H. (1983) Off-flavors from packaging materials in food products – Some case studies. *Instrum. Anal. Foods*, **1**, 413–421.

Piringer, O. (1981) Qualitätsmindernde flüchtige Spurenkomponenten in Verpackungen—Analytik und Eigenschaften. *Das Papier*, **356** (10A), V92–V94.

Raamshaw, E.H. (1985) Off-flavor in packaged foods. *CSIRO Food Res. Quart.*, **44** (4), 83–88.

Risch, S.J. (1988) Migration of toxicants, flavors, and odor-active substances from flexible packaging materials to foods. *Food Technol.*, **42** (7), 95–102.

Rüter, M. (1992) *Verpack.-Rundsch.* 43, techn.-wissensch, Beilage.

Saxby, M.J. February(1985) Identifying the taint. *Food Manufac.*, (2), pp. 57–59.

Seifert, R.M., Buttery, R.G. and Guadagni, D.G. (1975) Volatile components of jute sacks. *J. Sci. Food Agric.*, **26**, 1839–1845.

Sjostrom, L.B. (1950) Paper-package odors. *Modern Packaging*, (8), pp. 118–120.

Söderhjelm, L. and Eskelinen, S.E. (1984) Chemical properties of fiber-based food packages, in: *Euro Food Pack Conference*, 19–21 September, pp. 4246.

Söderhjelm, L. and Pärssinen, M. (1985) The use of descriptors for the characterization of odor in packaging materials. *Paperi ja Puu*, **67** (8), 412–416.

Steiner, I., Eberhartinger, S., Washüttl, J. and Kroyer, G. (1991) Migration of acetaldehyde from PET bottles, in: *Proceedings of 6th European Conference on Food Chemistry*, Hamburg, Vol. 2, 678–682.

Stepek, J., Duchacek, V., Curda, D., Horacek, J. and Sipek, M. (1987) *Polymers as Materials for Packaging*, Ellis Horwood, Chichester, p. 461.

Swoboda, P.A.T. and Peers, K.E. (1977) Metallic odor caused by vinyl ketones formed in the oxidation of butterfat. The identification of octa-1, cis-5-dien-3-one. *J. Sci. Food Agric.*, **28**, 1019–1024.

Thompson, L.J., Deniston, D.J. and Hoyer, C.W. (1994) Method for evaluating package-related flavours. *Food Tech.*, **1** (1), 90–94.

Tice, P. (1996) Packaging material as a source of taints, in: *Food Taints and Off-Flavors*, (ed. M.J. Saxby), Blackie Academic & Professional, London, pp. 226–260.

Tice, P.A. and Offen, C.P. (1994) Odors and taints from paperboard food packaging. *Tappi J.*, **77** (12), 149–154.

Tindale, C.R. and Whitfield, F.B. (1989) Production of chlorophenols by the reaction of fiberboard and timber components with chlorine-based cleaning agents. *Chem. Ind.*, (12), 835–836.

vom Bruck, C. G. and Hammerschmidt, W. (1976) Ermittlung der Fremdgeschmackschwelle in Lebensmitteln und ihre Bedeutung für die Auswahl von Verpackungs – materialien, in: *Internationale Konferenz über den Schutz verderblicher Güter durch Verpackung*, München, 9–11 June, 201–209.

vom Bruck, C.G. and Hammerschmidt, W. (1977) Ermittlung der Fremdgeschmacksschwelle in Lebensmitteln und ihre Bedeutung für die Auswahl von Verpackungsmaterialien, *Verpackungs-Rundshau*, 1, 1–4.

Whitfield, F.B. (1983) Some flavors which industry could well do without: Case studies of industrial problems. *CSIRO Food Res. Quart.*, **43**, 96–106.

Whitfield, F.B., Tindale, C.R., Shaw, K.J. and Stanley, G.(11) (1984) Contamination of cocoa powder by chlorophenols and chloroanisoles adsorbed from packaging materials. *Chem., Ind.*, pp. 772–774.

Ziegleder, G. (1998) Volatile and odorous compounds in unprinted paperboard. *Packag. Tech. Sci.*, **11**, 231–239.

Ziegleder, G., Stojacic, E. and Lustenberger, M. (1995) Cause and detection of off-odors in unprinted paperboard. *Packag. Tech. Sci.*, **8**, 219–228.

15
Possibilities and Limitations of Migration Modeling
Peter Mercea and Otto Piringer

As shown in Chapter 1, special emphasis is given in this book for interactions between plastic materials and packed goods and their consequences for quality assurance and legislation. The term interaction encompasses the sum of all mass transports from the package into the product as well as mass transport in the opposite direction. The mass transfers, often coupled with chemical reactions, lead to quality changes in the product and packaging material. Mass transport is understood as the molecular diffusion in, out, and through plastic materials like that shown schematically in Figure 1.3. The mass transport of package components into the product is also known as migration.

For the mathematical description and understanding of migration many analytical and numerical solutions of the diffusion equation, (7.12) are discussed in Chapters 7 and 8. The selected solutions of the diffusion equation included: diffusion from films or sheets (hollow bodies) into liquids and solids as well as diffusion in the reverse direction, diffusion controlled evaporation from a surface, influence of barrier layers and diffusion through laminates, influence of swelling and heterogeneity of packaging materials, and coupling of diffusion with chemical reactions. Due to the complex structure of many systems, only computer-supported treatments, as described and exemplified in Chapter 9, offer practical solutions for many problems.

Whereas 50 years ago most analytical solutions of the diffusion equation were known, the use of plastics was at a very beginning stage (Robinson-Görnhardt, 1957). Consequently, neither the problems of interactions with their consequences for the human health were known, nor the values of diffusion coefficients and partition coefficients needed for the solutions of the equations were available. But worldwide investigations over the last 50 years have demonstrated that interactions between plastics and foodstuffs or other products occur as foreseeable physical processes, as shown in a few selected publications during the last 30 years (Figge and Rudolph, 1979; Figge, 1980; Reid et al., 1980; Bieber et al., 1985; Till et al., 1987; Goydan et al., 1990; Castle et al., 1991; Baner et al., 1996; Hamdani et al., 1997;

Plastic Packaging. Second Edition. Edited by O.G. Piringer and A.L. Baner
Copyright © 2008 WILEY-VCH Verlag GmbH & Co. KGaA, Weinheim
ISBN: 978-3-527-31455-3

Lickly et al., 1997; O'Brien et al., 1997, 1999; Reynier et al., 1999; O'Brien and Cooper 2000, 2002). Standardization of migration measurements is based on this knowledge. However, the variety of substances occurring in interaction processes and the necessary time and cost requirements to carry out all the analysis for a complete quality assurance for consumer safety (Chapter 11), necessitate additional tools in order to fulfill this task. Modeling of potential migration is already used by the US Food and Drug Administration (FDA) as an additional tool to assist in making regulatory decisions (Chapter 12). The EU has recently introduced this option to use generally recognized migration models in the Directive 2002/72/EC (Chapter 13) as a novel conformity and quality assurance tool with the following statement in Article 8 (4):

> *"The verification of compliance with the specific migration limits provided for in paragraph 1 may be ensured by the determination of the quantity of a substance in the finished material or article, provided that a relationship between that quantity and the value of the specific migration of the substance has been established either by an adequate experimentation or by the application of generally recognized diffusion models based on scientific evidence. To demonstrate the noncompliance of a material or article, confirmation of the estimated migration value by experimental testing is obligatory."*

The realization of the above requirement has been recently achieved within the EU Project SMT4-CT98-7513 under the 5th Framework Program "Growth Evaluation of Migration Models in Support of Directive 2002/72/EC." The major objectives of this project were as follows:

- To demonstrate that a correspondence between the specific migration limit (SML) and a permitted maximum initial concentration (MIC) of a substance in the finished product can be established.
- To establish documentation that demonstrates the validity of underlying migration models for compliance purposes. Consequently, parameters used in the migration model have been selected in a way that a "worst case" estimate of migration is generated.

The final report (Hinrichs and Piringer, 2002) has been compiled and a summary of the results has been published recently (Begley et al., 2005).

The research project has established the mathematical equations to be applied and the conditions for their appropriate application with regards to plastics in contact with food. All these conditions and equations have been published in detail in the Practical Guide of the EU Commission in Annex 1, Mathematical Models, as well as in Chapter 7 and in previous publications cited in the reference list at the end of this chapter.

Beyond the characterization of the polymer and food (simulant), the key input parameters for the use of a migration model are the diffusion coefficient, D_P, of the migrant in the plastic material P, as well as the partition coefficient, $K_{P/L}$, of the migrant between P and the product (e.g., food, liquid) L.

15.1
Correlation of Diffusion Coefficients with Plastic Properties

The literature reports a series of sophisticated models for the theoretical estimation of diffusion coefficients in polymers (Chapter 5) but these models are, at least today, too complicated for practical applications. Therefore, a simpler approach was developed. A first approximation to estimate D_P was to correlate this coefficient with the relative molecular mass, M_r, of the migrant, with a matrix-specific (polymer) parameter, A_P, and the absolute temperature T, based on empirical data (Piringer 1993, 1994):

$$D_P = \exp(A_P - 0.008 \cdot M_r - 10450)/T \text{ (m}^2/\text{s)} \tag{15.1}$$

A similar approach has been proposed (Limm and Hollifield, 1996) to calculate D_P for migration estimation purposes from polyolefins

$$D_P = D_0 \cdot \exp(\alpha \cdot M_r^{1/2} - K \cdot M_r^{1/3})/T \text{ (cm}^2/\text{s)} \tag{15.2}$$

with the following values for D_0, K, and α

Polymer	K	α	ln(D_0)
PP	1335.7	0.597	−2.10
HDPE	1760.7	0.819	0.90
LDPE	1140.5	0.555	−4.16

To pursue the goal of obtaining a simple formula for the estimation of D_P, a refined equation for polyolefins and some other plastic materials has been developed (Brandsch et al., 2001) from Eq. (15.1):

$$D_P = 10^4 \cdot \exp(A_P - 0.1351 \cdot M_r^{2/3} + 0.003 \cdot M_r - 10454)/T \text{ (cm}^2/\text{s)} \tag{15.3}$$

with $A_P = A'_P - \tau/T$.

With this equation, a polymer-specific upper-bound diffusion coefficient, D_P^*, can be estimated and used instead of the actual diffusion coefficient, $D_P \leq D_P^*$, of a migrant in the polymer matrix. The parameter, A_P, is linked to the polymer and describes the basic diffusion behavior or a "conductance" of the polymer matrix towards the diffusion of migrants. In Eq. (15.3), $A_P \leq A_P^* = A'^*_P - \tau/T$, the parameter A_P^* depending on the temperature and the athermal constant A'^*_P should now be regarded as an "upper-bound" conductance of the polymer. The parameter τ, together with the constant 10454 in Eq. (15.3), both with the formal dimension of temperature, contribute to the diffusion activation energy, $E_A = (10454 + \tau)R$, where $R = 8.31451$ J/(mol K) is the gas constant. By analyzing from literature E_A data for a large series of migrants in many polymer matrices, it was concluded that one can take $\tau = 0$ for many polymers. Thus, taking $\tau = 0$ for LDPE one obtains $E_A = 86.92$ kJ/mol, which is in good agreement with the mean of $E_A = 87$ kJ/mol found from many literature data. For other important groups of plastics relevant to food packaging, e.g., HDPE and PET, a higher activation energy is generally observed. A good mean for these matrices is obtained with $E_A = 100\,\text{kJ/mol}$, which requires $\tau = 1577$.

To keep Eq. (15.3) functional and to work only with a minimum number of specific variables, to a first approximation τ was fixed at 0 and 1577, which lead to corresponding activation energies of $E_A = 87$ and $100\,\text{kJ/mol}$, respectively. It is known that in a given polymer and temperature range each migrant has different diffusion activation energy. Therefore, each migrant has a small specific contribution to E_A and thus influences also A_P. However, analyzing the available experimental data, one finds out that the main contribution to these values come from the specific structure of the polymer matrix and thus the influence of the migrant on E_A and, respectively, A_P may be neglected in a first approximation.

To validate the migration model in the context with the EU project, migration rates were collected from different sources (Begley et al. 2005) for LDPE, HDPE, PP, PET, PEN, PS, HIPS, and PA. All data were obtained from recent measurements using additives from the positive list of substances permitted under Directive 2002/72/EC. The migration measurements were carried out by following the conditions of Directive 97/48/EC for fatty food (simulants), in most cases olive oil, with good solubility for the additives. By using the migration software MIGRATEST Lite (Chapter 9) for each migration value the corresponding diffusion coefficient and athermal term A'_P has been calculated. In addition, to the migration data some recent experimental diffusion coefficients, obtained with up-to-date experimental methods, were available for PET, PEN, and PA and covered the temperature range of interest for food packaging materials. With all these data a representation of the characteristic migration behavior of a polymeric matrix was possible from a collection of experimental data obtained under very different conditions, at different temperatures including migrants of very different structures and molecular weights. From the mean value $\bar{A}'_P$ and standard deviation s for a specific polymer matrix, e.g., LDPE, an "upper bound" value, $A'^{*}_P = \bar{A}'_P + s \times t$, results by using the Student t-factor for a one (right)-side 95% confidence level with N samples. All these values are listed in Table 15.1 The A'^{*}_P values obtained in this way were very close to the previous values listed in the above-mentioned Practical Guide.

Table 15.1 Statistical evaluation of A_P values for migration modeling under "worst case" conditions.

Polymer	$\bar{A}'_P$	s	A'_P (max)	A'_P (min)	N	t	A'^{*}_P	τ
LDPE	10.0	1.0	11	7.0	27	1.7	11.7	0
HDPE	10.0	1.9	12.6	5.0	49	1.68	13.2	1577
PP	9.4	1.8	12.9	6.2	53	1.68	12.4	1577
PET	2.2	2.5	7.2	−4.3	58	1.67	6.35	1577
PEN	−0.34	2.4	3.8	−5.5	38	1.7	3.7	1577
PS	−2.8	1.25	0.0	−6.5	32	1.7	−0.7	0
HIPS	−2.7	1.67	0	−6.2	33	1.7	0.1	0
PA (6,6)	−1.54	2.0	2.3	−7.7	31	1.7	1.9	0

In addition to the above results, a few recently findings can be mentioned about the migration behavior of different plastics.

The migration of additives in 17 PP-samples has been measured (Begley et al., 2007). These samples cover the major types of PP used in food packaging. The diffusion coefficients with relative small molecular masses, $M_r = 136$ (limonene), as well as the migration of typical antioxidants used in PP up to M_r 1178 (Irganox 1010) have been measured at different temperatures. In addition, the diffusion data and the percentages of xylene soluble fractions have been correlated. This allows to predict the migration behavior of a PP sample by testing its "isotactic index" with xylene. The results clearly indicate that polypropylene can be subdivided from the migration point of view into the monophasic homopolymer (h-PP), monophasic random copolymer (r-PP), and heterophasic copolymer (heco-PP). The diffusion coefficients of r-PP are at least one order of magnitude higher than those of h-PP and comparable to the values for heco-PP. In Figure 15.1, the polymer-specific A_P values for the investigated h-PP and r-PP samples obtained with three additives at 40 °C is represented in dependence of the xylene-soluble amount.

This is a useful example of a correlation between specific migration property, A_P, and an easily measurable property of the polymer matrix.

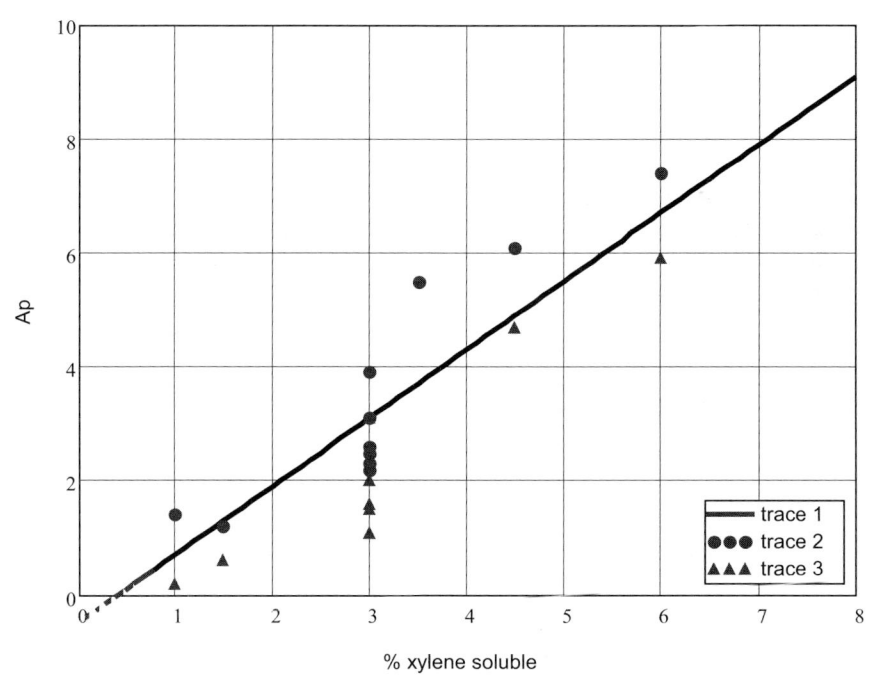

Figure 15.1 The polymer specific A_P-values for the investigated h-PP and r-PP samples obtained with three additives at 40°C in dependence of the %-w/w xylene-soluble amount (X). Trace 1: $A_P = -0.5 + 1.2X$; trace 2: Irganox 1010; trace 3: Irgafos 168 and Irgafos P-EPQ.

An influence on the migration speed is observed in all cases where concentrations >1% of additives with relative molecular masses $M_r < 1000$ are used in the polymer matrix. Roughly a linear increase of the A_P values is observed. In PVC, as an example, the A_P value is very small ($A_P \leq -4$ at 25°C) for rigid-PVC and increases until A_P 12–14 with typical plasticizer concentrations of 30%.

An important fact is the difference observed between the migration amount measured with HIPS samples by full immersion compared with one-sided migration cells (Lickly et al., 1997). Due to the two-phase structure of the plastic matrix, the normally homogeneous polystyrene-phase near the interface in a real food contact material is destroyed in the material edges after cutting. The consequence is an enhanced migration through the rubbery phase in these regions. An aging effect of the polymeric samples can also produce significant over-estimations in modeling, especially in the case of low molecular migrants, e. g., styrene. During long storage periods of packaging materials in the open atmosphere, considerable loss of the migrant occurs near the interface and consequently, the migrant is no longer homogeneously distributed in the plastic, as assumed in theory.

A relative strong decrease of the A_P values, due to crosslinking effects occur in many polymers, for example, for some printing inks ($A_P \approx -2$) and in some coatings ($A_P \approx 7$).

The diffusion in paper and board samples occurs principally in a different way than in plastics. Whereas a plastic material behaves as a solvent for the diffusing compound, in a paper matrix most of the diffusing substance is adsorbed on the surface of the fiber. In many foods, the diffusing compounds are partially solved in the matrix as in plastics and partially adsorbed as in a fiber material. The same complex process takes place in Tenax, a very powerful simulant for migration into dry foods. Despite the very complex structure of the considered matrix, the apparent diffusion values obtained from migration measurements often follow the diffusion law and the migration behavior of such matrices can also be characterized with specific A_P values. These values are in the range $6 < A_P < 13$, comparable with the polyolefins.

The above results should be regarded only as a first step in a longer process of refinement of the methods of migration measurement and the estimation models. Precise measurements of diffusion coefficients in the specific layers of a multilayer laminate are possible only in combination with the corresponding theoretical evaluation of the results (Chapters 8 and 9).

As already shown in Chapter 6, a considerable improved estimation of diffusion coefficients is now possible. For some well-defined polymer matrices and additives a full theoretical calculation of diffusion coefficients is possible.

In the following nine Figures 15.2(a)–15.4(c), the diffusion coefficients D_p (cm^2/s) of all compounds in polyolefins extracted from Appendix I at room temperature (23–25°C), 40 and 70°C are represented as functions of their relative molecular weights M_r, together with the corresponding curves obtained from Eq. (6.28) for two extreme values of the melting temperatures, $T_{m,p}$, and molecular masses, $M_{r,p}$, of the

15.1 Correlation of Diffusion Coefficients with Plastic Properties

corresponding polymers. Despite the fact that for a given M_r of the diffusing molecule the D_P values scattered often over three orders of magnitude, due to very different polymer samples and test methods used, the main stream of the values follows the trace prescribed from Eq. (6.28).

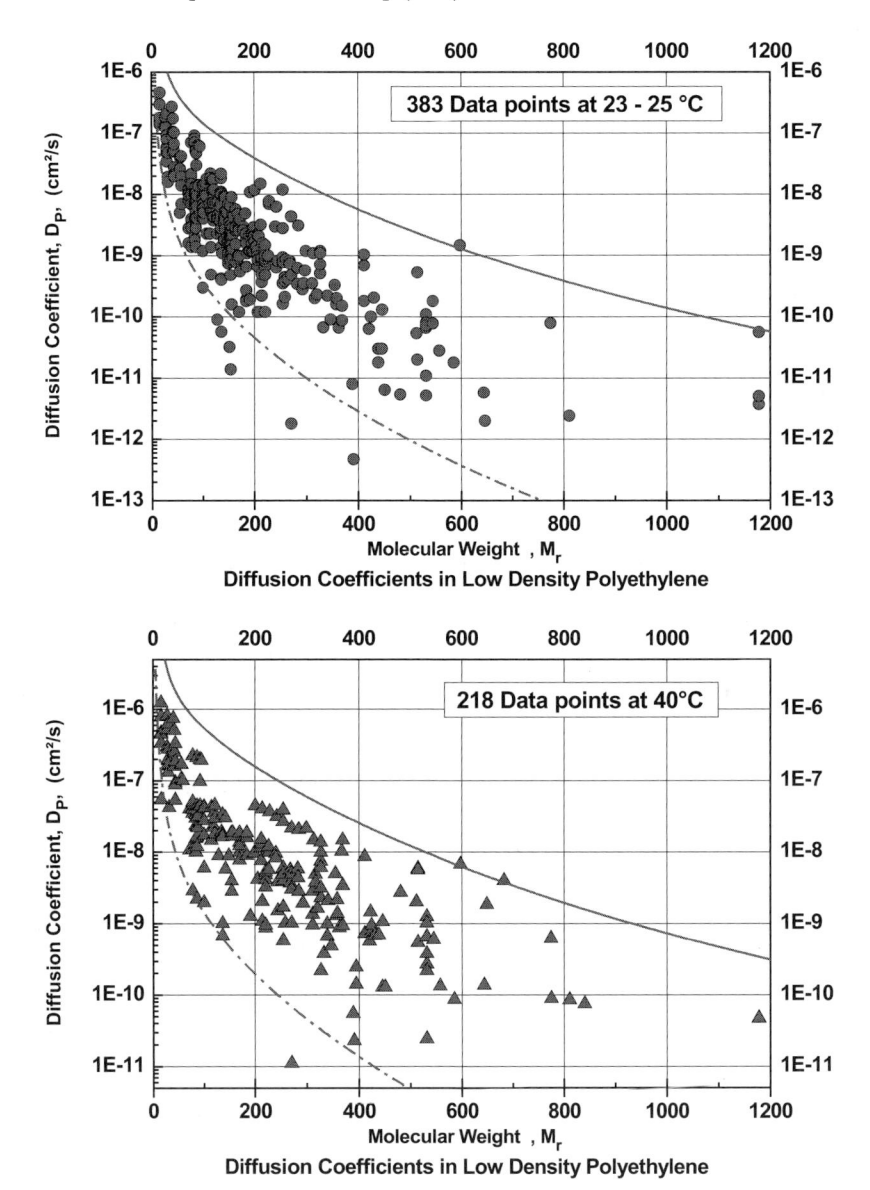

Figure 15.2 The diffusion coefficients D_p (cm²/s) of all compounds from Appendix I (LDPE). The upper curves are calculated with $T_{m,p} = 353$ K (80°C) and $M_{r,p} = 3000$ and the lower curves with $T_{m,p} = 393$ K (120°C) and $M_{r,p} = 100{,}000$ in Eq. (6.28).
a) At room temperature (23–25°C);.b) At 40°C; c) At 70 °C.

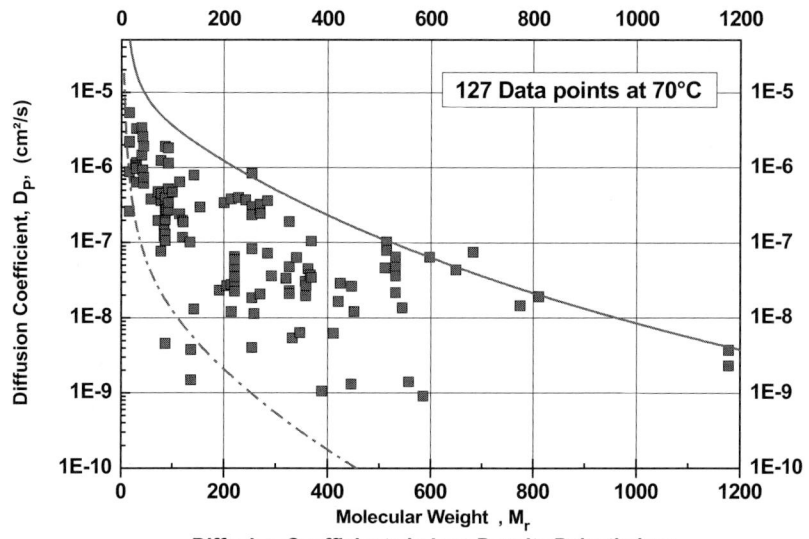

Diffusion Coefficients in Low Density Polyethylene

Figure 15.2 (*Continued*).

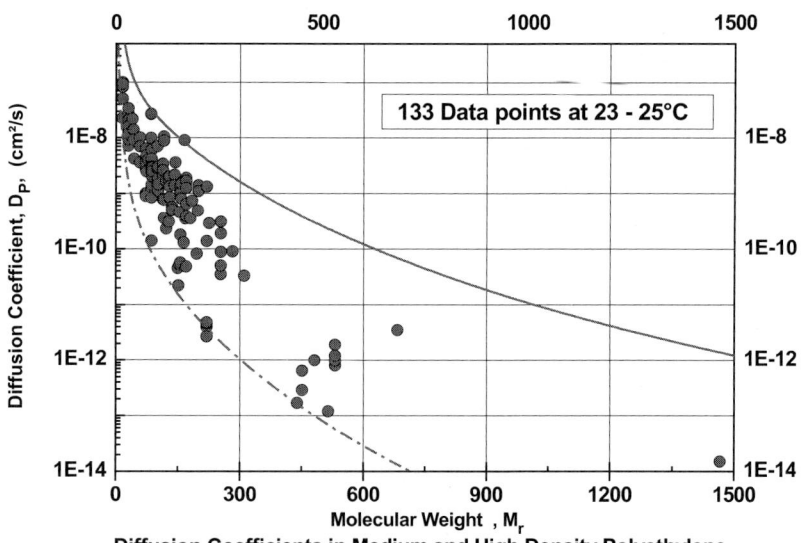

Diffusion Coefficients in Medium and High Density Polyethylene

Figure 15.3 The diffusion coefficients D_p (cm^2/s) of all compounds from Appendix I (HDPE). The upper curves are calculated with $T_{m,p} = 373$ K (100°C) and $M_{r,p} = 5000$ and the lower curves with $T_{m,p} = 403$ K (130°C) and $M_{r,p} = 500{,}000$ in Eq. (6.28). a) At room temperature (23–25°C);.b) At 40°C; c) At 70°C.

15.1 Correlation of Diffusion Coefficients with Plastic Properties

Diffusion Coefficients in Medium and High Density Polyethylene (98 Data points at 40°C)

Diffusion Coefficients in Medium and High Density Polyethylene (72 Data points at 70°C)

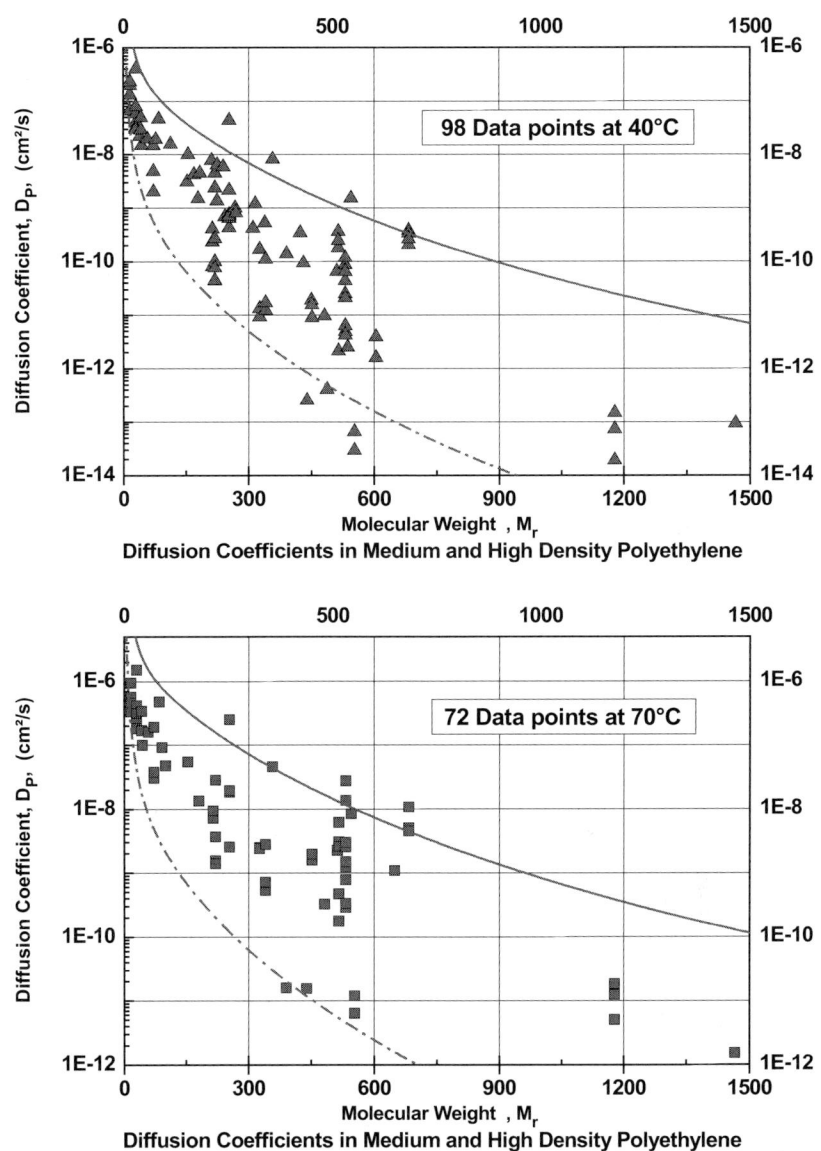

Figure 15.3 (Continued).

In Figure 15.5 the measured diffusion coefficients of components from two polymeric additives incorporated in two samples of the same LDPE matrix are represented as a function of M_r. These values obtained at 40°C cover the mass range $200 < M_r < 1000$ and correlate well with the D_p-curve obtained with Eq. (6.28) with $T_{m,p} = 363$ K $= 90$°C and $M_{r,p} = 15{,}000$. This agreement further support the assumption made for the development of the diffusion model in Chapter 6.

Figure 15.6 shows the diffusion coefficients of all alkanes listed in Appendix I for LDPE at 40 °C together with the corresponding curves obtained with Eqs. (15.2), (15.3), and (6.28). From Eq. (6.28), with $T_{m,p} = 90\,°C$ and $M_{r,p} = 25{,}000$, a stronger and, therefore, better decrease of the D_P values results in the low molecular mass region in comparison with Eq. (15.3) with $A_P = 10$. Consequently, D_p values calculated with Eq. (15.3) using an appropriate A_P value for small masses, are too big for migrants with higher molecular masses. The curve obtained with Eq. (15.2) is similar with the curve from Eq. (6.28), but with systematic higher, "upper-limit" values due to the set of constants used in combination with Eq. (15.2).

A strong differentiation of the values obtained with the three above equations results at high molecular masses as shown in Figure 15.7 Here, both semiempirical equations (15.2) and (15.3) fail because producing an unrealistic increase of the D_p values at high M_r values.

Among the improvements of the new diffusion model presented in Chapter 6 the following must emphasized:

- The activation energy of diffusion is determined from both the melting or glass temperature of the polymer matrix and the molecular mass of the diffusing

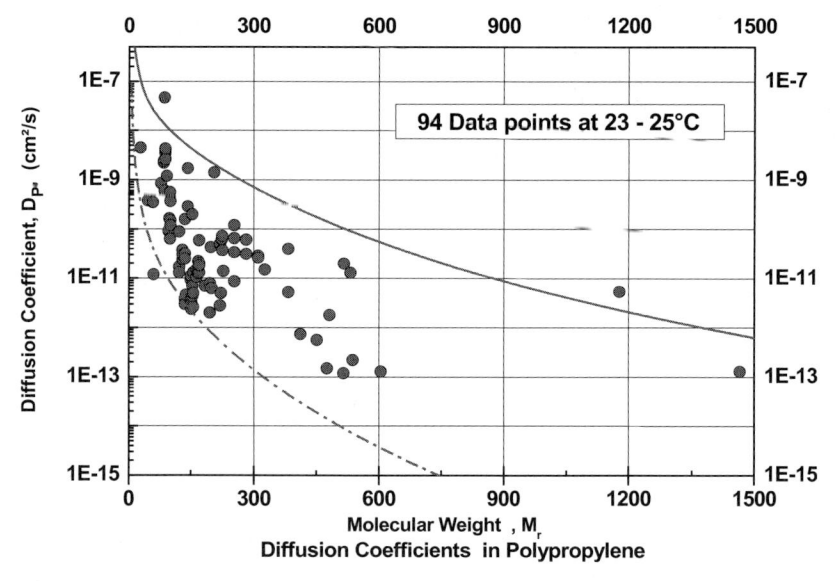

Figure 15.4 The diffusion coefficients D_p (cm^2/s) of all compounds from Appendix I (PP). The upper curves are calculated with $T_{m,p} = 403\,K\,(130°C)$ and $M_{r,p} = 1500$ and the lower curves with $T_{m,p} = 449\,K\,(176°C)$ and $M_{r,p} = 100{,}000$ in Eq. (6.28). a) At room temperature (23–25°C);.b) At 40°C; c) At 70 °C.

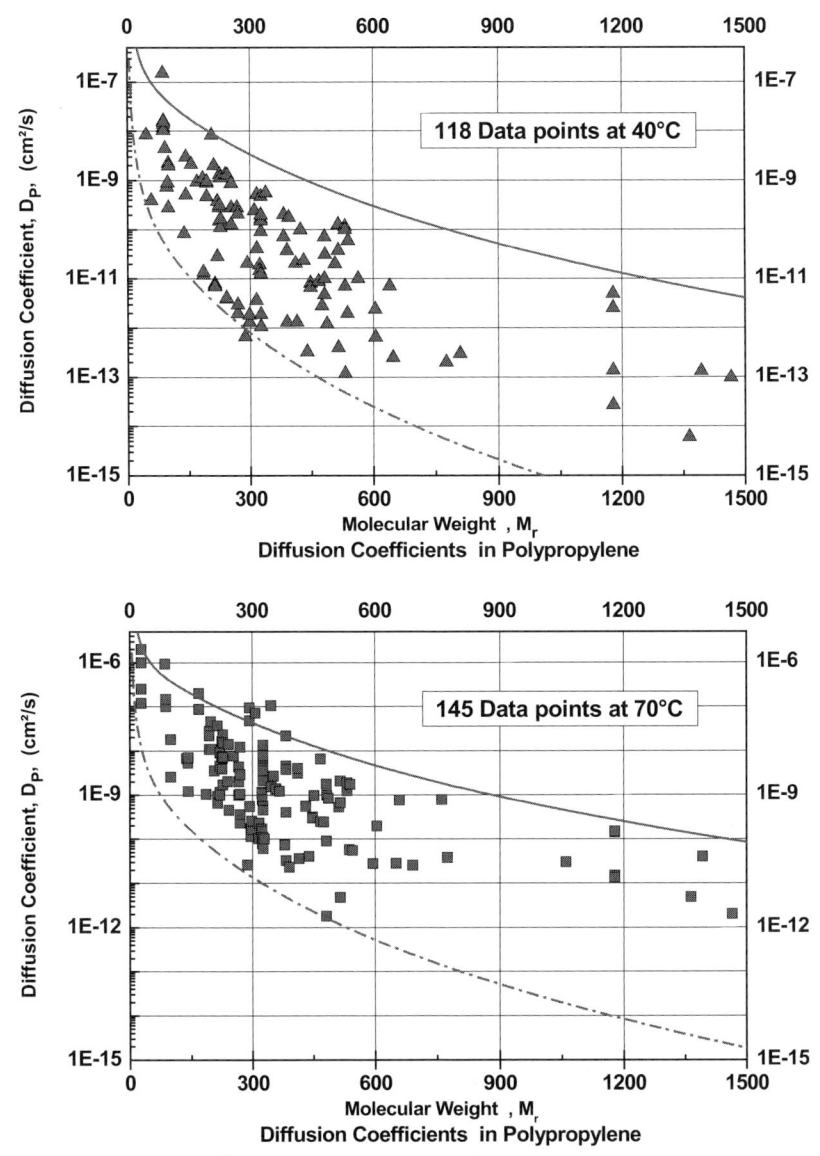

Figure 15.4 (Continued).

molecule. The somewhat arbitrariness used by selection of only two activation energies for Eq. (15.3) is eliminated in this way.

- A further important parameter is the molecular mass, $M_{r,p}$, of the polymer matrix. A polymer-like LDPE is polydisperse, in the sense that a sample spans a large range of molar masses (Chapter 2). From the diffusion point of view, the smaller molecules play a relatively more important role in comparison with the bigger molecules due to the easier movement of the diffusing solute through a matrix with

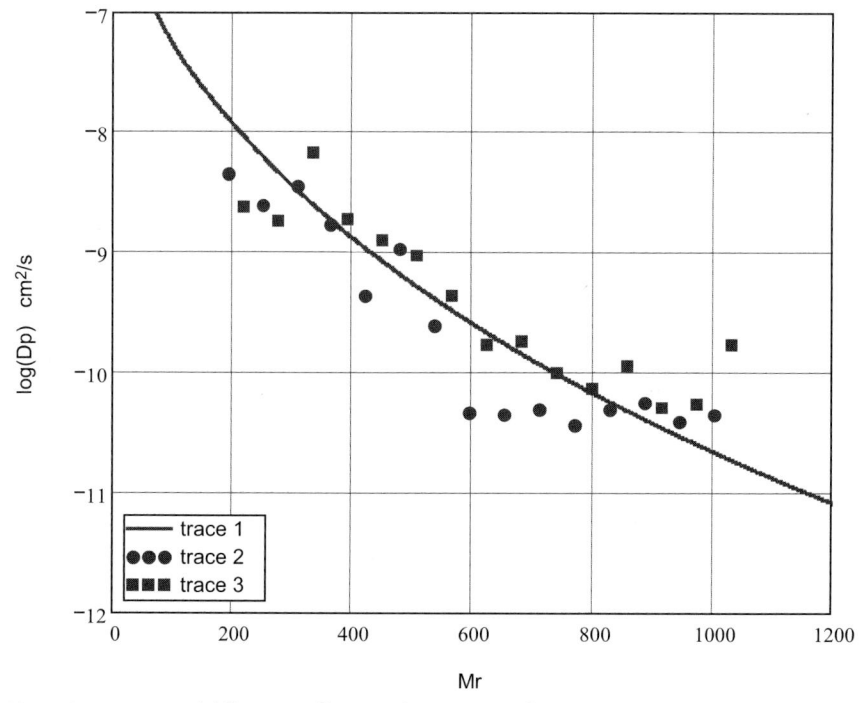

Figure 15.5 Measured diffusion coefficients of components from two polymeric additives (trace 2 and trace 3) in two samples of the same LDPE matrix at 40°C. The curve (trace 1) is calculated with Eq. 6.28 with $T_{m,p} = 363\,K = 90°C$ and $M_{r,p} = 15{,}000$.

a large portion of small molecules. As a good approximation, the number-average molar mass, $M_n = M_{r,p}$ is used in Eq. (6.28).

- Last but not least the diffusion model developed from the reference series of the *n*-alkanes can also be applied to other polymer matrices, e.g., PET as shown in Chapter 6. From the data used in the above-mentioned EU-project, Figures 15.8 and 15.9 show a correlation of diffusion coefficients measured for PS and PA with the calculated values using Eq. 6.28.

In PS as well as in PA the diffusion coefficients calculated with Eq. (6.28) decrease faster with M_r in comparison with the values calculated with Eq. (15.3) and the corresponding mean values $\bar{A}'_P$ from Table 15.1.

In the following Figures. 15.10(a)–15.12(c), the $\log(D_p)$ curves resulting from Eq. (6.28) for polyolefins, with the corresponding mean values for $T_{m,p}$ and $M_{r,p}$ at three temperatures, are compared with the corresponding curves obtained with Eq. (15.3), with the corresponding $\bar{A}'_P$ values from Table 15.1 and the upper limt values, $A_P^{'*}$ from Table 9.1. In all the cases, a relative faster decrease of the D_p values results with Eq. (6.28) in comparison with Eq. (15.3), with increasing M_r. This behavior is supported from the experimental results as shown above. Nevertheless,

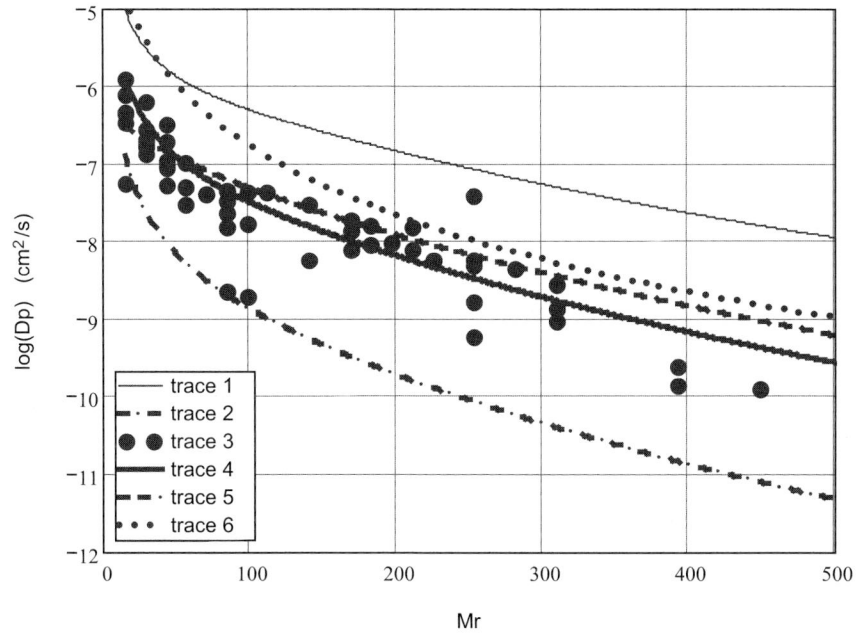

Figure 15.6 The diffusion coefficients of all alkanes (trace 3) listed in Appendix I for LDPE at 40°C together with the corresponding curves obtained with: Eq. (6.28) with $T_{m,p} = 353$ K (80°C) and $M_{r,p} = 3000$ (trace 1), $T_{m,p} = 393$ K (120°C) and $M_{r,p} = 100\,000$ (trace 2) and $T_{m,p} = 363$ K (90°C) and $M_{r,p} = 25\,000$ (trace 4). Eq (15.2) for LDPE (trace 6) and Eq. (15.3) with $A_P = 10$ (trace 5).

the upper limit values D_p^* calculated with $A_P'^*$ in Eq. (15.3) can still be used. But with the upper limit values for $T_{m,p}$ and $M_{r,p}$ in Eq. (6.28) even higher upper values, as shown in Figures 15.2–15.4, are obtained in comparison with the corresponding values from Eq. (15.3).

A general conclusion from the above results obtained with the theoretical improved modeling of diffusion coefficients is the possibility to correlate much better the diffusion behavior with well-defined properties of the polymer matrix. A still actual limitation of the modeling is the lack of experimental data obtained with samples with well-defined properties. To fill this gap is a challenge for the next future.

15.2
The Partition Coefficient

The second key input parameter needed in migration models is the partition coefficient, $K_{P/L}$, of the migrant between P and the product (e.g., food, liquid) L. The values of this dimensionless number cover many orders of magnitude,

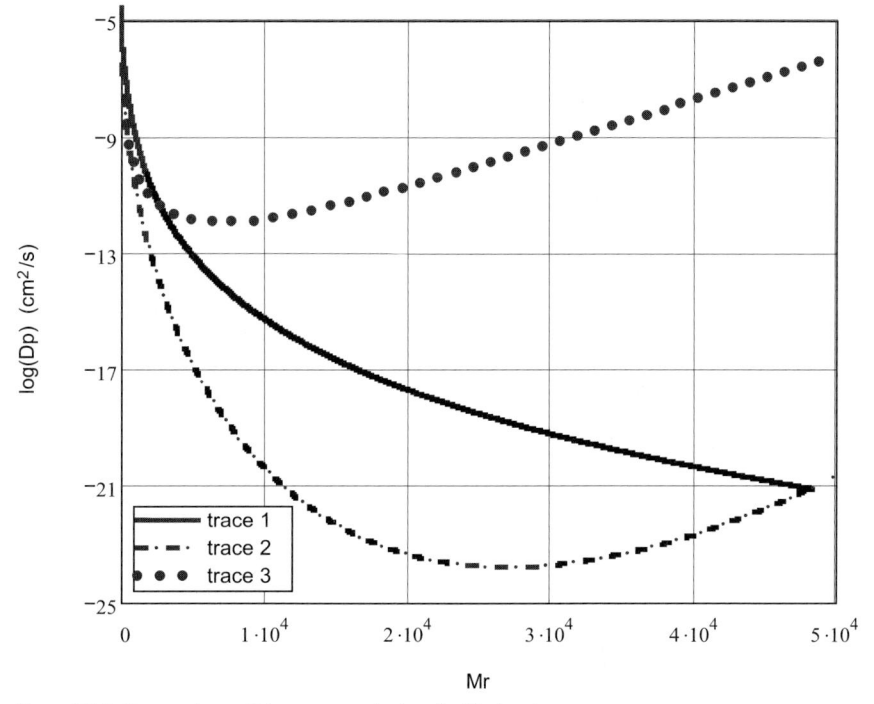

Figure 15.7 Comparison of the curves calculated with the three equations in Eq. (6.28) with $T_{m,p} = 363$ K (90 °C) and $M_{r,p} = 25{,}000$ (trace 1), Eq. (15.3) with $A_P = 10$ (trace 2) and Eq. (15.2) for LDPE (trace 3) at 100 °C.

depending on the polarities, structures and sizes of the migrant, the polymer and the product (Chapter 4). In order to simplify as much as possible the theoretical treatment of migration processes from the regulatory point of view, in the above-mentioned EU modeling project, an upper limit of $K_{P/L} = 1$ has been used for all fatty products and fatty food simulants, assuming a good solubility of the migrant in these phases. Due to a generally very low solubility of most organic compounds in water ($K_{P/L} \gg 1$) the value $K_{P/L} = 1000$ has been used for aqueous products as a sufficient good approximation for an upper limit. But this situation is, of course, not satisfactory for a more precise evaluation of product quality and safety.

Many results obtained from mass transport studies between a variety of systems during the last years (Chapters 4 and 11) have emphasized the necessity to consider partitioning in a much higher differentiation in all further treatments. As shown in many examples (Chapters 4 and 11) the actual used simulants for aqueous foods fail in many comparisons with migration tests in real foods. Consequently, it is no longer admissible to use such simulants without a case-by-case judgment of their reliability. It is also necessary to introduce in the modeling equations more realistic values for

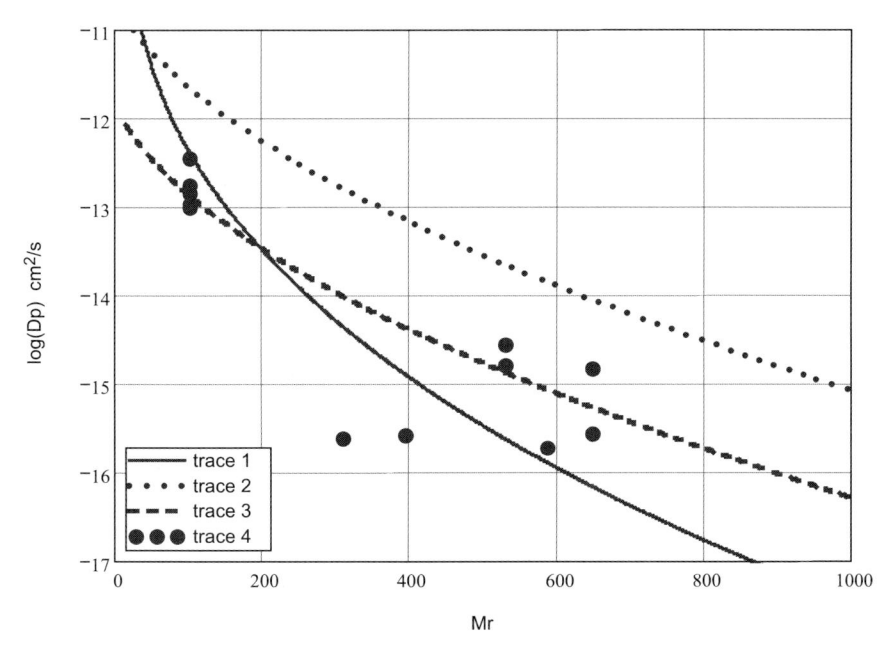

Figure 15.8 Measured diffusion coefficients (trace 4) in PS in comparison with calculated values with Eq. (6.28) with $T_{m,p} = 523$ K (250 °C) and $M_{r,p} = 75\,000$ (trace 1) and with Eq. (15.3) with $A_P = -2.8$ in Table 15.1 (trace 3) and $A_P = 0$ in Table 9.1 (trace 2) at 40 °C.

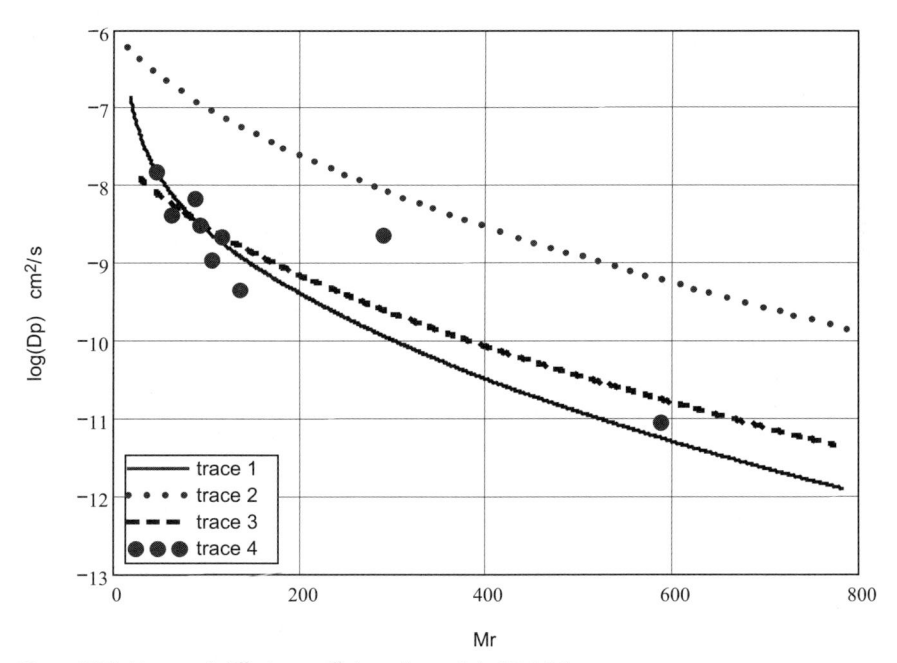

Figure 15.9 Measured diffusion coefficients (trace 4) in PA 6.6 in comparison with calculated values with Eq. (6.28) with $T_{m,p} = 535$ K (262 °C) and $M_{r,p} = 40\,000$ (trace 1) and with Eq. (15.3) with $A_P = -1.54$ in Table 15.1 (trace 3) and $A_P = 2$ in Table 9.1 (trace 2) at 40 °C.

the partition coefficients, obtained either from measurements or with available prediction methods. In Chapter 4, a few prediction methods are mentioned. Particularly important is the partition of a substance between a plastic matrix and an aqueous system. For such cases the simplest way is to equate the partition coefficient $K_{P/L}$ with the partition coefficient $P_{O/W}$ between octanol and water. Many $P_{O/W}$ values can be extracted from different databases. But this procedure can be used only as a first approximation.

The vapor pressure index method introduced in Chapter 4 has been found especially useful for migration modeling of additives from plastics into various products. In Appendix III, the vapor pressure index W ($W = Wa$) for a series of additives a is listed. This dimensionless number depends on the structure and polarity of the diffusing substance. Also in Appendix III are listed the dimensionless vapor pressure indexes G_W of the additives in water. Together with the relative molecular mass M_r the Henry constant h_w for water can be calculated with Eq. (4.57), with $j = (M_r + W + G_W - 2)/14$. The corresponding Henry constant h_p for polyolefins can be estimated in a similar

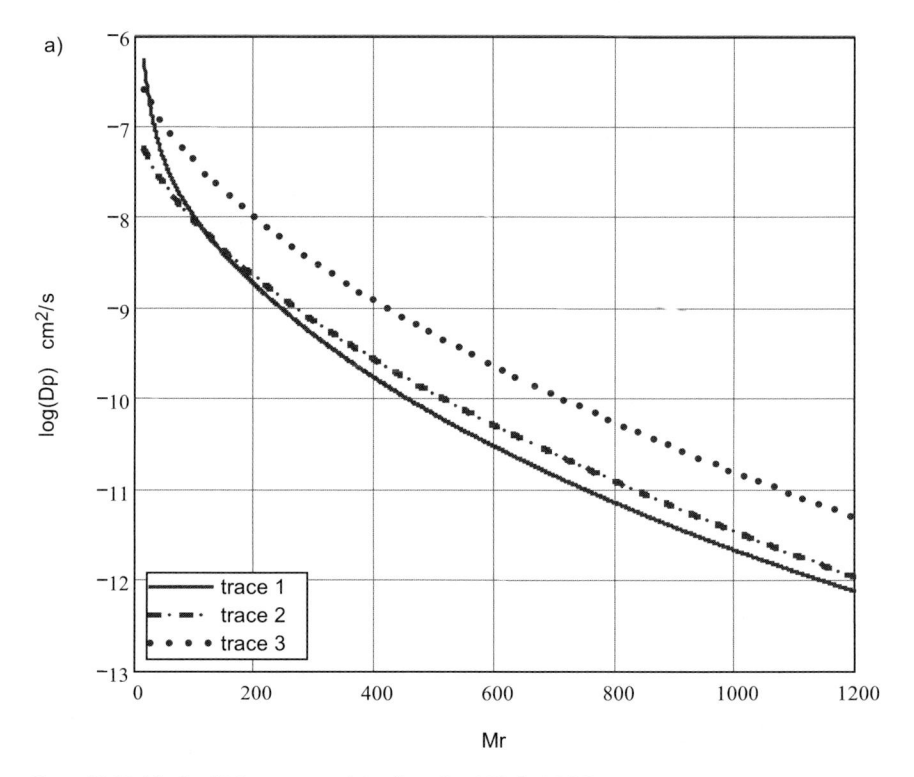

Figure 15.10 The $\log(D_p)$-curves resulting from Eq. 6.28 for LDPE with mean values $T_{m,p} = 363$ K $= 90\,°C$ and $M_{r,p}$ 25000 (trace 1) are compared with the corresponding curves obtained with Eq. (15.3) with $A'_P = 10$ from Table 15.1 (trace 2) and A'_P from Table 9.1 (trace 3). a) At 25 °C, b) at 40 °C, c) at 80 °C.

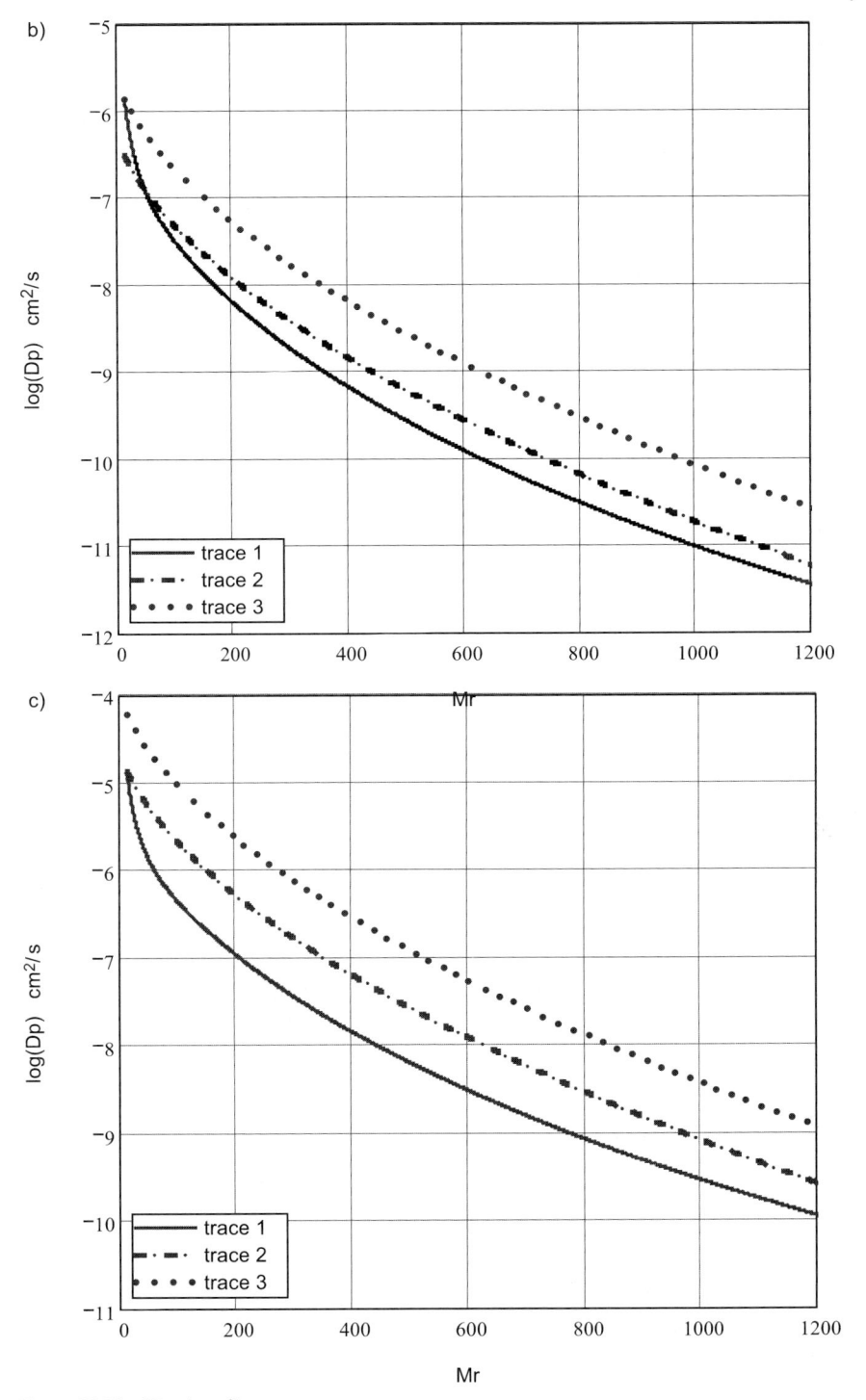

Figure 15.10 (*Continued*).

way, by using Eq. (4.57) with $j = (M_r + W + G_P\text{-}2)/14$, where the vapor pressure index of the additives for polyolefins are estimated with $G_P = -14\text{–}0.24\ M_r$. The partition coefficient $K_{P/L}$ of the additive between a polyolefin and water (L = W) can now be calculated with Eq. (4.58) at temperature T.

In the same way the vapor pressure indexes of additives can be estimated for other liquids, e.g., for ethanol, $G_L = G_E = -20\text{–}0.2\ M_r$ and any mixture of ethanol with water:

$G_L = (G_E - G_W) \cdot \frac{c_E}{100} + G_W$, where c_E is the concentration (%, w/w) of ethanol.

One important consideration should always be directed to the solubility of a migrant in connection with the toxicological important specific migration limit SML.

Let be n-octadecyl-3-(4-hydroxy-3,5-di-*tert*-butylphenyl)propionate [Irganox 1076] an example for estimating the partition coefficient (Chapter 11). This substance is a very widespread antioxidant used for the production of plastics and which has a SML restriction of 6 mg/kg food. The solubility of this additive in

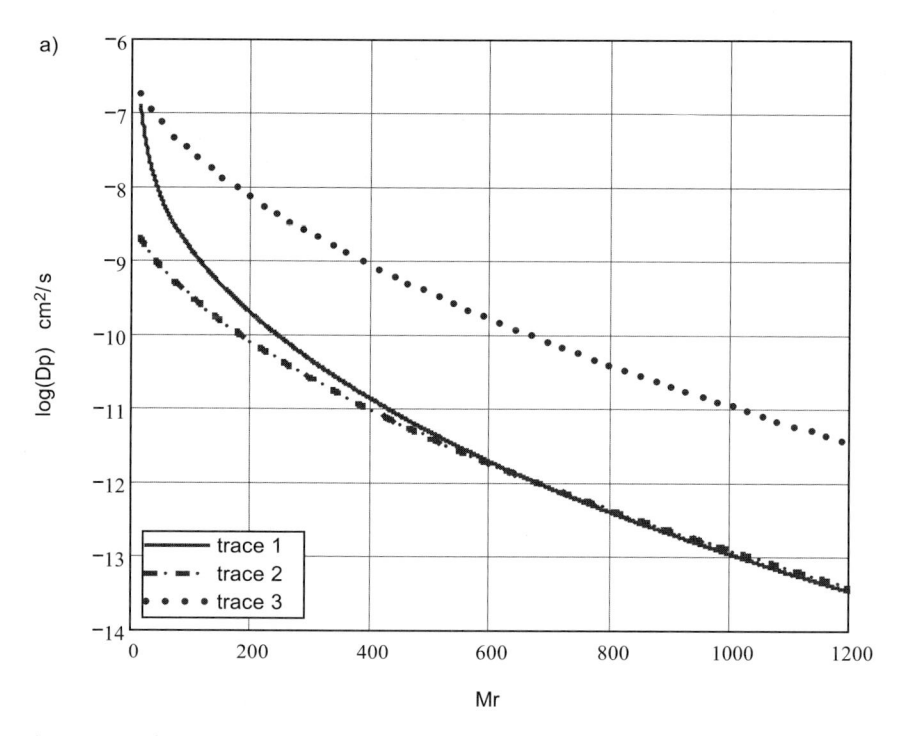

Figure 15.11 The $\log(D_p)$-curves resulting from Eq. (6-28) for HDPE with mean values $T_{m,p} = 393\ K = 120°C$ and $M_{r,p}$ 10 0000 (trace 1) are compared with the corresponding curves obtained with Eq. (15.3) with $A'_P = 10$ and $\tau = 1577$ from Table 15.1 (trace 2) and $A'_P = 14.5$ and $\tau = 1577$ from Table 9.1 (trace 3). a) At 40°C, b) at 80°C and c) at 120°C.

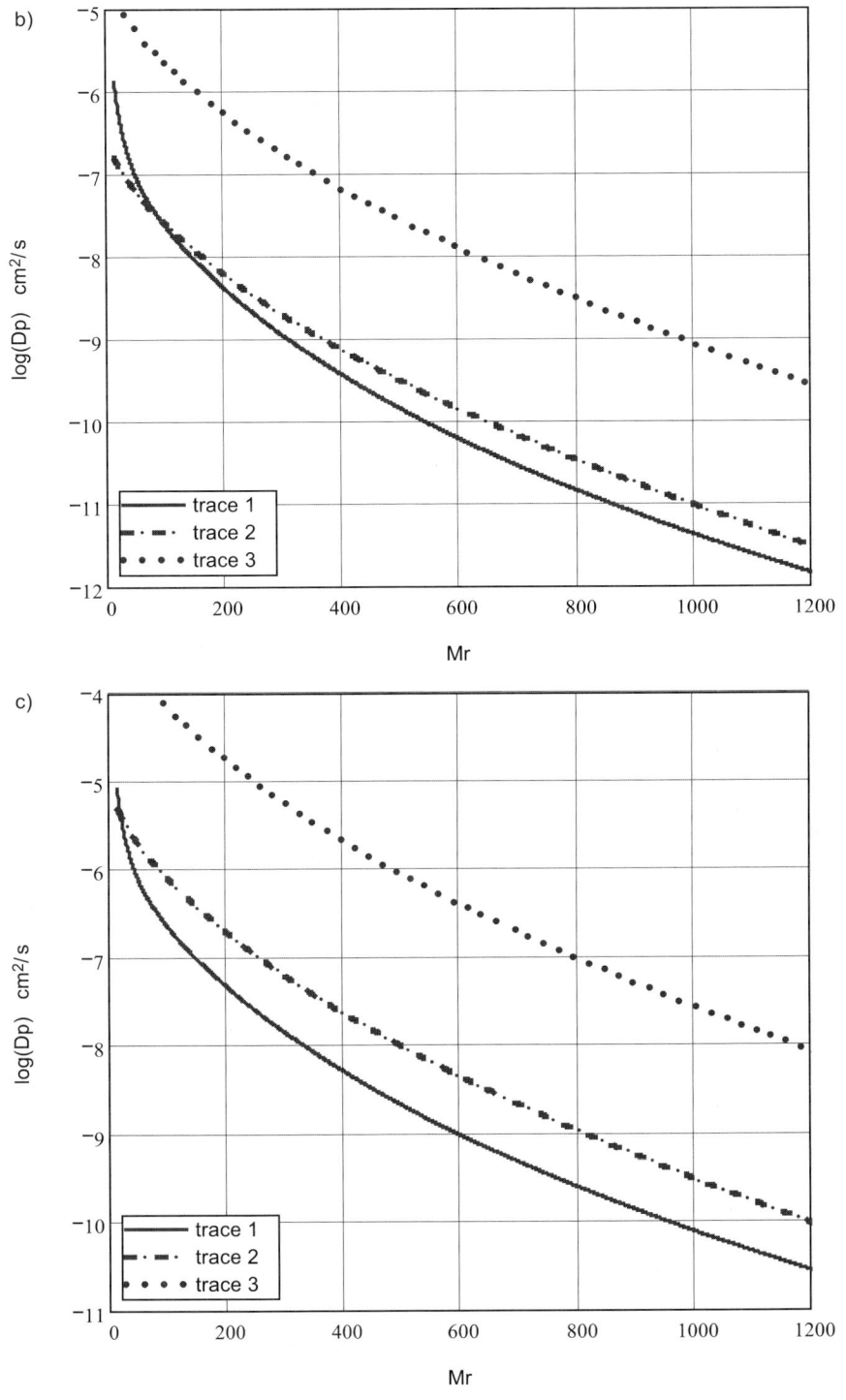

Figure 15.11 (*Continued*).

water is in the ppt [ng/L] range and can be expected to be also extremely low [ppb to ppm] in highly aqueous systems, such as the official EU simulants 3% acetic acid (B) and 10% ethanol (C), and even at higher ethanol contents up to 50%. Considering the $\log(P_{O/W}) = 13$ as a rough measure for the expected order of magnitude for the partition coefficient $K_{P/F}$ between a lipophilic polymer, such as a polyolefin and aqueous simulants of this additive, it results that any attempt to measure this concentration in an aqueous simulant must fail. By using the vapour pressure index method one finds for Irganox 1076 in Appendix III, $M_r = 531$, $W = -110$ and $G_W = -364$ and with the above relation, $G_P = -141$ and Eq. (4.57) and Eq. (4.58) $K_{P/L} = 3.8 \times 10^7$ results at 25°C. Although this value is much smaller in comparison with the $P_{O/W}$-value, the concentration of Irganox 1076 in water remains <0.05 ppb, assuming an equilibrium with a concentration of 1500 ppm in the plastic sample.

From an additional estimation for 50% ethanol/water one gets $G_L = -245$ and the partition coefficient decreases to $K_{P/L} = 4130$. Consequently, the concentration of Irganox 1076 in the liquid increases to 0.4 ppm at equilibrium with 1500 ppm in LDPE. Although even this concentration is of no concern in this

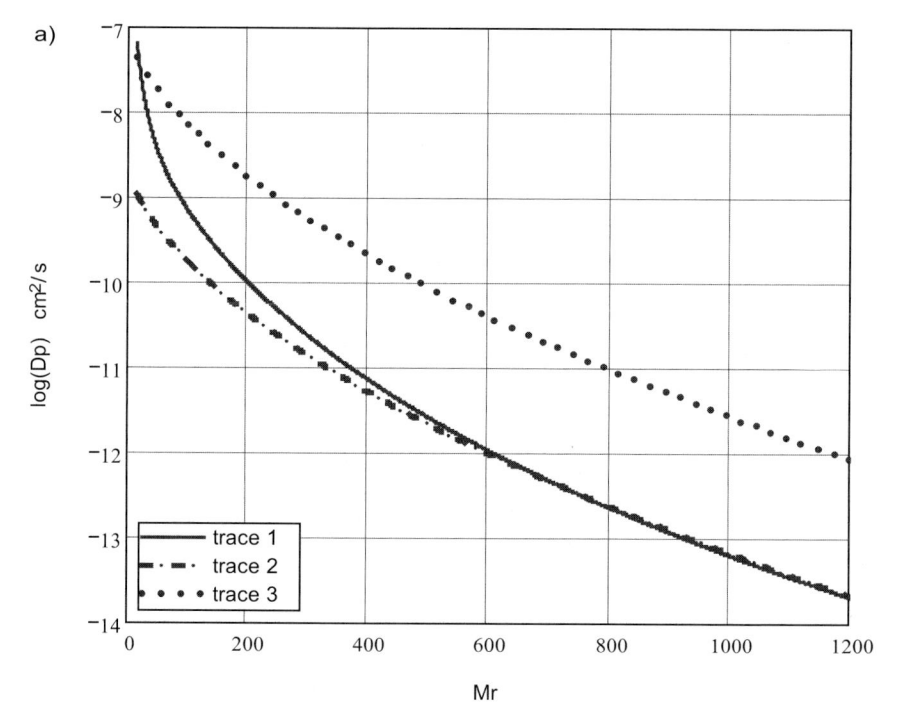

Figure 15.12 The $\log(D_p)$-curves resulting from Eq. (6-28) for PP with mean values $T_{m,p} = 418\,\text{K} = 145°\text{C}$ and $M_{r,p}$ 35000 (trace 1) are compared with the corresponding curves obtained with Eq. (15.3) with $A'_P = 9.4$ and $\tau = 1577$ from Table 15.1 (trace 2) and $A'_P = 13.1$ and $\tau = 1577$ from Table 9.1 (trace 3). a) At 40°C, b) at 80°C and c) at 120 °C.

b)

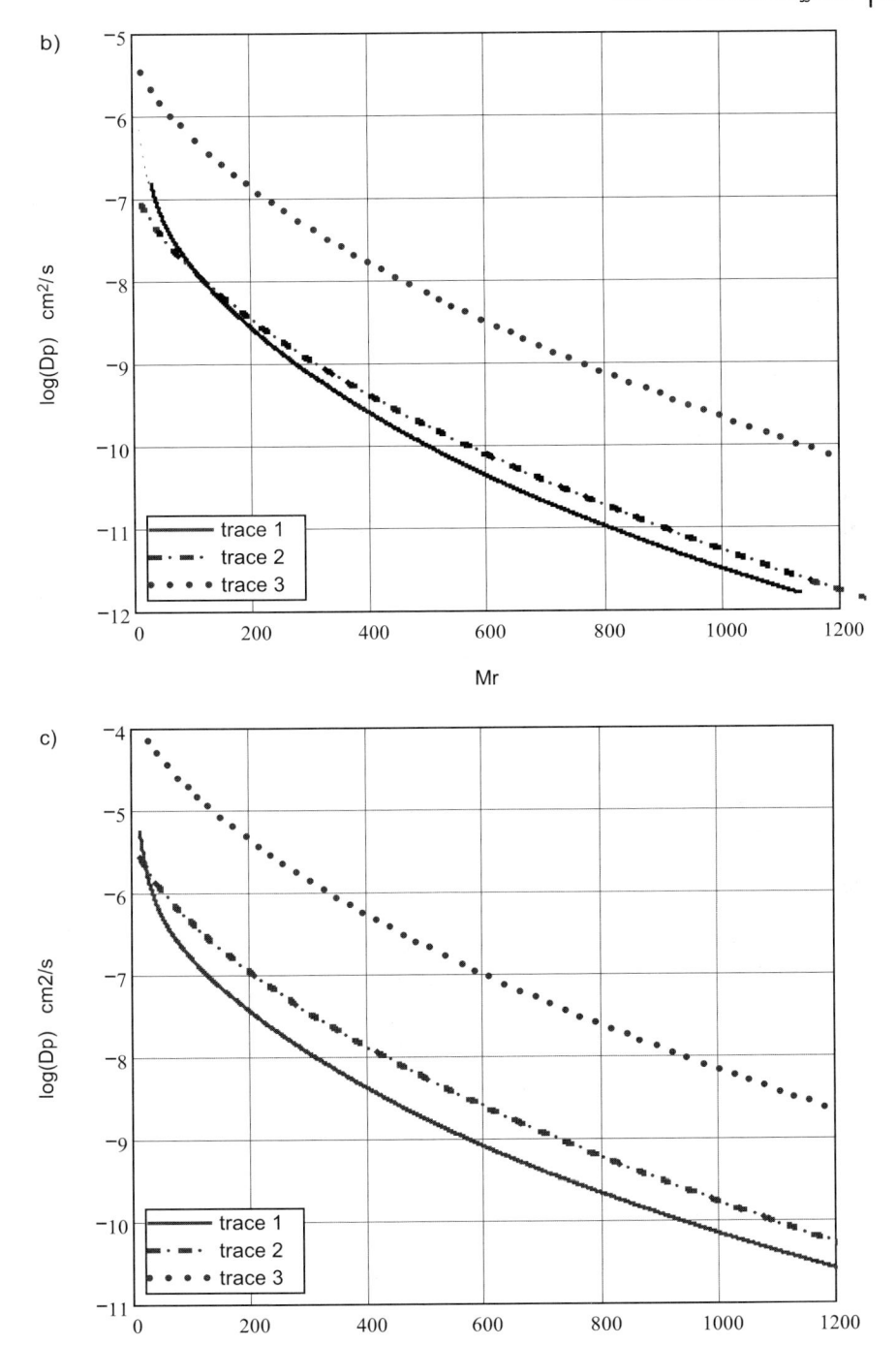

c)

Figure 15.12 (*Continued*).

example, it shows how drastically the partition coefficients can change with increasing the ethanol concentration. In many practical cases this may have important consequences.

The use of certain solvents as food simulants for controlling migration from plastics is widely practiced and allowed by food regulations. Using these simulants instead of real foods makes testing much easier, more sensitive and as a result much less expensive. Nevertheless, there is a real danger in some cases when the food simulant has a significantly lower migration value compared to the food. A well-known recent example was the presence of the photoinitiator ITX (isopropylthioxanthone) found in several milk products and beverages even though the migration test with the allowed food simulant (10% ethanol in water) showed that the migration limit values were not exceeded. For many similar migration testing problems finding the correct food simulant is of great practical importance. In principal, ethanol/water mixtures are very appropriate for this application (Chapter 4).

In Table 15.2, partition coefficients between LDPE and ethanol/water are given for six aroma compounds having similar sized molecules but very different polarities between LDPE and six different ethanol/water systems. Here one can also see the strong influence of the substance's polarity on the partition coefficient. The practical consequence of this behavior is that it is necessary to select the proper food simulant. The values found in parenthesis in Table 15.2 are the partition coefficients for the corresponding component between LDPE and whole milk. As one can see in this example, a 50% ethanol/water food simulant is not sufficient for nonpolar migrants like limonene but rather 75% ethanol/water is necessary to simulate the correct partitioning.

In Appendix II, experimental values of partition coefficients are collected from different sources, which may help to estimate the partitioning under different conditions.

The general conclusion drawn above for modeling diffusion coefficients is valid also for the modeling partition coefficients. A still actual limitation of the modeling is the lack of experimental data obtained in a systematic way, which allows to establish precise values for structure increments needed in the corresponding equations. To fill this gap is also a challenge for the next years.

Table 15.2 Partition coefficients, $K_{P/L}$, of aroma compounds between LDPE and ethanol/water mixtures and between LDPE/whole milk at 25°C.

ethanol/water	100%	75%	50%	25%	10%	0%
limonene	0.33	1.9 (4.8)	14.2	179	1156	5883
diphenylmethane	0.081	0.44	2.8 (2.5)	20	79	210
linalylacetate	0.043	0.32	2.9 (2.4)	36.5	222	894
camphor	0.038	0.14	0.58	2.8 (3.5)	8.3	17.6
phenylethylalcohol	0.051	0.058	0.074	0.097	0.12 (0.11)	0.14
cis 3-hexenol	0.0078	0.020	0.054	0.17 (0.16)	0.35	0.59

References

Baner, A.L., Brandsch, J., Franz, R. and Piringer, O. (1996) *Food Add. Contam.* **13**, 587–601.

Begley, T., Castle, L., Feigenbaum, A., Franz, R., Hinrichs, K., Lickly, T., Mercea, P., Milana, M., O'Brien, A., Rebre, S., Rijk, R. and Piringer, O. (2005) Evaluation of migration models that might be used in support of regulations for food-contact plastics. *Food Add. Contam.*, **22** (1), 73–90.

Begley, T., Brandsch, J., Limm, W., Siebert, H. and Piringer, O. (2007) The diffusion behavior of additives in polypropylene in correlation with polymer properties, submitted.

Bieber, W.D., Figge, K. and Koch, J. (1985) Interaction between plastics packaging materials and foodstuffs with different fat release properties. *Food Add. Contam.* **2**, 113–124.

Brandsch, J., Mercea, P., Rüter, M., Tosa, V. and Piringer, O. (2001) Migration modelling as a tool for quality assurance of food packaging. *Food Add. Contam.***19**, (Suppl.),29–41.

Castle, J., Mercer, A.J. and Gilbert, J. (1991) Migration from plasticized films into food. 5.Identification of individual species in a polymeric plasticizer and their, migration to food. *Food Add. Contam.***8**, 565–576.

Crank, J. (1975) *Mathematics of Diffusion*, 2nd edn, Oxford Science Publication, Oxford.

EU Commission .(1999) *Practical Guide*, Internet,http://cpf.jrc.it/webpack.

Figge, K. (1980) Migration of components from plastics packaging materials into packed goods – test methods and diffusion models. *Prog. Polym. Sci.*, **6**, 187–252.

Figge, K. and Rudolph, F. (1979) Diffusion im System Kunststoffverpackung/Füllgut. *Angewandte Makromolekulare Chemie* **78** (Nr. 1169), 157–180.

Goydan, R., Schwope, A.D., Reid, R.C. and Cramer, G. (1990) High-temperature migration of antioxidants from polyolefins. *Food Add. Contam.* **7**, 323–338.

Hamdani, M., Feigenbaum, A. and Vergnaud, J.M. (1997) Prediction of worst-case migration from packaging to food using mathematical models. *Food Add. Contam.* **14**, 499–506.

Hinrichs, K. and Piringer, O. (eds.) (2002) Evaluation of migration models to be used under Directive 90/128/EEC; Final Report Contract SMT4-CT98-7513. EUR 20604 EN; European Commission, Directorate General for Research, Brussels.

Lickly, T.D., Rainey, M.L., Burgert, L.C., Breder, C.V. and Borodinski, L. (1997) *Food Add. Contam.* **14**, 65–74.

Limm, W. and Hollifield, H.C. (1996) Modelling of additive diffusion in polyolefins. *Food Add. Contam.* **13**, 949–967.

O'Brien, A., Cooper, I. and Tice, P.A. (1997) Correlation of specific migration (C_f) of plastics additives with their initial concentration in the polymer (C_p); *Food Add. Contam.* **14**, 705–719.

O'Brien, A. and Cooper, I. (2001) Polymer additive migration to food – a direct comparison of experimental data and values calculated from migration models for polypropylene. *Food Add. Contam.*, **18**, 343–355.

O'Brien, A. and Cooper, I. (2002) Practical experience in the use of mathematical models to predict migration of additives from food contact polymers. *Food Add. Contam.*, **19** (Suppl.), 63–72.

O'Brien, A., Goodson, A. and Cooper, I. (1999) Polymer additive migration to food – a direct comparison of experimental data and values calculated from migration models for high density polyethylene (HDPE). *Food Add. Contam.*, **16** (No. 9), 367–380.

Piringer, O. (1993) *Verpackungen für Lebensmittel*, VCH-Verlag, Weinheim.

Piringer, O. (1994) Evaluation of plastics for food packaging. *Food Add. Contam.*, **11**, 221–230.

Reid, R.C., Sidman, K.R., Schwope, A.D. and Till, D.E. (1980) Loss of adjuvants from polymer films to foods or food simulants. Effect of the external phase. *Ind. Eng. Chem. Prod. Res. Dev.*, **19**, 580–587.

Reynier, A. and Dole, P. and Feigenbaum, A. (1999) Prediction of worst-case migration: presentation of a rigorous methodology. *Food Add. Contam.*, **16**, 137–152.

Robinson-Görnhardt, L. (1957) Kunststoffe in der Lebensmittelverpackung *Kunststoffe*. **47** (5), 265–267.

Till, D., Schwope, A.D., Ehntholt, D.J., Sidman, K.R., Whelan, R.H., Schwartz, P.S. and Reid, R.C. (1987) Indirect food additive migration from polymeric food packaging materials. *CRC Crit. Rev. Toxicol.*, **18**, 215–243.

Appendices

Appendix I

Peter Mercea

This section reviews most of the literature on the kinetics of small organic compounds (migrants) in polyethylenes (PEs) and polypropylenes (PPs) samples. An attempt was made to avoid a selection of data based on subjective criteria as to the quality of the experimental methods and/or model used to describe the diffusion process. However, in cases where a larger number of data were available for one and the same migrant, only the more recent experiments were cited. Then data was collected from experimental reports, which can be considered to be relevant to the topic of migration from polymeric packaging into foods and/or food simulants. In this respect the tables given below do not report diffusion data in swollen polymers because such situations are not suitable for food packaging.

There are a series of problems involved in an attempt to compare and interpret experimentally determined diffusion coefficients, D_p, in polymers.

First, problems result from the complex physical and chemical interactions between the migrants and the host polymer matrix. It is well documented that these interactions have a significant influence on the magnitude of D_p.

Then, problems are generated by the dependence of the morphological character of the polymer matrix upon its physical and chemical as well as manufacturing history. The D_p of a given migrant also depends on the molecular type and morphology/structure, density, and crystallinity of the polymer.

D_p depend on the temperature, T, too. The rate of this dependence often changes when the polymer undergoes a transition from a rubbery to a glassy state.

It is well known that the magnitude of D_p are also influenced by the initial amount of the migrant, $c_{p,0}$, formulated into the polymeric sample. High $c_{p,0}$ may have a plasticizing effect, which strongly influences the properties of a polymer matrix.

At last but not at least the D_p of a certain migrant may be influenced by the experimental setup used. There are numerous reports of migration experiments in which the polymeric sample is immersed into a liquid/solvent. It is logical to assume that such a liquid/solvent – which usually has relatively small molecules – diffuses into the polymer during the migration experiment. This process may influence to a certain degree not only the free volume available for the diffusional motion of the migrant, but even influence the interactions between the migrants and the host polymer matrix.

Plastic Packaging. Second Edition. Edited by O.G. Piringer and A.L. Baner
Copyright © 2008 WILEY-VCH Verlag GmbH & Co. KGaA, Weinheim
ISBN: 978-3-527-31455-3

Thus, taking into account those mentioned above, the sometimes large spread of D_p values reported in this appendix for one and the same migrant should not be considered as unusual or incorrect.

The rationale for the format of the tables given in the Appendix is the following.

The first column of the tables lists the chemical name of the migrant, as given in the publication cited. Because of that sometimes for one and the same organic migrant more than one chemical name appears in the tables. In column two the molecular weight, M_w, of the migrant is given.

Columns three and four give information about the polymer; namely about its density, crystallinity and/or morphology. In the case of polypropylenes, PPs, there are a series of abbreviations for the type of polymer, namely:

aT	– atactic PP,
iT	– isotactic PP,
HO	– homo polymer,
CO	– copolymer,
BO	– biaxially oriented PP
UO	– uniaxially oriented PP and
SB	– stereo block polymer,

Diffusion of a noninteracting migrant through an isotropic polymer matrix due to its random motion can be described by Fick's first law (Eq. 7.5) in which the rate constant D is defined as the diffusion coefficient. In the fifth column of the tables the diffusion coefficients, as reported in the cited publications, are given. There are several types of D_p, namely:

D	– concentration independent average diffusion coefficient,
$D_{c \to 0}$	– diffusion coefficient at "zero" diffusant concentration,
$D_{g.c.}$	– diffusion coefficient determined from inverse gas chromatography, and
D_s	– diffusion coefficient in a polymeric sample in contact with a solvent or food simulant.

The sixth column gives information about the single temperature, T (°C), at or temperature range in which the migration experiments were performed.

Columns 7 to 9 summarize the diffusion parameters for each migrant. In the seventh column the diffusion coefficients, units cm^2/s, are given. In order to make a comparison of the reported D values as easy as possible an attempt was made to give as many as possible D's for $T = 23$ °C (room temperature). In some cases this was possible only by extrapolating D's from measurements made at lower or higher temperatures. These situations are marked with (*. In those cases where a citation at 23 °C was not possible D's at other T's were given and marked with (** – T corresponding to the temperature given in column 6 or at another temperature given in the superscript – (^{70} for example.

In most cases the dependence of the diffusion in polymers on temperature is of Arrhenius type and described by the equation

$$D = D_0 \exp(-E_d/RT)$$

Table AI.1 Diffusion data for low molecular weight organic substances in Low Density Polyethylene (LDPE) and Linear Low Density Polyethylene (LLDPE) (Densities up to 0.930 g/cm^3 @ room temperature).

Diffusing Species		Polymer		Experiment			Diffusion parameters			Ref.
Name	Molec. weight M_w (g/mol)	Density @ (°C) ρ_P (g/cm^3)	Cristallinity (%)	Type of diffusion coefficient	Temp. range of experiment (°C)	Diffusion coefficient @ (23 °C) D (cm^2/s)		Pre-exponential coefficient $l_g D_0$	Activation energy E_D (kJ/mol)	
Methane	16.0	0.894 (25)	29.0	D	15; 45	4.64e-7		1.421	43.94	[1]
Methane	16.0	0.914 (25)	43.0	D	5; 55	1.7e-7		1.282	45.62	[1]
Methane	16.0	0.916 (25)	54.0	D	15; 50	1.94e-7		0.556	46.86	[2]
Methane	16.0	0.915 (25)	44.0	D	5; 50	1.58e-7		1.546	47.30	[3]
Methane	16.0	0.918	45.0	$D_{c\to 0}$	5; 35	2.98e-7		1.226	43.93	[4]
Methane	16.0	0.920 (23)	—	D	35; 50	1.79e-7$^{(*)}$		−1.712	28.52	[5]
Methane	16.0	—	—	D	30	2.28e-7$^{(**)}$		—	—	[6]
Ethylene	28.1	0.918	45.0	$D_{c\to 0}$	5; 35	1.34e-7		3.913	61.11	[4]
Ethylene	28.1	0.918	45.0	D	5; 35	1.81e-7		5.191	67.62	[4]
Ethylene	28.1	0.923 (25)	48.0	D	23; 73	1.25e-7		0.01023	39.18	[7]
Ethylene	28.1	—	—	D	30	1.01e-7$^{(**)}$		—	—	[6]
Ethane	30.1	0.894 (25)	29.0	D	5; 55	2.1e-7		2.036	49.38	[1]
Ethane	30.1	0.914 (25)	43.0	D	5; 55	5.87e-8		2.222	53.57	[1]
Ethane	30.1	0.920 (23)	—	D	33; 48	9.3e-8$^{(*)}$		0.948	34.46	[5]
Ethane	30.1	0.918	48.0	D	−26; 25	7.9e-8		2.874	56.52	[8]
Ethane	30.1	0.910 (25)	—	D	0; 50	4.98e-8		2.101	53.29	[9]
Ethane	30.1	0.918 (25)	—	D	25; 50	4.8e-8$^{(*)}$		2.505	55.66	[9]
Ethane	30.1	0.921 (25)	—	D	25; 50	3.48e-8$^{(*)}$		3.194	60.36	[9]
Ethane	30.1	0.924 (25)	—	D	20; 60	5.38e-8		1.888	51.89	[10]
Ethane	30.1	0.916 (25)	—	D	25	5.4e-8$^{(**)}$		—	—	[11]
Ethane	30.1	—	—	D	30	6.6e-8$^{(**)}$		—	—	[6]

Table AI.1 (Continued)

Diffusing Species		Polymer			Experiment			Diffusion parameters			Ref.
Name	Molec. weight M_w (g/mol)	Density @ (°C) ρ_P (g/cm³)	Cristallinity (%)	Type of diffusion coefficient	Temp. range of experiment (°C)	Diffusion coefficient @ (23°C) D (cm²/s)		Pre-exponential coefficient $l_g D_0$	Activation energy E_D (kJ/mol)		
Methanol	32.0	0.918 (23)	–	D_s	23	4.8e-8		–	–		[12]
Methanol	32.0	0.917 (23)	–	D	23	1.94e-8		–	–		[13]
Methanol	32.0	0.920 (25)	50.0	D	15;35	1.6e-8		0.031	44.36		[14]
Methanol	32.0	0.919	–	D	30	3.3e-8(**		–	–		[15]
Allene	40.1	0.894 (25)	29.0	D	10; 50	2.74e-7		1.414	45.20		[1]
Allene	40.1	0.914 (25)	43.0	D	10; 50	9.16e-8		1.750	49.80		[1]
Propylene	42.1	0.894 (25)	29.0	D	10; 50	1.75e-7		1.737	48.13		[1]
Propylene	42.1	0.914 (25)	43.0	D	10; 50	5.0e-8		1.933	52.31		[1]
Propylene	42.1	0.920 (25)	–	D	0; 22	1.07e-7(*		−0.260	38.0		[16]
Propane	44.1	0.894 (25)	29.0	D	10; 50	1.04e-7		2.248	52.31		[1]
Propane	44.1	0.914 (25)	43.0	D	10; 50	2.76e-8		2.264	55.66		[1]
Propane	44.1	0.915 (25)	44.0	D	25; 55	2.1e-8(*		3.491	63.19		[3]
Propane	44.1	0.918	45.0	$D_{c \to 0}$	5; 35	5.2e-8		2.929	57.86		[4]
Propane	44.1	0.920	–	D	30;48	1.98e-8(*		0.2735	45.20		[5]
Propane	44.1	0.920 (25)	–	D	0; 25	6.65e-8		3.607	23.3		[16]
Propane	44.1	–	–	D	30	3.1e-8(**		−7.176	–		[6]
Ethanol	46.1	–	–	$D_{c \to 0}$	49.1	2.7e-10(**		–	–		[17]
Propionitrile	55.1	–	–	D	25	5.0e-9(**		–	–		[18]
Isobutylene	56.1	0.922 (25)	60.0	D	−8; 30	4.0e-8		4.102	65.16		[19]
Isobutylene	56.1	0.922 (25)	60.0	$D_{c \to 0}$	−8; 30	2.6e-8		3.607	63.41		[19]
Acetone	58.1	–	–	D	25	7.0e-9(**		–	–		[18]
Butane	58.1	0.924 (25)	–	D	30; 60	4.2e-8(*		−0.4437	39.25		[10]
Butane	58.1	0.922 (25)	50.3	D	25	1.95e-8(**		–	–		[20]
Butane	58.1	0.924 (25)	51.0	D	25	1.4e-8(**		–	–		[20]
Neopentane	72.1	0.918 (25)	–	D	25; 50	1.6e-9(*		6.503	86.73		[9]

Compound								Ref
n-Pentane	72.1	0.918 (25)	—	D	25; 50	8.05e-9[*]	4.836	73.27	[9]
n-Butylaldehyde	72.1	0.922 (25)	50.0	D	25	1.18e-8[**]	—	—	[20]
n-Butylaldehyde	72.1	0.924 (25)	51.0	D	25	1.05e-8[**]	—	—	[20]
Butanal	72.1	0.919 (25)	—	D	25	2.8e-9[**]	—	—	[21]
Butylalcohol	74.1	0.922 (25)	50.0	D	25	1.08e-8[**]	—	—	[20]
Butylalcohol	74.1	0.922 (25)	51.0	D	25	9.0e-9[**]	—	—	[20]
Benzene	78.1	0.922 (25)	60.0	$D_{c \leftrightarrow 0}$	0	1.9e-9[**]	—	—	[19]
Benzene	78.1	—	70.0	$D_{c \leftrightarrow 0}$	25; 50	9.9e-9[*]	3.375	64.45	[22]
Benzene	78.1	0.916 (25)	54.0	$D_{c \leftrightarrow 0}$	25; 45	1.08e-8[*]	3.841	66.90	[23]
Benzene	78.1	0.918 (25)	54.0	$D_{c \leftrightarrow 0}$	25	1.98e-8[*]	—	—	[24]
Benzene	78.1	0.920 (25)	45.0	$D_{c \leftrightarrow 0}$	30; 40	1.41e-8[*]	0.4730	47.16	[25]
Benzene	78.1	0.915 (23)	42.0	$D_{c \leftrightarrow 0}$	23	3.8e-9	—	—	[26]
Benzene	78.1	0.918 (25)	—	$D_{c \leftrightarrow 0}$	25; 45	1.05e-8[*]	4.603	71.30	[27]
Benzene	78.1	0.918 (25)	45.0	$D_{c \leftrightarrow 0}$	25	4.0e-9[*]	—	—	[28]
Benzene	78.1	0.916 (25)	—	$D_{c \leftrightarrow 0}$	25; 35	1.4e-9[*]	−3.309	31.34	[29]
Benzene	78.1	0.917 (25)	—	D_{gc}	25	8.2e-9[**]	—	—	[30]
Benzene	78.1	0.918 (25)	—	$D_{c \leftrightarrow 0}$	25	2.15e-8[**]	—	—	[31]
Benzene	78.1	0.921 (25)	—	$D_{c \leftrightarrow 0}$	25	1.48e-8[**]	—	—	[31]
Benzene	78.1	0.922 (25)	—	$D_{c \leftrightarrow c \leftrightarrow 0}$	15; 35	9.9e-9	2.187	57.75	[31]
Benzene	78.1	0.928 (25)	—	$D_{c \leftrightarrow 0}$	25	6.9e-9[**]	—	—	[31]
Benzene	78.1	0.916 (25)	—	$D_{c \leftrightarrow 0}$	25; 45	7.1e-8[*]	1.876	51.1	[32]
Benzene	78.1	0.915 (25)	44.0	D	30; 45	2.9e-9[*]	1.768	58.35	[33]
Dimethylsulfoxide (DMSO)	78.1	—	70.0	$D_{c \leftrightarrow 0}$	25; 50	7.1e-9[*]	1.452	54.40	[22]
2-Hexene	84.2	—	70.0	$D_{c \leftrightarrow 0}$	25; 50	4.1e-9[*]	2.388	61.10	[22]
Cyclohexane	84.2	0.918 (25)	54.0	$D_{c \leftrightarrow c \leftrightarrow 0}$	25	6.1e-9[**]	—	—	[24]
Cyclohexane	84.2	0.915 (23)	42.0	$D_{c \leftrightarrow c \leftrightarrow 0}$	23	2.0e-9	—	—	[26]
Cyclohexane	84.2	0.921 (25)	—	$D_{c \leftrightarrow c \leftrightarrow 0}$	15; 35	1.04e-8	0.5264	48.20	[31]
Cyclohexane	84.2	0.922	—	$D_{c \leftrightarrow c \leftrightarrow 0}$	25; 30	1.8e-9[**]	4.848	77.00	[34]
Cyclohexane	84.2	0.922	—	D_S	40	3.4e-8[**]	—	—	[35]
Methylenechloride	84.9	0.912 (25)	47.0	$D_{c \leftrightarrow c \leftrightarrow 0}$	25	9.2e-8[**]	—	—	[36]
Methylenechloride	84.9	0.917 (25)	50.2	$D_{c \leftrightarrow 0}$	25	8.2e-8[**]	—	—	[36]

530 | Appendix I

Table AI.1 (Continued)

Diffusing Species		Polymer			Experiment		Diffusion parameters			Ref.
Name	Molec. weight M_w (g/mol)	Density @ (°C) ρ_P (g/cm³)	Cristallinity (%)	Type of diffusion coefficient	Temp. range of experiment (°C)	Diffusion coefficient @ (23 °C) D (cm²/s)	Pre-exponential coefficient $\lg D_0$	Activation energy E_D (kJ/mol)		
Methylenechloride	84.9	0.924 (25)	52.0	D_{c-c-0}	25	7.0e-8(**	–	–		[36]
Methylenechloride	84.9	0.924 (25)	–	D_{c-c-0}	25	7.1e-8(**	–	–		[37]
Pentanal	86.1	0.919 (25)	–	D	25	7.0e-8(**	–	–		[21]
n-Hexane	86.2	0.922 (25)	50.0	D	25	1.05e-8(**	–	–		[20]
n-Hexane	86.2	0.924 (25)	51.0	D	25	9.0e-9(**	–	–		[20]
n-Hexane	86.2	–	70.0	D_{c-c-0}	25; 50	5.3e-9(*	3.249	65.29		[22]
n-Hexane	86.2	0.918 (25)	54.0	D_{c-c-0}	25; 45	1.05e-8(*	3.563	65.40		[24]
n-Hexane	86.2	0.918 (25)	–	D_{c-c-0}	25; 45	8.4e-9(*	2.706	61.10		[27]
n-Hexane	86.2	0.918 (25)	45.0	D_{c-c-0}	25	3.0e-9(**	–	–		[28]
n-Hexane	86.2	0.918 (25)	–	D_{c-c-0}	25; 35	1.4e-9(*	−5.034	21.71		[29]
n-Hexane	86.2	0.918 (25)	–	D_{c-c-0}	25; 50	1.04e-8(*	2.703	60.56		[38]
n-Hexane	86.2	0.915 (25)	43.2	D_{c-c-0}	25	1.32e-8(**	–	–		[39]
n-Hexane	86.2	0.928 (25)	51.7	D_{c-c-0}	25	7.8e-9(**	–	–		[39]
n-Hexane	86.2	0.916 (25)	–	D_{c-c-0}	25; 45	5.1e-8(*	4.240	65.4		[32]
3-Methylpentane	86.2	–	70.0	D_{c-c-0}	25; 50	4.1e-9(*	2.547	61.94		[22]
Neohexane	86.2	–	70.0	D_{c-c-0}	25; 50	2.8e-9(*	2.900	64.86		[22]
Tetrafluormethane	88.0	0.918 (25)	45.0	D_{c-c-0}	20; 50	7.9e-9	3.043	63.15		[40]
Ethylacetate	88.1	0.922 (25)	50.0	D	25	5.3e-8(**	–	–		[20]
Ethylacetate	88.1	0.924 (25)	51.0	D	25	4.7e-8(**	–	–		[20]
Ethylacetate	88.1	0.906 (30)	35.8	D	30	3.0e-8(**	–	–		[41]
Ethylacetate	88.1	–	–	D	25	1.1e-8(**	–	–		[42]
p-Dioxane	88.1	–	70.0	D_{c-c-0}	25; 50	4.1e-9(*	4.912	75.33		[22]
1-Pentanol	88.2	0.919 (25)	–	D	25	6.4e-9(**	–	–		[21]

Compound	MW	Density (T)		Type	T	Value		Ref
2-Pentanol	88.2	0.919 (25)	—	D	25	9.7e-9(**	—	[23]
Toluene	92.1	0.918 (25)	54.0	$D_{c \to c \to 0}$	25; 45	1.43e-8(*	7.508	[24]
Toluene	92.1	0.920	45.0	$D_{c \to 0}$	30; 50	1.37e-8(*	1.604	[25]
Toluene	92.1	0.918 (30)	47.3	$D_{c \to c \to 0}$	30	4.1e-8(**	—	[43]
Toluene	92.1	0.919 (30)	48.0	$D_{c \to c \to 0}$	30	3.6e-8(**	—	[43]
Toluene	92.1	0.918 (30)	—	D	30	4.3e-8(**	—	[44]
Toluene	92.1	0.891 (70)	35.0	$D_{c \to c \to 0}$	70	5.22e-7(**	—	[45]
Toluene	92.1	0.910 (70)	45.0	$D_{c \to 0}$	70	3.42e-7(**	—	[45]
Toluene	92.1	0.918 (23)	40.6	D	30	2.13e-8(**	—	[46]
Toluene	92.1	0.916 (25)	—	$D_{c \to c \to 0}$	25; 45	6.3e-8(*	1.976	[32]
Phenole	94.1	0.918 (23)	—	D_S	23	4.5e-9	—	[12]
Methyl bromide	95.0	0.919 (25)	58.0	D	0; 30	6.05e-8	2.061	[19]
Methylcyclohexane	98.2	0.918 (23)	40.6	D	30	5.8e-9(**	—	[46]
n-Heptane	100.2	0.919	—	D	30	1.1e-8(**	—	[15]
n-Heptane	100.2	0.918	—	$D_{c \to c \to 0}$	25; 35	1.2e-9(*	−5.122	[29]
n-Heptane	100.2	0.922 (25)	—	$D_{c \to 0}$	25; 30	4.4e-9(*	2.480	[34]
n-Heptane	100.2	0.918	—	$D_{c \to 0}$	25; 50	9.0e-9(*	4.484	[38]
n-Heptane	100.2	0.918 (23)	40.6	D	30	7.9e-9(**	—	[46]
n-Heptane	100.2	0.922	—	D_S	40	2.4e-8(**	—	[35]
n-Heptane	100.2	0.922	—	D_S	40	6.0e-9(**	—	[35]
n-Heptane	100.2	0.916 (25)	—	$D_{c \to 0}$	25; 45	8.6e-9(*	4.622	[32]
n-Hexylaldehyde	100.2	0.922 (25)	50.0	D	25	8.0e-9(**	—	[20]
n-Hexylaldehyde	100.2	0.925 (25)	51.0	D	25	6.0e-9(**	—	[20]
cis-3-Hexen-1-ol	100.2	0.918 (23)	—	D_S	23	1.4e-8	—	[47]
Hexanal	100.2	0.919 (25)	—	D	25	3.0e-10(**	—	[21]
Ethylpropionate	102.1	—	—	$D_{c \to 0}$	30	1.22e-8(**	—	[48]
Hexylalcohol	102.2	0.922 (25)	50.0	D	25	7.0e-9(**	—	[20]
Hexylalcohol	102.2	0.924 (25)	51.0	D	25	5.0e-9(**	—	[20]
1-Hexanol	102.2	0.919 (25)	—	D	25	1.44e-8(**	—	[21]

Table AI.1 (Continued)

Diffusing Species	Polymer			Experiment		Diffusion parameters			Ref.
Name	Molec. weight M_w (g/mol)	Density @ (°C) ρ_P (g/cm^3)	Cristallinity (%)	Type of diffusion coefficient	Temp. range of experiment (°C)	Diffusion coefficient @ (23°C) D (cm^2/s)	Pre-exponential coefficient $l_g D_0$	Activation energy E_D (kJ/mol)	
2-Hexanol	102.2	0.919 (25)	–	D	25	4.05e-9$^{(**)}$	–	–	[21]
Hexylalcohol	102.2	0.922 (25)	–	D	25	6.0e-9$^{(**)}$	–	–	[20]
Hexylalcohol	102.2	0.924 (25)	–	D	25	5.0e-9$^{(**)}$	–	–	[20]
o-Xylene	106.2	0.918 (25)	–	$D_{c \to 0}$	25	9.4e-9$^{(**)}$	–	–	[24]
m-Xylene	106.2	0.918 (25)	–	$D_{c \to 0}$	25	1.46e-8$^{(**)}$	–	–	[24]
p-Xylene	106.2	0.918 (25)	–	$D_{c \to 0}$	25	1.57e-8$^{(**)}$	–	–	[24]
N-Methylaniline	107.1	0.924 (25)	–	D	50	4.2e-8$^{(**)}$	–	–	[49]
p-Cresole	108.1	0.918 (23)	–	D_s	23	2.3e-9	–	–	[12]
Anisole	108.1	0.918 (25)	42.0	D_s	25	1.8e-8$^{(**)}$	–	–	[50]
n-Octane	114.2	0.922 (25)	50.0	D	25	6.8e-9$^{(**)}$	–	–	[20]
n-Octane	114.2	0.924 (25)	51.0	D	25	6.0e-9$^{(**)}$	–	–	[20]
n-Octane	114.2	0.918	–	$D_{c \to 0}$	25; 50	7.1e-9$^{(*)}$	6.115	80.81	[38]
iso-Octane	114.2	0.918 (23)	40.6	D	30	5.2e-9$^{(**)}$	–	–	[46]
2,2,4-Trimethylpentane	114.2	0.918 (25)	–	$D_{c \to 0}$	25; 50	2.3e-9$^{(*)}$	6.113	83.59	[38]
2,2,4-Trimethylpentane (Isooctane)	114.2	0.922	–	D_s	40	5.2e-9$^{(**)}$	–	–	[35]
Ethylbutyrate	116.2	0.922 (25)	50.0	D	25	2.2e-8$^{(**)}$	–	–	[20]
Ethylbutyrate	116.2	0.924 (25)	51.0	D	25	1.75e-8$^{(**)}$	–	–	[20]
Ethylbutyrate	116.2	–	–	D	23	1.79e-8	–	–	[51]
Ethylbutyrate	116.2	–	–	D	20; 40	1.86e-8	–6.291	8.16	[52]
Ethylbutyrate	116.2	–	–	D	30	2.1e-8$^{(**)}$	–	–	[53]
Heptanol	116.2	0.918 (23)	–	D_s	23	5.3e-9	–	–	[54]
Heptanol	116.2	0.918 (23)	–	D_s	23	5.5e-9	–	–	[12]
1-Heptanol	116.2	0.919 (25)	–	D	25	4.9e-10$^{(**)}$	–	–	[20]

Appendix I | 533

Compound	MW	density (T)		Type	T (°C)	D		Ref
2-Heptanol	116.2	0.919 (25)	–	D	25	1.39e-9$^{(**}$	–	[21]
2,3-Benzopyrrole (Indole)	117.1	0.918 (23)	–	D_s	23	5.5e-9	–	[47]
Chlorophorm	119.4	0.918 (25)	50.0	$D_{c\to o}$	25	1.78e-8$^{(**}$	–	[24]
Phenylmethylketone (Acetophenone)	120.1	0.918 (23)	–	D_s	23	1.10e-8	–	[47]
Mesitylene	120.2	0.920 (25)	45.0	$D_{c\to o}$	30; 50	7.4e-9$^{(*}$	0.6019	49.47	[25]
n-Propylbenzene	120.2	0.920 (25)	45.0	$D_{c\to o}$	30; 50	1.16e-8$^{(*}$	1.083	51.10	[25]
N,N-Dimethylaniline (DMA)	121.2	0.920 (25)	50.0	D_s	15; 35	1.73e-8	−0.070	43.58	[14]
N,N-Dimethylaniline (DMA)	121.2	0.916 (25)	29.0	D	25	5.5e-9$^{(**}$	–	–	[55]
N,N-Dimethylaniline (DMA)	121.2	0.917 (25)	31.0	D	25	8.7e-9$^{(**}$	–	–	[55]
N,N-Dimethylaniline (DMA)	121.2	0.918 (25)	42.0	D	25	7.2e-9$^{(**}$	–	–	[56]
N,N-Dimethylaniline (DMA)	121.2	0.918 (25)	42.0	D_s	25; 45	3.8e-9$^{(*}$	3.934	69.99	[56]
N,N-Dimethylaniline (DMA)	121.2	0.920 (25)	50.0	D_s	15;34	7.41e-9	3.021	63.19	[56]
Cresylmethylether	122.2	0.918 (23)	–	D_s	23	1.2e-8	–	[47]
2-Phenylethylalcohol	122.2	0.918 (23)	–	D_s	23	4.3e-9	–	[47]
3-Octen-2-one (Methylheptenone)	126.2	0.918 (23)	–	D_s	23	7.3e-9	–	[47]
n-Octylaldehyde	128.2	0.922 (25)	50.0	D	25	4.3e-9$^{(**}$	–	[20]
n-Octylaldehyde	128.2	0.924 (25)	51.0	D	25	4.0e-9$^{(**}$	–	[20]
Octanal	128.2	0.919 (25)	–	D	25	9.0e-11$^{(**}$	–	[21]
n-Octanal (Aldehyde C$_8$)	128.2	0.918 (23)	–	D_s	23	2.3e-9	–	[47]
Octanal	128.2	–	–	D	20: 40	5.4e-9	−4.312	22.39	[52]
Ethylvalerate	130.2	–	–	$D_{c\to o}$	30	1.0e-8$^{(**}$	–	[48]
Octylalcohol	130.2	0.922 (25)	–	D	25	4.7e-9$^{(**}$	–	[20]
Octylalcohol	130.2	0.924 (25)	–	D	25	4.0e-9$^{(**}$	–	[20]
Amylaceticester (Isoamylacetate)	130.2	0.918 (23)	–	D_s	23	7.7e-9	–	[47]
p-isopropyltoluene (p-Cymene)	134.2	0.918 (25)	–	D_s	23	5.4e-9	–	[47]
2-(2-Ethoxyethoxy) ethanol	134.2	0.918 (23)	–	D_s	23	3.8e-9	–	[47]
n-Butylbenzene	134.2	0.920 (25)	45.0	$D_{c\to o}$	30; 60	6.4e-9$^{(*}$	0.514	49.33	[25]

Table A1.1 (Continued)

Diffusing Species	Polymer			Experiment		Diffusion parameters			Ref.
Name	Molec. weight M_w (g/mol)	Density @ (°C) ρ_P (g/cm³)	Cristallinity (%)	Type of diffusion coefficient	Temp. range of experiment (°C)	Diffusion coefficient @ (23°C) D (cm²/s)	Pre-exponential coefficient $lg D_0$	Activation energy E_D (kJ/mol)	
2,4,6 Trimethylphenol	136.2	0.918 (23)	–	D_s	23	2.3e-9	–	–	[12]
4-Isopropenyl-1-methyl-1-cyclohexene (Limonene)	136.2	0.918 (23)	–	D_s	23	4.3e-9	–	–	[47]
4-Isopropenyl-1-methyl-1-cyclohexene (Limonene)	136.2	0.923 (25)	50.4	D	25; 45	4.2e-10(*	−2.436	39.31	[57]
4-Isopropenyl-1-methyl-1-cyclohexene (Limonene)	136.2	0.930 (25)	55.3	D	25; 45	4.0e-10(*	−5.243	23.54	[57]
4-Isopropenyl-1-methyl-1-cyclohexene (Limonene)	136.2	–	–	$D_c \rightarrow_0$	23	5.71e-11	–	–	[51]
4-Isopropenyl-1-methyl-1-cyclohexene (Limonene)	136.2	–	–	D	20; 40	1.10e-8	−4.546	19.25	[52]
4-Isopropenyl-1-methyl-1-cyclohexene (Limonene)	136.2	–	–	D	23	1.85e-8	–	–	[58]
7-Methyl-3-methylene-1,6-octadiene (Myrcene)	136.2	0.918 (23)	–	D_s	23	7.0e-9	–	–	[47]
7-Methyl-3-methylene-1,6-octadiene (Myrcene)	136.2	–	–	D	20; 40	1.04e-8	−4.320	20.76	[52]
2-Methyl-benzoicacid (Phenylacetate)	136.2	0.918 (23)	–	D_s	23	2.5e-9	–	–	[47]
3-Phenyl-1-propanol	136.2	0.918 (23)	–	D_s	23	2.8e-9	–	–	[47]
2,6,6-Trimethylbicyclo (3,1,1) hept-2-ene (alpha-Pinene)	136.2	0.918 (23)	–	D_s	23	1.4e-9	–	–	[47]

Compound	MW	ρ (T)		D type	T (°C)	D			Ref
2,6,6-Trimethylbicyclo (3,1,1) hept-2-ene (alpha-Pinene)	136.2	–	–	D	20; 40	2.18e-8	–4.169	19.79	[52]
2,6,6-Trimethylbicyclo (3,1,1) hept-2-ene (alpha-Pinene)	136.2	–	–	D	23	9.7e-9	–	–	[58]
6,6-Dimethyl-2-methylenebicyclo (3,1,1) heptan-ropinene (Beta-Pinene)	136.2	0.918 (23)	–	D_s	23	1.4e-9	–	–	[47]
3,7,7-Trimethyl-bicyclo[4.1.0] hept-2-ene (Carene)	136.2	–	–	D_s	23	1.0e-8	–	–	[47]
2,4-Dimethyl-3-cyclohexen-1-carboxaldehyde	138.2	0.918 (23)	–	D_s	23	1.1e-9	–	–	[47]
n-Nonanal (Aldehyde C_9)	142.2	0.918 (23)	–	D_s	23	1.8e-9	–	–	[47]
n-Decane	142.2	0.922 (25)	50.0	D	25	4.2e-9(**)	–	–	[20]
n-Decane	142.2	0.924 (25)	51.0	D	25	3.7e-9(**)	–	–	[20]
n-Decane	142.2	–	–	$D_{g.c.}$	30; 80	3.4e-9(*)	–4.213	24.12	[30]
n-Decane	142.2	0.918 (25)	–	$D_{c\rightarrow 0}$	25; 50	3.6e-9(*)	8.644	96.83	[28]
cis-3-Hexen-1-yl-acetate	142.2	0.918 (25)	–	D_s	23	9.3e-9	–	–	[47]
7-Methylchinoline	143.2	0.918 (23)	–	D_s	23	4.3e-9	–	–	[47]
Ethylhexanoate	144.2	0.922 (25)	–	D	25	9.0e-9(**)	–	–	[20]
Ethylhexanoate	144.2	0.924 (25)	–	D	25	2.7e-9(**)	–	–	[20]
Ethylhexanoate	144.2	–	–	$D_{c\rightarrow 0}$	30	8.1e-9(**)	–	–	[48]
Nonanol	144.3	0.918 (23)	–	D_s	23	4.0e-9	–	–	[12]
1,2-Benzopyrone (Cumarin)	146.2	0.918 (23)	–	D_s	23	5.4e-9	–	–	[47]
1-Methoxy-4-(1-propenyl)benzene (Anethol)	148.2	0.918 (23)	–	D_s	23	5.0e-9	–	–	[47]
cis,trans 3,7-Dimethyl-2,6-octadien-1-nitrile (Citralva)	149.2	0.918 (23)	–	D_s	23	1.7e-9	–	–	[47]
N,N-Di-ethylaniline (DEA)	149.2	0.920 (25)	50.0	D_s	20; 39	2.1e-9	2.635	64.06	[59]
3,4-Methylen-dioxybenzaldehyde (Heliotropine)	150.1	0.918 (23)	–	D_s	23	8.7e-10	–	–	[47]

Table AI.1 (Continued)

Diffusing Species		Polymer		Experiment		Diffusion parameters			Ref.
Name	Molec. weight M_w (g/mol)	Density @ (°C) ρ_P (g/cm³)	Cristallinity (%)	Type of diffusion coefficient	Temp. range of experiment (°C)	Diffusion coefficient @ (23°C) D (cm²/s)	Pre-exponential coefficient lgD_0	Activation energy E_D (kJ/mol)	
Benzylacetate	150.2	0.918 (23)	–	D_s	23	7.0e-9	–	–	[47]
2,3,5,6 Tetramethylphenol	150.2	0.918 (23)	–	D_s	23	1.6e-9	–	–	[12]
Dimethylbenzylcarbinol	150.2	0.918 (23)	–	D_s	23	7.5e-10	–	–	[47]
Ethylbenzoate	150.2	0.918 (23)	–	D_s	23	1.1e-9	–	–	[47]
cis,trans 3,7-Dimethyl-2, 6-octadienal (Citral)	152.2	0.918 (23)	–	D_s	23	3.6e-9	–	–	[47]
cis,trans 3,7-Dimethyl-2, 6-octadienal (Citral)	152.2	–	–	$D_{c \to 0}$	23	3.2e-11	–	–	[51]
cis,trans 3,7-Dimethyl-2, 6-octadienal (Citral)	152.2	–	–	D	20; 40	2.22e-9	−0.899	43.94	[52]
cis,trans 3,7-Dimethyl-2, 6-octadienal (Citral)	152.2	–	–	D	23	3.5e-9	–	–	[58]
cis,trans 3,7-Dimethyl-2, 6-octadienal (Citral)	152.2	0.918 (23)	–	D_s	23	1.5e-9	–	–	[47]
1,7,7-Trimethyl- 2,2,1 heptane-2-one (Campher)	153.8	0.919	–	D	30	6.9e-9(**)	–	–	[15]
Carbontetrachloride	153.8	–	70.0	$D_{c \to 0}$	25; 50	2.9e-9(*)	6.092	82.86	[22]
Carbontetrachloride	153.8	0.918 (25)	50.0	$D_{c \to 0}$	25	6.6e-9(**)	–	–	[24]
Carbontetrachloride	153.8	0.918 (25)	45.0	$D_{c \to 0}$	25	8.0e-10(**)	–	–	[28]
1,7,7-Trimethylbicyclo 2,2,1 heptane-2-one (Borneol)	154.2	0.918 (23)	–	D_s	23	4.8e-10	–	–	[47]
3,7-Dimethyl-6-octene-1-al (Citronellal)	154.2	0.918 (23)	–	D_s	23	1.0e-9	–	–	[47]
1,8 Epoxy-p-Mentone (Eukalyptol)	154.2	0.918 (23)	–	D_s	23	1.0 e-9	–	–	[47]

Compound	MW	Density (T)		Type	T	Value			Ref
3,7-Dimethyl-1,6-octadiene-3-ylacetate (Linalool)	154.2	0.918 (23)	–	D_s	23	1.9e-9	–	–	[47]
3,7-Dimethyl-1,6-octadiene-3-ylacetate (Linalool)	154.2	–	–	$D_c \rightarrow_0$	23	1.39e-11	–	–	[51]
3,7-Dimethyl-1,6-octadiene-3-ylacetate (Linalool)	154.2	–	–	D	20; 40	2.7e-9	−5.391	18.04	[52]
2-cis-3,7-Dimethyl-2,6-octadiene-1-ole (Nerol)	154.2	0.918 (23)	–	D_s	23	2.1e-9	–	–	[47]
2-trans-3,7-dimethyl-2,6-octadiene-8-ole (Geraniol)	154.2	0.918 (23)	–	D_s	23	1.5e-9	–	–	[47]
2-Isopropyl-5-methylhexanone (Menthon)	154.2	0.918 (23)	–	D_s	23	2.1e-9	–	–	[47]
cis-2[2-Methyl-1-propenyl]-4-methyl-tetrahydropyran (Roseoxyde L)	154.2	0.918 (23)	–	D_s	23	3.1e-9	–	–	[47]
1-Methyl-4-isopropyl-1-cyclohene-1-ole (Terpineol)	154.2	0.918 (23)	–	D_s	23	2.2e-9	–	–	[47]
1-Methyl-4-isopropyl-1-cyclohene-1-ole (alpha-Terpineol)	154.2	–	–	D	20; 40	8.6e-10	0.418	53.73	[52]
n-Decylaldehyde	156.3	0.922 (25)	50.0	D	25	2.4e-9[**]	–	–	[20]
n-Decylaldehyde	156.3	0.924 (25)	51.0	D	25	2.1e-9[**]	–	–	[20]
3,7-Dimethyl-6-octene-1-ol (Citronellol)	156.3	0.918 (23)	–	D_s	23	2.6e-9	–	–	[47]
n-Decanal (Aldehyd C_{10})	156.3	0.918 (23)	–	D_s	23	1.4e-9	–	–	[47]
Decanal	156.3	0.919 (25)	–	D	25	1.6e-10[**]	–	–	[21]
Undecane	156.3	0.918 (23)	–	D_s	23	9.0e-9	–	–	[54]
Undecane	156.3	–	39.0	D	40	1.66e-8[**]	–	–	[60]
Undecane	156.3	–	44.0	D	40	1.9e-8[**]	–	–	[60]
2,6-Dimethyl-7-octene-2-ol (Dihydromyrcenol)	156.3	0.918 (23)	–	D_s	23	1.5e-9	–	–	[47]

Table A1.1 (Continued)

Diffusing Species	Polymer			Experiment		Diffusion parameters			Ref.
Name	Molec. weight M_w (g/mol)	Density @ (°C) ρ_P (g/cm³)	Cristallinity (%)	Type of diffusion coefficient	Temp. range of experiment (°C)	Diffusion coefficient @ (23°C) D (cm²/s)	Pre-exponential coefficient I_gD_0	Activation energy E_D (kJ/mol)	
2-Isopropyl-5-methylcy- clohexanole (Menthol)	156.3	0.918 (23)	–	D_S	23	1.2e-9	–	–	[47]
2-Mthoxynaphthalene (Yara Yara)	158.2	0.918 (23)	–	D_S	23	4.7e-9	–	–	[47]
Ethylheptanoate	158.2	–	–	$D_{c\to 0}$	30	6.9e-9(**	–	–	[48]
Decylalcohol	158.3	0.922 (25)	50.0	D	25	2.9e-9(**	–	–	[20]
Decylalcohol	158.3	0.924 (25)	51.0	D	25	2.2e-9(**	–	–	[20]
3,7-Dimethyl-1-octanol	158.3	0.918 (23)	–	D_S	23	9.2e-10	–	–	[47]
3,7-Dimethyl-octane-3-ol	158.3	0.918 (23)	–	D_S	23	1.3e-9	–	–	[47]
Diethylmalonate	160.2	0.918 (23)	–	D_S	23	4.4e-9	–	–	[47]
Dimethylphenylethylcarbinole	164.2	0.918 (23)	–	D_S	23	7.9e-10	–	–	[47]
Methoxy-4(2-propenyl)phenol (Eugenol)	164.2	0.918 (23)	–	D_S	23	2.6e-9	–	–	[47]
2-Methoxy-4-prophenylphenol (Isoeugenol)	164.2	0.918 (23)	–	D_S	23	1.55e-9	–	–	[47]
1-Phenylethylacetate (Styrolacetate)	164.2	0.918 (23)	–	D_S	23	2.9e-9	–	–	[47]
2-Phenylethylacetate	164.2	0.918 (23)	–	D_S	23	5.7e-9	–	–	[47]
n-Undecene-2-al (Aldehyd C_{11})	168.3	0.918 (23)	–	D_S	23	9.6e-10	–	–	[47]
cis-Undecene-8-al (Aldehyd C_{11} inter)	168.3	0.918 (23)	–	D_S	23	9.0e-10	–	–	[47]
Diphenylmethane	168.3	0.918 (23)	–	D_S	23	4.8e-9	–	–	[47]
Diphenyloxide	170.2	0.918 (23)	–	D_S	23	3.7e-9	–	–	[47]

Compound	MW	Density (T)			T	Value			Ref
Ethyloctanoate	170.3	0.922 (25)	50.0	D	25	3.2e-9(**)	—	—	[20]
Ethyloctanoate	170.3	0.924 (25)	51.0	D	25	2.8e-9(**)	—	—	[20]
n-Undecylaldehyde (Aldehyde C_{11})	170.3	0.918 (23)	—	D_s	23	1.0e-9	—	—	[47]
n-Dodecane	170.3	0.922 (25)	50.0	D	25	3.3e-9(**)	—	—	[20]
n-Dodecane	170.3	0.924 (25)	51.0	D	25	2.9e-9(**)	—	—	[20]
n-Dodecane	170.3	0.918 (23)	—	D_s	23	2.7e-9	—	—	[61]
Dodecane (Alcane C_{12})	170.3	0.918 (23)	—	D_s	6; 40	2.6e-9	4.729	75.45	[61]
Dodecane	170.3	—	39.0	D	40	1.86e-8(**)	—	—	[60]
Dodecane	170.3	—	39.0	D	40	9.97e-9(**)	—	—	[60]
Dodecane	170.3	—	44.0	D	40	1.23e-8(**)	—	—	[60]
Dodecane	170.3	—	44.0	D	40	7.76e-9(**)	—	—	[60]
2,4-Di-t-butylphenol	170.5	0.918 (23)	—	D_s	23	1.2e-10	—	—	[12]
2,6-Di-t-butylphenol	170.5	0.918 (23)	—	D_s	23	9.8e-10	—	—	[12]
Ethyl-Naphtylether (Bromelia)	172.2	0.918 (23)	—	D_s	23	3.9e-9	—	—	[47]
3,7-Dimethyl-8-hydroxyoctanal (Hydroxycitronellal)	172.3	0.918 (23)	—	D_s	23	5.5e-10	—	—	[47]
Methyleugenol	178.2	0.918 (23)	—	D_s	23	3.0e-9	—	—	[47]
2-Methoxy-4-propenylanisol (Methylisoeugenol)	178.2	0.918 (23)	—	D_s	23	2.6e-9	—	—	[47]
Butyrated hydroanisole (BHA)	180.2	0.912 (31)	—	D	31	3.4e-9(**)	—	—	[62]
Butyrated hydroanisole (BHA)	180.2	0.927 (31)	—	D	31	3.8e-9(**)	—	—	[62]
2-Methoxy-4-propenylanisol (Methylisoeugenol)	178.2	0.918 (23)	—	D_s	23	2.6e-9	—	—	[47]
Diphenylmethanone (Benzophenone)	182.2	0.918 (23)	—	D_s	23	4.9e-9	—	—	[47]
n-Dodecylaldehyde (Aldehyde C_{12})	184.3	0.918 (23)	—	D_s	23	1.9e-10	—	—	[47]
2-Methyl-undecanal (Aldehyde C_{12} MNA)	184.3	0.918 (23)	—	D_s	23	1.8e-10	—	—	[47]

540 | Appendix I

Table AI.1 (Continued)

Diffusing Species	Polymer			Experiment			Diffusion parameters		Ref.
Name	Molec. weight M_w (g/mol)	Density ρ_P @ (°C) (g/cm^3)	Cristallinity (%)	Type of diffusion coefficient	Temp. range of experiment (°C)	Diffusion coefficient @ (23°C) D (cm^2/s)	Pre-exponential coefficient l_gD_0	Activation energy E_D (kJ/mol)	
n-Undecalacton (Aldehyde C$_{14}$)	184.3	0.918 (23)	–	D_s	23	2.7e-10	–	–	[47]
n-Dodecylaldehyde	184.3	0.922 (25)	50.0	D	25	1.9e-9$^{(**}$	–	–	[20]
n-Dodecylaldehyde	184.3	0.924 (25)	51.0	D	25	1.6e-9$^{(**}$	–	–	[20]
Citronellylformiate	184.3	0.918 (23)	–	D_s	23	2.3e-9	–	–	[47]
Tridecane	184.4	–	39.0	D	40	1.6e-8$^{(**}$	–	–	[60]
Tridecane	184.4	–	39.0	D	40	9.0e-9$^{(**}$	–	–	[60]
Tridecane	184.4	–	44.0	D	40	1.86e-8$^{(**}$	–	–	[60]
Tridecane	184.4	–	44.0	D	40	8.98e-9$^{(**}$	–	–	[60]
2,6-Di-t-butyl-4-methylphenol	184.6	0.918 (23)	–	D_s	23	6.6e-10	–	–	[12]
Dodecanol	186.4	0.918 (23)	–	D_s	23	1.1e-9	–	–	[12]
3-Methoxy-4-hydroxy-benzaldehyde (Verdylacetate)	190.2	0.918 (23)	–	D_s	23	2.1e-9	–	–	[47]
2-Methyl-3-(4-isopropyl) phenylpropanal (Cyclamen aldehyde)	190.3	0.918 (23)	–	D_s	23	1.2e-9	–	–	[47]
Triisopropanolamine (TIPA)	191.0	0.920	–	D	40; 100	1.9e-10	5.52	86.4	[63]
Dimethylbenzylcarbinylacetate (DMBCA)	192.3	0.918 (23)	–	D_s	23	1.09e-8	–	–	[47]
4-[2,6,6-Trimethy-2-cyclohexene-1-yle]-3-butene-2-one (Ionone)	192.3	0.918 (23)	–	D_s	23	1.2e-9	–	–	[47]
Dimethylphthalate (DMP)	194.2	0.918 (23)	–	D_s	23	1.9e-9	–	–	[47]

Appendix I | 541

Compound	MW	Density (T)			T	Value 1	Value 2	Ref
Allyl-3-cyclohexylpropionate	196.3	0.918 (23)	–	D_s	23	2.4e-9	–	[47]
2,6-Dimethyl-2,6-octadiene-8-yl-acetate (Geranylacetate)	196.3	0.918 (23)	–	D_s	23	1.6e-9	–	[47]
1,7,7-Trimethylbicyclo-1,2,2-haphtanyl-2-acetate (Isobromylacetate)	196.3	0.918 (23)	–	D_s	23	2.3e-9	–	[47]
3,7-Dimethyl-1,6-octadiene-3-ylacetate (Linalylacetate)	196.3	0.918 (23)	–	D_s	23	1.2e-9	–	[47]
1-Methyl-4-isopropyl-1-cyclohexene-4-yl-acetate (Terpinylacetate)	196.3	0.918 (23)	–	D_s	23	1.2e-9	–	[47]
Tetradecane (Alcane C_{14})	198.4	0.918 (23)	–	D_s	23	2.5e-9	–	[61]
Tetradecane (Alcane C_{14})	198.4	0.918 (23)	–	D_s	6; 40	1.9e-9	4.282	[61]
3,7-Dimethyl-6-octene-1-ylacetate	198.3	0.918 (23)	–	D_s	23	2.9e-9	–	[47]
p-tert.-Butylcyclohexylacetate (Oriclene extra)	198.3	0.918 (23)	–	D_s	23	1.2e-9	–	[47]
Ethyldecanoate	200.3	0.922 (25)	50.0	D	25	2.1e-9 [**]	–	[20]
Ethyldecanoate	200.3	0.924 (25)	51.0	D	25	1.7e-9 [**]	–	[20]
Methylundecanoate	200.3	0.918	44.0	D	20; 70	1.18e-8	60.39	[64]
Amylcinnamicaldehyde	202.3	0.918 (23)	–	D_s	23	1.4e-9	–	[47]
3-[4-tert.-Buthylphenyl]-2-methylpropanale (Lilial)	204.3	0.918 (23)	–	D_s	23	1.4e-9	–	[47]
N,N-Di-n-Butyl-aniline (DBA)	205.3	0.920 (25)	50.0	D_s	24; 43	1.2e-9 [*]	0.8007	[59]
2,4-Di-tert-butylphenole	206.3	0.918 (23)	–	D_s	23	1.2e-10	55.01	[12]
2,6-Di-tert-butylphenole	206.3	0.918 (23)	–	D_s	23	9.8e-10	–	[12]
3-Methyl-3-phenylglycidate (Aldehyde C_{16})	206.3	0.918 (23)	–	D_s	23	2.2e-9	–	[47]
5-(2,6,6-Trimethyl-2-cyclohexene-1-yle)-3-methyl-3-butene-2-one (Methyljonone-alpha)	206.3	0.918 (23)	–	D_s	23	6.6e-10	–	[47]

Table AI.1 (Continued)

Diffusing Species	Polymer			Experiment			Diffusion parameters		Ref.
Name	Molec. weight M_W (g/mol)	Density ρ_P (g/cm³) @ (°C)	Cristallinity (%)	Type of diffusion coefficient	Temp. range of experiment (°C)	Diffusion coefficient @ (23°C) D (cm²/s)	Pre-exponential coefficient $I_g D_0$	Activation energy E_D (kJ/mol)	
4-(2,6,6-Trimethyl-2-cyclohexene-1-yle)-3-methyl-3-butene-2-one (Methylionone-gamma)	206.3	0.918 (23)	–	D_s	23	8.6e-10	–	–	[47]
Iso-amylsalicilate	208.3	0.918 (23)	–	D_s	23	2.1e-9	–	–	[47]
4-[4-Methyl-4-hydroxyamyl]-3-cyclohexene-carboxaldehyde (Lyral)	210.3	0.918 (23)	–	D_s	23	1.2e-10	–	–	[47]
Benzylbenzoate	212.3	0.918 (23)	–	D_s	23	3.2e-9	–	–	[47]
Pentadecane	212.4	–	39.0	D	40	1.15e-8[**]	–	–	[60]
Pentadecane	212.4	–	39.0	D	40	7.66e-9[**]	–	–	[60]
Pentadecane	212.4	–	44.0	D	40	1.3e-8[**]	–	–	[60]
Pentadecane	212.4	–	44.0	D	40	1.5e-8[**]	–	–	[60]
2,4-Dihydroxybenzophenone	214.2	–	–	D	5;100	2.2e-10	2.995	71.67	[65]
2,4-Dihydroxybenzophenone	214.2	0.920 (25)	43.0	D	35; 75	3.7e-10[*]	4.210	77.30	[66]
2,4-Dihydroxybenzophenone (DHB)	214.2	–	–	$D_c \rightarrow 0$	25	5.0e-10[**]	–	–	[67]
Tetradecanol	214.4	0.918 (23)	–	D_s	23	8.2e-10	–	–	[12]
Methyl Laureate	214.4	0.918 (25)	44.0	D	20; 70	9.3e-10	3.760	66.83	[64]
2-Hexyl-3-phenylpropenal (Jasmonal)	216.3	0.918 (23)	–	D_s	23	1.6e-9	–	–	[47]
2,6-Di-tert-butyl-4-methylphenol	220.4	0.918 (23)	–	D_s	23	6.6e-10	–	–	[12]
2,6-Di-tert-butyl-4-methylphenol (BHT)	220.4	–	–	D	25	1.5e-9[**]	–	–	[67]
2,6-Di-Tert-butyl-4-methylphenol (Ionol)	220.4	–	–	D	65; 95	1.7e-8[(G)]	0.832	54.82	[68]
2,6-Di-tert-butyl-4-methylphenol (Ionol)	220.4	0.918 (25)	44.0	D	23; 74	8.1e-10	3.842	73.28	[69]

Compound	MW	density						Ref
2,6-Di-*tert*-butyl-4-methylphenol (Ionol)	220.4	0.918 (25)	44.0	D	75; 90	3.45e-8[70]	9.260	109.8	[69]
2,6-Di-*tert*-butyl-4-methylphenol (BHT)	220.4	–	–	D	25	9.0e-10[**]	–	–	[70]
2,6-Di-*tert*-butyl-4-methylphenol (BHT)	220.4	0.918 (25)	44.0	D	10; 50	1.0e-9	4.150	74.47	[71]
2,6-Di-*tert*-butyl-4-methylphenol (BHT)	220.4	0.920 (25)	45.0	D_s	5; 60	4.8e-10	5.902	86.23	[72]
2,6-Di-*tert*-butyl-4-methylphenol (BHT)	220.4	0.917	–	D_s	30: 60	1.2e-10[*]	6.639	93.83	[73]
Hexadecane (Alcane C_{16})	226.4	0.918 (23)	–	D_s	6; 40	1.0e-9	4.773	78.01	[61]
Hexadecane (Alcane C_{16})	226.4	–	39.0	D	40	1.0e-8[**]	–	–	[74]
Hexadecane (Alcane C_{16})	226.4	–	44.0	D	40	1.2e-8[**]	–	–	[74]
2-Hydroxy-4-methoxybenzophenone (Chimasorb 90)	228.2	–	–	D	25	7.0e-9[**]	–	–	[70]
Methyl Tridecanoate	228.4	0.918 (25)	44.0	D	20; 70	7.8e-9	4.350	70.60	[64]
Phenylethylphenylacetate	240.3	0.918 (23)	–	D_s	23	3.0e-9	–	–	[47]
Hexadecanol	242.3	0.918 (23)	–	D_s	23	6.4e-10	–	–	[12]
Heptadecane	240.4	–	39.0	D	40	9.6e-9[**]	–	–	[74]
Heptadecane	240.4	–	44.0	D	40	8.3e-9[**]	–	–	[74]
Methylmiristate	242.4	0.918 (25)	44.0	D	20; 70	6.3e-9	4.710	73.15	[64]
Nonane-1,3-dioldiacetate (Jasmelia)	244.3	0.918 (23)	–	D_s	23	8.2e-10	–	–	[47]
Triphenylmethane	244.3	–	–	D	40	3.85e-9[**]	–	–	[60]
Triphenylmethane	244.3	–	–	D	40	1.52e-9[**]	–	–	[60]
Methylmiristate	242.4	0.918 (25)	44.0	D	20; 70	6.3e-9	4.710	73.15	[64]
Hexadecanone	254.4	0.918 (25)	44.0	D	30; 45	2.8e-9[*]	9.550	102.6	[64]
Hexadecanone	254.4	0.918 (25)	44.0	D	48; 70	3.68e-8[40]	2.770	61.0	[64]
n-Octadecane (Alcane C_{18})	254.5	0.918 (23)	–	D_s	6; 40	9.5e-10	5.416	81.82	[61]
Octadecane	254.4	–	39.0	D	40	6.0e-9[**]	–	–	[60]
Ostadecane	254.4	–	44.0	D	40	4.0e-9[**]	–	–	[60]

Table AI.1 (Continued)

Name	Molec. weight M_w (g/mol)	Density ρ_P @ (°C) (g/cm³)	Cristallinity (%)	Type of diffusion coefficient	Temp. range of experiment (°C)	Diffusion coefficient @ (23°C) D (cm²/s)	Pre-exponential coefficient lgD_0	Activation energy E_D (kJ/mol)	Ref.
n-Octadecane	254.4	0.917	—	D_s	30; 60	7.8e-10[(*]	5.678	83.80	[73]
n-Octadecane (Alcane C_{18})	254.4	0.917	—	D_s	30; 60	3.5e-10[(*]	3.034	70.74	[75]
n-Octadecane	254.4	0.914	—	D	40; 90	1.19e-8[(*]	1.491	53.3	[76]
n-Octadecane	254.4	0.917	—	D_s	30; 60	1.63e-10[(*]	0.389	57.67	[77]
2-Hydroxy-4-ethanediolbenzophenone	258.3	—	—	D	5;100	2.1e-10	2.949	71.56	[65]
1,3,4,6,7,8-hexahydro-4,6,6,7,8,8-haxamethyl-cyclopenta-2-benzopyrane (Galaxolid)	258.4	0.918 (23)	—	D_s	23	4.4e-10	—	—	[47]
7-Acetyl-1,1,3,4,4,6-Hexamethyl-tetra-hydronaphgthaline (Tonalid)	258.4	0.918 (23)	—	D_s	23	3.8e-10	—	—	[47]
N,N-Diphenyl-p-phenylene-diamine (DPPD)	260.3	—	—	D_s	22	8.7e-10[(**	—	—	[78]
Cedrylacetate	264.4	0.918 (23)	—	D_s	23	4.1e-10	—	—	[47]
Trichlormethylphenylcarbinylacetate (Roseacetol)	267.5	0.918 (23)	—	D_s	23	8.2e-10	—	—	[47]
2,6-Dinitro-1-methyl-3-methoxy-4-tert.-butylbenzole (Moschus Ambrette)	268.3	0.918 (23)	—	D_s	23	7.4e-10	—	—	[47]
Tetramethylpentadecane	268.3	—	—	D	40	5.6e-9[(**	—	—	[60]
Tetramethylpentadecane	268.3	—	—	D	40	3.7e-9[(**	—	—	[60]
Dicumilperoxyde	270.2	0.929 (25)	32.0	D	40: 70	1.02e-8[(30]	12.45	124.4	[79]
Tetramethylpentadecane	270.2	0.929 (25)	32.0	D	70	3.2e-7[(**	—	—	[80]
2-Hydroxy-4-n-butoxybenzophenone	270.3	—	59.0	D	70: 90	2.04e-8[(70]	2.763	68.63	[81]

Compound	MW	ρ (T)			T (°C)	D			Ref
Octadecanol	270.5	0.918 (23)	–		23	4.8e-10	–	–	[12]
Methyl Palmitate	270.5	0.918 (25)	44.0	D_s	30; 70	4.4e-9	4.410	72.36	[64]
Stearylalcohol	270.5	–	–	D	40	3.04e-9(**)	–	–	[60]
Stearylalcohol	270.5	–	–	D	40	1.09e-9(**)	–	–	[60]
Di-butyl-phthalate (DBP)	270.5	–	–	$D_{c\to 0}$	20; 40	1.8e-12	2.635	81.45	[82]
Trans-9-octanacide	282.5	0.918 (25)	44.0	D	20; 40	3.4e-10	13.14	128.1	[64]
Eicosane (Alcane C$_{20}$)	282.6	0.918 (23)	–	D_s	6; 40	6.3e-10	6.306	87.88	[61]
Stearic Acid	284.3	0.918 (25)	44.0	D	40; 70	8.2e-10(40	7.530	96.38	[64]
Heptadecanoate	284.5	0.918 (25)	44.0	D	30; 80	3.1e-9(*	6.580	85.50	[64]
Methylester 3-(3,5-di-tert.-butyl-4-hydroxy-phenyl) propionic acid	292.2	0.918	48.0	D	30; 60	2.8e-10(*	5.803	87.00	[83]
2,6-Dinitro-3,5-dimethyl-1-acetyl-4-tert.-butylbenzene (Moschus Ketone)	294.3	0.918 (23)	–	D_s	23	3.5e-10	–	–	[47]
2,4,6-Trinitro-1,3-dimethyl-5-tert.-butylbenzene (Moschus Xylol)	297.3	0.918 (23)	–	D_s	23	5.7e-10	–	–	[47]
Metyl Stearate	298.5	0.918 (25)	44.0	D	20; 38	1.2e-9	13.940	129.5	[64]
Docosane (Alcane C$_{22}$)	310.6	0.918 (23)	–	D_s	6; 40	3.5e-10	7.227	94.54	[61]
Docosane	310.6	–	39.0	D	40	1.35e-9(**	–	–	[60]
Docosane	310.6	–	44.0	D	40	9.6e-10(**	–	–	[60]
Methylnonadecanoate	312.5	0.918 (25)	44.0	D	20; 40	1.1e-9	12.091	119.4	[64]
2-(2-hydroxy-3-t-butyl-5-methylphenyl)-5-chloro-benztriazol (Tinuvin 326)	315.8	0.918 (23)	–	D_s	23	2.0e-10	–	–	[12]
Heptadecylbenzene	316.4	–	–	D	40	4.9e-9(**	–	–	[60]
Heptadecylbenzene	316.4	–	–	D	40	3.8e-9(**	–	–	[60]
Propylester 3(3,5-di-tert.-butyl-4-hydroxyphenyl) propionic acid	320.2	0.918	48.0	D	30; 60	2.3e-10(*	6.167	89.60	[83]
2-Hydroxy-4-octoxybenzophenone	326.4	–	–	D	5;100	5.1e-10	2.754	68.28	[65]
2-Hydroxy-4-octoxybenzophenone (HOB)	326.4	–	–	D	25	1.2e-9(**	–	–	[67]

Table AI.1 (Continued)

Diffusing Species	Polymer			Experiment			Diffusion parameters		Ref.
Name	Molec. weight M_w (g/mol)	Density ρ_P (g/cm³) @ (°C)	Cristallinity (%)	Type of diffusion coefficient	Temp. range of experiment (°C)	Diffusion coefficient @ (23°C) D (cm²/s)	Pre-exponential coefficient l_gD_0	Activation energy E_D (kJ/mol)	
2-Hydroxy-4-octoxybenzophenone (Cyasorb UV 531)	326.4	0.918 (25)	44.0	D	5; 40	1.1e-9	5.950	84.37	[69]
2-Hydroxy-4-n-octoxybenzophenone	326.4	–	59.0	D	70; 90	2.08e-8$^{(70}$	0.924	56.50	[81]
2-Hydroxy-4-octoxybenzophenone	326.4	0.919	43.0	D	35; 50	5.1e-10$^{(30}$	19.88	88.14	[84]
2-Hydroxy-4-octoxybenzophenone	326.4	0.919	43.0	D	60; 75	2.48e-8$^{(60}$	2.002	61.23	[84]
Methyl Eicosanate	326.5	0.918 (25)	44.0	D	20; 40	7.2e-10	11.77	118.5	[64]
Methyl Eicosanate	326.5	0.918 (25)	44.0	D	40; 80	1.37e-8$^{(40}$	5.221	78.4	[64]
Behenyl alcohol	326.6	–	–	D	40	2.2e-10$^{(**}$	–	–	[60]
2-Hydroxy-4-ethandiol-thioacetic acid ester	332.4	–	–	D	5; 100	6.7e-11	3.725	78.75	[65]
Tetracosane	338.7	–	39.0	D	40	9.9e-10$^{(**}$	–	–	[74]
Tetracosane	338.7	–	44.0	D	40	6.7e-10$^{(**}$	–	–	[74]
2-2-Methylene bis (4-methyl-6-t.-butylphenol) (Plastanox 2246)	340.4	0.918 (25)	44.0	D	10; 70	2.2e-10	8.240	101.4	[69]
2-Hydroxy- 4-ethandiol methylthioacetic acid ester	346.4	–	–	D	5; 100	9.0e-11	3.461	76.54	[65]
Methyl Docosanate	354.5	0.918 (25)	44.0	D	20; 40	3.3e-10	12.22	123.0	[64]
2(2-hydroxy-3-5-di-tert.-butyl-phenyl)-5-chloro-benzotriazole (Tinuvin 327)	357.5	0.918 (25)	44.0	D	5; 70	1.6e-10	6.770	93.83	[69]
4,4'-thio-bis-(3-methyl-6tert-butylphenol)	358.0	0.917 (25)	48.0	D	45; 70	2.24e-9$^{(40}$	2.236	65.32	[85]
4-4-Thio-bis-(6-t.-butyl-metacresol) (Santonox)	358.5	0.918 (25)	44.0	D	10; 70	2.0e-10	5.84	88.09	[69]

Compound	MW	density			T	D			Ref
4-4-Thio-bis-(6-t.-butyl-metacresol) (Santonox)	358.5	0.918 (25)	44.0	D	10	3.8e-11[**]	—	—	[86]
Hexylester of 3(3,5-di-tert.-butyl-4-hydroxyphenyl)	362.4	0.918	48.0	D	30; 60	6.6e-11[*]	10.45	116.9	[83]
N-amido bis(2,2,6,6-tertamethyl-4-piperidinyl)–β–amionpropioamide	366.6	0.921	0	D	49; 80	1.02e-8[40]	−1.60	38.30	[87]
2-2-Methylene-bis(4-ethyl-6-t.-butyl-phenol) (Plastanox 425)	368.5	0.918 (25)	44.0	D	5; 70	8.7e-11	8.89	107.4	[69]
Methyl Tricosanate	368.5	0.918 (25)	44.0	D	20; 40	1.5e-10	14.99	140.6	[64]
Methyl Tricosanate	368.5	0.918 (25)	44.0	D	42; 80	1.47e-8[40]	1.93	58.50	[64]
2-Hydroxy-4 ethandiol t-butylthioacetic acid ester	388.5	—	—	D	5; 100	8.0e-12	4.400	87.85	[65]
Di-octyl-phthalate (DOP)	390.6	—	—	$D_c \to 0$	20; 40	4.7e-13	18.66	175.6	[82]
Octacosane	394.6	—	39.0	D	40	2.46e-10[**]	—	—	[60]
Octacosane	394.6	—	44.0	D	40	1.41e-10[**]	—	—	[60]
2,2,6,6 -Tetramethyl-4-piperidinol (Dastib 845)	411.2	0.921	—	D	23	6.9e-10	—	—	[87]
2,2,6,6 -Tetramethyl-4-piperidinol (Dastib 845)	411.2	0.921	—	D_s	20; 40	1.04e-9	7.902	95.68	[88]
2,2,6,6 -Tetramethyl-4-piperidinol (Dastib 845)	411.2	0.917 (25)	23.0	D	25; 60	1.8e-10[*]	1.54	64.0	[69]
Squalane	422.6	—	—	D	40	1.46e-9[**]	—	—	[60]
Squalane	422.6	—	—	D	40	7.3e-10[**]	—	—	[60]
2-2-Methylene-bis-(4-methyl-6-methyl-cyclohexyl-phenole) (Novox WSP)	420.5	0.918 (25)	44.0	D	5; 70	6.3e-11	7.42	99.85	[69]
4-4-Methylene-bis-(2-6 di-tert. butyl-phenole) (Ionox 220)	424.5	0.918 (25)	44.0	D	5; 70	1.0e-10	7.96	101.8	[69]
2,5 di(5-tert-butyl-2-benzoxazolyl)-thiophene (Uvitex OB)	430.5	—	—	D_s	22	2.04e-10[**]	—	—	[78]

Table AI.1 (Continued)

Diffusing Species	Polymer			Experiment			Diffusion parameters			Ref.
Name	Molec. weight M_w (g/mol)	Density ρ_P @ (°C) (g/cm³)	Cristallinity (%)	Type of diffusion coefficient	Temp. range of experiment (°C)	Diffusion coefficient @ (23°C) D (cm²/s)	Pre-exponential coefficient l_gD_0	Activation energy E_D (kJ/mol)		
D,l-α-Tocopherol (Irganox E 201)	431.0	0.930	–	D	40	6.9e-10[**]	–	–		[63]
3,5-di-tert.-butyl-4-hydroxy-benzoic acid-(2,4-di-tert-butyl-phenyl) ester (Tinuvin 120)	438.6	0.918 (23)	–	D_s	23	1.8e-11	–	–		[12]
Methyl Octacosanate	438.6	0.918 (25)	44.0	D	20; 40	3.0e-11	14.65	142.7		[64]
2-Hydroxy-4 ethandiol n-octylthioacetic acid ester	445.5	–	–	D	5;100	3.0e-11	1.611	68.92		[65]
Dodecylester- 3(3,5-di-tert.-butyl-4-hydroxyphenyl) propionic acid	446.3	0.918	48.0	D	30; 60	1.3e-10[*]	6.891	95.06		[83]
n-Dotriacontane	450.9	0.917	–	D_s	30; 60	6.44e-12[*]	12.69	135.32		[73]
bis[2,2,6,6-tetramethyl-4-piperidinyl]-sebacate (Tinuvin 770)	480.7	0.921	–	D_s	20; 40	5.4e-12	3.587	72.83		[88]
bis[2,2,6,6-tetramethyl-4-piperidinyl-1-oxyl]-sebacate	511.3	0.917	23.0	D	40; 80	2.0e-9(40)	6.840	93.10		[89]
1,1,3-tris(2-methyl-4-hydroxy-5-butyl phenyl) butane (Topanol)	512.6	0.918 (23)	–	D_s	23	5.4e-11	–	–		[12]
Didodecyl-3-3-thiodipropionate (DLTDP)	514.4	0.918 (25)	44.0	D	5; 40	5.3e-10	9.940	108.9		[69]
Didodecyl-3-3-thiodipropionate (DLTDP)	514.4	0.918 (25)	44.0	D	40; 70	5.5e-9(40)	6.230	86.8		[69]
Didodecyl-3-3-thiodipropionate (DLTDP)	514.4	0.916 (25)	43.0	D	20; 50	2.0e-11	15.40	147.8		[90]

Name	MW	density (T)		type	conditions	value			ref
Octadecylester 3-(3,5-di-*tert*-butyl-4-hydroxyphenyl) propionate (Irganox 1076)	531.4	0.918 (25)	48.0	D	30; 60	6.6e-11[*]	8.170	104.0	[83]
3-(3,5-di-*tert*-butyl-4-hydroxyphenyl) propionate (Irganox 1076)	531.4	0.918 (25)	44.0	D	5; 40	1.1e-10	9.15	108.2	[69]
3-(3,5-di-*tert*-butyl-4-hydroxyphenyl) propionate (Irganox 1076)	531.4	0.916 (25)	43.0	D	30; 50	8.0e-11[*]	10.18	114.9	[90]
3-(3,5-di-*tert*-butyl-4-hydroxyphenyl) propionate (Irganox 1076)	531.4	0.924 (25)	48.0	D	30; 50	1.1e-11[*]	14.71	145.5	[90]
3-(3,5-di-*tert*-butyl-4-hydroxyphenyl) propionate (Irganox 1076)	531.4	0.928 (25)	51.0	D	30; 50	5.2e-12[*]	18.54	169.0	[90]
3-(3,5-di-*tert*-butyl-4-hydroxyphenyl) propionate (Irganox 1076)	531.4	–	40.5	D	25	1.1e-11[**]	–	–	[91]
3-(3,5-di-*tert*-butyl-4-hydroxyphenyl) propionate (Irganox 1076)	531.4	0.900	–	D_S	30; 60	7.2e-11[*]	5.58	113.0	[92]
3-(3,5-di-*tert*-butyl-4-hydroxyphenyl) propionate (Irganox 1076)	531.4	0.910	–	D	40	2.43e-11[**]	–	–	[63]
3-(3,5-di-*tert*-butyl-4-hydroxyphenyl) propionate (Irganox 1076)	531.4	0.918	–	D	40	3.8e-10[*]	–	–	[63]
1-1-3-tris(2-methyl-4-hydroxy-5-*tert*.-butyl-phenyl)butane (Topanol CA)	544.5	0.918	44.0	D	5; 70	7.8e-11	6.230	92.6	[69]
1-1-3-tris(2-methyl-4-hydroxy-5-*tert*.-butyl-phenyl)butane (Topanol CA)	544.5	–	40.5	D	25	1.8e-10[**]	–	–	[91]
2-Hydroxy-4-ethandiol *n*-dodecylthioacetic acid ester	557.5	–	–	D	5;100	2.8e-11	1.807	70.01	[65]
2-Hydroxy-4-ethandiol *n*-octadecylthio acetic acid ester	585.5	–	–	D	5;100	1.8e-11	1.602	69.92	[65]
N,*N*-Dioctadecyl-aniline (DODA)	597.6	0.918 (25)	42.0	D	25; 45	1.5e-9[*]	3.002	66.96	[56]
2,2'-Thiodiethyl-bis-[3-(3,5-di-*tert*-butyl-4-hydroxy phenyl)-propionat] (Irganox 1035)	643.4	0.918 (25)	44.0	D	5; 40	5.8e-12	14.00	143.0	[69]

Table AI.1 (Continued)

Diffusing Species		Polymer		Experiment		Diffusion parameters			Ref.
Name	Molec. weight M_w (g/mol)	Density @ (°C) ρ_P (g/cm³)	Cristallinity (%)	Type of diffusion coefficient	Temp. range of experiment (°C)	Diffusion coefficient @ (23 °C) D (cm²/s)	Pre-exponential coefficient lgD_0	Activation energy E_D (kJ/mol)	
Phosphorous acid, tris((2,4-di-tert-butylphenyl)ester (Irgafos 168)	646.0	0.905	–	D	80	2.4e-8[**]	–	–	[63]
Phosphorous acid, tris((2,4-di-tert-butylphenyl)ester (Irgafos 168)	646.0	0.918	–	D	80	1.1e-8[**]	–	–	[63]
Phosphorous acid, tris ((2,4-di-tert-butylphenyl)ester (Irgafos 168)	646.0	0.903	–	D	23	2.0e-12	–	–	[92]
Docosanyl Docosanate	649.1	–	–	D	40; 80	1.8e-9[40]	6.903	93.70	[93]
Distearyl-thio-dipropionate (DSTDP)	682.5	0.916 (25)	43.0	D	40; 80	4.0e-9[40]	6.12	87.0	[90]
1-3-5-Trimethyl-2-4-6-tri (3-5-di-tert-butyl-4-hydroxy benzyl)benzene (Ionox 330)	774.6	0.918 (25)	44.0	D	5; 70	7.9e-11	6.43	93.7	[69]
1-3-5-Trimethyl-2-4-6-tris (3-5-di-tert-butyl-4-hydroxy benzyl)benzene (Irganox 1330)	775.0	0.918	–	D	40	8.9e-11[**]	–	–	[63]
Terephthalate-2-2-methyleme-bis-(4-methyl-6-tert-butyl) phenole (HMP12)	810.6	0.918 (25)	44.0	D	5; 70	2.4e-12	16.78	160.9	[69]

Compound						
Polymer of 2,2,4,4-tetramethyl-7-oxa-3,20-diaza-20-(2,3-epoxypropyl)dispiro[5.1.11.2]-heneicosane-21-one (Hostavin N 30)	840.0	0.930	–	D	49	1.61e-10$^{(**)}$	–	–	[63]
Tertakis[3-(3,5-di-*tert*-butyl-4-hydroxy-phenyl)propionyloxymethyl]-methane (Irganox 1010)	1177.8	0.918 (25)	44.0	D	5;70	3.7e-12	8.95	115.5	[69]
Tertakis[3-(3,5-di-*tert*-butyl-4-hydroxy-phenyl)propionyloxymethyl]-methane (Irganox 1010)	1177.8	0.921 (23)	43.0	D	45;80	4.6e-11$^{(40}$	11.46	130.6	[94]
Tertakis[3-(3,5-di-*tert*-butyl-4-hydroxy-phenyl)propionyloxymethyl]-methane (Irganox 1010)	1177.8	–	–	D	25	5.02e-12$^{(**)}$	–	–	[70]
Tertakis[3-(3,5-di-*tert*-butyl-4-hydroxy-phenyl)propionyloxymethyl]-methane (Irganox 1010)	1177.8	0.917 (23)	23.0	D	25	5.6e-11$^{(**)}$	–	–	[95]
Tertakis[3-(3,5-di-*tert*-butyl-4-hydroxy-phenyl)propionyloxymethyl]-methane (Irganox 1010)	1177.8	0.922	45.0	D_S	49	1.56e-11$^{(**)}$	–	–	[96]

where $D_0 - cm^2/s$ – is a pre-exponential factor, E_d – kJ/mol – the energy of activation, R – 8.31 J/mol grd. – the gas constant, and T – K – the absolute temperature.

In the eigth and ninth columns D_0 and E_d are summarized for the experiments in which diffusion data were collected at different temperatures. Using these parameters with the above equation allows one to calculate the diffusion coefficient at any T point within the corresponding T range given in column 6. Moreover, one can calculate by extrapolation D's even beyond such a T range. However when doing so it is recommended not to exceed the extrapolation to far from the T range in which the experimental data were collected – column 6. Extrapolations up to $+/-$ 25% of the T range given in column 6 would most likely be on the safe side. Even when doing such a conservative extrapolation one should take care that for the new T the polymer did not change from rubbery to glassy or vice versa. It is known that for most of the polymers D_0 and E_d are usually differing in these two phases.

References

1 Michaels, A.S. and Bixler, H.J. (1961) *J. Polym. Sci.*, **50**, 413.
2 Kanitz, P.J.F. and Huang, R.Y.M. (1970) *J. Appl. Polym. Sci.*, **14**, 2739.
3 Bixler, H.J., Michaels, A.S. and Salame, S. (1963) *J. Polym. Sci., A*, **1**, 895.
4 Kulkarni, S.S. and Stern, A.S. (1983) *J. Polym. Sci., Polym. Phys. Ed.*, **21**, 441.
5 MacDonald, R.W. and Huang, R.Y.M. (1981) *J. Appl. Polym. Sci.*, **26**, 2239.
6 Costamagna, V., Strumia, M., Lopez-Gonzales, M. and Riande, E. (2006) *J. Polym. Sci., PartB: Polym. Phys.*, **44**, 2828.
7 Beret, S. and Hager, S.L. (1979) *J. of Appl. Polym. Sci.*, **24**, 1787.
8 Evnochides, S.K. and Henley, E.J. (1970) *J. Polym. Sci., A-2*, **8**, 1987.(1971) *AIChE Journal*, **17**, 880.
9 Brandt, W.W. (1959) *J. Polym. Sci.*, **41**, 403 and 415.
10 Robeson, L.M. and Smith, T.M. (1967) *J. Appl. Polym. Sci.*, **11**, 2007.(1968) **12**, 2083.
11 Yasuda, H., Stannett, V.T., Frisch, H.L. and Peterlin, A. (1964) *Die Makromolekulare Chem.*, **73**, 188.
12 Koszinowski, J. (1986) *J. Appl. Polym. Sci.*, **31**, 2711.
13 Podkowka, J. and Puchalik, A. (1982) *J. Appl. Polym. Sci.*, **27**, 1471.
14 He, Z., Hammond, G.S. and Weiss, R.G. (1992) *Macromolecules*, **25**, 1568.
15 Izydorczyk, J. and Salwinski, J. (1984) *J. Appl. Polym. Sci.*, **29**, 3663.
16 Dos Santos, M.L. and Leitao, D.M. (1971) *J. Polym. Sci., A-2*, **10**, 769.(1972) **10**, 1.
17 Thalmann, W.R. (1990) *Packaging Tech. Sci.*, **3**, 67.
18 Cuttler, J.A., Kaplan, E., McLaren, A.D. and Mark, H. (1951) *TAPPI*, **34**, 404. (1953) **36**, 423.
19 Rogers, C.E., Stannett, V. and Szwarc, M. (1960) *J. Polym. Sci.*, **45**, 61.
20 Shimoda, M., Matsui, T. and Osajima, Y. (1987) *Nippon Shokuhin Koyo Gakkaishi*, **34**, 402 and 535.
21 Johanson, F. and Leufvèn, A. (1994) *J. Food Sci.*, **59**, 1328.
22 McCall, D.W. and Schlichter, W.P. (1958) *J. Am. Chem. Soc*, **80**, 1861.
23 Fels, M. and Huang, R.Y.M. (1970) *J. Appl. Polym. Sci.*, **14**, 523 and 537.
24 Saleem, M., Asfour, A.-F.A., De Kee, D. and Harrison, B. (1989) *J. Appl. Polym. Sci.*, **37**, 617.
25 Doong, S.J. and Ho, W.S.W. (1992) *Ind. Eng. Chem. Res.*, **31**, 1050.
26 Fleischer, G. (1985) *Polym. Comm.*, **25**, 20.
27 Huang, R.Y.M. and Rhim, J.-W. (1990) *J. Appl. Polym. Sci.*, **41**, 535.
28 Kreituss, A. and Frisch, H.L. (1981) *J. Polym. Sci., Polym. Phys. Ed.*, **19**, 889.

29 Aboul-Nasr, O.T. and Huang, R.Y.M. (1979) *J. Appl. Polym. Sci.*, **23**, 1819. 1833 and 1851.

30 Gray, D.G. and Guillet, J.E. (1973) *Macromolecules*, **6**, 223.

31 Takeuchi, Y. and Okamura, H. (1976) *J. Chem. Eng. Japan*, **9**, 136.

32 Yeom, C.K. and Huang, R.Y.M. (1992) *J. Membrane Sci.*, **68**, 11.

33 Stern, A.S. and Britton, G.W. (1972) *J. Polym. Sci., Part A-2*, **10**, 295.

34 Fels, M. (1970) *AIChE J. Symposium Series*, **120**, 49.

35 Helmroth, I.E., Dekker, M. and Hankemeier, Th. (2003) *J. Appl. Polym. Sci.*, **90**, 1609.

36 Araimo, L., De Candia, F., Vittoria, V. and Peterlin, A. (1978) *J. Polym. Sci., Polym. Phys. Ed.*, **16**, 2087.

37 De Candia, Russo, R.F., Vittoria, V. and Peterlin, A. (1982) *J. Polym. Sci., Polym. Phys. Ed.*, **20**, 269.

38 Asfour, A.-F.A., Saleem, M., De Kee, D. and Harrison, B. (1989) *J. Appl. Polym. Sci.*, **38**, 1503.

39 Hedenqvist, M., Angelstok, A., Edsberg, L., Larsson, P.T. and Gedde, U.W. (1996) *Polymer*, **37**, 2887.

40 Stern, A.S., Sampat, S.R. and Kulkarni, S.S. (1986) *J. Polym. Sci., Part B Polym. Phys.*, **24**, 2149.

41 Phillips, J.C. and Peterlin, A. (1983) *Polym. Eng. Sci.*, **23**, 734.

42 Vahdat, N. and Sullivan, V.D. (2001) *J. Appl. Polym. Sci.*, **79**, 1265.

43 Ng, H.C., Leung, W.P. and Choy, C.L. (1985) *J. Polym. Sci., Polym. Phys. Ed.*, **23**, 973.

44 Markevich, M.A., Stogova, V.N. and Gorenberg, A.Ya. (1991) *Vysokomol. Soed.*, **A33**, 132.

45 Lützow, N., Tihminlioglu, A., Danner, R.P., Duda, J.L., DeHaan, A., Warnier, G. and Zielinski, J.M. (1999) *Polymer*, **40**, 2797.

46 Ghosh, S.K. (1982) *J. Appl. Polym. Sci.*, **27**, 331.

47 Koszinowski, J. and Piringer, O. (1990) *Verpackungs Rundschau*, **41**, 15.

48 Strandburg, G., De Lassus, P.T. and Howell, B.A. (1990) Barrier Polymers and Packaging (ed. W.J. Koros), ACS Symposium Series, Nr., 423, Washington DC, pp. 333.

49 Serota, D.G., Meyer, M.C. and Autian, J. (1972) *J. Pharm. Sci.*, **61**, 416.

50 He, Z., Hammond, G.S. and Weiss, R.G. (1992) *Macromolecules*, **25**, 501.

51 Theodorou, E. and Paik, J.S. (1992) *Packaging. Techn. Sci.*, **5**, 21.

52 Sadler, G.D. and Braddock, R.J. (1991) *J. Food Sci.*, **56**, 35.

53 Balik, C.M. (1996) *Macromolecules*, **29**, 3025.

54 Becker, K., Koszinowski, J. and Piringer, O. (1983) *Deutsche Lebensmittel-Rundsch.*, **79**, 257.

55 Zimerman, O.E., Cui, C., Wang, X., Atvars, T.D. and Weiss, R.G. (1998) *Polymer*, **39**, 1177.

56 Taraszka, J.A. and Weiss, (1997) *Macromolecules*, **30**, 2467.

57 Kobayashi, M., Kanno, T., Hanada, K. and Osanai, S.-I. (1995) *J. Food Sci.*, **60**, 205.

58 Lagaron, J.M., Cava, D., Gimenez, E., Hernandez-Munoz, P., Catala, R. and Gavara, R. (2004) *Macromol. Symposia*, **205**, 225.

59 Lu, L. and Weiss, J.A. (1994) *Macromolecules*, **27**, 219.

60 Reynier, A., Dole, P. and Feigenbaum, A. (1999) *Food Add. Contamin.*, **16**, 137.

61 Koszinowski, J. (1986) *J. Appl. Polym. Sci.*, **31**, 1805.

62 Howell, B.F., McCrakin, F.L. and Wang, F.W. (1985) *Polymer*, **26**, 433.

63 Begley, T., Castle, L., Feigenbaum, A., Franz, R., Hinrichs, K., Lickly, Ti., Mercea, P., Milana, M., O'Brien, A., Rebre, S., Rijk, R. and Piringer, O. (2005) *Food Add. Contamin.*, **22**, 73.

64 Moisan, J.Y. (1981) *Eur. Polym. J.*, **17**, 857.

65 Scott, G. (1988) *Food Additives Contam.*, **5**, 421.

66 Westlake, J.F. and Johnson, M. (1975) *J. Appl. Polym. Sci.*, **19**, 319.

67 Calvert, P.D. and Billingham, N.C. (1979) *J. Appl. Polym. Sci.*, **24**, 357.

68 Yushkevichyute, S.S., Shlyapnikov and Yu. A. (1965) *Vysokomol. soed.*, **A7**, 2094.

69 Moisan, J.Y. (1980) *Eur. Polym. J.*, **16**, 979.

70 Billingham, N.C. (1989) *Oxidation Inhibition in Organic Materials* (eds J. Pospisil and P.P. Klemchuk), CRC Press, Boca Ratton, FLA, Vol II, pp. 272.
71 Moisan, J.Y. (1985) *Polymer Permeability* (ed. J.W. Comyn), Applied Science Press, London, pp. 119.
72 Gnadek, T.P., Hatton, T.A. and Reid, R.C. (1989) *Ind. Eng. Chem. Res.*, **28**, 1030 and 1036
73 Chang, S.S., Pummer, W.J. and Maurey, J.R. (1984) *Polymer*, **24**, 1267.
74 Reynier, A., Dole, P. and Feigenbaum, A. (2001) *J. Appl. Polym. Sci.*, **82**, 2434.
75 National Bureau of Standards(NBS), Final Project Report NBS 82-2472,1982.
76 Auerbach, I., Miller, W.R. and Kuryla, W.C. (1958) *J. Polym. Sci.*, **28**, 129.
77 Chan, S.-S. (1984) *Polymer*, **25**, 209.
78 Wang, F.W. and Howell, B.F. (1984) *Polymer*, **25**, 1626.
79 Gupta, J.P. and Sefton, M.V. (1984) *J. Appl. Polym. Sci.*, **29**, 2383.
80 Gupta, J.P. and Sefton, M.V. (1986) *J. Appl. Polym. Sci.*, **31**, 2195.
81 Cichetti, O., Dubini, M., Parrini, P., Vicario, G.P. and Bua, E. (1968) *Eur. Polym. J.*, **4**, 419.
82 Thinius, K. and Schröder, E. (1964) *Plaste Kautschuk*, **11**, 390.
83 Möller, K. and Gevert, T. (1994) *J. Appl. Polym. Sci.*, **51**, 895.
84 Johnson, M. and Westlake, F.J. (1975) *J. Appl. Polym. Sci.*, **19**, 1745.
85 Roe, R.-J., Bair, H.E. and Gieniewski, C. (1974) *J. Appl. Polym. Sci.*, **18**, 843.
86 Moisan, J.Y. and Lever, R. (1982) *Eur. Polym. J.*, **18**, 411.
87 Malik, J., Hrivik, A. and Tomova, E. (1992) *Polym. Degrad. Stab.*, **35**, 61.
88 Malik, J., Hrivik, A. and Tomova, E. (1992) *Polym. Degrad. Stab.*, **35**, 125.
89 Dudler, V. (1993) *Polym. Degrad. Stab.*, **42**, 205.
90 Moisan, J.Y. (1980) *Eur. Polym. J.*, **16**, 989.
91 Földes, E. (1998) *Die Angewandte Makromol. Chem.*, **261/262**, 65.
92 Spatafore, R., Schultz, K., Thompson, T. and Pearson, L.T. (1992) *Polym. Bulletin*, **28**, 467.
93 Klein, J. and Briscoe, B.J. (1979) *Proc. Roy. Soc. Lond., A*, **365**, 53. (1975) *Nature*, **257**, 386.
94 Földes, E. and Turcsanyi, B. (1992) *J. Appl. Polym. Sci.*, **46**, 507.
95 Zweifel, H. (1998) *Stabilization of Polymeric Materials*, Springer, Berlin, Chap. 5.
96 Till, D., Schwope, A.D., Ehnholt, D.J., Sidman, K.R., Wehlen, R.H., Scwartz, P.S. and Reid, C.R. (1987) *CRC Critical Reviews in Toxicology*, **18** (3), 161.
97 Tochin, V.A., Shlyakhov, R.A. and Sapozhnikov, D.N. (1980) *Vysokomol. Soed. A*, **22** (4), 752.
98 Von Solms, N., Nielsen, J.K., Hassager, O., Rubin, A., Dandekar, A.Y., Andersen, S.I. and Stenby, E.H. (2004) *J. Appl. Polym. Sci.*, **91**, 1476.
99 Eby, R.K. (1964) *J. Appl. Phys.*, **35**, 2720.
100 Michaels, A.S., Bixler, J.H. and Fein, H.L. (1964) *J. Appl. Phys.*, **35**, 3165.
101 Lowell, P.N. and McCrum, N.G. (1935) *J. Polym. Sci. A-2*, **95**, (1971).
102 McCall, D.W. (1957) *J. Polym. Sci.*, **26**, 151.
103 Xiao, S., Moresoli, C., Bovenkamp, J. and De Kee, D. (1997) *J. Appl. Polym. Sci.*, **65**, 1833.
104 Williams, J.L. and Peterlin, A. (1970) *Die Makromol. Chem.*, **135**, 41.
105 Williams, J.L. and Peterlin, A. (1971) *J. Polym. Sci.*, **9**, 1483.
106 Peterlin, A., Williams, J.L. and Stannett, V.T. (1967) *J. Polym. Sci.*, **5**, 957.
107 Kwei, T.K. and Wang, T.T. (1972) *Macromolecules*, **5**, 128.
108 Blackadder, D.A. and Keniry, J.S. (1973) *J. Appl. Polym. Sci.*, **17**, 351.
109 Balik, C.M. and Simendinger, W.H., III (1998) *Polymer*, **39**, 4723.
110 Mohney, S.M., Hernandez, R.J., Giacin, J.R., Harte, B.R. and Miltz, J. (1988) *J. Food Sci.*, **53**, 253.
111 Peterlin, A. and Williams, J.L. (1972) *Br. Polym. J.*, **4**, 271.
112 Han, J.K., Miltz, J., Harte, B.R. and Giacin, J.R. (1987) *Polym. Eng. Sci.*, **27**, 934.
113 Jackson, R.A., Oldland S.R.D. and Pajaczkowski, A. (1968) *J. Appl. Polym. Sci.*, **12**, 1297.

114 Figge, K. and Rudolph, F. (1979) *Die Angew. Makromol. Chem.*, **78**, 157.
115 Klein, J. and Briscoe, B.J. (2065) *J. Polym. Sci., Polym. Phys. Ed.*, **15** (1977).(1977) *Nature*, **266**, 43.
116 Limm, W. and Hollifield, H.C. (1996) *Food Additives Contamin.*, **13**, 949.
117 Goydan, R., Schwope, A.D., Reid, R.C. and Cramer, G. (1990) *Food Additives Contamin.*, **7**, 323.
118 Limm, W. and Hollifield, H.C. (1995) *Food Additives Contamin.*, **12**, 609.
119 Lickly, T.D., Bell, C.D. and Lehr, K.M. (1990) *Food Additives Contamin.*, **7**, 805.
120 Li, N. and N., I.E. (1969) *Chem. Prod. Res. Dev.*, **8**, 281.
121 Sliepcevich, A., Stroti, G. and Morbidelli, M. (2000) *J. Appl. Polym. Sci.*, **78**, 464.
122 Franz, R. (1991) *Packaging Tech. Sci.*, **6**, 91.
123 Barson, C.A. and Dong, Y.M. (1990) *Eur. Polym. J.*, **26**, 329 and 453.
124 Niebergall, H. and Kutzki, R. (1982) *Deutsche Lebensmittel Rundsch.*, **78**, 323 and 428.
125 Koszinowski, J. and Piringer, O. (1987) *J. Plastic Film Sheeting*, **5**, 96.
126 Frensdorff, H.K. (1957) *J. Polym. Sci., A*, **2**, 341.
127 Vittoria, V. and Riva, F. (1986) *Macromolecules*, **19**, 1975.
128 Vittoria, V. (1985) *Polym. Comm.*, **26**, 213.
129 Wachler, T. and Göritz, D. (1995) *Macromol. Chem. Phys.*, **196**, 429.
130 Wieth, W.R. and Wuerth, W.F. (1969) *J. Appl. Polym. Sci.*, **13**, 685.
131 Long, R.B. (1965) *I. Eng. Chem. Fundam.*, **4**, 445.
132 Choy, C.L., Leung, W.P. and Ma, T.L. (1984) *J. Polym. Sci., Polym. Phys. Ed.*, **22**, 707.
133 Hjermstad, H.P. (1979) *J. Appl. Polym. Sci.*, **24**, 1885.
134 Ódor, L. and Geleji F. (1967) *Makromol. Chemie*, **100**, 11.
135 Moaddeb, M. and Koros, J.W. (1995) *J. Appl. Polym. Sci.*, **57**, 687.
136 Apostoloupoulos, D. (1991) *Winters, Packaging Techn. Sci.*, **4**, 131.
137 Semwal, R.P., Banerjee, S., Chauhan, L.R., Bhattacharya, A. and Rao, N.B.S.N. (1996) *J. Appl. Polym. Sci.*, **60**, 29.
138 Kolesnikova, N.N., Kiryushkin, S.G., and Shlyapnikov, Yu.A. ,(1985) *Vysokomol. soed.*, **A27**, 1880.
139 Hayashi, H., Sakai, H. and Matsuzawa, S. (1994) *J. Appl. Polym. Sci.*, **51**, 2165.
140 Gromov, B.A., Miller, V.B., Neiman, M.B. and Shlyapnikov, Yu.A. ,(1962) *J. Appl. Radiation Isotopes*, **13**, 281.
141 Dubini, M., Cichetti, O., Vicario, G.P. and Bua, E. (1967) *Eur. Polym. J.*, **3**, 473.
142 Holcik, J., Karvas, M., Kassovicova, D. and Durmis, J. (1976) *Eur. Poly. J.*, **12**, 173.
143 Hayashi, H. and Matsuzawa, S. (1992) *J. Appl. Polym. Sci.*, **46**, 499.
144 Hayashi, H. and Matsuzawa, S. (1993) *J. Appl. Polym. Sci.*, **49**, 1825.
145 Durmis, J., Karvas, M., Caucik, P. and Holcik, J. (1975) *Eur. Polym. J.*, **11**, 219.
146 Dudler, V. and Muinos, C.(1995)Polymer Durability: Degradation, Stabilization and Lifetime Predictions (eds L.R. Clough, N.C. Billingham and K.T. Gillen), ACS Advances in Chemistry Series No. 249, Washington D.C, pp. 441.
147 Billingham, N.C., Calvert, P.D. and Uzuner, A. (1989) *Eur. Polym. J.*, **25**, 839.
148 Billingham, N.C. (1989) *Makromol. Chem. Macromol. Symp.*, **27**, 187.
149 Luston, J., Pastusakova, V. and Vass, F. (1993) *J. Appl. Polym. Sci.*, **47**, 555.
150 Barson, C.A. and Dong, Y.M. (1990) *Eur. Polym. J.*, **26**, 449.
151 Okajima, S., Sato, N. and Tasaka, M. (1970) *J. Appl. Polym. Sci.*, **14**, 1563.
152 Reynier, A., Dole, P., Humbel, S. and Feigenbaum, A. (2001) *J. Appl. Polym. Sci.*, **82**, 2422.
153 Hsu, S.C., Lin-Vien, D. and French, R.N. (1992) *Appl. Spectroscopy*, **46**, 225.
154 Schwarz, T., Steiner, G. and Koppelmann, J. (1989) *J. Appl. Polym. Sci.*, **37**, 3335.
155 Mar'in, A.P., Borzatta, V., Bonora, M. and Greci, L. (2000) *J. Appl. Polym. Sci.*, **75**, 890.

Appendix II

Table AII.1 Partition coefficient data ($K_{P/L}$) for organic substances between LDPE and LLDPE and 100% ethanol.

Reference		[1]	[3]	[3]	[3]	[3]	[3]	[7]	[9]	[25]
Polymer phase		LDPE ρ=0.918	LDPE ρ=0.918	LDPE ρ=0.918	LDPE ρ=0.918	LDPE ρ=0.918	Average HDPE/LDPE	LDPE ρ=0.918	LDPE ρ=0.918	LDPE ρ=0.924
Temperature (°C)		23	Average 10–40	10	25	40	Average 10–40	23	23	40
Liquid phase substance	Cas #	100% Ethanol	100% Ethanol	100% Ethanol	100% Ethanol	100% Ethanol	100% Ethanol	100% Ethanol	100% Ethanol	100% Ethanol
Limonene	5989-27-5	0.42	0.29	0.29	0.21	0.2	0.28			
Diphenylmethane	101-81-5	0.25	0.084	0.084	0.066	0.073	0.099			
Diphenyloxide	101-84-8	0.14	0.11	0.11	0.1	0.11	0.15			
Isoamylacetate	123-92-2	0.046	0.029	0.029	0.067	0.049	0.046			
gamma-Undelactone (Aldehyde C14)		0.017	0.012	0.012	0.016	0.031	0.027			
Linalylacetate	115-95-7	0.065	0.047	0.047	0.045	0.047	0.055			
Camphor	76-22-2	0.064	0.043	0.043	0.033	0.044	0.040			
Citronellol	106-22-9	0.026								
Eugenol	97-53-0	0.013			0.023	0.044	0.018			
Dimethylbenzylcarbinol	100-86-7	0.0062	0.0091	0.0091	0.0064	0.011	0.011			
Menthol	89-78-1	0.0087	0.01	0.01	0.007	0.018	0.023			
2-Phenylethylalcohol	60-12-8	0.0028			0.029	0.033	0.017			
cis-3-Hexenol	623-37-0	0.0024				0.023	0.014			
Octane	111-65-9		0.83		0.81	0.84	1.2			
Nonane	111-84-2		1.5	2.0	1.2	1.3	2.0			
Decane	124-18-5	0.55	3.6	3.9	3.5	3.5	3.6			
Dodecane	112-40-3	0.97	3.3	6.0	2.0	2.0	4.4			1.3
Tetradecane	629-59-	1.3	5.4	9.0	3.5	3.2	8.0			2.0
Hexadecane	544-76-3	1.6	9.3	17	5.5	5.0	15			2.3
Octadecane	593-45-3	2.5	35	33	42	31	39			4.5
Eicosane	112-95-8	4.2	43	69	31	27	62			
Docosane	629-97-0	8.1	70	163	26	20	94			
Phenol	108-95-2							0.0026	0.0039	
p-Cresol	106-44-5							0.0056	0.0050	
2,4,6-Trimethylphenol	527-60-6							0.019		
2,3,5,6-Tetramethylphenol	527-35-5							0.030	0.019	

Reference		[7]	[25]	[31]
Polymer phase		LDPE $\rho=0.918$	LDPE $\rho=0.924$	LDPE $\rho=0.93$
Temperature (°C)		23	40	20
Liquid phase substance	Cas #	100% Ethanol	100% Ethanol	100% Ethanol
2,4-Di-t-butylphenol	96-76-4	0.016	0.0085	
2,6-Di-t-butylphenol	128-39-2	0.13	0.069	
2,6-Di-t-butyl-4-methylphenol (BHT)	128-37-0	0.19	0.099	1.1
3,5-Di-t-butyl-4-hydroxy-benzoic acid-(2,4-di-t-butyl-phenyl)-ester (Tinuvin 120)	4221-80-1	0.045		
2-(2-Hydroxy-3-t-butyl-5-methylphenyl)-5-chlorobenzztriazol (Tinuvin 326)	3896-11-5	0.71		
1,1,3-Tris(2-methyl-4-hydroxy-5-t-butylphenyl)butane (Topanol)	1843-03-4	0.00031		
Octadecyl 3-(3,5-di-tert-butyl-4-hydroxy phenyl) proprionate (Irganox 1076)	2082-79-3			1.6
Laurophenone	1674-38-0		1.8	
Triphenylene	217-59-4		0.63	
Heptanol	111-70-6	0.0047		
Dodecanol	112-53-8	0.021	0.63	0.03
Tetradecanol	112-72-1	0.029	0.75	
Hexadecanol	36653-82-4	0.033	0.83	
Octadecanol	112-92-5	0.053	1.1	
n-Dodecanal	112-54-9			0.03

Table AII.2 Partition coefficient data ($K_{P/L}$) for organic substances between LDPE and aqueous ethanol.

Reference		[2]	[1]	[3]	[3]	[3]	[3] Average	[2]	[1]	[3]	[3]
Polymer phase		LDPE	LDPE	LDPE	LDPE	LDPE	HDPE/LDPE	LDPE	LDPE	LDPE	LDPE
		$\rho = 0.918$	$\rho = 0.918$	$\rho = 0.918$	$\rho = 0.918$	$\rho = 0.918$	$\rho = 0.918$	$\rho = 0.918$	$\rho = 0.918$	$\rho = 0.918$	$\rho = 0.918$
Temperature (°C)		23	23	10	25	40	10–40	23	23	10	25
Liquid phase substance	CAS #	80% (w/w) Aqueous ethanol	80% (w/w) Aqueous ethanol	75% (w/w) Aqueous ethanol	75% (w/w) Aqueous ethanol	75% (w/w) Aqueous ethanol	75% (w/w) Aqueous ethanol	60% (w/w) Aqueous ethanol	60% (w/w) Aqueous ethanol	50% (w/w) Aqueous ethanol	50% (w/w) Aqueous ethanol
Limonene	5989-27-5	1.99		1.8	0.67	0.64	1.8	6.5		27	24
Diphenylmethane	101-81-5	0.82		0.26	0.17	0.16	0.34	3.7		2.5	2.4
Diphenyloxide	101-84-8	0.51		0.44	0.27	0.25	0.53	1.7		4.4	4.3
Isoamylacetate	123-92-2	0.069		0.036	0.31	0.24	0.14	0.16		0.33	0.27
gamma-Undelactone (Aldehyde C14)		0.03		0.01	0.03	0.03	0.032	0.073		0.096	0.085
Linalylacetate	115-95-7	0.15		0.12	0.065	0.067	0.15	0.59		1	0.96
Camphor	76-22-2	0.089		0.67	0.034	0.045	0.058	0.15		0.34	0.47
Citronellol	106-22-9	0.011						0.052			
Eugenol	97-53-0	0.023			0.016	0.012	0.024	0.047		0.038	0.03
Dimethylbenzyl-carbinol	100-86-7	0.0096		0.011	0.016	0.011	0.023	0.017		0.042	0.13
Menthol	89-78-1	0.02		0.021	0.013	0.011	0.056	0.053		0.12	0.13
Phenylethylalcohol	60-12-8	0.0045			0.021	0.017	0.016	0.0059		0.0095	0.14
cis-3-Hexenol	623-37-0	0.0034					0.0087	0.0058		0.012	0.028
Octane	111-65-9			1.6	1.5	1.1				270	28
Nonane	111-84-2			3.2	3	2.2				260	160
Decane	124-18-5		3.5	19	10	10			160	490	350
Dodecane	112-40-3		6.5	10	9	6			370	1100	1400
Tetradecane	629-59-		11	20	20	12			900	2900	3700
Hexadecane	544-76-3		20	43	47	25			7600	7400	9400
Octadecane	593-45-3		35	120	120	57			34000	21000	28000
Eicosane	112-95-8		59	410	370	140			160000	22000	600000
Docosane	629-97-0			2300	1700	510					

Reference		[3]	[3]	[2]	[1]	[3]	[3]	[3]	[3]	[2]	[1]
Polymer phase		LDPE	Average HDPE/LDPE 10–40	LDPE	LDPE	LDPE	LDPE	LDPE	Average HDPE/LDPE 10–40	LDPE	LDPE
		$\rho=0.918$		$\rho=0.918$	$\rho=0.918$	$\rho=0.918$	$\rho=0.918$	$\rho=0.918$		$\rho=0.918$	$\rho=0.918$
Temperature (°C)		40		23	23	10	25	40		23	23
Liquid phase	CAS #	50% (w/w) Aqueous ethanol	50% (w/w) Aqueous ethanol	40% (w/w) Aqueous ethanol	40% (w/w) Aqueous ethanol	35% (w/w) Aqueous ethanol	35% (w/w) Aqueous ethanol	35% (w/w) Aqueous ethanol	35% (w/w) Aqueous ethanol	20% (w/w) Aqueous ethanol	20% (w/w) Aqueous ethanol
Limonene	5989-27-5	28	32	9.7		130	120	210	160	275	
Diphenylmethane	101-81-5	2.7	3.9	6.6		10	7.7	18	9	560	
Diphenyloxide	101-84-8	4.7	6.5	20		19	21	33	28	180	
Isoamylacetate	123-92-2	0.29	0.42	0.99		0.55	0.69	0.91	0.81	4.1	
gamma-Undelactone (Aldehyde C14)		0.091	0.16	0.72		0.31	0.39	0.44	0.39	10	
Linalylacetate	115-95-7	1.2	1.3	0.86		3.8	5.5	0.092	3.1	–	
Camphor	76-22-2	0.38	0.53	0.44		0.39	0.86	6.4	2.4	5	
Citronellol	106-22-9			0.27					–	3.1	
Eugenol	97-53-0	0.26	0.14	0.26		0.082	0.095	0.093	0.11	1.5	
Dimethylbenzylcarbinol	100-86-7	0.18	0.11	0.075		0.24	0.41	1.1	0.58	0.32	
Menthol	89-78-1	0.13	0.18	0.45		0.32	0.45	0.62	0.51	5.4	
Phenylethylalcohol	60-12-8	0.23	0.12	0.019		0.35	0.25	0.18	0.16	0.098	
cis-3-Hexenol	623-37-0	–	0.022	0.025		0.25	–	0.15	0.088	0.077	
Octane	111-65-9	14									
Nonane	111-84-2	23									
Decane	124-18-5	43			4200						20000
Dodecane	112-40-3	110			27000						80000
Tetradecane	629-59-0	300			120000						280000
Hexadecane	544-76-3	600			780000						1100000
Octadecane	593-45-3	1300			330000						
Eicosane	112-95-8	3000			830000						

Table AII.2 (*Continued*)

Reference		[9]	[31]	[9]	[31]	[9]	[31]	[31]	[31]	[6]	[6]	[6]
Polymer phase		LDPE	LDPE	LDPE	LDPE	LDPE	LDPE	LDPE	LDPE	LDPE	Ionomer (Surlyn 1652)	Ionomer (Surlyn 1702)
		$\rho=0.918$	$\rho=0.93$	$\rho=0.918$	$\rho=0.93$	$\rho=0.918$	$\rho=0.93$	$\rho=0.93$	$\rho=0.93$			
Temperature (°C)		23	20	23	20	23	20	20	20	23	23	23
Liquid phase substance	CAS #	80% (w/w) Aqueous ethanol	80% Aqueous ethanol (v/v)	60% (w/w/) Ethanol	60% Aqueous ethanol (v/v)	40% (w/w) Ethanol	40% (v/v) Aqueous ethanol	20% Aqueous ethanol (v/v)	10% (v/v) Aqueous ethanol	Water + 10% (w/w) ethanol	Water + 10% (w/w) ethanol	Water + 10% (w/w) ethanol
Phenol	108-95-2	0.0046		0.0077		0.022						
p-Cresol	106-44-5	0.0079		0.0081		0.025						
2,3,5,6-Tetramethylphenol	527-35-5	0.050		0.11		0.64						
2,4-Di-t-butylphenol	96-76-4	0.032		0.105		17						
2,6-Di-t-butylphenol	128-39-2	0.55		2.72		33						
2,6-Di-t-butyl-4-methylphenol	128-37-0	0.92		5.7								
Ethyl butyrate	105-54-4							0.11	0.12	5	4	3
Ethyl hexanoate	123-66-0							0.7	1.28			
Ethyl octanoate	106-32-1						0.04					
Ethyl decanoate	110-38-3						0.1					
n-Hexanal	66-25-1		0.07		0.16		0.19					
n-Octanal	124-13-0						1.02					
n-Decanal	112-31-2				0.45		8.41					
n-Dodecanal	112-54-9		0.03		0.8		1.61	0.07	0.1			
n-Decanol	112-30-1				0.01		0.02					
n-Dodecanol	112-54-9		0.02		0.03		0.07	0.67	1.8			
Benzaldehyde	100-52-7									1.5		
Neral	5392-40-5										4	
Geranial	106-26-3											
Carvone	6485-40-1											6

Appendix II | 563

Reference		[31]	[17]	[6]	[6]	[6]	[17]
Polymer phase		LDPE ρ = 0.0.93	LDPE ρ = 0.916	LDPE	Ionomer (Surlyn 1652)	Ionomer (Surlyn 1702)	LDPE ρ = 0.916
Temperature (°C)		20	23	23	23	23	23
Liquid phase	CAS #	5% (v/v) Aqueous ethanol	Water (2.5% (v/v) EtOH, 0.2% (v/v) Tween 80)	Water + 2.5% (w/w) ethanol	Water + 2.5% (w/w) ethanol	Water + 2.5% (w/w) ethanol	Water (2.5% (v/v) EtOH, 0.2% (v/v) Tween 80)
substance							
Phenol	108-95-2						
p-Cresol	106-44-5						
2,3,5,6-Tetramethylphenol	527-35-5						
2,4-Di-t-butylphenol	96-76-4						
2,6-Di-t-butylphenol	128-39-2						
2,6-Di-t-butyl-4-methylphenol	128-37-0						
Ethyl butyrate	105-54-4	0.16					
Ethyl hexanoate	123-66-0	0.17					
Ethyl octanoate	106-32-1	1.66					
Ethyl decanoate	110-38-3						
n-Hexanal	66-25-1						
n-Octanal	124-13-0						
n-Decanal	112-31-2						
n-Dodecanal	112-54-9						
n-Decanol	112-30-1	0.17					
n-Dodecanol	112-54-9	2.67					
Benzaldehyde	100-52-7						
Limonene	5989-27-5		27	4800	4500	2700	79
Citral	5392-40-5			15	23	14	
Neral	106-26-3			21	38	25	
Geranial	141-27-5						
Carvone	6485-40-1						

Table AII.3 Partition coefficient data ($K_{P/L}$) for organic substances between LDPE and water.

Reference		[1]	[6]	[6]	[6]	[17]	[6]	[6]	[6]	[21]	[21]	[31]
Polymer phase		LDPE ρ = 0.918	LDPE	Lonomer (Surlyn 1652)	Lonomer (Surlyn 1702)	LDPE ρ = 0.916	LDPE	Lonomer (Surlyn 1652)	Lonomer (Surlyn 1702)	LDPE ρ = 0.922	LLDPE ρ = 0.922	LLDPE ρ = 0.93
Temperature (°C)		23	23	23	23	23	23	23	23	25	25	20
Liquid phase	CAS #	Water	Water + 10% (w/w) ethanol	Water + 10% (w/w) ethanol	Water + 10% (w/w) ethanol	Water (2.5% (v/v) EtOH, 0.2% (v/v) Tween 80)	Water + 2.5% (w/w) ethanol	Water + 2.5% (w/w) ethanol	Water + 2.5% (w/w) ethanol	Water	Water	Water
substance												
Limonene	5989-27-5	5300	4800	4500	2700							
Diphenylmethane	101-81-5	2850										
Diphenyloxide	101-84-8	1430										
Isoamylacetate	123-92-2	9										
gamma-Undelactone (Aldehyde C14)		35										
Camphor	76-22-2	211										
Citronellol	106-22-9	21										
Eugenol	97-53-0	5										
Dimethylbenzyl-carbinol	100-86-7	1										
Menthol	89-78-1	21										
2-Phenylethylalcohol	60-12-8	0.38										
cis-3-Hexenol	623-37-0	0.33										
Benzaldehyde	100-52-7		1.5	4	6							
Neral	5392-40-5						15	23	14			
Geranial	106-26-3						21	38	25			
Carvone	6485-40-1					27						
Ethyl butyrate	105-54-4		5	4	3					2.3	1.6	2.76
Butyl acetate	123-86-4									2.5	1.7	
Isopentyl acetate	123-92-2									5.3	3.6	

Ethyl-2-methyl butyrate	7452-79-1	6.4	5.2
Butyl propanoate	590-01-2	7.0	4.2
Hexyl acetate	142-92-7	24.8	19.0
Hexanal	66-25-1	3.7	1.5
trans-2-Hexenal	6728-26-3	11.6	11.8
Isopentanol	123-51-3	0.6	0.4
Hexanol	104-76-7	0.4	0.3
Ethyl hexanoate	123-66-0		3.61
Ethyl octanoate	106-32-1		0.15
n-Heptanol	111-70-6	1.6	1.76
n-Octanol	118-87-5		0.1
n-Decanol	112-30-1		0.14
n-Dodecanol	112-54-9		1.58

Table AII.4 Partition coefficient data ($K_{P/L}$) for organic substances between LDPE and other solvents.

Reference		[8]	[2]	[4]	[10]	[1]	[10]	[4]	[4]	[2]	[2]	[2]
Polymer phase		LDPE $\rho = 0.918$	LDPE $\rho = 0.918$	LDPE $\rho = 0.918$	LDPE – 3.5% vinyl acetate copolymer	LDPE $\rho = 0.918$	LDPE $\rho = 0.918$	LDPE $\rho = 0.918$	LDPE $\rho = 0.918$	LDPE $\rho = 0.918$	LDPE $\rho = 0.918$	LDPE $\rho = 0.918$
Temperature (°C)		23	23	23	23	23	23	23	23	23	23	23
Liquid phase substance	CAS #	100% Methanol	100% Methanol	100% Methanol	100% Methanol	Octanol	Propanol	Acetone	n-Hexane	Acetone	Benzene	n-Hexane
Limonene	5989-27-5	0.67	0.67		0.52	0.15	0.25		0.42			
Diphenylmethane	101-81-5	0.27	0.27			0.15			0.25			
Diphenyloxide	101-84-8	0.23	0.23		0.14	0.13	0.11					
Isoamylacetate	123-92-2	0.043	0.046		0.039	0.05	0.037					
gamma-Undelactone (Aldehyde C14)		0.052	0.052			0.022						
Linalylacetate	115-95-7	0.058	0.058		0.052		0.038					
Camphor	76-22-2	0.062	0.062			0.062						
Citronellol	106-22-9	0.046			0.002	0.0093	0.0030					
Eugenol	97-53-0	0.012	0.012			0.021						
Dimethylbenzylcarbinol	100-86-7	0.012	0.012			0.014						
Menthol	89-78-1	0.020	0.020		0.006	0.013	0.0060					
2-Phenylethylalcohol	60-12-8	0.0029	0.0097			0.013						
cis-3-Hexenol	623-37-0	0.0024	0.0024		0.002	0.0096	0.0030					
Heptanol	111-70-6		0.0044							0.025	0.068	0.16
Decane	124-18-5			2.8				0.58	0.24			
Dodecane	112-40-3			3.8				0.69	0.22			
Tetradecane	629-59-			5.9				1.00	0.24			
Hexadecane	544-76-3			11				1.8	0.30			
Octadecane	593-45-3			23				3.2	0.32			
Eicosane	112-95-8			–				6.2	0.41			
Docosane	629-97-0								0.39			
Tetracosane	646-31-1								0.48			
Hexacosane	630-01-3								0.66			
Octacosane	630-02-4								0.97			
Triacontane	638-68-6								1.3			

Reference		[8]			[8]
Polymer phase		LDPE			LDPE
		ρ = 0.918			ρ = 0.918
Temperature (°C)		23			23
Liquid phase	CAS #	100%	Liquid phase	CAS #	100%
substance		Methanol	substance		Methanol
Phenyl methyl ketone	98-86-2	0.030	Benzyl acetate	140-11-4	0.040	2,4-Dimethyl-3-cyclohexene-1-carboxaldehyde	68039-49-6	0.026
Agrunitrile	51566-62-2	0.070	Benzyl benzoate	120-51-4	0.094	3,7-Dimethyl octanol	106-21-8	
2-tert-Butylcyclohexyl acetate	88-41-5	0.092	β-Naphthylethyl ether		0.25	Dimethylphenylethyl carbinol	103-05-9	0.013
n-Octanal	124-13-0	0.020	Borneol	464-45-9	0.014	Dimethylbenzylcarbinyl acetate	151-05-3	0.049
n-Nonanal	124-19-6	0.026	Cedryl acetate	77-54-3	0.29	Dimethyl phthalate	84-66-2	0.012
n-Decanal	112-31-2	0.034	Citral (Neral, Geranial) cis	5392-40-5	0.016	Ethyl benzoate	93-89-0	0.07
n-Undecanal	112-44-7	0.043	Citral (Neral, Geranial) trans	5392-40-5	0.011	Eugenolmethyl ether	93-15-2	0.043
n-Undecene-2-al		0.043	Citralva (Geranyl nitrile) cis	5146-66-7	0.050	Eucalyptol (1,8-cineol)	470-82-6	0.10
cis-Undecene-8-al		0.024	Citralva (Geranyl nitrile) trans	5146-66-7	0.040	Galaxoid 50 (1,3,4,6,7,8-hexa-hydro-4,6,6,7,8,8-hexamethyl-cyclopenta-2-benzopyrane		0.22
n-Dodecanal	112-54-9	0.060	Citronellal	106-23-0	0.019	Geraniol	106-24-1	0.0066
2-Methylundecanal	110-41-8	0.059	Citronellyl acetate	105-87-3	0.071	Geranyl acetate	105-87-3	0.068
3-Methyl-3-phenyl glycidic acid ethyl ester	77-83-8	0.027	Citronellyl formate	105-85-1	0.086	Grisalva	68611-23-4	0.44
Allyl-3-cyclohexyl propionate	2705-87-5	0.14	Coumarin (Benzpyrone)	91-64-5	0.024	Heliotropin	120-57-0	0.019
Isoamylsalicylate	87-20-7	0.17	Cyclamenaldehyde	103-95-7	0.005	cis-3-Hexenyl acetate	3681-71-8	0.041
2-Pentyl-3-phenyl-2-propanal	122-40-7	0.064	p-Cymene	99-87-6	0.34	Alpha hexyl cinnamic aldehyde	101-86-0	0.10
Anethole	104-46-1	0.24	Diethyl phthalate	84-66-2	0.019	Hydroxycitronellal	107-75-5	0.0009
Diethylene gylcol monoethyl ether	111-90-0	0.0023	Diethyl malonate	105-53-3	0.011	Indol	120-72-9	0.0093
Benzophenone	119-61-9	0.046	Dihydro myrcenol	18479-58-8	0.0056	Isopropyl myristate	110-27-0	0.33
						Isobornyl acetate	125-12-2	0.086
						Isoeugenol	97-54-1	0.019

Table AII.4 (Continued)

Reference	[8]
Polymer phase	LDPE
	ρ = 0.918
Temperature (°C)	
Liquid phase substance	CAS #	100% Methanol

Liquid phase substance	CAS #	100% Methanol
Jasmelia	38285-49-3	0.047
Ionone	79-77-6	0.039
Cresyl methyl ether	104-93-8	0.15
Lilial	80-54-6	0.014
Linalool	78-70-6	0.0086
Lyral	31906-04-4	0.0018
Menthone	14073-97-3	0.082
Methyl quinoline	91-63-4	0.027
3-Octene-2-one		0.019
Methyl isoeugenol	93-16-3	0.055
A-iso-Methyl ionone	127-51-5	0.071
alpha-Methyl ionone	7779-30-8	0.060
gamma-Methyl ionone	127-51-5	0.066
Musk Ambrette	83-66-9	0.038
Musk Ketone	81-14-1	0.018
Musk Tibetene	145-39-1	0.050
Musk Xylol	81-15-2	0.045
Myrcene	123-35-3	0.36
Nerol	106-25-2	0.0047
Oryclene extra (p-tert-butylcyclohexyl acetate)	32210-23-4	0.11
Phenyl acetic acid	103-82-2	0.0010
2-Phenyl ethyl acetate	103-45-7	0.036
2-Phenylethylphenyl acetate	102-20-5	0.033
3-Phenylpropyl alcohol	122-97-4	0.0031
α-Pinene	80-56-8	0.64
β-Pinene	127-91-3	0.63
Rose acetol (trichloromethyl phenyl carbinyl acetate)	90-17-5	0.036
Rose oxide CO	3033-23-6	0.080
Rose oxide L-(cis-2[2-methyl-1-propenyl]-4-methyltetrahydro pyrane)	16409-43-1	0.12
1-Phenylethyl acetate	93-92-5	0.040
γ-Terpinene	99-85-4	0.62
Terpineol	98-55-5	0.012
Terpinyl acetate	80-26-2	0.088
Tetrahydro linalool (3,7-dimethyloctane-3-ol)	78-69-3	0.011
Tonalid	1506-02-1	0.10
Verdylacetate	5413-60-5	0.089
Yara Yara	93-04-9	0.29

Table AII.5 Partition coefficient data ($K_{P/L}$) for organic substances between HDPE and various liquids.

Reference Polymer phase Temperature (°C) Liquid phase substance	CAS #	[1] HDPE ρ=0.956 23 100% Ethanol	[3] HDPE ρ=0.956 10 100% Ethanol	[3] HDPE ρ=0.956 25 100% Ethanol	[3] HDPE ρ=0.956 40 100% Ethanol	[3] Average HDPE/LDPE 10–40 100% Ethanol	[4] HDPE ρ=0.956 23 100% Ethanol	[3] HDPE ρ=0.956 10 75% (w/w) Aqueous ethanol	[3] HDPE ρ=0.956 25 75% (w/w) Aqueous ethanol	[3] HDPE ρ=0.956 40 75% (w/w) Aqueous ethanol
Limonene	5989-27-5	0.36	0.53	0.23	0.22	0.28		3.9	2.1	1.6
Diphenylmethane	101-81-5	0.24	0.22	0.078	0.077	0.1		0.53	0.5	0.42
Diphenyloxide	101-84-8	0.16	0.28	0.14	0.14	0.15		0.92	0.7	0.58
Isoamylacetate	123-92-2		0.072	0.03	0.033	0.046		0.096	0.096	0.071
gamma-Undelactone (Aldehyde C14)		0.035	0.053	0.018	0.031	0.027		0.032	0.039	0.049
Linalylacetate	115-95-7	0.067	0.1	0.046	0.043	0.055		0.21	0.24	0.18
Camphor	76-22-2	0.057	0.047	0.035	0.036	0.4		0.045	0.079	0.079
Citronellol	106-22-9	0.016			0.002					
Eugenol	97-53-0	0.024		0.016	0.026	0.018		0.0078	0.0046	0.063
Dimethylbenzylcarbinol	100-86-7	0.017	0.017	0.011	0.013	0.011		0.02	0.045	0.034
Menthol	89-78-1	0.019	0.021	0.051	0.028	0.01		0.029	0.16	0.1
2-Phenylethylalcohol	60-12-8	0.011		0.021	0.022	0.017		0.011	0.029	0.019
cis-3-Hexenol	623-37-0	0.014		0.027	0.033	0.014		0.0076	0.0045	
Carvone	6485-40-1									
Octane	111-65-9		2.1	1.3	1.2	1.5		3	2.6	1.1
Nonane	111-84-2		3.6	2.1	1.8	2.5		4.2	4	2.2
Decane	124-18-5		5.8	2.7	2.4	3.6		6	14	10
Dodecane	112-40-3		11	3.3	2.5	5.5	0.94	13	9.8	6
Tetradecane	629-59-		22	6.2	4.1	11	1.3	32	22	12
Hexadecane	544-76-3		45	11	6.9	21	1.6	81	54	25
Octadecane	593-45-3		83	24	24	43	2.6	240	150	57
Eicosane	112-95-8		145	59	40	82	5.9	800	530	140
Docosane	629-97-0		215	65	41	120	13	2200	2300	510

Table AII.5 (Continued)

Reference		[3]	[3]	[3]	[3]	[3]	[3]	[17]	[2]	[4]
Polymer phase		HDPE	HDPE	HDPE	HDPE	HDPE	HDPE	HDPE	HDPE	HDPE
		$\rho = 0.956$	$\rho = 0.956$	$\rho = 0.956$	$\rho = 0.956$	$\rho = 0.956$	$\rho = 0.956$	$\rho = 0.941$	$\rho = 0.956$	$\rho = 0.956$
Temperature (°C)		10	25	40	10	25	40	23	23	23
Liquid phase substance	CAS #	50% (w/w) Aqueous ethanol	50% (w/w) Aqueous ethanol	50% (w/w) Aqueous ethanol	35% (w/w) Aqueous ethanol	35% (w/w) Aqueous ethanol	35% (w/w) Aqueous ethanol	Water (2.5% (v/v) EtOH, 0.2% (v/v) Tween 80)	Methanol	Methanol
Limonene	5989-27-5	52	33	31	230	150	140	35	0.74	
Diphenylmethane	101-81-5	7.3	5.3	3	24	15	13		0.27	
Diphenyloxide	101-84-8	13	9.1	3.1	45	28	23		0.23	
Isoamylacetate	123-92-2	0.68	0.55	0.36	1.2	0.89	0.68		0.043	
gamma-Undelactone (Aldehyde C14)		0.36	0.25	0.077	0.58	0.36	0.32		0.043	
Linalylacetate	115-95-7	2.6	1.7	0.45	7.4	2.7	0.088		0.059	
Camphor	76-22-2	0.45	0.56	1	0.87	1.1	4.5		0.070	
Citronellol	106-22-9								0.018	
Eugenol	97-53-0	0.24	0.2	0.059	0.16	0.11	0.094		0.020	
Dimethylbenzylcarbinol	100-86-7	0.061	0.052	0.22	0.35	0.55	0.84		0.012	
Menthol	89-78-1	0.23	0.21	0.26	0.59	0.47	0.6		0.014	
2-Phenylethylalcohol	60-12-8	0.17	0.1	0.087	0.045	0.032	0.04		0.0070	
cis-3-Hexenol	623-37-0			0.094	0.049	0.048	0.038		0.0030	
Carvone	6485-40-1							16		
Octane	111-65-9	43	82	67						
Nonane	111-84-2	71	230	94						
Decane	124-18-5	130	414	150						
Dodecane	112-40-3	390	1500	400						2.5
Tetradecane	629-59-	1200	4400	1200						3.5
Hexadecane	544-76-3	3000	14000	2800						5.8
Octadecane	593-45-3	10000	35000	5200						11
Eicosane	112-95-8	9400	400000	15000						27
Docosane	629-97-0									

Reference		[4]	[4]	[9]	[9]	[9]	[9]
Polymer phase		HDPE $\rho = 0.956$	HDPE $\rho = 0.956$	HDPE $\rho = 0.956$	HDPE $\rho = 0.956$	HDPE $\rho = 0.956$	HDPE $\rho = 0.956$
Temperature (°C)		23	23	23	23	23	23
Liquid phase substance	CAS #	n-Hexane	Acetone	100% Ethanol	80% (w/w) Aqueous ethanol	60% (w/w) Aqueous ethanol	40% (w/w) Aqueous ethanol
Dodecane	112-40-3	0.15	0.52				
Tetradecane	629-59-	0.15	0.67				
Hexadecane	544-76-3	0.19	1.1				
Octadecane	593-45-3	0.29	2.1				
Eicosane	112-95-8	0.42	4.3				
Docosane	629-97-0	0.70	9.4				
Tetracosane	646-31-1	0.54					
Hexacosane	630-01-3	0.63					
Octacosane	630-02-4	0.99					
Triacontane	638-68-6	1.4					
Dotriacontane	544-85-4	2.1					
Phenol	108-95-2			0.0053	0.0060	0.012	
p-Cresol	106-44-5			0.0078	0.0080		0.14
2,3,5,6-Tetramethylphenol	527-35-5			0.017	0.050	0.15	0.76
2,4-Di-t-butylphenol	96-76-4			0.012	0.029	0.090	0.20
2,6-Di-t-butylphenol	128-39-2			0.083	0.65	2.9	45.5
2,6-Di-t-butyl-4-methylphenol	128-37-0			0.084	1.11	5.4	

Table AII.6 Partition coefficient data ($K_{P/L}$) for organic substances between PP and various liquids.

Reference		[1]	[2]	[2]	[10]	[5]	[5]	[5]	[17]
Polymer phase		homo-PP $\rho = 0.902$	homo-PP $\rho = 0.902$	co-PP $\rho = 0.900$	OPP $\rho = 0.91$	PP $\rho = 0.9213$ (100% isotactic) weight fraction crystallinity = .734	PP $\rho = 0.9139$ (90% isotactic) weight fraction crystallinity = .655	PP $\rho = 0.883$ (80% isotactic) weight fraction crystallinity = .323	Isotactic homopolymer PP $\rho = 0.902$
Temperature (°C)		23	23	23	23	23	23	23	23
Liquid phase substance	CAS #	Ethanol	Methanol	Methanol	Methanol	Water with Tween 80 emulsifier	Water with Tween 80 emulsifier	Water with Tween 80 emulsifier	Water (2.5% (v/v) EtOH, 0.2% (v/v) Tween 80)
Limonene	5989-27-5	0.45	1	1.1	0.69	83	109	145	98
Diphenylmethane	101-81-5	0.20	0.35	0.34					
Diphenyloxide	101-84-8	0.22	0.27	0.29	0.20				
Isoamylacetate	123-92-2	0.10	0.11	0.12	0.084				
gamma-Undelactone (Aldehyde C14)		0.03	0.034	0.034					
Linalylacetate	115-95-7	0.066	0.055	0.060	0.10				
Camphor	76-22-2	0.042	0.12	0.13					
Citronellol	106-22-9	0.013	0.07	0.018	0.013				
Eugenol	97-53-0	0.020	0.05	0.016					
Dimethylbenzylcarbinol	100-86-7	0.013	0.0088	0.0093	0.017				
Menthol	89-78-1	0.018	0.018	0.019					
2-Phenylethylalcohol	60-12-8	0.00087	0.0074	0.0090	0.0090				
cis-3-Hexenol	623-37-0	0.00062	0.0037	0.0050		5	8	11	17
1-Carvone	6485-40-1								

co-PP = ethylene propylene copolymer

substance	CAS #	[19] co-PP 23 Aqueous vegetable-based beverage	[21] PP ρ = 0.904 25 Citrate buffer 0.976 g/L and 0.103 g/L sodium citrate in water	[23] Ethylene–propylene copolymer PP 18 Water	[9] homo-PP ρ = 0.902 23 100% Ethanol	[9] homo-PP ρ = 0.902 23 80% (w/w) Aqueous ethanol	[9] homo-PP ρ = 0.902 23 60% (w/w) Aqueous ethanol	[9] homo-PP ρ = 0.902 23 40% (w/w) Aqueous ethanol
Limonene	5989-27-5	270–310		40800				
Ethyl butyrate	105-54-4		4.4					
Butyl acetate	123-86-4		5.5					
Isopentyl acetate	123-92-2		9.3					
Ethyl-2-methyl butyrate	7452-79-1		11.3					
Butyl propanoate	590-01-2		18.8					
Hexyl acetate	142-92-7		45.1					
Hexanal	66-25-1		6.4					
trans-2-Hexenal	6728-26-3		26.2					
Isopentanol	123-51-3		1.1					
Hexanol	104-76-7		0.6					
2-Methylbutyrate				31				
Octanal	124-13-0			87				
Decanal	112-31-2			1490				
Linalool (nonan-2-one)	78-70-6			11				
Neral	5392-40-5			42				
Geranial	106-26-3 141-27-5			46				
Phenol	108-95-2				0.0061	0.0038		
p-Cresol	106-44-5				0.0061	0.0043	0.0090	
2,3,5,6-Tetramethylphenol	527-35-5				0.017	0.041	0.12	0.67
2,4-Di-t-butylphenol	96-76-4				0.0056	0.015	0.097	0.12
2,6-Di-t-butylphenol	128-39-2				0.062	0.47	2.9	35
2,6-Di-t-butyl-4-methylphenol	128-37-0				0.10	0.68	6.1	

Table AII.6 (Continued)

Reference Polymer phase Temperature (°C) Liquid phase substance	CAS #	[4] co-PP ρ = 0.900 23 Ethanol	[4] homo-PP ρ = 0.902 23 Ethanol	[4] co-PP ρ = 0.900 23 Methanol	[4] homo-PP ρ = 0.902 23 Methanol	[4] co-PP ρ = 0.900 23 Acetone	[4] homo-PP ρ = 0.902 23 Acetone	[4] co-PP ρ = 0.900 23 n-Hexane	[4] homo-PP ρ = 0.902 23 n-Hexane
Dodecane	112-40-3	1.5	1.3	4.4	3.5	0.79	0.72	0.30	0.21
Tetradecane	629-59-	1.7	1.5	5.6	4.7	0.87	0.77	0.28	0.19
Hexadecane	544-76-3	1.7	1.5	7.3	5.9	1.1	0.97	0.29	0.18
Octadecane	593-45-3	1.9	1.7	10	8.0	1.5	1.3	0.30	0.18
Eicosane	112-95-8	2.2	1.9	14	11	1.8	1.5	0.30	0.17
Docosane	629-97-0	2.7	2.3			2.3	1.8	0.32	0.18
Tetracosane	646-31-1							0.28	0.16
Hexacosane	630-01-3							0.28	0.16
Octacosane	630-02-4							0.28	0.14
Triacontane	638-68-6							0.29	0.13
Dotriacontane	544-85-4							0.36	0.15

Table AII.7 Partition coefficient data ($K_{P/L}$) for organic substances between PVC and various liquids.

Reference Polymer phase Temperature (°C) Liquid phase substance	CAS #	[1] PVC $\rho = 1.36$ 23 100% Methanol	[1] PVC $\rho = 1.36$ 23 100% Ethanol	[1] PVC $\rho = 1.36$ 23 40% (w/w) Aqueous ethanol	[1] PVC $\rho = 1.36$ 23 20% (w/w) Aqueous ethanol
Limonene	5989-27-5		0.60	30	110
Diphenylmethane	101-81-5	2.2	2.3	69	
Diphenyloxide	101-84-8	1.8	17	150	
Isoamylacetate	123-92-2	0.51	3.3	5.2	11
gamma-Undelactone (Aldehyde C14)		0.96	11	31	
Linalylacetate	115-95-7	0.17	0.20	0.46	0.93
Camphor	76-22-2	0.21	0.40	15	
Citronellol	106-22-9	0.084	1.2	2.1	9.1
Eugenol	97-53-0	0.40	3.6	4.9	9.8
Dimethylbenzylcarbinol	100-86-7	0.15	0.15	0.23	0.15
Menthol	89-78-1	0.025	0.050	0.43	0.32
Phenylethylalcohol	60-12-8	0.19	0.21	0.37	1.5
cis-3-Hexenol	623-37-0	0.11	0.07	0.16	0.33

Table AII.8 Partition coefficient data ($K_{P/L}$) for organic substances between PET and various liquids.

Reference		[10]	[10]	[10]	[21]
Polymer phase		PET	PET	PET	PET $\rho = 1.4$
Temperature (°C)		23	23	23	25
Liquid phase substance	CAS #	100% Methanol	100% Propanol	50% (w/w) Aqueous ethanol	Water
Limonene	5989-27-5	0.072	0.02	2.1	
Diphenylmethane	101-81-5				
Diphenyloxide	101-84-8	0.15	0.068	3.3	
Isoamylacetate	123-92-2	0.051	0.028	0.29	
gamma-Undelactone (Aldehyde C14)					
Linalylacetate	115-95-7	0.0090	0.0027	0.086	
Camphor	76-22-2				
Citronellol	106-22-9	0.015	0.0024	0.063	
Eugenol	97-53-0				
Dimethylbenzylcarbinol	100-86-7				
Menthol	89-78-1	0.0070	0.0028	0.048	
Phenylethylalcohol	60-12-8				
cis-3-Hexenol	623-37-0	0.038	0.022	0.16	
Ethyl butyrate	105-54-4				0.6
Butyl acetate	123-86-4				0.7
Isopentyl acetate	123-92-2				0.5
Ethyl-2-methyl butyrate	7452-79-1				0.3
Butyl propanoate	590-01-2				1.3
Hexyl acetate	142-92-7				3.8
Hexanal	66-25-1				1.1
trans-2-Hexenal	6728-26-3				0.9
Isopentanol	123-51-3				0.3
Hexanol	104-76-7				0.3

Table AII.9 Partition coefficient data ($K_{P/L}$) for organic substances between PA and various liquids.

Reference Polymer phase Temperature (°C) Liquid phase substance	CAS #	[10] Polyamide 23 100% Methanol[a]	[10] Nylon 6 23 100% Propanol	[21] Nylon 6 $\rho = 1.14$ 25 Water
Limonene	5989-27-5	0.030	0.038	
Diphenylmethane	101-81-5			
Diphenyloxide	101-84-8	0.072	0.096	
Isoamylacetate	123-92-2		0.024	
gamma-Undelactone (Aldehyde C14)				
Linalylacetate	115-95-7	0.051	0.012	
Camphor	76-22-2			
Citronellol	106-22-9	0.046	0.033	
Eugenol	97-53-0			
Dimethylbenzylcarbinol	100-86-7			
Menthol	89-78-1	0.062	0.030	
Phenylethylalcohol	60-12-8			
cis-3-Hexenol	623-37-0	0.025	0.048	
Ethyl butyrate	105-54-4			0.5
Butyl acetate	123-86-4			0.6
Isopentyl acetate	123-92-2			0.6
Ethyl-2-methyl butyrate	7452-79-1			0.5
Butyl propanoate	590-01-2			1.1
Hexyl acetate	142-92-7			2.0
Hexanal	66-25-1			1.6
trans-2-Hexenal	6728-26-3			1.0
Isopentanol	123-51-3			0.3
Hexanol	104-76-7			0.8

[a]Some swelling of polymer by MeOH occuring.

Table AII.10 Partition coefficient data ($K_{P/L}$) for organic substances between PS and various liquids.

Reference		[6]	[16]	[16]	[16]	[16]	[18]	[18]	[18]	[18]	[18]	[18]	[22]
Polymer phase		Crystalline PS	Crystalline PS	Crystalline PS	Crystalline PS	Crystalline PS	GPPS	GPPS	GPPS	HIPS	HIPS	HIPS	PS
Temperature (°C)		40	20	4	40	40	40	49	66	40	49	66	23
Liquid phase substance	CAS #	Water	Water	Water	3% (w/w) Acetic acid in water	8% (w/w) Aqueous ethanol	8% (w/w) Aqueous ethanol	8% (w/w) Aqueous ethanol	8% (w/w) Aqueous ethanol	8% (w/w) Aqueous ethanol	8% (w/w) Aqueous ethanol	8% (w/w) Aqueous ethanol	Water
Styrene	100-42-5	0.00033	0.00016	0.00008	0.00042	0.0006	0.000356	0.000556	0.00117	0.000425	0.000559	0.00124	
Toluene	108-88-3												800
D-Limonene	5989-27-5												5100
Ethyl acetate	141-78-6												1.9

HIPS = high impact polystyrene (rubber modified polystyrene). GPPS = general purpose polystyrene.

Table AII.11 Partition coefficient data ($K_{P/L}$) for pharmaceutical substances between polymers and water.

Reference		[12]	[12]	[12]	[12]	[12]	[12]	[12]
Polymer phase		Octanol	Polydimethyl-siloxane	LDPE	Polyethylene-co-vinyl acetate (40%)	Poly-ε-caprolactone	Poly-ε-caprolactam-co-ε-caprolatone (40/60)	Poly 2-hydoxyethyl methacrylate
Temperature (°C)		37	37	37	37	37	37	37
Liquid phase substance	CAS #	Water	Water	Water	Water	Water	Water	Water
Codeine	76-57-3	13.7	0.0472		2.00	3.26	26.1	11.9
Cortisone	53-06-5	29.5			1.71	6.86	37.9	
Hydrocortisone	50-23-7	40.7						27
Naltrexone	465-65-6	83.3	0.236		18.3	38.4	88.2	25.5
Corticosterone	50-22-6	87.1	0.569		9.79	21.3	75.9	
Amobarbital	57-43-2	117	0.0936		18.7	27.5	97.6	
Meperidine	57-42-1	528	16.8	5.83	95.5	48.5	140	
Androst-4-ene-3,17-dione	63-05-8	562	11.1	2.30	152	89.9	230	62.7
Norethindrone	68-22-4	817						70
17-α-Hyroxyprogesterone	68-96-2	1150						83
3-Ketodesogestrel	54048-10-1	1430			506	226		
Testosterone	58-22-0	2090	49.7	2.00	165	135	490	77.7
Estra-3,17β-diol		2090						177
Progesterone	57-83-0	7410	133	52.3	1620	1250	1650	129
L-Methadone	125-58-6	15300	1920	524	7800	4360	7860	
L-α-Aceylmethadol	509-74-0	20300		208				

Table AII.12 Partition coefficient data ($K_{W/L}$) for organic substances between liquids and water.

Reference Partition phases	CAS #	[1] Octanol/water	[11] Paraffin oil/water	[24] 1:1 Corn oil/water	[28] Sunflower oil emulsion/water	[32] Octanol/water	[32] Miglyol/water	[32] Miglyol/water + 3% β-lactoglobulin
Temperature (°C) Compound		23	28	23	25	25	25	25
Limonene	5989-27-5	35,000						
Diphenylmethane	101-81-5	19,000						
Diphenyloxide	101-84-8	11,000						
Isoamylacetate	123-92-2	175				160	140	110
gamma-Undelactone (Aldehyde C14)		1600						
Camphor	76-22-2	180						
Citronellol	106-22-9	2600						
Eugenol	97-53-0	243						
Dimethylbenzylcarbinol	100-86-7	73						
Menthol	89-78-1	1600						
2-Phenylethylalcohol	60-12-8	29						
cis-3-Hexenol	623-37-0	34			63			
Methanol	67-56-1		0.0114					
Ethanol	64-17-5		0.0332					
Propanol	71-23-8		0.209					
Butanol	71-36-3		2.39		2			
Pentanol	71-41-0		34.5					
Isopropanol	67-63-0		0.124					
Isobutanol	78-83-1		1.44					
Isopentanol	123-51-3		7.64					
2-Butanol	78-92-2		0.267					
Methyl acetate	79-20-9		2.58					
Ethyl acetate	141-78-6		11.8					
Butyl acetate	123-86-4		167					
Amyl acetate	628-63-7		248					

Propyl acetate	109-60-4	77.2	
Acetaldeyde	75-07-0	0.238	
Propanal	123-38-6	2.10	
Butanal	123-72-8	9.03	
Pentanal	110-62-3	65.1	
Hexanal	66-25-1	224	160
Isovaleraldehyde	590-86-3	42.4	
Furfural	98-01-1	3.0	
Acetone	67-64-1	0.345	
2-Butanone	78-93-3	3.21	
2-Pentanone	107-87-9	20.7	
Methyl carnosate			23.4
Carnosol	5957-80-2		15.9
Carnosic acid	3650-09-7		10.2

Table AII.12 (Continued)

Reference Partition phases Temperature (°C) Compound	Cas #	[28] Sunflower oil emulsion/water 25 400	[32] Octanol/ water 25	[32] Miglyol/ water 25	[32] Miglyol/water + 3% β-lactoglobulin 25	[34] Decane/ water 25	[34] Trichloroethylene/ water 25	[34] Benzene/ water 25	[34] Toluene/ water 25
Ethyl hexanoate	123-66-0	790							
Carvone	6485-40-1	250							
Anethole	104-46-1	13000							
Methyl acetate	79-20-9	4							
Octanal	124-13-0	1000							
2-Heptanone	110-43-0	160							
Ethyl butyrate	105-54-4	100							
Benzaldehyde	100-52-7	100	32	44	42				
Furfural	98-01-1	5							
Methyl butyrate	623-42-7	20							
Hexanol	111-27-3	79							
Linalool	78-70-6	400	3470	220	200				
3-Octanone	106-68-3	500							
Menthone	89-80-5	1300							
Ethyl octanoate	106-32-1	13,000							
Safranal	116-26-7	1000							
2-Nonanone	821-55-6		760	770	230				
i-Butanol	78-83-1					0.2	0.8	1.2	0.6
3-Methyl-1-pentanol	589-35-5					1.9	10	7.9	9.4
4-Methyl-1-pentanol	1320-98-5					2.7	13	9.8	11
2-Ethyl-1-butanol	97-95-0					2.4	13	12	10
5-Methyl-2-hexanol	627-59-8					7.2	55	42	35

Table AII.13 Partition coefficient data ($K_{G/L}$) for organic substances between gases and various food and liquid phases.

Reference				[7]	[13]	[13]	[13]	[13]
Gas/liquid phases Temperature (°C) Compound	Food phase		CAS #	Gas/food 23	N2/100% ethanol 25	N2/75% (w/w) aqueous ethanol 25	N2/50% (w/w) aqueous ethanol 25	N2/35% (w/w) aqueous ethanol 25
Ethyl acetate (CAS # 141-78-6)	Cocoa fat	Limonene	5989-27-5	0.0015	6.3E-05	8.9E-05	4.3E-03	2.9E-02
	Process cheese	Diphenylmethane	101-81-5	0.004	1.9E-06	4.1E-06	3.7E-05	2.4E-04
	Butter cookies	Diphenyloxide	101-84-8	0.015	1.8E-06	4.2E-06	4.1E-05	2.9E-04
	Water	Isoamylacetate	123-92-2	0.0053	7.5E-05	1.8E-04	8.5E-04	2.7E-03
Methyl ethyl ketone	Cocoa fat	gamma-Undelactone (Aldehyde C14)		0.0013	–	9.0E-07	1.1E-06	3.2E-06
(CAS # 78-93-3)	Process cheese	Linalylacetate	115-95-7	0.0019	3.1E-06	7.6E-06	5.9E-05	4.3E-04
	Butter cookies	Camphor	76-22-2	0.012	6.8E-06	1.0E-05	4.3E-05	1.7E-04
	Marmelade	Citronellol	106-22-9	0.0077	3.0E-06	4.7E-06	7.0E-06	1.4E-05
	Water	Eugenol	97-53-0	0.0017	1.4E-06	1.6E-06	1.2E-06	4.1E-06
Ethanol	Cocoa fat	Dimethylbenzylcarbinol	100-86-7	0.0077	1.6E-06	1.5E-06	3.8E-06	1.0E-05
(CAS # 64-17-5)	Process cheese	Menthol	89-78-1	0.00059	1.4E-06	1.5E-06	8.4E-06	4.6E-05
	Butter cookies	2-Phenylethylalcohol	60-12-8	0.0091	1.5E-06	1.7E-06	2.0E-06	4.9E-06
	Marmelade	cis-3-Hexenol	623-37-0	0.00091	7.4E-06	1.1E-05	2.9E-05	7.9E-05
	Water			0.00029				
Ethylene glycol	Cocoa fat			0.0023				
(CAS # 107-21-1)	Process cheese			0.00002				
	Butter cookies			0.0022				
	Marmelade			0.00053				
	Water			6.00E-06				

Table AII.13 (Continued)

Reference Gas/liquid phases Temperature (°C) Compound	CAS #	[13] N2/100% aqueous ethanol 25	[13] N2/66% (w/w) aqueous ethanol 25	[13] N2/33% (w/w) aqueous ethanol 25	[14] N2/water 25	[15] N2/water 25
Pentane	109-6-0	0.013	0.11	4.2		
Hexane	110-54-3	0.005	0.057	5.2		
Heptane	142-82-5	0.0019	0.028	2.6		
Octane	111-65-9	6.80E-04	0.011	1.8		
Nonane	111-84-2	2.1E-04	0.008	0.83		
Decane	124-18-5	5.9E-05	0.0024	0.67		
Dodecane	112-40-3	8.5E-06	5.0E-04	0.13		
Tetradecane	629-59-	2.0E-06	1.3E-04	0.06		
Hexadecane	544-76-3	6.6E-07	1.0E-04	0.02		
Polychlorinated dibenzodioxin 2,7-DiCDD	33857-26-0				2.40E-03	
Polychlorinated dibenzodioxin 2,7-DiCDD	30746-58-8				1.47E-03	
Polychlorinated dibenzodioxin 1,2,3,4-TCDD					8.16E-04	
2,2'-Dichloro biphenyl	13029-08-8					0.01
2,3-Dichloro biphenyl	16605-91-7					0.016
4,4'-Dichloro biphenyl	2050-68-2					0.014
2,2',3-Trichloro biphenyl	38444-77-8					0.035
2,3,4'-Trichloro biphenyl	38444-85-8					0.018

Appendix II | 585

Reference Gas/liquid phases	CAS #	[26] Air/water	[26] Air/5 g/L aqueous sodium caseinate	[26] Air/triolein	[27] Air/water	[27] Air/model sugar and starch fruit preparation	[33] Air/water	[33] Air/water	[33] Air/water	[33] Air/water	[32] Air/water	[32] Air/miglyol	[15] N2/water
Temperature (°C) Compound		25	25	25	30	30	25	60	70	80	25	25	25
Ethyl butyrate	105-54-4	0.0135	0.0125	0.00022	0.0183	0.031							
Ethyl hexanoate	123-66-0	0.034	0.022	0.00003	0.0302								
Ethyl acetate	141-78-6				0.0162								
Diacetyl	431-03-8				0.00126	0.00099							
Limonene	5989-27-5				0.251	0.215							
cis-3-Hexenol	623-37-0				0.00054								
Ethyl octanoate	106-32-1				0.0344								
Linalool	78-70-6				0.00387	0.00322					2.3	0.00022	
Benzaldehyde	100-52-7										1.7	0.0015	
Isoamylacetate	123-92-2										36	0.0052	
2-Nonanone	821-55-6										34	0.00064	
Hexanal	66-25-1				0.0116		0.011						
2-Octanone	111-13-7						0.0062	0.083	0.13	0.19			
Ethyl butanoate	105-54-4						0.018	0.075	0.12	0.19			
2-Heptanone	110-43-0						0.0057	0.08	0.12	0.17			
2-Hexanone	591-78-6							0.039	0.071	0.1			
t-2-Hexenal	6728-26-3							0.029	0.044	0.06			
2-Butanone	78-93-3							0.021	0.028	0.033			
1-Hexanol	4798-44-1							0.01	0.016	0.024			0.0007
Heptanol	111-70-6							0.0085	0.013	0.02			0.0009
Pentane-2-one	107-87-9												0.0032
Octane-2, 4-dione	140090-87-0												0.0088
Pentane-2, 4-dione	123-54-6												0.002
Oct-trans-2-enal													0.004
Methyl pentanoate	624-24-8												0.0125
Pentyl acetate	628-63-7												0.015
tert-Butyl propionate	20487-40-5												0.054

Table AII.14 Partition coefficient data ($K_{P/L}$) for organic substances between polymers and various foods.

Reference Polymer phase Temperature (°C) Liquid phase substance	CAS #	[1] LDPE 23 Skim milk (1.5% fat)	[1] HDPE 23 Skim milk (1.5% fat)	[1] LDPE 23 Whole milk (3% fat)	[1] HDPE 23 Whole milk (3% fat)	[1] LDPE 23 Wine (5.3% ethanol)	[1] HDPE 23 Wine (5.3% ethanol)	[1] LDPE 23 Wine (8% ethanol)	[1] HDPE 23 Wine (8% ethanol)
Limonene	5989-27-5	10	12	4.8	7.2	1100	–	1000	5000
Diphenylmethane	101-81-5	5.7	6	2.5	3.8	2100	2300	830	1600
Diphenyloxide	101-84-8	7.6	9.6	2.3	2.1	74	125	28	61
Isoamylacetate	123-92-2	2.3	3.1	1.8	1.5	4.5	5.4	4.7	5.3
gamma-Undelactone (Aldehyde C14)		2	3.5	1.3	1.1	34	77	12	190
Linalylacetate	115-95-7	5.7		2.4					
Camphor	76-22-2	6.8		3.5					
Citronellol	106-22-9	2.3	2	1.8	5.9	5.9	13	5.1	12
Eugenol	97-53-0	1.3	1.5	1.1	0.96	2.4	2.8	18	4.7
Dimethylbenzylcarbinol	100-86-7	3.1		1.2	1.2	0.34	0.67	0.44	0.65
Menthol	89-78-1					4.6	18	7.1	18
2-Phenylethylalcohol	60-12-8	0.18	0.21	0.11	0.12	0.15	0.3	0.2	0.22
cis-3-Hexenol	623-37-0	0.18	0.2	0.16	0.12	0.08	0.2	0.09	0.2

Reference		[9]	[9]	[9]	[9]	[9]	[9]	[9]	[9]	[9]	[9]	[9]	[9]	[9]
Polymer phase		LDPE	LDPE	LDPE	LDPE	LDPE	LDPE	LDPE	LDPE	LDPE	LDPE	HDPE	LDPE	HDPE
Temperature (°C)		23	23	23	23	23	23	23	23	23	23	23	23	23
Liquid phase substance	CAS #	Banana juice	Pineapple juice	Pear juice	Grapefruit juice	Orange juice	Orange drink 1	Orange drink 2	Fresh squeezed orange juice	Water	Skim milk (1.5% fat)	Skim milk (1.5% fat)	Wine (5.3% ethanol)	Wine (5.3% ethanol)
Limonene	5989-27-5	450	400	660						1170				
Diphenylmethane	101-81-5	310	310	320	230	180	300	190	160	1200				
Diphenyloxide	101-84-8							64	120	12				
Isoamylacetate	123-92-2							5.1	5.4	4				
gamma-Undelactone (Aldehyde C14)								11	13	21				
Linalylacetate	115-95-7	37	27	45		35	19	38	30	34				
Camphor	76-22-2	9	17	14		12	10	12	15	15				
Citronellol	106-22-9							3.5	6.4	5.2				
Eugenol	97-53-0							3.6	4.2	4.6				
Dimethylbenzylcarbinol	100-86-7							1.0	0.55	0.46				
Menthol	89-78-1							23	6.4	4.6				
2-Phenylethylalcohol	60-12-8							0.08	0.18	0.14				
cis-3-Hexenol	623-37-0							0.086	0.13	0.11				
Phenol	108-95-2										9.0	18		12
p-Cresol	106-44-5										5.0	7.0		5.1
2,4-Di-t-butylphenol	96-76-4	46	35	41	24	33.5				340			16	130
2,6-Di-t-butylphenol	128-39-2	450	440	480	220	255					12	200	1500	930
2,6-Di-t-butyl-4-methylphenol	128-37-0	580	580	620	240	285				6900	10	200	7000	2000

Table AII.15 Partition coefficient data ($K_{Paper/G}$) for organic substances between paper and air.

Reference Paper/air		[29] Paper/air	[29] Paper/air	[29] Paper/air	[29] Paper/air	[29] Paper/air	[29] Paper/air	[29] Paper/air
Paper phase		Testliner: 100% recycled 128 g/m2 191 μm	Testliner: 100% recycled 128 g/m2 191 μm	Testliner: 100% recycled 128 g/m2 191 μm	Fluting neutral sulfite semichemical pulp + 30% recycled 107 g/m2 209 μm	Fluting neutral sulfite semichemical pulp + 30% recycled 107 g/m2 209 μm	Kitchen towel: 100% recycled 46.7 g/m2 188 μm	Kitchen towel: 100% recycled 46.7 g/m2 188 μm
Temperature (°C)	Compound	100	70	40	100	70	40	100	70	40
	CAS #									
o-Xylene	95-47-6	24	106	213	21	166	418	16	30	23
Dodecane	112-40-3	5	3243		7	22		4	43	
Napthalene	91-20-3	190	359		93	235		48	62	
Diphenyl oxide	101-84-8	54	78		189	197		28	80	
2,6-Diisopropylnapthalene	24157-81-1	20			132	80		46	18	

Reference Paper/air		[30] Paper/air	[30] Paper/air	[30] Paper/air	[30] Paper/air
Paper phase		Testliner: 100% recycled 128 g/m2 191 μm	Testliner: 100% recycled 128 g/m2 191 μm	Liquid board triplex: bleached kraft + chemithermomechanical pulp 267 g/m2 478 μm	Liquid board triplex: bleached kraft + chemithermomechanical pulp 267 g/m2 478 μm
Temperature (°C)	Compound	100	70	100	70
	CAS #				
Acetophenone	98-86-2	568	778	138	169
Napthalene	91-20-3	201	213	72	56
Benzophenone	119-61-9	396	360	230	271
Dibutyl phthalate	84-74-2	1050	912	561	597
Methyl stearate	112-61-8	730	984	520	935

References

1 Kozinowski, J. and Piringer, O. (1989) The influence of partition processes between packaging and foodstuffs or cosmetics on the quality of packed products. *Verpackungs-Rundschau*, **40** (5), 39–44.

2 Becker, K., Koszinowski, J. and Piringer, O. (1983) Permeation von Reich- und Aromastoffen durch Polyolefine. *Deutsche Lebensmittel-Rundschau*, **79** (8), 257–266.

3 Baner, A.L. (1992) *Partition coefficients of aroma compounds between polyethylene and aqueous ethanol and their estimation using UNIFAC and GCFEOS*, PhD dissertation, Michigan State University, E. Lansing, MI.

4 Koszinowski, J. (1986) Diffusion and solubility of n-alkanes in polyolefines. *J. Appl. Polym. Sci.*, **31**, 1805–1826.

5 Letinski, J. and Halek, G.W. (1992) Interactions of citrus flavor compounds with polypropylene films of varying crystallinities. *J. Food Sci.*, **57** (2), 481–484.

6 Kwapong, O.Y. and Hotchkiss, J.H. (1987) Comparative sorption of aroma compounds by polyethylene and ionomer food-contact plastics. *J. Food Sci.*, **52** (3), 761–7763, 785.

7 Koszinowski, J. (1986) Diffusion and solubility of hydroxy compounds in polyolefines. *J. Appl. Polym. Sci.*, **31**, 2711–2720.

8 Koszinowski, J. and Piringe, O. (1990) Diffusion and relative solubility of flavour- and aroma-compounds in polyolefins, Verpackungs-Rundschau: Technisch-wissenschaftliche Beilage, 41 (3), 15–17.

9 Piringer, O. (1990) unpublished data.

10 Franz, R. (1990) unpublished data. Fraunhofer-Institut IVV, Freising, Germany.

11 Nelson, P.E. and Hoff, J.E. (1968) Food volatiles: gas chromatographic determination of partition coefficients in water-lipid systems. *J. Food Sci.*, **33**, 479–482.

12 Bao, Y.T., Samuel, N.K.P. and Pitt, C.G. (1988) The prediction of drug solubilities in polymers. *J. Polym. Sci.: Part C: Polym. Lett.*, **26**, 41–46.

13 Baner, A.L. and Piringer, O. (1994) Liquid/gas partition coefficients of aroma compounds and n-alkanes between aqueous ethanol mixtures and nitrogen. *J. Chem. Eng. Data*, **39**, 341–348.

14 Santl, H., Brandsch, R. and Gruber, L. (1994) Experimental determination of Henry's law constant (HLC) for some lower chlorinated dibenzodioxins. *Chemosphere*, **29** (9–11), 2209–2214.

15 Piringer, O. and Sköries, H. (1984) Selective enrichment of volatiles by gas-water partition in con- and countercurrent columns, in *Analysis of Volatiles* (ed. P. Schreier), Walter d Gruyter, Berlin, pp. 49–60.

16 Till, D.E., Ehntholt, D.J., Reid, R.C., Schwartz, P.S., Schwope, A.D., Sidman, K.R. and Whelan, R.H. (1982) Migration of styrene monomer from crystal polystyrene to foods and food simulating liquids. *Ind. Eng. Chem. Fundam.*, **21**, 161–168.

17 Halek, G.W. and Luttmann, J.P. (1991) Sorption behavior of citrus-flavor compounds in polyethylenes and polypropylenes, chapter 18, in *Food and Packaging Interactions II* (eds S.J. Risch and J.H. Hotchkiss), American Chemical Society, Washingoton DC.

18 Murphy, P.G., MacDonald, D.A. and Lickly, T.D. (1992) Styrene migration from general-purpose and high impact polystyrene into food-simulating solvents. *Food Chem. Toxic.*, **30** (3), 225–232.

19 Jabarin, S.A. and Kollen, W.J. (1988) Polyolefin properties for rigid food packaging. *Polym. Eng. Sci.*, **28** (18), 1156–1161.

20 Baner, A.L., Kalyankar, V. and Shoun, L.H. (1991) Aroma sorption evaluation of aseptic packaging. *J. Food Sci.*, **56** (4), 1051–1054.

21 Nielsen, T.J., Jägerstad, M.I., Öste, R.E. and Wesslén, B.O. (1992) Comparative

absorption of low molecular aroma compounds into commonly used food packaging polymer films. *J. Food Sci.*, **57** (2), 490–492.

22. Gavara, R., Hernandez, R.J. and Giacin, J. (1996) Methods to determine partition coefficient of organic compounds in water/polystyrene systems. *J. Agric. Food. Chem.*, **44**, 2810–2813.

23. Lebossé, R., Ducruet, V. and Feigenbaum, A. (1997) Interactions between reactive aroma compounds from model citrus juice with polypropylene packaging film. *J. Agric. Food Chem.*, **45**, 2836–2842.

24. Huang, S., Frankel, E.N., Aeschbach, R. and German, B.J. (1997) Partition of selected antioxidants in corn oil-water model systems. *J. Agric. Food Chem.*, **45** (6), 1991–1994.

25. Vitrac, O., Mougharbel, A. and Feigenbaum, A. (2007) Interfacial mass transport properties which control the migration of packaging constituents into foodstuffs. *J. Food Eng.*, **79** (3), 1048–1064.

26. Voilley, A. (2006) Flavour retention and release from the food matrix: an overview. Chapter 6, in *Flavour in Food* (eds A. Voilley and P. Etiévant), Woodhead Pub. Ltd., Cambridge.

27. Savary, G., Guichard, E., Doublier, J.-L. and Cayot, N. (2006) Mixture of aroma compounds: determination of partition coefficients in complex semi-solid matrices. *Food Res. International*, **39**, 372–379.

28. Shojaie, Z.A., Linforth, R.S.T. and Taylor, A.J. (2007) Estimation of the oil water partition coefficient, experimental and theoretical approaches related to volatile behavior in milk. *Fd. Chem.*, **103**, 689–694.

29. Nerín, C. and Asensio, E. (2004) Behaviour of organic pollutants in paper and board samples intended to be in contact with food. *Analytica Chimica Acta.*, **508**, 185–191.

30. Triantafyllou, V.I., Akrida-Demertzi, K. and Demertzis, P.G. (2005) Determination of partition behavior of organic surrogates between paperboard packaging materials and air. *J. Chromatography A.*, **1077**, 74–79.

31. Hwang, Y., Matsui, T., Hanada, T., Shimoda, M., Masumoto, K. and Osajima, Y. (2000) Desorption behavior of sorbed flavor compounds from packaging films with ethanol solution. *J. Agric. Food Chem.*, **48**, 4310–4313.

32. Seuvre, A.M., Espinosa Díaz, M.A. and Voilley, A. (2000) Influence of the food matrix structure on the retention of aroma compounds. *J. Agric. Food Chem.*, **48**, 4296–4300.

33. Jouquand, C., Ducruet, V. and Giampaoli, P. (2004) Partition coefficients of aroma compounds in polysaccharide solutions by the phase ratio variation method. *Food Chem.*, **85**, 467–474.

34. Wang, P., Dwarakanath, V., Rouse, B., Pope, G.A. and Sepehrnoori, K. (1998) Partition coefficients for alcohol tracers between nonaqueuos-phase liquids and water from UNIFAC-solubility method. *Adv. Water Res.*, **21** (2), 171–181.

Appendix III

A Selection of Additives Used in Many Plastic Materials

The Ref Nr-, restrictions and remarks are from the "Synoptic Document" (2005) of the European Commission, SANCO D3. In addition to the relative molecular masses, M_r, the corresponding structure increments, W and G_W (Chapters 4 and 15) are given.

Table AIII.1 Antioxidants and Primary Antioxidants.

CAS Nr Ref Nr	(Trade) Name	Structure	Restrictions and/or specifications	Remarks SCF-List
10191-41-0 93520	Irganox E201 α-Tocopherol	$M_r = 430.72$ $W = -92$ $G_W = -333$	OML: 60 mg/kg	Acceptable 1
128-37-0 46640	Ionol Lowinox Naugard BHT Dalpac 4 Topanol OC 2,6-Di-*tert*-butyl-4-methylphenol 2,6-Di-*tert*-butyl-p-cresol	$M_r = 220.36$ $W = -9$ $G_W = -172$	SML: 3 mg/kg	ADI: 0.05 mg/kg b.w. 1
2082-79-3 68320	Irganox 1076 Anox PP18 Lowinox PO35 Naugard 76 Octaethyl-3-(3,5-di-*tert*-butyl-4-hydroxyphenyl)propionate	$M_r = 530.88$ $W = -110$ $G_W = -364$	SML: 6 mg/kg	TDI: 0.1 mg/kg b.w. 2

119-47-1 66480	Cyanox 2246 Irganox 2246 Lowinox 22M46 Oxi-Chek 114 Vanox 2246 2,2′-Methylene-bis (4-methyl-6-*tert*-butyl-phenol)	$M_r = 340.51$ $W = 12$ $G_w = -191$	SML(T): 1.5 mg/kg (20)	Group TDI: 0.025 (with 66400) 2
35074-77-2 59200	Irganox 259 1,6-Hexamethylene-bis[3-(3,5-di-*tert*-butyl-4-hydroxyphenyl) propionate]	$M_r = 638.94$ $W = -149$ $G_w = -296$	SML: 6 mg/kg	TDI: 0.1 mg/kg b.w. 2
23128-74-7 59120	Irganox 1098 1,6-Hexamethylene-bis[3-(3,5-di-*tert*-butyl-4-hydroxyphenyl) propionamide]	$M_r = 636.97$ $W = -54$ $G_w = -218$	SML: 45 mg/kg	TDI: 0.75 mg/kg b.w. 2

Table AIII.1 (Continued)

CAS Nr / Ref Nr	(Trade) Name	Structure	Restrictions and/or specifications	Remarks SCF-List
36443-68-2 / 94400	Irganox 245 Triethyleneglycol-bis[3-(3-tert-butyl-4-hydroxy-5-methylphenyl)propionate]	$M_r = 586.77$; $W = -109$	SML: 9 mg/kg	TDI: 0.15 mg/kg b.w. 2
1709-70-2 / 95200	Ethanox 330 Irganox 1330 Alvanox 100 1,3,5-Trimethyl-2,4,6-tris(3,5-di-tert-butyl-4-hydroxybenzyl)benzene	$M_r = 775.22$; $W = -208$; $G_W = -216$	OML: 60 mg/kg	t-TDI: 1 mg/kg b.w. 2
1843-03-4 / 95600	Topanol CA 1,1,3-Tris(2-methyl-4-hydroxy-5-tert-butylphenyl)butane	$M_r = 544.82$; $W = -68$; $G_W = -395$	SML: 5 mg/kg	R: 5 mg/kg of food 3

$G_W = -309$

27676-62-6	Goodrite 3114 Irganox 3114 1,3,5-Tris (3,5-di-*tert*-butyl-4-hydroxybenzyl)-1,3,5-triazine-2,4,6-(1H,3H,5H)-trione	$M_r = 784.10$ $W = -103$ $G_W = -334$	SML: 5 mg/kg	R: 5 mg/kg of food 3
95360				
40601-76-1	Cyanox 1790 Irganox 170 1,3,5-Tris (4-*tert*-butyl-3-hydroxy-2,6-dimethylbenzyl)-1,3,5-triazine-2,4,6-(1H,3H,5H)-trione	$M_r = 699.94$ $W = -46$ $G_W = -320$	SML: 6 mg/kg	t-TDI: 0.1 mg/kg b.w. 2
95280				
6683-19-8	Adeskastab AO-60 Anox 20 Irganox 1010 Pentaerythritol tetrakis[3-(3,5-di-*tert*-butyl-4-hydroxyphenyl) propionate]	$M_r = 1177.67$ $W = -360$ $G_W = -547$	OML: 60 mg/kg	TDI: 3 mg/kg b.w. 2
71680				

Table AIII.1 (Continued)

CAS Nr Ref Nr	(Trade) Name	Structure	Restrictions and/ or specifications	Remarks SCF-List
32509-66-3 53670	Hostanox 03 Ethyleneglycol bis[3,3-bis(3-tert-butyl-4-hydroxyphenyl) butyrate]	$M_r = 795.08$ $W = -192$ $G_W = -308$	OML: 60 mg/kg	TDI: 0.1 mg/kg b.w. 2
96-69-5 92800	Irganox 415 Lowinox 44S36 Santonox R 4,4′-Thiobis(6-tert-butyl-3-methyl-phenol)	$M_r = 358.54$ $W = 21.5$ $G_W = -229$	SML: 0.48 mg/kg	t-TDI: 0.008 mg/kg 2
110553-27-0 40020	Irganox 1520 2,4-Bis(octylthiomethyl)-6-methyl-phenol	$M_r = 424.78$ $W = -41$ $G_W = -296$	SML(T): 5 mg/kg (41) with 38940 (2,4)	TDI: 0.1 mg/kg b.w. 2
41484-35-9 92880	Irganox 1035 Thiodiethanol-bis[3-(3,5-di-tert-butyl-4-hydroxyphenyl) propionate]	$M_r = 642.94$ $W = -146$ $G_W = -260$	SML: 2,4 mg/kg	TDI: 0.04 mg/kg b.w. 2

991-84-4 40000	Irganox 565 2,4-Bis(octylmercapto)-6-(4-hydroxy-3,5-di-*tert* butylanilino)-1,3,5-triazine	$M_r = 588.96$ $W = -110$	$G_W = -347$	SML: 30 mg/kg 2	TDI: 0.5 mg/kg b.w.
61167-58-6 31520	Irganox 3052 Sumilizer GM Acrylic acid. 2-*tert*-butyl-6-(3-*tert*-butyl-2-hydroxy-5-methylbenzyl)-4-methylphenyl ester	$M_r = 394.56$ $W = -26$	$G_W = -249$	SML: 6 mg/kg	TDI: 0.1 mg/kg b.w. 2
Secondary antioxidants (Phosphites/Phosphonites)					
26523-78-4 74400	Irgafos TNPP TNPP Phosphorous acid. tris(nonyl- and/or dinonylphenyl)ester	$M_r = 689.02$ $W = -199$	$G_W = -485$	SML: 30 mg/kg 2	TDI: 0.5 mg/kg b.w. 2
31570-04-4 74240	Irgafos 168 Phosphorous acid. tris(2,4-di-*tert*-butylphenyl)ester	$M_r = 646.94$ $W = -210$	$G_W = -430$	OML: 60 mg/kg	TDI: 1 mg/kg b.w. 2

Table AIII.1 (Continued)

CAS Nr Ref Nr	(Trade) Name	Structure	Restrictions and/ or specifications	Remarks SCF-List
26741-53-7 38820	Ultranox 626 Bis(2,4-di-*tert*-butylphenyl)pentaerythritol diphosphite	$M_r = 604.71$ $W = -183$ $G_w = -336$	SML: 0.6 mg/kg	TDI: 0.01 mg/kg b.w. 2
38613-77-3 83595	Irgafos P-EPQ Sandostab P-EPQ Reaction product of di-*tert*-butyl-phosphonite with biphenyl. obtained by condensation of 2,4-di-*tert*-butylphenol with Friedel–Craft reaction product of phosphorus trichloride and biphenyl	$M_r = 1035.44$ $W = -417$ $G_w = -610$	SML: 18 mg/kg	TDI: 0.3 mg/kg b.w. 2
118337-09-0 54300	Ethanox 398 2,2′-Ethylidene-bis (4,6-di-*tert*-butyl-phenyl)fluorophosphonite	$M_r = 486.66$ $W = -124$ $G_w = -360$	SML: 6 mg/kg	TDI: 0.1 mg/kg b.w. 2

Secondary antioxidants (sulfur compounds)

CAS	Name/Trade names	Structure	M_r / G_W / W	SML	Notes
123-28-4 93120	Argus DLTDP Cyanox LTDP Evanastab 12 Irganox PS 800 Thiodipropionic acid. didodecyl ester	$[H_{12}C_{25}-O-C(=O)-CH_2-CH_2-S-]_2$ $M_r = 514.85$ $W = -110$		SML(T): 5 mg/kg (21)	Group R: 5 mg/kg food (with Thiodipropionic acid. dioctadecyl ester 93280) 3
693-36-7 93280	Argus DSTDP Cyanox STDP Evanastab 18 Irganox PS 802 Lowinox DSTDP Thiodipropionic acid. dioctadecyl ester	$[H_{37}C_{18}-O-C(=O)-CH_2-CH_2-S-]_2$ $M_r = 683.18$ $W = -230$	$G_W = -348$	SML(T): 5 mg/kg (21)	Group R: 5 mg/kg food (with Thiodipropionic acid. didodecyl ester 93120) 3

Metal deactivators

CAS	Name/Trade names	Structure	M_r / G_W / W	SML	Notes
32687-78-8 38800	Irganox MD-1024 N,N'-Bis[3-(3,5-di-tert-butyl-4-hydroxyphenyl) propionyl]hydrazide	[t-Bu, HO, t-Bu phenyl –(CH$_2$)$_2$–C(=O)–NH–]$_2$ $M_r = 552.80$ $W = -11$	$G_W = -680$	SML: 15 mg/kg	TDI: 0.25 mg/kg b.w. 2

UV-absorber

CAS	Name/Trade names	Structure	M_r / G_W / W	SML	Notes
2440-22-4 61440	Eversorb 71 Lowilite 55 Mark LA 32 Tinuvin P Uvasorb SV 2-(2'-Hydroxy-5'-methylphenyl)benzotriazole	benzotriazole-N=N linked to HO-phenyl-CH$_3$ $M_r = 225.25$ $W = 90$	$G_W = -101$ $G_W = -138$	SML(T): 30 mg/kg (19)	Group TDI: 0.5 mg/kg b.w. (for 61440. 60400 and 60480) 2

Table AIII.1 (Continued)

CAS Nr Ref Nr	(Trade) Name	Structure	Restrictions and/ or specifications	Remarks SCF-List
3896-11-5 60400	Eversorb 73 Lowilite 26 Mark LA 36 Tinuvin 326 2-(2'-Hydroxy- 3'-tert-butyl-5'- methylphenyl)- 5-chlorobenzotriazole	$M_r = 315.81$ $G_W = -167$ $W = 41$	SML(T): 30 mg/kg (19)	Group TDI: 0.5 mg/kg b.w. (for 60400, 60480 and 61440) 2
3864-99-1 60480	Eversorb 75 Lowilite 27 Mark LA 34 Tinuvin 327 2-(2'-Hydroxy-3'. 5'-di-tert-butylphenyl)- 5-chlorobenzotriazole	$M_r = 357.89$ $G_W = -94$ $W = 12$	SML(T): 30 mg/kg (19)	Group TDI: 0.5 mg/kg b.w. (for 60480, 60400 and 61440) 2
70321-86-7 60320	Eversorb 76 Tinuvin 234 2-[2-Hydroxy-3,5-bis (1,1-dimethylbenzyl) phenyl]benzo-triazol	$M_r = 447.58$ $G_W = -198$ $W = -5$	SML: 1.5 mg/kg	TDI: 0.025 mg/ kg b.w. 2
131-57-7 61360	Cyasorb UV-9 Lowilite 20 Uvasorb MET Uvinul 3040 2-Hydroxy- 4-methoxybenzophenone	$M_r = 228.25$ $G_W = -137$ $W = 54$	SML(T): 6 mg/kg (15)	Group TDI: 0.1 mg/kg b.w. 2

131-56-6 48640	Uvinul 400 2,4-Dihydroxybenzophenone	$M_r = 214.22$ $W = 104$ $G_W = -104$	SML(T): 6 mg/kg (15)	Group TDI: 0.1 mg/kg b.w. 2
131-53-3 48880	Cyasorb UV-24 2,2'-Dihydroxy-4-methoxybenzophenone	$M_r = 244.25$ $W = 71$ $G_W = -131$	SML(T): 6 mg/kg (15)	Group TDI: 0.1 mg/kg b.w. 2
1843-05-6 61600	Chimasorb 81 Cyasorb UV-531 Eversorb 12 Hostavin ARO 8 Lowilite 22 Sumisorb 130 UV-Chek 301 Uvinul 3008 2-Hydroxy-4-n-octyloxybenzophenone	$M_r = 326.44$ $W = -2$ $G_W = -221$	SML(T): 6 mg/kg (15)	Group TDI: 0.1 mg/kg b.w. 2
23949-66-8 53200	Sanduvor VSU Tinuvin 312 2-Ethoxy-2'-ethyloxanilide	$M_r = 312.37$ $W = 83$ $G_W = -127$	SML: 30 mg/kg	TDI: 0.5 mg/kg b.w. 2

Table AIII.1 (Continued)

CAS Nr Ref Nr	(Trade) Name	Structure	Restrictions and/ or specifications	Remarks SCF-List
Hal(l)s				
52829-07-9 85280	Eversorb 90 Lowilite 77 Sanol LS-770 Tinuvin 770 Sebacic acid. bis(2,2,6,6-tetra- methyl-4-piperiyl)ester	$M_r = 480.74$, $W = -113$, $G_W = -86$	OML: 60 mg/kg	7
71878-19-8 81200	Chimassorb 944 Poly[6-[(1,1,3,3- tetramethylbutyl)- amino]-1,3,5- triazine-2,4-diyl]-[(2,2,6, 6-tetramethyl-4-piperidyl) imino]-hexamethylene- [(2,2,6,6-tetramethyl- 4-piperidyl)imino]	$M_r = 600.99$, $W = -156$, $G_W = -245$	SML: 3 mg/kg	TDI: 0.05 mg/kg b.w. 2
Lubricants				
010332-32-8 71660	Pentaerythritol monooleate 9-Octadecenoic acid (9Z).- 3-hydroxy-2, 2-bis(hydroxymethyl)- propyl ester	$M_r = 400.60$, $W = 35$, $G_W = -191$	SML: mg/kg	7
115-83-3 71695	Pentaerythritol tetrastearate Octadecanoic acid. 2,2- bis[[(1-oxooctadecyl) oxy]methyl]-1,3- propanediyl ester	$M_r = 1202.03$, $W = -124$, $G_W = -1070$		D

CAS / Code	Name	Structure	M_r, W, G_W	Notes
25637-84-7 56080	Glycerol 1,3-dioleate 9-Octadecenoic acid (Z)-, 2-hydroxy-1,3-propanediyl ester	(structure)	$M_r = 621.01$; $W = -139$; $G_W = -371$	ADI: not specified 1 Glycerol esters: also used as antifoaming and antistatic agents
1323-83-7 56320	Glycerol 1,3-distearate	(structure)	$M_r = 625.04$; $W = -146$; $G_W = -385$	ADI: not specified 1
27215-38-9 56780	Glycerol 1-monolaurate	(structure)	$M_r = 274.40$; $W = 55$; $G_W = -144$	Toxicologically acceptable 3
27214-38-6 56840	Glycerol 1-monomyristate	(structure)	$M_r = 302.46$; $W = 45$; $G_W = -165$	Toxicologically acceptable 3 W
25496-72-4 56960	Glycerol 1-monooleate 1,2,3-Propanetriol mono((Z)-9-octadecenoate)	(structure)	$M_r = 356.55$; $W = 0$; $G_W = -215$	ADI: not specified 1
26657-96-5 57150	Glycerol 1-monopalmitate	(structure)	$M_r = 330.51$; $W = 33$; $G_W = -187$	Toxicologically acceptable 3

Table AIII.1 (Continued)

CAS Nr / Ref Nr	(Trade) Name	Structure	Restrictions and/or specifications	Remarks SCF-List
31566-31-1 / 57520	Glycerol 1-monostearate / Octadecanoic acid, 2,3-dihydroxy-propyl ester	H$_3$C(H$_2$C)$_{16}$—C(=O)—O—CH$_2$—CH(OH)—CH$_2$OH $M_r = 358.57$ $W = 0$		ADI: not specified 1
120-40-1 / 63560	Lauric diethanolamide / N,N-bis(2-hydroxyethyl)lauramide	H$_3$C(H$_2$C)$_{10}$—C(=O)—N(CH$_2$CH$_2$OH)$_2$ $M_r = 287.45$ $W = 57$	$G_W = -220$	D
000120-40-1 / 39280				
57-11-4 / 89040	Stearic acid	H$_3$C(H$_2$C)$_{16}$—COOH $M_r = 284.49$ $W = 19$	$G_W = -123$ $G_W = -252$	ADI: not specified 1

Antiblocking/slip additives

CAS Nr / Ref Nr	(Trade) Name	Structure	Restrictions and/or specifications	Remarks SCF-List
124-26-5 / 88960	Stearamide / Octadecanamide	H$_3$C(H$_2$C)$_{16}$—C(=O)NH$_2$ $M_r = 283.50$ $W = 21$	$G_W = -230$	3
301-02-0 / 68960	Oleamide / (Z)-9-Octadecenamide	H$_3$C(H$_2$C)$_7$—CH=CH—(CH$_2$)$_7$—C(=O)NH$_2$ $M_r = 281.49$ $W = 36$	$G_W = -220$	3
112-84-5 / 52720	Erucamide / (Z)-13-Docosenamide	H$_3$C(H$_2$C)$_7$—CH=CH—(CH$_2$)$_{11}$—C(=O)NH$_2$ $M_r = 337.59$ $W = -15$	$G_W = -272$	3 (also used as external lubricant)

CAS# / Ref#	Name	Structure	Notes
110-30-5 53520	N,N'-Ethylenebisstearamide 1,2-bis(octadecanamido)ethane	H₃C(H₂C)₁₆—C(O)—NH—CH₂CH₂—NH—C(O)—(CH₂)₁₆CH₃ $M_r = 593.04$ $W = -97$ $G_W = -325$	3
3061-75-4 36960	Behenamide Docosanamide	H₃C(H₂C)₂₀—C(O)—NH₂ $M_r = 339.61$ $W = 5$ $G_W = -272$	Metabolized to ammonia and behenic acid 3

Antifogging additives

Glycerol esters: see lubricants

CAS# / Ref#	Name	Structure	Notes
29116-98-1 87280	Sorbitan dioleate Sorbitan di-(9Z)-9-octadecenoate	(sorbitan ring with R substituent, R = oleate ester —O—C(O)—(CH₂)₇—CH=CH—(CH₂)₇CH₃) $M_r = 693$ $W = -139$ $G_W = -356$	Group TDI: 5 mg/kg b.w. based on the group ADI 5 mg/kg b.w. for sorbitan esters of lauric and oleic acids 2
1338-39-2/ 63480 1338-39-2 87600	Lauric acid monoester with sorbitan/Sorbitan monolaurate	R = —O—C(O)—(CH₂)₁₀CH₃ $M_r = 346.47$ $W = 51$ $G_W = -122$	Group ADI: 5 mg/kg b.w.; also used as lubricant 1 D
1338-43-8 87680	Sorbitan monooleate Sorbitan mono-(9Z)-9-octadecenoate	R = —O—C(O)—(CH₂)₇—CH=CH—(CH₂)₇CH₃ $M_r = 428.61$ $W = 18$ $G_W = -171$	Group ADI: 5 mg/kg b.w. 1

Table AIII.1 (Continued)

CAS Nr Ref Nr	(Trade) Name	Structure	Restrictions and/or specifications	Remarks SCF-List
000123-77-3 36640	Azodicarbonamide	$M_r = 116.08$		3
Photoinitiators				
000119-61-9 38240	Darocur BP Benzophenone	$M_r = 182.22$ $W = 162$ $G_W = -19$	SML = 0.6 mg/kg	Group TDI: 0.01 mg/kg b.w. 2
		$W = 40$ $G_W = -140$		
000102-71-6 25480 000102-71-6 94000 000102-71-6	Triethanolamine Tris(2-hydroxyethyl)amine	$M_r = 149.19$ $W = 139$ $G_W = -21$		8 Coinitiator
25922				

Index

a

accelerators 20
acceptable daily intake (ADI) 427, 450
acetylacetone 478
achievable barrier effect 320
acid scavengers 71
acrylates 472
– relative threshold levels 472
acrylic acid 34, 38, 41, 44
acrylonitrile-butadiene rubber 49
acrylonitrile-butadiene-styrene (ABS) polymer 39
additive mole constants 101
additive molecular properties 7
additive structural increments 101
adsorption 350
afore-described barrier 326
alcohol soluble propionates (ASP) 59
aliphatic carboxylic acid 40
aliphatic diamines ethylenediamine 395
aliphatic oligomeric esters 69
aliphatic phosphites 74
n-alkanes 165, 168, 170
– critical temperatures 168
– homologous series 165
– melting temperatures 170
alkylhydroperoxides 24
O-alkylhydroxylamine group 78
aluminum-coated polymer 326
aluminum oxide layers 330
β-aminocrotonic acid 76
amorphous polymer phase 165
amylopectin 54
amylose 54
analogous weight fractions 96
analytical methods 380
– interpretation 380
– validation 380
antiacids, see acid scavengers
antifogging agents 64
antimicrobial effect 72
antistatic agents 65
Antoine equation 110
area-related migration value 369
area-to-volume ratio 273
aromatic amines 399
aromatic polyamide MXD-6 336
Arrhenius relation 316
Arrhenius-type equation 282
atom clusters 173
– melting temperatures 173
atomic force microscopy (AFM) 323
Avogadro constant 179

b

barrier improvement factor 311
barrier systems 298
BASIC computer program 107
batching oil 487
– formulation 484
benzoate-based phenols 77
benzyl alcohol 20
BET adsorption studies 331
biaxially orientated polypropylene (BOPP) 35
– substrate film 312, 340
biaxial orientation 30
biodegradable plastic materials 54
biodegradable polymer 55
bisphenol A-diglycidyl ether (BADGE) 20, 375, 459
bishphenol F-diglycidyl ether (BFDGE) 375, 459
block copolymer (BCP) 19
blow-molded containers 40
Boltzmann distribution 177
British Standards Institution (BSI) 467
built-in biocidal moieties 72

Plastic Packaging. Second Edition. Edited by O.G. Piringer and A.L. Baner
Copyright © 2008 WILEY-VCH Verlag GmbH & Co. KGaA, Weinheim
ISBN: 978-3-527-31455-3

tert-butylesters 24
tert-butyl-hydroxyanisol 408

c
camphorated off-flavours 479
carbon chain polymers 76
carcinogenic assessments 429
cardboard baking dishes 42
catalysts 25
– heterogeneous 25
– homogeneous 25
cationactive agents 65
ccp system 174
cellulose acetate butyrate (CAB) 59
cellulose acetate propionate (CAP) 59
chain-breaking antioxidants 73
chemical compounds 167
– homologous series 167
chemical interactions 4
– pharmaceuticals 4
– plastics 4
chemical modeling 99
chemical potential 90
– definition 91
chloroanisoles 486
chlorophenols 481, 482
chocolate products 488
chromatographic instrument 381
chromatographic system, *see* detection system
classification and regression tree (CART) algorithms 156
– calculations 156
– diagram 156
– structure 156
cobalt naphthenate 47
colorants 66
combinatorial activity coefficient contribution 107
Community Reference Laboratory (CRL) 445
Council on Environmental Quality (CEQ) 421
Crank–Nicolson method 251
crosslinked polyurethanes 48
crystalline polymers 31
– properties 31
crystallization 30
cumulative estimated daily intake (CEDI) 427
cyclic olefin copolymers (COC) 37

d
defect model 324
defect size distribution 315
degradation-sensitive moieties 70
dehydrating agent 75

detection system 381
diacylperoxide 24
dialkylperoxides 24
dicarboxylic acid 19
dicumylperoxide 24
diethylene glycol (DEG) 351, 459
different parameters 307
– units 307
diffusion coefficient 8–9, 164, 168, 178, 501
– *n*-alkanes 188
– correlation 501
– critical state 181
– gases 165, 178
– liquid 165, 184
– models 164
– plastic materials 188
– prerequisites 168
– solids 181
diffusion equation 499
diffusion model 7, 510
diffusion process 197
– differential equations 197
dimensionless group-contribution element 111
dimensionless partition coefficient 96
diphenyl butadiene migration 404
dipropylene glycol diacrylate (DPGDA) 60
direct experimental measurement 99
dispersion systems 25
distance–time scale 124
dynamic headspace technique 468

e
Einstein–Smoluchowski equation 152, 163
elastomer–modified thermoplastics 36
Elbro free volume model (ELBRO-FV) 108, 109
electron beam irradiation 37
empirical approximation method 100
environmental assessment (EA) 432, 434
equilibrium saturated vapor pressure 91
ether alcohols 493
ethoxylated fatty alkylamines 65
ethyl acetate 472
ethylene–propylene–diene rubber 49
ethylene–propylene elastomer 35
ethylene–vinyl acetate (EVA) 34–35
EU food packaging legislation 374
European commission 445, 500
– Joint Research Centre (JRC) 445
– practical guide 500
European Committee for Standardization 374

European Food Safety Authority (EFSA)   376
European harmonization process   351
European norms (EN)   374
excess free energy   94
excited state intramolecular proton transfer (ESIPT) mechanism   77

f
FABES formula   265, 282
FABES GmbH   9
– user-friendly programs   9
face-centered unit cell   173
Fat reduction factor (FRF)   455
favored polymeric multilayers   10
Federal Food, Drug, and Cosmetic Act (FFDCA)   417
finite difference (FD) method   9, 249
– numerical methods   9, 249
finite element method (FEM)   249
Flexographic inks   59
Food and Drug Administration (FDA)   417, 500
– regulation policy   352
food contact material (FCM)   10, 351
food contact notification (FCN)   418
food contact substances (FCS)   10, 417
food industry   465
– off-flavors   465
food-packaging materials   404
– safety   404
fossil-based raw materials   17
Fourier trigonometric series   210
Fraunhofer-Institute of Process Engineering and Packaging   374
free hydroxyl groups   49
free-radical fragments   80
free volume concept   103
free-volume model(s)   131, 133
free-volume parameters   139
frozen-food temperatures   45

g
gas chromatography-mass spectrometry (GC-MS)   468
Gaussian error   390
glassy polymer-solvent systems   138
glycerol propoxylate triacrylate (GPTA)   60
good manufacturing practice (GMP)   445
graft copolymers (GCP)   19
group-contribution Flory equation-of-state (GCFLORY)   108
– models   109
– polymer activity coefficient   109
group-contribution parameters   104

group-contribution thermodynamic polymer partition coefficient estimation methods   102
guide of uncertainty of measurements (GUM)   389

h
halogenated hydrocarbons   39
hard core volume, *see* van der Waals volume
HAS-functionalized PO   78
heat stabilizers (HS)   75, 85
– transformation products   85
Henry's constant   98
Henry's law coefficients   92–93
heptane matrix   185
heteroatoms   26
heterophasic copolymer (heco-PP)   503
heuristic analytical formula   310
high aspect ratio   337
high density polyethylene (HDPE)   18, 63, 108, 189
high-sensitive spectral methods   85
high-temperature migration   365
Hildebrand rule   176
hindered amine stabilizers (HAS)   66, 84
Holten–Anderson model   108
homologous series   100
– example   100
homopolymer   22, 35
– melt blending   22,
– plastic processing   22
HPLC methods   383
hydrochloric acid   52
hydrogen bonds   26
hydrogen peroxide   24
– derivatives
hydroperoxide deactivating antioxidants   74, 83
– transformation products   83
hydrophilic-hydrophobic balance   102

i
ideal gas law equation   97
ideal solution   92
– partitioning   92
indirect migration assessment   354
inorganic barrier layers   323
inorganic foils   9
inorganic layers   302, 320, 323
inorganic silicates   51
interaction energy   167, 176
– relative density   176
interaction model   166
– assumptions   166
intermediate polymeric layer   320

intermolecular binding relationships   101
intermolecular condensation reaction   17
internal food reactions   465
ionic addition polymerization reactions   18
isocyanates   399
– chemical degradation   399
isophthalic acid   395
isopropylisothioxanthone (ITX)   458
isotactic polypropylene (PP)   63

k

kinetic curve   364
kinetic theory   165
– gases   165
Kurderna–Danish evaporation apparatus   469

l

lamellae   170
– thickness   170
laminate theory   320
laminating agents   58
Laplace transforms   210
light screening pigments 76
light stabilizers   76
Likens–Nickerson steam distillation extraction apparatus   469
linear-low density polyethylene (L-LDPE)   32, 63
linear relationships   111
– advantage   111
linseed oil   484
longer shelf-lives   4
– factors leading   4
long-term heat aging (LTHA)   72
long-term migration, *see* high-temperature migration
long-term model experiments   83
low density polyethylene (LDPE)   18, 63, 108, 189
low-discoloring properties   73
lowest unoccupied molecular orbital (LUMO)   102
low molecular components   6
– diffusion coefficients   6
low molecular weight component systems   103
low molecular weight compounds   63
low molecular weight polyisobutylene   36
low-temperature extraction measurements   365
lubricant residues   486
lubricants   67

m

macroscopic alkane crystal samples   171
– melting points   171
macroscopic *n*-alkane sample   165
macroscopic defects   323
– properties   165
macroscopic particle systems   165, 167
– characteristic   167
– properties   165
maleic acid anhydride   38
mass spectroscopy (MS)   383
mass transport   5, 499
– external influences   5
mass transport equation   8, 263
mass transport phenomena   63
material-specific permeability   311
material-specific properties   99
mathematical models   500
maximum initial concentration (MIC)   500
mean-square displacement (MSD)   143
medium-forming substances   23
melamine-aldehyde resin   56
melamine resins   47
mercaptan-derived compounds   480
metallic off-flavors   485
metallocene random copolymers   35
methacrylic acid   38–39, 41, 44, 57
methylacrylamide–methylol ether   38
methyl acrylates   472
– relative threshold levels   472
4-methyl-4-mercaptopentan-2-one   481
microcrystalline waxes   57
MIGRATEST *EXP* software   9, 281–282
– migration estimations   281
MIGRATEST Lite   266, 273
– basic features   266
– estimation   276
– output information   278
migration   349, 366, 499
– alternative   366
– extraction   366
– process   350
– measurements   12, 500
modeling diffusion coefficients   520
modern food packaging migration testing   10
molal activity coefficient   97
– relationship   97
molar activation energy   179
molar activity coefficient   97
– relationship   97
molar concentration partition coefficients   96
molar fraction partition coefficient   95
– relationship   95
molar free enthalpy   90

moldy off-flavors 486
molecular descriptors 102
molecular models 126
mole fraction partition coefficient 95
– relationship 95
monoethylene glycol (MEG) 351, 459, 447
monolayer films 338
– permeation 338
monolayer polymer films 303
– substance transport 303
monomeric quinone methide 81
monophasic homopolymer 503
monophasic random copolymer 503
more than one inorganic barrier layer 332
– combinations 332
motor oil off-flavor 486
multilayer films 338
– permeation 338
multilayer (ML)-food system 259, 260
multilayer (ML) materials 8, 247
multilayer (ML) packaging 248
multilayer polymer films 305
– substance transport 305
mushroom-like off-flavors 486
musty off-flavors 486, 487
mutual recognition 461
– principle 461

n
naphthalene 487
National Environmental Policy Act (NEPA) 417, 421
Nernst's Law 92
nonadhesive rubber-covered rollers 51
nonbranched hydrocarbon 165
nonideal solutions 93–94
– partition coefficients 94
non-intentionally added substances (NIAS) 351
nonionic agents 65
nonpolar isooctane, *see* polar ethanol
nonpolar plastics 352
nonpolymeric liquids 98
– molar volume 98
no observed effect level (NOEL) 427
norbornene 37
novolac glycidyl ethers (NOGE) 459
n-pentane–polystyrene system 138
nucleating agents 67
nucleation 31
– heterogeneous 31
– homogeneous 31
number-average molar mass 28, 510

o
odor limit value concentration 475
off-flavor compounds 468
– identification 468
off-flavor contamination 12
off-flavor sensitive foods 468
olfactometer 470
oligomer 6
one-dimensional (1D) diffusion problem 8
one-dimensional P–F system 264
one inorganic barrier layer 313, 322
open-chain hydrocarbon 165
open-chain products 85
OPP films 35
optical brighteners 68
organic peroxides 24
organic polymers 51
organo-metal compounds 23
overall migration limit (OML) 350, 351, 450
oxidation-drying process 484
oxidation reactions 18
oxygen-triggered degradation 72
α-oxyperoxides 25

p
packaging 1, 4
– minimization 4
– plastics 349
packaging materials 465
particle system 168
– characteristic 168
particulate contaminations 324
partition coefficient 9, 89, 98–99, 102, 111, 510
– estimation 102, 111
– liquids 99
– polymers 99
– relationship 98
– thermodynamics 90
pencil-like off-flavor 488
penetrant-intramolecular energy 127
penetrant-polymer system(s) 127, 129, 135, 137, 140, 154–155, 157,
pentaerythritol tri-tetraacrylate (PETA) 60
permeability 6
permeation coefficient(s) 304, 316, 318
permeation processes 9
peroxide-controlled polymerizations 25
peroxide crosslinked ionomer 37
peroxyaliphatic fatty acids 24
peroxy-benzoic acids 24
peroxycarbonic acid 24
PET-inorganic layer 314
PET-packed foodstuff 407

phase ratio variation (PRV)   89
phenolic antioxidants   79
– transformation products   79
phosphoric acid   58
photoantioxidants   77
photocatalytic effect   74
photoinitiators   59
phthalate-based polycondensate resins   48
physicochemical parameters   89
plaque sorption method   89
plastic additives   6
– characteristic functions   6
– representative structures   6
plastic additives producers   349
plastic applications   1
– today's multitude
plastic-food system.   264
plasticizer   68–69,
– key role   69
plasticizer diethylhexyl adipate   369
plasticizer-free blends   41
plasticizer-functionalized polymers   69
plastic material barrier layer   5
plastic materials   15
– characteristics   15
plastic matrix   504
– two-phase structure   504
plastic processing   22
– melt blending   22,
plastic properties, see diffusion coefficients
plastic stabilizers   78
– transformation products   78
plastic-packaging material   353, 361
plastomers   32
platelets   337
– good parallel alignment   337
platelet-shaped particles   301, 322, 336
polar ethanol   363
polyaddition polymers   48
polyamide (PA) materials   43, 108, 352, 478
polybutylene tere-phthalate (PBT)   41
polycarbonate (PC)   42
polycondensation products   56
polycrystallinity   31
Polydimethyl siloxane networks   21, 52
polyester networks   21
polyetherpolyole   49
polyethylene (PE)   6, 17, 32–33, 107
– oxidation   33
polyethylene terephthalate (PET)   33, 108, 191, 301, 308, 314, 357
– bottles   292, 371, 477
– layer   340
– substrate film   312, 324

polymer-based packaging   297
– barrier function   297
polymer blends   43
polymer-borne free-radical intermediates   79
polymer-bound oxygenated groups   70
polymer-bound UV light-absorbing impurities   76
polymer chemistry   22
polymer films   332
– combinations   332
polymeric materials   302
– permeation   302
polymeric multilayered structures   9
polymerization   22
– chain reaction   25
– process   17, 32
– synthesis   19
polymer-like LDPE   509
polymer matrix   71, 265, 501
– diffusion behavior   501
polymer molecule   15
polymer-penetrant systems   142, 149, 152
polymer polycaprolactone (PCL)   55
polymer reactions   20
polymer-solvent systems   134
polymer substrates   302, 313, 323–324
– multilayer   313
– single layer   313
polyolefin packaging materials   475
polyoxyethylpentaerythritol tetraacrylate (PPTTA)   60
polyoxymethylene (POM) plastics   45
polyphenylene ether (PPE)   45
polyphenylene oxide   373
polyterephthalic acid diol ester   26
polytetrafluorethylene (PTFE)   46, 58
polyurethane (PUR) elastomers   48
polyurethane networks   21
polyvinyl chloride (PVC)   34, 39, 56, 63, 494
– outdoor application   75
polyvinylether   46
– polymerization   46
polyvinylidene chloride (PVDC)   41
– dispersion coatings   41
– films   41
polyvinylmethyl ether   46
polyvinyloctadecyl ether   46
predefined data banks   282
P-regenerated cellulose foil   54
present state-of-the-art technology   472
primary aromatic amines   398
printing inks   59
procaryotic microorganisms   55

processing aids 15
product mummification 4
propionate-type phenolic moiety 83
propylene oxide 49

q

quality assurance system 4, 6, 445
quality preservation 3
– packaging 3
quantitative structure activity relationship (QSAR) 7, 99, 102
– parameters 102
– statistical linear regression methods 102
quantitative structure property relationship (QSPR) 7, 102
quasi-chemical lattice model 107
quasi-homogenous material 303
quinone methide (QM) methods 81, 383–384

r

random walk 163
Raoult's law 92–94, 103
– accuracy 104
reaction injection molding (RIM) 21, 48
reactive extrusion (REX) 22
reactive polymer process 21
ready-prepared migration solution 390
real diffusion coefficient 265
regenerated cellulose film (RCF) 446
Registration, Evaluation and Authorisation of Chemicals (REACH) 63
regular solution theory (RST) 103
– scope 109
reinforcing agents 66
relative molecular mass 100, 101
relative threshold values 472
resin transfer molding (RTM) 21
retention index 110
– definition 110

s

Sackur–Tetrode equation 177
sacrificial stabilizer consumption 79
salty tastes 491
SAN copolymers 39
Scientific Committee for Food (SCF) 443
sealable films 370
sebacic acid 395
self-diffusion coefficient 165, 166, 331, 181, 184, 185
– n-alkanes 184, 185
– liquid alkanes 166
– metal 165, 181, 184

– salt 183
– semiconductor 183
– solid alkanes 166
self-installing executable program
self-service stores 1
semipermeable packaging materials 474
sensitized photolysis 79
sensorial properties 6
sensory-active compounds 474
– threshold concentrations 474
sensory evaluation 466, 472
– packaging materials 466
sewer-like off-flavors 491
shelf-life 3
short-term extraction measurements, see low-temperature extraction measurements
silicon-containing siloxane units 51
single branched molecules 111
single-component gas 91
single-phase homogeneous system 203
single-sided contact 371
SML methods 382, 384
sodium-cellulose-xanthogenate 54
solid phase microextraction (SPME) technique 468, 469
solubility coefficient 98, 303
– relationship 98
sophisticated package 4
– barrier layer 4
sorption coefficient 303
specific food-contact polymers 423
specific geometry factors 10
specific migration limit (SML) 265, 390, 350, 351, 450
stabilizers 70
standard cold seal formulation 480
standard olfactory methods 467
standard vapor phase 176
– definition 176
state-of-the-art hardening technology 20
static headspace technique 469
steam distillation 469
stimulant-based migration test 404
stirred-port version 22
Stokes–Einstein equation 164
structure–activity relationship (SAR) 427
– analysis 430
styrene homopolymers 39
styrene-acrylonitrile copolymer 39
styrene-butadiene rubber (SBR) 49
styrene rubber solution 38
– polymerization 38
substance transport 318
sulfur containing stabilizers 74, 78

surface-to-volume ratio 369–370
synthetic macromolecules 15
synthetic rubber 16

t

taint transfer test 467
technological development 3
– goal 3
temperature-resistant coatings 58
terephthalic acid 17, 358
– polycondensation reaction 17
tertiarybutylphenol disulfide 46
thermo-degrading PVC 71
thermodynamic group-contribution partition coefficient estimation methods 103–104, 108, 109
– comparison 108
thermodynamic models 99
thermoplastic polymethylmethacrylate (PMMA) 44
thermoplastics 17
– thermoset coatings 17
– thermoset plastic 28
thin polymer layers 318
threshold of regulation (TOR) 418
time-dependent evolution 287
time–temperature conditions 367, 384
time–temperature migration test 397
tolerable daily intake (TDI) 450, 457
toluene 493
2,6-toluene diisocyanate 48
tortuosity factor(τ) 321
total mass transfer 355
transition-state theory (TST) 150
translational partition function 177
transmittance 304
transport equations 9, 195, 200
triangle corner foods 378
tri-isopropanol amine 49
4,4,6-trimethyl-1,3-dioxane 487
trioxymethylene oligomers (trioxane) 45
tripropylene glycol diacrylate (TPGDA) 59
tris(2-ethylhexyl) phosphate 69
Trouton's rule 175, 176
two-component liquid phase 91
two-layer laminates 306
two-phase system 38, 91

u

ultrabarrier multilayer stacks 335
ultraviolet (UV) light 33
undesirable foreign substances 2
UNIFAC group-contribution method 7, 104, 107
unified quasi chemical (UNIQUAC) theory 104
– liquid mixtures 104
unsaturated polyester (UP) 47
urea-formaldehyde resins (UF) 46
user-friendly software 203
UV absorbers 79
– transformation products 79
UV diode array detection systems 383

v

vacuum web coater 324
van der Waals attractions 26
– equation 175
– forces 26, 171
– volume 104, 107, 108
vapor pressure index method (VPIM) 7, 110, 514
– partition coefficient estimation 109
vapor pressure unit contributions, see linear relationships
vinyl chloride (VC) 37, 40, 351, 441
vinyl chloride monomer (VCM) 459
vinylidene chloride (VDC) 37, 41, 44
volatile organic compounds 489
volume fractions 97

w

water vapor
– transport mechanisms 331
wax-like consistency 46
worst case
– assumption 355
– scenario 265

x

X-regenerated cellulose foil 54

z

zero permeated amount 339
Ziegler–Natta catalysts 18, 27, 32, 71